A&A LIBRARY

ASTRONOMY AND ASTROPHYSICS LIBRARY

More information about this series at http://www.springer.com/series/848

David Branch • J. Craig Wheeler

Supernova Explosions

Springer

David Branch
Homer L. Dodge Department of Physics and Astronomy
The University of Oklahoma
Norman, OK, USA

J. Craig Wheeler
Department of Astronomy
The University of Texas at Austin
Austin, TX, USA

ISSN 0941-7834 ISSN 2196-9698 (electronic)
Astronomy and Astrophysics Library
ISBN 978-3-662-57223-8 ISBN 978-3-662-55054-0 (eBook)
DOI 10.1007/978-3-662-55054-0

Cover figure: A false color image of Cassiopeia A using observations from both the Hubble and Spitzer telescopes as well as the Chandra X-ray Observatory
Credit: X-ray: NASA/CXC/SAO; Optical: NASA/STScI; Infrared: NASA/JPL-Caltech/Steward/O. Krause et al.

Printed on acid-free paper

This Springer imprint is published by Springer Nature
The registered company is Springer-Verlag GmbH Germany
The registered company address is: Heidelberger Platz 3, 14197 Berlin, Germany

Preface

JCW had long wanted to write a new monograph on supernovae. Despite an immense explosion in the literature and many review articles, there had not been one since that of Iosef Shklovsky in 1961. In 1981, JCW signed a contract to write such a monograph and even got an advance from a publisher who has long forgotten the contract and does not even publish that sort of book anymore. Along the way, each review article was intended for the monograph, but decades slipped by with only occasional work on the book itself. Major commitments to astro-politics later in life seemed to be the death knell of this long-held wish.

DB's longtime professional obsession has been the study of supernovae, but he never expected to write a book about it. For one thing, he thought that as long as he continued to work, he would want to spend as much time as possible trying to contribute to ongoing supernova research; for another, putting together a book looked like a heap of work. Thus, in the summer of 2008, when then *Praxis* editor Clive Horwood invited him to write one, he politely declined and cheerfully nominated a colleague at another institution for the honor. He then resumed planning for the rare opportunity to teach a small-enrollment graduate seminar course on supernovae in the fall semester. Such a course, entailing no particular syllabus, allows the instructor a great deal of freedom. For a "broad overview of current research" component of the course, he decided to have one student be responsible, each week, for gathering all astro-ph[1] abstracts directly relevant to supernovae (at that time typically ten per week). The following Monday, the student would distribute copies of the abstracts to the whole class. On Wednesday, after all students had read the abstracts, they would devote the hour to a discussion of the papers, led by the responsible student who would have read at least the introductions and conclusions. At about the time that DB began to be enthusiastic about broadening his own view of supernova research in this way, Clive informed him that his nominee had declined. Influenced by the prospective astro-ph experiment, DB found the nerve to undertake this task.

[1] A widely used electronic preprint service: arxiv.org/archive/astro-ph.

The astro-ph exercise did indeed prove to be an eye-opener to some interesting supernova research issues to which DB had been paying little attention. With the endorsement of the students (Michele Benesh, Tom Boone, Erin Cooper, and Leann Chau Dang—DB thanks them all), the class stayed with it for the entire semester.

The nominee unnamed above was JCW. By the time the possibility of collaboration arose, he had long since forgotten (but later recovered) his own mail from Clive Horwood in 2006, to which he replied that he thought DB would do an excellent job on such a project. (Return to paragraph two and iterate.)

The two of us were in communication on an entirely different matter in the spring of 2010, when the fact that DB had undertaken this book arose. After some tentative to and fro, we acknowledged our long mutual admiration (with some disagreement as to who is more admirable than whom) and compatible styles. A joint project was born.

Owing especially to (1) the appearance of Supernova 1987A in the nearby Large Magellanic Cloud, (2) the spectacularly successful use of supernovae as distance indicators for cosmology, and (3) the association of some supernovae with the enigmatic cosmic gamma-ray bursts, the pace of supernova research had been increasing sharply. Topical review articles, conference proceedings, and a steady stream of journal articles were available to those entering the field, but still no monograph presented a broad survey of modern supernova research and a guide to the current literature.[2] Our goal is to offer a modern account of what we think we know about the causes and consequences of supernovae, as well as of issues that remain to be resolved. The writing is aimed not so much at the supernova pundits of the world, who need no book from us, but for advanced undergraduate and graduate students of astronomy and physics and for astronomers and physicists contemplating research on supernovae or related fields. With this goal in mind, we have attempted to provide a modicum of pedagogy and rigor, but nothing like all the operating details used by experts in the vast array of required topics.

The emphasis is on the explosive phases of supernovae, with some attention to their consequences for and applications to astrophysics and cosmology. Some important related topics are necessarily given short shrift. We make no attempt at all to discuss the often difficult and painstaking process of obtaining and reducing observational data. The discussion of supernova remnants is largely confined to how they inform us about the supernovae that produced them rather than, for example, extending to particle acceleration processes and the origin of cosmic rays. Similarly, our interest in gamma-ray bursts is largely confined to their connections with supernovae. Supernova remnants and gamma-ray bursts need books of their own.

---

[2] *Supernovae and Nucleosynthesis: An Investigation of the History of Matter, from the Big Bang to the Present* by W. David Arnett (Princeton Un. Press, 1996) is a masterly treatise on the role of supernovae in the nuclear evolution of the universe, and *Stellar Explosions: Hydrodynamics and Nucleosynthesis* by Jordi José (CRC Press, 2016) is an excellent, broad survey of the physics of stars and their outbursts, including, but not limited to, supernovae.

Part I is devoted to a survey of the kinds of observations that inform us about supernovae, some basic interpretations of such data, and an overview of the evolution of stars that brings them to an explosive endpoint. Part II goes into more detail on core-collapse events in all their variety of photometric and spectroscopic properties: Which kinds of stars produce them, and how do they do it? Part II also discusses the recently-defined category of superluminous supernovae, although their explosion mechanisms remain unclear and may include thermonuclear explosions of a certain kind. Part III is concerned with identifying the stellar progenitors and explosion mechanisms of "traditional" thermonuclear events—Type Ia supernovae. Part IV is about consequences of supernovae and some applications—how they can be exploited to inform astrophysics and physics more generally. In the final chapter, we attempt to summarize what is known, what is not, and the exciting prospects for advancing our knowledge in the coming decades. An appendix provides a list of abbreviations.

References are provided in sufficient number to help the reader enter the literature, although inevitably our list will quickly become dated. We have included some history where it seemed especially appropriate, but this is not a history book, and many seminal papers are not cited.[3] With the exponential growth of the number of supernova papers, most of the references are, perforce, to recent publications. We attempted to consider those that were available to us by the end of June 2015. Only a small fraction of those that became available later are cited. We refrain from preceding references with "e.g." and appending "and references therein" because seeing these repeatedly would become tiresome.

Supernova research will continue to advance rapidly, which probably accounts for the lack of a recent monograph on the subject. The best that can be hoped for this book is that it will be of use to those who consult it—for a while.

The material in this book is the result of the ideas, work, and writings of countless colleagues around the planet. JCW is grateful to the Aspen Center for Physics where he worked on the book over the course of several summers. We are especially indebted to Bruno Leibundgut for reading the manuscript and offering numerous helpful comments and to J. Robinson Wheeler for a careful reading of the proofs.

Norman, OK, USA David Branch
Austin, TX, USA J. Craig Wheeler

[3]For an informed account of the early development of supernova research, see articles by Virginia Trimble in Rev. Mod. Phys., 54, 1183 (1982) and 55, 511 (1983).

Contents

Part I
Observational Overview and General Interpretations

Chapter 1
Overview

1.1 Introduction

Why study supernovae? At a basic level, one's attention is captured by anything new and unusual, and a new point of light in the sky is a novelty that inspires curiosity. Further study shows that these events are of gigantic power, and the imagination is fired with the awesome spectacle of stellar catastrophe. The maximum luminosity can equal that of a billion suns. Matter is thrown into space at a few percent of the speed of light with a kinetic energy of 10^{51} ergs—the energy equivalent of 10^{28} megatons of TNT. No terrestrial example compares. An energy of 10^{51} ergs is defined to be 1 Bethe, 10^{51} ergs $\equiv$ 1 B, a terminology adopted by the supernova community to honor Hans Bethe for his pioneering work on stellar nucleosynthesis and later work on supernova theory.

Besides this innate appeal, a host of scientific questions surround the supernova phenomenon. A key question is the nature of the physical mechanisms that lead to the explosions. Some supernovae are thought to be thermonuclear disruptions of entire stars; others result from gravitational collapse. The latter eject only their outer envelopes as their cores become compact objects: neutron stars and black holes.

Supernovae also allow us to do stellar autopsies to examine the innards of the stars that explode. The interior of a star is normally shielded from view, but explosion causes the star to expand and thin out. If we are clever enough, we can read the signals that flood from successive layers and thereby probe the nature of the pre- and post-supernova star. Supernovae also represent the endpoint of normal stellar evolution. By studying them and the evolutionary pathways that lead to them, we can gain new insights into the course of stellar evolution.

Supernovae have far-reaching consequences for astrophysics and cosmology. The ejected matter is enriched in heavy elements (in astronomy, "heavy" means carbon and elements of higher atomic number), some that were synthesized by nuclear reactions during the slow pre-explosion evolution of the star and some that were synthesized during the explosion itself. Thus supernovae are the main drivers of the

D. Branch, J.C. Wheeler, *Supernova Explosions*, Astronomy and Astrophysics Library, DOI 10.1007/978-3-662-55054-0_1

nuclear evolution (in astronomy, this is usually referred to as "chemical evolution") of the Universe.

Supernova ejecta sweeps up ambient circumstellar and interstellar matter (CSM and ISM) to form conspicuous long-lived supernova remnants (SNRs), such as the famous Crab Nebula and Cassiopeia A (Cas A). The energy released by supernovae in some circumstances stimulates and in others inhibits the formation of stars. Supernovae may drive galactic "fountains" and eject matter from galaxies in supernova-heated galactic winds. Thus supernovae are an important factor in the physical evolution of galaxies.

Finally, because they are so luminous, supernovae also serve as exquisite indicators of extragalactic distances. They are used to measure the cosmic distance scale (the value of the Hubble constant, H_0) and to determine the history of the cosmic expansion: deceleration by gravity at first, followed by acceleration driven by an entity called *dark energy* that may correspond to Einstein's cosmological constant.

This chapter presents an introduction to supernovae, a brief survey of the observations that inform us about them, and the basics of the mechanisms by which they are thought to explode. Subsequent chapters will delve into these issues in more detail.

1.2 Discovery

The transient nature of supernovae makes them difficult to study. No specific progenitor star or host galaxy can be yet anticipated, and because of their great distances, usually more than 10 megaparsecs (Mpc), they soon fade below detectability. Although hundreds of extended SNRs and hundreds of pulsars—rotating, magnetic, neutron stars—have been discovered in our Milky Way Galaxy, no contemporary supernova has been witnessed in the Galaxy for more than 400 years.

Our solar system as well as most supernovae that occur in the Galaxy are located within the dusty disk that constitutes the plane of the Galaxy. Owing to severe extinction of optical light by interstellar dust grains, only a small fraction of Galactic supernovae during the last two millenia has been detected by observers on Earth. The star visible to Chinese astronomers for 8 months in 185 AD probably was the supernova associated with the young SNR RCW 86. Other "new stars" that temporarily appeared in the sky in 1006, 1054, 1181, 1572, and 1604 are now known to have been supernovae. Some of these historical Galactic supernovae were visible even in daylight (see Stephenson and Green (2002) for a thorough summary of the historical record). Emission from their SNRs is now detected across the electromagnetic spectrum. The SNR Cas A, the brightest radio source in the sky apart from objects in the solar system, evidently was produced by a supernova that occurred around 1680, but no definite record of its sighting has been found. Other supernovae surely have erupted in the Galaxy since then. The SNR G1.9+0.3, near the Galactic center, has an age of only about 100 years, making it the youngest known remnant in the Galaxy. Nearly 300 other remnants of explosions in the Galaxy within the last $\sim 10^5$ years also are recognized.

On February 23, 1987, an extraordinary astronomical event took place when SN 1987A appeared in the Large Magellanic Cloud (LMC), a small irregular satellite galaxy of our Galaxy at a distance of only 50 kpc. At its brightest, SN 1987A reached apparent visual magnitude $m_V \simeq 3$, easily visible to the naked eye from sufficiently southern latitudes. SN 1987A has become the most well-observed supernova, and its aftermath will be observed far into the future, perhaps as long as there are astronomers on Earth.

Until another outburst is seen in our Galaxy or one of its satellites, investigations of the explosive phases must be based on much more distant events (Fig. 1.1). The study of extragalactic supernovae began in 1885, when a new star near the naked-eye limit of $m_V \simeq 6$ appeared close to the center of the Andromeda galaxy (M31),

Fig. 1.1 SN 2011fe in Messier 101. Credit: B.J. Fulton, Las Cumbres Observatory Global Telescope Network

the nearest large galaxy at a distance of 0.8 Mpc. Understanding of the titanic scale of the 1885 event came only in the 1920s with the recognition of the existence of other galaxies, including M31, at great distances from the Milky Way. Astronomers adopted the term *supernova* to contrast these distant, powerful events with classical novae that proved to be thermonuclear eruptions on the surfaces of white dwarfs in binary-star systems. These developments spurred Fritz Zwicky to begin systematic photographic searches for extragalactic supernovae (Zwicky 1938).

1.3 Spectral Classification

The primary classification of supernovae is based on the appearance of their optical spectra near maximum light, when the ejecta are optically thick. The spectra contain Doppler-broadened features that form above the photosphere with velocities that are typically about 10,000 km s^{-1}, or 0.03c. Spectral features commonly have *P Cygni* profiles characterized by emission around the rest wavelengths of the lines and blueshifted absorption that reflects formation of the lines in an expanding medium.

Modern techniques for discovering and studying extragalactic supernovae were developed by Zwicky, Walter Baade, and Rudolph Minkowski in the 1930s. The first few supernovae discovered by serendipity included examples corresponding to the two basic types of event now recognized, but the first 36 that Zwicky found himself were of only one type (Trimble 1982). For decades, the spectra of this latter type were not interpretable even though they were similar from event to event. They showed broad features that could not readily be associated with any known element. The light curve (luminosity as a function of time) had a predictable regularity. It rose to peak brightness in about 3 weeks, and then after declining for several more weeks it faded nearly linearly in magnitudes (a logarithmic measure of flux), or exponentially in luminosity with a half-life of about 50 days. Other observations identified a different kind of explosive event. Their maximum luminosities were dimmer by a factor of a few and they had a variety of light-curve shapes. The most important aspect of this type of supernova was that the spectra contained at least a few recognizable features—the Balmer lines of hydrogen. Minkowski (1941) classified the known supernovae as Type I (SN I) and Type II (SN II); spectra of SN II contain conspicuous hydrogen lines while spectra of SN I do not. Almost all extragalactic supernovae have readily fit into one of these two broad categories. The history of supernova spectral classifications and interpretations was reviewed by Branch (1990).

The classification scheme was subsequently extended according to additional spectral characteristics (Fig. 1.2). SN Ia have a distinctive spectral evolution that includes a strong absorption feature near 6100 Å, produced by singly-ionized silicon (Si II). In spectra of SN Ib, the Si II absorption is weak or absent, and strong lines of neutral helium (He I) develop around the time of maximum brightness. He I absorptions are not conspicuous in spectra of SN Ic, but a strong absorption of O I

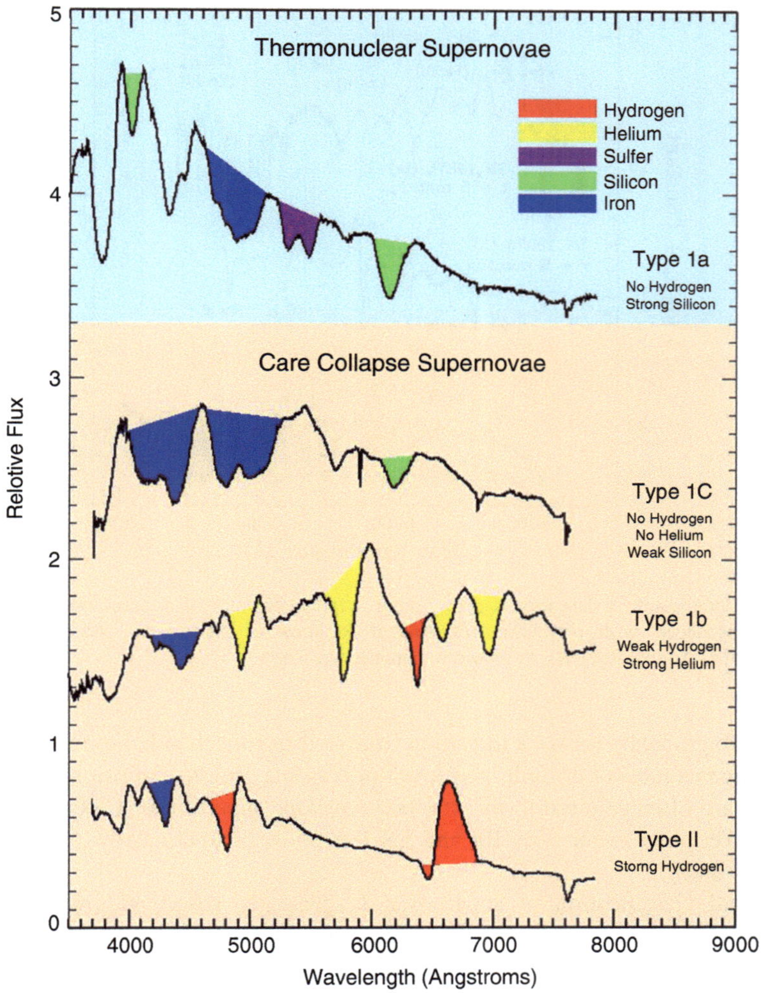

Fig. 1.2 Comparison of early-time spectra. Figure courtesy of D. Kasen

often is. Months after maximum, when the ejecta begin to become optically thin, SN Ia show forbidden lines of iron ions, while SN Ib and SN Ic show characteristic features of semi-forbidden Mg I] and forbidden [O I] and [Ca II] (Fig. 1.3).[1] SN Ib and SN Ic are sometimes referred to collectively as SN Ib/c, although in Chaps. 16 and 17 we will discuss the case for considering them as separate physical categories.

[1]Most of the spectra displayed in this book are in F_λ units, erg s^{-1} cm^{-2} Å^{-1}, as in Fig. 1.2, but some are in F_ν units, erg s^{-1} cm^{-2} hz^{-1}, as in Fig. 1.3. Because $F_\nu \propto \nu^2 F_\lambda$, a spectrum in F_ν units appears redder than it would in F_λ units. If a spectrum is not specified in a figure or its caption to be in F_ν units, it is in F_λ units.

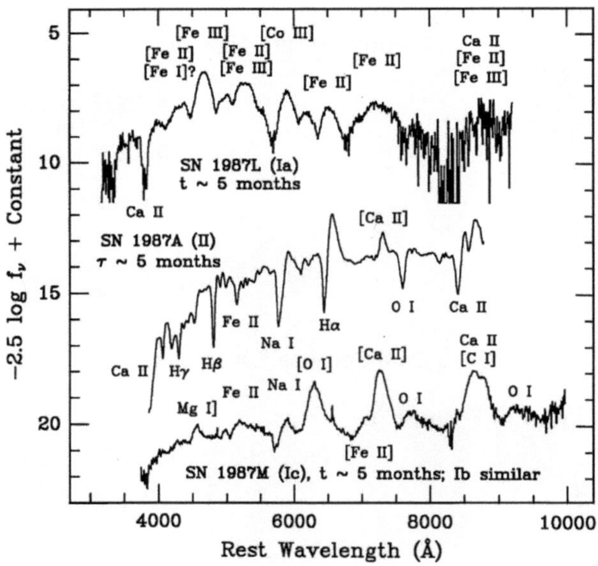

Fig. 1.3 Comparison of late-time spectra. From "Optical Spectra of Supernovae" (Filippenko 1997), reproduced with permission of Annual Reviews of Astronomy and Astrophysics, Volume 35, © Annual Reviews, http://www.annualreviews.org

An SN IIb gradually makes a transition from hydrogen-rich to helium-rich spectra, indicating that most of its hydrogen envelope was lost prior to explosion. SN IIb and SN Ib/c are often referred to as *stripped-envelope supernovae* (Clocchiatti et al. 1996a). Yet other events, SN IIn and SN Ibn, show distinct narrow circumstellar emission lines.

Spectral classification of supernovae is generally based on optical spectra taken near maximum light because for a long time spectra were restricted to the optical, and to the phase of highest flux and signal-to-noise ratio. Supernovae do, however, have characteristic features in the near-UV and near-IR that can extend or complement optical spectra.

1.4 Photometry

Light curves are used to constrain fundamental physical quantities such as the total ejected mass, the mass of synthesized, unstable ^{56}Ni, the kinetic energy of the ejecta, and the radius of the progenitor star. Most photometric measurements of supernovae are made in standard broad bands: U, B, V, R, I or u, g, r, i, z in the optical part of the spectrum (Fig. 1.4) and often J, H, and K in the near-IR. A *bolometric* light curve, tracking the total luminosity of a supernova, is fundamentally related to the physics of the explosion. In principle, the term bolometric implies an integration over all

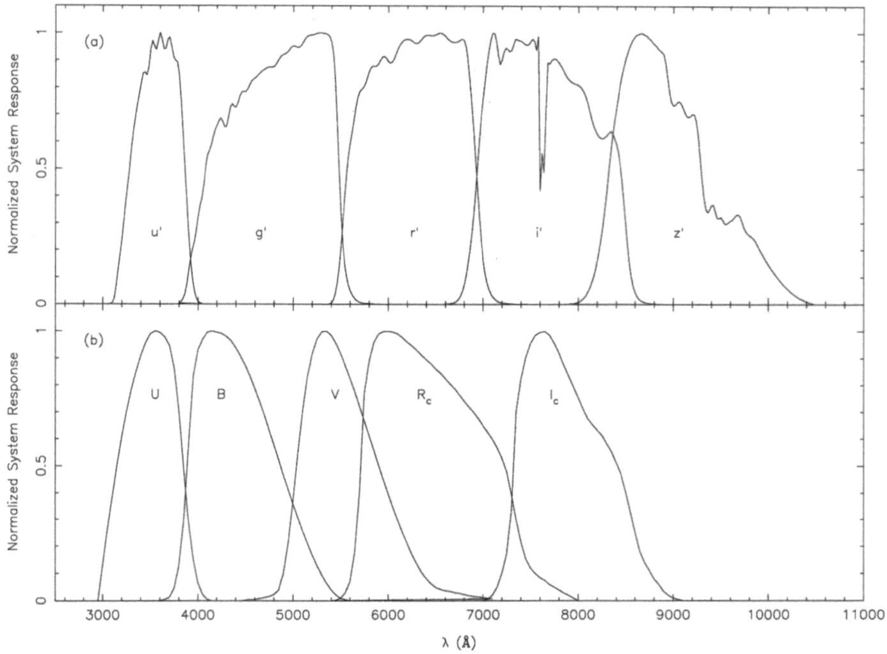

Fig. 1.4 Normalized responses of the ugriz (*top*, **a**) and UBVRI (*bottom*, **b**) bandpasses. The primes on the ugriz bands and the subscripts on the R and I bands are often omitted. From "The $u'g'r'i'z'$ Standard-star System" (Smith et al. 2002). © AAS. Reproduced with permission

wavelengths, but in the context of supernovae, the term often is synonymous with *UVOIR*—just near-UV, optical, and near-IR, combined. This limited integration is sometimes called "quasi-bolometric." The UVOIR wavelength regions encompass the thermal emission at temperatures of order 10^4 K that typically characterize supernova emission. A UVOIR light curve is constructed by suitably combining broad-band measurements. Emission at far-UV, X-ray, γ-ray, mid- and far-IR, and radio wavelengths is produced in other ways, and usually is small (although informative) compared to the UVOIR emission. This non-UVOIR flux is not always small, however, so we will maintain the distinction between bolometric and UVOIR. With estimates of distance and interstellar extinction, the apparent magnitude at the time of maximum brightness can be converted to an absolute magnitude.

A classic study of SN I light curves (Barbon et al. 1973) was influential in leading to the eventual recognition of different subclasses of SN I and to the diversity among SN Ia alone. An SN Ia reaches its maximum brightness some 20 days after explosion when it has a luminosity $\sim 2 \times 10^{43}$ erg s^{-1}, approaching 10^{10} times that of the Sun. The energy radiated in a few months exceeds 10^{49} ergs, about what the Sun radiates in 10^8 years. The majority of SN Ia are quite homogeneous with respect to absolute brightness, light-curve shape, and other properties, but there are outliers, and small systematic variations among the properties even of typical SN Ia are critical for

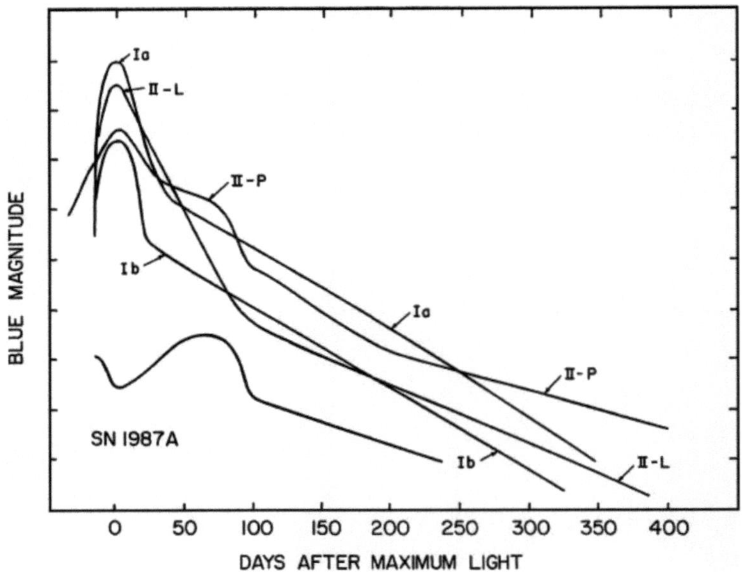

Fig. 1.5 A comparison of characteristic light curves. From "Introduction to Supernovae" (Wheeler 1990)

quantitative cosmology, including discovering and probing the nature of the dark energy that drives the cosmic acceleration.

Supernovae of other types display more observational diversity. Most of them are less luminous at their peaks than SN Ia although very bright exceptions have been found. A study of the absolute-magnitude distributions of the various types (Richardson et al. 2014) yielded the following mean values of M_B at maximum light (for $H_0 = 70\,\mathrm{km\,s^{-1}\,Mpc^{-1}}$): SN Ia: -19.3; SN IIP: -16.7; SN IIL: -17.9; SN IIn: -18.7; SN IIb: -17.0; SN Ib: -17.4; SN Ic: -17.3. These values have characteristic uncertainties of 0.3 magnitudes.

Characteristic light-curve shapes of the various supernova types are compared in Fig. 1.5. The light curve of an SN Ia consists of an initial rise and fall that lasts some 40 days, followed by a more slowly-fading tail. Light curves of SN Ib/c are qualitatively similar although usually dimmer than those of SN Ia. SN Ia and SN Ib/c light curves are powered by the radioactive-decay chain ^{56}Ni (half-life 6.1 days) through ^{56}Co (77 days) to stable ^{56}Fe. An early suggestion that a supernova light curve could be powered by this chain was by Pankey (1963).

A subclassification of SN II is based on light-curve shape. Light curves of most SN II have a postpeak *plateau* phase of nearly constant brightness for up to 3 months; these are designated SN IIP. Other SN II show an initial decline from peak that is nearly *linear* in magnitudes, with little or no plateau; these are SN IIL. Whether SN IIP and SN IIL are physically distinct or two extremes of some continuous distribution is not clear. Yet other SN II, particularly SN IIn, decline very slowly and often have a substantial contribution to their luminosity from the collision of their ejecta with CSM. Most SN II eventually undergo a linear (in magnitudes) tail phase, usually corresponding closely to the 77-day half-life of ^{56}Co.

SN 1987A had strong hydrogen lines and thus was an SN II, but it was subluminous and its light curve was of unusual shape. At discovery, although the optical brightness was increasing, the UVOIR luminosity was decreasing. After sudden shock heating, the ejecta were expanding and cooling adiabatically in what is sometimes called the *fireball* phase, by analogy to terrestrial nuclear explosions. The subsequent main peak was powered by radioactive decay. The decline from maximum was like the decline of some SN IIP from their plateaus, hence it has a connection to that type of event. SN 1987A was observationally conspicuous only because of its proximity; in more remote galaxies such subluminous events are difficult to detect. This illustrates the importance of a strong selection effect: in most observational samples, subluminous supernovae are under-represented relative to more luminous ones.

As searches have expanded, supernovae have been found that extend the limits of the usual photometric ranges (Fig. 1.6). Some are rare but exceptionally luminous, 10–100 times brighter than normal supernovae. Most of these are probably explosions that are interacting strongly with previously expelled CSM, but in some

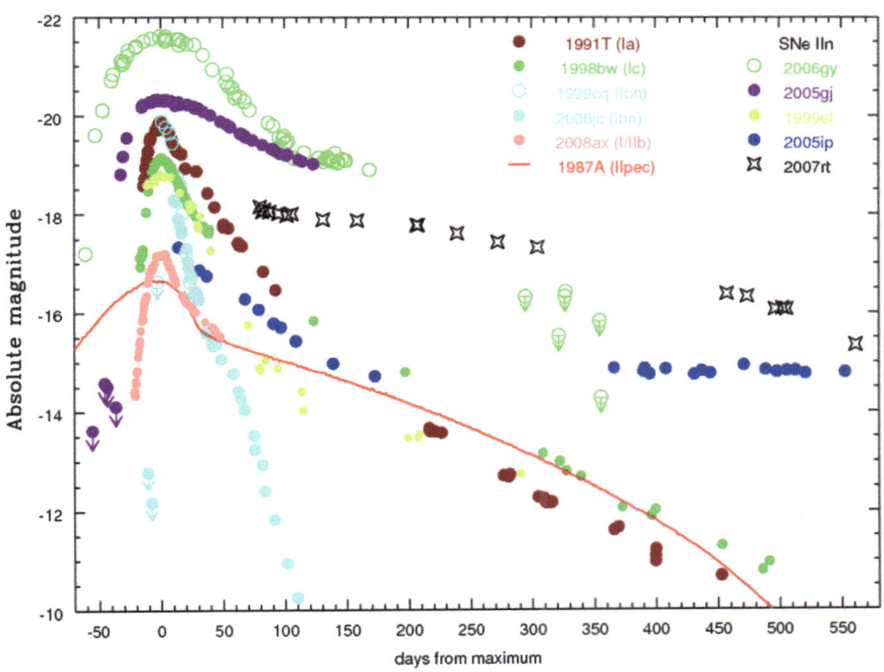

Fig. 1.6 A comparison of supernova light curves. From "Possible evidence of asymmetry in SN 2007rt, a type IIn supernova" (Trundle et al. 2009), reproduced with permission, © ESO

cases more exotic processes may be involved. Others are exceptionally dim and have been attributed to the formation of black holes with little ejection of radioactive isotopes to power the light curve, or to episodic expulsions of matter that are not true supernova explosions, or to other variations on the usual themes.

A photometric *color* is the difference between two broad-band magnitudes and thus a measure of the ratio of the fluxes in the two bands. Colors provide coarse information on the *spectral energy distribution* (SED). Typically SN Ia and SN II have $B - V \simeq 0.0$ near maximum light, similar to the color of a blackbody of temperature $\sim 10^4$ K. SN Ib/c are somewhat redder because their photospheres are cooler. For some time after maximum light, supernova colors become redder as the ejected matter cools.

1.5 Spectral Evolution

Spectra provide otherwise unattainable information on the composition and velocity of the ejected matter. While spectral classification is usually based on features observed near maximum light, the spectra of supernovae vary with physical conditions and hence with time. This spectral evolution is a valuable tool employed in the understanding of supernovae. Figure 1.7 shows the evolution of the spectra of a Type IIP supernova. Examples of other types will appear in appropriate chapters. As with light curves, a common practice is to consider UVOIR and other wavelength regions separately.

At early times, within months of explosion when the ejecta are optically thick, supernovae are in the *photospheric* phase. The ambient radiation field dominates the ionization and excitation state of the gas. The spectrum consists of a thermal continuum with superimposed line features having P Cygni profiles. The Doppler-broadened P Cygni features vary with time and provide a wealth of information.

The premaximum phase of supernovae is under-observed because many supernovae are still not discovered until near maximum light. Premaximum spectra of SN Ia show many of the same spectral features that become more distinct near maximum light, and the premaximum phase can also reveal high-velocity ($\sim$20,000 km s^{-1}) features that often disappear by maximum light. Premaximum spectra of SN II tend to show thermal continua with only weak hydrogen lines. They sometimes show transient features caused by interaction with CSM. Relatively few premaximum spectra of SN Ib/c have been obtained, but this is changing as more thorough searches allow more early discoveries.

Most supernova spectra are obtained near and somewhat after maximum light, the phase that is the primary basis for spectral typing. At maximum light, the photosphere is in outer layers of the ejecta. As the photosphere recedes deeper into the ejecta, crucial information on the composition structure (relative abundances versus ejection velocity) is revealed. During photospheric phases, the distinction between the velocity *at* and *of* the photosphere must be maintained. A single spectrum can reveal the velocity of matter currently *at*, not *of* the photosphere. As

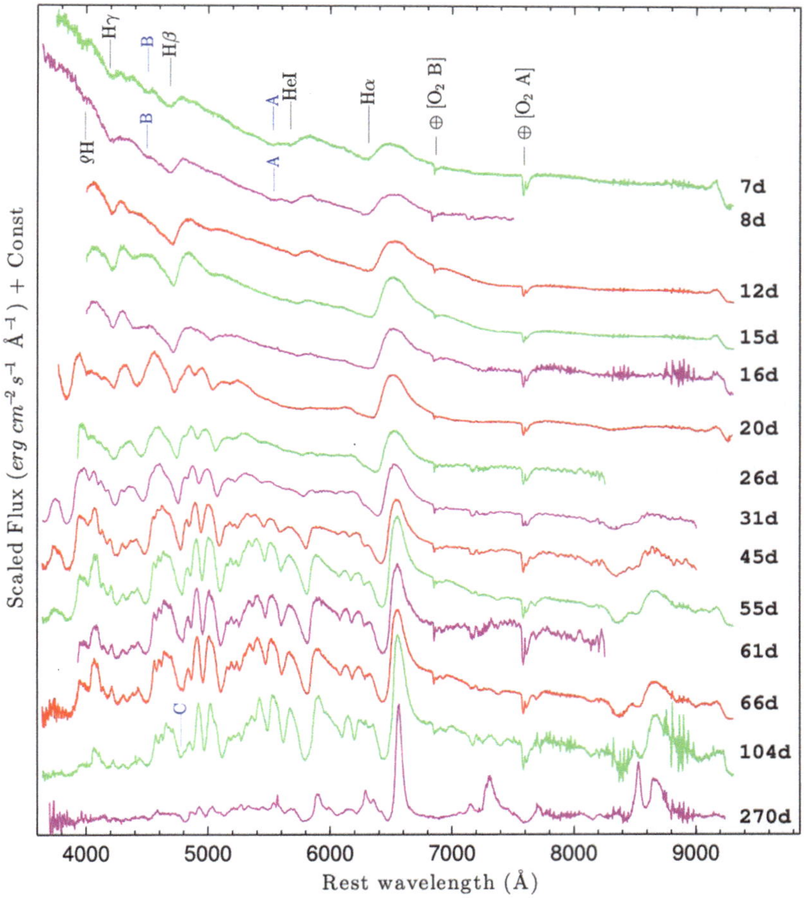

Fig. 1.7 Spectral evolution of the Type IIP SN 2012aw. Epochs are with respect to the estimated time of explosion. The $[O_2]$ absorptions form in Earth's atmosphere, not in the supernova. The characteristic line shapes evolve from broad P Cygni profiles to narrow, nearly symmetric emission lines. Figure 10 "Supernova 2012aw—a high-energy clone of archetypal Type IIP SN 1999em" from Bose et al. (2013), by permission of Oxford University Press on behalf of the Royal Astronomical Society

the ejecta expand, the column density and hence optical depth decline. When the ejecta become optically thin, they are in the *nebular* phase. The continuum is weak, and the spectral features are typically collisionally-excited forbidden emission lines centered on the rest wavelengths of the atomic transitions. As remarked above, SN Ia show distinctly different nebular-phase spectra from SN II and SN Ib/c, which tend to be similar apart from the permitted hydrogen lines in SN II. With their strong circumstellar interactions, SN IIn can display emission lines throughout their evolution. In the limits of early and late times, the terms photospheric and nebular are apt, but of course the transition from one phase to the other is gradual.

1.6 Explosion Mechanisms

Theoretical understanding of the processes leading to supernova explosions have proceeded in parallel with observational studies. Shortly after the discovery of the neutron, the potentially important role of neutron-star formation in triggering a supernova explosion was perceived by Baade and Zwicky (1934). Two of the basic processes that are still at the heart of the subject nearly 60 years later were elucidated by Hoyle and Fowler (1960).

One explosion mechanism operates when the core of a star reaches conditions of high electron degeneracy, for which thermonuclear burning is thermally and dynamically unstable. The star is consumed by the resulting supersonic shock wave, a process known as *detonation*, or by a rapid, but subsonic, combustion termed *deflagration*, or perhaps a mix of the two. Such an explosion tears the star apart, leaving no compact remnant. The environment usually considered for such an explosion is a dense evolved stellar core, a *white dwarf*. Stars of zero-age main-sequence (ZAMS) mass lower than about 8 $M_\odot$ become white dwarfs. In a binary stellar system, the hydrogen envelope can be lost by transfer to the companion star or by loss of mass entirely from the system. Later, if mass is added back onto the white dwarf, the white dwarf may increase in mass and shrink in volume until it reaches a critical condition of density and temperature and explodes. Since the white dwarf may be old, and will have hardly any hydrogen on its surface when it explodes, this type of model has long been suggested as the origin of SN Ia. Traditional models of SN Ia are accreting white dwarfs composed of carbon and oxygen that approach 1.4 $M_\odot$, when ignition of carbon-burning initiates the instability. Variations on this theme, including mergers of binary white dwarfs, are under consideration.

The other major mechanism to create a supernova explosion is core collapse. A star of ZAMS mass in excess of about 8 $M_\odot$ develops a core that implodes rather than forming a white dwarf. The best-studied cases involve models with ZAMS mass in excess of about 12 $M_\odot$, which develop iron cores. Iron has the highest binding energy per nucleon of any common element, so it can undergo only endoergic reactions that remove energy and pressure support from the star, leading to its collapse. An important variation on this theme is predicted to occur in stars of more modest ZAMS mass, about 8–12 $M_\odot$, which undergo non-explosive burning of nondegenerate carbon, but then form degenerate cores of oxygen, neon, and magnesium. These ONeMg cores are susceptible to electron capture, which leads to core collapse but with quantitative differences from iron-core collapse. A variation on this theme is binary-star mass accretion onto an ONeMg white dwarf, which is expected to lead to accretion-induced collapse (AIC).

A core collapse produces copious neutrino emission. Stirling Colgate recognized the critical role played by the neutrinos, one that we are still attempting to understand 50 years later (Colgate and White 1966). The collapsing sphere of neutrons halts and bounces at nuclear densities, unless it is too massive and collapses directly to a black hole. Although the details are not yet resolved, the process of neutron-star or black-hole formation liberates ∼100 B of gravitational energy, a hundred times more than necessary to eject the outer layers of the star as a supernova. Most of the energy is carried away by the neutrinos, to which the outer stellar layers are transparent; only about 1% needs to be deposited in the star to cause a healthy explosion. The ZAMS mass of the SN 1987A progenitor was about 15–20 $M_{\odot}$, well within the mass range predicted to undergo iron-core collapse. Detection of the burst of neutrinos from SN 1987A represented the birth of extragalactic neutrino astronomy and gave a remarkable confirmation of the basic theoretical picture of core collapse.

A number of lines of evidence show that SN Ib/c result from massive stars, so their evolution and the mechanism of their explosion is quite distinct from the white dwarf evolution and thermonuclear explosion thought to power SN Ia. The primary argument is that SN Ib/c are associated with regions of recent star formation, statistically suggesting that they arise in short-lived and hence initially massive stars. Model fits to light curves of some SN Ib/c demand more mass than can be readily attributed to a white dwarf explosion. The similarity of the nebular spectra of SN Ib/c to those of SN IIP is another strong argument. Similar arguments pertain to SN IIb. With the exception of SN 1987A, no direct evidence for core collapse, for instance in the form of a new-born neutron star, has been obtained for any extragalactic supernova, but the circumstantial evidence that SN IIb and SN Ib/c result from core collapse is strong.

For some rare events it is uncertain whether core collapse or explosion of or on a white dwarf is the operative mechanism. For example, the origin of SN IIn is somewhat ambiguous. For some of these, large masses seem to be involved. Other SN IIn, however, have underlying broad-line spectra that resemble those of SN Ia and might be thermonuclear explosions of white dwarfs inside dense circumstellar shells.

A third explosion mechanism may pertain to very massive stars, of ZAMS mass $\gtrsim$100 $M_{\odot}$. In these, helium burning produces an oxygen-rich nondegenerate core that becomes so hot that electron–positron pairs are spontaneously generated, effectively softening the equation of state and leading to contraction and heating of the oxygen core. Explosive burning of oxygen is predicted to lead to a strong thermonuclear explosion with total disruption of the star—a *pair instability* explosion. This may have been the fate of some the first stars to have been born after the end of the cosmological *dark ages*, stars that may have tended to be especially massive. A few of the contemporary superluminous events have also displayed some characteristics expected of pair-formation explosions.

Other mechanisms have been proposed to power supernovae, such as energy input from a new-born pulsar or a transition from a neutron star to a quark star. We will not pursue these in detail, but will mention and reference them as appropriate and encourage further work by those so committed.

1.7 Asymmetries and Polarization

Asymmetries are revealed by direct imaging of the ejecta of SN 1987A and some young Galactic SNRs, although issues of dust absorption and aspherical X-ray illumination complicate interpretation of the data. From the ring structure surrounding SN 1987A, we also know that the progenitor star's presupernova mass loss manifested substantial asymmetries.

A new window into the nature of supernova asymmetries was opened in the late twentieth century with the study of linear polarization of supernova light, especially spectropolarimetry. Light from a spherical explosion would not show intrinsic net polarization because polarized intensity beams from different parts of the projected surface would cancel, but asymmetry on the plane of the sky, whether global shape asymmetry of the ejecta or smaller-scale "clumping" of ejecta (composition or density clumping, or both) does produce net polarization. The power of spectropolarimetry is the capacity to measure asymmetry as a function of wavelength and hence to obtain information about the shape of the ejecta as a function of its composition.

Linear polarization in both lines and continuum is typically observed, but only at a level of a percent or less. Such observations are informative, but require substantial investments of telescope time and at present are feasible only for observationally bright supernovae, typically brighter than 18th magnitude even when using 8 to 10-m class telescopes. These data are more difficult to acquire and to interpret than ordinary flux spectroscopy, but certain trends have been revealed that will be outlined here. There has not yet been any useful circular polarimetry of supernovae.

The major trends are that SN Ia tend to show asymmetric features during the premaximum phase, but little or no detectable asymmetry near and after maximum light. This suggests that the outer layers of the ejecta contain significant asymmetries, but the inner layers less so. Asymmetries of nebular lines in the flux spectra of SN Ia do show some evidence for asymmetry in the deeper layers. Core-collapse events tend to behave the opposite way, with the degree of polarization increasing with time, suggesting that the inner layers of ejecta are more asymmetric. There is a tendency for polarization to imply a favored axis or direction to the ejecta, but there also is evidence for departures from axial symmetry. Some gamma-ray bursts are associated with core-collapse explosions. The energetic flow that drives emission of γ-rays is collimated, so the underlying process is presumed to be strongly aspherical.

1.8 Sites, Environments, and Rates

After a supernova has been observed, it is sometimes possible to find deep *archival* images of the site, serendipitously obtained prior to explosion. These have been used in some cases to detect and characterize the star that exploded, e.g., the progenitor of the Type II SN 1987A was a blue supergiant and that of the Type IIb SN 1993J was a yellow supergiant. More often, pre-explosion images provide only upper limits to the luminosity and therefore, by means of stellar-evolution theory, to the ZAMS masses of the progenitors. Pre-explosion and post-explosion images are also used to search for a former companion star (e.g., a companion has been detected in post-explosion images of SN 1993J) and to characterize the stellar population in the general vicinity of the explosion. Such images also are used to improve the accuracy of the final supernova photometry by allowing accurate subtraction of background light.

Because direct detections of progenitors are infrequent, statistical information about supernova sites is used to make inferences about the progenitors of the various types. Spiral and irregular galaxies contain both young and old stars, while elliptical galaxies have only old ones. SN II, SN Ib, and SN Ic are found almost exclusively in spirals and irregulars, and in spirals they tend to occur near the spiral arms, where young stars are concentrated. This is a clear sign that these types are produced by stars that are initially massive. Those formed with masses $\gtrsim 8\ M_\odot$ have nuclear lifetimes so short, $\lesssim 3 \times 10^7$ years, that their cores collapse before drifting far from where they were born. SN Ia are found in all kinds of galaxies and those in spirals show little if any tendency to concentrate to the arms. This indicates that they are produced by stars with ZAMS masses $\lesssim 8\ M_\odot$.

To estimate the relative occurrence rates of the different supernova types in the various types of galaxies, several observational effects must be taken into account. Most important is the discovery bias in favor of events of high luminosity and broad light curves. For example, because of their high peak luminosities, SN Ia are the majority type in magnitude-limited observational samples, but they actually occur less frequently than SN II. Another significant bias is the difficulty of discovering supernovae in spiral galaxies the disks of which are highly inclined to our line of sight. When these and additional effects are taken into account, it is found that in a volume-limited sample of a range of galaxy types, roughly half of all supernovae are SN II, a quarter are SN Ib/c, and a quarter are SN Ia. In spirals, the rates of all types are correlated with galaxy color: the bluer the galaxy (and therefore the higher the star-formation rate per unit galaxian mass during the last 10^9 years), the higher the supernova rate per unit mass.

Elliptical galaxies produce SN Ia at a lower rate than spirals do. Ellipticals mostly stopped forming stars more than a billion years ago and now contain few stars much more massive than the Sun, so SN Ia in ellipticals are produced by stars of relatively low ZAMS mass. The correlation between the SN Ia rate and the blueness of spirals implies, however, that most SN Ia in spirals are produced by stars that are moderately short-lived and have ZAMS masses considerably more massive than the Sun (but $\lesssim 8\ M_\odot$). Single stars of ZAMS mass less than $8\ M_\odot$ are expected to become white dwarfs and never explode. An SN Ia is thought to result when a white dwarf accretes matter from a binary companion star or merges with it. This can account for both the low but nonzero SN Ia rate in ellipticals and the correlation of the SN Ia rate with galaxy color in spirals.

Absolute rates are more uncertain than relative rates because the relevant selection effects are still more difficult to evaluate. When rates of supernovae in other galaxies are interpolated to our Galaxy, assumed to be of type Sbc, the total rate of Galactic supernovae is about 2.8 per century, for a mean interval between supernovae of 36 years.

1.9 Circumstellar Interaction

For core-collapse supernovae, circumstellar interaction (CSI) can be a significant source of radiation in the UVOIR range and is a major source of non-UVOIR radiation. The prime examples of CSI are SN IIn, some of which fall in the superluminous category. Circumstellar interactions are *radiative* when supernova photons are absorbed or scattered by CSM, and *hydrodynamical* when supernova ejecta collide with CSM. Hydrodynamical interaction produces a hot interface that can generate radiation over a wide range of wavelengths; this in turn produces further radiative interactions and non-UVOIR emissions. X-ray thermal Bremsstrahlung and inverse-Compton emission has been detected from more than a dozen supernovae, and synchrotron radio radiation from several dozen; enhanced optical and ultraviolet emission also can occur. In the photospheric and nebular phases, emission from CSI usually is weak compared to UVOIR emission from the ejecta, but in extreme cases the opposite is true. In all cases, observation of CSI is informative because it constrains the distribution of CSM and hence provides the best way to probe the mass-loss history of progenitor stars in their final phases of evolution. For a few radio emitters, very-long-baseline radio interferometry has spatially resolved the interaction regions and directly measured expansion rates.

UVOIR emission includes near-IR thermal emission from the photosphere, but dust grains far outside the photosphere, warmed by absorbing the dilute UVOIR radiation, can emit in the mid- and far-IR as well as at submillimeter wavelengths. This radiation is valuable for probing the amount of circumstellar dust that was present at the time of explosion and the amount of dust that condensed in the ejecta. Dust also scatters a fraction of the incident radiation to produce *light echoes*, which help to probe the distribution of CSM and ISM.

Emission from some decades-old, circumstellar-interacting core-collapse supernovae of the twentieth century, e.g., SN 1923A and SN 1957D, is still observable. These are sometimes referred to as historical extragalactic supernovae.

Evidence of CSI in SN Ia is much more elusive, but has been found in some cases. This evidence is especially crucial since it can help to identify the nature and evolution of the presumed underlying binary progenitor system.

1.10 Supernova Remnants

Matter ejected by a supernova sweeps up CSM and then ISM. When the mass of the swept matter becomes comparable to the mass of the ejecta, the ejecta begin to decelerate and kinetic energy is converted to other forms of energy. A hot region of interaction develops, and an extended *shell*-type SNR forms (Fig. 1.8). Cas A and the remnants of SN 1006, SN 1572 (Tycho Brahe's supernova), and SN 1604 (Johannes Kepler's) are well-observed examples of young shell-type SNRs, and most older Galactic remnants have shells. Radio emission from shells is synchrotron radiation produced by relativistic electrons accelerated in the hot interaction region. Almost half of the nearly 300 known Galactic SNRs are observed in X-rays, and almost a quarter at optical wavelengths. Nondetections are primarily caused by efficient absorption of X-rays and optical light by intervening ISM. Measurements of γ-rays at TeV energies verify that particle acceleration processes are occurring at the shock fronts of young SNRs. There is evidence that hadrons as well as electrons are accelerated to cosmic-ray energies by this process.

Fig. 1.8 An ultraviolet image of the Cygnus loop, the shell-type remnant of a supernova that occurred in the Galaxy about 6000 years ago. Credit: NASA/JPL-Caltech

The Crab Nebula, the remains of SN 1054, is a less common type of SNR. Synchrotron emission from the central regions, extending from radio to X-rays, requires continual injection of highly relativistic electrons. The source is a pulsar. The Crab neutron star rotates 30 times per second. Crab-like remnants are called *filled center*, or *plerions* (from the Greek word for full). *Composite* SNRs have both shells and filled centers. More than 10% of Galactic SNRs are composite while only a few percent are filled-center without associated shells.

Multiband studies of spatially-resolved SNRs in our Galaxy and its satellite galaxies provide information about the kinematics of the explosion, the composition structure of the ejecta, the CSM, the nature of the progenitor star or its companion, and the presence or absence of a collapsed central object. Unfortunately, the exact type of the supernova that produced the SNR is usually not known. Remarkably, modern observations of optical light echoes—supernova light that reflected off interstellar dust and reached Earth delayed by centuries in time owing to the longer path length—have revealed that the time-integrated spectrum of the Cas A supernova was similar to that of the Type IIb SN 1993J, and that Tycho's supernova was a typical Type Ia.

Young remnants of core-collapse supernovae are generally observed to be asymmetric. An observation of Cas A with the *Chandra* observatory revealed a bipolar structure in the silicon-rich ejecta and *Chandra* X-ray images of the Crab Nebula reveal asymmetries associated with the ejecta and the pulsar wind nebula. Young remnants of SN Ia also show departures from spherical symmetry. The overall morphology of SN 1572 and SN 1006, two remnants of SN Ia, is roughly circular, but both show evidence of a turbulent "broccoli/cauliflower" flow in the ejecta.

1.11 Gamma-Ray Bursts

Cosmic gamma-ray bursts are usually divided into those of short ($\lesssim 2$ s) and long ($\gtrsim 2$ s) duration. Sometimes they are called long-soft and short-hard, because the short bursts have higher mean photon energies. In long-soft bursts, γ-rays are inferred to be generated by highly-collimated relativistic flows—*jets*. The jets may be bipolar, but Lorentz beaming prevents the direct observation of any component propagating away from us. Interaction of the jets with ambient matter produces an *afterglow* at X-ray and longer wavelengths. Superimposed on the afterglow light curve, in some cases, is the light curve of a supernova. When spectroscopy has been available, the supernova has proven to be a member of a broad-lined subcategory of Type Ic, SN Ic-bl. Understanding the context of SN Ic, SN Ic-bl, and the long-soft gamma-ray bursts is a major enterprise in contemporary supernova research. The short-hard bursts may arise in mergers of binary neutron stars or magnetic jets associated with accretion-induced collapse, phenomena that are closely, if not directly, related to supernova research.

1.12 Summary

SN II, Ib, and Ic result from the gravitational collapse of highly-evolved cores of massive stars. SN Ia are thermonuclear explosions of carbon–oxygen white dwarfs. The nature of the progenitors of superluminous supernovae is not yet known. Advancing our understanding of supernovae relies primarily on interpretation of their photometric, spectroscopic, and polarimetric characteristics. Additional valuable information comes from the environments and occurrence rates of the various types of supernovae, and from SNRs.

Chapter 2
Search and Discovery

2.1 Introduction

Before supernovae can be studied, they must be found. Within redshift $z \lesssim 1$, supernovae occur at a rate exceeding one per second, but so far astronomers find only a small fraction of them, $\sim 10^{-5}$.

From 1885 to the mid-1930s, observations of nearby galaxies happened to reveal a few supernovae. Then came Fritz Zwicky. For an engaging account of early systematic searches begun by him in the 1930s that involved Schmidt telescopes, films and binoculars, glass plates and blink comparators, see Zwicky (1974).[1] Zwicky, members of an international search team organized by him, and their successors made many of the discoveries of the following four decades. Amateur astronomers also made, and continue to make, valuable discoveries. Of nearly mythological status in this area is the Reverend Robert Evans, who relied on his near-photographic memory of the appearance of thousands of bright galaxies to visually spot new supernovae, often before the time of maximum light (Evans 1994).

Ever more productive modern searches (Fig. 2.1) have turned from plates and films to CCD detectors and from labor-intensive search techniques to software pipelines and even robotic telescopes. Most discoveries continue to be made at optical wavelengths. Many searches have in common the goal of discovering supernovae before the time of maximum brightness, to obtain information on the outer layers of the ejecta as well as evidence of early CSI. Modern searches commonly employ the technique of accurate image subtraction. An image that contains the supernova is compared to an archival (presupernova) image of the site; the difference is the supernova.

[1]This is an opportunity to make a minor correction to the printed record in Zwicky (1974). Following his oral presentation, one question was asked. According to the record, Zwicky's response was followed by a statement by the questioner (D. Branch), but all words following the question were spoken by Zwicky.

D. Branch, J.C. Wheeler, *Supernova Explosions*, Astronomy and Astrophysics Library, DOI 10.1007/978-3-662-55054-0_2

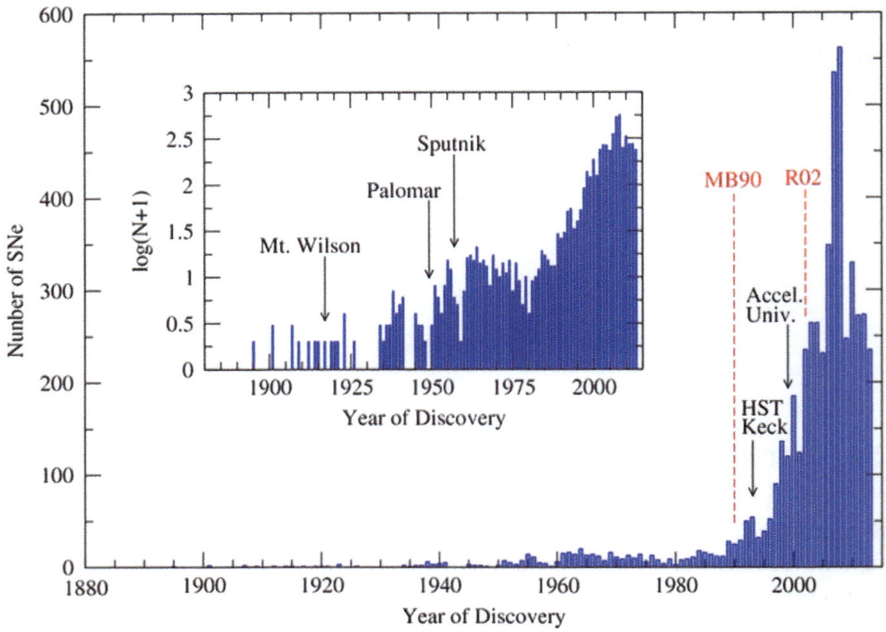

Fig. 2.1 Number of supernovae discovered versus time. The *inset* shows the same data on a log scale. From "Absolute-magnitude Distributions of Supernovae" (Richardson et al. 2014). © AAS. Reproduced with permission

Different equipment and strategies are appropriate for different redshift ranges. Considerations include where to look—the *fields* and the field of view (FOV); the time interval between successive visits to the fields—the *cadence*; and the exposure time per visit, which together with telescope aperture determines the limiting magnitude—the *depth*. For fixed resources (equipment and observing time) tradeoffs must be made. The number of discoveries, N, depends strongly on the limiting magnitude m_{lim}. For the idealized case of supernovae distributed randomly in transparent Euclidean space, $N \propto 10^{0.6m_{lim}}$, i.e., each additional magnitude yields a factor of about four increase in N. Similarly, for a fixed m_{lim} and a sample of supernovae having the same light-curve shapes but different absolute magnitudes M, $N \propto 10^{-0.6M}$. The bias against supernovae of low luminosity is severe. Extremely luminous events also are problematic to discover, because they are rare, but see McCrum et al. (2015) on finding such events among samples of apparently hostless supernovae.

For all but the most remote supernovae, discovery does not require the largest of telescopes. An important component of a search is prompt photometric and, when feasible, spectroscopic followup—otherwise little will be known about a supernova other than that it occurred. Effort has been made over the last several decades to obtain at least one identifying spectrum, usually near maximum light, for most discovered supernovae (Tomasella et al. 2014; Smartt et al. 2015), but information on the premaximum and later nebular phases is inadequate. Deciding which discoveries to follow up is another important consideration. Followup may be performed with the discovery instrument or, especially for spectroscopy, with larger telescopes.

Traditionally, supernovae were named in order of discovery as, e.g., 2000A to 2000Z followed by 2000aa to 2000az, then 2000ba to 2000bz, and so on. In this book, such names are used when they are available; however many supernovae discovered in recent searches do not receive such designations. In these cases, we use the names assigned by the discovery teams, e.g., SN PTF11iqb (often referred to in the literature as just PTF11iqb), discovered and named by the intermediate Palomar Transient Factory (Sect. 2.3).

Complete lists of extragalactic supernovae are maintained at Harvard,[2] Padova[3] (Barbon et al. 1999), and Sternberg.[4] Amateur astronomer David Bishop maintains the useful Bright Supernova List.[5] The Padova and Sternberg lists provide information on the properties of the host galaxies in addition to those of the supernovae. A unification of these lists and a brief discussion of statistical properties of supernova discoveries was provided by Lennarz et al. (2012). For a more extensive discussion of discovery statistics for the particular years 2010 and 2011, see Gal-Yam et al. (2013). A very extensive collection of supernova data, the Open Supernova Catalog (Guillochon et al. 2017) is available online.[6]

In the following sections, several of the noteworthy supernova searches are briefly described.

2.2 Nearby Supernovae

Supernovae in nearby galaxies are bright enough that some are found soon after explosion, some are observed for a long period of time (e.g., a year), and some are detected at non-UVOIR wavelengths. Thus nearby events have historically been among the most valuable ones for supernova science. These include some discovered by Rev. Evans, Tom Boles, Tim Puckett and other amateur astronomers, often using CCD detectors on telescopes of impressive aperture.[7] To obtain a good crop of discoveries, a common practice is to conduct a *targeted* search of a preselected list of galaxies.

[2]http://www.cbat.eps.harvard.edu/lists/Supernovae.html.

[3]http://graspa.oapd.inaf.it/asnc.html.

[4]http://www.sai.msu.su/sn/sncat/.

[5]http://www.rochesterastronomy.org/snimages/.

[6]https://sne.space.

[7]See, for example, the Puckett Observatory World Supernova Search (http://www.cometwatch.com).

The Lick Observatory Supernova Search[8] (*LOSS*; Filippenko et al. 2001) has performed a successful search of nearby galaxies in the northern sky since 1997. *LOSS* employs the fully robotic 0.76-m Katzman Automatic Imaging Telescope with a CCD camera having a FOV of 46 square arcminutes (large enough to include a nearby galaxy) to survey some 5000 galaxies with radial velocities less than about 5000 km s^{-1} at a cadence of 3–4 days. During a clear winter night, *LOSS* obtains more than 1200 images. The telescope control system checks the weather, opens the dome, points to the desired galaxies, finds and acquires guide stars, exposes, stores the data, and manipulates it to search for new supernova candidates. Only then do humans examine the images. The search limit is an unfiltered $m_{lim} \simeq 19$. In some years, *LOSS* has found nearly 100 bright supernovae at a mean $z \simeq 0.024$. For selected supernovae, the search telescope also obtains followup multiband photometry.

The CHilean Automatic Supernova sEarch (*CHASE*; Pignata et al. 2009) conducts a complementary search for supernovae in nearby galaxies in the southern sky using four fully robotic 40-cm telescopes. The FOV is 100 square arcminutes. *CHASE* surveys about 250 galaxies within a radial velocity of 3000 km s^{-1} with a cadence of 3–4 days. *CHASE* discovers dozens of supernovae per year, at a mean $z \simeq 0.023$. The search telescope also obtains photometry. The Carnegie Supernova Project (*CSP*; Hamuy et al. 2006; Folatelli et al. 2010) and other initiatives obtain additional optical photometry as well as near-IR photometry, spectroscopy, and polarimetry.

Some supernovae found by the exoplanet-finding Kepler mission and its successor mission K2[9] are discovered shortly after explosion and photometry is obtained at an extremely high cadence of 30 min (Olling et al. 2015).

Targeted searches generally focus on luminous galaxies because within a fixed galaxy type larger galaxies are expected to produce more supernovae, and observers want to look where supernovae are likely to be found. The price to be paid is a bias in favor of the kinds of supernovae that occur in luminous galaxies. There are growing indications for interesting populations of supernovae in star-forming dwarf galaxies, which are underrepresented in traditional targeted searches. There is also a lingering question of whether intergalactic supernovae exist (Zinn et al. 2012); targeted searches are basically mute on this issue. Targeted searches tend to include other biases, sometimes deliberately, e.g., excluding central regions of galaxies to avoid false alarms due to variable AGNs. Another limitation of targeted searches is that with their traditionally modest FOVs they cannot efficiently search the nearest rich clusters of galaxies. The clusters, especially the Virgo cluster, are too spread out on the sky to effectively monitor many galaxies at once, and there are too many galaxies to be targeted individually.

---

[8] w.astro.berkeley.edu/bait/kait.html.

[9] http://keplerscience.arc.nasa.gov/K2/.

The *unbiased* (non-targeted) All-Sky Automated Survey for Supernovae[10] (*ASAS-SN*; Shappee et al. 2014) begun in 2014, has, unlike most searches, the virtue of prompt data release rather than proprietary delay. The entire extragalactic sky is surveyed to $m_{lim} \simeq 16$ every two to three nights. By mid-2016 ASAS-SN had discovered about 400 supernovae in nearby galaxies.

2.3 Hubble-Flow Supernovae

Relative distances of supernovae in nearby galaxies usually have significant uncertainties because galaxy peculiar velocities, $\sim$200 km s^{-1}, are not negligible compared to the low galaxy recession velocities caused by cosmic expansion. For these galaxies, the use of redshift and the Hubble law does not provide accurate distances. To obtain a sample of supernovae that do have accurate relative distances, optimal for determining absolute-magnitude distributions of the supernova types and other purposes, one wants to reach into the *Hubble flow* ($cz \gtrsim 3000$ km s^{-1}, $z \gtrsim 0.01$). Searches at these and higher redshifts have been motivated mainly by the use of SN Ia as distance indicators for cosmology. In the Hubble flow, the exposures need not be extremely deep, but a wide FOV is required to survey a sufficient number of galaxies and achieve a satisfactory discovery rate.

The Calán/Tololo Supernova Search (Hamuy et al. 1993) was the last important search to use photographic plates for discovery observations. The search employed the Curtis Schmidt Camera at the Cerro Tololo Inter-American Observatory, which provided a very wide FOV of five by five degrees. The search surveyed several dozen fields that contained large numbers of galaxies. From 1990 to 1993, 32 supernovae were discovered in the range $0.01 \lesssim z \lesssim 0.1$, and CCD photometric coverage was obtained (Hamuy et al. 1996b). The historical significance of this search for the discovery of the acceleration of the cosmic expansion was discussed by Hamuy and Suntzeff (2015).

Unbiased searches can avoid some of the issues of targeted searches. One of the first efforts in this direction was the Texas Supernova Search (TSS; Quimby et al. 2005), which used the *ROTSE-IIIb* telescope at McDonald Observatory, originally designed as one of four robotic telescopes for the rapid discovery and monitoring of gamma-ray bursts (Akerlof et al. 2003). Each telescope had a FOV of 1.85 by 1.85 degrees. The wide FOV provided the capacity to image the entire nearby Virgo, Ursa Major, and Coma clusters of galaxies. The Virgo cluster could be covered in about 50 60-s images in a single night. Candidates were followed up with the Hobby–Eberly Telescope for spectral confirmation. The original goal was to discover supernovae in the nearby clusters at very early times, but serendipity brought distant supernovae of types never seen before. Beginning in 2004, the TSS conducted a nearly unbiased optical census of the transient sky down to unfiltered

[10]http://www.astronomy.ohio-state.edu/assassin/index.shtml.

$m_{lim} \simeq 18.5$. Within the first few years of operation, the project spanned an effective volume of about $10^9\,\mathrm{Mpc}^3$. In 2007, the project was extended as the ROTSE Supernova Verification Project (*RSVP*).[11] This search is unbiased with respect to the luminosity of supernova host galaxies and the central regions of galaxies. The project discovered about 20 supernovae per year at a mean $z \simeq 0.05$, a modest total contribution, but about 10% of the discoveries were among the most peculiar ever found, including five of the first six superluminous supernovae (SLSN; Chap. 18). SN 2006gy, the prototype of the new category of SLSN, appeared near the center of a mildly active galactic nucleus. Other discoveries were in dwarf galaxies that would have been overlooked by traditional targeted searches.

The unbiased search of the Nearby Supernova Factory[12] (NSF; Aldering et al. 2002; Copin et al. 2006) concentrated on the range $0.03 \lesssim z \lesssim 0.08$. The NSF relied on images obtained by the Near Earth Asteroid Tracking Team of NASA's Jet Propulsion Laboratory, which uses wide-field CCD mosaic imagers on two 1.2-m telescopes at Palomar Observatory. Each field included about 100 galaxies, the cadence was about 6 days, and $m_{lim} \simeq 22$. From 2003 to 2008 NSF found more than 600 supernovae. Followup of selected events (usually SN Ia) with a two-channel integral-field-unit optical spectrograph imager on a 2.2-m telescope gave accurate spectrophotometry from 3200 to 10,000 Å, from which broadband magnitudes could be synthesized without performing separate photometry.

The intermediate Palomar Transient Factory[13] (iPTF; Rau et al. 2009; Law et al. 2009), successor to the Palomar Transient Factory (PTF), is a fully automated, wide-field, unbiased survey aimed at systematic exploration of the optical transient sky with a focus on, but not a limitation to (Kulkarni 2012), supernovae. The survey uses an 8.1 square degree camera installed on the 48-in. Samuel Oschin telescope at Palomar Observatory. The automated Palomar 60-in. telescope obtains photometry. An exposure of 60 s reaches $m_{lim} \simeq 21.3$. The supernova search has a cadence of 5 days. Deep and wide-field imaging extends this sort of unbiased search in a major way. By mid-2015, the iPTF had discovered 2500 supernovae at mean $z \simeq 0.09$.

The Catalina Real-time Transient Survey[14] (*CRTS*; Drake et al. 2009, 2012; Mahabal et al. 2011) uses data obtained by the Catalina Sky Survey (*CSS*), the chief goal of which is to search for Near Earth Objects. The *CSS* 0.6-m Schmidt telescope is located on Mt. Bigelow in the Catalina Mountains of Arizona. The *CSS* typically covers about 800 square degrees of sky in a single night. By mid-2014, the *CRTS* had discovered more than 2000 supernovae at a mean $z \simeq 0.06$.

For 9 months between 2005 and 2007, the Sloan Digital Sky Survey[15] (*SDSS*; Frieman et al. 2008; Kessler et al. 2009) used the dedicated 2.5-m telescope at Apache Point, New Mexico, to conduct an unbiased search for supernovae in the

[11] http://www.rotse.net/rsvp/.

[12] http://snfactory.lbl.gov/.

[13] http://www.ptf.caltech.edu/iptf.

[14] http://crts.caltech.edu.

[15] http://www.sdss.org/supernova/aboutsupernova.html.

range $0.05 \lesssim z \lesssim 0.35$. The search region was a 300 square-degree southern equatorial field and the cadence varied from 2 to 5 days, with $m_{lim} \simeq 23.5$. More than 500 supernovae were found. *SDSS* obtained multiband photometry at every epoch of observation. Other telescopes carried out prompt spectroscopic followup of about half of the discoveries.

Traditional supernovae searches are based on imaging, but it is also possible to recognize supernova spectral signatures among large samples of galaxy spectra (Madgwick et al. 2003). Among 740,000 galaxy spectra obtained by the SDSS, 107 supernovae at a median $z \simeq 0.1$ were recognized, including 23 that had not been otherwise discovered (Graur et al. 2015).

2.4 High-Redshift Supernovae

An efficient technique pioneered by Norgaard-Nielsen et al. (1989) for discovering and following up batches of high-redshift ($z \gtrsim 0.2$) supernovae was developed in the late 1990s by the Supernova Cosmology Project[16] (*SCP*; Perlmutter et al. 1997, 1999). The procedure was to image many fields, some unbiased and some containing known clusters of galaxies, with a wide-field CCD camera on moderately large (2.5 to 4-m) telescopes shortly after a new Moon. Deep exposures resulted in "pencil beam" search images, each containing hundreds of high-redshift galaxies. Shortly before the next new Moon, the fields were imaged again. Image subtraction then revealed a dozen or so new supernovae, most of which had not yet reached peak brightness. (It helped that supernova light curves are time dilated by a factor $1 + z$.) This batch technique allowed advanced scheduling of followup observations with both the discovery telescopes and others. The *SCP* and the High-z Supernova Search team (*HZSS*; Schmidt et al. 1998; Riess et al. 1998)[17] used this technique to find the first dozens of high-redshift supernovae (typically $z \simeq 0.5$) and to discover the apparent acceleration of the cosmic expansion (Sect. 25.2.1).

Subsequently, from 2003 to 2008, the Supernova Legacy Survey[18] (*SNLS*; Astier et al. 2006; Conley et al. 2011) used the Megaprime camera on the Canada–France Hawaiian Telescope to conduct a *rolling* search—high-cadence repeated visits of selected fields, resulting in new discoveries during the followup of previous discoveries. Between 2003 and 2008, *SNLS* discovered about 1000 supernovae in the range $0.3 \lesssim z \lesssim 1$. The cadence of the photometry was 2 days and spectroscopic followup was performed with other telescopes. The Equation of State SupErNova Cosmology Experiment (*ESSENCE*; Miknaitis et al. 2007; Foley et al. 2009b) used the *CTIO* Blanco telescope and the MOSAIC camera between 2002 and 2008 to

[16]http://www-supernova.lbl.gov/.

[17]http://www.cfa.harvard.edu/supernova//HighZ.html.

[18]http://cfht.hawaii.edu/SNLS.

conduct a rolling search for supernovae in the range $0.1 \lesssim z \lesssim 0.8$ and discovered about 200 SN Ia.

The Panoramic Survey Telescope & Rapid Response System[19] (*Pan-STARRS*; Kaiser et al. 2002, 2010), located on Haleakala, Maui, employs a prototype 1.8-m telescope with a 10^9-pixel camera having a FOV of seven square degrees. Its Medium Deep Survey consists of ten pointings spaced approximately uniformly around the sky. The cadence and depth are optimized for studies of SN Ia. Single exposures should reach SN Ia at $z \simeq 0.5$. In 2010 *Pan-STARRS* began to discover a significant number of supernovae at a mean $z \simeq 0.24$ (Rest et al. 2014). When fully operational with four mirrors, the project is expected to detect from 10^4 to 10^5 supernovae per year at up to $z \simeq 1$.

The optical component of the *Subaru/XMM-Newton* Deep Survey carried out with the 8.2-m Subaru telescope discovered 39 SN Ia out to $z = 1.4$ (Okumura et al. 2014).

The searches mentioned above were carried out from the ground, occasionally supplemented by followup with the *Hubble Space Telescope (HST)*. For *very* high-redshift discoveries, $z \gtrsim 1$ (very high by current supernova standards, but not, of course, by the standards of quasars, galaxies, and gamma-ray bursts), the *HST* has been the instrument of choice, in spite of its narrow FOV. The *Hubble* Higher-z Supernova Search (*HHZSS*; Strolger et al. 2004; Riess et al. 2007) was the first space-based open-field supernova search, with an effective summed FOV of 300 square arcminutes, $m_{lim} \simeq 26$, and a cadence of 45 days. *HHZSS* discovered two dozen supernovae at $z \gtrsim 1$.

By targeting massive galaxy clusters at $0.9 \lesssim z \lesssim 1.5$, the *HST* Cluster Supernova Survey[20] (Dawson et al. 2009) is more efficient than surveying unbiased fields, and since many cluster galaxies are relatively dust-free, extinction does not strongly affect the supernova sample. The Cluster Lensing and Supernova survey with *Hubble*[21] (*CLASH*; Postman et al. 2012) exploits strong lensing by clusters to discover magnified background supernovae. *CLASH* discovered SN Ia candidates out to $z \simeq 1.8$ (Graur et al. 2014b). The Cosmic Assembly near-IR Deep Extragalactic Legacy Survey[22] (*CANDELS*; Grogin et al. 2011; Koekemoer et al. 2011), a sister *HST* program of *CLASH*, discovered 65 supernovae out to $z \simeq 2.5$ (Rodney et al. 2014).

During its 5-year duration, the Dark Energy Survey Supernova program (*DES SN*)[23] (Bernstein et al. 2012; D'Andrea 2014), begun in 2012, is expected to discover about 3500 SN Ia and provide well-sampled light curves in the range $0.05 < z < 1.2$.

[19] http://ps1sc.org/transients/.

[20] http://supernova.lbl.gov/2009ClusterSurvey.

[21] http://www.stsci.edu/~postman/CLASH/Home.html.

[22] http://candels.ucolick.org.

[23] http://www.darkenergysurvey.org.

2.5 Summary

Targeted searches of bright galaxies primarily find nearby supernovae. In the future, the discovery rate of such events will increase substantially, but not dramatically, simply because of the limited number of nearby galaxies. Nevertheless, nearby supernovae will remain the prime source of very early discoveries and the targets of spectropolarimetric and nebular-phase observations. Wide-field unbiased searches and pencil-beam surveys find significant numbers of supernovae in the Hubble flow and at high redshift, respectively. The number of such discoveries will increase tremendously in the near future (Sect. 26.4.1).

Chapter 3
Environments and Rates of Supernovae

3.1 Introduction

Valuable information on supernova progenitors is obtained from observations of the *environments* of supernovae. At a coarse level this means determining the Hubble morphological type or an integrated color of the host galaxy. Characterizing the stellar population near the site of a supernova is more useful, but because this requires a well-resolved galaxy image, it is largely restricted to supernovae in the local, low-redshift Universe, $z \lesssim 0.1$. The most direct and informative approach is to search for the actual supernova progenitor on a deep archival image obtained before the star exploded. This is feasible only for supernovae in nearby galaxies, within $\sim$20 Mpc, except in rare cases when the progenitor is extremely luminous. Ideally, the disappearance of the putative supernova progenitor is confirmed with a postsupernova image. Identifying, on archival or postsupernova images, a binary companion to the supernova progenitor or a cluster or association to which the progenitor belonged, also is informative.

Information on the *rates* at which the various types of supernovae occur in the various types of galaxies, both in the local Universe and as a function of redshift, also is valuable. Rates are useful for constraining the nature of supernova progenitors, planning supernova searches, and advancing our understanding of galaxy and chemical evolution.

3.2 Direct Detection of Progenitors and Companion Stars

Direct archival detection of supernova progenitors in external galaxies was reviewed by Smartt (2009, 2015) and Leonard (2011), which present updated lists of direct detections. The importance of searching for binary companions and confirming the disappearance of progenitors on postsupernova images were emphasized by

D. Branch, J.C. Wheeler, *Supernova Explosions*, Astronomy and Astrophysics Library, DOI 10.1007/978-3-662-55054-0_3

Kochanek (2009) and Maund et al. (2014b), respectively. Archival *HST* images have been an important source of data, but large ground-based telescopes with adaptive optics have extended this capacity. The first clear direct detection of a progenitor star was in the special case of the very nearby (0.05 Mpc), peculiar Type II SN 1987A in the LMC. The blue-supergiant progenitor that had appeared on numerous archival images was absent from postsupernova images. The ZAMS mass of the progenitor was estimated to have been about 20 $M_{\odot}$.

Direct detection of more distant progenitors has been successful mainly for SN IIP, yielding more than a dozen confirmed detections of red-supergiant progenitors as well as more than a dozen useful upper limits to the presupernova luminosity. Comparison of the luminosity ($\lesssim 10^5\ L_{\odot}$ in all cases) and colors of the progenitor with model stellar-evolution tracks gives an estimate of, or upper limit to, the ZAMS mass of the progenitor star, because the luminosity of the immediate presupernova depends mainly on the ZAMS mass. An early result was that the inferred ZAMS masses of the progenitors of SN IIP ranged from about 8 to 16 $M_{\odot}$. The lower value was consistent with the lower limit of stars expected to develop degenerate ONeMg cores that undergo core collapse (Sect. 8.5.3). The upper value was surprisingly low, because single stars of ZAMS masses up to 25 or 30 $M_{\odot}$ were expected to undergo core collapse inside red supergiants (RSG), thereby producing SN IIP. The apparent dearth of SN IIP progenitors in the range 16 to 25 or 30 $M_{\odot}$ was termed the "red-supergiant problem" (Smartt 2009). Subsequent direct detections of progenitors of SN 2008bk (Maund et al. 2014a), SN 2009md (Fraser et al. 2011), SN 2012A (Tomasella et al. 2013), SN 2012aw (Kochanek et al. 2012a), SN 2012ec (Maund et al. 2013a), and SN 2013ej (Fraser et al. 2014) continued to indicate relatively low masses (Smartt 2015).

Various explanations for the red-supergiant problem have been proposed. Perhaps stars in the missing mass range collapse directly to produce black holes with little optical display or, for other reasons related to the physics of core collapse, make only weak, dim supernovae that tend not to be discovered. The progenitors may be dimmed by the presence of circumstellar dust produced in their red-supergiant winds, causing the progenitor luminosities and masses to be underestimated (Walmswell and Eldridge 2012; Fraser et al. 2012; Van Dyk et al. 2012); if the dust is vaporized by the supernova radiation, the fact that it was present around the progenitor may not be recognized. Another possibility is that stars in the missing mass range evolve before exploding to become other types of core-collapse events (Yoon and Cantiello 2010; Sect. 3.6).

There are few direct constraints on the progenitors of SN IIL. Pre-explosion images produced an estimate of about 18 $M_{\odot}$ for the presupernova mass of SN 1979C (Van Dyk et al. 1999); this may be a lower limit depending on the coevality of the surrounding star cluster (Elias-Rosa et al. 2010). An upper limit of 20 $M_{\odot}$ was estimated for SN 1980K (Thompson 1982; Smartt et al. 2009). If it is a single object, a source in pre-explosion images of SN 2009kr appeared to be a yellow supergiant of about 18 to 25 $M_{\odot}$ (Elias-Rosa et al. 2010) or 15 $M_{\odot}$ (Fraser et al. 2010). The highly-extincted event SN 2009hd was a possible SN IIL with an upper limit to the progenitor mass of about 20 $M_\odot$ (Elias-Rosa et al. 2011). These values suggest that on average the progenitors of SN IIL may have higher masses than those of SN IIP.

The progenitor of the normal-luminosity Type IIn SN 2005gl was a luminous blue variable (LBV) star of very high mass, about 50 $M_\odot$ (Gal-Yam et al. 2007), which disappeared after the explosion (Gal-Yam and Leonard 2009). The LBV phase had been thought to be an immediate post-main-sequence stage of evolution, not a stage of evolution near to core collapse (Sect. 8.4.3), but the progenitor of SN 2005gl is part of accumulating evidence that some LBV stars are immediate supernova progenitors. The putative progenitor of the bright ($M_B \simeq -20$) Type IIn SN 2010jl may have had a ZAMS mass greater than 30 $M_\odot$ (Smith et al. 2011b). For a discussion of the possible Type IIn SN 2009ip, which may have had a ZAMS mass of about 70 $M_\odot$ (Smartt 2015), see Sect. 14.4. Most SN IIn, however, are not explosions of very massive stars (Habergham et al. 2014).

The best-explored case for SN IIb is SN 1993J. A K0 Ia supergiant progenitor of ZAMS mass in the range 13–20 $M_\odot$ was discovered in archival images, and its subsequent disappearance was confirmed (Maund and Smartt 2009). Postsupernova observations identified a probable former companion, a B2 supergiant (Maund et al. 2004; Fox et al. 2014). Assuming this identification to be correct, the nature of the presupernova system could be more closely defined from the blend of the progenitor and the companion in pre-explosion photometry; the star that exploded was a K0 supergiant (K0 Ia) with log $L/L_\odot \simeq 5.1 \pm 0.3$ and $T_{eff} \simeq 4270 \pm 500$ K; the B2 Ia companion had log $L/L_\odot \simeq 5.0 \pm 0.3$ and $T_{eff} \simeq 20{,}000 \pm 500$ K (Maund et al. 2004). If a "kick" velocity (Sect. 24.2.1) did not play an important role, the companion would have remained bound to the compact remnant of the supernova (Podsiadlowski et al. 1993; Woosley et al. 1994a). The velocity inferred from the narrow lines of the companion and the close blending of the companion in late images of the supernova are consistent with a low kick velocity.

Several other progenitors of SN IIb have been detected. That of SN 2011dh was a yellow supergiant of ZAMS mass about 13 $M_\odot$ (Maund et al. 2011). This star was present in archival observations, but not in postsupernova observations (Van Dyk et al. 2013). A blue point source in *HST* images obtained 3 years postexplosion may (Folatelli et al. 2014a) or may not (Maund et al. 2015) be a former binary companion. The progenitor of SN 2013df was a cool supergiant of ZAMS mass perhaps 13–17 $M_\odot$ (Van Dyk et al. 2014). A supergiant was excluded as the progenitor of SN 2008ax, while a Wolf–Rayet star of ZAMS mass about 28 $M_\odot$ or a stripped-envelope star of ZAMS mass 10–14 $M_\odot$ may be consistent with the data (Crockett et al. 2008). No surviving companion of the Type IIb supernova (Krause et al. 2008a) that produced the Cas A SNR has been found.

A source in an archival *HST* image (Cao et al. 2013) at the site of the Type Ib SN iPTF13bvn is likely to have been the progenitor star. The properties of the source (Eldridge et al. 2015; Kim et al. 2015b) and the supernova (Bersten et al. 2014; Fremling et al. 2014; Kuncarayakti et al. 2015) suggest that the progenitor was a 10–20 $M_\odot$ primary component of a binary system, not a Wolf–Rayet star. Deep archival images for about a dozen other SN Ib/c have not revealed a progenitor. This may be because the compact, hydrogen-deficient progenitors are expected to emit mainly in the blue and UV and hence, because of interstellar reddening and instrument response functions, to be difficult to detect, and because the progenitor may be difficult to detect against the light from a companion star. For any individual SN Ib/c event, a classical massive (ZAMS mass $\gtrsim$35 $M_\odot$) Wolf–Rayet star progenitor is difficult to exclude on the basis of archival observations, but the general lack of detections indicates that most SN Ib/c do not come from such stars (Smartt 2009; Eldridge et al. 2013). Most SN Ib/c probably come from stars of lower ZAMS mass that lose their hydrogen envelopes by means of binary interactions before their cores collapse.

SN Ibn are a rare form of SN Ib, the spectra of which include narrow emission lines of He I that form in helium-rich CSM (Sect. 16.5). An outburst observed about 2 years prior to SN 2006jc (Pastorello et al. 2007) suggested an LBV-like progenitor. There have been no other direct detections of SN Ibn progenitors.

The rare heterogenous class of superluminous ($m_B \lesssim -21$) supernovae tend to be discovered beyond the local Universe. There has been no direct detection of a progenitor.

Despite a longstanding conviction that SN Ia must be explosions of white dwarfs, direct evidence has remained elusive. The very early light curve of SN 2011fe strongly implied that the exploding star was more compact than any normal star, leaving a white dwarf as the only plausible progenitor (Sect. 20.2.1). Another longstanding conviction has been that the white dwarf is in a binary system, either accreting from a nondegenerate companion (the *single-degenerate* scenario) or eventually merging with a white dwarf companion (the *double-degenerate* scenario), but evidence for the nature of the putative companions also is elusive. Archival images at the sites of SN 2011fe and SN 2014J (Sect. 20.2) allowed significant constraints to be placed on a companion star, e.g., luminous red giants were excluded. This and additional constraints, some from young SNRs, on companions of SN Ia are discussed in Sect. 21.6.

3.3 Environments in the Local Universe

Practically all SN II, IIn, IIb, and Ib/c occur in spiral and irregular galaxies (van den Bergh et al. 2005), and many are clearly associated with star formation. In spirals, they have a tendency to be associated with the spiral arms, where star formation is enhanced. This is one reason why they are thought to come from core collapse.

Statistical studies of supernova environments were reviewed by Anderson et al. (2015b). In star-forming galaxies, the radial distributions of core-collapse supernovae and Hα emission are similar, suggesting that both are good tracers of recent star formation (Anderson and James 2009). It is more challenging to establish that different kinds of core-collapse supernovae are associated with different stellar populations or different degrees of star formation. There is some evidence that the environments of SN IIL are younger and of lower metallicity than those of SN IIP (Kuncarayakti et al. 2013a). It appears that SN Ic are more closely associated with ongoing star-formation environments than SN Ib, and that SN Ib may be more closely associated than SN II (Anderson et al. 2012). On average, the ZAMS masses of SN Ic progenitors exceed those of SN Ib (Kangas et al. 2013). SN Ic also come from higher-metallicity environments than do SN Ib (Kuncarayakti et al. 2013b), and SN Ib come from higher-metallicity environments than do SN II (Modjaz 2012). SN Ib and SN Ic be more centrally concentrated than the light of their host galaxies, but unlike SN Ic, SN Ib do not appear to concentrate to the locally brightest regions of their hosts. The environments of SN IIb have bluer colors and lower surface brightnesses than those of SN Ib, SN Ic, and SN IIP (Kelly and Kirshner 2012).

An exception to these trends is the subtype of broad-line SN Ic (SN Ic-bl; Sect. 17.4). Like the gamma-ray bursts with which some of them are associated, SN Ic-bl tend to occur in small star-forming galaxies of low metallicity (Wang and Dai 2014; Niino et al. 2015). Low metallicity may be required to avoid strong progenitor winds that would remove angular momentum necessary to the formation of gamma-ray bursts. Superluminous supernovae, especially those that are hydrogen-poor, also tend to appear in star-forming galaxies of low mass and low metallicity (Neill et al. 2011; Lunnan et al. 2014; Leloudas et al. 2015).

SN Ia occur in all types of galaxies (Hakobyan et al. 2012), and statistically their hosts are representative of normal field galaxies (Childress et al. 2013a). In star-forming galaxies, SN Ia trace the B-band light distribution, but not Hα emission or UV light, implying that the population in these galaxies is dominated by a relatively young (but $\gtrsim 10^8$ years) population (Anderson et al. 2015a). Early-type galaxies—ellipticals and lenticulars (galaxy type S0)—are primarily composed of old stars, but some such galaxies do have localized regions of recent star formation. In ellipticals, SN Ia "follow the light," i.e., their radial distribution is like that of the starlight (Förster and Schawinski 2008), and some SN Ia occur in early-type galaxies that are devoid of young populations (Schawinski 2009). Thus most SN Ia in early-type galaxies are produced by a relatively old population rather than only in the regions of recent star formation. On the other hand, in early-type galaxies that contain stellar populations younger than 3 Gyr, the SN Ia rate is enhanced compared to galaxies having no such populations (Gallagher et al. 2008). Therefore, although SN Ia in early-type galaxies are produced by a relatively old population (Gyrs), most are not from a *very* old (10 Gyr) population. In ellipticals, SN Ia are not predominantly in globular clusters, even though dynamical evolution in the crowded environment enhances the number of close binaries (Washabaugh and Bregman 2013). The SN Ia environments have long been regarded as a clue that their progenitors are long lived and hence of modest ZAMS mass, although the notion that some SN Ia arise in somewhat shorter-lived systems also has a long and respected history (Oemler and Tinsley 1979). The basic uniformity of explosion properties and modest ZAMS masses of the progenitors in turn suggest that the explosion is of a white dwarf of nearly the Chandrasekhar mass. Decades of work comparing observations to models have shown that white dwarfs composed of carbon and oxygen, the expected progeny of stars of ZAMS mass $\leq 8\ M_\odot$, are especially viable candidates. The need to bring a white dwarf to an explosive endpoint seems to demand evolution with a binary companion.

3.4 Rates in the Local Universe

Even in the local Universe, $z \lesssim 0.1$, supernova rates are difficult to determine, owing to the limited number of supernova discoveries in controlled searches that are suitable for rate determinations, to uncertain influences of interstellar extinction that depend in a complicated way on the inclination of a galaxy to our line of sight, and observational biases. In the local Universe, rates are based on targeted searching of preselected galaxies. The classical procedure for estimating supernova rates (Zwicky 1942; van den Bergh and Tammann 1991) begins by adopting a peak absolute magnitude and a light-curve shape for each supernova type. Then, for every observation of the search, the "surveillance time"—the length of time during which each type of supernova would be brighter than the limiting magnitude of the observation—is determined. For each type, the sum of the surveillance times is the "control time" of the search—the total time during which such a supernova would have been bright enough, in all the observations, to have been discovered. The rate of the supernova type is the number of events discovered divided by the corresponding control time.

Traditionally, local supernova rates are expressed in "supernova units" (SNu); e.g., 1 SNuB = 1 SN $(100\ \mathrm{yr})^{-1}\ (10^{10}\mathrm{L}_\odot^B)^{-1}$, where $\mathrm{L}_\odot^B$ is the solar luminosity in the B band. For example, if the B-band luminosity of our Galaxy is $2 \times 10^{10}\mathrm{L}_\odot^B$, 1 SNu corresponds to 2 Galactic supernovae per century. The blue luminosity of a galaxy is a measure of the star-formation rate averaged over several Gyr, characteristic of the main-sequence lifetime of an A-type star of a few solar masses; it is not a good measure of either the current rate of star formation or of galaxy mass. Luminosity in the near-IR, e.g., the K band at 2.1 μm, is a better measure of galaxy mass, so supernova rates are also expressed in units of SNuK, where $\mathrm{L}_\odot^B$ is replaced by $\mathrm{L}_\odot^K$, as well as in units of SNuM = 1 SN $(100\ \mathrm{yr})^{-1}\ (10^{10}\mathrm{M}_\odot)^{-1}$ (where galaxy mass is estimated from an empirical relation between luminosity and mass for each galaxy type).

A major study of supernova rates in a volume-limited sample of local galaxies, using the results of the LOSS supernova search, was based on about 1000 supernovae. A generalization of the control-time method allowed for a luminosity function and a family of light-curve shapes for each supernova subtype (Leaman et al. 2011; Li et al. 2011c). An important new finding was the "rate-size" relationship: within a given galaxy type, the supernova rate *per unit mass* is not constant but instead decreases with increasing galaxy mass. The results of Li et al. (2011b) for rates of the main supernova types—SN II, Ib/c, and Ia—in galaxies of different

Hubble types (and of a fiducial size) are presented in their Table 6 and Fig. 3.1 below. SN II and SN Ib/c rates, whether expressed in SNuB, SNuK, or SNuM, increase strongly from early-type to late-type galaxies; for example, in SNuM, the SN II and SN Ib/c rates increase from near zero in E and S0 galaxies to 0.6 and 0.2, respectively, in Sc galaxies. The SN Ia rate in units of SNuB decreases from early- to late-type, while in units of SNuK or SNuM it does not depend strongly on galaxy type; in SNuM the SN Ia rate is about 0.14.

Supernova rates as a function of integrated galaxy color, e.g., $B - K$, also are informative, because colors are related to the star-formation rate per unit mass or unit luminosity. On average, galaxies of different morphological types have different colors, but within each type galaxies have a range of colors and star-formation rates, so galaxy colors may be preferable to galaxy types to characterize host galaxies and their supernova progeny. Using galaxy colors rather than types also has the advantage that for more distant, poorly-resolved galaxies, colors are more readily observable than galaxy types. The results of Li et al. (2011b) for rates of SN II, Ib/c, and Ia in galaxies of different color (and fiducial size) are presented in their

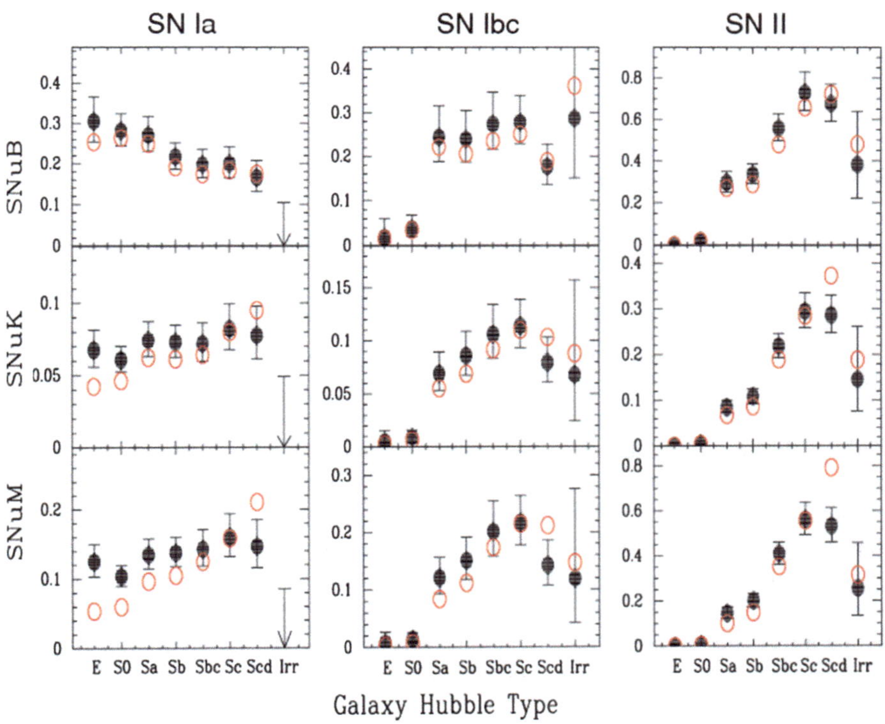

Fig. 3.1 Supernova rates (*solid symbols*) are plotted against galaxy type. Figure 13 "Nearby supernova rates from the Lick Observatory Supernova Search—III. The rate-size relation, and the rates as a function of galaxy Hubble type and colour" from Li et al. (2011b), by permission of Oxford University Press on behalf of the Royal Astronomical Society

Fig. 13 and Table 7. As expected, the SN II and SN Ib/c rates are much higher in blue galaxies than in red galaxies; in SNuM, the rates of SN II and SN Ib/c increase from near zero at galaxy $B - K = 4$ to 0.7 and 0.4, respectively, at $B - K = 2.5$. In units of SNuB, the SN Ia rate decreases from red to blue galaxies, while in units of SNuK or SNuM, it increases (Li et al. 2011b); in SNuM, it increases from about 0.05 at $B - K = 4$ to about 0.25 at $B - K = 2.5$.

The rates of SN Ia in galaxies of different Hubble type or color indicate that in blue, late-type galaxies a significant fraction of SN Ia come from rather young ($\sim 10^8$ years) populations. In the "A+B" parameterization of SN Ia rates (Scannapieco and Bildsten 2005; Mannucci et al. 2005), the SN Ia rate is expressed as the sum of two terms: the A term, accounting for "old" SN Ia, being proportional to galaxy mass and the B term, accounting for "young" SN Ia, being proportional to galaxy star-formation rate. For example, Li et al. (2011b) found[1]

$$SNuM(Ia) = (0.036 \pm 0.022) + (0.220 \pm 0.067) SNuM(CC),$$

where *SNuM(CC)*, the core-collapse rate, serves as a proxy for galaxy star-formation rate. The A+B parameterization is unable to account for all data on SN Ia rates; at least one additional parameter, such as host-galaxy mass, is required (Gao and Pritchet 2013).

3.5 The Galactic Rate

Direct estimates of Galactic supernova rates based on historical Galactic supernovae suffer from severe incompleteness (although allowance for incompleteness can be attempted; Adams et al. 2013). Estimates based on supernova remnants suffer from both incompleteness and uncertain remnant ages. Estimates from pulsars can give a lower limit to the rate of core-collapse supernovae, and observations of ^{26}Al γ-ray emission can provide a useful estimate of the core-collapse rate (Diehl et al. 2006). Perhaps the best estimates of Galactic rates are those inferred from rates determined for similar galaxies in the local Universe. For various assumptions about the properties of the Galaxy (Hubble type and luminosity), Li et al. (2011b) estimated that the Galaxy produces 2.84 supernovae per century (to within a factor of two, owing mainly to uncertainties in the properties of the Galaxy). The distribution by main supernova types was estimated to be: 0.54 SN Ia, 0.76 SN Ib/c, and 2.54 SN II per century. The corresponding mean intervals between SN Ia, SN Ib/c, SN II, and all supernovae are 185, 132, 65, and 35 years, respectively.

[1]Rates in SNu scale as h^2, where h is the Hubble constant in units of $100\,\mathrm{km\,s^{-1}\,Mpc^{-1}}$; Li et al. adopted $h = 0.73$.

3.6 Inferences from Rates About Progenitors of Core-Collapse Supernovae

Inferences about the progenitors of the various kinds of core-collapse supernovae can be drawn from environments, rates, and an adopted initial mass function (Smith et al. 2011a). Based on detailed results of Li et al. (2011b), Smith et al. adopted the following fractions of the types of core-collapse supernovae in a volume-limited sample (rounded here, owing to typical uncertainties of 0.04): 0.48 SN IIP, 0.06 SN IIL, 0.09 SN IIn, 0.11 SN IIb, 0.07 SN Ib, 0.15 SN Ic, and 0.04 peculiar SN Ib/c. They argued that the solution to the red-supergiant problem (Sect. 3.2) is that before exploding, stars in the range 17–25 $M_{\odot}$ evolve beyond the RSG phase to become other kinds of core-collapse supernovae; otherwise the number of progenitors of these other kinds would be too small.

The progenitors of some SN Ib/c may be single (or effectively single, in wide binaries), massive ($\gtrsim$35 $M_{\odot}$) stars that become classical Wolf–Rayet stars by losing most or all of their hydrogen envelopes, and perhaps also their helium, by means of stellar winds. Mass-loss rates are greater in higher mass stars and are enhanced at high metallicity by high radiative opacity. Smith et al. (2011a) argued that the number of such stars is sufficient to account for less than half of the observed SN Ib/c, so that most SN Ib/c must come from stars of lower ZAMS mass that experience Roche-lobe overflow in binaries. The observed tendency of SN Ic to come from more massive stars and more metal-rich environments than do SN Ib was attributed to especially high mass-loss rates *after* the hydrogen has been lost, because both high mass and high metallicity promote additional loss of helium.

In the favored scenario of Smith et al. (2011a; see their Fig. 7), SN IIP come from core-collapse in stars of ZAMS mass less than about 19 $M_{\odot}$; SN IIL from 19 to 23 $M_{\odot}$; and SN IIn above 23 $M_{\odot}$. SN IIb, SN Ib, and some SN Ic come from the full core-collapse mass range, with SN Ic coming preferentially from above 23 $M_{\odot}$ because of the enhanced post-hydrogen mass loss mentioned above. Whether the supernova types are cleanly separated in ZAMS mass in this way, when all observational constraints are taken into account, remains to be seen. For example, most SN IIn arise from the same population as SN IIP (Anderson et al. 2012). Binary evolution, including common-envelope episodes and stellar mergers, may prove to play a larger role than previously recognized, and other variables such as metallicity and rotation may be significant. If stars occasionally emit gouts of mass loss, their appearance may depend on the timing of core collapse with respect to processes occurring in the outer envelope.

3.7 Rates Versus Redshift

In the past, the mean metallicity of the Universe was lower and star-formation rates were generally higher. Thus measurements of supernova rates versus redshift can shed light on the nature of supernova progenitors as well as the evolution of the stellar population of the universe.

Most of the searches for nonlocal supernovae (redshift $z > 0.1$) described in Chap. 2 were conducted primarily to yield good crops of low-extinction SN Ia discovered prior to or near maximum light, to be used for cosmology. Variable sources unlikely to meet these criteria often were not followed up spectroscopically. This, together with the fact that most core-collapse supernovae are several magnitudes less luminous than SN Ia, produced a severe bias in the supernova sample, making it especially difficult to estimate the core-collapse rate. Many rate estimates were based on small numbers of supernovae, including some of uncertain classification owing to lack of spectra and well-sampled light curves.

Because nonlocal searches are not usually based on a list of target galaxies, rates are commonly expressed in volumetric terms, e.g., the number of supernovae $\mathrm{yr}^{-1}\,\mathrm{Mpc}^{-3}\,(h/0.7)^3$ (restframe year, comoving volume). Li et al. (2011b) estimated local volumetric rates, in these units (i.e., scaled from $h = 0.73$ to $h = 0.70$), for SN Ia, Ib/c, and II to be 2.7×10^{-5}, 2.3×10^{-5}, and 3.9×10^{-5}, respectively (with considerable uncertainty depending on the local luminosity densities and the luminosity functions for galaxies of different Hubble types; see their Table 10). The rate of superluminous supernovae at $z \simeq 0.16$ has been estimated to be about 2×10^{-7} in the above units (Quimby et al. 2013).

Given the short lifetimes of the massive progenitors of core-collapse supernovae, the variation of the core-collapse rate with redshift might be expected to track the rate of cosmic star formation, which can be estimated by other (nonsupernova) means (Hopkins and Beacom 2006). The star-formation rate increases by a factor of about 10 from $z = 0$ to $z \gtrsim 1$. Observations of the rates of core-collapse supernovae at various redshifts are generally consistent with this. The rate of core-collapse supernovae within 11 Mpc may be consistent with or somewhat greater than expectation based on the local rate of star formation (Horiuchi et al. 2011; Botticella et al. 2012; Xiao and Eldridge 2015). At intermediate redshifts $0.1 \lesssim z \lesssim 0.4$, core-collapse rates may be smaller than expected from the star formation rates by a factor of about two (Botticella et al. 2008; Horiuchi et al. 2014), probably because of observational bias against dim supernovae and/or black-hole formation. When allowance is made for bias against the discovery of dim supernovae (whether intrinsically dim or heavily obscured by dust) the core-collapse rate in the range $0.4 \lesssim z \lesssim 1.1$ is consistent with the elevated rate of star formation in that range (Fig. 3.2).

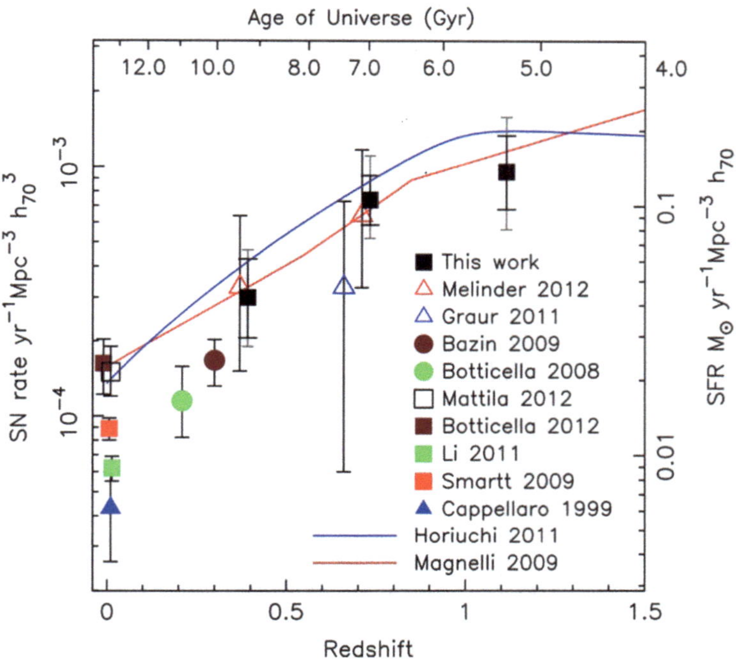

Fig. 3.2 Volumetric rates of core-collapse supernovae (*symbols*) are plotted against redshift. The *lines* are estimates of the cosmic star-formation rate. From "The Extended Hubble Space Telescope Supernova Survey: The Rate of Core-collapse Supernovae to $z \sim 1$" (Dahlen et al. 2012). © AAS. Reproduced with permission

SN Ia rates at various redshifts have been estimated based on several searches (Fig. 3.3). The volumetric SN Ia rate appears to rise to a maximum at $z \simeq 1.5$. As expected, the rise is less steep than that of the cosmic star-formation rate and the core-collapse supernova rate. Beyond $z \simeq 1.5$, the SN Ia rate appears to flatten or decrease. Although SN Ia rates do not closely track the star-formation rate, future SN Ia rates may still give useful constraints on star-formation rates at high redshifts, where rates of less luminous core-collapse supernovae are more difficult to estimate (Gal-Yam and Maoz 2004).

3.8 The SN Ia Delay-Time Distribution

A crucial relationship for inferring the nature of SN Ia progenitors is the *delay-time distribution* (DTD)—the SN Ia rate versus time that would follow a hypothetical instantaneous burst of star formation. The actual SN Ia rate versus time is the convolution of the DTD with the star-formation rate versus time. The SN Ia DTD was reviewed by Maoz and Mannucci (2012).

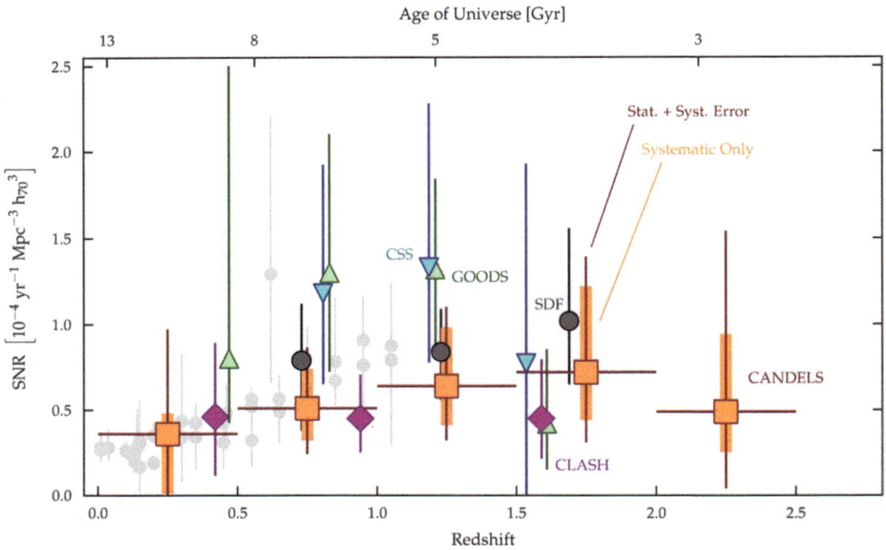

Fig. 3.3 Volumetric rates of SN Ia are plotted against redshift. From "Type Ia Supernova Rate Measurements to Redshift 2.5 from *CANDELS*: Searching for Prompt Explosions in the Early Universe" (Rodney et al. 2014). © AAS. Reproduced with permission

In general terms, the DTD is determined by measuring the SN Ia rate in stellar populations of different mean ages. This can be done in a number of ways. Perhaps the most general (but not necessarily the most accurate) way is, given the overall cosmic star-formation history and the observed dependence of the volumetric SN Ia rate on redshift, to invert the convolution (Horiuchi and Beacom 2010; Kistler et al. 2011; Perrett et al. 2012; Graur et al. 2014b). Another approach is to determine the SN Ia rate per unit mass in galaxies of different characteristic stellar-population ages; for example, in early-type field galaxies that have different redshifts. Because early-type galaxies formed their stars early in the history of the Universe, the characteristic age of their populations versus redshift can be estimated (Totani et al. 2008; Thomson and Chary 2011). Still another approach, which involves less averaging, is to look at the number of SN Ia hosted by individual galaxies or even subunits of galaxies for which star-formation histories have been determined (Brandt et al. 2010; Maoz et al. 2011, 2012). A novel application of this approach was based on SNRs of SN Ia in the Magellanic Clouds (Maoz and Badenes 2010), where star-formation histories are available as a function of position. Except for very short delays, the DTD can also be inferred from SN Ia rates in clusters of galaxies at different redshifts (Sand et al. 2012; Barbary et al. 2012). Clusters consist mainly of early-type galaxies that formed most of their stars near $z \simeq 3$, about 3 Gyr after the big bang, so the ages of their stellar populations as a function of redshift are well determined. Moreover, owing to their deep gravitational potential wells, clusters have retained all of the iron ever formed in them; this, after allowing for

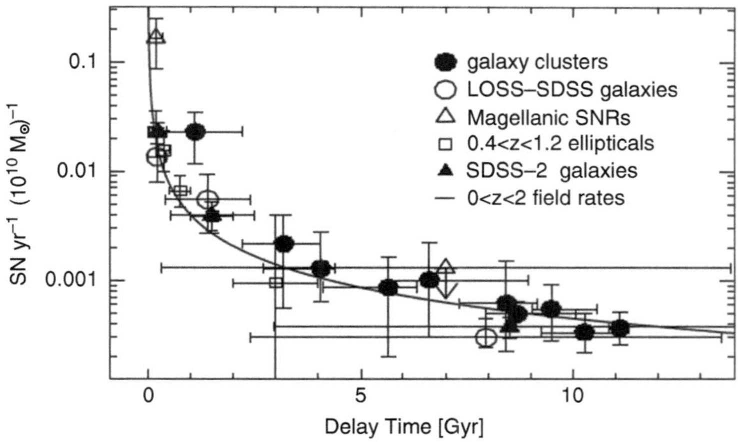

Fig. 3.4 The SN Ia delay-time distribution, estimated by various methods. The *curved line* is proportional to t^{-1}. From "Type Ia Supernova Rates and the Progenitor Problem: A Review," (Maoz and Mannucci 2012), reproduced by permission

a contribution of iron from core-collapse supernovae, permits a good estimate not only of the shape of the DTD but also of its absolute level. The various methods give consistent results (Maoz and Mannucci 2012; Graur et al. 2014b): the DTD inferred from observations is consistent with a power law, $\propto t^{-s}$ with $s \simeq 1$, from about 0.5 to 11 Gyr (Fig. 3.4). The nature of the DTD at ages less than 0.5 Gyr is uncertain (Rodney et al. 2014), but about half of all SN Ia appear to occur within 1 Gyr of star formation. This is why the SN Ia rate per unit mass increases from early- to late-type galaxies, and from red to blue galaxies.

3.9 Summary

Archival images at supernova sites have revealed an apparent red-supergiant problem for SN IIP and some information on the progenitors of a few supernovae of other types. The environments in which the various supernova types occur provide less precise statistical information on progenitor masses as well as the ages and metallicities of the stellar populations in which they occur. Like environments, supernova rates in relatively nearby galaxies provide statistical information on progenitors, and a basis for predicting supernova rates in our Galaxy. Knowledge of rates versus redshift provides additional information, such as the extent to which the core-collapse rate tracks the cosmic star-formation rate, and the shape of the SN Ia delay-time distribution.

Chapter 4
Spectra

4.1 Introduction

In a sense, a supernova *is* its composition and density structure. Only spectra can reveal relative abundances of elements versus velocity and hence depth within the ejecta. In principle, spectroscopic analysis can reveal the entire composition structure because as the ejecta expand, ever deeper layers are exposed.

During the first months after explosion, a supernova is optically thick and the spectrum forms outside a photosphere. The photosphere recedes with respect to the matter as time marches on, revealing deeper layers of ejecta. It is important to distinguish between the velocity of the photosphere and the velocity of the matter that, at a given instant, finds itself at the photosphere. The latter is the quantity directly determined by spectroscopy. During the photospheric phase, the main challenge to interpretation is that radiative-transfer effects are paramount; the emergent spectrum is the product of a complex series of photon emissions, absorptions, and scatterings. Fortunately, matter and radiation are sufficiently closely coupled that the assumption of *local thermodynamic equilibrium* (LTE) atomic level populations (occupation numbers) is servicable as a first-order approximation. In this approximation, the excitation and ionization states of the gas are determined by the local electron temperature and electron density.

Even when the radiation is decoupled from the electrons so that the radiation has a different effective temperature than the electrons, the assumption of LTE for the atomic level populations usually is not too bad. There are conditions where these levels tend to come to equilibrium with the radiation field, not the local electron temperature. LTE can be a decent approximation as long as the appropriate temperature is adopted. It is also true that level populations of some ions may be nearly in LTE, while those of other ions depart strongly from that condition. As a

D. Branch, J.C. Wheeler, *Supernova Explosions*, Astronomy and Astrophysics Library, DOI 10.1007/978-3-662-55054-0_4

rough rule of thumb, atoms with lots of levels and transitions, such as iron, may have excitations closer to LTE than simpler atoms like neutral hydrogen and neutral and ionized helium, for which the levels are few and widely spaced. A rough analogy is a bag of springs that is struck with a hammer. The energy will be more nearly equilibrated if the bag has many small springs rather than just a few large, stiff ones.

Eventually, the expanding ejecta become optically thin to continuum photons and the supernova becomes a nebula, self-excited by radioactive decay and perhaps CSI (Chap. 6). In the nebular limit of complete transparency, even in lines, radiative-transfer effects are absent because photons escape directly. The spectrum consists of emission lines, the profiles of which directly reflect the radial, and to some extent angular, distribution of level populations. Ionization fractions, however, are nowhere near LTE; thus conditions are extremely *non-local thermodynamic equilibrium* (NLTE) in the nebular phase.

There is, of course, no sharp division between the photospheric and nebular phases, but modeling in the two limits traditionally exploits different techniques and approximations. An outstanding difficulty in both phases is the large characteristic width of the Doppler-broadened spectral features, such that only about a dozen features appear in an optical spectrum at any one time.

Some of the information carried by supernova spectra can be inferred directly from observed line profiles, and observers routinely do some of this. Because spectral features are usually blended, however, deeper interpretation often involves comparing observed spectra with synthetic (calculated) spectra. Spectrum calculations may be highly simplified and rapid, or much more physically self-consistent and computationally demanding. Similarly, spectra are calculated for both simply parameterized input models of density and composition structures and for detailed numerical models based on nuclear-hydrodynamical simulations.

When light scatters through the ejecta of a supernova, it retains information about the orientation of the scattering layers. With few exceptions, extragalactic supernovae cannot be spatially resolved by direct imaging, and the total flux spectrum is rather insensitive to asymmetries. Polarization (Sect. 4.6) is the most powerful tool to explore the morphology of the ejecta. For a supernova with a typical photospheric radius of $\sim 10^{15}$ cm, the effective spatial resolution attained by polarimetry at 10 Mpc is $\sim 10\,\mu$as. Spectropolarimetry measures both the global shape of the emitting region as well as the shapes of smaller regions composed of particular chemical elements.

4.2 Elements of Line Formation in the Photospheric Phase

The fundamentals of radiative transfer in moving atmospheres were presented by Hummer (1968), Castor et al. (1975) and the book by Mihalas and Mihalas (1984), among others. In this section, we resort to simplifying approximations, including spherical symmetry, nonrelativistic flow, and the absence of CSI, in order to offer an intuitive understanding of the geometrical aspects of spectral-line formation in the photospheric phase.

4.2.1 The Velocity Law

At the onset of explosion, energy is suddenly released deep inside the star, followed by a brief hydrodynamical phase of duration, t_h, during which forces on the matter (pressure gradients pushing out, gravity pulling in) are substantial. The hydrodynamic phase ends when such forces have become negligible. Subsequently, each element of matter coasts at constant velocity so the more rapidly-moving elements get farther from the center of the explosion. When the time since explosion, t, has become much greater than t_h (ordinarily the case by the time the supernova is observed), $r = vt$, or $v = r/t$: at a given time, velocity is proportional to radius.

This simple velocity law usually is a good approximation, and it has nice consequences:

1. Velocity is a convenient comoving coordinate.
2. The ejecta are in *homologous expansion*, so the density structure evolves in a self-similar manner. The density at any comoving point scales as t^{-3}, and the shape of the density distribution does not change. This holds also for the full, complex, 3D structure. The density may vary with angle at a given radius, but it still scales as t^{-3} during homologous expansion.
3. The velocity gradient is homogeneous and isotropic. Not only is $dv/dr = 1/t$, but also $dv/ds = 1/t$ where ds is a displacement at any place and in any direction. As a photon moves, *at any place and in any direction*, it redshifts in the comoving frame (i.e., with respect to the matter) as $d\lambda/ds = (\lambda/c)dv/ds = \lambda/ct$.
4. For a distant observer, surfaces of constant line-of-sight velocity (*common-direction velocity surfaces*) are planes perpendicular to the line of sight (Fig. 4.1).

4.2.2 Resonant-Scattering Line Profile: Qualitative Overview

Assume spherical symmetry, that a perfectly sharp photosphere emits a continuous spectrum, and that a spectral line forms outside the photosphere by resonant scattering, i.e., a photon absorbed in an atomic transition is immediately emitted in the same transition,[1] in a random direction. There is no continuum opacity and no photon creation or destruction in the line-forming region. Consider an atomic transition of rest wavelength, λ_0. A photon emitted by the photosphere at wavelength $\lambda < \lambda_0$ redshifts as it moves into the line-forming region and may be scattered if it redshifts to λ_0. The observer receives scattered photons of wavelength λ_0 (in the

[1] This is not as bad as it sounds. Even though a particular absorption of a nonresonance line may often be followed by emission to a different lower level, statistically the emissivity of a line is often constrained to be not very different from the rate of absorption in the line (Chugai 1980; Abbott and Lucy 1985).

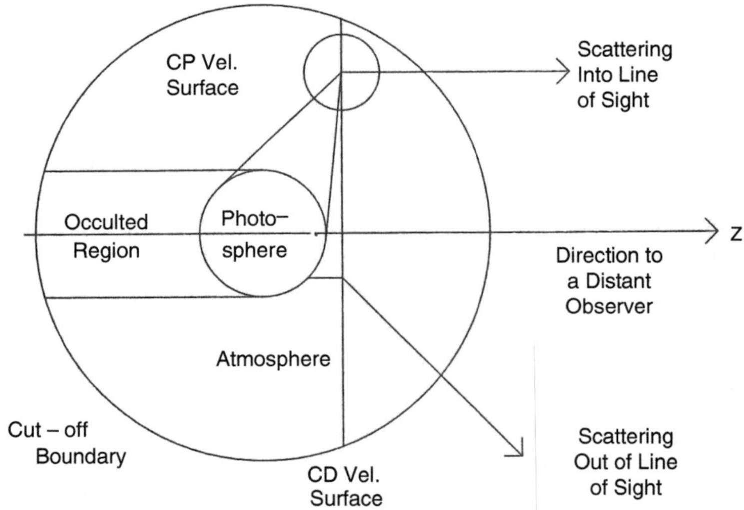

Fig. 4.1 Schematic representation of the line-forming layers of a supernova. From "Analysis of Supernova Spectra" (Jeffery and Branch 1990)

supernova restframe) only from the supernova midplane. Photons scattered from planes on the observer's side of the midplane have $\lambda < \lambda_0$ and those from planes on the far side of the midplane have $\lambda > \lambda_0$ (Fig. 4.1). Photons scattered towards the observer from behind the photosphere are occulted.

The line-forming region outside the photodisk (the projected photosphere) contributes an emission feature that is symmetric with respect to wavelength about λ_0. In the line-forming layer in front of the photodisk, some of the continuum photons that were directed towards the observer are scattered out of the line of sight, causing absorption at $\lambda < \lambda_0$ (Fig. 4.1). Because of occultation, the observer sees net absorption in the line profile.

In the flux spectrum, the net result is a line profile that has an emission peak at the rest wavelength and a blueshifted absorption. The combination of emission and blueshifted absorption, characteristic of an expanding atmosphere, is for historical reasons called a P Cygni profile[2] (Fig. 4.2).

[2]P Cygni, one of the intrinsically brightest stars in the Galaxy, is a luminous blue variable (LBV) of spectral type B1 Ia+. It sheds a nearly spherical wind that produces the famous line profile consisting of restframe emission and blueshifted absorption.

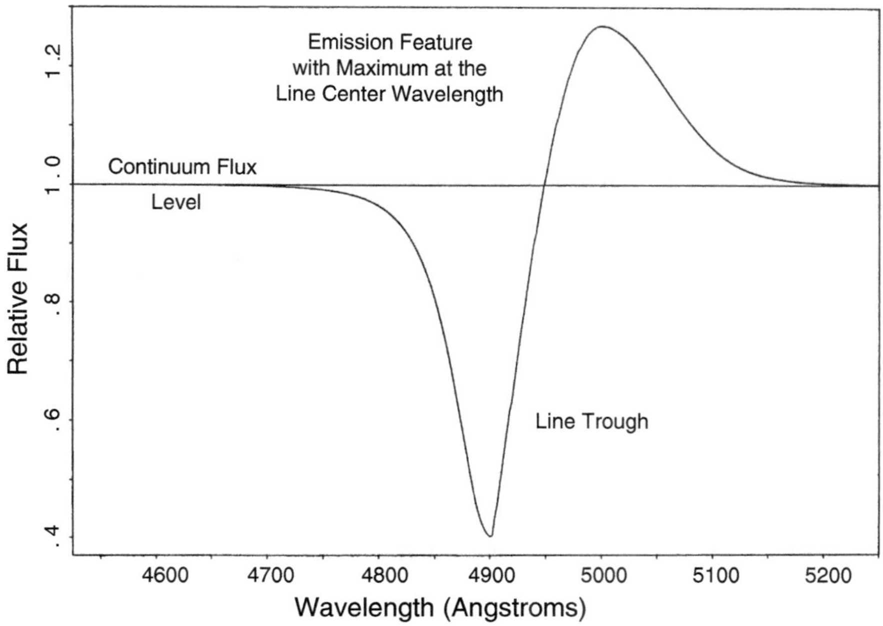

Fig. 4.2 A characteristic P Cygni line profile for a line of rest-wavelength 5000 Å and a velocity at the photosphere of 10,000 km s^{-1}. From "Analysis of Supernova Spectra" (Jeffery and Branch 1990)

4.2.3 Resonant-Scattering Line Profile: More Quantitative

To calculate a line profile, the line optical depth, $\tau(v)$, and the line source function, $S(v)$, in the line-forming region must be specified. In the comoving frame, the cross section of a spectral line is nonzero only in a narrow interval near the rest wavelength; if the line width is determined by thermal Doppler broadening, the width in velocity units is $v_{th} = 13\sqrt{(T/A10^4)}\,\mathrm{km\,s^{-1}}$, where A is the atomic weight of the atom. When a redshifting photon comes into resonance with an atomic transition, it remains in resonance as it crosses a region of *Sobolev length*, s, where $s dv/ds = v_{th}$, or $s = v_{th}t$. Because the ratio of the velocity-width of the resonance region and the supernova expansion velocity, v_{th}/v_{SN}, is small, a major simplification is possible: the *Sobolev approximation* (Sobolev 1960; Castor 1970; Lucy 1971), in which level populations are independent of position within the resonance region. Thus only the wavelength-integrated line cross section is needed, not its variation with wavelength. Because the resonance region is small, it is referred to as the *resonance point* and the line opacity is treated as a delta function at that point. (This may not be applicable for lines that have significant optical depth in intrinsically broad wings, such as Balmer lines in hydrogen-rich matter.)

The Sobolev approximation leads to an expression for the line optical depth:

$$\tau(v) = \frac{\pi \mathrm{e}^2}{\mathrm{m_e c}} f \lambda_o t n_l(v)[1 - \frac{g_l n_u(v)}{g_u n_l(v)}], \tag{4.1}$$

where f is the oscillator strength of the atomic transition, g_l and g_u are the statistical weights of the lower and upper levels, and the term in square brackets is the correction for stimulated emission. The radial dependence is only in the populations of the lower level, $n_l(v)$, and the upper level, $n_u(v)$, but these depend in turn on the radial dependences of the density, composition structure, ionization, and excitation. For simplicity $\tau(v)$ may be parameterized, e.g., as $\tau_{phot}(v/v_{phot})^{-n}$ or $\tau_{phot}\exp^{-(v-v_{phot})/v_e}$, where τ_{phot} and v_{phot} are the line optical depth and the velocity at the photosphere, respectively. (It can be useful to recognize that the power law has the same slope as the exponential at $v = nv_e$.)

The *source function*, with units of intensity, is a measure of the ability of the matter to emit radiation. In the Sobolev approximation, the line source function is

$$S(v) = \frac{2\mathrm{hc}}{\lambda^3}[\frac{g_u n_l(v)}{g_l n_u(v)} - 1]^{-1}. \tag{4.2}$$

For resonant scattering, the source function at any point must equal the mean intensity (intensity averaged over all incoming directions) at that point; i.e., what goes out is equal to what comes in. If the photosphere emits intensity I_{phot} in all outward directions, then $S(v) = W(v)I_{phot}$, where $W(v) = 1/2[1 - (1 - v_{phot}/v)^2)^{1/2}]$ is the geometrical dilution factor—the fraction of the sky subtended by the photosphere for an ion moving with velocity v. With these approximations, there are no free parameters for $S(v)$.

Now consider the wavelength dependence of intensity beams, I_λ, emitted along the observer's line of sight. The intensity of a beam with impact parameter greater than the radius of the photodisk is symmetric with respect to wavelength about λ_0 with (from the formal solution of the radiative-transfer equation)

$$I_\lambda = S(v)(1 - e^{-\tau(v)}), \tag{4.3}$$

where $\tau(v)$ and $S(v)$ are evaluated at the resonance point that is appropriate for wavelength λ. A beam with impact parameter smaller than the radius of the photodisk originates at the photosphere and has intensity

$$I_\lambda = I_{phot}e^{-\tau(v)} + S(v)(1 - e^{-\tau(v}), \tag{4.4}$$

where the first term is the continuum intensity that passes through the resonance point without being scattered out of the line of sight and the second term accounts for photons scattered into the line of sight at the resonance point. Since $S(v) < I_{phot}$, the scattering at the resonance point results in a net loss of intensity for the observer. Each such beam consists of the photospheric continuum with a superimposed blueshifted absorption.

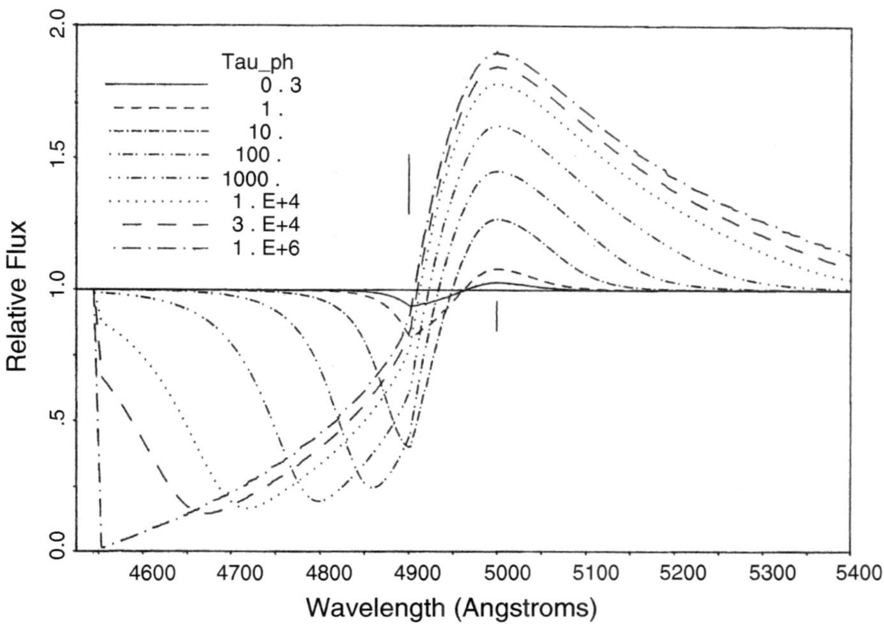

Fig. 4.3 Line profiles for line optical depth $\propto v^{-7}$, with various values of optical depth at the photosphere, Tau_ph. The *vertical line beneath* the continuum is at the rest wavelength of the line and the *vertical line above* the continuum is blueshifted to correspond to the velocity at the photosphere, 6000 km s^{-1}. The outer boundary of the line-forming layer is five times the radius of the photosphere. From "Analysis of Supernova Spectra" (Jeffery and Branch 1990)

All that remains is to appropriately sum the intensity beams to obtain the flux profile:

$$F_\lambda = 2\pi \int_{p_{min}}^{p_{max}} I_\lambda(p) p dp, \tag{4.5}$$

where p is the beam impact parameter in units of the photospheric radius, and the limits of integration vary from the front to the back of the line-forming region.[3]

Figure 4.3 shows that for sufficiently weak lines, the blueshift of the absorption minimum corresponds to the velocity at the photosphere. Thus the recession of the photosphere to deeper layers of ejecta can be inferred from a time series of spectra. For strong lines, since the line optical depth is high on common-direction surfaces well in front of the photodisk, the absorption minimum is more blueshifted.

[3]With additional simplifying assumptions it can be shown that if the line-forming region also is illuminated externally by emission from CSI, the line profile is "muted," i.e., the profile retains its shape, but with reduced amplitude of emission and absorption, or in the extreme case even "flipped," with absorption at the rest wavelength and blueshifted, net emission (Branch et al. 2000).

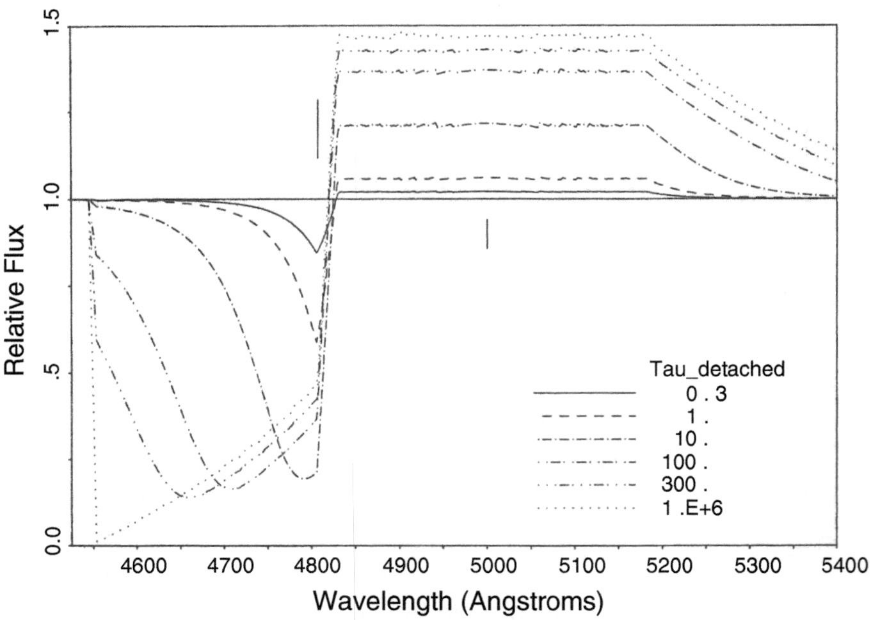

Fig. 4.4 Similar to Fig. 4.3, but for lines detached by one photospheric radius and various line optical depths at the detachment velocity, Tau_detached. From "Analysis of Supernova Spectra" (Jeffery and Branch 1990)

While it is common to assume a monotonically-declining density profile for supernova ejecta, the Sobolev approximation does not require this and functions equally well in a density distribution of arbitrary complexity. A common application is for a density distribution with a secondary peak that may result from the explosion dynamics or interaction with CSM. A different P Cygni shape is produced by a line that is "detached" from the photosphere, i.e., a line that has significant optical depth only above some detachment velocity that exceeds the velocity at the photosphere. This detachment can result from a nonmonotonic density or composition profile. The resulting line profile consists of a flat emission peak and an absorption that has a sharp red edge, blueshifted according to the detachment velocity (Fig. 4.4).

4.2.4 Multiple Scattering: Line Blending

If a photon scattered by one transition (B, for bluer) redshifts into resonance with another transition (R, for redder), the photon may be scattered again. This physical multiple scattering corresponds to the observational nuisance of line blending.

The source function of transition B, S_B (for clarity, we no longer indicate explicitly that source functions and optical depths are functions of v) is the same as above, but the source function of R is altered. To evaluate S_R at a resonance point, consider the sphere (a *common-point velocity surface*) centered on the resonance point (Fig. 4.1), the radius of which is determined by the velocity (wavelength) difference between B and R. Photons that are scattered by B somewhere on the sphere and directed toward the resonance point of R will arrive at the resonance point in resonance with R, increasing the mean intensity at the resonance point. On the other hand, continuum photons that would be in resonance with R at the resonance point are in resonance with B when they cross the sphere between the photosphere and the resonance point, and may be scattered into different directions, reducing the mean intensity at the resonance point. These considerations give

$$S_R = \frac{I_{phot}}{2} \int_{\mu(phot)}^{1} e^{-\tau_B} d\mu + \frac{1}{2} \int_{-1}^{1} S_B(1 - e^{-\tau_B}) d\mu, \quad (4.6)$$

where μ is the cosine of the angle with respect to the radial direction, $\mu(phot)$ is the cosine of the photospheric angular radius as seen from the resonance point, and S_B and τ_B are evaluated on the sphere. If the wavelength difference is sufficiently large, the sphere may cross the outer boundary of the line-forming layer, if one is imposed, and/or intersect or pass behind the photosphere; in these cases the limits of integration must be modified accordingly, which requires some bookkeeping.

To an extent, the two terms in the expression for S_R compensate, so simply using the single-line source function $S_R = W(v)I_{phot}$ would not be a bad approximation. The effects of multiple scattering, however, are essential for calculating the intensity beams and the flux profile. The expression for the intensity of a beam of impact parameter larger than the radius of the photodisk becomes

$$I_\lambda = S_B(1 - e^{-\tau_B})e^{-\tau_R} + S_R(1 - e^{-\tau_R}), \quad (4.7)$$

and that for a beam with impact parameter smaller than the radius of the photodisk becomes

$$I_\lambda = I_{phot} e^{-\tau_B} e^{-\tau_R} + S_B(1 - e^{-\tau_B})e^{-\tau_R} + S_R(1 - e^{-\tau_R}). \quad (4.8)$$

These expressions assume that both resonance points are within the line-forming region; when that is not the case, the expressions must be modified accordingly. By now, the reasoning behind these expressions for source functions and intensities, and the way to extend them to an arbitrary number of blending transitions, should be clear.

Figure 4.5 shows how the blends of a bluer line and a weaker redder line change as their wavelength separation decreases. Even though the red line is weaker, when the lines strongly overlap we see the absorption minimum of the red line rather than the emission peak of the blue line. Thus "absorptions trump emissions," i.e., in crowded spectra, absorption minima tend to appear at their expected blueshifted positions, while flux peaks may not correspond to line rest wavelengths. This is

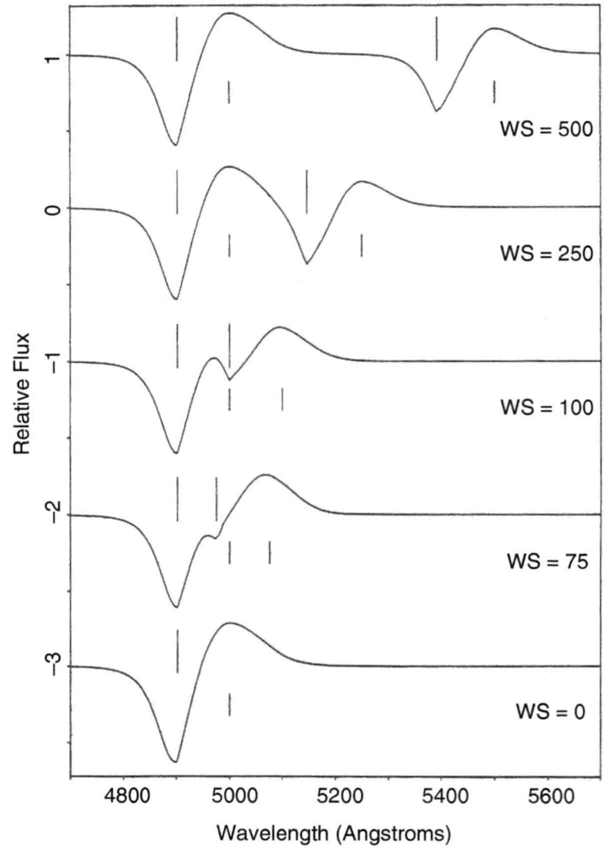

Fig. 4.5 Blends of a blue line and a weaker red line for various values of wavelength separation, WS. From "Analysis of Supernova Spectra" (Jeffery and Branch 1990)

why, when making line identifications in photospheric-phase spectra, one usually concentrates on absorption minima.

A more elaborate analytical and simple numerical treatment of the Sobolev approximation including nonscattering contributions to the source function and examples of further computations of line profiles in various density distributions was given by Jeffery and Branch (1990).

4.3 Lines To Be Considered

Supernova and stellar photospheric temperatures cover similar ranges, but the electron densities at supernova photospheres are lower, supernova compositions may differ strongly from the usual hydrogen-rich composition of stellar surfaces, and the severe supernova Doppler broadening means that only the strongest lines can be recognized. The LTE approximation for excitation and ionization can provide a useful guide to which lines may need to be considered when making supernova line identifications. Such a survey was carried out by Hatano et al. (1999b) for six different compositions: hydrogen-rich (solar); helium-rich (with hydrogen converted to helium); carbon–oxygen rich (with helium converted to equal parts carbon and oxygen); carbon-burned (especially abundant in oxygen, silicon, sulfur, and magnesium); oxygen-burned (especially abundant in silicon, sulfur, argon, calcium, and iron); and a time-dependent mixture of nickel, cobalt, and iron resulting from initially pure ^{56}Ni. For each composition and at a fixed characteristic electron density ($5 \times 10^9\,\mathrm{cm}^{-3}$), the Sobolev optical depths of the strongest lines of each ion are functions of temperature, as illustrated in Fig. 4.6 for the hydrogen-rich case. The figure shows, in agreement with observations of SN II, that at high temperature the spectrum is predicted to be nearly featureless, then lines of H, Ca II, and Fe II appear as the temperature falls. For each ion having a line strong enough to be of potential interest, for any of the six compositions, a single-ion synthetic spectrum produced by the SYNOW code (Sect. 4.4.1) was presented in Hatano et al. (1999b).

4.4 Synthetic Spectra for the Photospheric Phase

4.4.1 SYNOW

The SYNOW code (Fisher 2000; Branch et al. 2003) is the simplest of the widely-used supernova synthetic-spectrum codes. The basic assumptions are homologous expansion, spherical symmetry, a sharp blackbody photosphere, and resonant-scattering line formation treated in the Sobolev approximation. The code does not do continuum transport and it does not solve equations of statistical equilibrium (Sect. 4.4.3), so it is not suitable for inferring quantitative abundance ratios or for rigorously testing explosion models. The main function of SYNOW is to take line multiple scattering into account as described in Sect. 4.2.4, so that it can be used in an empirical spirit to make line identifications and determine the velocity at the photosphere as well as the velocity intervals in which each ion is detected. For each ion that is introduced, the optical depth of a reference line, either at the photosphere or at some detachment velocity, is a free parameter, and the corresponding optical depths of the other lines of the ion are calculated assuming LTE excitation at a chosen temperature. The radial distribution of line optical depths can be a power law, exponential, or Gaussian. When fitting to an observed spectrum, the most important parameters are the optical depths of the reference lines, the velocity at the photosphere, and whatever maximum or minimum (detachment) velocities may be imposed on each ion. Most of the established supernova line identifications in the photospheric phase have been made by comparing observed spectra with SYNOW synthetic spectra.

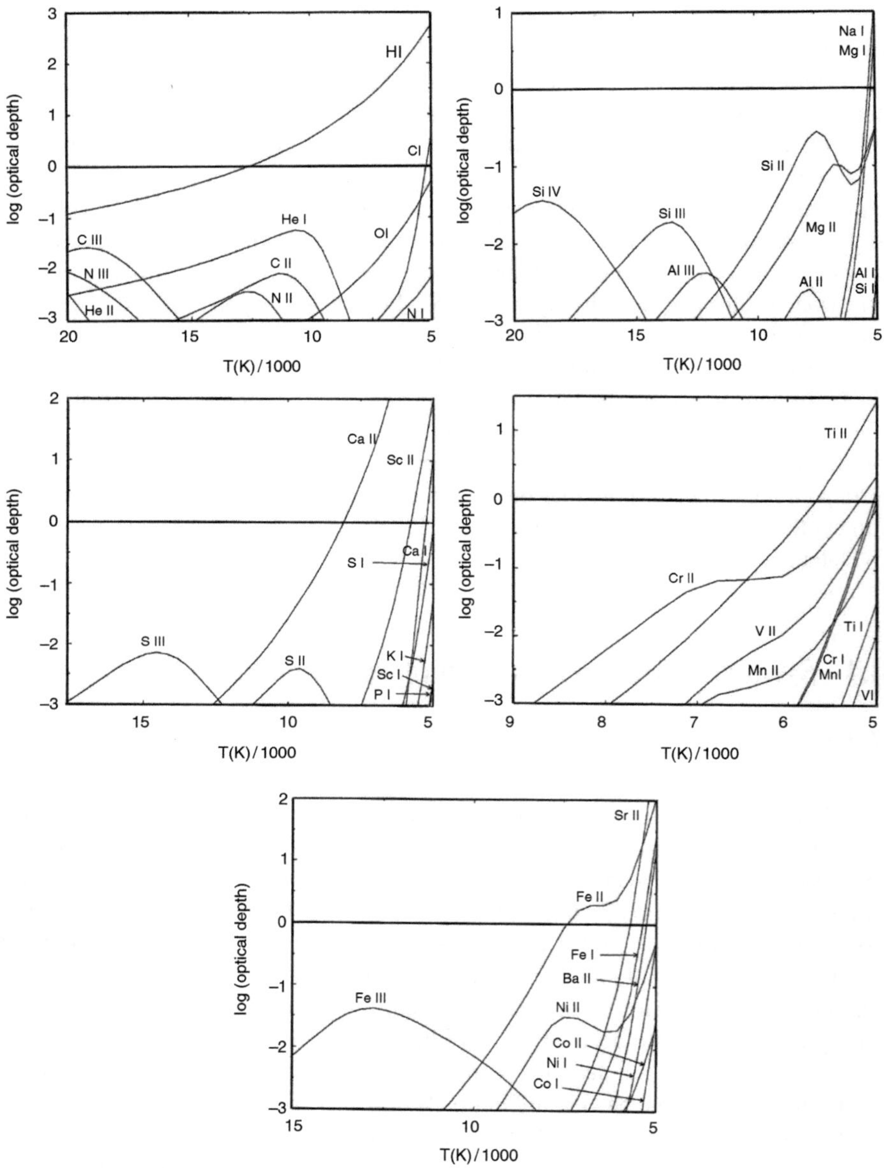

Fig. 4.6 Logarithm of Sobolev LTE optical depths of ion reference lines at the photosphere plotted against temperature for a hydrogen-rich composition. From "Ion Signatures in Supernova Spectra" (Hatano et al. 1999b). © AAS. Reproduced with permission

All that SYNOW requires from atomic physics are ionic partition functions and a very large list of lines, with wavelengths, log(gf) values, and excitation potentials. Small errors in log(gf) values inevitably have some effect on SYNOW spectra, but otherwise the atomic data are essentially known. Because SYNOW is iteration-free, a synthetic spectrum can be generated in seconds, and fitting to an observed spectrum can be done interactively (the human iterates).

The SYN++ code, a successor to the original SYNOW fortran code, is written in modern C++ and offers further enhancements. Since homogeneous fitting of very large sets of spectra by humans is not feasible, the SYNAPPS code, a combination of SYNOW and APPSPACK (Asynchronous Parallel Pattern Search, a minimizer PACKage) automatically performs SYNOW fits to observed spectra. SYN++ and SYNAPPS were described by Thomas et al. (2011) and Nugent et al. (2008), and both are publicly available.[4] Figure 4.7 shows a SYN++ fit to a spectrum of a SN Ia.

4.4.2 Elementary Monte Carlo

Monte Carlo (probabilistic) methods offer a flexible and intuitive approach to calculating supernova spectra. Packets of radiant energy, each packet representing many photons, are injected into the model ejecta and tracked through randomized scatterings and absorptions until they escape the line-forming region. A 1D code described by Mazzali and Lucy (1993) is frequently used for all types of supernovae (e.g., Altavilla et al. 2007; Hachinger et al. 2013). This code assumes a sharp blackbody photosphere, makes the Sobolev approximation, and does not solve rate equations. The main differences with respect to SYNOW are (1) for input composition and density structures, an approximate radiative-equilibrium temperature distribution and internally-consistent ionization ratios are calculated iteratively; (2) electron scattering (but not continuous absorption) in the line-forming region is taken into account; and (3) instead of invoking resonant scattering, an approximate treatment of photon branching (Lucy 1999; Mazzali 2000) allows an absorption in one nonresonant transition to be followed by emission in another. To reduce statistical noise, the emergent spectrum is not determined by binning and counting escaping packets according to wavelength, but by means of the formal solution for intensity beams (Sect. 4.2.3), using line and continuum source functions derived from the packet statistics (Lucy 1999).

4.4.3 Detailed Calculations

At considerable computational expense, much more physically realistic calculations of synthetic spectra are carried out for parameterized density and composition structures as well as for detailed nuclear-hydrodynamical explosion models having numerical density and composition structures. For explosion models, time series of

[4]https://c3.lbl.gov/es/.

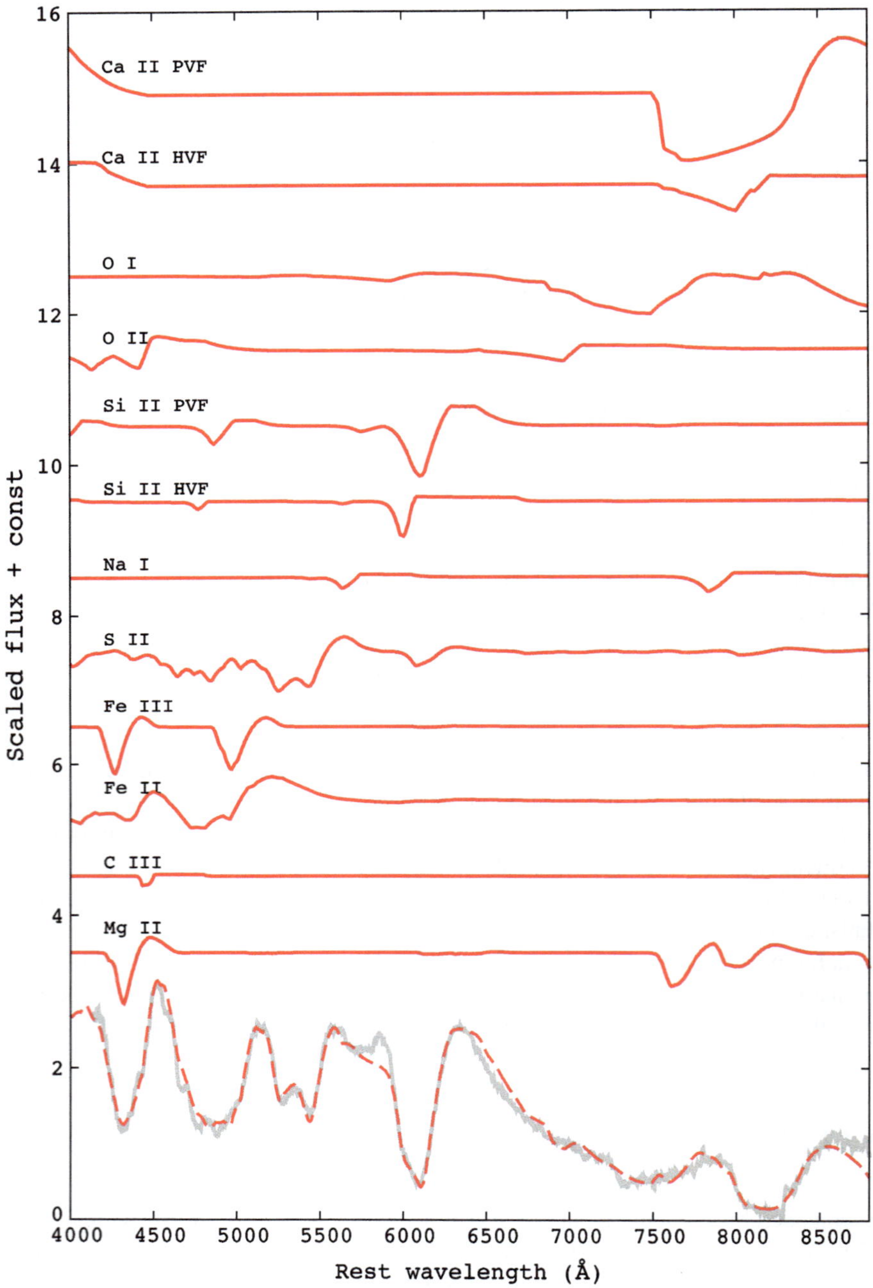

Fig. 4.7 At the bottom, a SYN++ fit to a maximum-light spectrum of the Type Ia SN 2010kg. Single-ion spectra of the ions used in the fit also are shown. PVF and HVF refer to photospheric-velocity features and high-velocity features (Sect. 20.4), respectively. Figure 1 "Possible detection of singly ionized oxygen in the Type Ia SN 2010kg" from Barna et al. (2016), by permission of Oxford University Press on behalf of the Royal Astronomical Society

synthetic spectra provide a stringent confrontation with observations. Homologous expansion is sufficient for many supernova applications, but ideally all other approximations discussed above should be abandoned, making self-consistent synthetic-spectrum calculations a formidable problem; to include CSI, even more so.

Worst of all, real supernovae are three-dimensional.

4.4.3.1 Non-Local Thermodynamic Equilibrium (NLTE)

Even in 1D, the ideal calculation involves the following. (1) Atomic level populations in each radial zone are obtained by solving coupled NLTE rate equations (equations of statistical equilibrium), taking the rates of all relevant radiative and collisional transitions between levels into account. For each ion, there is one equation for every level treated in NLTE, and the equations for all ions are coupled by the constraint of charge conservation. In practice, the number of levels that can be treated in NLTE is limited; the populations of a much larger number of additional levels, which contribute to the opacity, may be calculated in LTE. (2) The wavelength-dependent mean intensity in each zone is obtained by solving the special-relativistic radiation-transfer equation, taking into account all sources of radiation in other zones and the intervening opacity. (3) The kinetic temperature in each zone is obtained by enforcing energy conservation, with nonlocal energy deposition of the products of nuclear decay taken into account. The level populations, mean intensities, and temperatures are mutually dependent, so final solutions are found by means of nested sets of iterations.

The number of radial zones may be $\sim 10^2$. The number of ions treated in NLTE may be $\sim 10^2$. The number of atomic levels treated in NLTE may be $\sim 10^5$. The number of atomic transitions treated in NLTE and LTE may be $\sim 10^5$ and $\sim 10^7$, respectively. If the Sobolev approximation is dropped, then the cross section across each bound–bound transition should be resolved in the comoving frame, requiring $\sim 10^5$ wavelength points for the radiation field. These numbers add (multiply) up.

The equations for level populations, mean intensities, and temperatures all contain time-dependent terms because the physical state of a supernova at one time depends on its physical state at previous times. In some calculations, some or all of the time-dependent terms are neglected, so that the supernova is modeled in "snapshot" mode, with the speed of light effectively being taken to be infinite. Undesirable consequences of this approximation were discussed by Hillier and Dessart (2012).

4.4.3.2 Nonthermal Excitation and Ionization

For supernovae, nonthermal excitation and ionization caused by the products of nuclear decay should be taken into account. When a γ-ray from a nuclear decay undergoes Compton scattering off a free or bound electron, a substantial fraction of its energy is lost, producing a primary nonthermal electron that, in addition to undergoing Coulomb interactions with thermal electrons, can cause excitation and ionization. When the primary electron causes ionization, a secondary nonthermal electron is produced with an energy that is several times larger than the relevant ionization potentials and is capable of causing further excitation and ionization (Fano 1963; Gould 1972; Lucy 1991; Swartz 1991). In this way, each original γ-ray can lead to numerous excitations and ionizations. In general, the deposition of γ-ray energy is nonlocal (Sect. 5.4.2), but the mean free path of a nonthermal electron is short enough that its energy is effectively deposited "on the spot."

Generally, nonthermal excitation and ionization processes are not important for excitation and ionization levels that can be significantly populated thermally. The outstanding manifestation of the importance of nonthermal excitation in the photospheric phase is the appearance of He I lines in the optical spectra of SN Ib. These lines have lower-level excitation potentials of about 20 eV, which are not populated thermally to any significant extent. While nonthermal processes produce only a small excitation fraction, this fraction is orders of magnitude higher than that resulting from thermal excitation, and since the helium abundance is high, significant He I line optical depths are produced (Lucy 1991; Swartz et al. 1993b).

4.4.3.3 Multiple Electron Scattering

In some supernovae, especially SN IIn and superluminous supernovae, lines form in regions that are above the optically-thick photosphere where thermalization occurs, but still in regions that have an optical depth of a few to electron scattering. In this circumstance, lines formed above the photosphere can be subject to multiple electron scattering before emerging to the observer. The width of the lines can then be affected not only by the bulk expansion velocity of the material in the line-forming region, but also by the thermal distribution of the electrons. Lines broadened by electron scattering tend to have symmetric wings with Lorentzian profiles (Münch 1948; Auer and van Blerkom 1972; Hillier 1991; Chugai 2001; Laor 2006).

The thermal velocity of electrons at electron temperature, T_e, is $v_e \simeq 700 \sqrt{(T/10^4)}\,\mathrm{km\,s^{-1}}$. The number of scatterings scales as about the square of the electron-scattering optical depth, τ_e, so the full width half maximum (FWHM) of the line in velocity units is about $\tau_e v_e$. When observed, these symmetric scattering-line profiles typically show FWHM of $\sim$2000 km s^{-1} and are often described as being of "intermediate width," falling between the typical broad lines of supernova ejecta and the narrow lines of slow-moving CSM. The scattering profiles can also show superimposed narrow P Cygni profiles from line scattering in slow-moving matter beyond the electron scattering region and very narrow line peaks from especially slow-moving matter that has been excited by, for instance, breakout radiation, but still moves with speeds characteristic of the wind that created the CSM.

The details of the line profile can depend on whether photons are generated internally to the scattering region and on the density and velocity profiles of the

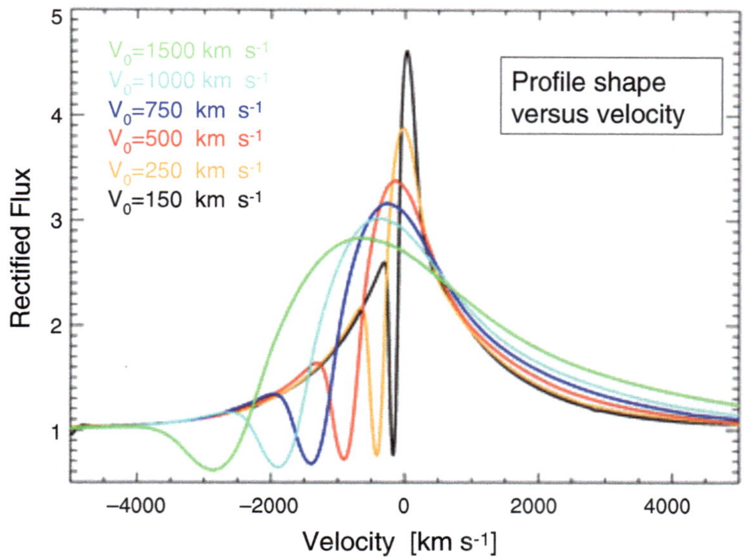

Fig. 4.8 The effect of multiple electron scattering on line profiles as the bulk expansion velocity is decreased. Figure 15 "SN 1994W: an interacting supernova or two interacting shells?" adapted from Dessart et al. (2009)

scattering matter. Compton scattering could affect the line profiles, but tends to be small in practice for typical temperatures and electron velocities. The line peak can be slightly redshifted for spherical geometry and expansion (Fransson and Chevalier 1989), but may have different behavior in asymmetric geometries.

Lines with multiple-scattering wings are observed in a variety of species but tend to be especially prominent in the Balmer lines. In the case where these lines are formed within the multiple-scattering region, Hα, Hβ, and Hγ form at increasing electron-scattering optical depth, which leads to wings of increasing width from Hα to Hβ to Hγ. If the Balmer lines form outside the electron-scattering region, as they may in some circumstances, the lines should have similar profiles and lack the scattering wings.

The effect of multiple electron scattering becomes especially apparent or even dominant for low expansion velocities. Figure 4.8 shows the transition from a P Cygni profile to a symmetric Lorentzian profile as the bulk expansion velocity is decreased (Dessart et al. 2009). In general, Monte Carlo techniques are most appropriate for computing line profiles subject to multiple electron scattering.

4.4.3.4 Advanced Spectrum Codes

Thanks to advances in (1) high-performance scientific computing including super-computer parallelization (2) the development of extremely efficient computational algorithms, and (3) laboratory atomic-data measurements, powerful synthetic-spectrum codes exist. Numerical stability and convergence are important issues to be considered. These codes also require as input an enormous amount of atomic (and in some cases molecular) data—for bound–bound and bound–free radiative and collisional transitions, dielectronic recombination, charge-exchange reactions, and more. Despite the rapid and continual improvements in measurements of atomic data, much further progress in this area is needed (Hillier 2011).

Detailed synthetic-spectrum codes may be "finite difference" (grid-based, deterministic), based on iteratively obtaining simultaneous solutions to differential and intego-differential equations, or Monte Carlo (probabilistic), or a hybrid involving both. Detailed codes differ in their computational approach to the problem; in practice, even a given code often makes different approximations for different applications. Because most codes undergo continual development, a few of them are mentioned, but not described, here. For details, the reader should consult the cited works, as well as future papers that will describe further code developments. Applications of these codes to specific supernovae will be discussed in later chapters.

Highly-developed grid-based codes include PHOENIX (Hauschildt and Baron 1999; Lentz et al. 2000; Baron et al. 2009; Jack et al. 2012b), CMFGEN (Hillier and Dessart 2012; Li et al. 2012), and RAGE/SPECTRUM (Frey et al. 2013). An example of a hybrid code is HYDRA (Höflich 2005, 2009; Höflich et al. 2006).

Several powerful Monte Carlo codes, e.g., SEDONA (Kasen et al. 2006) and ARTIS[5] (Kromer and Sim 2009; Bulla et al. 2015), have been developed. In principle, such codes will be able to take all relevant physical effects into account. They are implemented on massively parallel systems, and they are particularly efficient for treating 3D, time dependence, and polarization.

4.5 The Nebular Phase

The photospheric and nebular phases offer complementary insights and challenges for analysis. Since the continuum opacity is negligible, the nebular phase provides a direct view into the inner regions of the ejecta.

In the absence of CSI, the nebular phase is powered by radioactive decay, primarily that of ^{56}Co for several hundred days. At very late times (years), longer-lived species such as ^{57}Co and ^{44}Ti may become important. Gamma rays and positrons deposit energy by Compton scattering to produce nonthermal fast electrons, the energy of which quickly goes into atomic ionization and excitation as well as Coulomb heating of the thermal electrons. Optical emission lines are

[5]TARDIS (Kerzendorf and Sim 2014), a simplified, rapid version of Artis, is publically available at http://pypi.python.org/pypi/tardis.sn.

produced by recombination, collisional excitation, and fluorescence (Kozma and Fransson 1992).

Because the nebular phase is associated with especially low densities that result from expansion, the lines produced in this phase tend to be *forbidden* rather than the *permitted* lines that characterize the higher-density photospheric phase. The characteristic line profile of the nebular phase is an emission line that peaks near the rest wavelength. Because the lines tend to form in deep, relatively high-density, slow-moving layers (typically a few thousand $\mathrm{km\,s^{-1}}$) of the ejecta, in the late phases of some supernovae line blending is not too severe. Thus it is somewhat more feasible to extract information without making synthetic-spectrum calculations. Some line identifications are straightforward, e.g., [O I] $\lambda\lambda$6300,6364, Mg I] λ4571, and [Ca II] $\lambda\lambda$7291,7323 in core-collapse supernovae, as well as Balmer lines in SN II. In SN Ia, line identifications in optical spectra are less obvious, but some isolated, readily identifiable lines appear in the near-IR.

In the limit of complete transparency, the flux of an emission line as a function of wavelength is determined by the distribution of the upper-level populations in the ejecta. If the line emits all the way to the center of the ejecta, the flux profile peaks at the rest wavelength; if there is a central "hole" in the emissivity, then the center of the flux profile is flat. This is useful, e.g., in probing the radial distribution of various iron-group elements in the deep layers.

Photospheric-phase flux spectra are relatively insensitive to asymmetry, but nebular-phase flux spectra can provide some information on both small-scale clumping and global shape asymmetry. For example, wavelength-shifted peaks of emission lines can provide information on asymmetries along the observer's line of sight. This is complementary to spectropolarimetry, which measures asymmetry on the plane of the sky (Sect. 4.6).

In general, analysis of nebular spectra involves spectrum calculations, using line lists that include forbidden lines. As for the photospheric phase, nebular-phase synthetic-spectrum calculations of various levels of complexity prove to be useful. Relatively simple parameterized approaches that can be used as diagnostic tools were described by Bowers et al. (1997) and Fransson et al. (2002). A one-zone code that frequently has been used as a diagnostic tool for various types of supernovae (Mazzali et al. 1997, 2010) has as input parameters just the time since explosion and the mass, composition, and outer velocity of the ejecta. Heating is calculated from ^{56}Co decay and cooling is by radiative emission. The electron density, temperature, NLTE level populations, and emission-line fluxes are determined by enforcing statistical and thermal equilibrium. An example fit is shown in Fig. 4.9.

Modern multizone codes for calculating nebular-phase synthetic spectra of hydrodynamical explosion models include NERO (Maurer et al. 2011), a 1D steady-state code. The steady-state assumption is valid from months to years after explosion when all atomic processes are so fast that the supernova almost instantaneously emits almost all of the energy deposited by radioactivity. Time-dependence can be neglected because the output is determined by the instantaneous radioactive input, not on the history of the supernova.

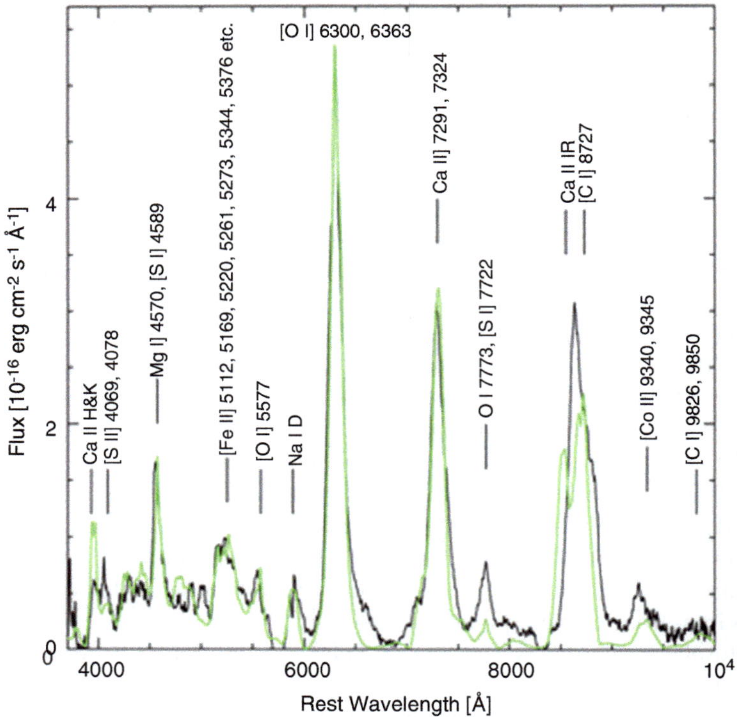

Fig. 4.9 A nebular-phase spectrum of the Type Ic SN 2007gr, obtained 172 days after explosion, is compared to a synthetic spectrum. Figure 4 "The Type Ic SN 2007gr: a census of the ejecta from late-time optical-infrared spectra" from Mazzali et al. (2010), by permission of Oxford University Press on behalf of the Royal Astronomical Society

After several years, owing to low density and temperature, atomic processes occur more slowly and the radiative output does not necessarily match the radioactivity input. The output now does depend on the history, and time-dependence must be taken into account. For example, an important time-dependent effect is *ionization freezeout*, which occurs when the recombination time is no longer short compared to the radioactivity timescale. During freezeout, the level of ionization remains nearly constant and a steady-state treatment would underestimate the ionization and the luminosity (Fransson and Kozma 1993). When the cooling timescale becomes longer than the expansion timescale, adiabatic cooling must be taken into account.

Although continuum opacity is negligible in the nebular phase, permitted lines in the UV may have large optical depths, in which case nonlocal radiative transport must be taken into account (Jerkstrand et al. 2011). Even in the optical, permitted lines (e.g., lines of Fe II) may have significant optical depths well into what is commonly regarded to be the nebular phase (Branch et al. 2008; Black et al. 2016). In the highly-idealized case of a permitted line forming inside a transparent, uniformly-emitting core, the P Cygni profile is qualitatively different from that of the photospheric phase (Friesen et al. 2012). The emission peak and the absorption dip are of low contrast with respect to the continuum and the emission peak is redshifted with respect to the rest wavelength.

4.6 Spectropolarimetry

Supernova explosions are intrinsically 3D events with significant departures from spherical symmetry. Understanding asymmetries in supernovae is important for phenomena that range from the origin of gamma-ray bursts to the cosmological applications of SN Ia.

In the photospheric phase, flux spectra of aspherical models can be closely reproduced by appropriate adjustment of the parameters of spherical models, so that it is difficult to deduce the geometry from the flux spectra (Höflich et al. 1996). Flux spectra give no information on important issues such as the different orientations for different chemical species. Nebular flux spectra may be more useful indicators of asymmetry since line profiles can reveal velocity asymmetries and small-scale clumping structure, but again the two techniques together can be a powerful probe of geometry since the spectropolarimetry yields information on the plane of the sky while nebular spectroscopy yields information along the line of sight by means of the Doppler shift.

Spectropolarimetry has shown that virtually all supernovae are significantly aspherical. Core-collapse supernovae behave differently than SN Ia. Polarization of core-collapse supernovae tends to increase with time, suggesting that asymmetry is stronger in the inner layers and that the explosion process itself is strongly aspherical. Core-collapse supernovae tend to establish a preferred direction of asymmetry. An axially-symmetric geometry can account for many basic features, but departures from axial symmetry are also inferred for most core-collapse events. In SN Ia, the asphericity is stronger in the outer layers, providing constraints on the burning process and perhaps on the circumstellar environment. A history of the early observations, those of SN 1987A, and the contemporary record was presented by Wang and Wheeler (2008).

For polarization in a scattering atmosphere, the electron density is more important than the total mass density, and optical depth and occultation effects also are important. The extended atmosphere of a supernova yields more polarization than would a plane-parallel atmosphere. Much of the basic understanding of the possible behavior of polarized supernovae, especially of the continuum, is elucidated by ellipsoidal models. To produce the same net polarization, prolate models tend to require larger distortion than oblate models (Höflich et al. 1996). While real supernova photospheres are undoubtedly more complex than ellipsoids, simple models remain a handy way to estimate the global asphericity from the magnitude of the continuum polarization (Hoflich 1991).

The situation is more complex for line polarization, which may depend on non-LTE effects and on the angular as well as the radial distribution of the radiation.

Lines that form P Cygni scattering profiles in asymmetric expanding atmospheres should show greater linear polarization than in the continuum. Polarization increases at the absorption minimum because the absorbing material blocks predominantly unpolarized forward-scattered flux from the portion of the photosphere along the line of sight. This increases the relative proportion of polarized flux that scatters from the asymmetric limb. The polarization tends to decrease at the emission peak because the emitted unpolarized flux tends to dilute the underlying polarized continuum flux. The result is an *inverse P Cygni* polarization profile (McCall 1984; Jeffery 1987).

The polarization of flux from a source is characterized by the four Stokes parameters, I, Q, U, and V (Chandrasekhar 1960), where I is the intensity, Q and U measure the linear polarization, and V measures the circular polarization. The Stokes vectors Q and U have an amplitude and a direction, but are *quasi-vectors* for which the directions 0° and 180° are identical since they are related to the intensity, which is the square of the amplitude of the electric vector.

Operationally, I is the total flux and Q and U are differences in flux determined with the electric vector oscillating in two orthogonal directions on the sky. The parameter U represents angles on the sky that are rotated by 45° with respect to Q. The degree of linear polarization, P, and the polarization angle, θ, are expressed in terms of the Stokes parameters as $P = \sqrt{Q^2 + U^2}/I$ and $\theta = \frac{1}{2}\arctan\frac{U}{Q}$, or

$$Q/I = P\cos 2\theta, \; U/I = P\sin 2\theta. \tag{4.9}$$

The astronomical convention is that $\theta = 0°$ points to north on the sky.

To measure the Stokes vectors, the incoming flux is split into *ordinary* and *extraordinary* rays, usually with a Wollaston prism. Because polarimetry involves the difference of the ordinary and extraordinary rays, errors per pixel are amplified, and estimates of polarization from Eq. (4.9) can be biased to large values by noise. Proper treatment of errors in data was discussed by Patat and Romaniello (2006). Interstellar polarization (ISP) introduced by dust in either the host galaxy or our Galaxy also complicates the interpretation of the data. Polarization from supernovae varies across spectral lines and with time; these factors help to separate out the fixed ISP.

Spectropolarimetry data are often presented as a function of wavelength in the Q/U plane. A major advantage to presenting the data in the Q/U plane is that one can then do vector analysis by eye to determine how the amplitude and polarization angle will respond to a change of the uncertain ISP. In addition, the tendency of the data to follow a single axis or to depart from a primary axis in loops or other structures is readily apparent, independent of the ISP.

Polarization data can be analyzed in terms of two principal components. In the Q/U plane, the two components correspond to projecting the polarization vectors onto a *dominant axis* and an *orthogonal axis*. The coordinates of the new system are given by rotating the original coordinates counterclockwise so that the Q axis is parallel to the dominant axis in the new coordinate system and the dominant axis points toward the center of the data as it clusters on the Q/U plane. The components parallel and perpendicular to the dominant axis are given by

$$P_d = ((Q - Q_{ISP})\cos\alpha + (U - U_{ISP})\sin\alpha)/I, \tag{4.10}$$

and

$$P_o = -((Q - Q_{ISP})\sin\alpha + (U - U_{ISP})\cos\alpha)/I, \tag{4.11}$$

where Q_{ISP} and U_{ISP} are the Stokes parameters of the ISP, $\alpha = 2\theta_d$ is the rotation angle in the Q/U plane, and θ_d is the polarization position angle of the dominant axis. The spectral profiles of P_d and P_o are insensitive to the choice of ISP, and hence have some advantage compared to calculating the degree of polarization using Eq. (4.9).

Although there are prominent and interesting exceptions, a dominant axis often can be identified. This represents the orientation on the sky of the preferred physical axis of the supernova ejecta. The spectropolarimetry projected onto the dominant axis represents global geometric deviations from spherical symmetry. In particular, if the ejecta are strictly axially symmetric, the data on the Q/U diagram fall on a straight line, defining the dominant axis. The components perpendicular to the dominant axis represent physical deviations of the ejecta distribution from the dominant axis. Thus, if the geometry departs significantly from axial symmetry, the data will scatter about the dominant axis in the direction of the orthogonal axis. These basic situations and their representation in the Q/U plane are shown schematically in Fig. 4.10.

The top left panel of Fig. 4.10 presents a smooth, axially and mirror-symmetric structure with the axis tilted on the sky and with respect to the line of sight. Sampling different wavelengths across a line profile corresponds to viewing the geometry along different slices normal to the line of sight. Different wavelengths thus show different values of Q and U, depending on the geometry and optical depth, and hence different polarization amplitudes, but at a fixed polarization angle, as illustrated in the top right panel. The lower left panel of Fig. 4.10 illustrates a case in which the axisymmetry is broken by "clumps" that vary in composition or optical depth and therefore in their wavelength-dependent structure. Clumps of high-opacity matter block underlying patches of photosphere and lead to a wavelength-dependent geometry on planes of constant velocity even if the underlying photosphere is symmetric. For this situation, the data do not fall on a straight line in the Q/U plane, as illustrated in the lower right panel. The underlying axisymmetric geometry may still be evident by producing a dominant axis, but the clumping causes an appreciable physically-significant distribution of the data along the orthogonal axis. Clumps blocking an asymmetric photosphere yield more complex structure, but no difference in principle.

This basic phenomenology serves to define spectropolarimetry types according to their distribution in the Q/U plane (Wang and Wheeler 2008). The classification scheme is insensitive to the value of the ISP. The spectropolarimetric (SP) types are:

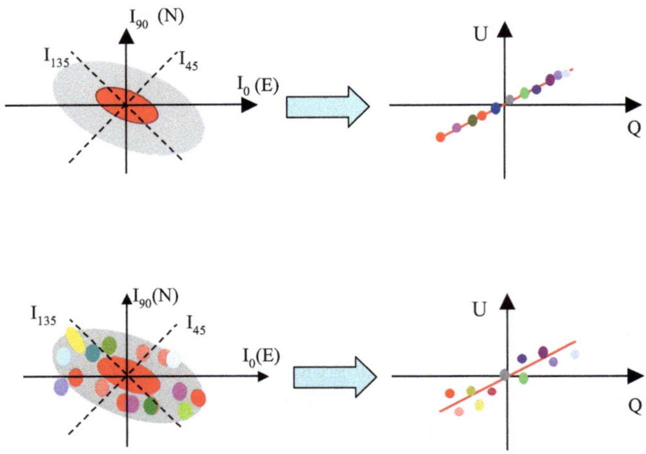

Fig. 4.10 *Top left*: A smooth, axisymmetric structure. The directions denoted by I represent the measurement of the flux at the angles needed to construct the Q and U polarization components. *Top right*: The resulting wavelength-dependent polarization amplitude plotted in the Q/U plane follows a straight line, the dominant axis. The *colored circles* represent the polarization measured at different wavelengths. *Lower left*: A case for which the matter is distributed in clumps of different composition and optical depth. *Lower right*: The polarization distribution in the Q/U plane is no longer along a single line. The departure from axial symmetry caused by the clumping yields a finite distribution along the orthogonal axis. The Q/U diagrams in the *upper and lower right panels* represent spectropolarimetry Types D0 and D1, respectively. From "Spectropolarimetry of Supernovae" (Wang and Wheeler 2008), reproduced with permission of Annual Reviews of Astronomy and Astrophysics, Volume 46, © Annual Reviews, http://www.annualreviews.org

SP Type N0: the data are consistent with observational error. The centroid of the distribution may be offset from zero polarization due to ISP, but there is no measurable supernova polarization.

SP Type N1: the data show no significant elongation in any preferred direction, but the distribution apparently is wider than would be consistent with observational error, indicating either intrinsic polarization or underestimation of observational error.

SP Type D0: the data form an elongated ellipse in a Q/U diagram, with the distribution orthogonal to the major axis of the ellipse being consistent with observational error. The locus of the data can be well approximated by a straight line, the position of which defines the dominant axis. (For strong ISP and broad wavelength coverage, the ISP may vary with wavelength, complicating the analysis.)

SP Type D1: the data form an elongated ellipse in a Q/U diagram so that a dominant axis can be identified, but deviations orthogonal to the dominant axis are significant.

SP Type L: the data show large changes in amplitude and position angle across strong spectral lines. These variations result in prominent loops in the Q/U diagram.

Because the spectropolarimetric properties of supernovae evolve with time and wavelength, the SP type also may evolve, and different portions of the spectrum may suggest different SP types.

Continuum polarization gives information on the global shape of the photosphere. Polarization across spectral lines gives information about small-scale opacity variations in the ejecta, due to geometry or composition, or both. For complex composition structure, both the degree and angle of polarization can vary across strong spectral lines. This produces loops in the Q/U plane. Since these loops are functions of the wavelength across the line feature, they are functions of the velocity and the depth of the portion of the structure that contributes to that wavelength. Because the loops result from changes in the amplitude and angle of polarization as a function of wavelength, velocity, and depth, they specifically demonstrate departures from axial symmetry. Loops and their implied nonaxisymmetric structures are common in both core-collapse supernovae and SN Ia. Figure 4.11 shows the wavelength-dependent data for the Type Ib SN 2008D (Sect. 16.2.2) in the Q/U plane. A complex loop structure is especially prominent in the Ca II near-IR triplet and somewhat less so in the Ca II H&K. Examples of how loops may be formed are given in Kasen et al. (2003), Hoffman et al. (2008), and Hole et al. (2010).

The discussion above of the decomposition of the polarization into dominant and orthogonal components implicitly assumes that there are only two components to the polarization, each of which has a fixed axis. One is due to the ISP and the other is intrinsic to the supernova ejecta. In common circumstances, the geometry is more complicated than this simple decomposition can treat.

A powerful method to illustrate the special attributes of spectropolarimetry is to plot the data in a polar diagram where the radial displacement is proportional to the line absorption minimum and the angle represents the orientation in the Q/U plane. Figure 4.12 gives an example (again for SN 2008D) where it is seen that some chemical elements (oxygen, calcium) are ejected at different angles with respect to other elements (helium, silicon, iron) and that the orientations vary with epoch. Total flux spectra in neither the photospheric nor nebular phase are able to give such perspective. Figure 4.12 is presented again in Chap. 16 (Fig. 16.13) where SN 2008D is discussed in more detail.

Supernova spectropolarimetry is still a young field, but some basic conclusions can be reached that have important implications for the future of supernova research. Spectropolarimetry has been obtained for every major spectral type of supernova of various luminosity classes and peculiarities. They are all polarized and hence aspherical in some significant way. Spectropolarimetry of supernovae can also reveal information about the properties of the interstellar dust in the host galaxy. Current data suggest that this dust frequently has properties different from those of dust in the Galaxy (Patat et al. 2015).

Spectropolarimetry is still in a data-driven phase, but much theoretical work has laid the groundwork for theoretical understanding (Jeffery 1987, 1991; Hoflich 1991; Chugai 1992b, 2006; Kasen et al. 2003). Exploring more complex structures requires more general models. Monte Carlo radiative-transfer models are the ones

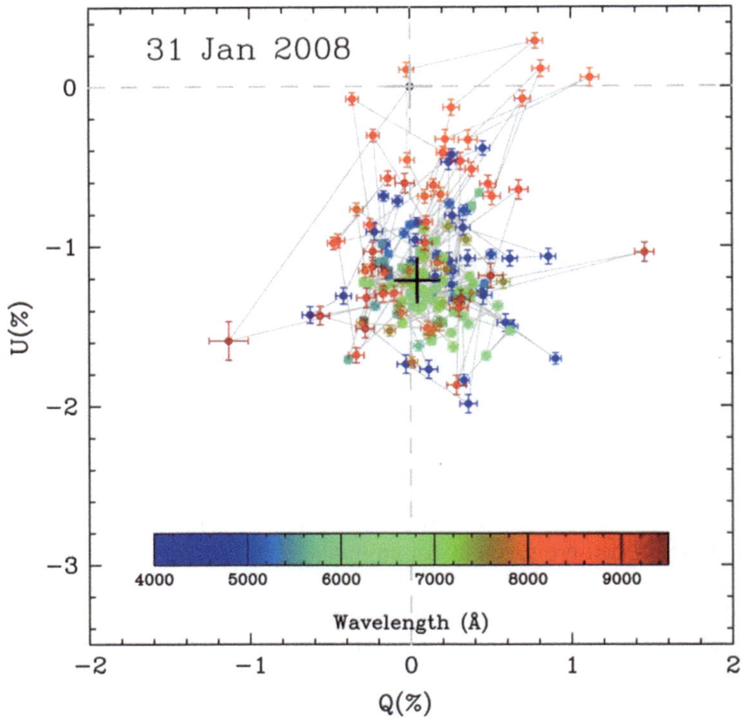

Fig. 4.11 Spectropolarimetry of the Type Ib SN 2008D near V-band maximum presented in the Stokes Q/U plane. The location of the ISP is indicated by the *black cross*. The *gray cross* and *dashed lines* denote null polarization. The observed data are color coded according to wavelength, as shown by the color bar. Note the excursion of *orange points* in the upper right corresponding to the Ca II near-IR triplet; this "loop" structure corresponds to a variation in polarization amplitude and angle with wavelength across the line and implies a departure from bipolar symmetry. From "The Early Asymmetries of Supernova 2008D/XRF 080109" (Maund et al. 2009). © AAS. Reproduced with permission

most likely to provide the needed generality of the underlying geometry and the capacity to handle the associated radiative transfer with a realistic amount of computer time. Contemporary examples include HYDRA (Höflich 2005), SEDONA (Kasen et al. 2006), and ARTIS (Kromer and Sim 2009). These codes treat aspherical geometries, γ-ray deposition, and line and continuum transfer. They can compute the spectral and polarimetric evolution as would be observed from arbitrary viewing angles.

Because there are so many variables associated with full 3D structure, analysis of supernova polarization will require the development of careful methodology (Kasen and Plewa 2007) and the acquisition of abundant data. The applications of spectropolarimetry to individual supernova spectral types will be discussed in subsequent chapters.

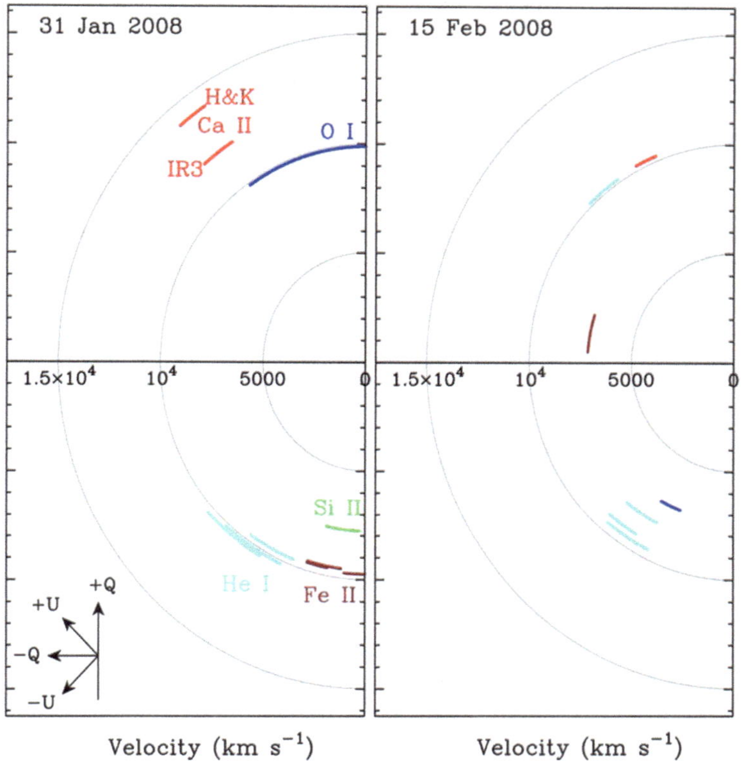

Fig. 4.12 Polar plots showing the velocity and polarization angle at absorption minimum for lines of helium, oxygen, silicon, calcium, and iron near maximum light and about 2 weeks later for the Type Ib SN 2008D. See also Fig. 16.13. From "Models for Type I supernovae—Partially incinerated white dwarfs" (Maund et al. 2009). © AAS. Reproduced with permission

4.7 Summary

The spectra of supernovae are typically divided into two phases: the photospheric phase when permitted lines form above an optically-thick photosphere and the nebular phase when the ejecta are optically thin in the continuum and forbidden lines dominate at the associated low densities. In the photospheric phase, lines typically show the P Cygni profiles characteristic of expanding atmospheres with peaks near the rest wavelength of the line and blueshifted absorptions characteristic of the velocity of the material above the photosphere. In the nebular phase, radiative-transfer effects may be small and line profiles reflect the expansion of the ejecta in which they form. The power to drive continuum and line emission comes from the original heat of explosion in some cases and from radioactive decay and CSI in other cases and at later epochs.

Spectra of the photospheric and nebular phases usually receive different treatments, but in principle the two phases involve the same problem of time-dependent radiative transfer, with different effects having different levels of importance. Eventually it will be feasible to treat both phases with a single advanced code. Full solution of the spectrum problem will require 3D and incorporation of the effects of CSI.

Departures from axisymmetry are ubiquitous in core-collapse and thermonuclear supernovae. Spectropolarimetry provides a potentially powerful tool to characterize and explore the physical nature of those asymmetries.

Chapter 5
Light Curves

5.1 Introduction

Characteristic light curves of the main supernova types were illustrated in Chap. 1. In this chapter, we consider the most important physical processes that affect light curves. We begin by presenting simple concepts that capture much of the essence of supernova light curves, including some useful rules of thumb that explain, for instance, why most events require about the same time to reach maximum light and have about the same photospheric radius at that time. We then consider the energy sources that power light curves, assuming that supernovae explode in a vacuum. Effects associated with the circumstellar and interstellar environment are reserved for later chapters; light curves of SN IIn, and perhaps also SN IIL, are strongly affected by CSI (Chap. 6). We then briefly describe how the various mechanisms come into play for the main supernova types and consider the processes that affect broadband light curves. Because the opacity is wavelength dependent and constant neither in time nor space, the evolution of the spectral energy distribution (SED), and therefore light curves in individual broad bands, is more complicated than bolometric or UVOIR light curves.

5.2 Physical Conditions

Consider a generic supernova that is 20 days past explosion and has both density and temperature independent of radius (a crude assumption, for simplicity). Ejection velocity is proportional to radius, with $v = 10{,}000\,\mathrm{km\,s^{-1}}$ at the surface, the photosphere. Then the radius is $R = vt = 1.7\times10^{15}$ cm. Assume that the luminosity

D. Branch, J.C. Wheeler, *Supernova Explosions*, Astronomy and Astrophysics Library, DOI 10.1007/978-3-662-55054-0_5

is $L = 2 \times 10^{43}\,\mathrm{erg\,s^{-1}}$. From the blackbody relation, and assuming the internal temperature and the effective temperature are the same,

$$T \simeq \mathrm{T}_\odot \left(\frac{L}{\mathrm{L}_\odot}\right)^{1/4} \left(\frac{R}{\mathrm{R}_\odot}\right)^{-1/2}, \tag{5.1}$$

so that $T \simeq 10{,}000\,\mathrm{K}$. The density is $\rho \simeq 10^{-13} m\,\mathrm{g\,cm^{-3}}$, where m is the ejected mass in units of solar masses. The number density of nuclei is

$$N_{nuc} = \frac{\rho}{\mathrm{m_H}\mu} = 6 \times 10^{10} \frac{m}{\mu}\ \text{nuclei cm}^{-3}, \tag{5.2}$$

where μ is the mean molecular weight (mean number of nucleons per nucleus). At this density and temperature, the matter is ionized. For single ionization, the electron density, N_e, is equal to N_{nuc}.

These properties enable us to rank the energies that are involved. The kinetic energy of our generic supernova is

$$E_k = \int \frac{1}{2} \rho v(r)^2 4\pi r^2 dr = \frac{3}{10} m v^2 = 6 \times 10^{50} m\,\text{ergs}. \tag{5.3}$$

The energy in the form of radiation is

$$E_{rad} = \mathrm{a} T^4 V = 1.5 \times 10^{48}\,\text{ergs}, \tag{5.4}$$

where V is the volume. The available recombination energy (a kind of potential energy) is

$$E_{rec} \gtrsim N_e I V \simeq 2.6 \times 10^{46} \frac{m}{\mu} \frac{I}{13.6}\,\text{ergs}, \tag{5.5}$$

where I is the mean ionization energy in eV, and for normalization we have used the hydrogen energy of 13.6 eV. The thermal energy in random particle motion is

$$E_{th} = 3 N_{nuc} \mathrm{k} T V = 5 \times 10^{45} \frac{m}{\mu}\,\text{ergs}. \tag{5.6}$$

In real supernovae the numbers differ, but this ordering of the energies is generally correct. Most of the energy is kinetic, the internal energy is dominated by that of radiation, not particles, and the recombination energy is almost negligible.

5.3 Understanding Basic Properties of Supernova Light Curves

A supernova does typically require several weeks to reach maximum light when the radius is $\sim 10^{15}$ cm, the photospheric temperature is $\sim 10^4$ K, and the velocity at the photosphere is $\sim 10^4$ km s^{-1}, as we assumed in Sect. 5.2. Here we outline the basic physics that determines these characteristic values.

Stars have masses of order 1 $M_\odot$. In their final stages, stars have cores, the regions that power the explosions, of radius comparable to that of Earth, $R_c \sim 10^8$ cm. The outer regions may be much more extended, lightly bound, and easy to expel, but a true explosion must disrupt that small, compact core. The binding energy of the core is $GM^2/R_c \sim 10^{51}$ erg or 1 B (for Bethe). Because this much energy must be released to unbind the core, this is the basic energy scale of most supernova explosions, whether by core collapse or thermonuclear disruption of a white dwarf. The energy per unit mass sets the scale of the associated explosion velocity, $v \sim \sqrt{2GM/R_c} \sim 10^4$ km s^{-1}, essentially the escape velocity from the core.

Simply blowing up the star will not necessarily make it bright. Shocked matter can be very hot, but also very optically thick, in which case it cannot radiate efficiently. The fundamental properties of the light curve depend on two timescales: the time for photons to diffuse out, and the time for the supernova to expand. Whatever the explosion mechanism, the shocked matter expands and becomes less optically thick with time. The luminosity of a supernova near maximum light depends on radiation diffusion and hence on an appropriate mean UVOIR opacity, κ, the cross section per unit mass. The opacity in ionized ejecta is typically $\kappa \sim 0.1\,\mathrm{cm^2\,g^{-1}}$.

In general, radiation diffuses out by photon random walk. The photon mean free path is $\ell \sim 1/\kappa\rho$, where ρ is the density, so the photon requires about $N \sim (R/\ell)^2$ steps to travel a distance R. Each step has time of flight ℓ/c so the time required to diffuse out of the ejecta is

$$t_{diff} \sim \frac{N\ell}{c} \sim \frac{R^2}{\ell c} \sim \frac{\tau R}{c}, \tag{5.7}$$

where the optical depth is $\tau \sim \kappa\rho R \sim R/\ell$. A more careful derivation of the diffusion time (Arnett 1996), accounting for the fact that in a star the photons diffuse in three dimensions, gives

$$t_{diff} \simeq \frac{3R^2}{\pi^2 \ell c} \simeq \frac{R^2}{3\ell c} \simeq \frac{\tau R}{3c}. \tag{5.8}$$

The other important timescale, the dynamic time for the supernova to expand, is $t_{dyn} \sim R/v$. Early in the expansion, when the ejecta are still very optically thick, little light can diffuse out. At very late times, the ejecta turn optically thin, heat is lost, and the luminosity is again low. Maximum luminosity occurs when the diffusion time is comparable to the dynamic time. By equating the two, this occurs when $\tau \sim 3c/v \sim 30$. Note the important implication that maximum light is not when the ejecta turn transparent, as one might naively expect, and that the optical depth at maximum light is roughly independent of the opacity.

The time at which maximum light is reached, t_{max}, does depend on the opacity and other parameters of the explosion. From the previous considerations,

$$t_{max} \sim R_{max}\tau_{max}/3c \sim \kappa\rho_{max}R_{max}^2/c \sim \kappa M/4\pi R_{max}c, \tag{5.9}$$

where we have used $\rho_{max} \sim 3M/(4\pi R_{max}^3)$ as the mean density of the ejecta near maximum light. The radius is $R_{max} \sim vt_{max}$, so $t_{max}^2 \sim \kappa M/4\pi vc$, and with $v^2 \sim 2E_k/M$, where E_k is the kinetic energy, we have

$$t_{max} \simeq \frac{1}{4}(\kappa/c)^{1/2}(M^3/E_k)^{1/4}. \tag{5.10}$$

For an opacity of $0.1\,\mathrm{cm^2\,g^{-1}}$, $M = 1\,\mathrm{M_\odot}$, and $E_k = 1$ B, the time to reach maximum light according to Eq. (5.10) is 9 days, and is not very strongly dependent on the specific choices of these parameters. The time would be somewhat longer for higher ejected mass.

The radius of the ejecta at which maximum light occurs is

$$R_{max} \sim vt_{max} \sim 1.7(\kappa/c)^{1/2}(ME_k)^{1/4}. \tag{5.11}$$

For fiducial parameters, Eq. (5.11) gives $R_{max} \sim 4\times10^{15}$ cm, again rather insensitive to the parameters. Expressions similar to Eqs. (5.10) and (5.11) for t_{max} and R_{max} often are inverted and used with observational estimates of t_{max} and v (hence R_{max}) and an assumed value of κ to make estimates of M and E_k.

The luminosity of a supernova near maximum light also can be roughly estimated. When a strong shock reaches the stellar surface it delivers half of its energy as kinetic energy, and half as thermal energy, E_{th}, that can, in principle, be radiated away. The fiducial luminosity would then be $L_{max} \sim E_{th}/t_{max}$. By the time of maximum light, however, the thermal energy has been diminished by adiabatic expansion which, if the postshock material is dominated by radiation pressure, scales as $(R_0/R_{max})^4$ where R_0 is the radius of the progenitor star. For an RSG progenitor of radius $\sim10^{14}$ cm, the thermal energy is diminished by a factor of $\sim10^4$ before it can be radiated. Then

$$L_{max} \sim E_{th}/t_{max}(R_0/R_{max})^4 \sim 10^{42}\,\mathrm{erg\,s^{-1}}, \tag{5.12}$$

comparable to observed values. This expression often is inverted and used with observational estimates of t_{max} and v to make an estimate of R_0. If the progenitor radius is much smaller than that of an RSG, the initial thermal energy is essentially depleted by the epoch of maximum light and an alternative source of luminosity is needed.

Students of supernovae should be familiar with the above rules of thumb. Although we have been casual in this section about factors of 2 and π, some of the estimates are reasonably robust. The estimates of the optical depth and radius near maximum light turn out to be reasonable over a wide range of conditions, being appropriate to SN IIP with massive, recombining, hydrogen envelopes as well as to superluminous supernovae that are caused by collisions of ejecta with dense circumstellar shells (Chap. 18). More serious estimates of the maximum luminosity require more careful consideration and additional physics.

More accurate semi-analytic light-curve models were presented by Arnett (1982, 1996) assuming any power source to be concentrated in the center of spherical, self-similar, homologously-expanding ejecta. A basic assumption is that the solution is separable into spatial and temporal components. Solution of the energy equation coupled with that for radiation diffusion under these assumptions then gives a diffusion time of $t_d = \kappa M/\beta c R$, where β is a constant of integration that is characteristically 13.8 for various reasonable density profiles. The hydrodynamic time is $t_h = R/v_{sc}$, where v_{sc} is a scale velocity. For times well in excess of the time necessary to settle into homologous expansion, the scale velocity can be the velocity of any element of matter at a given radius and time, but it is customary to think of it as the velocity of the outermost matter at radius R at a given epoch. The notion of the velocity of the outermost matter is sensible in the context of a constant-density model, but it is less well defined for more realistic situations with substantial density gradients. The velocity of matter at the photosphere at maximum light can be measured, and that velocity, v_{ph}, is often taken to represent v_{sc}. Since the location of the photosphere depends on the distribution of density and opacity, this assumption is reasonable, but not precise.

With the identification $v_{ph} = v_{sc}$, the diffusion and hydrodynamic timescales can be combined to construct an effective timescale

$$t_{eff} = \sqrt{2 t_d t_h} = \left(\frac{2\kappa M}{\beta c v_{ph}}\right)^{1/2}, \tag{5.13}$$

that is independent of R. Models show that this timescale is closely related to the rise time of the light curve, or t_{max}. Once again, this identification is not exact, even within the context of simple models where fits to observations can yield differences of a few days between t_{eff} and t_{max} (Chatzopoulos et al. 2013b). The expression in Eq. (5.13) can be inverted to give

$$M_{diff} = \frac{1}{2}\frac{\beta c}{\kappa} v_{ph} t_{max}^2, \tag{5.14}$$

enabling an estimate of the mass in terms of the observed quantities, v_{ph} and t_{max}. The mass determined in this way, M_{diff}, is specifically that component that affects the diffusion time. This characteristic mass may be only the ionized component that contributes to the opacity, κ. The mass as determined by Eq. (5.14) is often taken to be an estimate for the ejected mass, but if some of the ejecta are not ionized and hence of low opacity, Eq. (5.14) may only give a lower limit to the ejecta mass.

The kinetic energy of the ejecta can then be estimated from the mass and velocity. The energy is related to the mass through the mean square expansion velocity defined as $< v^2 >= 2E_k/M$. This velocity cannot be measured directly. Rather, one needs to relate this velocity to the measured quantity, v_{ph}. For a constant density sphere, $v_{ph}^2 = 5/3 < v^2 >$. With this relation, one can write from Eq. (5.13)

$$t_{max} = \left(\frac{\kappa}{\beta c}\right)^{1/2} \left(\frac{6}{5}\frac{M^3}{E_k}\right)^{1/4} \simeq 0.28 \left(\frac{\kappa}{c}\right)^{1/2} \left(\frac{M^3}{E_k}\right)^{1/4}. \tag{5.15}$$

The approximate derivation of Eq. (5.10) produced this same scaling and even, perhaps fortuitously, nearly the same numerical coefficient. Equations (5.14) and (5.15) can then be solved to give an estimate of the kinetic energy of the mass that contributes to the diffusion near peak:

$$E_k = \frac{3}{20}\frac{\beta c}{\kappa} v_{ph}^3 t_{max}^2. \tag{5.16}$$

Note from Eq. (5.14) that the mass depends linearly on the photospheric velocity for a given rise time. The kinetic energy, $E_k \propto Mv^2 \propto v_{ph}^3$ for a given rise time, is thus especially sensitive to the velocity. Care must be taken when using these relations to estimate ejected mass and kinetic energy from the rise time and the photospheric velocity, both of which vary from event to event and have observational errors associated with them. In addition, Eqs. (5.14) and (5.16) are based on a simple, constant-density, constant-opacity approximation. For a more realistic structure with a decreasing (approximately power-law) density profile, the observed v_{ph} will tend to be larger than the value of v_{sc} for a constant-density model with the same value of E_k/M (Vinkó et al. 2004). Use of the observed value of v_{ph} may then lead to overestimates of M and E_k. In some works, issues of appropriate numerical scalings in these relations are side-stepped by normalizing to a "well-understood" situation, for instance by comparing a particular stripped-envelope supernova to the rise time and photospheric velocity of a typical SN Ia. This may be a valid procedure, but it demands that the two underlying structures be homologous, which may not be the case. While these formulas can give a rough guideline to the properties of a given supernova, in some circumstances they can give estimates for the mass and energy that are incompatible with analogous estimates based on late-time light-curve data where the power derives from the trapping of γ-ray energy from radioactive decay. A possible resolution is that only a portion of the ejecta is ionized and contributes to the diffusive width of the peak (Sect. 5.4.2).

Even more accurate calculations of supernova light curves must address details of variable opacity, complex composition structures, radiative departures from LTE, recombination of hydrogen envelopes in the case of SN IIP, and other effects. More realistic models tune the whole physical structure to reproduce the rise time and photospheric velocity at a given epoch. The rise time is a function of the band chosen, and estimates of the photospheric velocity depend somewhat on the P Cygni absorption of the particular ion chosen; ideally, models would be consistent in that regard.

5.4 Energy Sources

Various energy sources and radiation mechanisms shape the light curves of supernovae.

5.4.1 *Shock Energy: Breakout, Fireball, and Plateau*

A supernova first shines, brightly but briefly, when the shock wave from the explosion approaches the surface of the star. For essentially all supernovae, the shock and postshock regions are radiation dominated. The supernova begins to brighten when the optical depth between the shock and the photosphere of the progenitor star becomes low enough that radiation can begin to escape. The criterion is similar to that for maximum light in the supernova: the diffusion time must be comparable to the dynamic time. In the case of breakout, this occurs when the shock is at an optical depth $\tau \simeq c/v \simeq 30$, before it has reached the photosphere. A radiation precursor raises the surface temperature to $\gtrsim 10^5$ K, and the supernova brightens in the UV and X-rays. Owing to the finite speed of light, an observer receives this breakout emission from different parts of the progenitor's surface at different times. For a spherical shock, the duration of breakout is related to the light crossing time, R_0/c—about an hour for an extended RSG progenitor. If the underlying explosion is strongly asymmetric, the shock approaches the surface at different times and the duration of breakout is related to the longer shock-crossing time, R_0/v_{shock} (Couch et al. 2009, 2011; Suzuki and Shigeyama 2010; Matzner et al. 2013)—about a day for an extended RSG. The total energy radiated during breakout is $E_{bo} \simeq 8\pi R_0^2 \kappa^{-1} c v_{bo}$, where v_{bo} is the shock velocity at breakout (Katz et al. 2012). For core-collapse supernovae and SN Ia, the radiated energies are $\gtrsim 10^{46}$ and $\lesssim 10^{42}$ ergs, respectively. Estimates of E_{bo} and v_{bo}, when observed in the future, could be used to estimate R_0.

When breakout begins, electrons and baryons swept up by the shock move with equal speeds with respect to postshock material, so a short time is required for the electrons and baryons each to thermalize among themselves (Nakar and Sari 2010). The baryons are hotter since they have the same velocity but greater mass, so electrons and baryons come into thermal equilibrium with each other during a somewhat longer (but still short) time interval. Since breakout begins as soon as radiation begins to flow freely, the radiation may not be in thermal equilibrium with the matter. When radiation can freely flow through the shock, the radiation-mediated shock disappears. (In the presence of CSM, the radiation can accelerate particles and cause a shock to reform, as a matter-mediated shock; Chap. 6.)

At this writing, one photospheric shock breakout, in the Type Ib SN 2008D (Sect. 16.2.2), may have been observed, but this is debated. Numerous detections of breakout from core-collapse events are expected in the coming decades. For SN Ia, the breakout flash from the small progenitors is expected to be hot enough to emit γ-rays. Very brief, bright γ-ray breakout flashes may be observable by future all-sky monitors (Höflich and Schaefer 2009; Piro et al. 2010).

After breakout, the hot, shocked ejecta are exposed. The ejecta expand and cool nearly adiabatically, which causes the luminosity to decline. Although this phase is sometimes referred to as breakout, or breakout tail, we prefer to make the distinction between the brief, bright breakout per se and the longer, dimmer phase of subsequent cooling, which we refer to as the *fireball* phase, by analogy to the similar early phase of terrestrial nuclear explosions. During the fireball phase, the UVOIR luminosity decreases monotonically, but peaks may appear in the UV and optical as the SED shifts to longer wavelengths.

Fireball radiation within hours and days of breakout has been observed in several core-collapse supernovae. Fireball radiation has not yet been seen in an SN Ia, but UV/optical fireball emission within an hour of breakout may be detectable in the future (Rabinak et al. 2012).

For SN IIP, the progenitors of which have massive, extended hydrogen envelopes, the fireball phase lasts for weeks, until matter behind the shock becomes cool enough for hydrogen to recombine. During the following several months, an SN IIP continues to radiate some of the energy deposited by the shock, at a nearly constant temperature, that of hydrogen recombination, about 5500 K. This is the diffusive *plateau* phase of approximately constant luminosity. As discussed above, when a progenitor of smaller radius explodes, adiabatic expansion severely reduces the temperature before the condition $\tau \simeq c/v$ is attained and any fireball phase is brief. This applies to SN Ia, SN Ib, and SN Ic, for which the progenitor radii, ranging from 10^8 cm for a white dwarf up to perhaps 10^{11} cm for an SN Ib/c progenitor, are much smaller than the fiducial radius, $\sim 10^{15}$ cm, of supernovae at maximum light. For these events to shine brightly for as long as they do, another source of energy must be available.

5.4.2 *Radioactive Decay of ^{56}Ni and ^{56}Co*

The other energy source is well known. During the explosion, material with equal numbers of protons, Z, and neutrons, N, is burned rapidly ($\sim$1 s), allowing little time for electron captures or beta decays to change the Z/N ratio. For the white dwarf progenitors of SN Ia, the fuel is primarily ^{12}C and ^{16}O, and for core-collapse supernovae it is primarily ^{28}Si, all three of which have equal Z and N. If this material is burned to near nuclear-statistical-equilibrium, the most tightly bound nucleus with $Z = N$, ^{56}Ni, is predominantly produced. The ^{56}Ni is unstable, $^{56}\mathrm{Ni} + e^- \rightarrow {}^{56}\mathrm{Co} + \gamma + \nu_e$, with a half-life of 6.1 days (e-folding lifetime $\tau_{Ni} = 6.1/ln2$ days = 8.8 days). In turn, ^{56}Co decays to stable ^{56}Fe, 81% of the time by electron capture, ^{56}Co + $e^- \rightarrow$ ^{56}Fe + γ + ν_e, and 19% by positron decay, ^{56}Co $\rightarrow$ ^{56}Fe + e^+ + $\gamma + \nu_e$, with half-life 77 days ($\tau_{Co} = 111.3$ days). The key point is that energy released by radioactivity cannot begin to be adiabatically degraded until the decays occur. Thus, the delayed release and deposition of radioactivity energy in the form of γ-rays and positrons (neutrinos escape without depositing energy) avoids most of the adiabatic conversion into kinetic energy, heats the ejecta, and leads by thermal emission to high UVOIR luminosity. The fact that the decay times of ^{56}Ni and ^{56}Co are roughly the same as the timescale on which a solar mass of material exploded with 1 B of energy becomes nearly optically thin is a coincidence of physics, one that allows SN Ia to shine so brightly.

More quantitatively, the energy released per ^{56}Ni decay is $Q_{Ni,tot} = Q_{Ni,\gamma} + Q_{Ni,\nu} = 2.16\,\text{MeV}$, where on average the energies released per ^{56}Ni decay are $Q_{Ni,\gamma} = 1.75\,\text{MeV}$ as γ-rays with $Q_{Ni,\nu} = 0.41\,\text{MeV}$ wasted on neutrinos (although some day these may be detectable). The most intense γ-ray line is at 0.812 MeV.

The total energy released per ^{56}Co decay is

$$\begin{aligned} Q_{Co,tot} &= 0.81\left(Q_{Co,\gamma,ec} + Q_{Co,\nu,ec}\right) \\ &\quad + 0.19\left(Q_{Co,ann,+} + Q_{Co,\gamma,+} + Q_{Co,kin,+} + Q_{Co,\nu,+}\right) \\ &= 4.57\,\text{MeV}, \end{aligned} \tag{5.17}$$

where $Q_{Co,ann,+} = 1.022\,\text{MeV}$ and on average $Q_{Co,\gamma,ec} = 3.73\,\text{MeV}$, $Q_{Co,\nu,ec} = 0.84\,\text{MeV}$, $Q_{Co,\gamma,+} = 2.09\,\text{MeV}$, $Q_{Co,kin,+} = 0.63\,\text{MeV}$, and $Q_{Co,\nu,+} = 0.83\,\text{MeV}$. Subscripts *ec* and + refer to the respective channels, and γ, ν, *ann*, and *kin* refer to particles or processes in the output channels. In the electron-capture channel, the most intense γ-ray line is at 0.847 MeV. In an average ^{56}Co decay, the energy released by direct γ-ray photons is $0.81Q_{Co,\gamma,ec} + 0.19Q_{Co,\gamma,+} = 3.42\,\text{MeV}$, by annihilation is $0.19Q_{Co,ann,+} = 0.19\,\text{MeV}$, by positron kinetic energy is $0.19Q_{Co,kin,+} = 0.12\,\text{MeV}$, and by neutrinos is $0.81Q_{Co,\nu,ec} + 0.19Q_{Co,\nu,+} = 0.81\,\text{MeV}$. The total average released energy potentially available to heat supernova ejecta (the total released minus the neutrino energy) is

$$< Q_{Co} >= 0.81\; Q_{Co,\gamma,ec} + 0.19\left(Q_{Co,ann,+} + Q_{Co,\gamma,+} + Q_{Co,kin,+}\right) = 3.73\,\text{MeV}, \tag{5.18}$$

3.61 MeV in γ-rays and 0.12 MeV in positron kinetic energy. Of the energy potentially available for heating, 92% comes from direct γ-rays and 8% from positron kinetic energy and annihilation.

These values for the average energy released per decay lead to an expression for the specific rate of production of energy per gram of initial ^{56}Ni, s_R, and hence to an expression for the *radioactivity luminosity*—the rate of energy released as γ-rays and positrons but not neutrinos:

$$L_R = s_R M_{Ni} = \left(\frac{1}{\tau_{Co} - \tau_{Ni}} \left[Q_{Ni,\gamma} (\frac{\tau_{Co}}{\tau_{Ni}} - 1) - < Q_{Co} > \right] e^{-t/\tau_{Ni}} + < Q_{Co} > e^{-t/\tau_{Co}} \right) \frac{M_{Ni}}{56 \mathrm{m}_\mu}, \tag{5.19}$$

where M_{Ni} is in grams, and m_μ is the unit of atomic mass, or

$$L_R = (6.45 \times 10^{43} e^{-t/8.800} + 1.45 \times 10^{43} e^{-t/111.3}) M_{Ni}\, \mathrm{erg\, s}^{-1}, \tag{5.20}$$

where t is in days and M_{Ni} is in solar masses. The time-integrated radioactivity luminosity is $1.89 \times 10^{50} M_{Ni}$ ergs. Derivations of these results and further details are in Nadyozhin (1994) and Arnett (1996).

The energy released according to Eq. (5.19) may not be the energy deposited in the ejecta because with time some of the γ-rays and some of the positrons may be emitted directly into space without depositing their energy. Determining the fraction of the energy produced in each process associated with the decay of ^{56}Ni and ^{56}Co that is ultimately deposited in the ejecta is not trivial. The sources of the γ-rays and positrons are likely to be distributed in a complex way in the ejecta; the locations of the sources of γ-rays and positrons are not necessarily where their energy is thermalized; the density and composition structures are likely to be complex; density clumping will reduce the average deposition of energy; helium has a smaller photoionization cross section than metal-rich matter allowing soft X-rays produced in the down-scatter to escape more easily and hence giving a smaller effective γ-ray thermalization opacity; the γ-rays are emitted as line features; Compton scattering of γ-rays is sensitive to the scattering angle; the secondary electrons undergo complex down-scattering before depositing their energy as thermal energy; and the outcome for positrons depends on the presence and distribution of magnetic fields. If the material is optically thin, all the detailed pathways must be followed to predict the optical spectra. The most accurate way to account for these manifest aspects is through a Monte Carlo calculation, but that can be extremely expensive computationally even in spherically-symmetric configurations, never mind 3D reality (Colgate et al. 1980; Ambwani and Sutherland 1988). Most Compton scattering is into the forward direction with little energy loss, and the largest energy loss tends to be in the rare right-angle scattering. An important approximation is then to assume the γ-ray proceeds in one direction until it undergoes right-angle scattering thus yielding its energy. In this approximation, one can treat the γ-ray transfer as an "absorption" with an effective gray absorptive opacity (Swartz et al. 1995).

To account for the transfer and loss of the ^{56}Ni and ^{56}Co decay products, one can define a dimensionless local deposition function, δ. This function determines the rate of energy deposition at a given point in the ejecta from sources at any point in the radioactive matter. A single element of matter in the radioactive material producing energy at a specific rate, s_R, would deposit energy in an element of mass dm at a remote location a distance ℓ away at the specific rate

$$ds_{dep} = \frac{\kappa_\gamma}{4\pi\ell^2} e^{-\tau} s_R dm, \tag{5.21}$$

where τ is the optical depth between the two locations. To get the total local deposition function, δ, at a given location, one must integrate s_{dep} at that location over the distribution of all the source material, $dM_{Ni} = X_{Ni}\rho dxdydz$, and normalize by the integrated specific rate of production of decay energy, which is the total rate of production of energy divided by the total mass contributing to the decay, M_{Ni}. The local deposition function defined in this way is a function of position in the ejecta. It can be nonzero even at locations that contain no radioactive species and is a function of time (Sutherland and Wheeler 1984; Swartz et al. 1995; see Dessart et al. (2012a) for an application of this method), as illustrated in Fig. 5.1. There can be separate local deposition functions for γ-rays and for positrons. If the source is confined to a small volume in the center of the ejecta, then the local deposition function in a mass element dm at distance r from the center is just $d\delta(r,t) = \kappa_\gamma/(4\pi r(t)^2)e^{-\tau(r)}dm$.

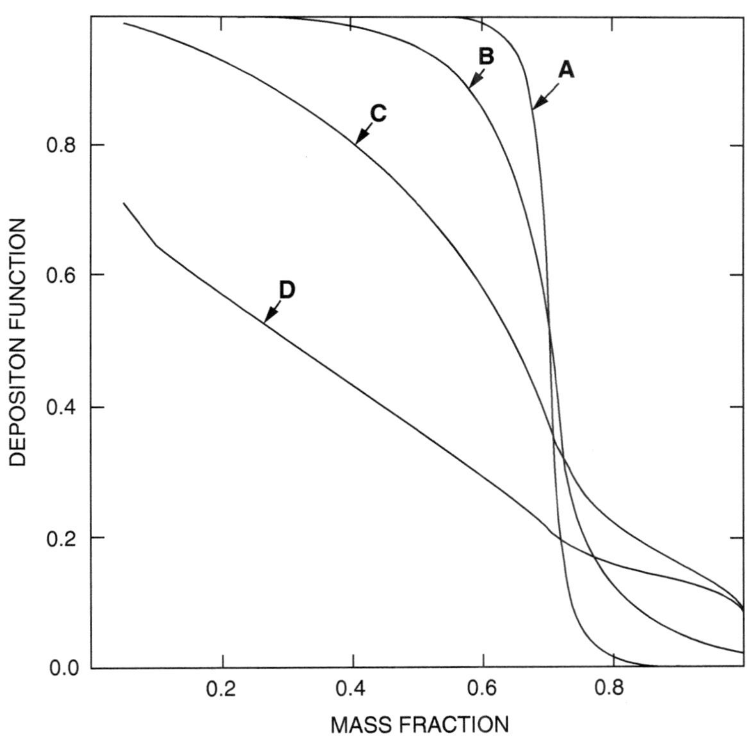

Fig. 5.1 Gamma-ray deposition functions versus mass fraction for an SN Ia model at times (in units of 10^6 s) of 0.5 (*line A*), 1 (*line B*), 2 (*line C*), and 4 (*line D*) after explosion. From Sutherland and Wheeler (1984)

If the ejecta are in homologous expansion, the appropriate integral over the source distribution needs to be done only once; the subsequent scaling with time follows the scaling of the optical depth, $\tau \propto t^{-2}$.

The net deposition function, D, expresses the fraction of the total radioactive energy liberated that is actually deposited somewhere in the ejecta, providing heat to power the light curve. In the simplest form, the luminosity produced by radioactive decay would be written $L(t) = D(t)L_R(t)$, where L_R is the total power produced by radioactive decay, as given by Eqs. (5.19) and (5.20). In practice, there are complications because there are, in principle, different deposition functions for the γ-rays and the positrons as determined by the local deposition functions.

To account for this possibility, Eq. (5.19) can be recast in the form

$$L_R = \left[\frac{Q^*_{Ni}}{\tau_{Ni}} e^{-t/\tau_{Ni}} + \frac{Q^*_{Co}}{\tau_{Co} - \tau_{Ni}} \left(e^{-t/\tau_{Co}} - e^{-t/\tau_{Ni}}\right)\right] \frac{M_{Ni}}{56\mathrm{m}_\mu}, \tag{5.22}$$

where $Q^*_{Ni} = Q_{Ni,\gamma} D_\gamma$ and

$$Q^*_{Co} = \left(0.81 Q_{Co,\gamma,ec} + 0.19 Q_{Co,\gamma,+}\right) D_\gamma + 0.19 \left(Q_{Co,ann,+} D_\gamma + Q_{Co,kin,+}\right) D_+, \tag{5.23}$$

where D_γ and D_+ are the net deposition functions for γ-rays and positrons, respectively. Note that the γ-rays produced by the annihilation of the positrons are subject to both deposition functions; the positrons must be stopped in the ejecta, but the resulting annihilation γ-rays are then subject to the same deposition as the directly-produced γ-rays.

For a given model, the net deposition function for γ-rays, D_γ, is most properly computed with a Monte Carlo treatment, or with the effective gray opacity approximation outlined above, to determine the local deposition functions. A simple analytic approximation that is frequently adopted is to assume the radioactive source is confined to the center of a spherical distribution of ejecta. In this case, the fraction of the γ-rays produced that escape the ejecta is $e^{-\tau}$, where τ is the total γ-ray optical depth from the center of the ejecta. The net deposition function, the fraction of the energy produced that is left behind somewhere in the ejecta, can then be written as $D_\gamma = 1 - e^{-\tau}$ (Swartz and Wheeler 1991). Assuming that γ-ray interaction can be treated as an absorptive process with an appropriate constant γ-ray opacity κ_γ, that the optical depth scales as t^{-2} in homologous expansion, and approximating the ejecta as a constant-density sphere with outer velocity $v \simeq v_{ph}$, the net deposition function can be expressed as $D_\gamma = 1 - e^{-A_\gamma t^{-2}}$, where the γ-ray optical depth is $\tau_\gamma = \kappa_\gamma \rho R = A_\gamma t^{-2}$ and $A_\gamma = (3\kappa_\gamma M_{ej})/(4\pi v_{\mathrm{ph}}^2)$. These expressions can easily be generalized to a power-law density profile. Fits to Monte Carlo calculations give $\kappa_\gamma \simeq 0.03\,\mathrm{cm}^2\,\mathrm{g}^{-1}$ (Colgate et al. 1980; Sutherland and Wheeler 1984). A more accurate polynomial fit for the net deposition function, $D_\gamma(\tau)$, for a central source in a sphere of constant density was presented by Colgate et al. (1980). This calculation accounts for the sensitivity to the input γ-ray spectrum, which is small, and the loss of soft X-rays that do not completely thermalize.

A similar treatment can be adopted for the positrons, with the complication that, unlike γ-rays, positrons can be affected by the presence of magnetic fields in the ejecta. If the magnetic fields are sufficiently strong and tangled, then the positron mean free path is short and the effective optical depth is large. In this case, the net positron deposition would be $D_+ = 1$. It is also possible that the magnetic field could be radially combed by the explosion so that while positrons would spiral along the field lines, they have a finite chance to escape (Colgate et al. 1980; Chan and Lingenfelter 1993; Milne et al. 1999). In this case, the net deposition function could be again approximated as $D_+ = 1 - e^{-A_+t^{-2}}$, but with an opacity and optical depth corresponding to the physics of the positrons. In one treatment, this was approximated with $\kappa_+ = 7\,\mathrm{cm^2\,g^{-1}}$ (Colgate et al. 1980; see also Penney and Höflich 2014).

The basic point is that the deposition function of the positrons is likely to be different than that of the γ-rays at later times when the γ-rays begin to leak. Taking this into account, the energy deposited can be written as

$$\begin{aligned} L_R = {} & [\left(6.57\times10^{43}e^{-t/8.800} + 1.32\times10^{43}e^{-t/111.3}\right)D_\gamma \\ & +4.64\times10^{41}\left(e^{-t/111.3} - e^{-t/8.800}\right)D_+ \\ & +7.52\times10^{41}\left(e^{-t/111.3} - e^{-t/8.800}\right)D_+D_\gamma]M_{Ni}\ \mathrm{erg\,s^{-1}}, \end{aligned} \tag{5.24}$$

where the time is in days and the mass in solar units. If D_γ and D_+ are unity, Eq. (5.24) reduces to Eq. (5.20). If the positrons are trapped, $D_+ \simeq 1$, for a substantial time, the energy deposited by positrons can begin to exceed that by γ-rays after 200–300 days for typical supernova conditions. There were indications in SN 1987A (Chap. 11) and in SN 1994I (Sect. 17.2) that positrons can partially leak.

This prescription for power input, Eq. (5.24), can be incorporated in semianalytic light-curve models that assume the radioactive power source to be concentrated in the center (Arnett 1982, 1996). Solution of the energy and radiation diffusion equations gives

$$L(t) = M_{Ni}e^{-x^2}[(s_{Ni} - s_{Co})\int_0^x 2ze^{z^2-2zy}dz + s_{Co}\int_0^x 2ze^{z^2-2yz+2zs}dz](1 - e^{-A_\gamma t^{-2}}), \tag{5.25}$$

where $x = t/t_{eff}$, t_{eff} is the effective diffusion time from Eq. (5.13), $y = t_{eff}/2\tau_{Ni}$ with $\tau_{Ni} = 8.8$ days, $s = t_{eff}(\tau_{Co} - \tau_{Ni})/2\tau_{Co}\tau_{Ni}$ with $\tau_{Co} = 111$ days, M_{Ni} is the initial mass of ^{56}Ni, and $s_{Ni} = 4.0\times10^{10}\,\mathrm{erg\,g^{-1}\,s^{-1}}$ and $s_{Co} = 6.8\times10^{9}\,\mathrm{erg\,g^{-1}\,s^{-1}}$ are the energy generation rates due to ^{56}Ni and ^{56}Co decay. The factor $(1 - e^{-A_\gamma t^{-2}})$ accounts for the increasing transparency to γ-rays.This simple prescription assumes that positrons are trapped, $D_+ = 1$, but neglects the deposition of positron kinetic energy, the last term on the RHS of Eq. (5.24). The solution in Eq. (5.25) can be generalized to account for a finite initial radius of the exploding star, other sources of power input, and leakage of both γ-rays and positrons (Arnett 1996; Cappellaro et al. 1997; Chatzopoulos et al. 2012; Szalai et al. 2016). At late times when the diffusion time is short, Eq. (5.25) reduces to Eq. (5.22).

The basic separable, similarity solution we have sketched here in Eq. (5.25) has several remarkable properties. At early times, corresponding to very small values of x in Eq. (5.25), the luminosity scales as $L \propto t^2$, a prescription sometimes used to extrapolate early light curves to determine the time of explosion. This early-time behavior does not depend explicitly on the nature of the radioactive input. Its origin is also more subtle than assuming $L = 4\pi(vt)^2\sigma T^4$ with constant v and T (Arnett 1996).

Another important property of the solution is the relation between the input power from Eq. (5.20) and the luminous output from Eq. (5.25). Prior to maximum light, the power input exceeds the luminous output. The excess energy is stored to diffuse out later. Shortly after maximum light, the luminosity radiated is more than the instantaneous input because of that stored, diffusing energy. According to the separable, self-similar solution, the bolometric power output is identically equal to the power input precisely at maximum light (established by setting the time derivative of Eq. (5.25) equal to zero). This property is known as "Arnett's law." At times somewhat after maximum light, the ejecta are predicted to become optically thin. Whatever power is injected quickly thermalizes and diffuses out, and the power input is then again identically equal to the bolometric power output. With the assumption that all the energy of radioactive decay is trapped, either the maximum-light condition or the later tail can be used to estimate M_{Ni}. In practice, γ-rays are probably efficiently trapped near maximum light, but less efficiently so at later epochs, so using the late-time tail to measure M_{Ni} can be problematic. Using Arnett's Law is a popular way to estimate M_{Ni} in a supernova thought to be powered by radioactive decay: the brighter the absolute maximum luminosity, the more ^{56}Ni.

Equations (5.14), (5.16), and (5.25) are thus widely used to estimate the mass in the ejecta, the kinetic energy, and the amount of ^{56}Ni in an explosion, based on the observed rise time, peak luminosity, photospheric velocity and an estimated opacity.

An analogous treatment can be applied to the late-time tail, assuming it to be powered by radioactive decay. The slope of the late-time light curve depends on the trapping efficiency of γ-rays. The mass that contributes to trapping of γ-rays may be nontrivial to compute accurately since the trapping may depend on clumping and different opacity for down-scattered X-rays as a function of composition, but for purposes of illustration, we will adopt the common assumption that all ejecta are equally effective at trapping γ-rays. In this case, the optical depth to the γ-rays scales approximately as

$$\tau_\gamma \sim \kappa_\gamma M_{ej}/v^2t^2 \sim \kappa_\gamma M_{ej}^2/E_k t^2, \tag{5.26}$$

where κ_γ is the opacity to γ-rays, M_{ej} is the ejecta mass, and E_k is the kinetic energy. If two supernovae have similar late-time light curves and hence γ-ray optical depth, then the mass and energy should scale as $M_{ej} \propto v^2$ and $E_k \propto M_{ej}^2 \propto v^4$. Note that the argument employed here is based on the slope of the tail. This argument is related to but different from that associated with the width of the peak, for which the property

$\tau_{opt} \sim c/v$ leads to a different constraint on the ejecta mass and energy (Eqs. (5.14) and (5.16)): $M_{diff} \propto v$ and $E_k \propto v^3$ for a constant optical opacity and a given light-curve rise time. Both of these constraints need to be at least approximately satisfied. Failure to satisfy them implies a breakdown in underlying assumptions, for instance spherical symmetry or constant opacity.

For a simple expanding-shell model with central γ-ray deposition from radioactive decay for which the γ-ray opacity scales as in Eq. (5.26) and the net deposition function is taken to be $1 - e^{-\tau_\gamma}$, a convenient parameter, T_γ, with units of time characterizes the late-time behavior:

$$T_\gamma = \left(\frac{C\kappa_\gamma M_{ej}^2}{E_k}\right)^{1/2}, \tag{5.27}$$

where C is a dimensionless structure constant that depends on the slope of the density profile (typically $C \simeq 0.05$). Note that $T_\gamma = A_\gamma^{1/2}$ where A_γ is a parameter introduced above. The parameter T_γ can be determined from observations and hence plays a similar role to the rise time in establishing constraints on the ejecta mass and energy when fitting the peak of the light curve. For large values of this parameter, the light curve remains parallel to the γ-ray deposition line; for small values of this parameter, γ-ray leakage sets in early and light curves may display considerable variation; for intermediate values, in practice $T_\gamma \simeq 75$ days, the light curves tend to have the same behavior (Clocchiatti and Wheeler 1997). More careful considerations must account for positron deposition, a variety of radioactivities, ionization freezeout, and other effects.

Using the expression for the rise time, $t_{max} \simeq t_{eff}$, from Eq. (5.13) and for M_{diff} from Eq. (5.14), the parameter T_γ can also be written

$$T_\gamma = \left[\left(\frac{5}{3}\beta C\right)\left(\frac{c}{v_{ph}}\right)\left(\frac{\kappa_\gamma}{\kappa}\right)\left(\frac{M_{ej}}{M_{diff}}\right)\right]^{1/2} t_{max}. \tag{5.28}$$

A determination of the value of T_γ from observations then allows constraints to be put on model parameters, for instance the ratio of the opacities or the ratio of the masses.

Using Eq. (5.27) with the relation $E_k = 1/2 M_{ej} < v^2 >= 3/10 M_{ej} v_{ph}^2$, one can write expressions for M_{ej} and E_k as

$$M_{ej} = \frac{3}{10}\frac{v_{ph}^2 T_\gamma^2}{C\kappa_\gamma} \tag{5.29}$$

and

$$E_k = \frac{9}{100}\frac{v_{ph}^4 T_\gamma^2}{C\kappa_\gamma}. \tag{5.30}$$

This method to estimate the ejecta mass and energy does not depend on the optical opacity, as do Eqs. (5.14) and (5.16).

Equation 5.28 can be inverted to write

$$\frac{M_{diff}}{M_{ej}} = \frac{5}{3}\beta C \frac{c}{v_{ph}} \frac{\kappa_\gamma}{\kappa} \left(\frac{t_{max}}{T_\gamma}\right)^2. \tag{5.31}$$

As we will discuss in Chaps. 16 and 17, the mass ratio determined in this way from the observed parameters and standard assumptions about the opacities is often considerably less than unity. This suggests that some of the mass that traps γ-rays does not participate in controlling the diffusion time, perhaps because a substantial portion of the ejecta is not ionized and hence does not generate a significant optical opacity.

The discussion of Eqs. (5.27)–(5.31) has focused on γ-ray deposition, but in practice, consideration must also be given to positron deposition. In this case, Eq. (5.24) should be employed to fit the late-time tail, accounting for the different deposition properties of γ-rays and positrons. In a basic model, global deposition functions of the form $D_\gamma = 1 - e^{(T_\gamma/t)^2}$ and $D_+ = 1 - e^{(T_+/t)^2}$ can be adopted. Separate values can then be obtained for $T_\gamma = A_\gamma^{1/2}$ and $T_+ = A_+^{1/2}$ by, for instance, multiparameter minimization techniques (Szalai et al. 2016). When allowing for positron deposition, the value measured for T_γ will be smaller than if the late-time tail were fit by γ-ray deposition alone.

In more sophisticated models, care must be taken about the radius, composition, recombination (not only in hydrogen, but in some cases in helium and even oxygen) and time and spatial dependence of the opacity, as well as the distribution of the radioactive species. Even assuming spherical symmetry, this requires more elaborate models and, ideally, radiation-hydrodynamical calculations (Sect. 5.6). More accurate models should also account for asymmetry in the overall morphology, the composition, and the distribution of radioactive species.

5.4.3 *Gamma-Ray Light Curves*

One astronomer's loss is another astronomer's gain. The hard photons from radioactive decay that are not trapped in the ejecta to cause heating and power the UVOIR light curve may be directly detectable and hence an important source of information on the relevant physics.

The fact that X-rays from SN 1987A were detected by *Ginga* before they were expected from estimates of down-scattering of γ-rays was one of the first clues that the radioactive ^{56}Ni was mixed further out in the ejecta than explosion models had anticipated. The 0.847 and 1.238 MeV lines of ^{56}Co were later directly detected. Detailed models investigated the line profiles and integrated fluxes of these lines in comparison to observations (Sect. 11.4.2).

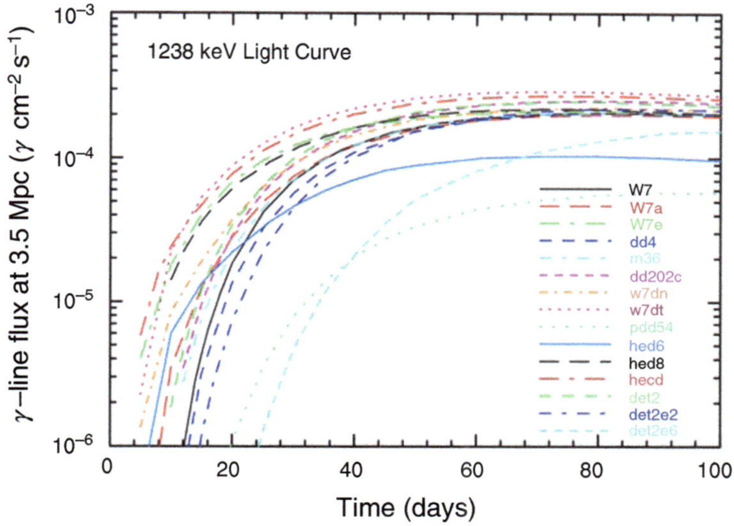

Fig. 5.2 Predicted light curves of the 1.238 MeV line of ^{56}Co for a selection of SN Ia explosion models and a distance of 3.5 Mpc. From "Expectations for the Hard X-Ray Continuum and Gamma-Ray Line Fluxes from the Type Ia Supernova 2014J in M82" (The and Burrows 2014). © AAS. Reproduced with permission

Similar models and expectations (Fig. 5.2) arose for SN Ia that produce even more ^{56}Ni, but it was not until the advent of SN 2014J (Sect. 20.2.2) that γ-rays were directly detected from a SN Ia. The ^{56}Co lines and a γ-ray continuum from SN 2014J were detected by the venerable *INTEGRAL* mission (Churazov et al. 2014).

5.4.4 *Buried Pulsar/Magnetar*

The effects of the formation of a compact object on the energetics and luminosity of a supernova were considered shortly after the discovery of pulsars as rotating, magnetized neutron stars (Gunn and Ostriker 1969). The energy input by magnetic multipole is given by

$$L_p(t) = \frac{E_p}{t_p}\frac{l-1}{(1+t/t_p)^{\ell}}, \tag{5.32}$$

where $E_p = I_{ns}\Omega_i^2/2$ is the initial rotational energy of the pulsar with $I_{ns} \sim 10^{45}\,\mathrm{g\,cm^{-2}}$ the neutron-star moment of inertia, Ω_i the initial angular velocity, and t_p the characteristic timescale for spin-down. This timescale is usually taken to be that of an ℓ =2 magnetic dipole of fiducial tilt (e.g. Kasen and Bildsten 2010)

$$t_p = \frac{6I_{ns}c^3}{B^2R_{ns}^6\Omega_i^2} \simeq 1.3 I_{ns,45}R_{ns,6}^{-6}B_{14}^{-2}P_{10}^2 \text{ years}, \tag{5.33}$$

where $R_{ns,6}$ is the radius of the neutron star in units of 10^6 cm, B_{14} is the dipole field in units of 10^{14} G, and the initial rotational period of the pulsar in units of 10 ms is $P_{10} = 1.4 I_{ns,45}^{1/2}E_{p,50}^{-1/2}$, where $E_{p,50}$ is the initial rotational energy in units of 10^{50} ergs $\equiv$ 0.1 B. For canonical values, the field is given by $B_{14} \simeq P_{10}/t_{p,yr}^{1/2}$. To provide the energy of a supernova explosion, the magnetic field must be very high, $\sim 10^{14}$ G, in the range of *magnetars*, but more modest fields could still produce appreciable luminosity. In the model based on separability and self-similarity, the first law of thermodynamics, and radiation diffusion as applied above to radioactive decay, the corresponding expression for the light curve of a supernova powered by a pulsar or magnetar is

$$L(t) = \frac{E_p}{t_p}e^{-x^2/2}\int_0^x e^{z^2/2}\frac{z}{(1+yz)^2}dz, \tag{5.34}$$

where $x = t/t_{eff}$ and $y = t_{eff}/t_p$, with t_{eff} again being the effective diffusion time of Eq. (5.13). This model assumes that the pulsar input power is thermalized near the center of the ejecta. Numerical MHD models suggest that thermalization may not occur, with the energy flowing in jets from within the ejecta (Bucciantini et al. 2012)

5.5 Application to Supernova Types

In this section we briefly describe the general behavior of the light curves of the main supernova types. More detailed discussions will appear in later chapters.

5.5.1 SN Ia

We begin with typical SN Ia because, apart from the so-far unobserved breakout and fireball emission and currently ambivalent evidence for collision with nearby CSM or a companion, only one energy source—radioactivity—is responsible for powering the light curve.

For a typical SN Ia, peak luminosity $L_{UVOIR} \gtrsim 10^{43}$ erg s^{-1} occurs about 20 days after explosion. From spectral line blueshifts and widths, the velocity at the photosphere is inferred to be $v \simeq$ 10,000 km s^{-1}. These are basic parameters needed to constrain the explosion, as outlined in Sect. 5.3. From v and the time to maximum light, the radius of the photosphere near maximum light is $R = vt \simeq 10^{15}$ cm. To account for the luminosity, the temperature at the photosphere, from $\sigma T_{ph}^4 = L/4\pi R^2$, must be $T_{ph} \sim 10^4$ K. The simple numerical example in Sect. 5.2 corresponds approximately to the properties of a SN Ia.

Suppose that an SN Ia is the thermonuclear disruption of a white dwarf that accretes enough matter from a companion star to approach the Chandrasekhar limit of 1.4 $M_\odot$. While accreting, the white dwarf contracts to a radius of $\sim 10^8$ cm. The central density increases to $\rho_c \gtrsim 10^9\,\mathrm{g\,cm^{-3}}$ and the central temperature to $T_c \gtrsim 10^8$ K, at which point unstable carbon fusion ignites and a nuclear burning front propagates through the white dwarf in about 1 s, releasing nuclear energy while synthesizing heavier elements, raising the temperature to $T \gtrsim 10^9$ K, and driving the white dwarf into expansion. Until the time of maximum light, the expansion is nearly adiabatic, with $T \propto R^{-1}$ for a radiation-dominated ($\Gamma_1 = 4/3$) gas, as internal energy is converted to kinetic energy. At maximum light, the radius of the photosphere is $\sim 10^7$ times larger than that of the white dwarf, so adiabatic expansion would reduce the internal temperature to $T \sim 100$ K, far too low to account for the peak luminosity. Thus the initial internal energy of the ejecta cannot lead to the large, hot photosphere that is needed to produce the UVOIR emission at maximum light. An explosion as just described would be dark, practically unobservable. Something is missing, namely the delayed input of radioactive decay.

As shown in Fig. 5.3, for a short time after maximum light, L_{Ia} slightly exceeds L_R as previously pent up energy escapes. Subsequently, L_{Ia} falls below L_R as the ejecta thin out and an increasing fraction of the γ-rays are able to escape as

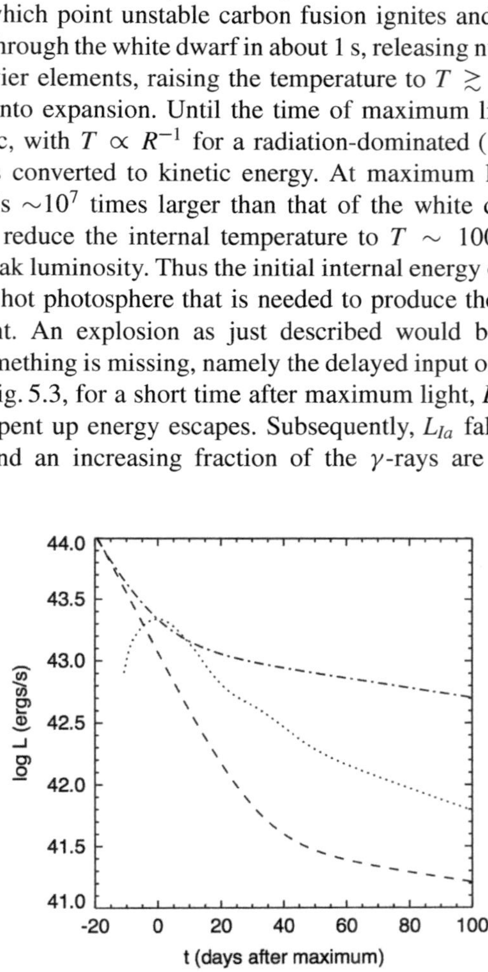

Fig. 5.3 The *dotted line* shows an observed UVOIR light curve of a typical SN Ia, L_{Ia}. The *dot-dashed curve* corresponds to complete trapping of decay products, i.e., it is the radioactivity luminosity, L_R (Eq. (5.20)). Using Arnett's Law, that the luminosity at maximum light equals the radioactive input at that time (Sect. 5.4.2), the characteristic SN Ia peak luminosity requires $M_{Ni} \simeq 0.6\ M_\odot$. The *dashed line* corresponds to complete γ-ray escape and complete positron trapping. The supernova light curve tracks neither complete γ-ray trapping nor complete γ-ray escape. From "Optical Light curves of Supernovae" (Leibundgut and Suntzeff 2003)

the column density ($g\,cm^{-2}$) and the γ-ray optical depth decrease as t^{-2}. For a characteristic γ-ray opacity of $0.03\,cm^2\,g^{-1}$ and standard SN Ia parameters, the γ-ray optical depth (radially from the center) remains above unity for about a month after maximum light.

Late-time light curves of SN Ia, although involving comparatively little light and challenging to observe, can be informative. By about 200 days, practically all γ-rays escape, and for the next several hundred days the light curve depends on the fate of the positrons. In the limit of a sufficiently strong and tangled magnetic field, positrons are trapped and deposit their kinetic energy by collisions with electrons before slowing down and annihilating with electrons to produce 0.511 MeV γ-rays, which escape. As long as energy deposition is dominated by completely-trapped positrons, the slope of the bolometric light curve should match the ^{56}Co decay rate. On the other hand, for a sufficiently weak magnetic field or one that has been "radially combed" by the explosion, an increasing fraction of the positrons may escape before depositing all of their kinetic energy, and the light curve should decline faster than the ^{56}Co decay rate (Sect. 5.4.2).

There are several possible complications to inferring the extent of positron trapping from the slope of the late-time light curve. One is a thermal instability called an *infrared catastrophe* (IRC; Axelrod 1980). More than a year after explosion, the ejecta cool to a critical temperature, about 2000 K. Then, excitation of fine-structure levels of abundant iron ions, followed by radiative de-excitation by mid- and far-IR emission lines, causes the temperature to drop catastrophically to $\sim$100 K. The optical and near-IR emission may decrease substantially as the mid- and far-IR emission rises, although redistribution of UV emissivity resulting from nonthermal excitation and ionization into the optical and near-IR may mitigate this decrease (Sect. 20.2.1). The mid- and far-IR emission, not normally included in the UVOIR luminosity, is difficult to observe, and the accuracy of observed late-time SN Ia light curves that do include mid- and far-IR emission is not yet sufficient to place strong constraints on the extent to which positron trapping occurs.

Somewhat later, depending on the small and uncertain amount of short-lived ^{57}Ni (half-life 1.5 days) that is synthesized, the decay of its daughter ^{57}Co to stable ^{57}Fe, $^{57}\mathrm{Co}+e^- \rightarrow\ ^{57}\mathrm{Fe}+\gamma + \nu_e$, with half-life 272 days, may slow the decline rate of the light curve. The relatively low energy of these γ-rays compared to those of ^{56}Co entails somewhat higher γ-ray opacity and therefore greater probability of energy deposition. Excited states of a stable daughter nucleus make transitions to lower states by γ-ray emission until the ground state is reached. In the case of ^{57}Fe, however, the transition to the ground state from a low-lying level usually is by ejection of an inner electron (*internal conversion*) carrying 14.4 keV of kinetic energy (Seitenzahl et al. 2009). Like positrons, these fast electrons may deposit some or all of their kinetic energy. Evidence for slowing of the late-time light curve of SN 2012cg by ^{57}Co decay was presented by Graur et al. (2016).

Another effect that may slow the decay of the light curve is *ionization freezeout* (Sect. 4.5). After the ejecta density becomes so low that the recombination time is longer than the ^{56}Co decay time, the ionization level remains almost constant, so the rate of energy release by delayed recombination is higher than would otherwise be the case. That is, energy is released by recombining electrons that were produced by ionization at earlier epochs.

The decay of ^{55}Fe $+ \ e^- \rightarrow$ ^{55}Mn $+ \ \nu_e$, with half-life 2.7 yr, may contribute significant energy. The nuclide ^{55}Fe is the daughter of short-lived (half-life 0.7 days) ^{55}Co. The decay is from ground state to ground state, so does not involve direct γ-ray emission, but nevertheless can lead to photon production. After a nucleus decays by capturing a K- or L-shell orbital electron, electronic de-excitation occurs with release of energy. This process is generally negligible compared to the energy of positrons and γ-rays from nuclear reactions, but at very late times, when γ-ray escape is practically complete, deposition by means of ^{55}Fe Auger electrons (electronic de-excitation by ejection of outer electrons) and X-rays may be the dominant source of power to the light curve. In practice, the contribution of ^{55}Fe may be overwhelmed by luminosity resulting from other effects including CSI and light echoes (Chap. 6).

In some supernovae (e.g., SN Ib/c and SN 1987A), the decay of ^{44}Ti (half-life 60 years) may play a role, but since SN Ia produce large quantities of iron-group isotopes, the decay of ^{55}Fe is likely to be more important for them.

5.5.2 SN Ib/c

Massive-star progenitors of core-collapse supernovae that have lost their hydrogen envelopes have radii in the range 0.1–1 $R_\odot$. When the core collapses, an outgoing shock wave forms at the base of the mantle heating and ejecting it. In minutes, the shock reaches the photosphere of the progenitor where breakout produces an X-ray flash (Sect. 5.4.1). SN Ib/c are observed to be substantially polarized, implying asymmetric structure (Sect. 4.6). Timescales of minutes are expected in models with substantial asymmetries. Especially for the case of an asymmetric shock, the spectrum is not expected to be a single-temperature blackbody, since the emitting regions will vary both in time and in depth at different frequencies and NLTE and nonthermal processes may play a role, but typical energies of keV are expected. The duration of the breakout is a measure of the radius of the progenitor, but the relation may not be straightforward when the breakout is asymmetric. Breakout may have been observed from the Type Ib SN 2008D, but this is uncertain (Sect. 16.2.2).

Owing to the substantial presupernova radius, adiabatic cooling is not nearly as severe for the progenitor of a SN Ib/c as for a white dwarf progenitor of a SN Ia, and the breakout flash can be followed by a fading UVOIR (mainly UV) fireball display for several days, owing to diffusive release of internal energy deposited by the shock. Fireball radiation has been observed in the case of the Type Ic 1999ex (Stritzinger et al. 2002).

After the fireball phase and in the absence of CSI, the light curve is powered by ^{56}Ni and ^{56}Co decay, as in SN Ia. The basic physical processes are similar, although differences from SN Ia light curves are produced by different ejected masses, kinetic energies, compositions, opacities, and asymmetries. Usually, SN Ib/c involve larger asymmetry than SN Ia and lower M_{Ni}, although in SN Ic-bl, some of which are associated with gamma-ray bursts (Sect. 17.4.3), M_{Ni} and/or E_k may be higher than in SN Ia.

The late-time considerations for SN Ib/c are much like those for SN Ia. For example, the decreasing decline rate of the Type Ic-bl SN 1998bw, after a phase of decay at the ^{56}Co rate, might be attributed to energy deposition by ^{57}Co (Seitenzahl et al. 2009). A difference with respect to SN Ia is that after decades the decay of long-lived ^{44}Ti (half-life 60 yr) may be important. The decay of ^{44}Ti produces hard X-rays and γ-rays, which escape, and positrons, which may be trapped. The expected amount of synthesized ^{44}Ti is model dependent and subject to nuclear reaction-rate uncertainties (Tur et al. 2010), but for SN Ib/c at late times ^{44}Ti decay is likely to be more important than the ^{55}Fe decay that may be significant for SN Ia.

5.5.3 SN IIb

SN IIb are explosions of massive stars that have lost most but not all of their hydrogen envelopes by means of a stellar wind or by processes associated with binary mass transfer. Even a modest ($\lesssim$1 M$_\odot$) remaining hydrogen envelope may allow the progenitor to have a large radius, so the fireball phase could last long enough to be observed, as in the case of the bright Type IIb SN 1993J (Sect. 15.2). In practice, some progenitors of SN IIb may have rather large radii, but others may be more compact such that the fireball phase is truncated (Sect. 15.3.1). As in SN Ib/c, the fireball phase is followed by a luminosity minimum, then a rise to a secondary maximum and a subsequent tail phase, both powered by ^{56}Co decay as in SN Ia and SN Ib/c. Recombination is likely to quantitatively affect the shape of the second maximum (Sect. 15.4.2).

5.5.4 SN IIP

An RSG progenitor has a massive hydrogen envelope, $\sim$10 M$_\odot$, of large radius, $100 \lesssim R_\odot \lesssim 1000$, which leads to the light-curve plateau of SN IIP.

Within the extended envelope, the shock generated by core collapse requires hours to days to reach the photosphere, at which point breakout produces a bright flash in the UV and X-rays. The characteristics of the flash depend mainly on the presupernova radius and on the temperature in the shock, where NLTE and nonthermal processes may be important. Breakout has not yet been observed for an SN IIP.

Early fireball emission in the UV was observed for SNLS-04D2dc (Schawinski et al. 2008; Gezari et al. 2008a) and modeled in spherical symmetry by Tominaga et al. (2009) and with nonspherical shocks by Couch et al. (2009). The fireball phase lasts for a few weeks, until the outer layers become cool enough for hydrogen to recombine. During this phase, the shifting SED of the fireball radiation may produce a hump in the optical.

When the fireball cools to below 10,000 K, hydrogen in the outermost layers begins to recombine. An extended cooling and recombination phase follows, during which hydrogen recombines from the outside in. Outside the recombination front, the opacity of the neutral gas is low and beneath it, the opacity is high, so the photosphere follows the recombination front. As the ejecta expand, the photosphere recedes with respect to mass, but maintains a roughly constant radius. The temperature remains roughly constant at the recombination temperature of hydrogen, about 5500 K at the relevant electron densities, so the luminosity also remains roughly constant, resulting in the light-curve plateau (Fig. 5.4). The duration of the plateau, typically $\sim$100 days, is determined by the mass of the hydrogen envelope, the kinetic energy, and the extent to which ^{56}Ni is mixed towards the surface. The total energy radiated in the typical plateau phase, $\sim 10^{42}$ erg s^{-1} for $\sim 10^{7}$ s, is $\sim 10^{49}$ ergs. The bulk of the energy radiated was deposited by the initial shock. This energy is released by the inward passage of the recombination front rather than by outward diffusion, but the recombination energy per se, $\sim 3 \times 10^{46}$ ergs per solar mass of hydrogen, is low.

During the plateau phase, the velocity *of* the photosphere (dR_{phot}/dt) is nearly zero, but the velocity of the matter *at* the photosphere ($v_{phot} = R_{phot}/t$, where t is the time since explosion) as measured by blueshifts of the absorption components of P Cygni profiles is, of course, not zero (Chap. 4). Given a nearly constant photospheric temperature governed by hydrogen recombination, the luminosity at a given time after explosion should scale as $L \propto v_{phot}^2$. This correlation has been measured and used to estimate the luminosity and thus distance of SN IIP (Sect. 25.3).

When the recombination front reaches the base of the hydrogen envelope, it recedes rapidly in both mass and radius into the inner, metal-rich ejected core material, causing a plunge in luminosity and the end of the plateau phase. The

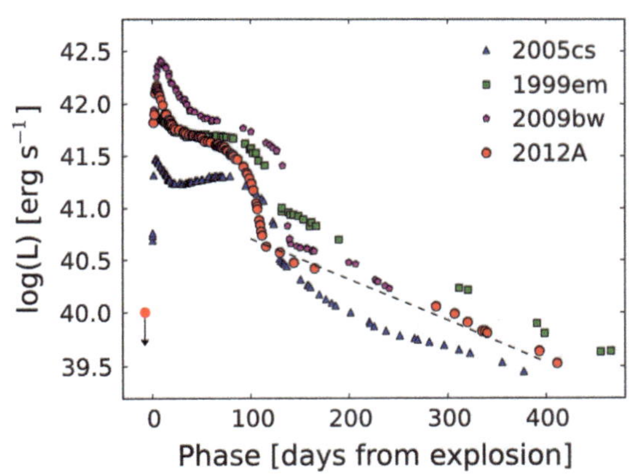

Fig. 5.4 UVOIR light curves of four SN IIP. The *dashed line* shows the ^{56}Co decay rate. Figure 5 "Comparison of progenitor mass estimates for the Type IIP SN 2012A" from Tomasella et al. (2013), by permission of Oxford University Press on behalf of the Royal Astronomical Society

plateau phase is followed by a *plateau-tail phase* (Utrobin 2007a; also known as the post-recombination plateau) during which radiation is released by the warm inner layers of the ejecta. This phase can be barely noticeable or extended, depending on the extent of mixing and whether M_{Ni} is low or high.

SN IIP eject radioactive ^{56}Ni, typically $\sim$0.1 M$_\odot$ but with a wide range. During the early plateau phase, the radioactivity luminosity is substantially less than the shock-deposited thermal energy that is released by the hydrogen recombination front, but if ^{56}Ni is mixed towards the surface, radioactive heating can extend the plateau. After the plateau and plateau-tail phases, the initial shock energy is depleted, but ^{56}Co decay continues (Fig. 5.4). As long as γ-rays are completely trapped, the light curve decays at the ^{56}Co rate and the luminosity (for known distance and corrected for extinction) can be used to measure M_{Ni}.

To first order, SN IIP light curves, at least through the ^{56}Co-powered tail, seem to be understood, and analytic and semi-analytic approaches can be useful (Zampieri et al. 2003; Nagy et al. 2014). At later times, additional effects, such as other radioactivities (^{57}Co, ^{22}Na, ^{44}Ti), CSI, ionization freezeout, dust formation, accretion onto a neutron star, and pulsar energy input, are possible. These are discussed in subsequent chapters.

5.5.5 SN 1987A

SN 1987A represents a special case. It had a massive envelope, giving it some properties of an SN IIP, but because the progenitor was blue and relatively compact, $R \simeq 3 \times 10^{12}$ cm, it suffered considerable adiabatic cooling and thus had some properties reminiscent of SN IIb. Breakout was not observed, but the early fireball phase was. The UVOIR luminosity of the fireball declined, but the optical light curve rose as emission shifted from the UV into the optical bands. The fireball emission declined for several weeks, before the UVOIR light curve began to rise to a secondary maximum powered by radioactive decay. The secondary maximum did not obey Arnett's Law (Sect. 5.4.2) because the large envelope mass imposed slower diffusion times than for, e.g., SN Ia. Consequently, there was more hysteresis between the input power and the output luminosity. After the second maximum, the bolometric (UVOIR plus far-IR) light curve closely followed the decay of ^{56}Co for about 3 years, while γ-rays from radioactive decay were efficiently trapped by the massive ejecta. At later times, the rate of decline of the light curve decreased as ionization freezeout and other radioactivities became important.

5.5.6 Light-Curve Extremes

Some supernovae show strong variations in their light-curve behavior compared to the traditional types. Some are overluminous, others are underluminous.

A large mass of ^{56}Ni can, in principle, produce a very bright light curve. Evolutionary models of pair-instability supernovae (Chap. 10) predict ^{56}Ni masses of about 10 $M_\odot$. Another mechanism to produce very bright light curves is CSI (Chap. 18).

Several models have been proposed to produce underluminous light curves. One invokes the electron-capture induced collapse of an ONeMg white dwarf or stellar core (Sect. 9.4.2). In spherically symmetric models, this process is expected to generate rather small amounts of ^{56}Ni and hence to produce only rather dim light curves if radioactive decay is the only energy source. Other processes, for instance the production of magnetic jets in the collapse, could alter this conclusion.

Another model to produce dim light curves is collapse of the core of a massive star to form a black hole. Much of the matter that would otherwise get shocked and ejected as ^{56}Ni could be swallowed by the black hole. Collapse to a black hole has been invoked to account for some supernovae that displayed plateaus but little or no radioactive tail. The extreme version of this process would be complete collapse of the entire star to form a black hole, with no supernova, ejection of matter, or radiative emission. The result would be the sudden disappearance of a previously bright, massive star (Kochanek et al. 2008). Searches for such disappearances have begun (Gerke et al. 2015). The formation of black holes does not guarantee dim or absent supernovae, however. In the collapsar model of γ-ray bursts, the formation of the black hole and the associated production of a relativistic jet is purported to generate very bright supernovae (Sect. 17.4.3). Even if a traditional supernova does not occur, the hydrodynamic response of an RSG envelope to the abrupt loss of mass of the core by neutrino emission might produce a weak, dim transient (Nadyozhin 1980; Piro 2013; Lovegrove and Woosley 2013).

5.6 Detailed Calculations

Detailed calculations of light curves are carried out for hydrodynamical models of supernovae, with various approximations and degrees of complexity (Höflich et al. 2010; Pumo and Zampieri 2011; Kasen et al. 2011; Jack et al. 2012b; Utrobin and Chugai 2013; Dessart et al. 2013a; Frey et al. 2013; Tominaga et al. 2013). Many of the comments regarding detailed spectrum calculations (Sect. 4.4.3) pertain also to light-curve calculations. Owing to the finite speed of light, the luminosity at one time depends on what happened at previous times, so light-curve calculations must include time dependence (Hillier and Dessart 2012).

5.7 Summary

In the absence of CSI, supernova light curves are predominantly powered by internal energy deposited by the shock from the explosion and by complete or partial trapping of the products of the ^{56}Ni decay chain. A key difference between these two energy sources is that the shock energy is deposited early and is subsequently adiabatically degraded by expansion. Radioactivity energy is deposited over an extended period of time and thus is less degraded by expansion. Many qualitative aspects of supernova light curves can be captured by simple dimensional analysis or analytic models, but detailed simulations require careful consideration of radiation hydrodynamics, time-varying opacities, asymmetries, and other complications. All supernovae must produce early brief bursts of breakout radiation as the shock approaches the stellar surface; this has proven difficult to capture in practice. Input power from a pulsar or magnetar is considered to account for some events. While the leakage of γ-rays will diminish the brightness of the optical light curve, there will be an associated γ-ray light curve that can provide diagnostics of the explosion.

The small radius ($\sim 10^8$ cm) of the white dwarf progenitor of a SN Ia leads to prompt degradation of any initial internal energy, and the observed light curve on the rise to maximum and later is powered by decay of ^{56}Ni and ^{56}Co. Radii of SN Ib and SN Ic progenitors ($\lesssim 10^{11}$ cm) are large enough to lead to an early shock-powered fireball phase that has been detected in a few cases. The envelopes of some SN IIb progenitors are extended, up to $\sim 10^{13}$ cm; this leads to more readily detectable fireball phases. Subsequently, as for SN Ia, SN Ib, SN Ic, and SN IIb are powered by ^{56}Ni and ^{56}Co decay and thus have a qualitatively similar shape with a pronounced peak followed by an extended exponential tail. SN Ia typically produce more ^{56}Ni and hence are brighter at peak than typical core-collapse events.

The light curves of SN IIP arise from RSG progenitors with massive hydrogen-rich envelopes (radius $\sim 10^{13}$ cm). SN IIP are dominated by shock-deposited energy until the end of the plateau phase. During this phase, a hydrogen recombination wave proceeds inward in mass while remaining at nearly constant radius. The plateau phase ends when the entire hydrogen envelope has expanded and recombined, thus dissipating the original shock energy. After the end of the plateau, ^{56}Co decay dominates.

At late phases of all types of supernovae, additional effects including radioactive decays of other nuclides, ionization freezeout, and CSI become significant or dominant.

Chapter 6
Circumstellar Interaction

6.1 Introduction

During their evolution stars, especially those of high luminosity such as the progenitors of core-collapse supernovae, produce CSM by means of winds. CSM also can be produced by impulsive ejections of matter and by processes associated with duplicity, such as sloppy mass transfer, common-envelope episodes, and novae. After a supernova occurs, hydrodynamic and radiative interactions with CSM can lead to observable emissions and absorptions across the electromagnetic spectrum. These features constrain some of the supernova properties and carry vital information on the density structure, velocity, and composition of the CSM. This is our best way to learn about the mass-loss behavior of stars in the late phases of their evolution, prior to explosion. That information, in turn, gives clues to the nature of the progenitor star and, perhaps, to the explosion mechanism.

Some supernovae, especially SN IIn, show evidence of strong CSI from the beginning. For core-collapse supernovae in general, emission from CSI tends to decline with time more slowly than other sources of emission. In most cases CSI eventually dominates the emission, an extended supernova remnant interacting with CSM being the long-term example.

In this chapter, the emphasis is on some of the ways in which supernovae and CSM interact. Inferences about individual supernovae from observational studies of CSI will be discussed in later chapters. We begin with the idealized case of spherical supernova ejecta interacting with CSM from a spherical, steady-state wind of the progenitor star. More detail on this case is provided in the review article by Chevalier and Fransson (2003). Most of this chapter applies to core-collapse supernovae. Evidence of SN Ia CSM, which is more elusive and tends be manifested in different ways, is discussed in Sect. 6.11.

D. Branch, J.C. Wheeler, *Supernova Explosions*, Astronomy and Astrophysics Library, DOI 10.1007/978-3-662-55054-0_6

6.2 Hydrodynamic Interaction

In a geometrically-thin region just above the photosphere of the progenitor star, wind matter can be accelerated by continuum radiation pressure and/or by absorption of radiation in spectral lines. Farther from the star, net forces on the wind matter are weak, so the matter reaches a terminal velocity at which it subsequently coasts. Beyond the narrow acceleration region, the CSM formed by a steady-state wind can thus be considered to have a velocity, v_w, that is constant in time and space and is on the order of the escape velocity from the star's photosphere.

For a constant wind mass-loss rate, $\dot{M}_w$, conservation of mass requires the density of the CSM to be

$$\rho_w = \frac{\dot{M}_w}{4\pi r^2 v_w} = 5 \times 10^{-17}\, \dot{M}_{-5}\, r_{15}^{-2}\, v_{10}^{-1}\ \ \mathrm{g\,cm^{-3}}, \tag{6.1}$$

where $\dot{M}_{-5} = \dot{M}_w/(10^{-5} M_\odot\ \mathrm{yr}^{-1})$, $r_{15} = r/(10^{15}\,\mathrm{cm})$, and $v_{10} = v_w/(10\,\mathrm{km\,s^{-1}})$. For a steady-state wind, the CSM density thus declines as r^{-2}. The ratio $w = \dot{M}_w/v_w$ is called the *wind parameter*, a measure of the density at a given radius. Assuming that the wind had a finite duration, the total circumstellar mass of $M_w = \dot{M}_{-5}\, t_5\, M_\odot$ is contained within an outer (or "cutoff") radius of $3 \times 10^{18}\ v_{10}\, t_5\,\mathrm{cm} \simeq v_{10}\, t_5\,\mathrm{pc}$, where t_5 is the wind duration in units of 10^5 years.

Because the velocity of the outer layers of the supernova ejecta ($\gtrsim$10,000 km s^{-1}) is much higher than v_w, the ejecta interact hydrodynamically with a comparatively stationary CSM. As a *forward shock*, or *blast wave*, propagates outwards, accelerating and heating the shocked CSM, a *reverse shock* propagates inwards (in mass), reheating the outer layers of the ejecta. The forward-shocked CSM and the reverse-shocked ejecta are separated by a *contact discontinuity* where the density peaks as shown in Fig. 6.1a. Between the contact discontinuity and the forward shock, the electron temperature can be lower than the ion temperature (Chevalier and Fransson 2003).

Prior to hydrodynamical interaction with the wind, the radial density profile of the outer layers of the ejecta usually can be described by a steep power law, $\rho_{SN} \propto v^{-n}$, with $n \simeq 10$. For $n > 5$ an analytic self-similar solution, or *similarity solution* (where radial profiles of physical quantities change in scale but not in shape), provides insight into the basic physical picture (Chevalier 1982a). When the interaction region between the two shocks is treated in a thin-shell approximation, where the radial distance between the two shocks is taken to be small compared to the radius of the contact discontinuity, the analytics are simplified. For example, the shell radius increases as $R_s \propto t^{(n-3)/(n-2)}$; the maximum ejecta velocity just inside the reverse shock is $v_{ej} = R_s/t \propto t^{-1/(n-2)}$; and the velocity of the shell is $v_s = dR_s/dt = v_{ej}(n-3)/(n-2)$. The ratio of the masses in the shocked ejecta and the shocked CSM is $(n-4)/2$ ($= 3$ for $n = 10$), and the corresponding ratio of the densities is $(n-4)(n-3)/2$ ($= 21$ for $n = 10$). The ratio of the electron temperatures of the shocked CSM and shocked ejecta is $(n-3)^2$ (neglecting radiative cooling,

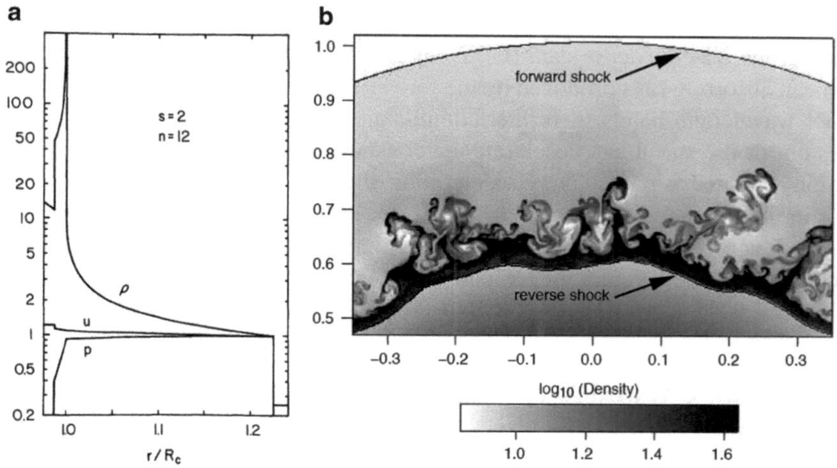

Fig. 6.1 (**a**) The variation of density, ρ, pressure, p, and gas velocity, u, in the interaction region between homologously expanding matter with density profile $\rho \propto r^{-12}$, as might characterize an RSG, and an external medium expanding at constant velocity with density profile $\rho \propto r^{-2}$, characteristic of a steady-state wind, as a function of radius normalized to the radius of the contact discontinuity, R_c. The physical variables have been normalized to their values at the outer shock wave. The forward and reverse shocks are at $r/R_c = 1.22$ and 0.98, respectively. A cool dense shell can form at the contact discontinuity. Note that the velocity profile is distinctly nonhomologous between the forward and reverse shocks. From Chevalier (1982a). (**b**) Two-dimensional calculation of the shock structure for supernova ejecta with $\rho \propto r^{-6}$ interacting with a steady-state wind with $\rho \propto r^{-2}$. Note the Rayleigh–Taylor instability at the contact discontinuity. The density is scaled to the value immediately behind the forward shock. From "Supernova Interaction with a Circumstellar Medium" (Chevalier and Fransson 2003), courtesy John Blondin

see below). Typically, the temperature of the shocked CSM is $\sim 10^9$ K and that of the shocked ejecta is $\gtrsim 10^7$ K, so both radiate in the X-ray band.

An important aspect of CSI, as illustrated in Fig. 6.1a, is that the velocity, while well behaved across the contact discontinuity, is distinctly nonhomologous in the entire region between the forward and reverse shocks. This means that the standard assumption of homology, $v \propto r$, which is hard-wired into many supernova spectral codes (Sect. 4.4), is not valid in this region. Strictly speaking, this will be true as long as the forward and reverse shocks exist, including the propagation of the forward shock into the ISM. The quantitative issue of when this nonhomology can be ignored has not been thoroughly explored.

In practice, of course, many complications arise. As lower-density, shocked CSM decelerates higher-density, shocked ejecta in the vicinity of the contact discontinuity, Rayleigh–Taylor instabilities develop at the interface, with narrow "fingers" or broader "mushrooms" of ejecta penetrating into the shocked CSM (Chevalier et al. 1992; Fig. 6.1b). Another effect is that the shocked ejecta may be able to radiate efficiently, thus lowering the temperature of the ejecta to $\lesssim 10^4$ K while maintaining the ram pressure. The density, already enhanced in the vicinity of the contact discontinuity, then rises even further, as the shocked ejecta are compressed into a *cool dense shell* (CDS). The CDS has its own radiative properties; e.g., it can absorb X-rays produced by the forward and reverse shocks and reradiate in other wavelength bands. A further complication is that when the radiation first breaks out of the star it can accelerate the constant-velocity wind. Since the burst radiation is diluted as r^{-2}, so is the acceleration, the result being an induced velocity profile in the innermost unshocked CSM of $v \propto r^{-2}$ (Fransson 1984). In addition, conduction across the contact discontinuity and across the shocks, if not inhibited by magnetic fields, may alter the temperature structure.

6.3 Optical, UV, and X-ray Emission from the Shocked Regions

In the case of a steady wind, the forward shock sweeps up mass at a rate $\dot{M}_{sh} = 4\pi r^2 \rho_w v_{sh}$, where v_{sh} is the velocity of the shock. If the energy produced in the shock is radiated away, the corresponding luminosity is

$$L = \dot{M}_{sh} v_{sh}^2/2 = 2\pi \rho_w r^2 v_{sh}^3 = \dot{M}_w v_{sh}^3/2v_w. \tag{6.2}$$

The luminosity is initially radiated in X-rays and UV. The reverse shock also produces hard radiation. In this and the next section we assume that at most wavelengths the CSM is optically thin. The optically thick case is discussed in Sect. 6.9.

The shocked regions of the ejecta and the CSM radiate by means of several processes. (1) Inverse-Compton scattering of photospheric photons by fast electrons in the shocked CSM produces a power-law continuum in the UV and X-rays and perhaps low-energy γ-rays. (2) Both the forward- and reverse-shocked regions emit X-rays by free–free radiation. (3) The reverse-shocked ejecta also contribute X-ray line emission from metals. (4) During the first months after the explosion, much of the emission from the reverse shock may be absorbed in the overlying shocked ejecta and reradiated in the UV and optical. This is especially likely if a CDS has formed. Some of the hard flux from the forward shock may also be radiated back into the CDS, absorbed there, and reradiated. (5) The CDS can cool to the point where atoms recombine and produce emission lines, especially Hα . These lines typically have intermediate width, $\sim$1000 km s^{-1}, compared to broader lines from high-velocity unshocked ejecta interior to the decelerated CDS and to narrower emission lines from unshocked CSM (Sect. 6.4).

Line profiles of the optical/UV radiation that results from reprocessing of X-rays in the CDS depend on the optical depth of the CDS. For low optical depths, the line profile is typically a Gaussian of intermediate width corresponding to the expansion velocity of the CDS. For higher optical depths, lines can develop blueshifted absorption components (Fransson 1984; Chugai et al. 2004; Dessart et al. 2009).

6.4 Optical, UV, and X-ray Emission from the Unshocked Regions

In addition to emission from the hot, shocked regions produced by the hydrodynamical interaction, and radiation due to absorption by and re-emission from a CDS, *radiative* interactions can cause emission from the unshocked regions. (1) The unshocked CSM can be ionized by radiation from photospheric shock breakout. Subsequent recombination leads to narrow emission lines in the optical, especially Hα, and in the UV, e.g., Lyman-α and lines of ions such as C III, C IV, N V, and Si IV. The width of these narrow lines provides a measure of v_w. A caveat is that the innermost wind material may be accelerated by breakout radiation (Sect. 6.2). (2) The unshocked CSM can be excited by X-rays from the interaction region, again producing narrow emission and absorption features in the optical and UV. This also can produce coronal emission lines, e.g., lines of [Fe VII] and [Fe X]. (3) The outermost layers of the unshocked ejecta, ionized and heated by radiation that is emitted inward from the interaction region, can contribute broad emission and absorption lines of high-ionization species such as C III, C IV, N V, and Si IV that would not otherwise form in the ejecta (Fransson et al. 1984).

6.5 Radio Emission

Another common manifestation of CSI is synchrotron radio emission, which requires relativistic electrons and a magnetic field. The former are thought to be produced by *diffusive shock acceleration* (Blandford and Ostriker 1978; Bell 1978), a subclass of *Fermi acceleration* (Fermi 1949) in which particles are accelerated by multiple reflections across shocks. The magnetic field may be the initial circumstellar field, compressed and further amplified in the interaction region, and/or it may be generated by random motions associated with the Rayleigh–Taylor instability. The radio luminosity depends on the uncertain small fractions of the postshock thermal energy that go into relativistic electrons and magnetic fields (Maeda 2013b).

An idealized model for the radio emission (Chevalier 1982b), based on synchrotron emission from the interaction region, wavelength-dependent free–free absorption in the unshocked CSM, and other assumptions, predicted that

$$F_\lambda \propto \left(\frac{\lambda}{20\,\mathrm{cm}}\right)^{(\gamma-1)/2} \left(\frac{t}{t_{20}}\right)^{-(\gamma+5-6m)/2} \exp\left[-\left(\frac{\lambda}{20\,\mathrm{cm}}\right)^2 \left(\frac{t}{t_{20}}\right)^{-3m}\right], \tag{6.3}$$

where γ is the power-law index of the electron energy spectrum, t_{20} is the time in days at which the optical depth to free–free absorption in the unshocked CSM at a wavelength of 20 cm falls to unity, and $m = (n-3)/(n-2)$, where n is the power-law index of the ejecta density profile. The emitted flux increases with wavelength (typically $\gamma \gtrsim 2$), but the exponential term that accounts for free–free absorption ($\propto \lambda^2$) external to the interaction region produces a cutoff of the emergent spectrum that shifts to longer wavelengths with time. Thus the spectrum is first observable at shorter wavelengths and later "turns on" rather abruptly at successively longer wavelengths. The time until the radiation is observable at a given wavelength provides a measure of the wind parameter, $w = \dot{M}/v_w$. The ratios of peak fluxes at different wavelengths (which occur at different times) provide a measure of m and hence n. After the CSM becomes optically thin to free–free absorption, the ratio of fluxes at different wavelengths (at a given time) provides a measure of γ. In some cases a rapid, achromatic decrease in the fluxes at a later time, compared to the power-law prediction, reveals an outer cutoff radius (an abrupt decrease in density) of the CSM. This leads to an estimate of the total mass of CSM, as well as, for an assumed value of v_w, the duration of the presupernova wind.

In reality, complications again may arise (Sramek and Weiler 2003). (1) $\dot{M}/v_w$ may have varied with time. Some supernova progenitors undergo enhanced wind mass loss during a restricted time, from $\sim 10^4$ years down to even just years prior to explosion, thus imposing an especially small radial extent of the dense CSM. (2) The density structure of the wind may be modified by the presence of a binary companion. (3) Free–free absorption may be by cool gas mixed into the interaction region, rather than by external unshocked gas (Weiler et al. 1990; Chandra et al. 2012a). (4) Synchrotron or inverse Compton cooling of the relativistic electrons may alter the radio light curve and spectrum (Chevalier et al. 2006).

The model described above was developed for and applied to the Type IIL SN 1979C and SN 1980K (Chevalier 1982b). Subsequently it was realized that conditions might be different, especially for relatively low values of w such as pertained to SN 1987A and usually pertain to SN Ib/c owing to high values of v_w. For small values of w, synchrotron self-absorption by the same relativistic electrons that emit the synchrotron radiation is significant or even dominant (Chevalier 1998; Fransson and Björnsson 1998; Weiler et al. 2011). The self-absorption has consequences for the early radio light curve and spectra. When self-absorption is dominant, analysis of a well-observed radio light curve yields an estimate of the radius, and therefore the velocity, of the interaction region. A useful diagnostic results from plotting lines of constant velocity, assuming synchrotron self-absorption to be dominant, in a plot of peak radio luminosity versus the time to reach the peak. Placing the parameters of an observed supernova in the diagram yields a prediction of the velocity of its interaction region; if the velocity is lower than the velocity inferred from the blue edge of spectral lines or from very-long-baseline interferometry, the luminosity predicted by the self-absorption model is too low, and free–free absorption is dominant (Chevalier 1998; Chevalier and Fransson 2003). A model fit to the radio light curves of the Type IIb SN 1993J is shown in Fig. 6.2.

In a few cases, e.g., the Type IIL SN 1979C, the Type IIn SN 1986J, SN 1987A, the Type IIb SN 1993J, and the Type IIP SN 2004et, the radio emission has been

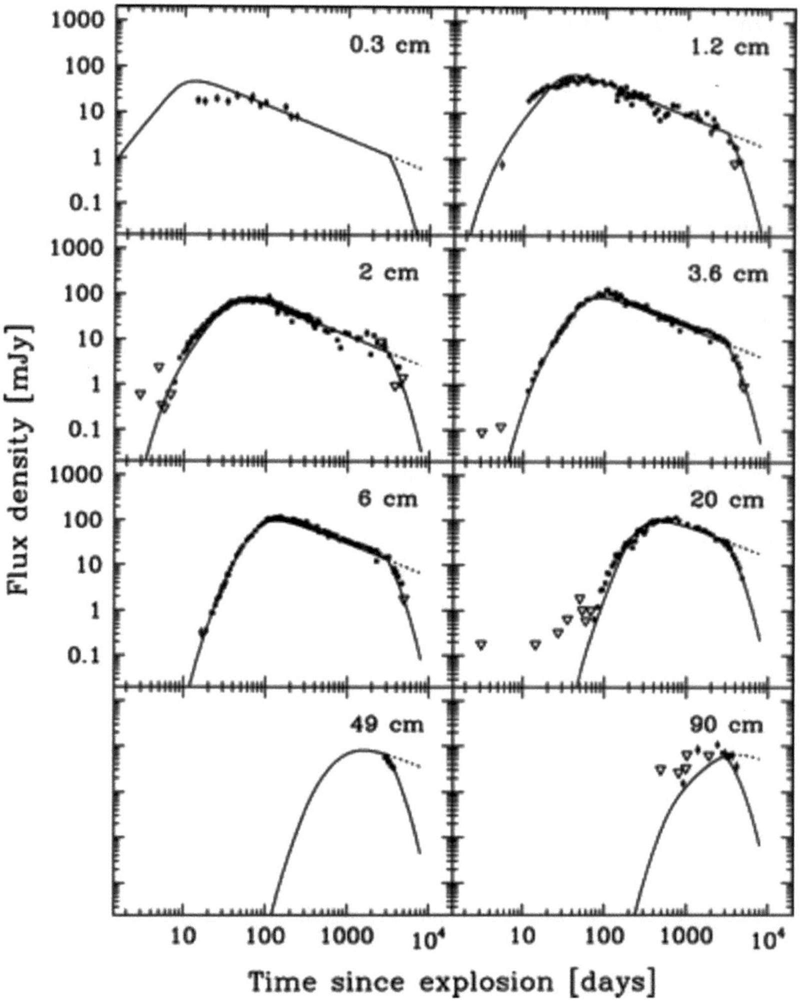

Fig. 6.2 Radio light curves of SN 1993J. *Solid lines* represent a combined synchrotron and thermal free–free absorption model described by "Long-Term Radio Monitoring of SN 1993J" (Weiler et al. 2007). © AAS. Reproduced with permission

imaged by means of very-long-baseline interferometry, revealing fairly to very symmetric shell structure and allowing the expansion of the interaction region to be observed directly. The radio-emitting shell of SN 1979C has been undergoing almost free expansion ($m \equiv (n-3)/(n-2) = 0.91 \pm 0.09$), where $R \propto t^m$; Marcaide et al. (2009b). The shell of the very well studied SN 1993J also has been almost freely expanding ($m \simeq 0.9$; Martí-Vidal et al. 2011a).

6.6 Dust and Infrared Emission

Circumstellar matter of RSGs includes dust grains. After a supernova occurs, the temperature of a circumstellar dust grain is determined by a balance between heating by absorption of supernova radiation and cooling by IR emission. Because typical dust grains evaporate at about 1500 K, depending on their composition (amorphous or graphitic carbon dust, or silicates), a core-collapse supernova can evaporate dust out to $r_e \simeq 10^{18}\, L_{45}^{0.5}$ cm, where L_{45} is the maximum supernova luminosity in units of $10^{45}\,\mathrm{erg\,s^{-1}}$; this maximum probably occurs at shock breakout. For $r > r_e$, the dust survives and subsequently cools (Fig. 6.3).

Evaporation of dust grains by supernova radiation can cause the circumstellar dust extinction of a supernova to be less than that of the progenitor star. If the dust density varies as r^{-2} from, say, ten times the radius of the progenitor, r_p, to an outer cutoff radius, r_0, then the ratio of optical depths to dust absorption after and before the supernova is $(r_o/r_e-1)/(r_o/10r_p-1)$, which may be small due to large $r_o/10r_p$. Evaporation does not necessarily destroy most of the dust, however, because the ratio of dust masses after and before the supernova is $(r_o - r_e)/(r_0 - 10r_p)$, which is not necessarily small. Much of the extinction of the progenitor occurs in the inner layers of the circumstellar dust, where the supernova evaporates the dust, while much of the dust mass may be in extended, outer portions of the CSM where dust is not evaporated.

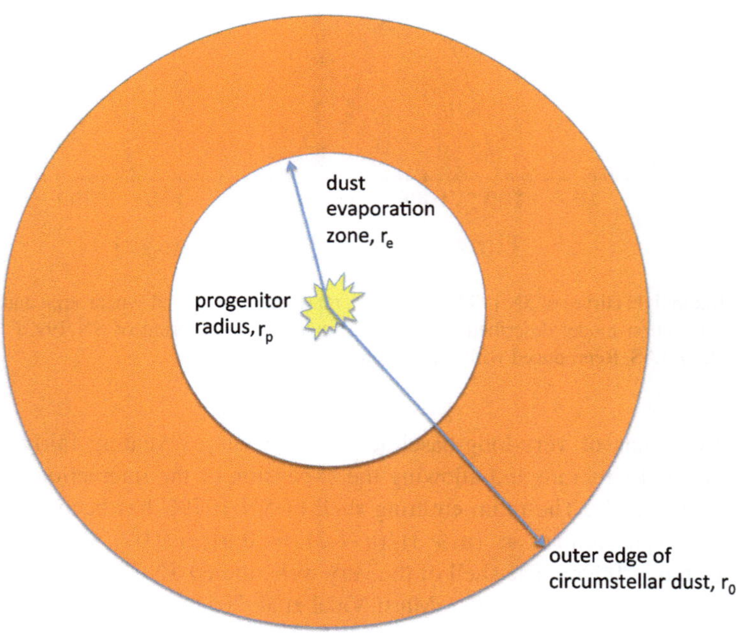

Fig. 6.3 Schematic of circumstellar dust shell following evaporation of inner dust

Surviving dust of temperature T emits in the IR, peaking at $\lambda_{max} = 6(T/500\,\mathrm{K})\,\mu\mathrm{m}$ according to the blackbody Wien law. This can produce an *infrared echo* if the postsupernova optical depth to dust absorption in the UV is significant. In some cases, the luminosity of the dust emission eventually exceeds the direct supernova emission. The emitting dust may be so spatially extended that the light curve of the echo is strongly affected by light travel-time effects. The ratio of the IR and total emitted energies provides an estimate of the dust optical depth, hence also the total mass of surviving circumstellar dust. For more details, see Dwek (1983, 1985). IR echoes from interstellar dust also have been observed (Meikle et al. 2011).

Radiation from the CSI region can heat surviving circumstellar dust and produce enhanced IR emission. After decades, when the forward shock reaches the dust, collisional heating of the dust may lead to further IR emission (Tanaka et al. 2012b). Eventually, the shock may destroy some or all of the dust by sputtering.

Because dust grains not only absorb but also scatter radiation, they can produce a direct echo of scattered supernova light in the optical and UV. This can be difficult to separate from the emission from the supernova photosphere and the shocked circumstellar regions, but it has been observed in a few cases at late times.

After the supernova occurs, dust may form in the CDS (Pozzo et al. 2004; Meikle et al. 2011) and eventually in the deeper metal-rich layers of the supernova ejecta; these topics are considered in some of the later chapters.

6.7 Clumps

Stellar winds commonly have density irregularities, often discussed in terms of density-enhanced "clumps" embedded in a lower-density interclump medium, sometimes with a large density contrast. When a shock of velocity, v_s, enters a clump, internal and external pressures tend to become comparable, so the shock slows down to a clump velocity $v_c = (\rho_s/\rho_c)^{1/2} v_s$, where ρ_s/ρ_c is the ratio of the densities external and internal to the clump. In the clump, the lower velocity and higher density can lead to enhanced emissivity and radiative cooling. Encounters of the forward shock with clumps in the CSM can produce X-rays as well as optical spectral lines of intermediate width owing to the reduced velocity of the shock in the clump (Chugai and Danziger 1994). The shock also propagates around the clump, which tends to lead to clump fragmentation (Klein et al. 1994).

Clumps are invoked to account for the shape of the radio light curves of some supernovae. When their presence is taken into account, the inferred value of the wind parameter is reduced. If circumstellar dust is in optically-thick clumps, standard estimates of the total mass of dust based on IR emission are likely to be too low.

6.8 Shells

Detached CSM shells can be produced in various ways. (1) The relatively fast winds of massive main-sequence stars produce low-density "bubbles" surrounded by shells of swept-up interstellar matter. (2) The fast ($\sim$1000 km s^{-1}) wind of a star that has returned to the blue following an RSG phase compresses the RSG wind into a shell. (3) A shell may form as a consequence of an episodic ejection of matter from the supernova progenitor or from a common-envelope episode of a binary system. Episodic mass ejection is not well understood, but is witnessed in planetary nebulae, novae, luminous blue variables (LBVs), and in all its asymmetric glory in η Carinae, an extremely luminous star that may itself be some version of an LBV. There is ample evidence that the ejecta of some supernovae do interact with a variety of shells. When the delayed interaction of supernova ejecta with a shell occurs, a mass of ejecta comparable to the mass of the shell is decelerated, and some of the kinetic energy of the ejecta is converted to radiation. For example, the interaction of ejecta expanding at 10,000 km s^{-1} with a 1 M$_\odot$ shell could in principle radiate almost 10^{51} ergs at a time $\sim$0.3r_{16} years after the explosion, where r_{16} is the radius of the shell in units of 10^{16} cm. If the shell happens to be at a radius $\sim$$10^{15}$ cm, the typical radius of a supernova photosphere, the conversion of kinetic energy to luminous energy is maximally efficient and the luminous output is especially bright.

For an optically-thin shell, the expression for the luminosity generated by a forward shock given in Sect. 6.3 applies. In principle, the run of $\rho_w(r)$ can be inferred from the evolution of the luminosity and the shock speed.

Shell ejection may produce episodes of high luminosity even before the supernova occurs. A shell may collide with dense CSM from a preceding wind phase (Smith et al. 2013b). A progenitor such as an LBV or a very massive star subject to pulsational pair instability (Sect. 10.7) may eject multiple shells, with the possibility of collisions between the shells. Such events can produce *supernova impostors*—luminous eruptions that may be mistaken for genuine supernovae (Van Dyk et al. 2000; Smith et al. 2011c; Sect. 14.4).

6.9 Optically-Thick CSM

Early work on CSI was done in the context of optically-thin CSM, but it became recognized that in some cases, especially SN IIn or superluminous supernovae, CSM produced by winds or by episodic ejections can be massive enough to be optically thick in the continuum, at least after being ionized by supernova radiation, and a photosphere can be present in the CSM.

For optically-thin CSM, the breakout flash from the supernova is brief ($\lesssim$1 h) and the emission is mainly in the UV and X-rays. This is followed by a fireball phase and then the main peak of the supernova optical light curve, powered by the diffusive release of shock-deposited energy or by radioactivity (Sect. 5.4). In contrast, for sufficiently optically-thick CSM (wind optical depth $\gtrsim c/v_{sh}$), breakout can be much more extended in time and the UV and X-ray emission can be degraded to the optical, such that breakout emission dominates the whole peak of the optical light curve.

Even for the case of a steady-state wind, for which the density is constrained to $\rho_w \propto r^{-2}$, a wide variety of peak luminosities and light-curve shapes can be produced, depending on the wind parameter and the outer extent of the thick wind (Chevalier and Irwin 2011; Ginzburg and Balberg 2012). After breakout, a collisionless shock is predicted to form and generate hard X-ray emission hundreds of days after explosion, as the optical emission declines (Svirski et al. 2012). Such late, hard X-ray emission has been detected from a few supernovae, suggesting that their light curves were powered by breakout in optically-thick winds (Ofek et al. 2013).

Some massive stars are surrounded by massive circumstellar shells. Very massive stars, $\gtrsim$40 $M_\odot$, which never become RSGs, may lose more mass by means of episodic ejections than by steady winds (Smith and Owocki 2006; Dwarkadas 2011; Shiode and Quataert 2014). Unlike in the case of a steady wind, the distance of the shell from the supernova progenitor at the time of explosion, and the density profile within the shell, are not a priori constrained. These properties as well as the mass of the shell undoubtedly cover wide ranges. Consequently a very wide variety of light curves can be produced by interactions between ejecta and shells (van Marle et al. 2010; Chatzopoulos et al. 2013b; Piro 2015b).

The widths of spectral lines are not always determined by Doppler broadening alone. If the CSM has a significant optical depth to electron scattering, an otherwise narrow emission line can be broadened by electron scattering to yield a line of intermediate width, the width being determined by the scattering conditions rather than by an expansion velocity (Sect. 4.4.3). When the photosphere is outside of a forward shock that is still at a large optical depth in dense CSM, the emission of, e.g., Hα, is from photoionized hydrogen ahead of the shock. The line would be narrow if it were not broadened by electron scattering. No absorption component is expected, and neither a CDS nor the supernova photosphere are directly observable.

As the shock proceeds through the shell, the photosphere can recede to beneath the forward shock Behavior similar to the optically-thin case described above then can occur, replete with a forward shock as well as reverse shocks in the shell and in the ejecta, formation of a CDS, and absorption and re-emission from the shell.

When the photosphere retreats beneath the forward shock but is within the CDS, the width of the Hα line is determined by the expansion velocity of the CDS, unless the electron-scattering optical depth in the CDS is significant. A blueshifted absorption component may form, depending on the line optical depth. The photosphere of the superluminous SN 2006gy, for example, may have retreated from outside the forward shock to within the CDS about 20 days after maximum light (Sect. 18.2.1).

6.10 Core-Collapse Supernovae

All core-collapse supernovae have CSI, but the strengths and observational manifestations of the interaction vary widely. For weak interactors, e.g., most SN IIP, early optical spectra may show no obvious manifestations of CSI. At the other extreme, the defining feature of an SN IIn is the presence of at least one narrow emission line, usually Hα, in the optical.

The progenitors of SN IIP are RSGs, which are observed to have moderate wind mass-loss rates ranging from $\lesssim 10^{-6}$ M$_\odot$ yr^{-1} for a star of ZAMS mass about 8 M$_\odot$ to $\gtrsim 10^{-5}$ M$_\odot$ yr^{-1} for a ZAMS mass of 25 M$_\odot$. Owing to the low wind velocity of $\gtrsim 10\,\mathrm{km\,s^{-1}}$, the wind parameter may be high enough to produce observable CSI for relatively nearby SN IIP. If $v_w \sim 10\,\mathrm{km\,s^{-1}}$ and $v_{SN} \sim 10{,}000\,\mathrm{km\,s^{-1}}$, then observing 1 year of CSI after the supernova probes the mass-loss history of the RSG during the 1000 years immediately preceding the supernova. Observations of thermal or inverse Compton scattering X-ray emission and synchrotron radio emission can be used to estimate presupernova mass-loss rates and ZAMS masses of individual SN IIP progenitors; these quantities are generally consistent with expectation (Chevalier et al. 2006). The progenitors of SN IIL and some SN IIb may be red or yellow supergiants that lose significantly more mass prior to explosion and therefore display stronger CSI.

Progenitors of SN Ib/c, whether classical Wolf–Rayet stars from massive stars or less massive stars, the envelopes of which have been stripped in binaries, have wind mass-loss rates of $\sim 10^{-5}$ M$_\odot$ yr^{-1} but high wind velocities $\sim 1000\,\mathrm{km\,s^{-1}}$. Observing 1 year of CSI therefore probes the progenitor mass-loss history only during the decades immediately preceding the supernova.

Owing to strong CSI, SN IIn show narrow spectral lines, and their light curves are typically brighter and broader than those of SN IIP. The CSM could be from steady winds or from shells produced by episodic ejection of mass from the progenitor. The result is a bright, slowly-declining optical light curve. The light curves of some, but perhaps not all, superluminous supernovae also are powered by strong CSI.

6.11 Type Ia Supernovae

The presence of CSM associated with SN Ia might confirm the expectation that duplicity is involved as well as provide clues to the nature of the progenitor, e.g., whether single- or double-degenerate systems or a mix of both are involved (Chap. 21). In most single-degenerate models, some CSM is expected to be associated with SN Ia, while in most double-degenerate models the amount of CSM might be very small.

A rare class of supernovae that undergo strong interaction with hydrogen-rich CSM, at least some of which are SN Ia, is discussed in Sect. 23.2. For typical SN Ia, radio and X-ray emission from CSI has not been detected. For many SN Ia, evidence of CSM is entirely lacking, and the circumstellar environments appear to be remarkably clean.

Narrow absorption lines of Na I D, Ca II, and a few other species are often observed in supernova spectra and are generally attributed to the ISM of our Galaxy or the host galaxy. A narrow circumstellar component would be difficult to distinguish from an interstellar absorption in the host. In a few SN Ia, however, narrow Na I absorption near the redshift of the host has been observed to vary with time. This is attributed to ionization of circumstellar sodium followed by recombination. A statistical tendency of narrow (nonvariable) Na I D absorptions in SN Ia to be more blueshifted than redshifted is attributed to outflowing CSM associated with at least a significant fraction of SN Ia (Sect. 21.6.5).

In addition to the broad photospheric P Cygni features in early-time SN Ia spectra, which typically form at velocities $\gtrsim$10,000 km s^{-1}, high-velocity absorptions of Ca II and some other ions, in the range 20,000–30,000 km s^{-1}, are common. One possibility for their origin is that they reflect the sweeping up of CSM, but other origins are not excluded.

6.12 Summary

The interaction of supernova ejecta with CSM is a rich subject. Hydrodynamic interaction between ejecta and CSM resulting from a presupernova wind or episodic ejection produces a forward shock that heats the CSM and a reverse shock that reheats outer layers of the ejecta. Optical, UV, and X-ray emission from both shocked and unshocked regions ensues, and acceleration of electrons to relativistic energies in the interaction regions leads to synchrotron radio emission. Circumstellar dust that escapes vaporization by the supernova can absorb supernova light and reradiate an IR echo. Observations of these emissions provide valuable information on the mass-loss histories of the supernova progenitors. Breakout of the supernova shock in optically-thick CSM can dominate the optical peak of the light curve, and produce a variety of light curves including extremely luminous ones.

Chapter 7
Supernova Remnants

7.1 Introduction

When does a supernova become a supernova remnant? A tidy rule would be that a supernova remnant (SNR), but not a supernova, is primarily interacting with ISM, but this would not work for some young SNRs such as *Cas A* that are still interacting primarily with CSM rather than ISM, and obviously not for pulsar-powered SNRs such as the *Crab*. Maintaining that the emission from an SNR, but not a supernova, is primarily from interaction with an ambient medium, whether CSM or ISM (and making exceptions of Crabs), would be no better because some SN IIn would have to be regarded as SNRs from the moment of their discovery; some would even evolve from SNRs to supernovae. One rule of thumb could be simply age: a supernova becomes an SNR at an age of about 100 years.

Contemporary supernovae (practically all beyond the Local Group of galaxies) are observationally resolved in time but not in angle, while SNRs (those in the Galaxy and nearby galaxies) are resolved in angle but in only a narrow slice of time. Extragalactic events extending back to SN 1923A that were discovered in the twentieth century and are still observable owing to CSI (Long et al. 2012) are sometimes called supernovae and sometimes called SNRs. SN 1885A in the Andromeda galaxy is now a borderline SNR. The radio source G1.9+0.3 is the youngest known Galactic SNR at an age $\gtrsim$100 years. Remnants of Galactic supernovae that were observed by eye before the invention of the telescope are called historical Galactic SNRs. As usual, SN 1987A is an exception to a rule, being resolved in both time *and* angle. In spite of its youth, some authors have been referring to "SNR 1987A" ever since the ejecta began to impact the inner circumstellar ring (Sect. 11.6). It may be appropriate to think of SN 1987A as in transition from supernova to SNR.

D. Branch, J.C. Wheeler, *Supernova Explosions*, Astronomy and Astrophysics Library, DOI 10.1007/978-3-662-55054-0_7

7.2 SNR Populations

7.2.1 Galactic SNRs

About 300 Galactic SNRs are known, most having been recognized as extended sources of synchrotron radio emission (Green 2014a). Because SNRs are concentrated to the central regions of the Galactic plane, they are often highly extincted in the optical, and even more in X-rays, where SNRs are intrinsically luminous (typically $10^{36}\,\mathrm{erg\,s^{-1}}$). Practically all of the known Galactic SNRs have been detected in the radio, where extinction by dust is not an issue, compared to more than 40% in X-rays, 30% in the optical, and 30% in the IR.

Even the radio sample of Galactic SNRs is seriously incomplete, for several reasons. Surveys generally are restricted to within a few degrees of the Galactic plane and it is difficult to recognize SNRs of low radio surface brightness in directions in which the diffuse Galactic radio emission is strong. In addition, owing to the limited angular resolution of radio surveys it is difficult to distinguish SNRs of small angular size ($\lesssim$10 arcminutes) from background radio sources. If the mean interval between Galactic supernovae is 35 years (Sect. 3.5) and all supernovae become SNRs, then the number of known SNRs suggests that they are detectable for more than 10^4 years, perhaps 10^5 years if the degree of SNR incompleteness is very high (Helfand et al. 1989; Li et al. 1991).

In the radio, about 80% of the known Galactic SNRs have a *shell* or partial shell morphology, a few percent are *pulsar-wind nebulae* (PWN) like the Crab (Sect. 7.4.2), and $\gtrsim$10% are *composite* (shell and PWN). About a dozen are *mixed morphology*—a shell and, instead of a PWN, a central, unresolved, thermal X-ray source that often is referred to as a *compact central object* (CCO). The lack of detected PWN in most SNRs may imply that the PWN phase is relatively short-lived or that many PWN are intrinsically much fainter than that of the Crab. Among the more than 1000 known Galactic pulsars, fewer than a hundred are associated with known SNRs, primarily because a pulsar is detectable long ($\sim 10^6$ yr) after a SNR is not.

Most known Galactic SNRs have angular sizes in the range of a few to $\sim$50 arcminutes. Distances and therefore linear diameters generally are uncertain, but most diameters are in the range of a few to $\lesssim$100 pc. Thus all but the youngest SNRs have swept up a substantial amount ($\sim 10^4\,M_\odot$) of ISM, so the SNR properties depend mainly on the density and structure of the local ISM, with the only supernova properties "remembered" being the initial kinetic energy and whether there is a pulsar. Owing to the incompleteness and poorly known distances, it is difficult to infer the Galactic distribution of SNRs. Some statistical studies of the known SNRs are useful, however; for example, the distribution of SNRs in Galactic longitude suggests that the distribution per unit area with respect to Galactic radius peaks around 3 kpc from the Galactic center (Green 2014b).

A plot of radio surface brightness, Σ, versus linear diameter, D, for shell SNRs shows a correlation (the *Σ–D relation*) that was originally interpreted in terms of SNR age—young SNRs having high Σ and small D, old ones having low Σ and large D. The relation is strongly affected by observational selection biases and is now recognized to contain significant intrinsic scatter owing to variations in supernova explosion energy and ejected mass, as well as in the density of the ambient ISM. Nevertheless, for most shell SNRs no other distance estimate is available, so the observed Σ is often used to make a rough estimate of D, and hence distance (Bandiera and Petruk 2010; Pavlović et al. 2013).

7.2.2 Extragalactic SNRs

The LMC at a distance of 50 kpc and the SMC at 64 kpc are close enough for radio discovery of SNRs, and because the extinction along the line of sight is low, optical and X-ray observations also are feasible. About 80 SNRs are known in the LMC and SMC, with about half having been detected in X-rays. Owing to less confusion with other sources than in the Galaxy, about 60% of the SNRs in the LMC have been detected in the IR (Seok et al. 2013). In an external galaxy that is not nearly edge-on to the line of sight, the spatial distribution of the known SNRs is revealed, and the similar distances of the SNRs facilitates statistical studies of their properties (Badenes et al. 2010).

Compared to Galactic and Magellanic-Cloud SNRs, SNRs in more distant galaxies are difficult to identify in radio and X-rays because of their lower fluxes and smaller angular sizes. Fortunately, the generally moderate extinction of SNRs allows candidates to be identified as extended sources of optical emission lines characterized by a relatively high ratio ($\gtrsim$0.4) of [S II] $\lambda\lambda$6717,6731 to Hα flux, caused by shock-heating of the ISM (Mathewson and Clarke 1973). In photoionized nebulae, sulfur is more highly ionized so this flux ratio is lower. The strength of the [Fe II] 1.644 μm line also is useful (Blair et al. 2014). Among large galaxies of the Local Group, more than 300 SNR candidates are catalogued in M31 (Lee and Lee 2014a) and nearly as many in M33 (Long et al. 2010; Lee and Lee 2014b). From a study of the stellar environments of a subsample of M31 and M33 SNRs with deep *HST* images, it has been inferred that only a small fraction, if any, of stars of ZAMS mass exceeding about 40 $M_\odot$ explode as SN (Jennings et al. 2014).

In nearby galaxies beyond the Local Group, more than 1000 SNRs and candidates are recognized (Vučetić et al. 2015), including more than 250 in M83 (Blair et al. 2014; Long et al. 2014a). The small, nearby (3.3 Mpc) starburst galaxy M82 is deduced to produce supernovae at a high rate, about eight per century, most of them highly extincted by dust. The several dozen known SNRs in M82 include a large number that are especially young (Fenech et al. 2008).

7.3 Evolution of Shell SNRs

7.3.1 Classical Theory

The classical theory of the evolution of shell SNRs (Woltjer 1972; Spitzer 1998; Draine 2011) applies to those that develop in a homogeneous, relatively high-density ISM rather than in a hot, low-density bubble. The evolution includes several phases.

Before the mass of the swept-up ambient medium exceeds that of the ejected mass, the (nearly) *free expansion*, or *ejecta dominated* phase typically lasts hundreds to thousands of years and achieves diameters of a few to tens of parsecs. The basic theory for this phase is that of Chevalier (1982a) as was discussed in the context of CSI in Sect. 6.2.

When the swept mass exceeds the ejected mass by a factor $\gtrsim$10, the interaction evolves towards a state that can be modeled as that of a point explosion in a uniform medium. During the ensuing *adiabatic* or *Sedov* phase, energy conservation is a good assumption because temperature is high ($T \gtrsim 10^7$K) and therefore radiative losses are small compared to the total energy. As swept mass is added to the shell, the velocity decreases. A similarity solution (Sedov 1959) for the case of constant ISM density, n, and energy, E, gives $R \propto (E/n)^{1/5}t^{2/5}$ and thus the shock velocity $v_{sh} = 2R/5t$. In this phase, the SNR diameter is $\lesssim$100 pc. Most observed SNRs are in this phase, which lasts tens of thousands of years (Bandiera and Petruk 2010). A caveat to the assumption of energy conservation during the Sedov phase is that SNRs may typically put about 10–30% of their energy into cosmic rays by means of diffusive shock acceleration. There may never be a truly adiabatic phase. During the Sedov phase the momentum increases by a factor ~10 owing to work done by the hot shocked gas.

After ~10^4–10^5 years, when $v_{sh} \lesssim 200\,\mathrm{km\,s^{-1}}$, temperature ($T \propto v_{sh}^2$) falls below 10^6 K, emissivity in UV and optical lines rises, and conservation of energy no longer applies. The *radiative*, *isothermal*, or *snowplow* phase of constant momentum begins. Because its energy is efficiently radiated away, the shell of swept-up ISM narrows and its density rises. The thin, cool shell contains most of the mass and kinetic energy, while the hot, low-density gas interior to the shell contains most of the thermal energy.

After ~10^6 years, when $v_{sh} \lesssim 10\,\mathrm{km\,s^{-1}}$, comparable to random motions in the ISM, the shell fragments, loses its identity, and merges with the ISM. The evolution of an SNR in an inhomogeneous ISM is rather more complicated (Cox and Smith 1974; McKee and Ostriker 1977; Korolev et al. 2015).

7.3.2 Presupernova Shells

The ambient medium may be expected to have a complicated radial (and probably angular) structure that may include one or more shells resulting from presupernova interactions between CSM (Sect. 6.8) and ISM. For an overview, see Lozinskaya (1992).

While on the main sequence, the fast wind of a massive progenitor star blows a low-density cavity, or bubble, surrounded by a thin, dense, cool "main-sequence shell" of wind and swept-up ISM. Depending on the ZAMS mass, the wind mass-loss rate is $\dot{M}_w \sim 10^{-6}$–10^{-5} $M_\odot$ yr^{-1}, the wind velocity $v_w \gtrsim 1000\,\mathrm{km\,s^{-1}}$, the bubble radius is of order parsecs, and the shell mass may be about 10–100 $M_\odot$.

A star that then spends part of its life as an RSG has a wind that is slower, $v_w \sim 10\,\mathrm{km\,s^{-1}}$, and more dense. For a steady-state wind, the density profile is $\rho \propto r^{-2}$. If the star explodes while it is an RSG, the initial interaction is with this dense wind, and a "red-supergiant shell" of ejecta and swept-up wind matter forms. Eventually this shell interacts with the main-sequence shell, driving a new forward shock into the main-sequence shell and a new reverse shock back into the RSG shell.

If the progenitor does not undergo an RSG phase, but evolves directly from the main sequence to explode as a Wolf–Rayet star, the fast Wolf–Rayet wind further compresses the main-sequence shell. Initially, the supernova interaction is with the low-density wind matter, producing a "Wolf–Rayet shell" of ejecta and swept-up wind matter that eventually interacts with the main-sequence shell.

When an RSG contracts to become a Wolf–Rayet star before exploding, the structure is even more complex. The fast Wolf–Rayet wind produces a low-density bubble surrounded by a dense shell of Wolf–Rayet and RSG wind matter. This shell, in turn, is surrounded by a low density cavity interior to the main-sequence shell.

Even in spherical symmetry, complicated interactions are possible (Dwarkadas 2005). Double X-ray-emitting shells are predicted in some cases, filled spheres of X-ray-emitting gas in others. In some cases, the SNR may skip the Sedov phase because the ejecta collide with a shell of CSM and go directly into a radiative cooling phase. The overall longterm evolution of the SNR depends primarily on the ratio of the mass initially in shells to the ejecta mass. Given the probability that massive-star progenitors have complex CSM, interpreting observations of SNRs on the assumption of a uniform ambient medium can lead to large errors.

7.4 Young SNRs

Young SNRs ($\lesssim 3 \times 10^3$ years) are not highly contaminated by swept-up ISM and therefore provide the most information about the supernovae that produced them. SNRs that are exceptionally useful are those for which the explosion date is at least approximately known, and especially for which the precise spectroscopic type of the supernova is known from observation of the supernova spectrum by means of time-delayed scattered-light echoes off interstellar dust. In this section, the discussion is restricted to Cas A, known to be a remnant of an SN IIb; the SNRs of five historical supernovae of the second millennium including the remnant of Tycho's SN 1572, known to have been an SN Ia; the probable SNR of the first historically-recorded supernova, SN 185; the youngest known Galactic SNR, G1.9+0.3; and the

SNR of SN 1885A in M31. Because the Galactic SNRs have angular diameters of arcminutes, they are well resolved from radio to X-rays.

7.4.1 Cassiopeia A

A light-echo spectrum of the supernova that produced the Cas A SNR closely resembled the spectrum of the Type IIb SN 1993J (Krause et al. 2008a). Furthermore, spectra reflected from several different sight lines showed expansion velocities differing by up to 4000 km s^{-1}, providing direct evidence for an asymmetric explosion (Rest et al. 2011b).

A suggestion that the Cas A supernova was glimpsed by Flamsteed in 1680 and catalogued as an ordinary variable star of 6th apparent magnitude (Ashworth 1980) is doubtful. An estimate of the explosion time, 1681 $\pm$ 19, based on proper motions and radial velocities of optically-emitting knots of ejecta (Fesen et al. 2006), is suggestive, but the 10 arcminute discrepancy between the positions of the star recorded by Flamsteed and the center of the SNR is much larger than Flamsteed's typical measurement error (Stephenson and Green 2002). The supernova possibly was seen by Gian Domenico Cassini on or shortly before 1671 as a star of $m_V \simeq 4$ (Soria et al. 2013). At a distance of 3.4 kpc (Reed et al. 1995), with visual extinction $A_V \simeq 6.2$ magnitudes (Eriksen et al. 2009), $m_V \simeq 4$ would correspond to $M_V \simeq -14.4$. This is dimmer than the typical $M_V \simeq -17$ for SN IIb, but the extinction is uncertain and, of course, if the supernova *was* seen it was not necessarily at maximum light. If the supernova had been brighter than $m_V \simeq 3$ it very likely *would* have been noticed, which implies (for $A_V = 6.2$) that at its peak the supernova was dimmer than $M_V = -15.4$. It seems safe to conclude only that the supernova was not a particularly bright one.

With an angular diameter of 4 arcminutes (for a linear diameter of $\gtrsim$4 pc) and an age of only about 330 years, Cas A (Fig. 7.1) provides our best view of a young, oxygen-rich remnant of a core-collapse supernova. In X-rays, radio, and IR, the most conspicuous feature of Cas A is a bright, patchy, approximately circular ring. Just outside the ring, a forward shock propagates into CSM at about 6000 km s^{-1}; on the inside, the ring is bounded by a reverse shock moving, in the Lagrangian sense, into the supernova ejecta at 4000 km s^{-1}. Low-density ejecta that have recently crossed the reverse shock provide most of the X-ray luminosity, while ejecta knots of higher density provide most of the optical and IR emission (Patnaude and Fesen 2014). Slower, cooler, low-luminosity matter inside the reverse shock is less conspicuous, but can be detected by means of IR emission lines (Isensee et al. 2012; Milisavljevic and Fesen 2015), IR and submillimeter continuum radiation from $\sim$1 $M_\odot$ of dust that has formed in the ejecta (De Looze et al. 2017; Bevan et al. 2017), and free–free absorption of radio synchrotron emission (DeLaney et al. 2014).

A "jet" expanding to the NE out to about 15,000 km s^{-1} and a fainter "counterjet" to the SW are seen in the optical (Fesen et al. 2006) and in X-rays (Hwang et al. 2004). The material appears to be silicon- and sulfur-rich ejecta flowing in a bipolar

Fig. 7.1 A false color image of Cassiopeia A using observations from both the *Hubble* and *Spitzer* telescopes as well as the Chandra X-ray Observatory. Credit: NASA/JPL-Caltech

fashion with opening half-angles of about 40° (Milisavljevic and Fesen 2013). Because there is no evidence for relativistic jets, a terminology such as "high-velocity silicon-rich ejecta streams" is sometimes used.

Optical emission is faint but informative. Numerous compact ($\lesssim$1 arcsecond) knots and filaments of emission are detected outside the forward shock. Some of them, called *quasistationary flocculi*, show strong Hα emission and are slow-moving ($\sim$200 km s^{-1}), comparatively dense clumps of CSM. Others, called *fast-moving knots* (FMKs), are clumps of ejecta that cooled after crossing the forward shock, but have been reheated by CSI. The projected distribution of the FMKs is approximately circular, except around the jet and counterjet where the FMKs tend to be farther out. Thousands of FMKs are detected, the majority being bright in S II $\lambda\lambda$6716,6731 emission lines and inferred to be silicon- and sulfur-rich ejecta, as in the jets. The ballistic motions of the knots allow them to move faster than the bulk of the ejecta; the typical outward velocity of a knot is 8000 km s^{-1}. About a hundred FMKs, typically at 10,000 km s^{-1}, are bright in N II $\lambda\lambda$6548,6583 lines and inferred to have a high N/H abundance ratio. Only a few FMKs, also near 10,000 km s^{-1}, have detectable Hα emission.

The "global" interpretation of Cas A (Young et al. 2006; Hwang and Laming 2012) seems reasonably clear. The remnant contains $\sim$3.5 M$_\odot$ of ejecta, including

$\sim$0.1 $M_\odot$ of iron, and the kinetic energy is about 3 B. The bulk of the iron probably was synthesized as ^{56}Ni. Allowing for the presence of a neutron star, the mass of the immediate supernova progenitor was about 5 $M_\odot$. The FMKs reveal that the progenitor contained a nitrogen-rich near-surface layer in which the CNO cycle had operated. This is not consistent with a single-star progenitor of ZAMS mass in the range 25–40 $M_\odot$ because if such a star had lost enough mass to become 5 $M_\odot$, then its nitrogen-rich layer would have been lost before explosion. A single-star progenitor in the range 12–25 $M_\odot$ would have retained too much of its hydrogen envelope to be an SN IIb. The progenitor probably was a star of ZAMS mass in the range 15–25 $M_\odot$ that lost most of its hydrogen envelope by interacting with a binary companion. The observed deceleration of the forward shock and the thermal X-ray emission from shocked gas along the outer boundary of the remnant are consistent with CSM from a steady-state wind, $\rho \propto r^{-2}$, and a high density consistent with an RSG having a wind velocity of $\sim$10 km s^{-1} and a mass-loss rate of $\gtrsim 10^{-5}$ $M_\odot$ yr^{-1} (Chevalier and Oishi 2003; Lee et al. 2014). If the wind was from the progenitor rather than from a companion, then the progenitor of this SN IIb was extended rather than compact (Sect. 15.3.1).

All of this global interpretation is consistent with SN IIb in general and SN 1993J in particular, except that the former companion star of SN 1993J is seen, while no candidate companion has been found in Cas A. One possibility is that a binary pair merged before exploding.

Several X-ray studies suggested that the iron-emitting ejecta tended to be exterior to the silicon-emitting ejecta (Hughes et al. 2000; Hwang et al. 2004; Hwang and Laming 2012), at odds with any simple "onion-skin model" of the supernova explosion. In addition, observations of hard X-rays from the decay chain $^{44}\mathrm{Ti}+e^- \rightarrow {}^{44}\mathrm{Sc}+\gamma+\nu_e$ (half-life 60 years) followed by $^{44}\mathrm{Sc}\rightarrow {}^{44}\mathrm{Ca}+e^+ +\gamma+\nu_e$ (half-life 3.9 h) tend to come from regions closer to the center of Cas A than the thermal X-rays from iron. The decay X-rays trace all of the ^{44}Ti, however, while the thermal X-rays trace only iron that has been reverse-shocked. There may be unobserved iron in the central regions of the ejecta. More than 10^{-4} $M_\odot$ of ^{44}Ti was synthesized in the explosion (Grefenstette et al. 2014; Siegert et al. 2015).

Interpretation of the richly asymmetric structures of Cas A is challenging. Radial velocities, proper motions, and variable emissivities across the remnant can be used to attempt 3D kinematic reconstructions, but there are ambiguities having to do, for example, with viewing the SNR in projection on the sky and with obscuration by dust that has formed in the ejecta. According to 3D reconstructions (DeLaney et al. 2010; Milisavljevic and Fesen 2013), the dominant morphological feature of optically bright ejecta is a torus-like structure, with an axis tilted about 30° from the plane of the sky. In and around the torus, much of the ejecta is arranged in circular ring-like structures that may be produced by the pressure of expanding bubbles of ejecta rich in ^{56}Ni (Milisavljevic and Fesen 2015). Such rings appear to be a common characteristic of young core-collapse SNRs (Vogt and Dopita 2010, 2011; Winkler et al. 2009; Milisavljevic et al. 2012). The jet and counterjet are streams of high-velocity (12,000 km s^{-1}) silicon and sulfur-rich ejecta protruding through the perimeter of the torus.

The first of a new class of unresolved X-ray sources, called central compact objects (CCOs), was found in Cas A in the first image obtained by *Chandra* (Tananbaum 1999). The CCO is 7 arcseconds southeast of the expansion center, which implies a CCO transverse velocity of 350 km s^{-1}, nearly perpendicular to the jet/counterjet structure (Fesen et al. 2006). The CCO is quite dim, with an X-ray luminosity of only a few solar luminosities. Neither a pulsar nor a PWN have been detected. A model of a cooling 1.65 $M_{\odot}$ neutron star having a low surface magnetic field ($\lesssim 10^{11}$ G) and a carbon-rich atmosphere provided a reasonable account of the X-ray spectrum and luminosity (Heinke and Ho 2010; Chang et al. 2010; Suleimanov et al. 2014). An inferred decrease of the surface temperature of 3.5% in 10 years (Elshamouty et al. 2013) is too fast for standard theories of neutron-star cooling. The temperature decrease, if it is real (Posselt et al. 2013), may signal a relatively recent transition to neutron superfluidity with associated cooling by enhanced neutrino emission (Page et al. 2011; Shternin et al. 2011; but see Bonanno et al. 2014).

Ouyed et al. (2015b) advanced the novel speculation that the inner layers of the ejecta of a standard supernova were impacted by relativistic ejecta of a subsequent explosion of the surface layers ($\sim 10^{-3}$ $M_{\odot}$, carrying a kinetic energy of 10 B) of a neutron star, as the neutron star made a transition to a more stable strange quark matter phase. For a judicious choice of delay time between core collapse and neutron-star explosion of several days, spallation reactions in the inner supernova ejecta may have led to a reduction of the ^{56}Ni abundance (and hence the current iron abundance), an increase in the abundance of ^{44}Ti, and the presence of carbon, which could have fallen back onto the CCO.

7.4.2 *SN 1054 and the Crab Nebula*

The Crab Nebula (Fig. 7.2), in Taurus, is the remnant of a core-collapse supernova of 1054 AD that was observed and recorded in China, Japan, and Korea. The supernova was visible in daytime for 3 weeks and at night for 21 months. The Crab is a filled-center remnant powered from within by a pulsar. With an angular diameter of 6 arcminutes at a distance of 2 kpc (Trimble 1973), the linear diameter is $\lesssim$4 pc and the mean expansion velocity has been about 1000 km s^{-1}. For reviews of the Crab, see Davidson and Fesen (1985) and Hester (2008).

At the heart of the nebula is the pulsar, detectable across the electromagnetic spectrum from radio to γ-rays. The pulsar is a rotating neutron star with $\gtrsim 10^{49}$ ergs remaining in rotational kinetic energy and a $\sim 10^{12}$ G magnetic field, with the dipole axis tilted with respect to the rotation axis. The pulsed emission from X-rays to radio is synchrotron radiation. The current pulse period of 33 ms is lengthening by 4×10^{-13} s/s, implying a current spindown luminosity of $\lesssim 10^{39}$ erg s^{-1}. The initial pulse period is estimated to have been about 20 ms. Only a small fraction of the spin-down luminosity is pulsed; most of it probably is carried from the neutron star as kinetic energy of an ultrarelativistic, highly-magnetized, pulsar wind. About 0.1 pc

Fig. 7.2 An *HST* composite image of the Crab Nebula. Credit: NASA, ESA and Allison Loll/Jeff Hester (Arizona State University). Acknowledgement: Davide De Martin (ESA/Hubble)

from the pulsar, interaction between the wind and the supernova ejecta produces a wind-termination shock that thermalizes the wind energy and accelerates electrons to relativistic velocities. Beyond this shock, these fast electrons power the PWN, or *synchrotron nebula*, that dominates the total radiative output of the Crab.

The PWN is bounded by another shock, of radius about 2 pc, caused by interaction between the PWN and surrounding gas. Beyond this shock is a cage-like system of elongated filaments of ejecta and CSM. The thermal line emission of the filaments over a wide range of ionization states, up to [Ne V], is primarily a result of photoionization of the filaments by the synchrotron emission. The average current outward motion of the filaments is about 1400 km s^{-1}. Relatively dense fingers of ejecta, caused by Rayleigh–Taylor instabilities, extend inwards from the filaments. The filaments and fingers also contain warm dust that formed in the ejecta (Loll et al. 2013); the mass of warm dust is uncertain, but probably $\lesssim 10^{-2}$ M$_\odot$ (Temim et al. 2012). Far-IR and submillimeter observations with *Herschel* indicated a mass of 0.02–0.2 M$_\odot$ of cold (30 K) dust (Temim and Dwek 2013; Gomez et al. 2012b;

Owen and Barlow 2015). The Crab appears to have condensed a substantial fraction of its refractory elements into dust.

Positional variations of emission-line strengths indicate extreme helium-richness in some places (MacAlpine and Satterfield 2008) and a wide range of heavier nucleosynthesis products (Charlebois et al. 2010; Satterfield et al. 2012). Features associated with the supernova ejecta, such as a high helium-to-hydrogen ratio in a band oriented east–west and ejecta extending to $2650\,\mathrm{km\,s^{-1}}$ toward the north (Black and Fesen 2015) suggest a north–south bipolar flow and an equatorial torus of ejecta. Thus the progenitor rotation axis, in projection on the sky, appears to have been oriented north–south. The proper motion of the pulsar, 13 milli-arcseconds per year nearly perpendicular to the north–south axis, corresponds to $120\,\mathrm{km\,s^{-1}}$ transverse velocity. The long axis of the asymmetric PWN runs from SW to NE and is tilted into the sky by about 30° giving the PWN its elliptical shape as projected on the sky. The pulsar spin axis and associated bipolar flow of non-thermal particles runs from the SE to the NW, along the long axis of the overall filamentary remnant structure. The kick of the neutron star is thus nearly normal to the bipolar flow in the ejecta, a property shared with Cas A, and the orientations of the PWN and the pulsar axis are about midway between.

It has been suggested that the Crab was a normal SN IIP (Hester 2008). If so, hydrogen-rich ejecta of substantially higher velocity than $1400\,\mathrm{km\,s^{-1}}$ should be present outside the filaments. Such matter has not been observed directly; it could be present, however, and it has been invoked to confine the sharp outer edge of the system of filaments. Ejecta expanding at $10^4\,\mathrm{km\,s^{-1}}$ would by now have traveled 10 pc. The Crab is expanding into a low-density cavity in the ISM of diameter about 100 pc, which could cause interactions between the forward and reverse shocks with the ISM to be weak and account for the nondetection of these shocks (Fesen et al. 1997; Lundqvist and Tziamtzis 2012).

For an estimated peak $m_V \simeq -5$ (Fesen et al. 2012), a distance of 2 kpc, and an extinction of $A_V = 1.6$ (Sollerman et al. 2001), the peak absolute magnitude would have been $M_V \simeq -18$, rather bright for a normal SN II. It has been suggested that SN 1054 was an SN IInP (Smith 2013; Sect. 14.3.1), a subenergetic subset of SN IIn that have high maximum luminosities owing to CSI (Mauerhan et al. 2013b; see also Woosley and Heger 2015). In this model, a large fraction of the kinetic energy of the ejecta was converted to radiation, and little high-velocity matter is expected. Several months after explosion, the decelerated ejecta and swept CSM are proposed to have formed a slowly expanding shell that was subsequently accelerated and fragmented by the PWN, producing the filaments of the Crab.

Estimates of the total mass of the observed nebula include $\lesssim 54\ \mathrm{M_\odot}$ (Fesen et al. 1997) and $\gtrsim 7\ \mathrm{M_\odot}$ (Owen and Barlow 2015). The associated kinetic energy is $\lesssim 0.1$ B. Parameters like these, together with estimates of element abundances in the ejecta, led to suggestions that the Crab supernova was an electron-capture explosion of a star of ZAMS mass from 8 to 10 $\mathrm{M_\odot}$ that lost most of its hydrogen envelope by presupernova winds (Nomoto et al. 1982; Wanajo et al. 2009; Yang and Chevalier 2015; Jerkstrand et al. 2015c). The predicted light curve of such an explosion includes a bright, short plateau that is consistent with early-time observations of

SN 1054 (Tominaga et al. 2013). Such an explosion is expected to eject little ^{56}Ni, $\gtrsim 10^{-3}$ $M_{\odot}$. If the late-time light emission was powered by ^{56}Ni decay, however, then about 0.06 $M_{\odot}$ of ^{56}Ni would have been required to allow the supernova to be visible to the naked eye for 21 months (Sollerman et al. 2001). Alternatively, the late-time emission may have been powered by CSI (Moriya et al. 2014). A pulsar contribution to the late-time emission also is possible.

7.4.3 SN 1181 and SNR 3C58

A bright star of 1181 that was recorded in China and Japan reached a peak brightness probably brighter than $m_V = 0$ and was visible at night for 6 months. It very likely was a supernova (Stephenson and Green 2002).

The filled-center SNR 3C58, in Cassiopeia, is much less luminous than the Crab, but otherwise there are similarities: 3C58 contains a 66 ms pulsar and a PWN, it exploded in a low-density ISM bubble, no high-velocity shell of ejecta has been detected, and it contains numerous knots and filaments. The distribution of the radial velocities of the knots is bimodal (Fesen et al. 2008). Members of the faster group, with mean velocity 770 km s^{-1}, are knots that apparently were ejected in a bipolar fashion. Members of the slower group, with velocity $<$ 250 km s^{-1}, are shocked circumstellar knots, many of which are bright in N II $\lambda\lambda$6548,6583 lines and appear to have enhanced N/H abundance ratios.

A strong argument for associating SN 1181 with 3C58 is that the positional agreement is excellent, and no other plausible SNR candidate has been found (Stephenson and Green 2002). Nevertheless, doubts have been expressed about the association, based on properties of the PWN (Chevalier 2005) as well as on the size of the remnant combined with observed motions of the knots (Fesen et al. 2008). At the traditionally adopted distance of 3.2 kpc, the 6 by 10 arcminute angular size of the SNR corresponds to a linear size of 6 by 10 pc. To achieve this size since 1181, the time-averaged expansion velocity would be 3500–5700 km s^{-1}, but measured radial velocities of numerous knots and filaments seldom exceed 1100 km s^{-1}. A redetermination of the distance yielded only 2 kpc (Kothes 2010), which would reduce the linear size and ease some of the problems. Still, observed proper motions of knots of only 0.02–0.07 arcseconds per year are much less than the time-averaged 0.2 arcseconds per year that is required if 3C58 is the remnant of SN 1181 (Fesen et al. 2008). Substantial deceleration of the knots would be indicated. The issue of association may be resolved by future X-ray measurements of the proper motion of the pulsar, which is offset from the center of the remnant by nearly 1 arcminute (Gotthelf et al. 2007).

With $m_V < 0$, a distance of 2 kpc, and $A_V = 2.1$ (Fesen et al. 2012), the peak absolute magnitude of SN 1181 would have been $M_V < -13.6$, thus possibly subluminous.

7.4.4 SN 1572 and the Tycho SNR

The Tycho SNR (G120+1.4; 3C10), the remnant of SN 1572, is unique among Galactic SNRs in that both its age and its supernova type are firmly known. A light-echo spectrum established that SN 1572 was a spectroscopically normal SN Ia (Krause et al. 2008b). The supernova, in Cassiopeia, was observed in China and Korea, and especially in Europe. Tycho Brahe made careful estimates of the brightness for more than a year. The peak apparent magnitude was $m_V \simeq -4.5$. Even before the light-echo spectrum was obtained, the light curve and colors already had been inferred to be those of a normal SN Ia (Ruiz-Lapuente 2004). For an extinction of $A_V \simeq 2$ magnitudes (Fesen et al. 2012) and a typical SN Ia $M_V \simeq -19.3$, the distance would be about 3.6 kpc, consistent with most estimates of the distance to the remnant (Hayato et al. 2010).

As expected for the remnant of an SN Ia, there is no evidence of a neutron star in the SNR. In radio and X-rays, the SNR has a limb-brightened, nearly circular shell structure (Fig. 7.3), about 8 arcminutes in angular diameter, about 9 pc in linear diameter. Radio and hard X-ray images show coincident sharp outer edges that delineate the nonradiative forward shock, moving at about 4600 km s^{-1}. Images of thermal X-rays from shocked ejecta show a fleecy or broccoli-like complexity. Detailed 3D hydrodynamical modeling showed that Rayleigh–Taylor instabilities at the interface between the ejecta and a uniform ambient medium are capable of generating such structure without invoking initial clumping of the ejecta or the ambient medium (Warren and Blondin 2013).

The only optical emission is from faint filaments, the spectra of which show Balmer lines that have both narrow and much broader (1800 km s^{-1}) components. The former are produced by cold neutral hydrogen atoms that are collisionally excited when they encounter the forward shock at the low preshock temperature. Such Balmer-dominated spectra are consistent with a neutral ISM and a Type Ia explosion. Breakout radiation from a core-collapse supernova would ionize the surrounding ISM. The broader component of the Balmer lines is produced by proton-neutral charge exchange that creates fast neutrals at the postshock temperature (Chevalier and Raymond 1978; Heng 2010).

Thermal X-rays from ejected matter include lines of elements ranging from carbon to iron. Doppler broadening of such lines indicates that silicon, sulfur and argon ejecta have expansion velocities about 4700 km s^{-1} while iron ejecta have about 4000 km s^{-1} (Hayato et al. 2010). Calculations of the X-ray emission based on a grid of SN Ia hydrodynamical models have shown that good agreement with observations is obtained for 1D Chandrasekhar-mass delayed-detonation models that synthesize 0.7 $M_\odot$ of ^{56}Ni, have a kinetic energy $\gtrsim$1 B, and explode into a uniform ambient medium having a density $\sim$1 cm^{-3} (Badenes et al. 2006). This analysis showed that in the SNR, iron must be generally inside the silicon, as it is in the explosion model. Observations of thermal X-rays indicate that titanium and chromium are spatially co-located with iron, as expected (Miceli et al. 2015),

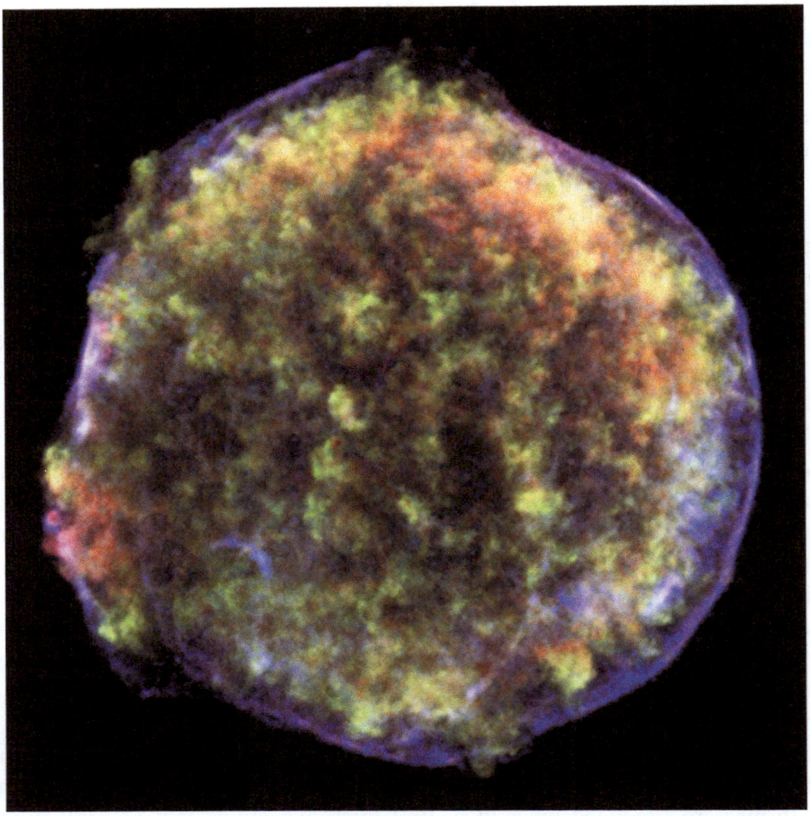

Fig. 7.3 A *Chandra* X-ray image of the remnant of Tycho's supernova of 1572. Credit: NASA/CXC/SAO

and observations of hard X-rays from the ^{44}Ti decay chain imply, for a distance of 3.6 kpc, that $\gtrsim 10^{-5}$ $M_{\odot}$ of ^{44}Ti was synthesized in the explosion (Troja et al. 2014).

X-ray proper-motion measurements of the forward shock and reverse-shocked ejecta suggest a lower density of the ambient medium, $\lesssim 0.2\,\mathrm{cm}^{-3}$ (Katsuda et al. 2010). Models in which the ejecta initially interact with an r^{-2} stellar-wind structure before interacting with a homogeneous ISM may resolve the discrepancy, while implying that the supernova progenitor was in a single-degenerate binary system (Chiotellis et al. 2013).

Herschel observations indicated the presence of $\sim 10^{-2}$ $M_{\odot}$ of cool (90 K) dust in swept-up ISM, but provided no evidence that dust formed in the ejecta (Gomez et al. 2012a). This is consistent with other evidence that SN Ia are not important dust producers.

A G-type subdwarf was suggested to be a former donor star in a SD system, but the identification is not conclusive (Bedin et al. 2014). Additional constraints on the presence of former donors in SN Ia SD systems are discussed in Sect. 21.6.

7.4.5 SN 1604 and the Kepler SNR

The supernova of 1604, in Ophiuchus, was visible at night for a year, and its evolving brightness was well observed by Johannes Kepler and others. The peak apparent visual magnitude was about −3 and the light curve was consistent with that of an SN Ia. The SNR does appear to be the remnant of an SN Ia, based on the low O/Fe ratio in the ejecta (Reynolds et al. 2007) and also on the centroid energy of the iron Kα emission from the ejecta. The centroid energy depends on the ionization level of iron; in SNRs of SN Ia, iron is less ionized than it is in SNRs of core-collapse events, owing to weaker CSI (Yamaguchi et al. 2014).

Much like Tycho's SNR, in the radio and X-rays Kepler shows a limb-brightened shell, in this case irregular with higher brightness in the north (Fig. 7.4). Optical emission from Kepler is largely confined to Balmer-dominated spectra with narrow and broader components. The Balmer emission indicates the presence of neutral ISM, which supports the Type Ia identification (Sect. 7.4.4).

The SNR is almost 7° out of the Galactic plane, which at an (uncertain) distance of 5 kpc would correspond to almost 600 pc. At this distance from the Galactic plane

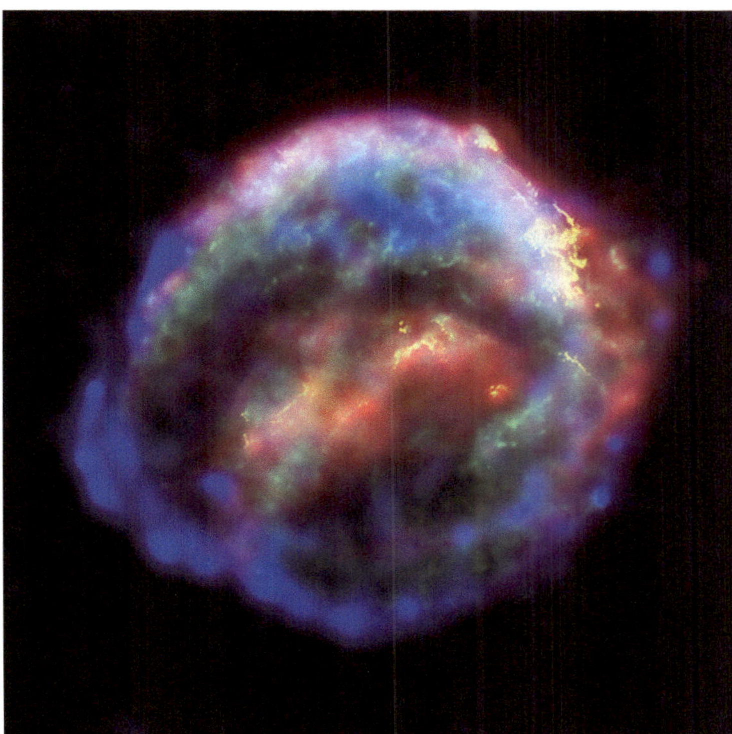

Fig. 7.4 A composite X-ray, optical, and IR image of the remnant of Kepler's supernova of 1604. Credit: NASA/ESA/JHU/R. Sankrit & W. Blair

the particle density of the ISM is typically $\lesssim 10^{-2}\,\text{cm}^{-3}$, yet the supernova ejecta is interacting, especially in the north, with a dense (1–100 cm^{-3}) nitrogen-rich ambient medium. This matter may be the stripped envelope of an oxygen-rich AGB star (Sect. 8.3.2)—a former donor in a single-degenerate binary system (Chiotellis et al. 2012). Neither a stripped donor star nor a neutron star has been identified.

The presence of dust in the SNR, inferred to have been present in the donor's circumstellar wind, supports the AGB-star hypothesis because such dust would not be expected to be present around a main-sequence donor. There is no evidence that dust formed in the ejecta (Williams et al. 2012).

Provided that the SNR is at a distance of at least 7 kpc, the observed thermal X-ray spectrum could be accounted for by an explosion model that ejected about 1 $M_\odot$ of ^{56}Ni (Patnaude et al. 2012). For an angular diameter of 4 arcminutes and a distance >7 kpc, the linear diameter is $\gtrsim 9$ pc, implying a mean expansion velocity since 1604 of $\gtrsim 10{,}000\,\text{km s}^{-1}$. For $m_V \simeq -3$ and $A_V \simeq 2.8$ (Reynolds et al. 2007), this distance would imply $M_V < -20$, unusually bright for an SN Ia, but perhaps consistent with a suggestion based on X-ray spectra that the supernova was an overluminous SN 1991T-like event (Katsuda et al. 2015).

7.4.6 SN 1006 and SNR G327.6+14.6

The supernova of 1006, in Lupus, was recorded in China and Japan and remained visible at night for more than 3 years. At its peak, the event seems to have been extremely bright, perhaps $m_V \simeq -8.5$ (Stephenson 2010). For a distance of 2.2 kpc (Winkler et al. 2003) and $A_V \simeq 0.3$, this would correspond to $M_V \simeq -20.5$, possibly overluminous although consistent with a typical SN Ia brightness within the uncertainty in m_V.

The SNR associated with SN 1006, G327.6+14.6 (Fig. 7.5), is known to be a remnant of an SN Ia, on the basis of the Kα energy centroid (Yamaguchi et al. 2014) and because it is more than 500 pc out of the Galactic plane and expanding into a low-density ($\sim 0.1\,\text{cm}^{-3}$) ISM (Katsuda et al. 2009). Like the Tycho remnant, the SNR shows a roughly circular limb-brightened shell in radio and X-rays, faint Balmer-dominated filaments in the optical, and its X-ray images have a fleecy appearance. With an angular diameter of 30 arcminutes, the SNR has a diameter of about 20 pc and a mean expansion velocity since 1006 of about $10{,}000\,\text{km s}^{-1}$. The current expansion velocity varies strongly with azimuth (Winkler et al. 2014), indicating asymmetry in the ejecta and/or in the ambient medium.

An OB subdwarf star within 3 arcminutes of the projected center of the SNR (Schweizer and Middleditch 1980) is behind the remnant and therefore useful for absorption spectroscopy of unshocked ejecta, material that is too cold to be readily detected in emission. UV spectra of this star and several background QSOs show broad Fe II, Si II, Si III, and Si IV absorptions. The Fe II absorptions are nearly symmetric about their rest wavelengths, but the silicon absorptions are redshifted,

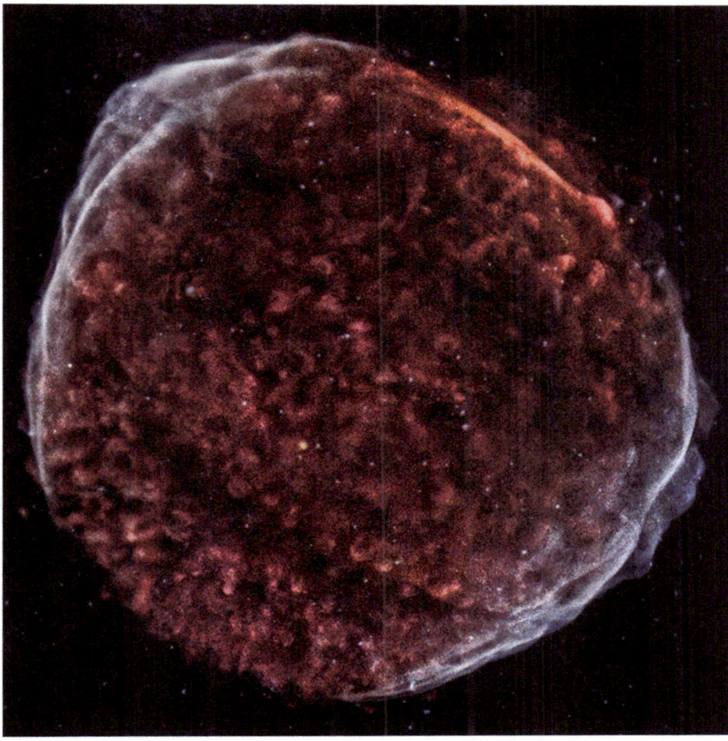

Fig. 7.5 A *Chandra* X-ray image of the remnant of SN 1006. Credit NASA/CXC/SAO

indicating a back-to-front asymmetry perhaps owing to a low density of ISM on the back side (Hamilton et al. 1997; Winkler et al. 2005). A sharp red edge of the Si II $\lambda 1260$ absorption, redshifted by $7000\,\mathrm{km\,s^{-1}}$, reveals the velocity of unshocked ejecta now encountering the reverse shock (Hamilton et al. 2007). An observed $44\,\mathrm{km\,s^{-1}}$ decrease in this velocity in a decade confirms the standard picture of a reverse shock working its way into the ejecta (Winkler et al. 2011). Abundances inferred from the thermal X-ray spectrum of the entire remnant are consistent with those expected of an SN Ia (Uchida et al. 2013).

In spite of deep searches, no former companion star to the supernova progenitor has been found (González Hernández et al. 2012; Kerzendorf et al. 2012).

7.4.7 SN 185 and SNR RCW 86

The supernova of 185, in Centaurus, was recorded in China. Estimates of the period of visibility at night range from 8 months to 2 years (Stephenson and Green 2002).

The shell-type SNR RCW 86 (G315.4–2.3) appears to be the remnant of SN 185 (Williams et al. 2011). If so, for a distance of about 2.5 kpc (Helder et al. 2013b), the angular radius of 21 arcminutes corresponds to a linear diameter of 35 pc and requires an average shock speed of $7800\,\mathrm{km\,s^{-1}}$, much higher than the presently measured speeds across most of the remnant. The SNR appears to have expanded rapidly into a wind-blown bubble, only recently encountering higher-density ISM to rapidly reduce the expansion speed (Vink et al. 2006). From analysis of thermal X-ray emission, the mass of iron and the total ejected mass were estimated to be about 1 $M_\odot$ and 1–2 $M_\odot$, respectively (Yamaguchi et al. 2011). These masses, the lack of evidence for a neutron star, the iron Kα energy centroid, and the Balmer-dominated shocks surrounding the remnant indicate that the supernova was of Type Ia, in which case the wind-blown bubble may have been produced by a single-degenerate progenitor binary system (Williams et al. 2011; Broersen et al. 2014). No former companion has been identified.

The supernova was observed only a few degrees above the horizon (Schaefer 1993). Nevertheless, within the large uncertainties of atmospheric and interstellar extinction, an SN Ia at its peak could have been easily visible (Williams et al. 2011).

7.4.8 G1.9+0.3: The Youngest Known Galactic SNR

If the Galaxy produces 3 supernovae per century then about 10 supernovae should have occurred since Cas A. G1.9+0.3, within 2° of the direction of the Galactic center, is the only known remnant of such a recent event.[1] In radio and X-rays, the SNR shows a somewhat asymmetric shell of irregular brightness (Borkowski et al. 2014) with an angular diameter of 1.7 arcminutes. The observed angular expansion rate of $\gtrsim$0.6% per year, together with an allowance for some past deceleration, gives an age of only $\gtrsim$100 years (Reynolds et al. 2008; Carlton et al. 2011). In view of heavy interstellar absorption of the X-ray spectrum, the SNR is assumed to be near the distance of the Galactic center, 8.5 kpc. If so, the current velocity of the forward shock is $14{,}000\,\mathrm{km\,s^{-1}}$, the linear diameter is 5 pc, and the average expansion velocity since 1900 has been almost $20{,}000\,\mathrm{km\,s^{-1}}$. The X-ray absorption implies many magnitudes of extinction in the optical, so the supernova could not have been seen from Earth and optical emission from the SNR will not be detected.

The supernova appears to have been an SN Ia (Reynolds et al. 2009; Borkowski et al. 2010; Yamaguchi et al. 2014). G1.9+0.3 is the only known Galactic SNR that is brightening in both radio and X-rays.

The radio emission and most of the X-rays are of synchrotron origin. Thermal X-ray emission lines are from reverse-shock heated ejecta that were initially

[1]The SNR G354.4+0.0 may have an age of 100–500 years (Roy and Pal 2013).

expanding at $\gtrsim$18,000 km s^{-1}. In some places, this high-velocity ejecta is enhanced in silicon and sulfur; in other places, it is enhanced in iron, indicating an asymmetric abundance distribution (Borkowski et al. 2013). On the basis of a probable detection of the 4.1 keV line from the ^{44}Ti decay chain, it has been inferred that $\sim 10^{-5}$ M$_\odot$ of ^{44}Ti was synthesized in the explosion (Borkowski et al. 2010).

7.4.9 *SN 1885A and Its SNR in the Andromeda Galaxy*

SN 1885A (also known by a variable-star designation, S Andromedae), which occurred within the central bulge of the Andromeda galaxy (M31) only 16 arc-seconds from the galaxy's center, reached a peak $m_V \lesssim 6$. The supernova type is uncertain. With an estimated $A_V \simeq 0.23$, the peak absolute magnitude was $M_V \simeq -18.7$. This brightness, together with the probable absence of hydrogen and helium lines (de Vaucouleurs and Corwin 1985), the rapidly decaying light curve, somewhat red colors, and the non-star-forming location in M31, led to suggestions (van den Bergh 1994) that SN 1885A may have been a relatively weak SN 1991bg-like SN Ia (Sect. 20.5). Alternatively, it may have been a member of a rare, peculiar class of SN 2002bj-like events (Perets et al. 2011) that appear to be explosions on old white dwarfs that eject not more than a few tenths of a solar mass (Sect. 23.4).

The SNR has not been definitively detected in emission, but in near-UV images it is seen in absorption against the light of the bulge of M31 as a circular dark spot of angular diameter 0.8 arcseconds (Fesen et al. 1989). The linear diameter of the spot is 3 pc and the mean expansion velocity since 1885 has been 12,000 km s^{-1}. Spectra obtained with the *HST* (Fesen et al. 1999b) revealed that the absorption was primarily by the Ca II H&K lines, with weaker absorptions of Ca I, Fe I, and Fe II. These features form in unshocked ejecta that is partially photoionized by UV emission from bulge stars. The total mass of iron was estimated to be 0.21 M$_\odot$, but with large uncertainties. Narrow passband *HST* images (Fesen et al. 2007, 2015) showed that Ca II absorption occurred mainly in a broad, roughly symmetric shell centered on 6000 km s^{-1} but extending out to 12,500 km s^{-1} while Fe II absorption occurred mainly in four plumes extending from the center out to 10,000 km s^{-1}.

7.5 Summary

The hundreds of known SNRs in the Galaxy and in nearby galaxies provide some useful statistical information such as their distributions within galaxies, but since most of them have swept up large amounts of ISM, the information provided on the nature of the supernovae that produced them is limited. Young SNRs are more informative. Young Galactic SNRs, which are well resolved in angle, generally are roughly circular in outline, but when viewed at high resolution they show complex 3D structures.

Cas A, known to be the remnant of an SN IIb, probably was an explosion of a star of ZAMS mass in the range 15–25 $M_{\odot}$ that lost most of its hydrogen envelope in a binary interaction, although no former companion to the supernova progenitor has been found. Cas A contains a rapidly cooling CCO. The nature of the core-collapse event that produced the Crab is debated. The radio source 3C58 contains a 66 ms pulsar and resembles the Crab in some other respects. Whether 3C58 is the remnant of SN 1181 remains uncertain, and all that is known about the type of SN 1181 is that it was some sort of core-collapse event.

The Tycho remnant is known to be that of a spectroscopically normal SN Ia, and the properties of the remnant are generally consistent with hydrodynamical models for SN Ia. Outstanding questions include whether an interaction of the ejecta with CSM, and/or the presence of a surviving donor star can be established; either or both would suggest a single-degenerate progenitor system. The Kepler remnant is that of an SN Ia, possibly an overluminous one. Interaction between ejecta and CSM provides possible evidence of a single-degenerate progenitor system. SN 1006 also was an SN Ia. No former companion to the supernova progenitor has been found in the remnant, casting doubt on a single-degenerate scenario in this case. SN 185 probably was an SN Ia, perhaps from a single-degenerate binary system.

At an age of $\gtrsim$100 years, G1.9+0.3 is the only Galactic SNR known to be younger than Cas A. Some evidence suggests that it is the remnant of an SN Ia that is interacting with uniform-density ISM. The remnant of SN 1885A in the Andromeda galaxy is detected in absorption against the light of the galaxy's bulge stars. The supernova may have been a subluminous SN Ia or a member of a separate class of explosions involving white dwarfs.

Chapter 8
Evolution to Catastrophe

8.1 Introduction

The principles and many details of stellar evolution are discussed in classic books, including those by Schwarzschild (1958), Cox and Giuli (1968), Clayton (1984), Arnett (1996), Hansen et al. (2004), Maeder (2009), Bisnovatyi-Kogan (2011), Kippenhahn et al. (2012) and Iben (2013). Details of the very late stages of evolution are not covered so thoroughly in text books, but are extensively discussed in literature that is too vast to be summarized here. Our goal in this Chapter is rather to give a synopsis of the relevant physics and stages of evolution that precede supernova explosions and a brief summary of observations that constrain the theory. Many of these topics are discussed in some depth in the reviews of Woosley et al. (2002) and Langer (2012) and in the series of articles by Paxton et al. (2011, 2013, 2015).

The most fundamental property of a single star is its ZAMS mass, although other factors such as metallicity and rotation also are important. Stellar ZAMS masses range from 0.08 to perhaps $\sim$300 $M_\odot$, although ever higher-mass stars are contemplated in the context of the rapid seeding and growth of supermassive black holes in the early Universe. Stars much less massive than the Sun ($M_\odot \simeq 2.0 \times 10^{33}$ g) do not complete the main-sequence core-hydrogen-fusion phase in a Hubble time. There are many of them, however, and they are important in a variety of contexts.

The evolution of the Sun is expected to be typical of stars of modest mass. The main-sequence lifetime of the Sun is $\gtrsim 10^{10}$ years, after which its helium core will contract, hydrogen will fuse in a surrounding shell, and the envelope will expand to become a red giant (Sect. 8.3) attaining $R \simeq 200$ $R_\odot$ ($R_\odot \simeq 7.0 \times 10^{10}$ cm) and $L \gtrsim 1000$ $L_\odot$ ($L_\odot \simeq 3.8 \times 10^{33}$ erg s^{-1}). The helium core will contract, heat, and fuse to a mixture of carbon and oxygen (C/O). The electron-degenerate C/O core, surrounded by helium- and hydrogen-burning shells, will never fuse because

D. Branch, J.C. Wheeler, *Supernova Explosions*, Astronomy and Astrophysics Library, DOI 10.1007/978-3-662-55054-0_8

enhanced mass loss during the red-giant phase (Sect. 8.4) will remove the bulk of the envelope and halt the evolution of the core. After $\sim 10^9$ years as a red giant, the Sun will eject its envelope as a planetary nebula and become a C/O white dwarf of $\gtrsim 0.5$ M$_\odot$. Stars of ZAMS mass up to at least 6 M$_\odot$ undergo similar evolution, although they do so more rapidly, become larger and brighter, and end up as more massive C/O white dwarfs (Sect. 8.2). Evolution in the ZAMS mass range of $\gtrsim 6$ to $\lesssim 13$ M$_\odot$, which is more uncertain, is discussed in some detail in Sects. 8.5.3 and 8.5.4. In the context of presupernova evolution, the prime characteristic of stars with ZAMS mass $\lesssim 13$ M$_\odot$ is that they form a degenerate core of some sort during the late stages of their evolution. Degenerate cores composed of C/O may explode. Those composed of oxygen, neon, and magnesium (ONeMg) may collapse.

After leaving the main sequence, stars in the ZAMS-mass range $\gtrsim 13$ to ~ 70 M$_\odot$ undergo burning of successively heavier fuels, first in the core and then in surrounding shells. A principle characteristic of the evolution in this mass range is that electron degeneracy remains small or negligible in the inner cores. In this case, the stars, and their inner cores, display a negative specific heat in accord with the virial theorem (Cox and Giuli 1968). If there is no nuclear input, the core loses thermal energy to the outer, cooler, portions of the star and contracts and heats until the threshold temperature to ignite a nuclear fuel is reached. This nondegenerate evolution guarantees that when a fuel is exhausted in the core, the core contracts and heats until the ash of one stage of nuclear burning becomes the fuel of the next stage. This behavior gives rise to the famous "onion-skin" structure of massive stars, shells of ever heavier elements. This behavior continues until the stars in this mass range form an iron core (Sect. 8.5.5). The nuclear properties of iron are endothermic, effectively giving the iron core an adiabatic index less than 4/3, the value required for stability. The iron core is thus doomed to collapse (Chap. 9). Stars with ZAMS masses above ~ 70 M$_\odot$ may undergo thermonuclear explosion and disruption following pair instability (Sect. 8.6).

Single stars with ZAMS mass in the range $\gtrsim 13$ to $\lesssim 30$ M$_\odot$ are expected to be RSGs (Sect. 8.3) when their cores collapse (Chap. 12; but see Sect. 12.5.4). Stars that are somewhat more massive may experience an RSG phase but lose mass to become more compact *Wolf–Rayet* stars (Sect. 8.4.1) before their cores collapse (see Chaps. 15–17 for discussion of collapse in stars stripped of their extended envelopes). Stars still more massive, exceeding the *Humphreys–Davidson limit* of ~ 50 M$_\odot$ (Humphreys and Davidson 1979), may go directly from the main-sequence to the Wolf–Rayet phase.

The more massive a ZAMS star is, the brighter, with $M \propto L^3$ for solar-mass stars with temperature-dependent opacities and $M \propto L$ for very massive, radiation-dominated stars for which the opacity is primarily electron scattering. In theoretical contexts, it is convenient to refer to the luminosity, L, but to relate to observations it is often convenient to refer to absolute magnitudes. The bolometric magnitude can be expressed as $M_{bol} = 4.72 - 2.5\log(L/L_\odot)$, where 4.72 is the absolute bolometric magnitude of the Sun. Observations of magnitudes observed in restricted bands require other bolometric corrections. Because massive stars burn hotter and brighter, they live a shorter time. For massive stars, the lifetime is roughly constant, independent of mass. The main-sequence lifetimes of 8, 15, 20, 50, and 100 $M_\odot$ stars are 30, 11, 8, 4, and 3×10^6 years, respectively (Paxton et al. 2011, 2013).

Important but uncertain aspects of stellar evolution are processes associated with the transport of energy and angular momentum and with composition mixing. On the main sequence, low-mass stars such as the Sun have radiative cores and convective envelopes, while the opposite is true of massive stars. The extended envelopes of post-main-sequence phases generally are convective. Owing to rapid mixing by mass motions, a convection zone maintains a homogenous composition. Complications include *convective overshoot* whereby plumes of material have sufficient momentum to proceed beyond the formal boundary between convective instability and stability and *semi-convection*, a process in which there is just enough mixing of composition to yield a composition gradient consistent with neutral convective instability (Kippenhahn and Weigert 1994). For all these processes, there will be differences depending on whether composition gradients are, for *Ledoux* convection, or are not, for *Schwarzschild* convection, included in the criterion for convective instability. For many applications, *mixing-length theory* may be adequate to capture the flux carried by mechanical motion, but convection is likely to be more complex, involving chaotic, turbulent flows that are characterized by intermittency (Arnett et al. 2009; Arnett and Meakin 2011b).

The classic means of characterizing stellar evolution is by means of the *Hertzsprung–Russell diagram* (H–R diagram or HRD) that plots brightness against temperature. In observational terms, the brightness is often the magnitude and the temperature is represented by a color. For theorists, the brightness is represented by the bolometric luminosity and the temperature by the effective temperature of the photosphere. In the upper panel, Fig. 8.1 gives a theoretical representation of the evolution of stellar models with ZAMS mass between 2 and 10 $M_\odot$ as they evolve as protostars onto the main sequence, rapidly cross the *Hertzsprung gap* after core hydrogen exhaustion, and then climb the giant branch. Stars around 10 $M_\odot$, display an excursion away from the giant branch to hotter temperatures, a *blue loop*, that is sensitive to issues like overshoot of convective regions. The lower panel of Fig. 8.1 presents the complementary information on the locus of the central temperature and density over the same course of evolution.

Figure 8.2 summarizes the evolution to various explosive endpoints of single, nonrotating, massive stars in the log T_c–log ρ_c plane, where T_c and ρ_c are central temperature and density. This parameterization, the stellar-interior theorist's version of the Hertzsprung–Russell diagram, allows one to track and categorize the interior evolution of stellar models rather than their directly observable properties. The next several sections outline the basic physics and evolution encapsulated in this figure. Figure 8.2 does not present the evolution of low-mass stars in any detail.

This chapter is laid out as follows. We first discuss the formation of white dwarfs in order to establish nomenclature that we employ throughout the chapter. We then discuss the processes that dictate the structure and evolution of red giants that control much of presupernova evolution. We next address various mechanisms of mass loss that are again crucial factors that determine the nature of presupernova

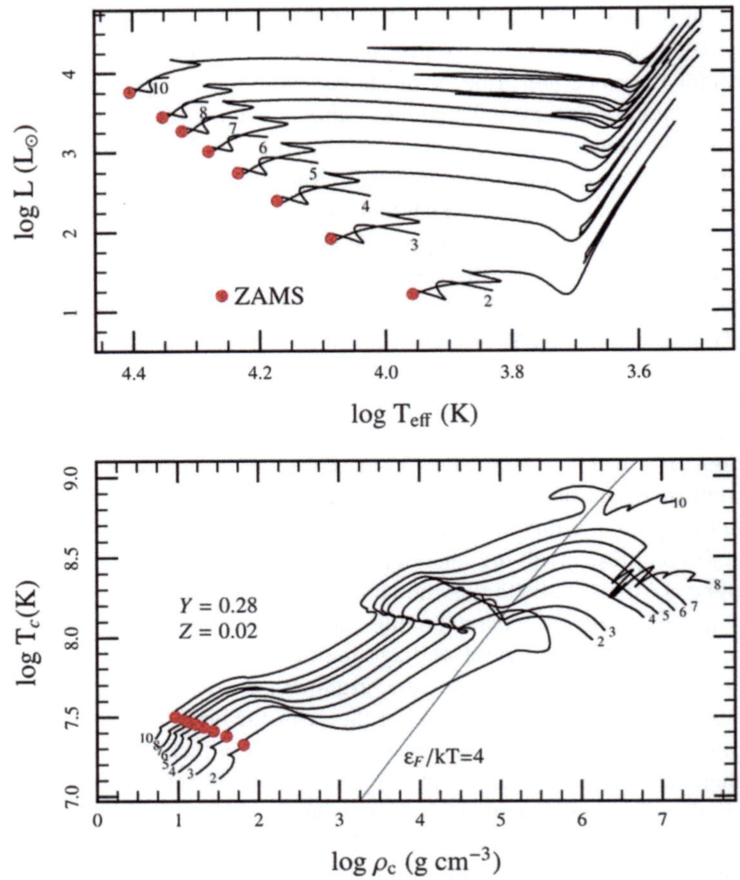

Fig. 8.1 *Upper panel*: evolutionary tracks in the Hertzsprung–Russell diagram of bolometric luminosity versus effective temperature for stellar models of solar metallicity and ZAMS mass from 2 to 10 $M_{\odot}$; the location of the ZAMS is given by the *red dots*. *Lower panel*: the evolution of the same models represented in the plane of central temperature and density; the *diagonal line*, $\epsilon_f/kT = 4$, represents the separation of nondegenerate matter on the left from degenerate matter on the right, where ϵ_f is the Fermi energy and kT is a measure of the thermal energy of a particle. From "Modules for Experiments in Stellar Astrophysics (MESA)" (Paxton et al. 2011). © AAS. Reproduced with permission

evolution. Having set the stage, we then discuss the inner workings of nonrotating, single-star models that directly lead to supernovae. Finally, we briefly outline how all these processes play out when rotation is incorporated and when stars are in binary systems.

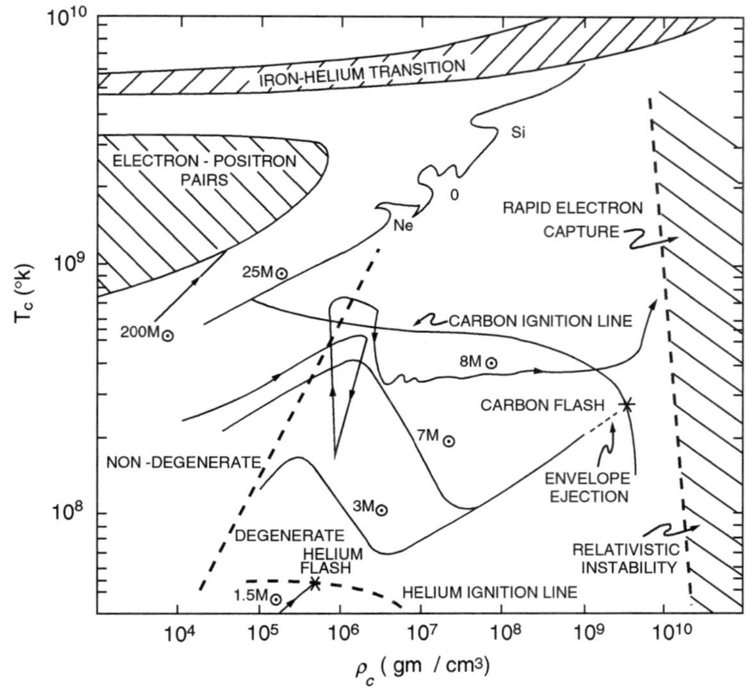

Fig. 8.2 Evolutionary tracks approaching a final catastrophe are depicted in terms of the locus of the central temperature and density. Stars of solar metallicity with ZAMS mass less than about 8 $M_\odot$ converge to a common track; as indicated, most (or all) lose their envelopes to leave stable C/O white dwarfs without igniting degenerate carbon. Most stars of mass about 8–13 $M_\odot$ probably develop degenerate ONeMg cores that collapse because of the combined effects of electron capture on the products of carbon burning and the destabilizing effects of general relativity, but some may eject their envelopes to produce white dwarfs (Sect. 8.5.3). The track shown represents a mass of 8 $M_\odot$ for which carbon ignites off-center, causing the wide excursion. At the upper end of this mass range, oxygen may ignite off-center and burn to completion before ignition of central oxygen or core collapse (Sect. 8.5.4). Stars of mass about 13 to ∼70 $M_\odot$ develop iron cores that collapse due to electron capture and photodisintegration into helium (Sect. 8.5.5). Ignition of neon, oxygen, and silicon in the core followed by respective shell-burning stages causes the excursions that are marked along the track corresponding to 25 $M_\odot$. Very massive stars encounter the region of electron–positron pair formation and instability (Sect. 8.6). From "Introduction to Supernovae" (Wheeler 1990)

8.2 Making White Dwarfs

Some supernovae are thought to come from exploding white dwarfs (Chap. 21). It is thus important to understand what stars make what sort of white dwarfs under various circumstances. The ZAMS mass below which a white dwarf is formed is sometimes labeled M_{up}, but in other contexts this label has been applied to the maximum ZAMS mass that forms a degenerate C/O core. In some of the literature there is an implicit or explicit assumption that these limits are the same, but this is not necessarily the case. Because we will try to carefully distinguish between these two limits, we use $M_{wd,max}$ to refer to the ZAMS mass below which a white dwarf forms and $M_{CO,max}$ to refer to the mass below which a star develops a degenerate C/O core. There is an analogous ZAMS mass, $M_{ONeMg,max}$, below which a star forms a degenerate core of ONeMg (Sect. 8.5.3). Above $M_{ONeMg,max}$, a star is expected to form an iron core (Sect. 8.5.5). As discussed below, the range of observational uncertainty in $M_{wd,max}$ overlaps the ranges of theoretical uncertainty in $M_{CO,max}$ and $M_{ONeMg,max}$. It is thus important to constrain these numbers independently.

There are theoretical estimates of $M_{wd,max}$, and this number can be constrained from observations; however, at this writing the quantities $M_{CO,max}$ and $M_{ONeMg,max}$ are purely parameter-dependent theoretical constructs. Estimates of $M_{CO,max}$ and $M_{ONeMg,max}$ vary by about 2 $M_{\odot}$, depending on model assumptions, as discussed below (Sects. 8.5.2 and 8.5.3). Models generally give $M_{ONeMg,max}$ greater than $M_{CO,max}$ by 2–3 $M_{\odot}$.

The critical aspect for supernovae is that if $M_{wd,max}$ is greater than $M_{CO,max}$, then no single stars ignite carbon in degenerate matter, but if $M_{wd,max}$ is less than $M_{CO,max}$ then some stars should undergo thermonuclear explosion within extended hydrogen envelopes. Whatever the processes are by which stellar envelopes are ejected to form planetary nebulae, these processes also may apply to stars with cores of ONeMg. If $M_{wd,max}$ is high enough, then even some single stars that evolve beyond carbon burning may avoid core collapse by becoming ONeMg white dwarfs (Sect. 8.5.3), and it becomes especially interesting to understand the value of $M_{wd,max}$ and the physics that determines it. If some stars that develop ONeMg cores cannot eject their envelopes, then these stars should explode by core collapse within extended envelopes.

The challenge is to determine what ZAMS masses give rise to white dwarfs and, among the white dwarfs, which are composed of C/O and which of ONeMg. The value of $M_{wd,max}$ can be constrained by studies of white dwarfs in open clusters. If white dwarfs are present in a cluster, then $M_{wd,max}$ is assumed to exceed the cluster main-sequence turn-off mass. In this way $M_{wd,max}$ has been estimated to be 6.6–8.8 $M_{\odot}$ (Dobbie et al. 2006), $\gtrsim 6$ $M_{\odot}$ (Rubin et al. 2008), and > 7.1 $M_{\odot}$ (Williams et al. 2009). These numbers are roughly consistent with theoretical predictions for $M_{CO,max}$, but they do not preclude $M_{CO,max} > M_{wd,max}$ so that some single stars with C/O cores explode within extended hydrogen envelopes, nor do they exclude $M_{ONeMg,max} > M_{wd,max}$ so that some single stars with ZAMS masses exceeding $M_{CO,max}$ form ONeMG white dwarfs and others undergo electron-capture core collapse within extended envelopes.

The relation between ZAMS masses and the masses of the white dwarfs that they form is called the *initial-final mass relation* (IFMR; Weidemann and Koester 1983). The IFMR is best constrained by studies of white dwarfs in open clusters of various ages. In a given cluster, an initial mass is inferred from the main-sequence turn-off and a final mass is inferred from measurements of the masses of white dwarfs in the cluster. An example is given in Fig. 8.3, which shows that stars in the ZAMS mass

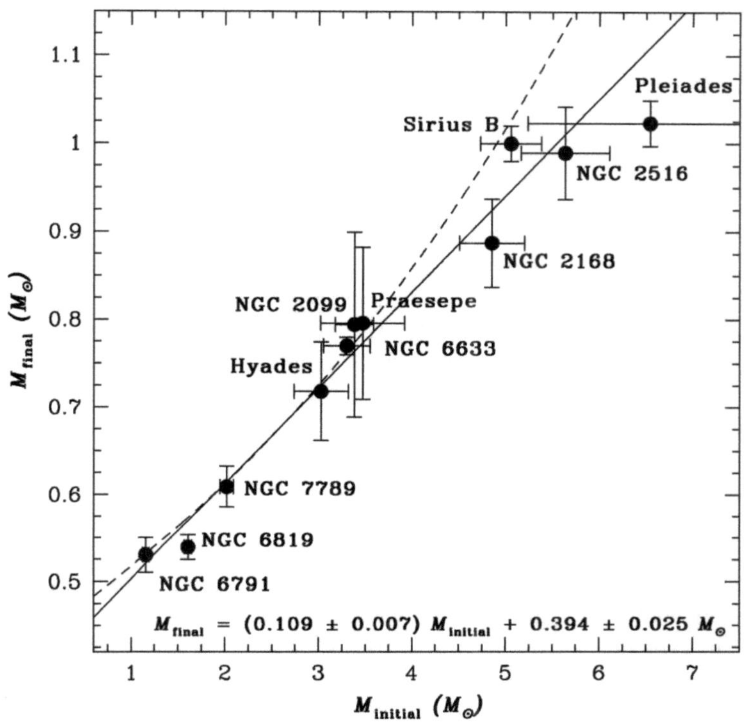

Fig. 8.3 The initial-final mass relation, inferred from white dwarfs in open clusters of various ages. From "The Initial-final Mass Relation: Direct Constraints at the Low-mass End" (Kalirai et al. 2008). © AAS. Reproduced with permission

range 1–7 $M_\odot$ lose increasingly large fractions of their mass before becoming white dwarfs.

The mass distribution of field (non-cluster) DA white dwarfs (the most common white dwarfs, those with hydrogen-rich atmospheres) was found to peak sharply at $0.593 \pm 0.016\ M_\odot$ (Kepler et al. 2007). About 19% of the white dwarfs have mass around 0.8 $M_\odot$ and 9% have mass around 1.12 $M_\odot$. A small tail of the distribution extends below 0.6 $M_\odot$. Members of the 0.593 $M_\odot$ component and probably the 0.8 $M_\odot$ component are C/O white dwarfs while members of the 1.12 $M_\odot$ component may be ONeMg white dwarfs. Entering 1.12 $M_\odot$ into an accurate IFMR would yield upper and lower limits to the value of $M_{CO,max}$ and $M_{ONeMg,max}$, respectively.

Observationally, there are issues with accurately determining white dwarf masses, and the steeply-declining mass function toward higher ZAMS masses means that the statistics are poor in the interesting mass range of $M_{CO,max}$, $M_{ONeMg,max}$, and $M_{wd,max}$. Theoretically, the first burning-shell thermal pulse (Sect. 8.3.2) tends to have an especially pronounced effect to increase the core mass and hence the eventual white dwarf mass. This effect is expected to be stronger at lower metallicity. On the other hand, hot-bottom burning (Sect. 8.3.2) proceeds more rapidly at lower metallicity. This may reduce the lifetime of the thermally-pulsing phase so that there is less change in the core mass before envelope expulsion. The bottom line is that the values of $M_{wd,max}$, $M_{CO,max}$, and $M_{ONeMg,max}$ cannot yet be accurately determined.

8.3 Red Giants and Red Supergiants

Red giants represent a post-main-sequence phase in which the luminosity has increased substantially, the radius has swollen, and the photospheric temperature has declined to less than about 5000 K. Stars in this stage are thought to come from ZAMS masses less than about 8 $M_{\odot}$. Stars of greater ZAMS mass are expected to become RSGs that are even brighter (absolute visual magnitude $M_V < -5$), larger in radius, and cooler. Betelgeuse is a prime example of an RSG visible to the naked eye.

The nature of red giants and their RSG kin is important for presupernova evolution. SN IIP are observed to explode as RSGs. Knowing which stars shed their envelopes (Sects. 8.2 and 8.4) and which do not is important for understanding which stars evolve to explosion and which become white dwarfs that do not, on their own, explode. Some RSGs may harbor degenerate C/O cores (Sect. 8.5.2), others cores of ONeMg (Sect. 8.5.3), and yet others will be evolving to produce iron cores (Sect. 8.5.5). An important goal of presupernova evolution is thus to determine which red giants or RSGs have C/O cores, which ONeMg cores, and which nondegenerate cores. All these structures could, in the right circumstances, explode, and the interior evolution that discriminates them is very complex. Clues to this internal evolution come from the luminosity, surface abundances, and, in principle, asteroseismology.

The basic computational understanding of the formation of red giants was established by Martin Schwarzschild and others in the 1950s. A complete fundamental understanding of the phenomenon remains somewhat elusive. Red giants first form after the phase of main-sequence hydrogen burning. The nondegenerate helium core radiates energy and contracts in accord with the virial theorem. Whether the envelope expands or contracts depends on the nature of shell burning; burning shells tend to lock in the temperature and serve as a node in the expansion and contraction of various portions of the star, within, between, and beyond the shells. The formation of a hydrogen-burning shell after core hydrogen burning serves as such a node. In this case, the energy of the contraction of the helium core prior to core helium ignition is transferred outward. This excess energy flows to the outer, unburned hydrogen envelope. The net energy gain, again in accord with the virial theorem, leads to expansion and cooling of the envelope, thus helping to promote the formation of a red giant. The envelope expands as the core contracts with the hydrogen-burning shell sitting at a node in the motion.

This simple picture is one aspect of the problem, but does not trivially account for the amplitude of the effect, the very large radii, $\sim 10^{13}$ to 10^{14} cm, to which the envelope can expand. Another aspect is that thermal stability breaks down and is only restored when the large radii are attained. Asteroseismological techniques are being developed that can reveal the inner structure of red giants, discriminating those having only hydrogen-burning shells from those undergoing core helium burning (Chaplin and Miglio 2013) and even gleaning hints of internal magnetic fields (Fuller et al. 2015a). These techniques will surely lead to deeper understanding of the processes involved in the advanced stages of stellar evolution (Aerts 2015; Paxton et al. 2015).

8.3.1 Dredge-Up

The formation of large, cool, convective envelopes results in several important stages of *dredge-up*. The *first dredge-up* is when a star evolves up the red-giant branch in the H–R diagram and the deep convective envelope mixes to the surface matter that has been subject to some hydrogen burning, especially by the CNO process. This matter is typically enriched in ^{13}C relative to ^{12}C and in nitrogen relative to carbon. Measures of these abundances give critical insight into the interior evolution.

The ZAMS mass below which degenerate C/O cores form, $M_{CO,max}$ (Sect. 8.2), is determined by interior evolutionary processes. In particular, the mass of the C/O core is determined by the mass of the helium core in which it grows. Evolutionary calculations have shown that the processes that determine the mass of the helium core in the relevant mass range are complex. Stars develop central cores of helium while in the main-sequence phase. Next, stars of several solar masses undergo central helium burning while hydrogen burns in a surrounding shell. When helium is exhausted in the center, there is a phase prior to helium-shell ignition in which the hydrogen shell burning is temporarily extinguished, and the convective hydrogen envelope penetrates the helium core. This is the *second dredge-up* (Becker and Iben 1979). A *Kippenhan diagram* (Kippenhahn et al. 1965) is a powerful graphical means to capture the evolving inner structure of the late stages of a stellar model in a manner that reveals the location of radiative and convective regions and of burning shells. Figure 8.4 illustrates the second dredge-up in a model of ZAMS mass 5 $M_\odot$ by means of a Kippenhahn diagram. The dredge-up is represented by the deep penetration of the outer convective envelope at $\sim 8 \times 10^7$ years that reduces the mass of the helium core from about 1.2 $M_\odot$ (m/M $\simeq$0.25) to about 0.9 $M_\odot$ (m/M $\simeq$ 0.18), thus guaranteeing that the subsequent C/O core is less than M_{Ch}. The *third dredge-up* is associated with the thermally-pulsing double-shell phase discussed in Sect. 8.3.2.

The final mass of the helium core, which determines the mass of the eventual C/O core, depends on the degree to which the helium core is reduced in mass as it is swept up into the encroaching hydrogen envelope during the second dredge-up. Evolutionary studies incorporating Schwarzschild convection, semiconvection, and no overshoot, found that the convective envelope penetrated and reduced the mass

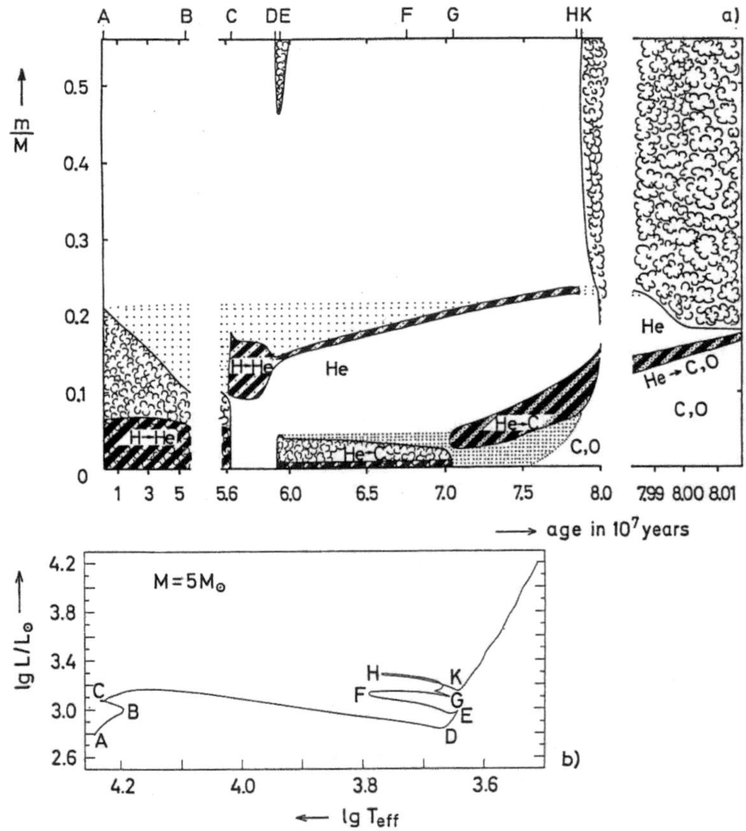

Fig. 8.4 A Kippenhahn diagram showing the convective, shell-burning structure of a star of ZAMS mass 5 $M_\odot$ (*top*, **a**), where the vertical axis is in units of the total mass, and the corresponding path in the Hertzsprung-Russell diagram (*bottom*, **b**) with common phases marked by the alphabetic letters. The second dredge-up is represented by the deep penetration of the hydrogen-rich envelope that reduces the mass of the helium core at phase K at $\sim 8 \times 10^7$ years as the model begins its final ascent up the supergiant branch. Note also the thinning of the hydrogen- and helium-burning shells that lead to the thermally-pulsing, double-shell phase, the third dredge-up, and the formation of the degenerate C/O core. From "Stellar Structure and Evolution" (Kippenhahn and Weigert 1994)

of the helium core in stars of ZAMS mass up to about 8 or 9 $M_\odot$, yielding in turn low-mass, degenerate C/O cores (Paczyński 1971; Becker and Iben 1980). This is the origin of the fact "known" by all astronomers that stars with ZAMS mass below 8 $M_\odot$ make degenerate C/O cores. This is reflected in Fig. 8.2 by portraying stars up to at least 7 $M_\odot$ as engaging in the common-core evolution.

This value of $M_{CO,max} \simeq 8$ $M_\odot$ is, however, model dependent. Other studies (Bertelli et al. 1985; Maeder and Meynet 1989) suggested that outward convective-core overshoot is important already on the main sequence, yielding more massive helium cores. The net effect would be that stars with ZAMS mass greater than $M_{CO,max} \simeq 6\ \mathrm{M}_\odot$ would have such massive helium cores that they do not develop degenerate C/O cores. Some computations have found an instability at the end of convective-core helium burning when fresh helium is mixed in, resulting in an increase in the burning rate and an expansion of the inner convective region. These "breathing pulses" may also result in more massive helium cores and a lower value of $M_{CO,max}$ (Castellani et al. 1985; Lattanzio 1986). Rotational and magnetically-induced mixing also can change $M_{CO,max}$ and other boundaries (Wheeler et al. 2015a).

8.3.2 AGB Stars, SAGB Stars, Thermal Pulses

Evolved low- to intermediate-mass stars are observed to occupy the *giant branch* on the H–R diagram where they are presumed to have helium cores and hydrogen-burning shells, and the *asymptotic giant branch* (AGB) where they are presumed to have proceeded beyond core helium burning to form degenerate cores of C/O or ONeMg. Physically, such stars have expanded to the *Hayashi track*, the boundary in the H–R diagram beyond which fully convective stars cannot maintain hydrodynamic and thermal equilibrium (Hayashi 1961). At this stage, stars with C/O cores have thin hydrogen- and helium-burning shells. Stars with ONeMg cores have in addition a carbon-burning shell. AGB stars typically show long-period pulsations as, for instance, *Mira variables*.

In standard evolution theory (Becker and Iben 1980), degenerate C/O cores form on the AGB with masses that increase monotonically with ZAMS mass. For some luminosity range above the base of the AGB, models obey a relation between core mass and luminosity (Paczyński 1970):

$$\frac{L}{\mathrm{L}_\odot} = 59{,}250\left(\frac{M_{core}}{\mathrm{M}_\odot} - 0.522\right), \tag{8.1}$$

where M_{core} is, strictly speaking, the mass of the hydrogen-exhausted core, slightly greater than that of the C/O core. In principle, stars on the AGB could exist up to the maximum luminosity according to Eq. (8.1), when the core reaches the Chandrasekhar mass and the luminosity becomes about 52,000 $\mathrm{L}_\odot$ or $M_{bol} \simeq -7.1$. The luminosity of a star on the AGB may thus give a clue to its interior structure and evolution.

In theoretical language, models with degenerate C/O cores are AGB models. The average star with a degenerate ONeMg core is expected to be brighter than the average AGB star. For this reason, models with ZAMS mass in the range $M_{CO,max}$ to $M_{ONeMg,max}$ are called *super-asymptotic giant-branch*, or SAGB models (Garcia-Berro and Iben 1994). The bright red-giant Antares might be an SAGB star, and an M-type star in the SMC with a pulsational period of 1749 days and an absolute magnitude of $M_{bol} = -8.0$ is a candidate SAGB star (Groenewegen et al. 2009). A caveat is that luminosity alone may not be a sufficient discriminator because AGB stars with ZAMS mass just below $M_{CO,max}$, as well as more massive RSGs, $M \gtrsim 15\ M_\odot$, are predicted to have similar luminosities to stars with ZAMS mass just greater than $M_{CO,max}$. While more massive stars have different effective temperatures, AGB stars with ZAMS mass just above and just below $M_{CO,max}$ may be difficult to tell apart.

While the luminosity is an important clue to the interior evolution, a significant complication is that the luminosity may not be a simple metric of the core mass as suggested by Eq. (8.1). A *hot-bottom envelope* is one in which the outer convective envelope penetrates sufficiently close to the hydrogen-burning shell that CNO burning can occur within the convective envelope on the AGB (Scalo et al. 1975). This is expected to occur in the more massive AGB stars and in SAGB stars. Stars with hot-bottom envelopes would not obey the same relation between core mass and luminosity as given in Eq. (8.1). With a hot-bottom envelope, a core can remain nearly constant in mass, but the surface luminosity increases because of increasing core degeneracy and burning within the convective envelope (Wagenhuber and Groenewegen 1998; Siess 2010).

Another significant feature of AGB and SAGB models is the occurrence of repetitive, thermally-pulsing shell burning. Models of AGB and SAGB stars have early (E-AGB) and later thermally-pulsing (TP-AGB) phases (Garcia-Berro and Iben 1994). After a finite, but usually small, growth of the core mass and luminosity in the E-AGB phase, models for which the base of the convective envelope is beyond the thin hydrogen and helium shells enter the double-shell-flash TP-AGB stage. The thin helium shell undergoes a thermal runaway (Schwarzschild and Härm 1965; Weigert 1966). Burning increases the temperature, but since the shell is geometrically thin, even if it expands it does little work on the overlying layers. The result is that the pressure does not decline substantially. The density declines, but the shell stays hot, and helium burning continues. The expanding helium shell does, however, extinguish the hydrogen shell. This allows the outer convective envelope to penetrate beneath the extinct hydrogen shell and sweep some freshly-produced carbon to the surface in the *third dredge-up*. The third dredge-up is more efficient at higher masses and lower metallicities.

The hydrogen shell eventually re-ignites and produces fresh helium. When the helium concentration becomes sufficient, the helium shell re-ignites and the cycle begins again. Models in this stage can have thousands of thermally-pulsing double-shell flashes, posing a severe challenge to numerical stellar-evolution models. Many calculations of AGB evolution invoke "synthetic" evolution to average over the complex thermally-unstable shell burning, but this may obscure physics involved in the pulses, nucleosynthesis, mixing, and mass loss. The evolution and nucleosynthesis of AGB stars was reviewed by Herwig (2005).

8.3.3 C, S, and M Stars

Although the AGB luminosity does not give an unambiguous measure of the interior evolution, surface abundances provide important complementary information. If carbon outnumbers oxygen by number, oxygen is locked up in CO molecules, which have no observable bands in the optical. The excess carbon is free to make other associations. The star then has strong carbon absorption bands (C_2, CN, CH) and is classified as a C-type AGB star (a "carbon star"). If carbon is less abundant than oxygen, it is locked up in CO molecules, and the excess oxygen is free to form other associations. AGB stars that have strong oxide bands (e.g., TiO, ZrO) in the optical are classified as M-type AGB stars. The first helium thin-shell flashes generate carbon and isotopes such as ^{13}C at ZAMS mass $M \gtrsim M_\odot$ and ^{22}Ne at $M \gtrsim 5\ M_\odot$ that later produce neutrons. In subsequent shell flashes, some of the neutron-rich species act as seeds to generate *s-process* nucleosynthesis (Iben 1975). The appearance of these nuclides at the stellar surface gives rise to an S-type AGB star. The brightest AGB stars show an excess of the s-process element rubidium (García-Hernández et al. 2006). Relevant data may also come from the rapidly evolving post-AGB stars, also known as proto-planetary nebulae, that have ejected their obscuring envelopes but have not yet cooled to form a white dwarf and a resolved planetary nebula. These stars have a complex and interesting array of abundance properties that give clues to their evolution (van Aarle et al. 2011; De Smedt et al. 2012; D'Ercole et al. 2012).

The luminosities, numbers, and ratios of C-type to M-type AGB stars are measures of the presence of carbon and oxygen in the atmosphere, hence related to the evolution of the core of the star. The distributions of C-type and M-type stars also depend on the metallicity (lower primordial oxygen more easily results in a C-type star when carbon is mixed to the surface) and on ejection of the envelope, that is, the value of $M_{wd,max}$. Which type, C, S, or M, is manifested by an AGB star depends on the ZAMS mass and evolutionary phase. A key question is the nature of these AGB stars and the constraints that they put on the ZAMS masses involved, the evolutionary state of the brightest AGB stars, and on the envelope ejection process (Marigo and Girardi 2007).

In principle, stars on the AGB could be C-type right up to the maximum luminosity on the AGB. This is not the case, however, since in both cluster and field stars the bolometric magnitude of the carbon stars in the Magellanic Clouds is truncated rather sharply at about $M_{bol} \simeq -6.1$, corresponding to a core mass of about 0.88 $M_\odot$ (Blanco et al. 1980). There are some bright M-type and S-type stars that can be assigned to the AGB in the luminosity range from the carbon-star cutoff to the maximum AGB luminosity at $M_{bol} \simeq -7$ to -8 (Wood et al. 1983). There are relatively few of these stars observed brighter than the carbon-star limit (Mould and Reid 1987), perhaps because of heavy mass loss (Sect. 8.4) associated with the formation of dust, making the stars observable only in the infrared when the stars become *OH/IR masers* (Wood et al. 1992).

The fact that some AGB stars exceed the limiting magnitude according to Eq. (8.1), $M_{bol} \simeq -7.1$, implies that envelope ejection does not automatically occur at more modest luminosities. The brightest, highly-obscured, long-period AGB stars seem to require rather large ZAMS masses, about 6 $M_\odot$ (Ventura et al. 2000). This suggests that $M_{wd,max}$ is at least this high, but does not clearly identify the composition of the degenerate core. AGB stars become carbon stars when the third dredge-up associated with thermal pulses brings carbon to the surface. AGB stars may survive to higher luminosities than the observed carbon-star limit by burning carbon to nitrogen in a hot-bottom envelope; the reduction of carbon in the atmosphere may convert these stars to M-type (Iben and Renzini 1983). Nearly all M-type AGB stars in the Magellanic Clouds are lithium rich (Smith and Lambert 1990).

Clues to distinguish AGB from SAGB stars may arise because thermal shell flashes in stars with C/O and ONeMg cores have quantitatively different nucleosynthesis. The thermal pulses in SAGB models tend to be shorter and weaker than in AGB models. Because of the high gravity of the relatively massive ONeMg cores, TP-SAGB models have higher temperatures that tend to promote the ^{22}Ne(α,n)^{25}Mg reaction at the base of the pulsing shell and a pronounced hot-bottom envelope, leading to enhanced abundances of ^{4}He, ^{7}Li, ^{14}N, ^{13}C, ^{17}O, ^{25}Mg, ^{26}Al, and ^{27}Al (Siess 2010). While these species may serve to distinguish typical AGB stars from typical SAGB stars, it remains unclear whether an appropriate abundance analysis would distinguish the most massive AGB stars, those with ZAMS mass just less than $M_{CO,max}$, from slightly more massive SAGB stars. For instance, rubidium can be produced by hot-bottom burning in both AGB and SAGB models with ZAMS mass $\gtrsim$4 $M_\odot$ (Karakas et al. 2012).

There is great potential in the study of C, S, and M stars to determine $M_{wd,max}$, $M_{CO,max}$ and $M_{ONeMg,max}$ through studies of luminosity and composition distributions of AGB stars. In practice, the existence of hot-bottom envelope evolution is a significant complication that alters compositions in important, potentially observable ways, but also makes the luminosity a more ambiguous measure of the evolutionary state. At this writing, it is not clear that there is any observational feature that allows observed AGB stars to be cleanly differentiated between those with C/O cores and those with ONeMg cores nor to determine the ZAMS mass above which envelopes are ejected.

8.4 Mass Loss

Mass loss is important during post-main-sequence phases of all stars and even during the main-sequence phases of massive stars. Stars with ZAMS mass less than $M_{wd,max}$ lose their extended red-giant envelopes. Stars with ZAMS mass in excess of about 8 $M_\odot$ are expected to die as RSGs that have suffered at least modest mass loss or as Wolf–Rayet stars that have suffered a lot. Owing to mass loss, stars with ZAMS mass between about 30–50 $M_\odot$ are no longer RSGs by the time their cores collapse. Stars with mass in excess of the Humphrey–Davidson limit, near 50 $M_\odot$, never become RSGs because they suffer substantial mass loss on the main sequence. Stars can also lose their hydrogen envelopes by mass transfer in binary systems (Sect. 8.8). As noted in Sect. 8.1, the Sun's final mass will be about half of its ZAMS mass. For most of the more massive stars, the mass of the final white dwarf or the mass of the core immediately prior to core collapse is an even smaller fraction of the ZAMS mass. A rule of thumb is that the core of a massive star composed of helium and heavier elements is about 1/3 of the ZAMS mass. The progenitors of SN Ia, SN Ib, SN Ic, and SN IIb all have had substantial mass loss and mass loss is implicated for various other supernova types.

8.4.1 Radiation-Driven Mass Loss in Massive Stars

Feebly for lower-mass stars and more powerfully for brighter, more massive stars, mass is driven in the form of winds from the surfaces of stars by the radiation flow from their surfaces. If the mass-loss rate, $\dot{M}$, depends on L, R, and M, then by dimensional analysis (Reimers 1975), the mass loss must scale as

$$\dot{M} \propto \frac{LR}{GM}. \tag{8.2}$$

The actual rate depends on the constant of proportionality (which might also depend on stellar parameters).

The radiation that drives winds exchanges momentum most effectively in spectral lines (Castor et al. 1975). Hydrogen and helium have few lines, so even at their typical lower abundance, metals provide most of the lines that drive the wind. This makes the winds dependent on metallicity and the effective temperature that determines the ionization and excitation states as well as luminosity (Vink et al. 2001). Rotation tends to enhance mass loss by effectively reducing the local gravity at the stellar surface. Mass loss is inhibited at low metallicity. This is especially important in the context of the presupernova evolution of the first stars of *Population III* at an epoch when the metallicity was low, or even null. Prescriptions for mass loss in various conditions and references to the original literature are given in Paxton et al. (2011, 2013) and Langer (2012).

Determination of mass-loss rates depends on knowing the density of the wind. This is often measured by free–free and recombination processes for which the emissivity is proportional to ρ^2. This makes these processes susceptible to clumping in the wind that artificially enhances the estimated mean density if a smooth wind is assumed. This consideration is thought to reduce line-driven mass-loss rates by factors of several compared to standard prescriptions employed in many current stellar-evolution codes (Smith 2014).

Line-driven mass loss is of rather little importance to presupernova evolution for stars of ZAMS mass of 10 $M_\odot$ or less. For more massive stars, O stars, mass loss even on the main sequence can be important by reducing the mass of the star by substantial amounts even before the end of hydrogen burning. If the hydrogen envelope is shed by winds or other processes the result can be a Wolf–Rayet star or another form of hydrogen-deficient star. These stars are also expected to be subject to strong line-driven winds. Mass-loss rates for Wolf–Rayet stars are estimated to be in the range 10^{-5} to 10^{-4} $M_{\odot}$ y^{-1}. At the highest rates, winds can also deplete surface helium, yielding bare, nondegenerate C/O cores that may be related to some SN Ic.

8.4.2 AGB Mass Loss

Mass loss on the AGB sets the limit $M_{wd,max}$. Determining this mass loss, both observationally and physically, is difficult. The Reimers formula, Eq. (8.2), and elaborations on it (de Jager et al. 1988) have been adopted for the mass-loss rate of models for which $L(t)$, $R(t)$, and $M(t)$ are known, but it is not clear that this is a proper usage of the formula. Used in this way, the Reimer's formula overpredicts mass-loss rates for some evolutionary states and greatly underpredicts mass-loss rates at the tip of the AGB. A factor omitted in the Reimers scaling is the effect of pulsations. AGB mass loss is usually, and perhaps always, preceded by a pulsational phase. Lower-mass Mira variables pulsate in the fundamental "breathing" mode. The pulsation changes the structure of the star. Both the structure of the envelope and the pulsation period itself are substantially different from estimates based solely on linear stability analysis of hydrostatic stellar-evolution models (Ya'Ari and Tuchman 1996, 1999). The transition of an AGB star to a carbon star is likely to change the mass loss by changing the character of the dust (Ventura et al. 2012), and the process of mass loss itself may affect when and how a given star becomes a carbon star. The fact that all this is occurring in a dynamically-pulsing star adds to the complexity of the problem.

Extensive mass loss leading to the exposure of the semi-degenerate core, the process that determines the limit $M_{wd,max}$, is by means of a *superwind* (Renzini 1981; Iben and Renzini 1983; Willson 2000). The superwind may not be a distinct physical mechanism, but a consequence of mass loss depending very sensitively on the physical state of the star. Mass-loss rates rise steeply with increasing luminosity and radius and decreasing effective temperature. Winds are likely to be exacerbated by dust formation, as well as by pulsation that expands the envelope in the binding potential and produces shocks and colder radiative pockets where dust can form. Theoretical models that match observational constraints can be used to extrapolate to other conditions. Such models suggest that a larger proportion of low-metallicity stars may produce core masses that reach the Chandrasekhar limit before expelling their envelopes. Thus at lower metallicity there may be an increased tendency to produce degenerate carbon explosions within hydrogen envelopes (which might be relevant to SN 2002ic-like events; Sect. 23.2). These models also suggest that white dwarfs from low-metallicity populations tend to be less frequent and more massive than those associated with higher-metallicity populations.

Empirical relations describe *which* AGB stars lose mass, that is, their mass, luminosity, and radius while doing so, but not *how* stars lose mass, that is, how the mass loss versus time depends on $M(t)$, $L(t)$, and $R(t)$ (Willson 2000). In addition, empirical relations between mass-loss rates and stellar parameters are dominated by selection effects. For AGB stars, it is very difficult to simultaneously measure L, R, and M for a large sample of stars.

An alternative prescription for AGB mass loss that is consistent with observations and models is that the stellar mass remains nearly constant until the mass-loss rate increases rapidly (Willson 2000). Empirically, the mass-loss rate increases exponentially with pulsation period up to a period of about 500 days for masses of 2 $M_\odot$, up to 750 days for masses of 5 $M_\odot$, and then remains constant at the superwind value of somewhat more than 10^{-5} $M_\odot$ y^{-1} (Vassiliadis and Wood 1993).

The question remains of the physics that directly determines the total expulsion of the envelope. This is the ultimate process that sets $M_{wd,max}$ and therefore whether some stars with degenerate C/O cores fail to eject their entire envelope and whether any or all single stars that correspond to model SAGB stars lose their entire envelopes to produce ONeMg white dwarfs.

Of possible relevance to this issue is that evolutionary models of TP-AGB and TP-SAGB stars that undergo mass loss are plagued by an instability that may play a role in real stars. The instability occurs near the base of the convective envelope when the radiation pressure builds and the local luminosity becomes greater than the *Eddington limit*, the condition where the pressure gradient exceeds the local gravity (for discussions of this issue, see Joss et al. 1973; Paxton et al. 2013). In such a situation, the density and gas pressure tend to zero, and codes typically crash, suggesting that the instability leads to failure of hydrostatic equilibrium and dynamic expulsion of the remaining envelope (Wood and Faulkner 1986). Lau et al. (2012) found this instability in models with hot-bottom envelopes when the mass of the envelope declined to about 2 $M_\odot$. The exact values of the envelope mass and the presence of the instability depend on the treatment of convection, but the general trend is that the envelope mass when the instability occurs increases with the ZAMS mass. This suggests that envelope ejection may be more effective at higher mass for any stars with degenerate cores and thus that $M_{wd,max}$ may exceed $M_{ONeMg,max}$.

A true understanding of this instability and its possible role in AGB envelope ejection probably will require multidimensional stellar evolution to better understand plume mixing (Scalo 1973) and related issues (Willson 2000). If the instability is real and ubiquitous in AGB and SAGB stars of large ZAMS mass, then it may be that all massive AGB and SAGB stars eject their envelopes; hence none may explode as SN II triggered by degenerate carbon ignition in a C/O core or by electron capture and collapse in a degenerate ONeMg core. Either of these processes could occur later, however, if the white dwarf forms in a binary system (Sect. 8.8). Careful studies of the IFMR (Sect. 8.2) may give clues to the mass loss and envelope ejection as functions of ZAMS mass and metallicity (Kalirai et al. 2014).

8.4.3 Mass Loss by Luminous Blue Variables

Luminous blue variables (LBVs) are bright, hot stars that lose mass in extensive bursts that are quite different from the steady mass loss expected by line-driven winds (Sect. 8.4.1). It was once thought that this phenomenon was associated only with the most massive stars during their young main-sequence phases. It has become clear, however, that this mode of mass loss may occur in lower-mass stars (still in excess of about 20 $M_\odot$) and late in their evolution. From the radii and expansion velocities of CSM, some stars must have ejected substantial gouts of matter only years or decades before exploding (Smith 2014; Owocki 2015).

Understanding the episodic mass loss associated with LBVs and SN IIn (Chap. 14) or SLSN-II (Chap. 18) remains a large challenge. Espisodic mass loss may be related to generic processes that beset massive stars in their late stages of evolution. Hysteresis associated with the finite-time response of convection may lead to the radiative flux temporarily exceeding the local Eddington limit (Guzik and Lovekin 2012). In addition, iron-core stars are predicted to reach a phase of complex, turbulent, semi-dynamical nuclear burning when, for instance, oxygen burning rates are comparable to convective-overturn timescales (Arnett and Meakin 2011b; Couch et al. 2015). This turbulent burning may drive gravity waves and acoustic modes that propagate out through the star. Dissipation of these acoustic waves has the potential to drive mass loss from the surface (Shiode and Quataert 2014). It remains to be seen how this mass loss depends on stellar mass, rotation, magnetic fields, metallicity and age. Pulsational pair instability has been explored in this context (Sect. 8.6).

8.5 Presupernova Evolution

8.5.1 Helium Flash

The *helium flash* is not a presupernova mechanism, but it gives a hint of the role of degeneracy and associated complex issues of mixing that dominate other stages of evolution. Because of the temperature insensitivity of the equation of state of degenerate matter, mechanical response to nuclear burning is muted, leading to thermal and even dynamic instability. The burning can then lead to changes in the stellar structure and associated mixing processes.

Stars with ZAMS mass $\lesssim 1.5\ M_\odot$ undergo central ignition of degenerate helium, leading to the helium flash (Fig. 8.2). Whether this is a semi-dynamical or only a thermal event is debated. Studies of the helium flash discriminate between metal-poor and solar-metallicity stars and explore both 2D and 3D (Dearborn et al. 2006; Mocák et al. 2010). The observed systematics of the tip of the red-giant branch and the *horizontal branch* are generally consistent with the predicted behavior of the helium flash. After the helium flash, stars are predicted to become less luminous at a rate of perhaps several thousandths of a magnitude per year, but this has not yet been observed directly.

An important issue is whether hydrogen is stirred from the outer envelope down into the helium core during the helium flash with the potential to amplify the strength of the nuclear runaway. Current results suggest that the helium flash is probably associated with some mass loss, but not with extensive mixing between the helium core and the hydrogen envelope. Stars of very high metallicity may lose enough mass on the red-giant branch to avoid the helium flash, and therefore never fuse helium into carbon and oxygen in their cores.

8.5.2 Evolution to Degenerate Carbon Ignition

In stars of relatively low mass, the first threat of catastrophe is ignition of degenerate carbon in stars with ZAMS masses below $M_{CO,max}$. Carbon burns at high temperature, $T \simeq 3 \times 10^8$ K. For stars of low to moderate mass this temperature is reached only at such high density that the matter is electron-degenerate and prone to rapid dynamical thermonuclear runaway (Hoyle and Fowler 1960).

Single stars that develop degenerate C/O cores have virtually identical internal structures, differing only in the envelope mass, and hence total mass (Paczyński 1970). The basic property of these cores is the unique density structure of degenerate matter. In addition, the temperature distribution is controlled by the balance of neutrino cooling and heating associated with accretion of mass onto the core from overlying hydrogen-burning and helium-burning shells (Barkat 1975). This convergence in central conditions is shown by the models of 3 and 7 $M_\odot$ in Fig. 8.2.

There may be some variation in the structure of these degenerate C/O cores because the fractions of carbon and oxygen produced by helium burning depend on temperature and hence could vary in radius in different ways for cores that grew in stars of different total mass. Because core helium burning leads to relatively oxygen-rich matter and subsequent shell burning leads to relatively carbon-rich matter, composition gradients are predicted to vary with mass (Domínguez et al. 2001). The composition structure is also likely to be somewhat different in white dwarfs that accrete in binary systems (Sect. 8.8). The effort to use SN Ia for cosmological applications (Sect. 25.2) has brought close attention to these subtle variations in composition and their effect on the spectra, colors, and light curves of SN Ia.

The ZAMS mass range of single stars that undergo this "common-core" evolution toward ignition of degenerate carbon is affected by two very different physical processes. One is the dredge-up (Sect. 8.3.1) that determines the ZAMS mass above which carbon is ignited in nondegenerate matter, i.e., the value of $M_{CO,max}$. The other is ejection of the envelope to become a planetary nebula (Sect. 8.4.2), as determined by $M_{wd,max}$ (Sects. 8.2 and 8.3.2).

In this context, carbon ignition is defined to take place when carbon-burning rates exceed neutrino-loss rates. Carbon ignites when the C/O core exceeds a minimum mass of about 1 $M_\odot$. Carbon burning is first controlled by convective transport. In the absence of loss of the envelope and quenching of shell-burning that feeds mass to the core, carbon burning eventually proceeds on a dynamical timescale, leading to explosion (Chap. 19).

8.5.3 Degenerate Oxygen–Neon–Magnesium Cores

For ZAMS masses above $M_{CO,max}$, carbon ignites off center in a nondegenerate core of $\lesssim$1.4 $M_\odot$ and burns out, leaving the oxygen basically unburned and with neon, magnesium, and some sodium as the burning products. Off-center carbon ignition occurs prior to strong neutrino losses that lead to the degenerate common-core phase (Sect. 8.5.2). This stalls the second dredge-up until after carbon burning is completed in the core. Figure 8.2 shows the evolutionary track of a star of $M \simeq 8\ M_\odot$ that ignites carbon off-center in a nondegenerate region. Subsequent off-center carbon shell burning consumes more of the C/O mass. At this point, models with $M > M_{CO,max}$ and $M < M_{ONeMg,max}$ undergo the second dredge-up that sweeps essentially all material remaining in the helium core into the outer convective envelope. Figure 8.5 shows the Kippenhahn diagram for a model of ZAMS mass $\simeq$10 $M_\odot$ with an initial helium-core mass of $\simeq$2.2 $M_\odot$. The carbon-burning shell ignites somewhat less than 10^5 years prior to the end of the second dredge-up phase and then moves both into the center and outward. The dredge-up ensures that the inner core of ONeMg cannot grow to exceed the Chandrasekhar mass, so the model develops a degenerate ONeMg core. When this core mass is less than the critical value for neon ignition, $M \simeq 1.37\ M_\odot$ (Nomoto 1984), the core undergoes convergence to a common evolutionary track. If the envelope is not ejected, these degenerate ONeMg cores are predicted to eventually undergo electron capture and gravitational collapse (Sect. 9.4.3). Cores with mass above $M_{ONeMg,max}$ ignite neon (Sect. 8.5.4) and proceed to form iron cores (Sect. 8.5.5).

Computation of the late evolution in this mass range is exceedingly complex and sensitive to the ZAMS mass, to the physical input, and probably to numerical procedures. The evolution depends on metallicity, overshooting, semi-convection, thermal pulses, hot-bottom burning, and other factors. An important feature in such models is that electron capture occurs on evolutionary time scales. This significantly alters the composition, the thermonuclear systematics, the equation of state, and the condition of dynamical equilibrium. Nuclear burning is semi-degenerate and rapid, so that the usual assumption of complete mixing on a nuclear timescale becomes invalid. This leads to problems associated with the treatment of time-dependent convection and composition gradients. The combination of semi-degenerate structures and neutrino losses tends to cause temperature inversions. This leads to off-center ignition, further complicating the final evolution, as also seen for the semi-degenerate helium flash (Sect. 8.5.1). These factors determine which

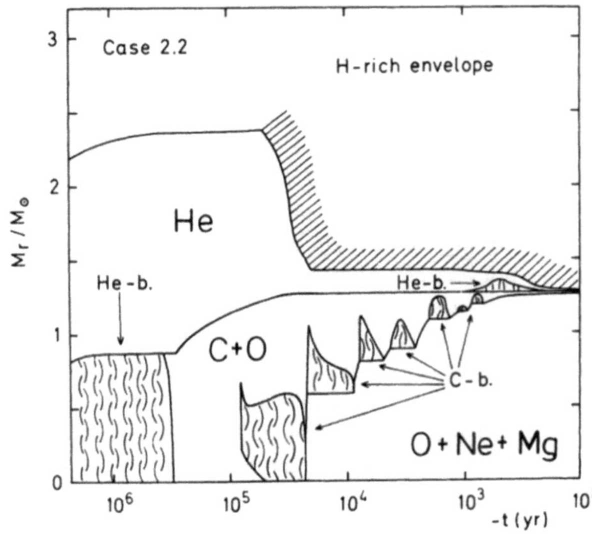

Fig. 8.5 A Kippenhahn diagram showing the convective, shell-burning structure of a model of ZAMS mass about 10 $M_\odot$. Time is measured from the end of the second dredge-up phase, which is represented by the deep penetration of the hydrogen-rich envelope that reduces the mass of the helium core. Note also the off-center ignition of carbon in a convective shell, the propagation of the convective shell to the center, the subsequent ignition of off-center carbon-burning shells, and the development of a core of degenerate ONeMg. From "Evolution of 8–10 solar-mass stars toward electron capture supernovae. II—Collapse of an O + Ne + Mg core" (Nomoto 1987). © AAS. Reproduced with permission

models produce ONeMg white dwarfs and which may possibly evolve to become electron-capture supernovae.

The complex evolution of stars of ZAMS mass in the range 7–12 $M_\odot$ has been discussed by Barkat et al. (1974), Nomoto (1984, 1987), Woosley and Weaver (1995), Ritossa et al. (1996), Eldridge and Tout (2004), Siess (2007, 2010), Poelarends et al. (2008), Ventura and D'Antona (2011), Paxton et al. (2011, 2013, 2015), Takahashi et al. (2013), Jones et al. (2013), Doherty et al. (2015), Schwab et al. (2015) and others. Estimates of $M_{ONeMg,max}$ range from 8 to 11 $M_\odot$.

8.5.4 *Off-center Oxygen and Neon Ignition*

For ZAMS masses somewhat higher than $M_{ONeMg,max}$, models show that the second dredge-up (Sect. 8.3.1) does not occur, the core exceeds the Chandrasekhar mass, and oxygen and neon ignite off-center. Some of the neon photodisintegrates to produce alpha particles, which combine with remaining neon. The result is that, although neon has a higher atomic number than oxygen, neon ignites before oxygen. The evolution of stars in the ZAMS range 10–15 $M_\odot$ through neon and oxygen ignition was studied by Woosley et al. (1980), Nomoto (1984), Nomoto and Hashimoto (1988), and Weaver and Woosley (1993). The models suggest the presence of another critical mass, $M_{NeO,max} \simeq 13\ M_{\odot}$, above which thermonuclear burning at each stage is initiated in the center and, with the exception of some thermal instabilities, the evolution proceeds directly toward the iron-core phase, as will be described in Sect. 8.5.5. Below $M_{NeO,max}$, neon ignites off-center, leaving a composition inversion in intermediate stages, with neon-burning products—silicon, sulfur, and argon—overlying the ONeMg core.

The evolution is again very complicated and quite sensitive to ZAMS mass in the range $M_{ONeMg,max} < M < M_{NeO,max}$, about 8–13 $M_{\odot}$. Most published models in this mass range are nonrotating and nonmagnetic; more realistic models may reveal at least quantitatively different behavior. One thing is clear: all models with ZAMS mass greater than $M_{CO,max}$ form cores that are expected, in the absence of total envelope ejection, to collapse. If they explode, and the statistics of SN II suggest that they do (Sects. 3.2 and 12.5.4), they must leave neutron stars or black holes.

Nomoto et al. (1982) and Tominaga et al. (2013) suggested that a star in the ZAMS mass range of 8–13 $M_{\odot}$ may have been the progenitor of the Crab nebula. They invoked a star of 9 $M_{\odot}$ to form a core of ONeMg surrounded by silicon, sulfur, and argon without having an appreciable layer of carbon, since the Crab is not observed to be enhanced in carbon. These models undergo some helium-core penetration resulting in contamination of the envelope with helium. By invoking a judicious amount of mass loss from the envelope before the second dredge-up, the model can provide an appreciable helium enrichment. High-resolution filter images of the Crab revealed evidence for CNO and for helium, carbon, and oxygen burning (Satterfield et al. 2012). Other analyses have suggested that the Crab could have been an ordinary SN IIP (Sect. 7.4.2).

8.5.5 *Iron-Core Evolution*

Evolving models of massive stars from the main sequence to their unstable endpoint represents a special challenge. The models must incorporate a wide range of reaction rates and nucleosynthesis and complex stellar structures. Models of the evolution of nonrotating, solar-metallicity massive stars to form iron cores that collapse were presented by Woosley and Weaver (1988), Nomoto and Hashimoto (1988), Woosley et al. (2002), Limongi and Chieffi (2006) and others. There are still important uncertainties in the mass and evolution of the cores of these stars associated with the rate of the $^{12}C(\alpha, \gamma)^{16}O$ reaction and in the effect of electron captures on the core. Current models suggest that Coulomb interactions in the equation of state play a significant role and that the cause of the instability of the iron core is electron capture, abetted by photodisintegration, rather than the latter process alone as earlier models indicated (Chap. 9).

For masses above $M_{NeO,max} \simeq 13\ M_\odot$, the early evolution becomes more complex due to problems of convective overshoot. The treatment of composition gradients during hydrogen burning affects the size of the helium core, which in turn affects the details of the final stages and the mass of the core to be associated with a given ZAMS mass. Main-sequence mass loss also begins to play an important role in stars with mass $\gtrsim 30\ M_\odot$ (Maeder and Meynet 1987, 1989; Vink et al. 2011; Ekström et al. 2012). The final core masses and evolution are dependent on this important, but rather uncertain, process (Sect. 8.4).

More massive stellar models avoid thermal instability associated with partial degeneracy, but the later stages of evolution are complex because of the presence of multiple burning shells. At the lower end of this mass range, models undergo central carbon burning followed by several stages of carbon shell burning. These models also typically have more than one shell-burning stage of neon, oxygen and silicon. A representative evolutionary track for a model of ZAMS mass of 25 $M_\odot$ is shown in Fig. 8.2. Successive fuels are burned, terminating in the development of an iron core, surrounded by burning shells separated by layers of inert fuel. The resulting core structure depends in part on the mass of the helium core, which can grow by 10–50% during central helium burning by addition of mass from the hydrogen-burning shell.

Another reason for the complexity of evolution in this mass range is that the results depend on details of the interaction of the burning shells and the entropy distribution of the overlying matter (Fig. 8.6). An important factor is whether a convective shell extends from less than to more than the Chandrasekhar mass. If a convective region spans the Chandrasekhar mass, the cores tend to be less degenerate. The result is that the evolution, the mass of the resulting iron core, and the density gradient in the surrounding layers as measured by the "compactness" of the inner layers (Sect. 9.4) can be strongly nonmonotonic with respect to ZAMS mass (Barkat and Marom 1990). The variation may even be chaotic, in the sense of being very sensitive to initial conditions (Sukhbold and Woosley 2014). Significant uncertainties are also associated with the fact that in the very late stages of oxygen and silicon burning, convection occurs on the time scale of nuclear burning. This means the evolving composition is not rapidly homogenized and may again lead to dynamic, chaotic burning characterized by intermittent bursts (Arnett and Meakin 2011b). Other major uncertainties involve the role of angular-momentum distribution and transport and associated circulation and mixing that may be affected by the coupled effects of rotation and magnetic fields (Sect. 8.7).

Above the Humphreys–Davidson limit of about 50 $M_\odot$, stars may lose their envelopes through heavy mass loss already on the main sequence (Sect. 8.4) and become Wolf–Rayet stars. The envelope also may be lost in the course of binary mass transfer (Sect. 8.8). The loss of the envelope may affect some quantitative aspects of the evolution, but, unlike stars with ZAMS mass less than $M_{ONeMg,max}$, the nondegenerate core evolution of massive stars proceeds in qualitatively the same manner even in the absence of the envelope since the controlling processes of radiative energy loss and core contraction are internal to the core and do not depend on the presence or absence of an extended envelope. Once the core mass

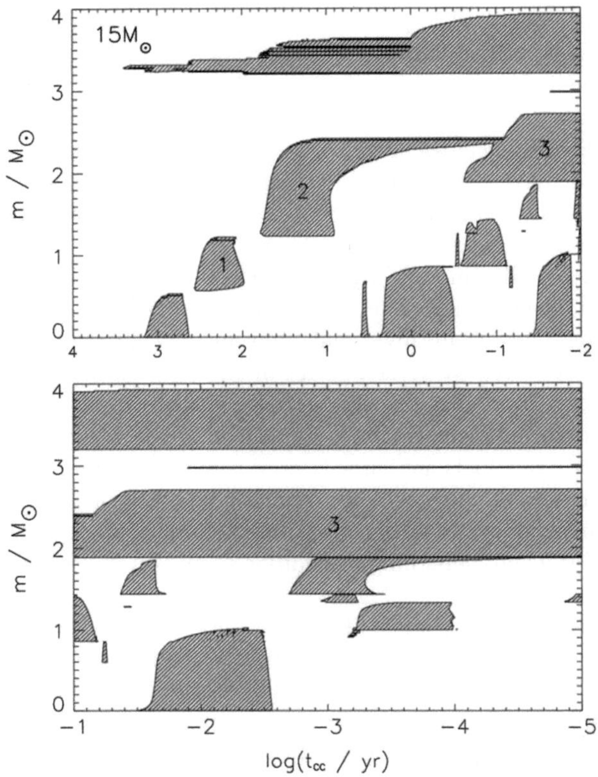

Fig. 8.6 A Kippenhahn diagram showing the convective, shell-burning structure of the inner 4 M$_\odot$ of a model star of ZAMS mass 15 M$_\odot$. Time is measured from the point of iron-core collapse. The *top panel* shows the phase from central carbon burning (with three subsequent shell-burning phases as labeled); the *bottom panel* shows the phase from the end of the first oxygen shell-burning phase until minutes before core collapse. *Cross-hatched* regions are convective. From "The Compactness of Presupernova Stellar Cores" (Sukhbold and Woosley 2014). © AAS. Reproduced with permission

is determined, perhaps by processes very sensitive to the precise ZAMS mass or the presence of a binary companion, the subsequent evolution is less sensitive to details. This may have implications for the progenitor evolution and the nature of the explosion of stripped-envelope supernovae: SN IIb, SN Ib, and SN Ic.

All stars with ZAMS mass above $M_{ONeMg,max}$ and below the limit discussed in the next section, about 100 M$_\odot$, develop iron cores and collapse (Chap. 9). For this reason it is thought that the late stages of evolution leading to SN IIP, SN IIb, SN Ib and SN Ic are all roughly similar. What is not clear is which stars leave behind neutron stars and which black holes. Because of the possibility of nonmonotonic evolution, iron-core masses, and core compactness, there may not be a single dividing line below which neutron stars are produced and above which black holes result. Likewise, it is not known what evolution leads to "normal" pulsars, what to highly-magnetized magnetars, and what to compact central objects as observed in Cas A (Sect. 7.4.1). Some magnetars may arise from especially high-mass stars that are otherwise thought to perhaps foster black-hole formation (Muno et al. 2006). Magnetars may represent the extreme in a multi-parameter continuum, rather than a physically separate population (Viganò et al. 2013). Another puzzle is the origin of gamma-ray bursts (Sect. 17.4.3), whether those are associated with black-hole or neutron-star formation, and to what degree the presupernova evolution is a controlling factor.

8.6 Dynamic Instability

Collapse is initiated when the core of a massive star becomes dynamically unstable. For a Newtonian gravitating object in hydrostatic equilibrium, the density scales as

$$\rho \propto \frac{M}{R^3} \tag{8.3}$$

and the pressure scales as

$$P \propto \frac{GM^2}{R^4}. \tag{8.4}$$

To be neutral to adiabatic compression (or expansion), with pressure neither rising nor falling compared to that required to maintain dynamic equilibrium, pressure must scale as $P \propto \rho^{4/3}$. The condition on the equation of state to yield neutral stability is thus $\Gamma_1 = 4/3$. This critical value is somewhat different (slightly larger) when general relativity rather than Newtonian gravity is employed. The response of P to an increase of ρ via the equation of state of a relativistic gas gives just that behavior, $P \propto \rho^{4/3}$, and so maintains a condition near to that of neutral stability. This applies to white dwarfs near $M_{\rm Ch}$, to radiation-dominated stars, and to the lepton-dominated fluid during core collapse.

For stars in which electron degeneracy makes a significant contribution to the pressure, electron capture contributes to softening the equation of state. With the removal of electrons, the pressure does not rise sufficiently upon a compressive perturbation to maintain hydrostatic equilibrium. For sufficiently strong electron capture, the effective value of Γ_1 is less than 4/3, and collapse ensues. This is the dominant process for degenerate cores of ONeMg for which the neon and magnesium are susceptible to electron capture (Sect. 8.5.3). Electron capture is also a contributing factor to the instability of the cores of somewhat more massive stars that develop iron cores when electron capture occurs on iron-group elements.

For more massive stars, another significant destabilizing effect is *photodisintegration*. As the iron core approaches $T \sim 10^{10}$ K, iron-group elements photodisintegrate into constituent helium nuclei, protons, and neutrons. As this condition is reached, many nuclear reactions enter the state of NSE in which the forward rates are exactly balanced by reverse rates (Hoyle 1946; Clifford and Tayler 1965). As for chemical equilibria, nuclear equilibria can be described by equating the chemical potentials of the reactants and products, and the solution described by the appropriate Saha equation. The resulting abundances do not depend on reaction rates (by assumption), but on T, ρ and Y_ℓ, the ratio of leptons to nucleons. Compression of a fluid in NSE does not lead only to adiabatic heating. Some of the work of compression goes into driving the disintegration and reducing the net nuclear binding energy. The result is, once again, that the pressure does not rise sufficiently under compression to maintain hydrostatic equilibrium, $\Gamma_1 < 4/3$, and dynamic contraction begins. As the ZAMS mass increases, photodisintegration becomes more dominant over electron capture as the cause of iron-core instability. Once the material in an iron core reaches NSE, instability and catastrophic collapse are only moments away.

There is evidence that very massive stars, $M > 100\ M_\odot$, exist in the Galaxy and the LMC and SMC (Panagia et al. 1983; Humphreys and McElroy 1984; Fitzpatrick and Garmany 1990; Hunter et al. 1995; Crowther et al. 2010). Stars of this mass may also predominate among the first stars to form at zero metallicity after the dark ages (Abel et al. 1998, 2000; Bromm et al. 2002; Bromm and Larson 2004). A caveat is that conditions of the first stars may be susceptible to fragmentation so that the first stars may have been of smaller mass than once thought (Stacy et al. 2010; Greif et al. 2011). Stars of $\gtrsim 100\ M_\odot$ are radiation-pressure dominated and nearly neutrally dynamically stable. Models of such stars show a tendency to pulsational instability leading to violent pulsations, but the pulsations are stabilized in the nonlinear pulsational regime and after the main-sequence phase (Appenzeller 1970; Ziebarth 1970; Baraffe et al. 2001).

In late stages of their internal evolution, stars that are sufficiently massive encounter another instability because they reach internal temperatures $\sim 10^9$ K at relatively low densities, conditions under which electron–positron pairs are produced copiously. The corresponding endothermic conversion of internal energy into rest mass associated with pair formation softens the equation of state so an appropriately mass-weighted average value of the adiabatic index $\Gamma_1 \equiv |dlnP/dln\rho|_{ad}$ is less than 4/3, thus destabilizing the stars (Rakavy and Shaviv 1967; Barkat et al. 1967; Fraley 1968). In the ρ—T plane (Fig. 8.2), the region for which pair formation leads to $\Gamma_1 \leq 4/3$ is bounded at low temperature (that depends on the density) where there is insufficient energy to create electron–positron pairs. The region is bounded above at $T \simeq 3 \times 10^9$ K nearly independent of density, because for higher temperatures relatively little energy is lost in the process of converting photons to rest mass. The region is bounded at density above $\sim 6 \times 10^5$ g cm^{-3} as degeneracy limits the phase space for newly-created electrons. The region of creation of electron–positron pairs is also a regime where there is substantial production of neutrinos by pair annihilation. The neutrino losses drive energy loss and contraction of the core more than photon losses from the outer boundary.

In these very massive stars, the products of helium burning tend to be enriched in oxygen because of the enhanced rate of the $^{12}C(\alpha, \gamma)^{16}O$ reaction at higher temperatures. Models of these stars thus tend to form oxygen cores. The pair-formation instability sets in prior to oxygen burning. The ZAMS mass that represents the transition from iron-core stars to pair-instability stars depends on the mass of the core, which is subject to uncertainties of convective overshoot, mass loss, and mixing (Ober et al. 1983; Maeder and Meynet 1989; Heger et al. 2003), but the basic criterion for pair instability is that the oxygen core exceeds about 60 $M_{\odot}$. Nonrotating models of low metallicity evolve to pair-instability for helium-core masses ranging from about 65 to 140 $M_{\odot}$ that correspond, in turn, to ZAMS masses of about 140–260 $M_{\odot}$ (Heger et al. 2003).

In general, lower metallicity suppresses mass loss and allows relatively lower-mass main-sequence stars to form more massive oxygen cores and to encounter the pair-instability regime. For a given ZAMS mass, higher metallicity leads to lower-mass oxygen cores, thus raising the required ZAMS mass to reach pair instability. Single-star models predict that no stars reach the condition of pair instability if the metallicity is greater than about 1/3 solar because of the strong, metallicity-dependent mass loss (Langer et al. 2007). Alternatively, very massive stars may form by mergers in dense stellar clusters more rapidly than winds can deplete their mass (Yungelson et al. 2008; Pan et al. 2012d). Other complications can affect the oxygen core mass and hence the critical mass for pair instability. One example is the potential for mixing of helium down into the oxygen core, which depends on the choice of convective model (Schwarzschild or Ledoux) and may result from multidimensional plume-like flows.

Stars in the mass range predicted to reach pair instability, being close to neutral dynamical instability, are susceptible to quasi-homogeneous evolution such that nearly all the main-sequence mass could be burned to heavier elements. In particular, models that rotate (Sect. 8.7) are subject to mixing that tends to decrease the ZAMS mass required to produce pair instability (Chatzopoulos and Wheeler 2012a; Yoon et al. 2012a). For very low metallicity that mutes mass loss, rotation near breakup can reduce the ZAMS mass that proceeds to pair instability to around 65 $M_{\odot}$ (Fig. 8.7).

The issue of whether either some first stars or contemporary stars undergo pair-instability explosions rather than iron-core collapse remains open. Dynamical models and candidates for pair-instability supernovae will be explored in more detail in Chaps. 10 and 18.

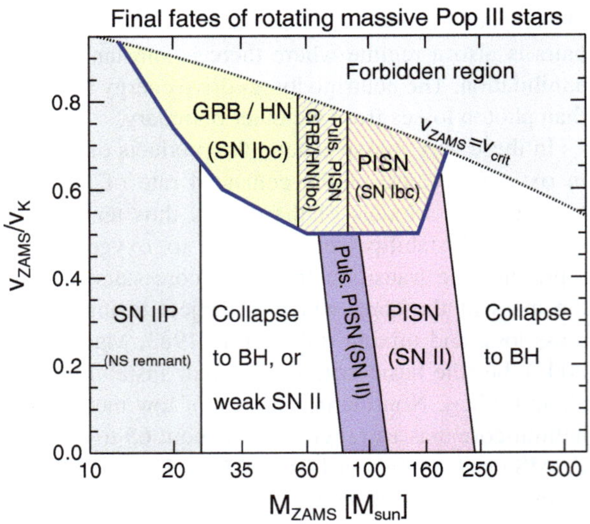

Fig. 8.7 A diagram of the theoretical end points of models of various ZAMS masses of zero metallicity as a function of the initial rate of rotation in units of the surface Keplerian velocity. From "Evolution of massive Population III stars with rotation and magnetic fields" (Yoon et al. 2012a) reproduced with permission, © ESO

8.7 Evolution with Rotation

All stars rotate. Even single massive stars that could lead to core collapse tend to show rapid surface rotation. Binary-star evolution (Sect. 8.8) is likely to be involved in many and maybe even all stripped-envelope supernovae, almost surely for SN Ia, and close binary evolution involves rotation. Stars in close binaries can be brought to modestly rapid rotation by tidal interaction and synchronization. Other evidence for the importance of rotation for supernovae comes from the study of compact objects, white dwarfs and neutron stars, all of which rotate at various speeds. The effect of rotation on the evolution of stars, theoretically with models and observationally with the techniques of asteroseismology, is of intense current interest. The basic physics and implications of rotation and rotationally-induced mixing are summarized in the book by Maeder (2009) and in the review of Langer (2012). Figure 8.7 illustrates how the end point of evolution of very low metallicity stars might depend on the initial rotation.

The contraction of stellar cores associated with different stages of nuclear burning leads to spin-up of the cores and significant differential rotation. On the other hand, there are a variety of dissipative processes that provide torque to slow rotation or otherwise redistribute angular momentum. Studies of evolution with such effects included show that evolution, particularly in the late stages, may be significantly altered (Endal and Sofia 1976, 1978). Rotation can also affect mass loss by amplifying the effects of radiatively-driven winds (Heger et al. 2000; Paxton et al. 2013). Since angular momentum is lost by line-driven stellar winds, mass-loss rates depend on the metallicity (Sect. 8.4.1).

Rotation drives processes that transport both angular momentum and chemical species. Rotation induces the *dynamical shear instability*, the *secular shear instability*, the *Solberg–Hoiland instability*, *Eddington–Sweet* or *meridional* circulation, the *baroclinic instability*, and the *Goldreich–Schubert–Fricke instability* (Heger et al. 2000; Maeder et al. 2013). Rotation, especially shear associated with differential rotation, can drive dynamos that generate magnetic fields by the *Spruit–Tayler mechanism* (Spruit 1999, 2002; Heger et al. 2005), and by the *magnetorotational instability* (Balbus and Hawley 1998; Wheeler et al. 2015a). These magnetic processes can also transport angular momentum and abet chemical mixing. Gradients in mean molecular weight inhibit many or all of these instabilities, thus limiting mixing, but other processes, such as doubly- or triply-diffusive instabilities (Acheson 1978; Menou et al. 2004), can overcome some of this inhibition. Properly incorporating all the associated physics remains a huge challenge.

The efficiency of rotationally-induced mixing can be calibrated from observations of nearby, rapidly rotating massive stars. Rotationally-induced mixing has been invoked to explain the nitrogen enhancement that is observed in some massive main-sequence stars. Mixing dredges up the products of CNO burning that would otherwise remain deep in the stars (Heger et al. 2000; Meynet and Maeder 2000; Maeder 2000), although binary mass transfer may also produce surface enhancements of nitrogen. Rotationally-induced mixing has been invoked to account for the relative frequencies of O-type and Wolf–Rayet stars over a range of metallicities (Meynet and Maeder 2005).

In a sufficiently close binary, tidal forces should lock the secondary star into rotational synchronism with the orbital period, with the resulting rapid rotation inducing rotational mixing. This mechanism is especially efficient for a secondary having a convective envelope. In this context, rotational mixing in high-mass systems has been studied by de Mink et al. (2009) and Brott et al. (2011a).

Above a critical rotation, evolutionary tracks go upwards and bluewards in the H–R diagram, very close to those of fully homogeneous evolution. Maeder (1987) discussed the effects of rotationally-induced mixing on the evolution of massive stars. He concluded that there would be substantial mixing produced by the small-scale 3D tail of the turbulent spectrum of the baroclinic instability, and that the effective diffusion coefficient could be sufficiently large to mix most massive stars during their main-sequence lifetimes. The notion that quasi-homogeneous evolution might lead to larger core masses has been applied in the context of the progenitors of gamma-ray bursts (Heger et al. 2005; Yoon and Langer 2005b) and explored in other work (Heger et al. 2000; Langer et al. 2007; Ekström et al. 2008; Brott et al. 2011a,b). The potential influence on the pair-instability mechanism is discussed in Sect. 8.6.

The rotation of compact objects give some constraints on the rotational evolution of the progenitor stars. There is little evidence that compact objects are born rapidly rotating. On the contrary, white dwarfs seem to rotate surprisingly slowly (Angel 1977; Kawaler 2004; Tayar and Pinsonneault 2013; Brinkworth et al. 2013). While this may reflect internal dissipation of rotation in the progenitor, considerable spin-down might occur during the process of mass ejection (Spruit 1998). Most pulsars rotate far from break-up speeds, although they also could have been spinning more rapidly at birth and slowed down during the explosion or later.

The internal rotational states of stars may be even more complex than currently understood. The simplest expectation is that convection both efficiently homogenizes compositions and alters the rotation such that angular momentum per unit mass is uniform. Coupled with meridional circulation, convection might lead to constant rotation on cylinders. Asteroseismological studies of the Sun have shown, however, that the rotation rate is neither constant on spherical shells nor on cylinders in the convective zone, but nearly on cones of constant latitude (Howe 2009). If such puzzles exist in the best-studied star, then surely others exist in the late stages of supernova progenitor stars that we have yet to pose, much less resolve.

8.8 Binary-Star Evolution

More than half of the stars massive enough to be core-collapse supernova progenitors are in binary systems, and a significant fraction of these are in systems close enough that their evolution is affected by their companions (Sana et al. 2013; Dunstall et al. 2015). Of order 20% even of apparently single stars may have undergone merger (de Mink et al. 2014). Thus, while some massive stars evolve as effectively single stars, the final outcome of a significant fraction are influenced by a close binary companion. Binary evolution is directly implicated or hypothesized in all stripped-envelope supernovae: SN IIb and SN Ib, and SN Ic. Virtually all scenarios for the progenitor evolution of SN Ia invoke binary evolution. The basic processes of binary evolution were reviewed by Iben and Tutukov (1984, 1985) and Bours et al. (2013). The outcome depends on the ZAMS masses and separations of the stars and probably on the metallicity.

A close binary system is one in which the evolution of at least one of the stars is affected by the presence of the companion. Even the theory of single-star evolution has significant uncertainties (e.g., those associated with convection), but these pale in comparison with the uncertainties associated with the evolution of stars in binaries. Binary evolution can involve tidal interaction, external irradiation, mass transfer from one star to the other, mass loss from the system owing to common-envelope episodes or sloppy transfer, and more.

The most important aspect of binary evolution is that mass can be transferred between the two stars. In principle, the combined mass may be conserved, but it is more likely that some mass, in some cases a great deal, is lost from the system. The evolution of such systems is uncertain because it depends in part on poorly understood processes of mass and angular-momentum loss that occur during mass transfer. These effects determine the separation and masses of the stars as they continue to evolve. Fortunately, there are observed systems such as cataclysmic variables and binary X-ray sources that constrain these processes (Warner 1995; Tanaka and Lewin 1995).

Each star in a binary system has its own gravitational domain. For the basic case of circular orbits, these domains are characterized by the *Roche lobes* surrounding the stars. Paczyński (1971) gave a simple analytic fit to the radius of the Roche lobe of a star of mass M_1 that is less massive than its companion, M_2, in a binary system with separation a and total mass $M_T = M_1 + M_2$ as

$$\frac{R_{RL,1}}{a} \approx \left(\frac{M_1}{M_T}\right)^{1/3}. \tag{8.5}$$

A more accurate fitting formula for all mass ratios, $q_1 = M_1/M_2$, was given by Eggleton (1983):

$$\frac{R_{RL,1}}{a} \approx \frac{0.49q_1^{2/3}}{0.6q_1^{2/3} + ln(1 + q_1^{1/3})}. \tag{8.6}$$

In the basic process of mass transfer, one of the stars fills its Roche lobe and transfers matter through the *inner Lagrangian point* into the gravitational domain of the other star. This matter may be directly accreted by the companion, but is more likely to first form an *accretion disk*, especially when the companion is a compact star.

Theoretical studies of binary stellar evolution have served to define several key phases of mass transfer. *Case A* mass transfer occurs during core hydrogen burning as helium accumulates in the center of the more massive star and it expands. If the two stars are so close that the more massive one fills its Roche lobe while still on the main sequence, the system is expected to evolve to a contact system. The outcome is uncertain, but may involve a merger of the two stars into one (Flannery 1976; Webbink 1976). *Case B* transfer occurs during helium-core contraction or at the beginning of core helium burning when models predict large envelope expansion and the star becomes a red giant. *Case C* mass transfer occurs during core helium burning or later stages when models predict RSGs of very large radius. At the wider separations involved in Case B and Case C, the primary, depending on its mass, leaves a degenerate stable core or a more massive, evolving core. The degenerate core could be composed of helium if the initial separation were close enough that transfer occurred during the hydrogen shell-burning phase as the star expanded toward the giant branch. At still wider separations, mass loss may be postponed until a degenerate C/O or ONeMg core is formed. In either case, the stripped degenerate cores cool to become stable white dwarfs, inert until the companion evolves. More massive nondegenerate cores continue evolving to catastrophe.

Mass and angular momentum can be lost from a binary system by winds from either star and by shedding matter into orbits beyond the Roche lobes of the stars where matter can orbit both stars but still be bound to the system. Some processes of mass and angular momentum loss are intrinsic to the mass-transfer process itself, particularly when mass is transferred at a rate so high that it cannot be assimilated by the mass-gaining star.

Many observed binary systems with compact companions are found in orbits that are so small that the stars are closer together than the expected size of the main-sequence progenitor of the compact component. This condition is thought to arise when binary stars form a common envelope (Ivanova et al. 2013). For rapid mass transfer, the energy of accretion cannot be radiated sufficiently rapidly and a giant-type envelope surrounds the accreting star and engulfs the mass-losing star. The common envelope is in hydrostatic and perhaps also thermal equilibrium. A common envelope typically forms when the mass loss is unstable and occurs on a dynamical timescale. The accreting star and the core of the mass-losing star spiral together within the common envelope and then, presumably, eject the envelope as their orbital kinetic energy grows and more thermal energy is deposited into the envelope.

The criterion for envelope ejection is frequently characterized by the binding energy of the common-envelope system expressed as

$$\frac{GM_dM_e}{\lambda R} = \alpha \left(\frac{GM_cM_2}{2a_f} - \frac{GM_dM_2}{2a_i} \right), \tag{8.7}$$

where M_d is the total mass of the donor star, M_e is the mass of its envelope, M_c is the mass of its core, and M_2 is the mass of the secondary star that is assumed to be constant. The dimensions a_i and a_f are the initial separation of the donor and secondary and the final separation of the core and secondary, respectively, and R is the radius of the donor star before mass transfer. The parameter λ is a structural parameter that determines the binding energy of the original envelope of the donor, and α is a parameter that sets the scale of the binding energy after the formation of the common envelope. For given parameters, this formula gives the final separation, a_i, as a function of the nature of the two stars in the binary. In this formulation, the problem can be parameterized with the product $\alpha\lambda$. For appropriate values of this parameter, envelope expulsion can be delayed until the donor core and the secondary are very close together.

An alternative prescription for loss of the common envelope is cast in terms of the angular momentum of the system

$$\frac{J_f - J_i}{J_i} = \gamma \frac{M_{1,f} - M_{1,i}}{M_{T,i}}, \tag{8.8}$$

where J_i and J_f are the initial and final angular momenta of the binary system, $M_{1,i}$ and $M_{1,f}$ the initial and final masses of the donor star, $M_{T,i}$ the initial total mass, and γ is an efficiency parameter (Nelemans et al. 2000). For reasons that are not completely clear, this prescription for the final separation seems to better account for systems containing two white dwarfs than the energy prescription.

The common-envelope process can thus lead to evolved cores and unevolved stars in close proximity. This process is thought to underlie the formation of cataclysmic variables and binary X-ray sources. Most single-degenerate models of SN Ia invoke the common-envelope process as a key step in the progenitor evolution.

The conditions for which double stars survive the common-envelope process versus when the donor core and the secondary merge is not clear (Deloye and Taam 2010; Taam and Ricker 2010). The evidence for strong breaking of spherical symmetry in favor of axial symmetry suggests that merging might have been the fate of the progenitor of SN 1987A, although there is no direct evidence for past duplicity.

If the two components survive, the subsequent evolution when the original secondary begins to lose mass is also rather uncertain, but is, if anything, more likely to involve a common envelope. In the second stage of mass transfer, the star receiving mass is compact so the accretion energy is high. The resulting thermal and radiation pressure tend to prevent matter from settling onto the compact star. A second phase of common-envelope evolution can lead to two compact objects orbiting one another. This evolution is invoked in the double-degenerate scenario of SN Ia progenitor evolution (Sect. 21.3).

Whereas common envelopes are proposed to be in hydrostatic equilibrium, the mass transfer, especially in the second stage of transfer onto a compact object, can lead to sufficient radiation pressure that hydrostatic equilibrium is not possible. This occurs when the luminosity from accretion exceeds the *Eddington limit*. For the basic case when the opacity to the radiation is due to electron scattering, the limiting luminosity can be expressed as

$$L_{Edd} = \frac{4\pi G c m_p}{\sigma_{es}} M_1 \simeq 3.3 \times 10^4 \left(\frac{M_1}{M_\odot}\right) L_\odot, \tag{8.9}$$

where m_p is the mass of a proton, σ_{es} is the cross section for electron scattering, and M_1 is the mass of the accreting star. If the accretion rate is too high, then the accreted matter cannot be in hydrostatic equilibrium and must be expelled from the system. This condition thus also gives rise to a limiting mass accretion rate

$$\dot{M}_{Edd} = \frac{4\pi c m_p}{\sigma_{es}} R_1 \simeq 1.5 \times 10^{-8} \left(\frac{R_1}{10\ \mathrm{km}}\right) M_\odot\ \mathrm{yr}^{-1}, \tag{8.10}$$

where R_1 is the radius of the accreting star, here normalized to be typical of a neutron star. While the Eddington limit is most frequently discussed in the context of binary evolution, similar situations can arise within the depths of massive stars where the opacity may exceed that of electron scattering and radiation pressure becomes large. This may lead to nonmonotonic density structures and even to instabilities (Sect. 8.4.2).

If the original primary is massive enough, it leaves a collapsed remnant, a neutron star or black hole. A spherically-symmetric impulsive explosion in a binary system that conserves linear and angular momentum disrupts the binary system if more than half of the pre-explosion mass of the system is ejected. In reality, the criterion to disrupt the binary is likely to be more complicated because the explosion may impose an impulse on any collapsed remnant of the explosion (Sutantyo 1978). In the case of neutron-star or black-hole formation, even if explosive mass ejection occurred, disruption of the binary system is inhibited if the progenitor had lost its hydrogen envelope, so that less than half of the pre-explosion mass of the system is ejected. Such a system probably becomes a hard X-ray source. SN Ib/c may be a crucial link in this connection. In some circumstances, both stars may produce neutron stars that are retained in a close binary. Merging of two neutron stars has been implicated in short, hard gamma-ray bursts and is discussed in the context of the production of r-process nuclides (Sect. 24.5).

8.9 Summary

The range of uncertainty in the mass below which white dwarfs are formed, $M_{WD,max}$, is similar to the ranges of uncertainty of the upper limit for degenerate central carbon ignition, $M_{CO,max}$, and the upper limit for degenerate ONeMg cores, $M_{ONeMg,max}$. Cluster and IFMR studies suggest that single stars with mass below about 6–8 $M_\odot$ lose their envelopes and become white dwarfs. Ambiguities remain because the white dwarf mass distribution does not specifically constrain the inner composition of DA white dwarfs. The highest mass component in the distribution of Kepler et al. (2007) may be ONeMg white dwarfs. Similar uncertainties apply to the IFMR. It may extend to high core masses, but are those cores C/O or ONeMg?

It is thus not clear either observationally or theoretically just where the upper limit lies for the production of white dwarfs, the value of the mass for the transition between stars that proceed to degenerate carbon ignition and those that do not, those that are subject to collapse by electron capture in degenerate cores of ONeMg, and those that proceed to iron-core collapse. There is still substantial uncertainty as to whether any single stars manage to retain their envelopes and proceed to degenerate carbon ignition or to electron-capture supernovae. Nor is there any convincing evidence that any observed class of supernova arises from degenerate carbon ignition or electron capture in single stars.

Studies of the progenitors of SN IIP show that they are of relatively low ZAMS mass, about 8–16 $M_\odot$, but do not undergo the second dredge-up and proceed directly to iron-core collapse (Chap. 12).

If any stars do evolve to degenerate carbon ignition as either single or binary stars, there are then a host of interesting problems associated with the physics of the thermal runaway. Among these are the convective Urca process and the question of whether the runaway leads to a subsonic deflagration or a supersonic detonation. These topics are explored in Chap. 19.

Single stars that form iron cores are doomed to collapse. SN 1987A agreed remarkably well with basic predictions of iron-core collapse in a star of ZAMS mass about 20 $M_\odot$. The circumstances under which iron-core collapse results in an explosion leaving behind a neutron star or a black hole with or without an explosion is still a matter of discussion.

Much of the literature of stellar evolution that has been summarized here is in the context of nonrotating, nonmagnetic models. Although progress is being made in understanding these complexities, much remains to be learned in this regard.

A long-standing challenge has been to relate the interior evolution to the external properties, especially surface abundances, or to the properties revealed after the star explodes. Recent advances in asteroseismology of stars over a large range of masses and evolutionary states promises a revolution in our ability to peer within stars as they evolve. Given our relative ignorance of the related effects of rotation and magnetic fields on stellar evolution, we may be in for some surprises. We are struggling to understand the Sun as revealed by detailed asteroseismology, and there is every reason to think that these problems will get worse in more complex, advanced stages of evolution where composition layering, differential rotation, and mixing are important. We have a good first order understanding of stellar evolution, but being able to peer within stars is likely to shake the mental image of spherically-symmetric stellar structure and evolution that we inherited from Fred Hoyle, Martin Schwarzschild, and others.

Part II
Massive-Star Supernovae

Chapter 9
Core Collapse

9.1 Introduction

The problem of core collapse is exceedingly complex. Few subjects cover such a broad range of physics in such a tightly coupled way: hydrodynamics and turbulence, magnetic fields and rotation, weak and strong particle interactions, neutrino transport, hot and catalyzed matter at nuclear densities and above, condensed matter and many-body physics, and the potential, at least, to produce exotic "strange" particles, never mind black holes. Recent work even suggests that collapse could radiate gravitational radiation into macroscopic higher dimensions. A thorough coverage of the vast literature on this topic is out of the question. Instead, we will present some of the basic physics as it pertains directly to the issue of supernova explosions.

As discussed in Chap. 8, the cores of massive stars collapse for a variety of reasons. At extreme densities and in the absence of substantial electron captures, instability induced by general relativistic effects increases the critical adiabatic index for stability. Theory suggests that electron capture dominates for the cores of stars of ZAMS mass about 8–13 $M_{\odot}$ that form degenerate cores of ONeMg, and by white dwarfs of this composition that accrete (Sects. 8.6 and 9.4.2). Such a fate may also meet C/O white dwarfs that have cooled substantially before accreting (Sects. 9.4.2 and 21.3.1).

For more massive stars $\gtrsim$12 $M_{\odot}$, which are expected to form iron cores (but see Sect. 8.7 for uncertainties), instability is initiated by a combination of photodisintegration of iron and electron capture in lower-mass semidegenerate cores and by photodisintegration in higher-mass nondegenerate cores (Fig. 8.2; Sects. 8.5.5 and 9.4). For some of the most massive stellar models, instability to electron–positron pair formation decreases the effective adiabatic index below the neutral stability limit of $\Gamma_1 = 4/3$ (Sect. 8.6) and may lead to collapse in some circumstances (and explosions with no compact remnant in others; Chap. 10).

D. Branch, J.C. Wheeler, *Supernova Explosions*, Astronomy and Astrophysics Library, DOI 10.1007/978-3-662-55054-0_9

The bottom line is that, with the exception of the possibility that isolated ONeMg white dwarfs may be stable, the great majority of stars with mass in excess of $\sim$8 $M_\odot$ are expected to collapse. Details will vary with initial ZAMS mass and perhaps initial composition and rotational state, but the outcome that a compact object, a neutron star or a black hole, must form is unavoidable. Section 9.7 will briefly discuss some even more arcane possibilities.

The discovery of pulsars showed that collapse involves rotation and magnetic fields. After an early spate of work involving these factors, they have been substantially neglected until recently. The reasons have been entirely rational and pragmatic. The 1D collapse problem, especially the coupling of dynamics with neutrino transport and the nuclear equation of state, has proved so difficult that it is prudent to "solve" that problem before including the complexities of rotation and magnetic fields.

Supernova 1987A (Chap. 11) showed that 1D models were inadequate to account for the mixing of elements revealed by observations. Cas A (Sect. 7.4.1) is rent with patterns of nonspherical flow. Other evidence has accumulated, including polarization data (Sect. 4.6), that collapse is exceedingly asymmetric and frequently bipolar, although substantial deviations from axial symmetry are also seen. It may be that collapse is intrinsically unstable to nonspherical perturbations related to the standing accretion shock instability or other effects (Sect. 9.4.7), and that motion in 3D amplifies and is critical to the explosion by neutrino processes. It is still far from clear which of these processes can account for the observed departures from spherical symmetry. Rotation and magnetic fields may yet prove to be important in creating or abetting the explosion (Sect. 9.5). Even given all these considerations, the standard physics of (nearly) spherical collapse is still critical to determine the environment in which collapse and explosion occur.

9.2 History of the Collapse Problem

A brief history is appropriate to illustrate the effort and ingenuity that has gone into the collapse problem. The notion that collapse to form a neutron star with a resultant release of binding energy to power a supernova explosion came only a few years after the discovery of the neutron (Baade and Zwicky 1934). Early discussions proposed the iron–helium phase transition as the triggering mechanism for iron-core collapse (Hoyle 1946) and raised the possibility that the collapse of an iron core would allow the outer layers of fuel to contract, heat, and explode (Hoyle and Fowler 1960). This thermonuclear process does not work because the burning blows inward, not outward, due to the low pressure induced by the collapsing inner core (Colgate and White 1966). The collapse does release a huge binding energy, $\sim$300 B, and most of that is carried off by neutrinos that could be the agent of explosion (Colgate and White 1966). Colgate and White made the simplifying assumption that a constant fraction of the outgoing neutrino flux would be deposited in proportion to mass in the outer envelope. This is regarded as too simple and optimistic by current standards, but it established a critical proof of principle that the role of the neutrinos must be considered in any model of collapse-induced explosion. The stage was thus set for nearly 50 years of effort to improve the physics of core collapse and neutrino transport. We have not yet seen the end of the game.

In early work on collapse, neutrinos were treated as free streaming. The result was a seeming, but erroneous, self-consistency. With the free-streaming assumption, the neutrino density was always low so that issues of Fermi statistics of neutrinos seemed unimportant. Driving this assumption was the knowledge that neutrino cross sections tend to scale as $\sigma \propto E_{\nu}^2$. This opens a low energy "window," and it was argued that neutrinos can leak out of this window and not fill a Fermi sea. The argument was analogous to looking at an old rusty bucket with holes in the bottom. It is intuitively clear that the longterm solution is an empty bucket. It is also clear, however, that a fireman with a hose can fill the bucket; this is a matter of the rate of input compared to the rate of leakage. Collapse provides a fire hose of neutrinos. Closer inspection showed that the free-streaming approximation implicitly allowed the occasional occupation of a single cell of phase space by more than one neutrino of the same spin.

Proper treatment of the neutrino statistics revealed that neutrinos are trapped because of phase-space blocking, and that at densities above $\sim 10^{11}\,\mathrm{g\,cm^{-3}}$ the free-streaming approximation breaks down completely (Mazurek 1974; Sato 1975). Several years were required, roughly 1975–1978, before this suggestion, regarded as radical at the time, became accepted wisdom. Careful studies of the neutrino and energy transport showed that this result held even in the case where statistical equilibrium was not assumed a priori in Boltzmann transport calculations with detailed cross sections. For most purposes, one can regard the neutrinos to be in thermal equilibrium and use Fermi–Dirac statistics and the concept of a neutrino chemical potential, μ_{ν}, in the bulk of the high density, infalling matter. The result was gross changes in the systematics of neutrino transport and hence the perceived nature of the collapse.

It is instructive to contrast the physics of collapse with and without neutrino trapping. Without trapping, the iron core would be rendered unstable by the iron–helium phase transition and electron capture (Sect. 9.4) at a temperature of about 10^{10} K. The adiabatic index, $\Gamma_1 \equiv (dlnP/dln\rho)_{ad}$, would at first be less than the nonrelativistic stability limit, $\Gamma_1 \lesssim 4/3$, and both density and temperature would rise during collapse. There would be rapid electron capture, so nuclei would become very neutron rich and quickly dissolve into free nucleons. The electron density would rapidly decline and the neutron density would go up. Nonrelativistic, nondegenerate neutrons would then dominate the subsequent collapse. From this point on, the collapse would proceed with $\Gamma_1 \simeq 5/3$ and $T \propto \rho^{2/3}$. The neutrons would provide higher pressure at a given density than with neutrino trapping so collapse would be halted at relatively low density, $\rho \simeq 4 \times 10^{13}\,\mathrm{g\,cm^{-3}}$, about 10% of nuclear density, $\rho_{nuc} \simeq 3 \times 10^{14}\,\mathrm{g\,cm^{-3}}$ (Bruenn 1985), and less binding energy would be available.

In contrast, if neutrinos are not free to escape their phase space is filled and electron capture is suppressed, e.g., in the reaction $p + e \rightarrow n + \nu_e$ there is no place to put the ν_e. Trapping occurs where the neutrino diffusion time is comparable to the dynamical collapse time, which occurs at a density of $\sim 10^{12}\,\mathrm{g\,cm^{-3}}$ (Sect. 9.4). At higher density, neutronization and neutron drip are suppressed and there is relatively little dissolution of nuclei, with their retention aided by the accessible degrees of freedom in the excited states. With $P_{ion} = \mathrm{n}_A \mathrm{k}T < An_n\mathrm{k}T$ (n_A is the Avogadro number, A is atomic weight, and n_n is the number density of neutrons), there are fewer particles and hence a lower ion pressure. The number of electrons and neutrinos is approximately constant within the trapping radius so the collapse is dominated by relativistic particles, for which the adiabatic index is $\Gamma_1 = 4/3$, just the neutral stability value, and $T \propto \rho^{1/3}$. Collapse proceeds in nearly neutral dynamical stability until nuclei interact and stiffen the equation of state, bringing collapse to a halt at about twice the nuclear density. The core then bounces and drives a strong shock back out into the infalling matter.

Around 1973, the theory of weak neutral currents incorporating the new Weinberg–Salaam–Glashow theory of weak interactions was incorporated in collapse models (Freedman 1974; Dicus et al. 1976). This major change in physics implied coherent scattering of neutrinos on nucleons, with the implication that the cross section would scale with atomic weight as $\sigma_{coh} \propto |A\Psi|^2$ rather than $A|\Psi|^2$, as would be the case for incoherent scattering from individual nucleons, where A is the atomic weight and $|\Psi|^2$ is the probability of interaction. With neutral-current interactions, $\sigma_{coh} \propto A^2\sigma_n$, where σ_n is the cross section for single-nucleon scattering. Since the atomic weight of nuclei is large, a much larger interaction between neutrinos and matter was predicted, raising hopes that a larger energy deposition would lead to a larger explosion energy. These hopes were undercut by the intrinsic nonlinearity and feedback of the collapse problem. Over the next couple of years, detailed calculations showed that if the neutrinos interact with the infalling flow of matter, the infall is cushioned. It is that infall, however, that delivers the binding energy to generate the neutrinos in the first place. The net effect was that the higher cross sections reduced the rate of neutrino production. The results again teetered on the edge of success, but models tended to predict complete collapse to black holes.

The 5 years from about 1978 to 1983 were the era of nuclear physics. This effort was triggered by the realization that neutrinos are trapped in the collapse. The trapping of neutrinos blocks excessive neutronization and hence *neutron drip* and the production of abundant free neutrons. Consequently, nuclei remain intact to high densities. Previous equations of state at nuclear densities were calculated in the cold, catalyzed limit corresponding to $T = 0$. With neutrino trapping and coherent neutrino scattering, the collapse is nearly adiabatic. The initial entropy is rather low, but finite (Sect. 9.3.1). The result is that matter at nuclear density is "hot." The implication of neutrino trapping was that the equation of state must be determined self-consistently with the collapse dynamics and with the neutrino generation and transport. In addition, the nature of weak interactions had to be reconsidered, allowing for electron capture on excited nuclear states, blocking at high neutron enrichment, and neutrino phase-space limitations.

In the mid 1980s there was considerable work with the new nuclear equations of state (Bethe et al. 1979; Burrows and Lattimer 1984) and some optimism that with this major change in understanding a solution was near. An important aspect was that with dynamical collapse to nuclear density and beyond, the proto-neutron star (PNS) overcompressed and underwent a violent rebound on the dynamical timescale of milliseconds that drove a strong shock wave back out into the infalling matter. This led to intense exploration of the *bounce shock* or *prompt* mechanism, investigating the possibility that this shock could proceed out of the core with the attendant violent ejection of outer material. There were hints of success for the prompt mechanism but, as for the inclusion of coherent scattering, feedback proved to be a critical limiting factor. The bounce shock must move outward against the ram pressure of infalling matter, and it loses energy to dissociation of iron. The shock thus stalls at a nearly fixed radius, $\sim 10^8$ cm, becoming, at least temporarily, a standing accretion shock well outside the PNS, but still well inside the collapsing core. One of the few points of agreement in this complex topic is that core collapse does generate an initially-strong bounce shock, but also that the bounce shock alone stalls and cannot create a successful explosion. The stage was thus set for the long, continuing search for alternatives. Principal among these has been the *delayed* neutrino-driven mechanism (Bethe and Wilson 1985). The key idea in this class of models is that subsequent neutrino loss from the PNS could reinvigorate the standing shock in about 1 s.

The era since SN 1987A (Chap. 11) has been dominated by efforts to improve the techniques of neutrino transport and by considerations of multidimensional effects in the context of the delayed mechanism. SN 1987A showed unanticipated outward mixing of matter rich in ^{56}Ni that spherical models could not accommodate. The seeds for some of the relevant physics had been planted earlier with the recognition that leakage of neutrinos from the newly-formed neutron star would cause instability to convection (Epstein 1979). This led to detailed multidimensional models. The hope was that convective overturn would enhance the neutrino transport and deposition and revive the stalled shock, the essence of the delayed mechanism. Variations on this theme involved currents of cold matter sinking from beneath the shock to be heated and then rising, carrying their excess pressure to enhance the shock (Herant et al. 1994). These delayed-mechanism models have also failed to provide unambiguous success, but work in this context continues apace. We have learned that the core-collapse ambiance tends to adjust to changes in microphysics through feedback on the equation of state, the weak interactions, the neutrino transport and the hydrodynamics so that the result remains just short of a robust explosion. The situation is somewhat like a pot on the stove; the lid rattles, releasing steam, but the lid does not get blasted up to the kitchen ceiling. This is known in some quarters as *Mazurek's law* (Lattimer and Prakash 2000).

Simulations of both collapse dynamics and neutrino transport have become more accurate with computational advances over the last two decades. There is hope that accurate computations in 3D, including multidimensional neutrino transport, will produce robust explosions. Recognition of the standing accretion shock instability (SASI; Sect. 9.4.7) showed that gross asymmetries may arise deep in the collapse structure. Polarization observations and the recognition that long gamma-ray bursts arise in massive stars from which the energy is highly collimated stimulated the idea that rotation and perhaps magnetic effects may play a role even in typical core-collapse events. Core collapse remains a daunting problem.

9.3 Collapse Physics

9.3.1 Entropy

The first stages of collapse are caused in part by photodissociation associated with *nuclear statistical equilibrium* (NSE; Sect. 9.4) and are governed by two key parameters, the entropy per baryon in units of the Boltzmann constant, s/k, and the electron number per baryon, Y_e, at the onset of collapse. These parameters couple neutrino transport, dynamics, and the equation of state (Bethe et al. 1979; Shapiro and Teukolsky 1986; their Chap. 18). Strong and electromagnetic interactions are nearly in equilibrium so changes in entropy are generated only by weak interactions. The initial value and evolution of Y_e are also critical in this context. The net entropy per baryon at the start of collapse in the electrons and nucleons is $s/\mathrm{k} \simeq 0.9$. For an iron composition, the initial value of Y_e is 0.42.

The change of entropy, s, can be written as (Shapiro and Teukolsky 1986; their Equation 18.6.15)

$$T\dot{s} = -(\mu_e + \mu_p - \mu_n - \mu_\nu)\dot{Y}_e + (\mu_\nu - < E_\nu >_{esc})\frac{Y_\nu}{\tau_{esc}}, \tag{9.1}$$

where the μ terms are the chemical potentials of the electrons, protons, neutrons, and neutrinos, respectively, Y_ν is the neutrino number per baryon, $< E_\nu >_{esc}$ is the mean energy of the escaping neutrinos, and τ_{esc} is the characteristic escape time for the neutrinos, a diffusion time in one limit and the free-streaming time, R/c, in another. The first term in Eq. (9.1) represents the entropy change due to departures from β equilibrium and the second due to loss of neutrinos. It is important to examine the difference in the chemical potentials of the electrons and the neutrinos, $\mu_e - \mu_\nu$. If the neutrino phase space fills up so that μ_ν increases and $\mu_e - \mu_\nu$ decreases, there is less energy to drive electron capture. The complementary quantity that arises in Eq. (9.1) is $\mu_n - \mu_p \equiv \hat{\mu}$. To remove protons and create neutrons, an amount of energy $-\hat{\mu}$ must be available to drive the electron capture, and the energy required becomes larger as the nuclei become more neutron rich. The electron capture must also overcome the excitation of the daughter nucleus, e_{ex}.

Detailed calculations of collapse account for electron capture on heavy nuclei and the effects of filled nuclear shells, which block electron capture on nuclei. Most of the captures occur just before neutrino trapping when phase space limitations restrict the electron-capture rate. For the initial value of $Y_e \simeq 0.42$, the final value at neutrino trapping is $Y_e \simeq 0.35$. The electron fraction in the center decreases to $Y_e \simeq 0.27$ (Janka 2012; Sect. 9.4.1).

Prior to collapse, the quasistatic evolution occurs on a timescale that is long compared to the weak-interaction timescale and excited states are unimportant, so $\mu_\nu = 0$. Statistical equilibrium in this situation can be expressed as $\mu_e = \hat{\mu}$. Between the onset of dynamical collapse and neutrino trapping, the electron chemical potential rises on the dynamical timescale as the density increases. The quantity $\hat{\mu}$ goes up only on the weak interaction timescale. As the rate of electron capture increases, more neutrons are created and $\mu_e - \hat{\mu} > 0$. The quantity $\mu_e - \hat{\mu}$ has been termed the "out-of-whackness" that determines the rate of electron capture and hence the rate of change of entropy. For the expected change in Y_e in conditions with few free protons and capture primarily on nuclei, the sum of the chemical potentials, which would be exactly zero in total β equilibrium, must be positive and large enough to overcome the excitation energy of the daughter nucleus. Since the weak interactions are out of equilibrium, electron captures increase the entropy. Neutrino losses, on the other hand, decrease the entropy. The net change in entropy during collapse results from the complex interplay of electron capture and neutrino losses.

Numerical calculations show that electron capture onto free protons dominates the change in Y_e during the early collapse. Considering only the capture on free protons, $\hat{\mu}$ is small and the energy lost in neutrinos, $\simeq \mu_e$, is large so the entropy decreases. In contrast, for capture on nuclei, $\hat{\mu}$ is large and the energy lost to neutrinos, $\simeq \mu_e - \hat{\mu} - e_{ex}$, is small, so the entropy increases. The low initial entropy promotes the retention of nuclei; electron capture is suppressed, but the entropy rises. As s rises, electron capture proceeds more rapidly on free protons, but the entropy declines. These effects tend to offset one another, and the net effect is only a mild increase in entropy before neutrino trapping, to $s/k \simeq 1.0$–1.5 (Bethe et al. 1979). Collapse up to neutrino trapping is thus nearly adiabatic.

Departures from adiabaticity depend on details of the treatment of the equation of state, the internal degrees of freedom of nuclei, and the neutrino transport. After neutrino trapping, weak interactions come nearly into equilibrium, expressed as $\mu_e - \mu_\nu = \mu_p - \mu_n = -\hat{\mu}$. Electron capture and neutronization continue, but the neutrinos are trapped and the number of leptons per baryon, $Y_l = Y_e + Y_\nu \simeq const.$

The weak-interaction equilibrium would be complete were it not for a window at small neutrino energy where the neutrino scattering and absorption cross sections are low, which allows neutrino leakage. The blockage of neutrinos also means that they cannot carry off substantial energy. After neutrino trapping, collapse is nearly exactly adiabatic until the equation of state stiffens at and above nuclear density, bounce occurs, and a bounce shock forms.

9.3.2 The Equation of State

The tendency for s/k and Y_e to remain constant during collapse means that to a good approximation the equation of state can be parameterized by s/k and Y_e and computed independently of the dynamics during collapse. After core bounce, $s/\mathrm{k} \gg 1$. Then the equation of state can be coupled to the dynamics in a numerical calculation to compute the changes in s/k and Y_e and determine whether an explosion occurs. The equation of state must treat the matter as a mixture of heavy nuclei with their own internal degrees of freedom, α particles, free neutrons and protons, and the lepton component: electrons and positrons, as well as the three flavors of neutrinos and antineutrinos (at least; some treatments also consider "sterile" neutrinos). The central temperature at nuclear density corresponds to about 10 MeV. At these energies, excited states of nuclei are relevant but substantially less than the nucleon Fermi energy ($\sim$40 MeV).

Proper treatment of the nuclei must include the internal shell structure, surface energy, the spin-dependent nucleon–nucleon potential (applicable in the nonrelativistic limit), and many-body interactions. A simple model of nucleon interactions in a nucleus would have the size of a nucleus be of order the range of the nucleon–nucleon potential, so that nuclear density would increase with A and, if two-body interactions dominate, the binding energy would be proportional to A^2. Experiments show, however, that the size of nuclei scales as $\sim A^{1/3}$, i.e., the nuclear density is approximately constant and the binding energy scales approximately as A. This property of nuclear matter, known as *saturation*, must be reflected in the properties of the nuclear potential. Saturation means that there must be a component of the nuclear force that is attractive for few aggregated nucleons but becomes repulsive for many. Saturation results from the Pauli exclusion principle for nucleons, *exchange forces* involved by swapping spin or isospin between two nucleons, and the repulsive core of the nuclear potential. The nuclear force does not obey a superposition principle so the total interaction in a many-body system of nucleons does not reduce to the sum of two-particle interactions.

Early treatments of the equation of state of hot nuclear matter characterized nuclei in terms of an equivalent single species of atomic number Z and atomic weight A and used a compressible liquid-drop model (Lamb et al. 1978, 1981; Bethe et al. 1979; Epstein and Pethick 1981). The composition was determined by assuming equilibrium among the free nucleons, nuclei, and leptons, and hence that the nucleon chemical potentials are the same for bound and free nucleons, subject to charge neutrality and baryon conservation. An analysis of the equation of state at zero temperature was given by Shapiro and Teukolsky (1986; their Chap. 8). The finite temperature equation of state in wide use today is that of Lattimer and Swesty (1991), which is based on a *Skyrme* nucleon–nucleon potential. Alternatives have been given by Shen et al. (2010a) and Mathews et al. (2013). The effect of these alternatives on core collapse have been compared by Couch (2013a).

Since neutrino trapping blocks phase space and inhibits electron capture, the electron and proton numbers remain nonzero during collapse. At the modest entropy that characterizes collapse, it remains energetically favorable for protons to remain bound in nuclei. With $Z/A \lesssim 0.35$, the resulting nuclei must be large and neutron rich to allow high proton number until the nuclei dissolve above nuclear density. Atomic weights of $A \lesssim 1000$ are possible. The systematics of neutron drip are different from the case of cold, catalyzed matter. A "vapor" of free nucleons forms outside the massive nuclei. At $\rho \simeq \rho_{nuc}/2 \simeq 1.5 \times 10^{14}\,\mathrm{g\,cm^{-3}}$, it becomes energetically favorable to form various geometries of vapor, and there is a phase transition from nuclei surrounded by a vapor of free nucleons to a sea of nuclei surrounding bubbles of free nucleon vapor. These geometries can involve spherical bubbles, rods, and layers variously known as "meatballs," "spaghetti," "lasagna," or generally the "pasta" phase of nuclear matter (Pethick and Ravenhall 1995).

9.3.3 *Weak Interactions*

The theory of weak interactions as they apply to cooling neutron stars was given by Shapiro and Teukolsky (1986; their Chap. 11); much of this theory applies in general to astrophysical weak interactions. The basic theory of electroweak interactions is as described by Weinberg, Salaam, and Glashow. In this theory, the electromagnetic force is mediated by chargeless, massless, vector bosons (photons) and the weak force by two charged massive vector bosons, W^+ and W^-, as well as the neutral intermediate vector boson, Z^0. The Universal Fermi coupling constant G_F, which expresses the strength of weak interactions, is given by the parameters of the WSG theory as determined from experiment. The existence of reactions mediated by the neutral Z^0 allow neutral current reactions that substantially alter the physics of the collapse problem. Charged currents allow reactions such as

$$n \rightarrow p + e + \bar{\nu_e} \tag{9.2}$$

and

$$p + e \rightarrow n + \nu_e, \tag{9.3}$$

the basic reactions of β decay and electron capture, and their analogs in nuclei. Neutral currents allow neutrino-nucleon scattering processes such as

$$\nu + n \rightarrow \nu + n \tag{9.4}$$

and

$$\nu + p \rightarrow \nu + p, \tag{9.5}$$

where ν can be any flavor of neutrino. Neutrino-lepton scattering reactions are mediated by both charged and neutral currents to allow reactions such as

$$\nu_e + e \rightarrow \nu_e + e \tag{9.6}$$

and

$$\nu_\mu + \mu^- \rightarrow \nu_\mu + \mu^-. \tag{9.7}$$

At low energies, typical cross sections scale as $G_F^2 E^2$ where E is the center of mass energy. In particular, neutrino cross sections typically scale as E_ν^2.

Energy released in typical β-decay reactions is small compared to the rest-mass energy of nucleons, so perturbation theory can be used to compute the decay rates in accord with *Fermi's golden rule*. Allowed transitions carry off no orbital angular momentum. The spin component of the combined lepton wave function in the exit channel can have either a total spin of zero, a singlet state, or of unity, a triplet state. In the singlet lepton state, the nucleon does not change either its spin or total angular momentum (e.g., as $n \rightarrow p$). These reactions are governed by the vector part of the weak interaction. For triplet transitions the nucleon wave function must undergo a spin flip. The total angular momentum can then be conserved if $\Delta J = \pm 1$ or 0, where the former is the *Gamow–Teller selection rule* and the latter the *Fermi selection rule*. The triplet reactions involving a spin flip are governed by the axial vector part of the weak interactions. Relevant rates for weak interactions were presented by Fuller et al. (1982) and associated references and updated by Thielemann et al. (2003).

Neutrino-interaction processes relevant to core collapse were summarized by Janka (2012):

- electron neutrino absorption on neutrons and electron capture ($\nu_e + n \longleftrightarrow e^- + p$)
- electron antineutrino absorption on protons ($\bar{\nu_e} + p \longleftrightarrow e^+ + n$)
- electron neutrino absorption on nuclei ($\nu_e + A(Z,N) \longleftrightarrow A(Z+1, N-1) + e^-$)
- ν_e, ν_μ, and ν_τ scattering on nuclei ($\nu + A(Z,N) \longleftrightarrow \nu' + A(Z,N)$)
- ν_e, ν_μ, and ν_τ scattering on neutrons ($\nu + n \longleftrightarrow \nu' + n$)
- ν_e, ν_μ, and ν_τ scattering on protons ($\nu + p \longleftrightarrow \nu' + p$)
- ν_e, ν_μ, and ν_τ scattering on electrons and positrons ($\nu + e^\pm \longleftrightarrow \nu' + e^\pm$)
- ν_e, ν_μ, and ν_τ neutrino production and absorption from pair production ($e^+ + e^- \longleftrightarrow \nu + \bar{\nu}$)
- nucleon–nucleon Bremsstrahlung ($\nu + \bar{\nu} + (n \text{ or } p) + (n \text{ or } p) \longleftrightarrow (n \text{ or } p) + (n \text{ or } p)$)
- neutrino–neutrino scattering ($\nu_{\mu,\tau} + \nu_{\bar{\mu},\tau} \longleftrightarrow \nu_e + \bar{\nu_e}; \nu + [\nu_e, \bar{\nu_e}] \longleftrightarrow \nu' + [\nu_e', \bar{\nu_e}']$)

With appropriate approximations, the moments of the reaction rates give the neutrino emissivity and the corresponding mean free path.

There are three categories of neutrino scattering processes: neutrino-nucleon scattering, neutrino-electron scattering, and neutrino–antineutrino scattering.

Because of the large mass of the nucleons, neutrino-nucleon scattering can be treated as conservative, although the most rigorous numerical treatments include effects of nucleon recoil as well as nucleon degeneracy and associated blocking factors. When the nucleons are in nuclei there are significant collective effects. Neutrino-electron scattering and neutrino-antineutrino scattering substantially effect the neutrino energy and tend to thermalize the neutrino spectrum. Neutrino-electron scattering also increases the neutrino flux and the rate of leptonization by causing high-energy neutrinos to scatter to lower energies where the mean free paths are longer. Neutrino-nucleon scattering is mediated by the exchange of the neutral Z boson. The reaction rate is the same for all flavors of neutrinos and antineutrinos.

9.3.4 Neutrino Transport

Proper handling of neutrino transport remains one of the biggest challenges in the core-collapse problem. In a sense, the neutrino-transport problem is easier than, for instance, a fully NLTE radiative-transport problem where the opacity of millions of lines is a sensitive nonlinear function of the radiation distribution for which one is trying to solve (Burrows 2013). The energy dependence in neutrino transport is far simpler. An important difference of neutrino transport in the context of core collapse compared to radiation transport in most stars is that the neutrino transport is a radiation-hydrodynamics problem, which is also an extremely challenging problem in photon transport.

The ideal of an adequately-resolved full Boltzmann transport calculation with general relativity in 3D remains beyond the capabilities of current computational facilities. Various approximations are thus invoked in order to make progress. At one limit, a central neutrino "light bulb" is used as a simple way to parametrize the outgoing neutrino flux. This approach is inadequate in many ways, but it did enable early studies of neutrino-driven convection and continues to be used to explore other aspects of the problem such as multidimensional dynamics. Another approach is to assume that neutrinos diffuse with opacities that are independent of energy, the gray approximation. Computationally-efficient parameterizations of the deleptonization, the entropy changes, and the neutrino stress for use in collapse simulations have been developed that capture much of the physics (Liebendörfer 2005).

A more sophisticated technique than a simple specified source of neutrino flux, is to employ *multi-energy group flux-limited diffusion* (MGFLD) in which the transport is appropriately diffusive at high "optical" depths and is free streaming in the limit of vanishing optical depths (Castor 1972; Mihalas and Weaver 1982). MGFLD techniques allow for the transport of neutrinos in energy as well as physical space and are relatively computationally tractable. The basic equation of MGFLD neutrino transport was given by Bruenn (1985). One term represents interaction between matter and neutrinos and incorporates all neutrino emission, absorption, and inelastic scattering processes, which redistribute neutrinos among energy zones. Another term represents the diffusion of neutrinos between mass shells, and a third term represents redistribution of neutrinos in energy by the relative Doppler shifts between zones. Inelasticity of neutrino-electron scattering makes this process difficult to model. Unlike elastic scattering, absorption, or emission processes, inelastic neutrino-electron scattering couples all energy groups. Calculating this process in detail is computationally intensive. In some circumstances, a *Fokker–Planck approximation* can be made, sacrificing physical accuracy for computational tractability. In this approximation, only the interactions between a specific neutrino energy group and its neighboring groups are computed.

In MGFLD, an interpolation scheme is used to connect the diffusive and free-streaming regimes. The interpolation scheme is called a *flux limiter* because it also serves to ensure that the energy is not effectively transported more rapidly than the speed of light. A widely-used flux limiter is that of Levermore and Pomraning (1981), although this particular scheme has also been criticized (Bludman and Cernohorsky 1995). Unfortunately, the most interesting transport regimes are often in the region of modest optical depth where the MGFLD technique is most uncertain and flux-limiter interpolations are most unreliable (Janka 1992).

Because μ and τ neutrinos and antineutrinos have almost identical cross sections they are, in some treatments, calculated using one set of neutrino groups in MGFLD. This results in three sets of neutrino groups, one for electron neutrinos, one for electron antineutrinos, and a combined group containing the remaining flavors of neutrinos. More sophisticated techniques have been developed to employ multiple energy groups for all three flavors of neutrinos. In principle, oscillations between the neutrino types may occur (Suwa et al. 2011; Zhang and Burrows 2013), but simulations show that the matter density associated with electrons dominates that of neutrinos. In this circumstance, the evolution of neutrinos to different flavors gets out of phase for neutrinos propagating on different trajectories, thus suppressing collective oscillations.

More sophisticated treatments of neutrino transport include velocity-dependent terms due to the motion of the stellar plasma. These require a Lorentz transformation between the comoving and observer frames of reference. Some of these simulations have directly integrated the Boltzmann transport equation, and others have used moment equations for the neutrino number, energy, and momentum using a variable-Eddington-factor closure scheme, in both Newtonian gravity and with general relativity. Neutrino–neutrino scattering couples regions that are, in principle, widely separated in both space and energy. This process is ignored in contemporary collapse calculations, but may be important for neutrino detectors.

The neutrino transport problem naturally gets more complex in multiple dimensions. MGFLD generically tends to underestimate angular variations of the radiated neutrinos and to make the neutrino flux more spherical than it should be. A technique employed in multidimensional simulations is the *ray-by-ray* approximation, in which the spherical transport problem is solved on each angular bin in polar coordinates. This approximation assumes that the neutrino intensity is axially symmetric around each radial direction and hence that the neutrino flux is purely radial. The ray-by-ray approximation generically artificially enhances angular variations since all fluxes are radial. This can lead to local "hot spots" in the emission near the neutrinosphere that radiate only in the radial direction. Yet another approach is called the *isotropic diffusion source approximation*, in which the neutrino distribution function is decomposed into components representing trapped and streaming neutrinos.

More advanced treatments of neutrino transport have explored Monte Carlo transport methods (Abdikamalov et al. 2012), general relativistic treatments (Müller et al. 2010 and subsequent papers in the series; Cardall et al. 2013) and Boltzmann transport (Liebendörfer et al. 2005; Sumiyoshi and Yamada 2012; Lentz et al. 2012; O'Connor and Ott 2013), increasingly often in multiple dimensions as computational facilities become more capable.

9.4 Collapse Dynamics

Overviews of collapse physics and dynamics were given by Bethe (1990), Arnett (1996), Mezzacappa (2005), Janka (2012), and Burrows (2013). The literature on the topic is large and growing steadily. Here we will sketch some of the basic aspects.

9.4.1 Initial Iron-Core Collapse

The most intensely-studied case of core collapse is in the context of the iron cores of massive stars that are rendered unstable by a combination of photodisintegration and electron capture (Sect. 8.6). Despite the complexities of the nuclear interactions (Sects. 9.3.1 and 9.3.2), the inhibition of weak interactions due to neutrino trapping means that the pressure during collapse is, to a good approximation, dominated at first by relativistically degenerate electrons and, at sufficiently high densities, by relativistically degenerate neutrinos, so $\Gamma_1 \simeq 4/3$. Variations in Γ_1 depend on electron capture and hence on details of the equation of state (Sect. 9.3.2) and neutrino transport (Sect. 9.3.4). In practice, the collapsing matter is near to neutral dynamic stability in realistic simulations.

For collapse under conditions of neutral stability, the density and hence mass distributions are preserved and collapse is *homologous* such that the mass, $M(x)$, within a given fractional radius, $x = r/r_0$, is constant. For homologous collapse, the velocity of a given mass point at given fractional radius is constant, $r(M) = v(M)t$, thus $v \propto r$ at a given instant, t (Fig. 9.1). At a given point of fixed interior mass, $M(x)$, $\rho(x) \propto t^3$, where t is measured from the onset of collapse. Because it is nearly in a condition of neutral stability, the portion of the structure that contracts homologously maintains the same density and pressure profiles without altering the binding energy (Goldreich and Weber 1980; Yahil 1983). In more realistic models, the velocity remains very closely proportional to radius in the homologous core, but $\rho(M)$ shifts somewhat and $v(M)$ is not precisely constant.

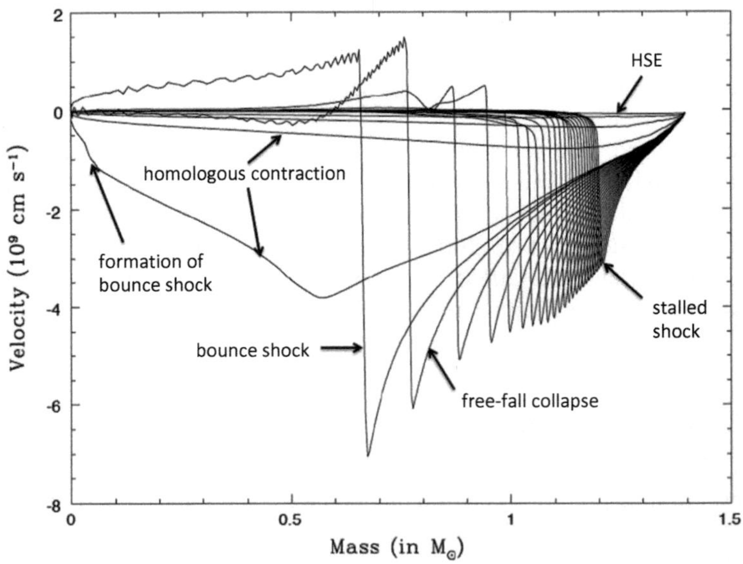

Fig. 9.1 The evolution of the velocity profile as a function of mass for the 1D collapse of the iron core of a model of ZAMS mass 11 $M_\odot$. The velocity is initially zero in hydrostatic equilibrium (HSE), but quickly develops a characteristic "v" shape, with the inner part representing the contraction of the homologous core and the outer part in free fall. Note that homologous contraction corresponds to $v \propto r$; the near linearity in this plot of v versus mass is an artifact of convolving the density and mass profiles. Bounce occurs about 1 s after the onset of collapse, resulting in the formation of a strong shock. The steepest shock corresponds to about 1.5 ms after bounce. The shock stalls at about 17 ms after bounce. Adapted and reprinted with permission from "Colloquium: Perspectives on core-collapse supernova theory" (Burrows 2013), RMP 85, 245. © 2013 by the American Physical Society

At the onset of core collapse, the material is effectively transparent to neutrinos, but as collapse proceeds to higher density and neutrino phase-space blocking becomes appreciable, the neutrinos are trapped in the infalling matter. Neutrino cross sections typically scale as the square of the neutrino energy (Sect. 9.3.3). The criterion for trapping is similar to that for the determination of the time of maximum light for a supernova: neutrino trapping occurs when the diffusion time for the neutrinos is comparable to the free-fall collapse timescale. For a neutrino mean free path of $\lambda_\nu \simeq 2\,\mathrm{km}(10\,\mathrm{MeV}/\epsilon_\nu)^2$, a typical neutrino energy of $\simeq$20 MeV, and a collapse time of $\simeq 1.6$ ms at the point of interest, this occurs at a density of about $\rho \sim 10^{12}\,\mathrm{g\,cm^{-3}}$ (Bethe 1990). Beyond this region of trapping, there is a region where neutrinos can diffuse outward in mass. The point where the neutrinos can begin to stream freely, the *neutrinosphere*, is typically at a density of $\sim 10^{11}\,\mathrm{g\,cm^{-3}}$. The location of the neutrinosphere is a function of neutrino energy, thus it is not at a single radius even in spherical models, but within a region of some breadth. In the absence of trapping, the emission of electron neutrinos would be radiated at rather

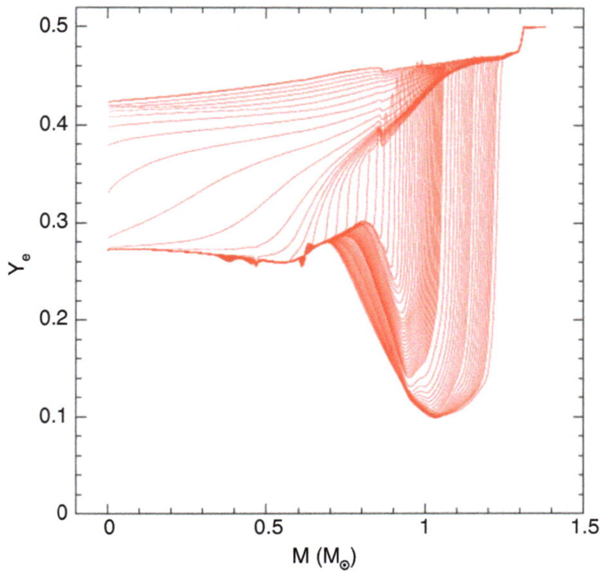

Fig. 9.2 The evolution of the electron number per baryon, Y_e, as a function of mass for the 1D collapse of the iron core of a model of ZAMS mass 11 $M_\odot$. At the beginning of collapse, the outer material has $Y_e \simeq 0.5$ and the inner iron core has $Y_e \simeq 0.42$. The inner portions have $Y_e \simeq 0.27$ at bounce. Reprinted with permission from "Colloquium: Perspectives on core-collapse supernova theory" (Burrows 2013), RMP 85, 245. © 2013 by the American Physical Society

high energies and for a brief time; the lower energies and extended time observed for SN 1987A represented direct evidence that trapping occurs (Burrows 2013).

Figure 9.2 shows the evolution of Y_e during collapse. The Chandrasekhar mass can be written as $M_{ch} = 5.76 < Y_e >^2 M_\odot$, where Y_e is the mean number of electrons per nucleon. For onset of collapse in an iron core, $Y_e \simeq 0.42$ so $M_{ch} \simeq 1.0\ M_\odot$, whereas a typical core has $M_{core} \simeq 1.5\ M_\odot$. The iron core exceeds the effective Chandrasekhar mass and thus, while perhaps semidegenerate, is not dominated by degeneracy pressure. The mass of the inner homologous core is $M_{hc} \simeq M_{ch}(Y_{e,trap}) \simeq 0.6$ to $0.8\ M_\odot$, again considerably smaller than the total collapsing iron core. At the surface of neutrino trapping for the case of an iron core, $Y_e \simeq 0.35$ so $M_{ch} \simeq 0.7\ M_\odot$. At trapping, the material within M_{ch} essentially constitutes the portion of the core that proceeds to collapse homologously.

With $v \propto r$ at a given t, at some radius the inward flow becomes supersonic. At larger radii, pressure waves cannot propagate to equilibrate pressure gradients, and matter must be in approximate free fall with

$$v(r) = \sqrt{\frac{2GM(r)}{r}}, \tag{9.8}$$

and hence $v \propto r^{-1/2}$, with $M(r)$ approximately constant beyond the homologous core. The velocity profile early in collapse thus shows a maximum infall velocity at the point where these two segments meet (Fig. 9.1).

To the degree to which neutral stability is maintained during collapse, the collapse of the inner core is homologous until nuclear density is attained, at which point nuclei merge to form a uniform degenerate gas of nucleons with interactions. The nuclear equation of state stiffens so pressure rises steeply with further increase in density and steep pressure gradients form. Collapse is halted when the central density is $\rho_c \simeq 2\rho_{nuc} \simeq 6 \times 10^{14}\,\mathrm{g\,cm^{-3}}$. The velocity is relatively low in the homologous core. As outer material collapses supersonically in free fall, it plummets onto the low entropy, quasistatic core, and a shock forms near the boundary of the homologous core. The shock-formation point defined by the location where the entropy first reaches a value of 3 k per nucleon occurs at 0.4–0.5 $M_\odot$ (Janka 2012). The inner core overshoots and overcompresses compared to its final equilibrium condition, then rebounds and drives a bounce shock outward, initially both in mass and radius (Fig. 9.1).

The potential for this bounce shock to immediately drive an explosion is known as the prompt mechanism. All current calculations of iron-core collapse show that while the bounce shock must occur, it is insufficient alone to cause ejection of the outer layers of the star. The prompt mechanism fails in this case. The shock is depleted by the inward momentum (ram pressure) of the free-falling material and by the energy required to dissociate infalling iron (Bethe 1990). The shock disintegrates iron to helium, and it must propagate through $\sim$1 $M_\odot$ of iron to get beyond the collapsing iron core into the outer parts of the star. This nuclear binding energy is lost, at least temporarily. It could be recovered in a successful explosion if the helium recombines to iron-group elements releasing that nuclear binding energy, but processes other than core bounce must drive the explosion.

9.4.2 *Electron Capture and Accretion-Induced Collapse*

While much work on core collapse has focused on iron cores, there has long been a parallel effort on the degenerate ONeMg cores of less massive stars that may collapse due to electron capture on ^{20}Ne and ^{24}Mg (Sect. 8.5.3). For an ONeMg core, $Y_e = 0.5$ and M_{ch} is close to the standard value of 1.44 $M_\odot$. Electron capture reduces Y_e and hence the effective Chandrasekhar mass. Rather than having the mass grow to approach or exceed M_{Ch}, as for many models of SN Ia, in this context, the mass is constant and M_{Ch} shrinks.

The slope of the stellar mass function suggests that relatively low-mass stars should dominate stellar death rates. Growing evidence confirms that most SN IIP involve ZAMS masses at the lower end of the mass function (Sect. 12.5.4). These lines of evidence have generated renewed interest in the possibility that a substantial fraction of SN IIP arise in electron-capture, rather than iron-core collapse (but note the evidence in Sect. 12.5.4 that no SN IIP progenitors arise in AGB stars with degenerate cores).

In addition to single-star evolution where electron capture may lead directly to degenerate core collapse, electron capture may also occur during accretion onto a white dwarf composed of ONeMg in a binary system, a process called *accretion-induced collapse* (AIC). AIC is also predicted to happen in some circumstances when a C/O white dwarf cools for a long time to low temperatures before accretion begins (Saio and Nomoto 1985; Nomoto and Kondo 1991; Schwab et al. 2015). AIC can also arise in the context of *merger-induced collapse* (Sect. 21.3). Evidence that some pulsars are only recently born with little "kick" in ancient globular clusters is consistent with their formation by AIC (Boyles et al. 2011).

Since collapse ensues prior to thermonuclear burning, during this form of collapse oxygen burns in a standing front that does not significantly impede the inward collapse (Hillebrandt et al. 1984). The outcome is a rather low-mass homologously-collapsing core that is conducive to explosion. Even the prompt mechanism may produce explosions in models with collapsing ONeMg cores since the overburden of mass is substantially less, resulting in smaller accretion rates and less ram pressure than for iron cores. In addition, the core is predicted to be quite distinct from the outer dilute envelope, so the binding energy of essentially all the matter beyond the collapsing core is small. The envelope is thus easier to eject than for the typical case involving iron-core collapse. For AIC, there is no envelope and the outcome might be quite different than for a typical supernova (Darbha et al. 2010).

Electron-capture induced collapse results in a failed, weak, or strong explosion depending on details of the nuclear equation of state and neutrino transport (Hillebrandt et al. 1984; Burrows and Lattimer 1985; Baron et al. 1987; Mayle and Wilson 1988; Woosley and Baron 1992; Kitaura et al. 2006; Fryer 2009; Wanajo et al. 2011; Moriya et al. 2014). As for iron-core collapse, it is important to treat this sort of collapse in multiple dimensions with rotation and magnetic fields (Dessart et al. 2007; Metzger et al. 2008; Abdikamalov et al. 2010; Sect. 9.5).

9.4.3 *Post-Bounce Dynamics*

There are some generic similarities between the dynamics of electron-capture induced collapse and iron-core collapse, but quantitative differences. Here we will primarily discuss the outcome of iron-core collapse where most of the work has been done.

State of the art core-collapse simulations cannot yet do all the physics at the necessary resolution, but Moore's law has allowed great computational progress and promises much more to come (Sarikas et al. 2012; Kuroda et al. 2012; Bruenn et al. 2013; Wongwathanarat et al. 2013; Dolence et al. 2013; Murphy et al. 2013; Ott et al. 2013; Sumiyoshi et al. 2015; Lentz et al. 2015). One point that has become increasingly clear is that while 1D models can capture some of the essence of the

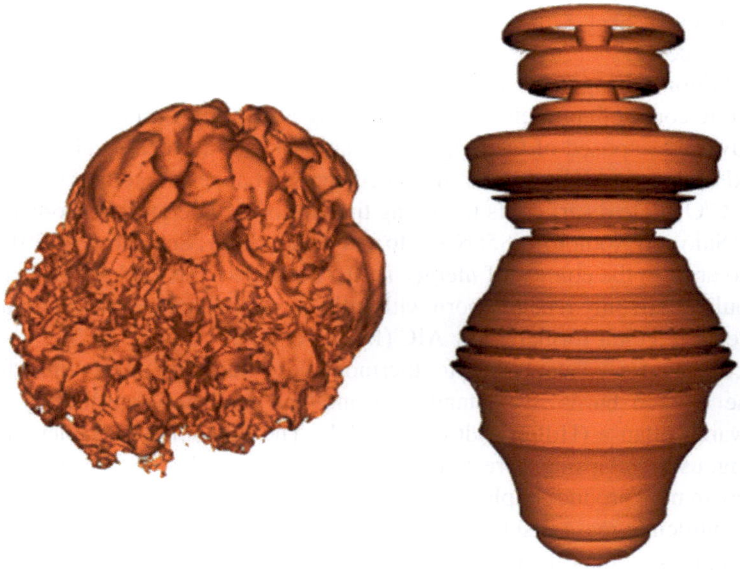

Fig. 9.3 A nonrotating, nonmagnetic 3D simulation of a core collapse (*left*) contrasted with a 2D simulation beginning from similar initial conditions (*right*) but then artificially rotated in the third dimension and visually rendered in a similar way. The plots represent contours of constant entropy 850 ms after bounce. The radial scale is about 400 km. The faux 3D plot differs qualitatively from the true 3D simulation. From "On the Impact of Three Dimensions in Simulations of Neutrino-driven Core-collapse Supernova Explosions" (Couch 2013b). © AAS. Reproduced with permission

problem and are computationally tractable for exploring physics, 2D models may be actively deceptive. Not only do they enforce an artificial axial symmetry, but they are susceptible to the well-known phenomenon that turbulence tends to cascade to large scales in 2D, whereas in 3D reality turbulence cascades to small scales. This makes 2D models prone to being qualitatively wrong (Fig. 9.3; Couch and O'Connor 2014). At this writing, the most sophisticated 3D collapse calculations fail to robustly explode, suggesting that some physics is still missing.

9.4.4 Standing Shock Phase

For nonrotating iron-core collapse, the failed bounce shock peters out to produce a standing *accretion shock* in a few to tens of milliseconds. The shock stalls at the radius, $\sim 2 \times 10^7$ cm, where the postshock pressure balances the ram pressure of the infalling matter, $P_{sh} \sim \rho_{in} v_{in}^2$ (Fig. 9.4). The accretion shock remains at roughly constant radius as freely-falling material collides with it, and mass accumulates beneath it. When the shock stalls, the preshock density still exceeds 10^{11} g cm^{-3}.

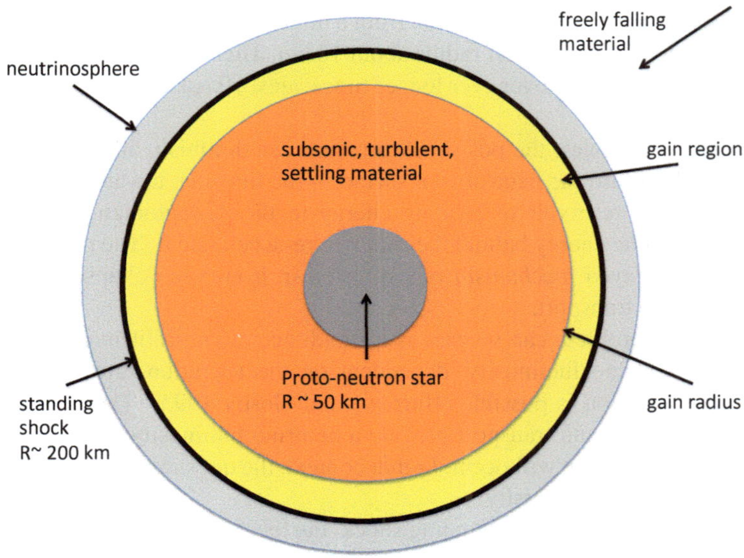

Fig. 9.4 Schematic diagram (not to scale) of the configuration of a collapsing, nonrotating, nonmagnetic, iron core about 100 ms after core bounce when the bounce shock stalls. The hot PNS formed from the collapsing homologous core has a mass of $\simeq 0.7\ M_\odot$ and a radius of $\simeq$50 km. The shock stalls at a radius of $\simeq$200 km, still inside the neutrinosphere from where neutrinos can stream freely outward. The outer freely-falling matter (beyond which the outer portions of the star will momentarily still be in hydrostatic equilibrium) is decelerated at the shock and then settles subsonically onto the PNS. This settling matter, and that within the PNS, is strongly turbulent. Between the gain radius and the standing shock is a thin gain region where the deposition of neutrinos is predicted to be greater than the loss. Deposition of neutrino energy in the gain region may re-accelerate the standing shock and generate an explosion by the delayed mechanism. The standing shock will be unstable to the SASI (Sect. 9.4.7). Rotation and magnetic fields may also play a role (Sect. 9.5)

The shock thus stagnates beneath the neutrinosphere (Fig. 9.4) and hence before a successful shock breakout of neutrino flux. Electron–positron pair annihilation is the dominant source of neutrinos in the postbounce shock wave, a process that produces neutrinos of all types (Sect. 9.3.3).

After its formation, the inner core, the remnant of the homologous collapse, is nearly in hydrostatic equilibrium for hundreds of milliseconds, but is predicted to be strongly convective (Sect. 9.4.6). Beyond the static core, postshock material is in sonic communication so that it feels pressure gradients as it settles onto the inner core (Fig. 9.4). This material is turbulent and subject to strong plume flows as neutrino-cooled matter sinks and neutrino-heated matter rises (Sect. 9.4.6). Outside the standing shock, matter is in free fall. The region within the standing shock remains a copious source of neutrinos with a total energy comparable to the binding energy of the inner core. As neutrinos are lost by diffusion and mechanical transport from the inner regions, the process of *de-leptonization*, the core undergoes thermal Kelvin–Helmholtz contraction from a hot PNS with a radius of ~50 km to a "cold," compact neutron star of radius about 10 km. This contraction provides even more binding energy, but over ever longer timescales, 10s of seconds (Burrows and Lattimer 1986).

On longer timescales, the postshock pressure can diminish, allowing the shock to shrink back toward the neutron-star surface—the first step toward total collapse. Alternatively, the reservoir of binding energy in the de-leptonizing neutron star might deposit some energy behind the shock to re-accelerate it. The principle focus of divining a *delayed mechanism* of explosion in iron-core collapse remains the study of neutrino transport.

The collapse problem can be cast as a Bondi accretion problem in which there is a critical neutrino luminosity, dependent on the accretion rate, above which no stationary solution is possible (Burrows and Goshy 1993). The implication is that if the dynamics and transport provide a neutrino luminosity above the critical luminosity, an explosion will occur. In that context, the question is whether collapse produces the critical luminosity.

In a region just beneath the standing shock, the local rate of energy deposition by neutrinos exceeds the intrinsic local energy-loss rate. The inner radius of this region is called the *gain radius* and the region between the gain radius and the standing shock is called the *gain region* (Fig. 9.4). For typical accretion rates, $\gtrsim 0.1\ M_\odot\ s^{-1}$, several percent of the neutrino luminosity from the neutrinosphere can be absorbed in the gain region. The duration of the exposure of matter to neutrino heating depends on the residence time of the matter in the gain region. The residence time is increased by nonradial, turbulent flows. The turbulent ram pressure helps to accelerate the shock, thus promoting explosion at a lower postshock thermal pressure (Couch and Ott 2015). These are intrinsically multidimensional effects. In principle, sufficient energy from the neutrino flux can be deposited in the gain region to increase the postshock pressure, thus accelerating the shock and producing an explosion. In practice, it has been difficult to show that this process is robust. Among other issues, these considerations demand a proper treatment of turbulence, which, like so many of the aspects of the complex collapse problem, remains a huge challenge computationally and even to the fundamental theory of fluid dynamics.

9.4.5 Explosion Phase

In the context of the delayed mechanism based on neutrino heating, the process may be conducive to more successful explosions with higher explosion energies if the explosion occurs early in the postbounce epoch. At later times, less mass tends to be resident in the gain region where it can be heated by neutrinos. Any delay in the explosion allows absorbed neutrinos to be reradiated, rendering them ineffective in driving the explosion. In models, energy does not accumulate in the material beneath the accretion shock until the explosion begins to succeed, moving matter outward. This suggests that a delay in the onset of the explosion is detrimental since

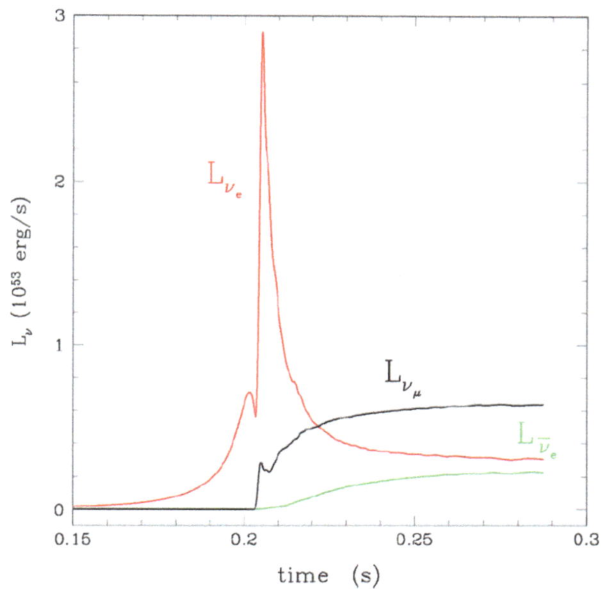

Fig. 9.5 The neutrino luminosity as a function of time after bounce for the 1D collapse of the iron core of a model of ZAMS mass 11 $M_\odot$. The burst in ν_e occurs as the shock passes through the neutrinosphere. The ν_μ luminosity represents the sum of the ν_μ and ν_τ neutrinos and their antineutrinos. Reprinted with permission from "Colloquium: Perspectives on core-collapse supernova theory" (Burrows 2013), RMP 85, 245. © 2013 by the American Physical Society

the energy of neutrinos radiated before the explosion begins is wasted. The time at which model explosions ensue after the shock stagnation depends on factors such as the equation of state and the intensity of postshock turbulence, as well as the intrinsic properties of neutrino physics.

While the bounce shock is predicted to stagnate within the neutrinosphere, a successful explosion moves the shock to lower densities and produces a distinct, prompt burst of neutrinos (Fig. 9.5). This prompt burst comprises mainly ν_e since the infalling matter at the neutrinosphere is still suffering rapid electron capture on protons. The luminosity in ν_e exceeds 10^{53} erg s^{-1} for $\sim$0.01 s in the spike and then reaches a typical value $\sim 10^{52}$ erg s^{-1} after about 0.1 s. Ongoing accretion can be detected, in principle, because during accretion the fluxes of ν_e and $\bar{\nu_e}$ are substantially higher than those of ν_μ and ν_τ and their antineutrinos. After accretion stops, the luminosities of all species of neutrinos are predicted to be about equal because the temperature distribution within the PNS is relatively flat and the neutrinospheres of all species of neutrinos are coincident.

Analysis suggests that lower-mass iron cores and larger Y_e favor explosion because there is less iron to photodisintegrate and the shock reaches the steep density gradient, where it accelerates, earlier. The steepness of the density gradient can be quantified by the measure of *bounce compactness*, defined as the ratio of interior mass to radius at a fiducial radius at bounce (O'Connor and Ott 2011). At face value, these criteria would favor smaller ZAMS masses. Because the mass of the iron core can be a nonmonotonic, even chaotic function of the ZAMS mass (Sect. 8.5.5), there may be *islands of explodability* where conditions are most conducive to a delayed explosion mechanism (Ugliano et al. 2012; O'Connor and Ott 2013; Sukhbold and Woosley 2014). Quantitative determination of the existence and location of these especially conducive conditions must take into account the rotation and magnetic evolution of the progenitor stars and in the subsequent collapse. An important consideration is that the outcome of the collapse may depend on the precise conditions in the progenitor, including the perturbations due to dynamical, chaotic burning in the final stages (Sect. 8.5.5; Couch and Ott 2013; Couch et al. 2015). One of the islands of explodability may be in the range of ZAMS masses that produces ONeMg cores (Sect. 9.4.2).

If the collapse successfully triggers an explosion, then a shock wave races out through the star, ejecting and partially transforming the matter. Most work on this phase, especially the computation of nucleosynthetic yields (Sect. 24.5), has been done in the context of 1D models. In the inner layers, the shock tends to reprise the nuclear evolution of the progenitor: the silicon shell is partially burned to iron-group elements, especially ^{56}Ni, the ONeMg layers to silicon, and some of the carbon to ONeMg. Helium tends to be at sufficiently low densities that it is not freshly burned in the explosion. The overall distribution of these bulk elements is changed quantitatively but not qualitatively in the explosion, yielding similar bulk composition before and after the explosion (Johnston and Yahil 1984).

There is growing observational evidence that the explosion energy of core-collapse supernovae increases with the ZAMS mass of the progenitor (Sect. 12.5.3). If this trend is borne out, it gives more impetus to consider core collapse not as a phenomenon restricted entirely to the inner iron core, but one closely coupled to the nature of the progenitor.

9.4.6 Fluid Instabilities

There are good reasons to think that multidimensional effects are critical to the core-collapse problem. For example, direct imaging of SN 1987A (Chap. 11), the Crab Nebula, and Cas A (Chap. 7) shows obvious departures from spherical symmetry, and spectropolarimetry (Sect. 4.6) has demonstrated that this is also true for distant core-collapse events. Physics also points in this direction.

Computational techniques are now sufficiently advanced that it is possible to simulate the initial iron-core collapse, the bounce, the stalling of the shock, the development of strong, neutrino-driven convection and of various instabilities, and then follow the resulting asymmetric shock to the surface of the star and beyond. These sorts of sophisticated and extensive simulations give promise for making direct connection to various observables. Even in the absence of the effects of rotation and magnetic fields (Sect. 9.5), the resulting complex dynamics may give insight into the apparent "mixing" by plumes and clumps of hydrogen inward and oxygen and ^{56}Ni outward in SN 1987A (Sect. 11.4.4). These simulations also give insight into the origin of neutron star "kicks" (Wongwathanarat et al. 2010; Nordhaus et al. 2012).

The postbounce structure can develop negative entropy gradients, making the structure susceptible to thermal convection. Neutrino leakage gives rise to composition gradients, as determined by the distribution of Y_e, that also can drive convection. For *neutron-finger* instabilities, analogous to *thermohaline, salt-finger* instabilities in the oceans, perturbations that would be formally stable dynamically are rendered unstable by heat diffusion so that instabilities proceed on thermal timescales (Wilson and Mayle 1988). Current understanding is that this process does not significantly enhance the neutrino flux because the lepton number equilibrates between fingers and stabilizes them more rapidly than thermal transport can destabilize them (Bruenn and Dineva 1996).

Multidimensional simulations show that various instabilities drive overturn, entailing plumes of sinking, colder, material and rising, warmer, material (Herant et al. 1994; Janka 2012). The neutrino transport then responds to the plume structure, with sinking fingers absorbing neutrinos and rising fingers radiating neutrinos. This may enhance the net neutrino deposition between the gain radius and the shock and re-accelerate the standing shock. The standing shock is subject to instabilities (Sect. 9.4.7), and there may be intrinsically large-scale transport of energy and composition by magnetorotational effects (Sect. 9.5). The convection beneath the standing shock may drive a *lepton-number emission self-sustained asymmetry* in which the dipolar mode of the standing shock leads to greater lepton flux from one hemisphere (Tamborra et al. 2014a). The structure both during collapse and upon successful explosion is expected to be complex and turbulent (Fig. 9.3; left panel).

As the explosion ensues, conditions arise where dense material is accelerated by lighter material, giving rise to Rayleigh–Taylor instabilities. This is prominently the case at the boundary between the denser helium core that is decelerated by the lower-density hydrogen envelope. Similar effects can occur with smaller amplitude at inner-core composition boundaries. Shear associated with Rayleigh–Taylor fingers or other effects triggers Kelvin–Helmholtz instabilities, and shocks propagating across steep density gradients near composition discontinuities induce Richtmyer–Meshkov instabilities. These instabilities tend to lead to small-scale mixing of composition in the ejecta. The strong turbulence beneath the shock can impose initial, rather large-scale asymmetries. These asymmetries can also trigger Rayleigh–Taylor instabilities that may produce large-scale plumes in the nonlinear limit. Figure 9.6 illustrates the nonlinear limit of the action of Rayleigh–Taylor instabilities in the dynamics of a nonrotating model. The plumes could contribute to observed asymmetries, but might be modified by rotation.

Collision of inner, denser regions with lighter, outer regions also produces reverse shocks. The reverse shocks can propagate back into the inner ejecta and promote fallback onto the inner compact object (Ugliano et al. 2012; Wong et al. 2014). Supernova progenitors that have been stripped of their hydrogen envelopes

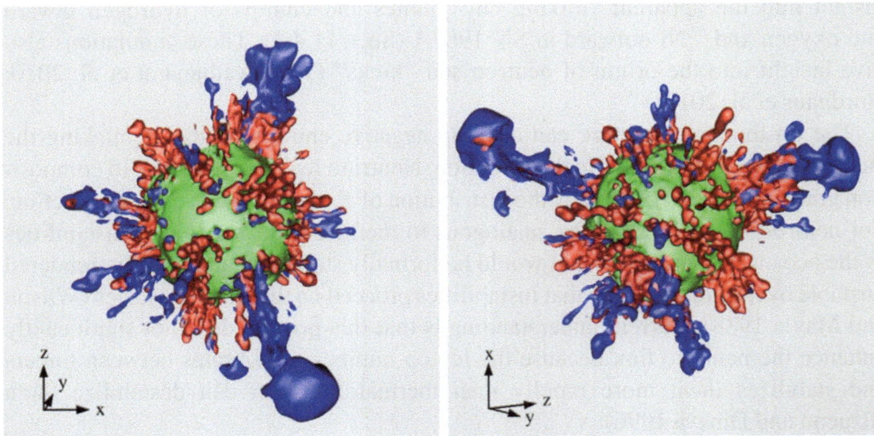

Fig. 9.6 Two different views of a nonrotating, nonmagnetic 3D simulation of core collapse in a model blue supergiant of ZAMS mass 15.5 $M_\odot$. The structure is illustrated at about 2 h after collapse, near the time of shock breakout from the surface. Convective perturbations have driven Rayleigh–Taylor instabilities at composition boundaries. Nickel-rich plumes (*purple*) proceed beyond oxygen-rich knots (*red*), with both extending beyond the carbon-rich layers (*green*). Each panel is 7.5×10^{12} cm on a side, twice the initial radius of the progenitor supergiant. From "Three-dimensional Simulations of Mixing Instabilities in Supernova Explosions" (Hammer et al. 2010). © AAS. Reproduced with permission

(Chaps. 15–17) tend to suffer less fallback because of the reduced power of the reverse shock.

Deposition of energy from radioactive decay can create *nickel bubbles* that may protrude as large-scale isolated "fingers" extending as far as the base of the helium shell (Fig. 9.6). This phenomenon may have been witnessed in the remnant of Cas A (Sect. 7.4.1). MHD processes may also lead to large-scale structures in the ejecta. All these phenomena are likely to lead to the shock emerging asymmetrically from the surface (Couch et al. 2011; Matzner et al. 2013).

9.4.7 The Standing Accretion Shock Instability: SASI

The *standing accretion shock instability* (SASI) is an important ingredient of the collapse problem. It was discovered in basic 2D hydrodynamic calculations with simplified structure and no neutrino transport (Blondin et al. 2003). It has been reproduced in simulations incorporating more accurate structures and neutrino transport, and in 3D. There is no question that the instability exists, but there is uncertainty as to its cause and effects. The basic idea is that a perturbation to a spherical standing shock creates nonspherical flow that feeds back on the shock to amplify the perturbations. There is debate as to whether the feedback process involves a primarily radial advective–acoustic cycle (Foglizzo et al. 2007) or a purely acoustic, primarily azimuthal, process (Blondin and Mezzacappa 2006). In the former, the perturbed shock sheds vortices that are advected down to the neutron star where they create acoustic waves that propagate back up to the shock, amplifying the perturbation. In the latter, the whole cycle is proposed to be driven by acoustic waves.

The earliest 2D calculations of collapse models with SASI showed dramatic oscillations in an $\ell = 1, m = 0$ dipole mode. These simulations immediately raised questions of whether the SASI could reproduce the "bipolar" structure of core collapse supernovae revealed by spectropolarimetry. This remains unclear for a number of reasons. The excitation of the $\ell = 1$ mode, while interesting, is clearly an artifact of the axial symmetry imposed in 2D calculations. A related "acoustic" mechanism arose in some simulations as a large-scale instability in which g-modes are excited on the PNS surface that drive a large acoustic flux (Burrows et al. 2006). These simulations even provided unipolar explosions in some cases, with attendant kicks to the neutron star. The acoustic mechanism driven by the excitation of g-modes has not been robustly reproduced. In the nonlinear limit, the SASI modes and g-modes may be susceptible to parasitic Rayleigh–Taylor and Kelvin–Helmholtz instabilities (Weinberg and Quataert 2008; Guilet et al. 2010), sensitive to the detailed physics such as nuclear dissociation and neutrino cooling (Fernández et al. 2014), and dominated by neutrino-induced convection (Burrows et al. 2012).

In addition, the behavior of the SASI in 3D is rather different (Blondin and Mezzacappa 2007; Hanke et al. 2013; Fernández 2015; Fig. 9.7). Models with no angular momentum in the original iron core developed spiral modes in response to the SASI such that a net angular momentum was imparted to the matter interior to the standing shock. The implication was that the neutron star (not treated explicitly in the first 3D simulations), would be counter-rotating in order to conserve angular momentum. This brings the prospect of relaxing the coupling between the rotational state of the original iron core and the spin rate of the neutron star, in violation of standard assumptions (and intuition). Blondin and Mezzacappa (2007) found that

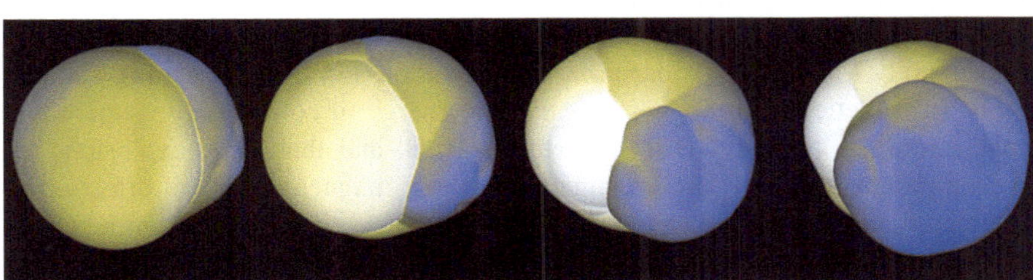

Fig. 9.7 Dynamics of the Standing Accretion Shock model. The unstable standing shock develops transverse spiral flows in this simulation, leading to intersecting shocks. The *blue regions* represent the leading edge of the spiral SASI wave that propagates from right to left across the principle standing shock. The propagation of the spiral wave results in colliding shocks and a shock triple point illustrated here by the discontinuity between the blue and white surfaces at the leading edge of the SASI wave. Reprinted by permission from Macmillan Publishers Ltd.: NATURE, Blondin and Mezzacappa (2007)

for low iron-core rotation there is no correlation of the spin axis of the neutron star and that of the iron core. For moderate rotation, they found that the outer matter follows the iron-core angular momentum, but because the SASI demands counter-rotation of the neutron star to conserve angular momentum, the neutron-star angular momentum tends to be independent of that of the iron core. If the progenitor core rotation is a few times faster, the SASI is not able to overcome the relic angular momentum of the core, and the neutron-star spin will be parallel to that of the iron core. The primary axis of the ejecta in SN 1987A is tilted with respect to the plane of the inner ring by about 15°, reminiscent of the result of Blondin and Mezzacappa (2007) for moderate rotation. Polarization studies of SN Ib/c also give suggestions of tilted plumes or jets (Sects. 16.3.3 and 17.3.3). These enticing results have not been robustly reproduced in other simulations (Fryer and Young 2007; Iwakami et al. 2009; Takiwaki et al. 2012; Burrows et al. 2012). The angular momentum transferred to the neutron star by SASI-driven flows as the explosion proceeds and during any subsequent accretion may be insufficient to account for substantial neutron-star spin. This process may supplement the angular momentum contributed by the progenitor star.

Independent of the effect on the rotation of the PNS, the SASI is expected to affect the collapse dynamics. Its presence tends to move the accretion shock outward in a way that might enhance the probability of explosion. The nonradial flows associated with the SASI may lead to a larger efficiency of neutrino energy deposition in the gain region. The associated secondary shocks will convert kinetic energy into thermal energy and higher entropy. This may enhance the overall turbulent plume flow and result in additional impetus to the shock to propagate outward.

9.5 Rotation and Magnetic Fields

While there have been many models of nonrotating collapse in 1D, and many that considered rotation in multiple dimensions in the absence of magnetic fields, it is almost surely true that all collapse involves rotation and that all rotating collapse involves magnetic fields. Suggestions that the energy of core-collapse supernovae ranges from the typical 1 B to an extreme 10 B, and that some explosions are "engine-driven," all call for investigation of how magnetorotational effects may in some circumstances add to or even supplant the energy derived from neutrino-transport processes.

The rotational energy of the neutron star can, in principle, provide a substantial amount of energy to a core-collapse explosion, but with a reasonably firm upper limit. The moment of inertia of the PNS is $I_{PNS} \simeq 3 \times 10^{46}\,\mathrm{g\,cm^2}$. The rotational energy of the PNS is then

$$E_{\mathrm{rot,PNS}} \simeq \frac{1}{2} I_{PNS} \Omega_{PNS}^2 \simeq 0.9\mathrm{B} \left(\frac{M_{NS}}{1.5 M_\odot}\right) \left(\frac{P_{PNS}}{25\ \mathrm{ms}}\right)^{-2} \left(\frac{R_{PNS}}{50\ \mathrm{km}}\right)^2 . \tag{9.9}$$

After deleptonization, the neutron star has a radius of about 10 km. For an extreme rotational period of 1 ms, the rotational energy would be ~20 B. This is a substantial energy, but essentially an upper limit to the energy available to power a magnetorotational explosion with a neutron star. If a given event—supernova, gamma-ray burst or other—requires more than this, a magnetorotational process can probably be ruled out. In practice, only a fraction of the energy, the free energy of differential rotation, could be tapped. Even for smaller energy input, the generation and release of magnetic energy could abet and shape the explosion.

Immediately after the discovery of pulsars, there were suggestions that rotation and magnetic fields could be a significant factor in the supernova explosion mechanism (Ostriker and Gunn 1971; Bisnovatyi-Kogan 1971; Kundt 1976). The characteristic power of nonrelativistic MHD outflow is given by (Blandford and Payne 1982)

$$L_{\rm MHD} = \frac{B^2 r^3 \Omega}{2} \simeq 3 \times 10^{45} {\rm erg\,s^{-1}} \left(\frac{B}{10^{12}\,{\rm G}}\right)^2 \left(\frac{R_{NS}}{10\,{\rm km}}\right)^3 \left(\frac{P_{NS}}{1\,{\rm ms}}\right)^{-1}. \quad (9.10)$$

Typical pulsar dipole fields of 10^{12} G and rotation periods of several to several tens of milliseconds yield electrodynamic power of $\sim 10^{44}$ to $10^{45}\,{\rm erg\,s^{-1}}$ that is insufficient to produce a strong explosion, but the recognition of *magnetars* (Duncan and Thompson 1992; Usov 1992) invited consideration of larger field strengths. For $B \sim 10^{15}$ G the MHD luminosity, $\gtrsim 10^{51}\,{\rm erg\,s^{-1}}$, might rival that of the neutrino flux. The discovery that core-collapse supernovae are routinely polarized and substantially bipolar (Sect. 4.6), and the nearly simultaneous discovery of the connection between supernovae and gamma-ray bursts (Sect. 17.4.3), brought new attention to the issues of rotation and magnetic fields and their possible effects on the energetics and dynamics of core collapse.

Early simulations investigated the role of magnetic fields in core collapse (LeBlanc and Wilson 1970; Meier et al. 1976; Müller and Hillebrandt 1979; Symbalisty 1984), but these models tended to require what appear to be unphysically large fields in the initial iron core to produce interesting dynamical effects. This issue still plagues many sophisticated multidimensional MHD models, but there are hints that fields of order 10^{12} G may exist prior to collapse in restricted regions at the boundary of the iron core (Wheeler et al. 2015b).

9.5.1 Basic Magnetorotational Physics: Length and Time Scales

There are a number of key length and time scales in the magnetorotational collapse problem, including those associated with the progenitor star: the radius of the star, $R_{\rm star} \sim 10^{13}$ cm for an RSG and $R_{\rm He} \sim 2 \times 10^{10}$ cm for a helium core. The inner iron core that collapses to form a neutron star has a typical radius, $R_{\rm Fe} \sim 4 \times 10^8$ cm. The dynamical or sound-crossing time for the helium core is

$$\tau_{\rm dyn} \simeq 20\, s \frac{R_{10}^{3/2}}{(M/{\rm M}_\odot)^{1/2}}, \qquad (9.11)$$

where R_{10} is the radius in units of 10^{10} cm. The dynamical time for a shock wave or jet to emerge from a red giant is from 10^3 to 10^5 s, and from a stripped helium core, the progenitor of an SN Ib or SN Ic, from 1 to 10 s is typical. Other relevant length scales are the radius of the speed of light cylinder:

$$R_{LC} = \frac{\rm c}{\Omega} = \frac{{\rm c}P}{2\pi} \simeq 50\,{\rm km}\left(\frac{P_{NS}}{\rm ms}\right), \qquad (9.12)$$

where Ω is the rotational frequency, and the Alfvén radius at which magnetic pressure is balanced by the ram pressure, e.g.,

$$\frac{1}{2}\rho v^2 \simeq \frac{1}{8\pi}B^2, \qquad (9.13)$$

which, for a dipole field with $B \simeq B_{NS}\left(\frac{R}{R_{NS}}\right)^{-3}$, is

$$R_A \simeq 3.0 R_{NS}\left(\frac{B}{10^{14}\,{\rm G}}\right)^{1/3}\left(\frac{\rho}{10^8\,{\rm g\,cm^{-3}}}\right)^{-1/6}\left(\frac{v}{10^8\,{\rm cm\,s^{-1}}}\right)^{-1/3}. \qquad (9.14)$$

After core bounce, a PNS forms with a radius of $R_{PNS} \simeq 50$ km, and the bounce shock stalls at a radius of about 200 km. Over a cooling time of about 5–10 s, the PNS deleptonizes by neutrino emission, cools, and contracts to form the final neutron star structure (Sect. 9.4.3). If the neutron star is rotating and magnetized, this cooling phase may be associated with contraction, spin-up, and amplification of the magnetic field. These changes can significantly alter the physics associated with the neutron star and its interaction with its surroundings and hence the explosion process itself (Wheeler and Akiyama 2007; Metzger et al. 2011).

During the thermal contraction of the PNS, P, R_{LC}, and R_A all change significantly. As a result, the light cylinder may contract from $R_{LC} \sim 10^3$ km, beyond the radius of the stalled shock and comparable in size to the original iron core, to $R_{LC} \simeq$ 50 km, comparable to the radius of the neutron star and well within the stalled shock. The light cylinder may also contract from beyond R_A to significantly less than R_A. This has implications for angular-momentum transport and the generation of radiation.

9.5.2 *Magnetic-Field Compression and Wrapping*

The fraction of the rotational energy of a PNS that can be tapped to drive an explosion depends on the behavior of the magnetic field. The magnetic field of the PNS is uncertain. Upon compression with flux freezing, the strength of the poloidal or toroidal component of the magnetic field scales as $B \propto R^{-2}$ or $\rho^{2/3}$ and hence would increase by a factor of about 10^3 by basic compression (Obergaulinger et al. 2014). If the precollapse core has a field strength comparable to that of a magnetized white dwarf, $\sim 10^6$ G, then a field of $\sim 10^9$ G could arise from flux freezing. An initial field of $\sim 10^9$ G could give the field typically associated with pulsars, and an initial field of order 10^{12} G would be required to give that of a magnetar. Even then, the field would be far weaker than that required to be dynamically important, $\gtrsim 10^{16}$ G.

If the rotation period is short compared to the convective timescales, corresponding to a *Rossby Number* $Ro \simeq P_{PNS}/\tau_{\rm conv} \simeq 1$, an α–Ω type dynamo might operate (Duncan and Thompson 1992). This is a severe requirement in the immediate PNS phase when the radius is rather large. An alternative magnetic-field amplification mechanism is linear amplification (Meier et al. 1976; Kluźniak and Ruderman 1998). In this process, the differentially-rotating neutron star could wrap the initial poloidal seed field, B_p, into strong toroidal fields that then might emerge from the star through buoyancy or as a magnetic jet. After n_ϕ revolutions of the neutron star, the initial field could be wrapped and amplified to produce a toroidal field

$$B_\phi \simeq 2\pi n_\phi B_p. \tag{9.15}$$

The number of revolutions to reach a dynamically significant $B_f \simeq 2 \times 10^{16}$ G is

$$n_f = \frac{B_f}{B_0}\frac{1}{2\pi} \simeq 3 \times 10^3 \left(\frac{B_0}{10^{12}\,\rm G}\right)^{-1}. \tag{9.16}$$

For $B_0 \sim 10^{12}$ G produced by field compression and a PNS rotation period of 25 ms, the amplification timescale to reach dynamically significant field strengths would be on the order of 100 s, rather long compared to most notions of the timescale to produce an explosion. This timescale would be decreased if the initial field were greater.

As the PNS contracts and spins up, the timescale for the linear field amplification, t_f, decreases and the rotational energy of the neutron star increases. If the contracted neutron star with a radius of $\sim$10 km spins with a period of 1 ms, t_f would be substantially less than the cooling, contraction time of $\sim$10 s. Thus sometime during the contraction a strong field could be produced.

9.5.3 The Magnetorotational Instability: MRI

While flux freezing seems unlikely to generate dynamically strong magnetic fields, and linear wrapping is likely to be too slow to produce dynamically strong fields on dynamical timescales, there are other processes that may generate significant magnetic fields on much shorter timescales. Principal among these is the magnetorotational instability (MRI; Balbus and Hawley 1998).

Ignoring entropy gradients, the condition for the instability of the slow magnetosonic waves in a magnetized, differentially-rotating plasma is (Balbus and Hawley 1998)

$$\frac{d\Omega^2}{d\ln r} + (\mathbf{k}\cdot\mathbf{v_A})^2 < 0, \tag{9.17}$$

where **k** is the wavenumber and

$$v_\mathrm{A} = \frac{B}{\sqrt{4\pi\rho}} \tag{9.18}$$

is the Alfvén velocity. When the initial magnetic field is small, $(\mathbf{k}\cdot\mathbf{v_A})^2$ is negligible, and the local instability criterion for the MRI is simply that the angular velocity gradient be negative (Balbus and Hawley 1998):

$$\frac{d\Omega^2}{d\ln r} < 0. \tag{9.19}$$

This instability operates when the angular velocity, not the angular momentum, has a negative gradient. A negative gradient in angular velocity is a natural result of the differential motion that accompanies the collapse of a rotating iron core to form a differentially-rotating PNS (Akiyama et al. 2003).

The growth of the magnetic field associated with the MRI is exponential with characteristic timescale of order Ω^{-1} (Balbus and Hawley 1998). The MRI thus tends to dominate the process of wrapping field lines, which grows only linearly in time. The MRI may also operate under conditions of moderate rotation that are not sufficient to compete with the PNS convective timescales to drive an α–Ω dynamo. The resulting unstable flow is expected to develop turbulence and drive a dynamo that amplifies and sustains the field.

The field grows due to the MRI dynamo action until it reaches a saturation field limit. An order of magnitude estimate for the saturation field can be obtained by equating the Alfvén velocity to the rotational velocity, $v_\mathrm{A} = r\Omega$, or

$$B_{sat} \sim \sqrt{4\pi\rho}r\Omega \simeq 10^{15}\mathrm{G}\left(\frac{R_{NS}}{10\,\mathrm{km}}\right)\left(\frac{\rho}{10^{10}\ \mathrm{gcm^{-3}}}\right)^{1/2}\left(\frac{P_{NS}}{1\ \mathrm{ms}}\right)^{-1}, \tag{9.20}$$

a field strength that arises naturally by this physics but is also interestingly close to the fields identified in magnetars. While it is common in various circumstances to associate dipole fields with neutron stars, the field structures associated with the MRI are likely to be substantially dominated by the toroidal component. The field can in principle grow quickly, in tens of milliseconds, to generate dynamically significant magnetic fields of order 10^{15} to 10^{17} G where negative shear is strongest, at the boundary between the PNS and the settling material interior to the standing shock.

While the likelihood of the MRI operating in the collapse ambience seems compelling, trying to assess its effect quantitatively remains a challenge. One of the principle problems is that the wavelength corresponding to the most rapidly growing mode is predicted to be much smaller than the macroscopic length scales of the problem, so the scales of the unstable modes cannot be directly resolved in any direct numerical simulation that addresses the scales of the iron core and the standing shock and the neutron star (Obergaulinger et al. 2006).

Some simulations attempt to emulate the effect of the MRI by adopting initial fields, such that compression and winding eventually produce the sorts of fields expected for the MRI at saturation. Simulations that adopt stronger initial magnetic fields can produce dramatic magnetic jets generated by the hoop stress of the wound-up fields (Burrows et al. 2007), but it is not clear that the effects of the rapid exponential growth of the MRI are the same as the slower geometric growth of the wrapping process, even if they have the same final field. Simulations are underway that do multidimensional and general relativistic MHD simulations of the MRI and related magnetic physics (Font et al. 2011; Endeve et al. 2012; Masada et al. 2012; Winteler et al. 2012; Sawai and Yamada 2014). Some progress may be made in the near future by adopting "subgrid" models of the MRI in global simulations.

Figure 9.8 shows the outcome of a highly resolved 3D simulation of core collapse in a star with modest initial magnetic field ($\sim 10^{10}$ G), but sufficient initial angular momentum that the PNS rotates with a period near the limit of 1 ms (Mösta et al. 2015). The field is amplified to $> 10^{14}$ G by compression and linear winding. Under these conditions, this simulation resolves the most rapidly growing modes of the MRI. At bounce, the structure is unstable to the MRI at the boundary of the PNS. The instability leads to channel flow, then turbulence. The field grows exponentially ($\tau \simeq 0.5$ ms) and becomes fully turbulent in about 3 ms, three rotations. The simulation then displays an inverse cascade to produce a well-ordered, large scale, toroidal field surprisingly rapidly, after only about six rotations. The toroidal field attains a maximum value $\sim 5 \times 10^{15}$ G, strong enough to be reminiscent of magnetars and to be of dynamical significance.

There have been speculations that the MRI or other magnetic effects could drive powerful jets in core-collapse supernovae (Wheeler et al. 2000, 2002; Lazzati et al. 2012), but it is far from clear that this is routinely the case. Even with the rapid field growth associated with the MRI, the conditions required to drive organized jets are stringent in terms of the rapid initial rotation that may be required. Driving coherent, long-lived jets may be difficult because the associated MHD flows may be unstable

Fig. 9.8 Magnetic field structure 30 ms after core bounce resulting from a 3D general-relativistic MHD model of the collapse of the iron core of a model of ZAMS mass 25 $M_{\odot}$ with an initial period of 2.25 s and a poloidal field of 10^{10} G. The simulation resolved the most rapidly growing modes of the MRI. The resulting structure is unstable to the MRI and rapidly grows a toroidal field of strength $\sim 10^{15}$ G. The rotation axis is vertical. *Yellow and blue regions* indicate field strengths of 10^{15} G. Reprinted by permission from Macmillan Publishers Ltd.: NATURE, Mösta et al. (2015)

in 3D (Mösta et al. 2014). Some polarization observations are consistent with the choking of jet-like flows that fail to penetrate the surface of the star (Sect. 17.3.3).

Another interesting possibility is that turbulence associated with the MRI may be robustly significant. Turbulence develops in the nonlinear phase of the MRI when the field strength approaches saturation. The turbulence acts as an effective viscosity, transporting angular momentum and depositing rotational energy as heat that could, for instance, supplement neutrino deposition in the gain region (Thompson et al. 2004, 2005; Ott et al. 2006) and may increase the residency time in the gain region and add turbulent ram pressure (Sect. 9.4.3). The effect of the MRI to grow strong, large-scale magnetic fields and to drive turbulence may complement and enhance basic neutrino-driven processes.

9.5.4 *Other MHD Processes*

There are thus many issues that arise concerning coupling of rotation to the magnetic field: the nonlinear growth, saturation, and subsequent evolution of the magnetic field; the circumstances in which MHD jets or other MHD flows form; whether jets are needed to explode stars as typical supernovae as well as those associated with gamma-ray bursts; and the fate of the compact objects—neutron star, pulsar, magnetar, or black hole. The spiral modes of the SASI in multiple dimensions (Sect. 9.4.7) intrinsically involve shear and hence the potential to drive some form of dynamo process (Endeve et al. 2012). This possibility demands a proper treatment of turbulence in addition to the magnetic properties.

The effects of rotation and field amplification raise issues not just of MHD processes, but of interesting problems of plasma physics. Alfvén waves could propagate outward and be dissipated and deposited specifically just behind the standing shock (Wheeler and Akiyama 2007; Suzuki et al. 2008). Large-amplitude electromagnetic waves might be produced that could drive shocks, generate pions by proton–proton collisions, or create electron–positron pairs depending on the circumstances (Wheeler et al. 2000). The dynamically significant magnetic fields predicted at MRI saturation, of order 10^{15} to 10^{16} G, are much stronger than the QED limit of 4×10^{13} G, so quantum effects may become important. Sufficiently strong magnetic fields might affect neutrino cross sections. Magnetorotational processes near the neutron star may disrupt the structure of any organized dipole field and promote the generation of outflows of various characteristics (Metzger et al. 2011) including relativistic MHD winds (Bucciantini et al. 2008; Metzger et al. 2014). Fallback onto magnetars might generate propellor-driven explosions (Piro and Ott 2011), and there have been suggestions that magnetar input may influence some superluminous supernovae (Chap. 18).

9.5.5 *Non-Axisymmetric Instabilities: NAXI*

It is not clear that neutron stars can be born with very high rotation rates, but if so, then other classes of instabilities may ensue that can affect the outcome of core collapse. Rotation with a ratio of rotational to binding energy $T/|W| > 0.01$ ($P \simeq$ 20–30 ms) may be subject to non-axisymmetric instability (NAXI; Andersson 1998;

Shibata et al. 2003; Ott et al. 2005). The thresholds, growth rates, and saturation depend on the degree of differential rotation (Ou et al. 2004; Wheeler and Akiyama 2007) and are affected by magnetic fields (Rezzolla et al. (2001a,b)). A dynamic instability to a bar-like mode can occur for $T/|W|$ as low as $\sim$0.2 (Shibata and Sekiguchi 2005). NAXI can be produced by the presence of a corotation point within a *resonant cavity* formed by an extremum in vorticity density in a differentially rotating structure that serves to reflect acoustic waves (Ou and Tohline 2006). While one-armed $m = 1$ spiral modes can be generated by this mechanism and by the SASI (Sect. 9.4.7), the azimuthal acoustic modes due to the SASI near the standing shock are not the same as those of the NAXI that occur within a differentially-rotating PNS.

Most work on NAXI has considered the neutron star to be in complete isolation and ignored the magnetic field, but both the supernova ambiance and magnetic fields are potentially significant. It is important to realize that for sufficiently rapidly rotating PNS, these instabilities can occur deep inside the supernova material, before the ejecta have time to expand and dissipate. The MRI grows magnetic field on a timescale more rapid than all but dynamical bar modes. This means that NAXI usually occurs in a magnetized medium. NAXI thus interacts with surrounding, magnetized matter, likely generating significant magnetoacoustic flux that propagates out into the ambient matter, whether or not an explosion has been initiated by some other mechanism (Wheeler and Akiyama 2007; Fu and Lai 2011). Even relatively slowly-rotating iron cores can potentially generate rotating neutron stars subject to NAXI, and hence to magnetoacoustic flux.

9.6 Black Hole Formation

This chapter has focused primarily on the production of neutron stars as a result of core collapse. Evidence that nature also produces black holes of stellar mass, while still largely circumstantial, is compelling. While it is not yet technically possible to directly observe event horizons or other strong-field effects (but see Sect. 24.2.2; Abbott et al. 2016) and isolated black holes remain elusive, there is ample evidence for the existence of stellar-mass black holes in binary systems. The first black hole candidates, e.g., Cyg X-1, were identified as the theory of massive-star evolution and core collapse was being developed in the late 1960s and early 1970s.

These black holes must arise by core collapse, but the progenitor evolution and the specifics of the collapse mechanism remain unclear. An early perspective was that there is a ZAMS mass cut in the neighborhood of 30 $M_{\odot}$, below which neutron stars and supernovae are formed and above which black holes are the end product. The general supposition was that higher-mass stars would form black holes but not supernovae, basically by the failure of the core-collapse explosion mechanism and hence direct collapse to a black hole.

Subsequent developments have added considerable complexity to this simple, bimodal view. One development was the formation of a neutron star in SN 1987A as indicated by the neutrinos (Chap. 11). The neutron star lasted for at least the 10 or so seconds that the neutrinos were detected, because neutrinos, like light, would be trapped if an event horizon formed. Astronomers have been searching for that neutron star ever since, with increasingly tight limits. Lack of evidence of a neutron star does not prove the existence of a black hole, but researchers began to think about the possibility that black-hole formation and supernova explosion could co-exist (Chevalier 1997). Some work has entertained the notion that stars with main-sequence masses as low as 20 $M_\odot$, including SN 1987A itself, might make black holes. On the other hand, neutron stars were found in binary systems and young clusters that pointed strongly to the ability of some very massive stars to form neutron stars (Muno et al. 2006).

This topic is also made more complex by the realization that the mass of iron cores and the surrounding density structures, which presumably play an important role in determining whether a neutron star or black hole results from core collapse, vary nonmonotonically with ZAMS mass—even chaotically in the sense of being sensitive to initial conditions (Sect. 8.5). We now recognize that the outcome may depend on rotation, metallicity, and binary evolution. There may be islands of both explodability (Sect. 9.4.3) and of collapsibility (Kochanek 2014; Pejcha and Thompson 2015) in a complicated parameter space (Sects. 8.5, 24.2.1, and 24.2.2).

Work on collapse physics illuminates these issues. Considering the physics of the collapse of cores of about $M_{\rm Ch}$ outlined in this chapter, it is unlikely that black holes can form by direct collapse of such cores. The matter within the stalled shock is sonically isolated from the supersonic infalling matter of the outer iron (or ONeMg) core. When the inner core bounces and produces the stalled shock, the amount of matter destined to fall onto the inner core cannot be communicated. The basic process of homologous infall, bounce, and bounce shock that results in the formation of a PNS must occur whether or not the core is doomed to ultimate collapse.

The basic process that differentiates the creation of a black hole rather than a neutron star is fallback, a notoriously uncertain, multidimensional problem. The distributions of neutron-star and black-hole masses contain clues to the physics of supernova explosions and the roles of neutrino-driven convection and fallback (Sekiguchi and Shibata 2011). Instabilities of various kinds may promote the growth of a convective region beneath the standing shock that can deliver energy to the gain region and revive the shock (Sect. 9.4.3). Higher rates of fallback tend to suppress the explosion. There are tradeoffs of density structure and the timescales of instabilities and fallback that could abet the formation of neutron stars or black holes, depending on circumstances. A rapidly-growing instability, e.g., Rayleigh–Taylor, might initiate convection behind the standing shock and build up enough energy in that region to yield a relatively rapid explosion on the timescale of 300 ms, leaving a neutron star. If insufficient energy is accumulated in the turbulent region, the SASI (Sect. 9.4.7) might build up a sufficient turbulent region on a longer timescale, 1000 ms. If neither mechanism works, then fallback could yield a black hole with only a weak explosion. The transition from little fallback yielding neutron stars to substantial fallback leading to black holes might be abrupt with respect to ZAMS mass (Fryer et al. 2012; Belczynski et al. 2012). As outlined, this basic scenario does not address issues of the nonmonotonic nature of the late stages of evolution, the possibility of islands of explodability, and magnetorotational effects, e.g., that the MRI (Sect. 9.5.3) may develop strong magnetic fields and turbulence on timescales of tens of milliseconds that could compete with Rayleigh–Taylor instabilities.

If the collapsing core is considerably greater than M_{Ch}, direct collapse to a black hole without the intermediate step of a PNS may be possible. This may occur for very massive stars depending on their metallicity, rotation, and mass loss (Heger et al. 2003). At even higher ZAMS masses, collapse to form black holes may occur directly from the main sequence due to general relativistic effects (Fowler and Hoyle 1964). For nonrotating models, this may occur for masses larger than 10^6 $M_\odot$, but with rotation this mass might be increased to 10^9 (Fowler 1966). Alternatively, even such high masses might be stabilized and powered by the annihilation of dark matter within them (Freese et al. 2008).

Further discussion of black holes is given in Sect. 24.2.2.

9.7 Quark and Strange Stars

While most work on collapse involves either hadronic (in practice baryonic, mostly neutrons and some protons) stars or black holes, there is the possibility that a collapsed object could be composed of more exotic species such as quarks. A transition of the composition from hadrons to quarks could release energy that might be observable and even power an explosion (Ouyed et al. 2002). The binding energy is already large. A phase transition might release a comparable amount of energy. If the core of a neutron star makes a transition to a phase composed of up and down quarks at much higher density at the point where strange quarks become the favored state of matter, considerable energy, $\sim$100 B, could be released. There might be a first-order phase transition from hadrons to quarks in the accretion phase that follows core bounce that could also, in principle, affect the neutrino flux or nucleosynthesis (Sagert et al. 2009; Fischer et al. 2011; Nishimura et al. 2012; Nakazato et al. 2013b; Claret 2014; Drago et al. 2014). Objects composed of strange quarks and Λ hyperons might have specialized, observable properties (Mannarelli et al. 2014).

Current limits on neutron-star masses constrain the equation of state of high-density matter and hence all these interesting scenarios.

9.8 Summary

The nature of core collapse surely depends on the composition and structure of the core, ONeMg or iron. The outcome of collapse may depend on the detailed structure of the inner regions of the progenitor, including dynamic, turbulent burning, at the moment of collapse. The outcome also will depend on the density gradient of the outer material: the compactness or overburden. There may be islands of explodability and of collapsibility in the mass distribution of progenitors. The magnetorotational state of the progenitor and the presence and action of a binary companion will also affect the collapse dynamics.

Proper treatment of the core-collapse problem must account for multidimensional neutrino transport of all flavors of neutrinos, neutrino-induced convection, SASI, and Rayleigh–Taylor and Richtmyer–Meshkov instabilities and associated "fingers" and plume-like flows. The most accurate simulations will employ general relativity. How these effects ultimately affect the outcome remains a work for the future as computational capacity increases.

Core collapse almost certainly happens in a rotating context. There will undoubtedly be seed magnetic fields, and there are a variety of processes—compression, field-line wrapping, the α–Ω dynamo and the MRI and subsequent nonlinear effects—that may produce a dynamo action to amplify magnetic fields to dynamically significant levels. No current simulations properly simultaneously resolve the most rapidly-growing modes of the MRI and the stellar length scales. Proper treatment of MHD effects requires full 3D simulations. Some progress in following the growth and saturation of magnetic fields generated by the MRI and other processes may be made in the near future by adopting subgrid models of the MRI.

Pulsars are not necessarily born rapidly rotating, but the original spin for an observer of neutron stars is effectively the final spin for the supernova process. We still do not sufficiently well understand how rotation and magnetic fields are amplified and dissipated before the shock hits the surface and the ejecta dissipate to reveal the neutron star. Rapidly-rotating PNS are subject to NAXI. These effects have been explored to some extent, but very little work has been done coupling the non-axisymmetric effects with the magnetic field. Much of the rotational energy could be dissipated by the production of intense Alfvén waves.

We do not yet have an adequate understanding of which core-collapse events yield typical pulsars and which yield magnetars, CCOs, black holes, and gamma-ray bursts.

Chapter 10
Pair-Instability Supernova Models

10.1 Introduction

Above some mass limit, the late stages of evolution of massive stars are predicted to change qualitatively (Sect. 8.6). Rather than forming iron cores that collapse, models reach conditions ($T \gtrsim 10^9$ K, $\rho \lesssim 10^5$ g cm^{-3}) where electron–positron pairs are copiously produced. This softens the equation of state and renders the star dynamically unstable (Rakavy and Shaviv 1967; Barkat et al. 1967; Fraley 1968). The resulting explosions are called pair-instability supernovae (PISN). Much effort has gone into understanding the formation of the first stars of the Universe (many of which may have been very massive), the evolution of low-metallicity stars that reach pair-instability conditions, and the predicted properties of their explosions. The discovery in the contemporary Universe, in which the metallicity is not extremely low, of superluminous supernovae that had some observational characteristics of PISN (Sect. 18.4) brought a new focus to the issue of which stars can undergo pair instability. This chapter summarizes the dynamics and outcome predicted for stars that encounter this instability.

10.2 Pre-Explosion Evolution

Evolutionary calculations predict that the pair instability occurs after models have undergone central helium burning and formed massive cores composed primarily of oxygen. In nonrotating models, the pair instability is predicted to occur for oxygen-core masses in the range 60–130 $M_\odot$, with slightly more mass within the outer boundary of the helium core. The ZAMS masses that lead to cores in this range depend on metallicity, rotation, and mass loss. For nonrotating models of very low metallicity, the ZAMS mass is in the range 140–260 $M_\odot$ (Heger and Woosley 2002; Chen et al. 2014a). For rapidly-rotating models of low metallicity, the lower ZAMS

D. Branch, J.C. Wheeler, *Supernova Explosions*, Astronomy and Astrophysics Library, DOI 10.1007/978-3-662-55054-0_10

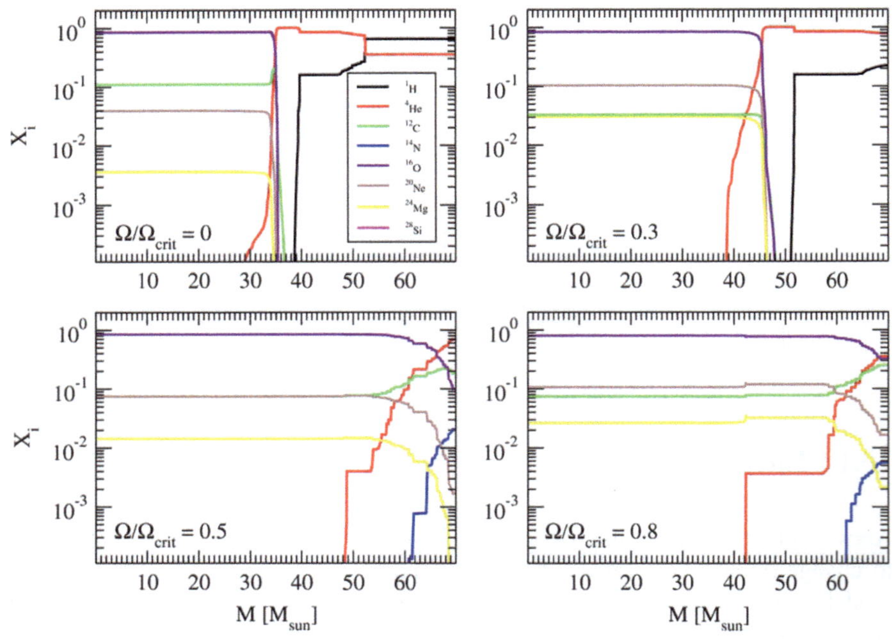

Fig. 10.1 Composition structure of models of a star of ZAMS mass 70 $M_\odot$ just prior to core oxygen ignition as a function of the rotation rate on the ZAMS in units of the critical Keplerian velocity. Note that the mass of the oxygen core (*purple line*) grows with rotation rate so that with $\Omega/\Omega_{crit} \gtrsim 0.5$, the oxygen core crosses the threshold, ~65 $M_\odot$ for the pair-instability and encompasses essentially the entire star. From "Effects of Rotation on the Minimum Mass of Primordial Progenitors of Pair-instability Supernovae" (Chatzopoulos and Wheeler 2012a). © AAS. Reproduced with permission

mass can be reduced to ~65 $M_\odot$ (Chatzopoulos and Wheeler 2012a; Yoon et al. 2012a; Fig. 10.1).

At higher metallicity, up to ~1/3 solar (above which PISN are not expected to occur), higher mass-loss rates cause the range of ZAMS masses leading to PISN to shift to higher values (Yusof et al. 2013). At these metallicities, PISN at the high end of the ZAMS-mass range (~500 $M_\odot$) are expected to be similar to those from low-metallicity stars; at the low end of the range (~150 $M_\odot$) PISN may resemble somewhat overluminous SN IIP (Kozyreva et al. 2014a; Whalen et al. 2014b).

An aspect of pair instability that has yet to be explored properly is the 3D nature of the evolution that precedes the explosion. This is a general problem in stellar evolution, but especially important in high-mass stars because they are so close to neutral stability and hence especially sensitive to displacement of matter with little buoyancy stability. There may, for instance, be plume mixing between the helium and oxygen cores that is not captured at all by standard spherically-symmetric stellar-evolution simulations invoking standard prescriptions for convective mixing. Inward mixing of helium might give rise to sporadic burning. The combined effects

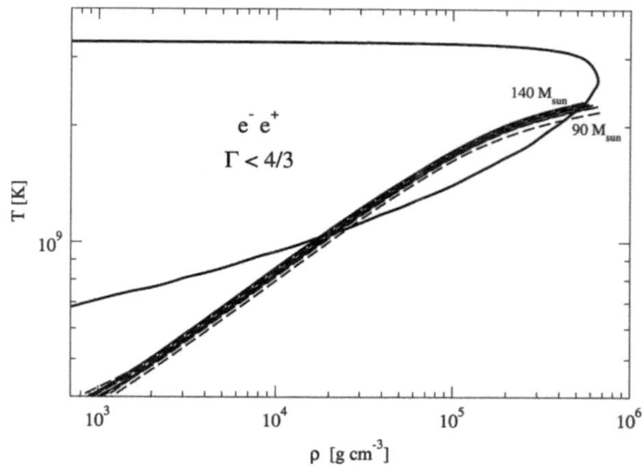

Fig. 10.2 Structure of rotating models with ZAMS mass from 90 to 140 $M_\odot$ at the point of explosion by the pair-instability mechanism. The *thick black line* denotes the boundary of the pair-formation region. The centers of the models evolve into and then out of the pair-formation region before the explosion ensues. The instability is triggered by the matter in an off-center shell where Γ_1 is a minimum. From "Finding the First Cosmic Explosions. IV. 90–140 $M_\odot$ Pair-instability Supernovae" (Smidt et al. 2015). © AAS. Reproduced with permission

of subsequent adiabatic expansion and neutrino cooling would shift the evolution to a lower adiabat, leading to higher density at a given temperature and shifting a given model away from pair instability and toward iron-core collapse.

10.3 Dynamics of Pair Instability

The pair instability sets in when an appropriately mass-averaged value of the adiabatic index Γ_1 becomes less than 4/3. In 1D models, this condition is met when the center of the star has already attained sufficiently high densities to suppress pair formation (Fig. 10.2). The pair instability thus occurs in an off-center shell where Γ_1 is a minimum. The rapid contraction of the shell drives it inward, leading to compression and heating of the inner core of oxygen. Unlike the case of iron-core collapse, the oxygen in these stars is subject to strong energy release by thermonuclear burning. For nonrotating models, the star is totally disrupted. At the upper end of the range of oxygen-core masses that explode as PISN, a large amount of oxygen is burned. This can yield a very large energy, $\sim$100 B, and tens of solar masses of ^{56}Ni. Throughout most of the mass range, the energy and ^{56}Ni production are considerably lower; at the low end of the range, very much lower. Demographics would argue that the lower-mass events are likely to dominate any observed sample of PISN.

Modest rotation does not change the picture qualitatively (Chatzopoulos et al. 2013a; Chen 2015). Rotating oxygen cores produce somewhat less energetic PISN and eject less ^{56}Ni than nonrotating cores of equal mass because rotational support suppresses the collapse, especially around the stellar equator, leading to a weaker shock. It is possible that with sufficient rotation, a star that would otherwise explode as a PISN could instead collapse to form a neutron star or black hole.

10.4 Nucleosynthesis

PISN models have rather characteristic nucleosynthetic yields (Heger and Woosley 2002; Ohkubo et al. 2003, 2009) that depend on the mass of the oxygen core and on the peak temperature reached during the contraction. The yields may be affected by rotation. Nonrotating models of PISN are predicted to yield approximately a solar distribution of intermediate-mass elements of even electric charge such as silicon, sulfur, and argon. There is no stage of stable nuclear burning after core helium burning to provide a significant neutron excess. At low metallicity, the models are thus strongly deficient in intermediate-mass elements with odd nuclear charge such as aluminum and scandium. At higher metallicity, this odd-even effect is less pronounced (Kozyreva et al. 2014b). A lack of conditions conducive to the s-process and the r-process yields virtually no elements heavier than zinc.

Searches for the predicted PISN abundance distribution in Galactic stars of very low metallicity have not revealed the pattern expected for PISN (Sect. 24.8). The nucleosynthetic imprint of PISN on stars in the early universe is, however, likely to be a complicated matter, depending on the degree of mixing during the explosion and on how PISN ejecta are mixed into the ambient CSM and ISM (Joggerst and Whalen 2011).

10.5 Predicted Observational Properties

The decay of tens of solar masses of ^{56}Ni can potentially yield a very bright explosion that could be observed at cosmological distances, but the large ejecta mass and associated diffusion time mean that a substantial portion of the ^{56}Ni can decay before maximum light. This yields a lower luminosity than the large ^{56}Ni mass might imply. The brightness and width of the light curve also depend on the degree to which the hydrogen envelope is retained as well as other factors (Scannapieco et al. 2006; Kasen et al. 2011; Pan et al. 2012c; Dessart et al. 2012b, 2013b; Smidt et al. 2015). The massive ejecta will trap γ-rays so the tail of the light curve should follow the rate of ^{56}Co decay for hundreds of days.

Stars that retain their hydrogen envelopes and hence have extended plateaus in their light curves are likely to be observable for long times (Fig. 10.3). At high redshift, the light curve is slowed even further by the time-dilation factor, $1+z$. This

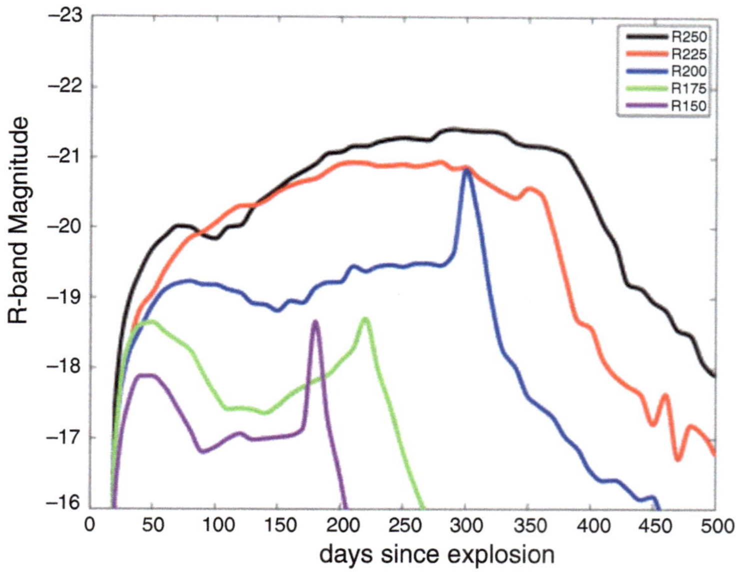

Fig. 10.3 Sample rest frame R-band light curves for nonrotating PISN models with ZAMS mass from 150 to 250 $M_\odot$ that explode as RSGs. Figure 1 "Pair-instability supernovae at the epoch of reionization" from Pan et al. (2012c), by permission of Oxford University Press on behalf of the Royal Astronomical Society

renders problematic the detection of such events at high redshift by their temporal variation. The shorter breakout shock and subsequent fireball might be very bright and observable as a transient even at cosmological distances, but then the relevant rest-frame timescales are only hours for extended hydrogen envelopes and even shorter for progenitors with less extended envelopes (Kasen et al. 2011).

It is possible that during the RSG phase, the hydrogen envelope of a PISN progenitor develops pulsational instabilities that remove most of the hydrogen without affecting the evolution of the core. If so, interaction between the PISN ejecta and dense hydrogen-rich CSM could strongly affect the observational properties (Moriya and Langer 2015; Sect. 18.2).

Models with bare helium cores can show a secondary maximum in the near-IR that is reminiscent of that seen in SN Ia (Sect. 5.5.1). The physics is similar in the two cases (Kasen et al. 2011). As the ejecta expand and cool, doubly ionized iron-group elements recombine to singly ionized species, which have much greater line emissivity in the near-IR, leading to an increase in the luminosity there. The secondary maximum might provide evidence for large amounts of ^{56}Ni.

Spectra on the rise to maximum brightness depend on the composition of the outer layers, and hence on the mass-loss and mixing history. Sample model spectra are shown in Fig. 10.4. The ejecta might contain a substantial amount of helium that is not sufficiently excited by γ-ray deposition to be detectable in spectra. After peak brightness, the deeper ejecta, which are rich in iron-group elements independent of

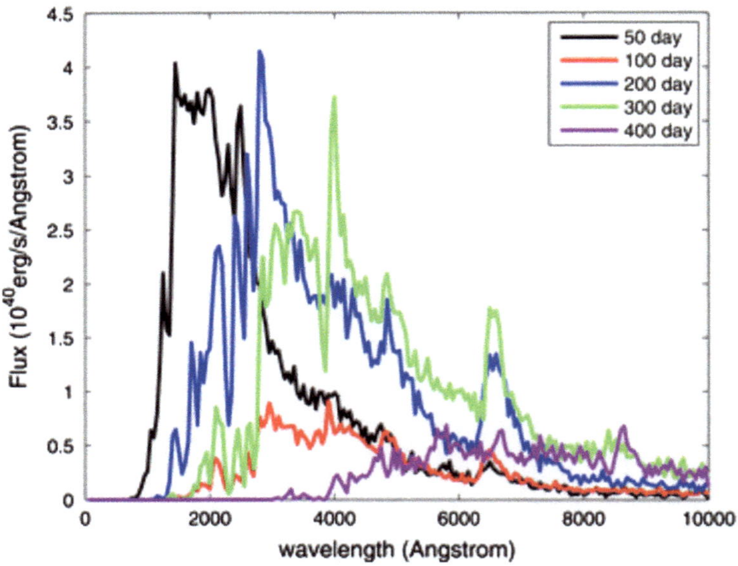

Fig. 10.4 Evolution in time of model spectra of a nonrotating model with ZAMS mass of 250 $M_\odot$ that explodes as an RSG by the pair-instability mechanism. Figure 3 of "Pair-instability supernovae at the epoch of reionization" from Pan et al. (2012c), by permission of Oxford University Press on behalf of the Royal Astronomical Society

the details of the outer envelope, will be subject to strong line blanketing by Fe I and Fe II lines and hence will tend to have rather low photospheric temperatures and red colors (Dessart et al. 2012b, 2013b; Chatzopoulos et al. 2015). It is possible that PISN could be recognized by deconvolving their spectra at a single epoch from that of the host galaxy, thus overcoming the problems of recognizing PISN from their slow light curves.

Because the *JWST* promises to view very early epochs in the Universe, when the mass function may have been weighted to high masses and hence some of the stellar explosions may have been PISN, much effort has gone toward predicting the detectability of PISN with *JWST* (Scannapieco et al. 2006; Hummel et al. 2012; Pan et al. 2012c; Smidt et al. 2015).

10.6 Upper Limit to PISN: Collapse to Black Holes

Not all PISN are expected to disrupt completely. Above some ZAMS mass, e.g., 500 $M_\odot$ at high metallicity (but still with an oxygen-core mass below 130 $M_\odot$), the collapse proceeds too far to be entirely reversed; some of the central matter becomes iron and collapses to form a massive ($\sim$100 $M_\odot$) black hole (Whalen et al. 2014b).

For oxygen-core masses above about 130 $M_{\odot}$, the collapse induced by pair formation is sufficiently robust that oxygen burning cannot reverse the implosion at all. The entire cores of such stars are predicted to collapse to form black holes (Fowler and Hoyle 1964; Wheeler 1977; Bond et al. 1984; Heger and Woosley 2002). This situation might lead to the direct collapse of the entire star to form a black hole with no substantial explosion and luminous output, but if the star were rotating sufficiently, an accretion disk might form around the black hole (Fowler and Hoyle 1964). That, in turn, could be conducive to the formation of a jet that would explode the star and affect the nucleosynthesis (Fryer et al. 2001; Ohkubo et al. 2006).

10.7 Pulsational Pair Instability: PPISN

In his unpublished 1967 Ph.D. thesis, Zalman Barkat noted that some models that skirted the pair-instability regime on the low-mass side ejected shells of matter, but survived the instability and continued to evolve. This pair-instability process ejects a few $M_{\odot}$ of the outer portion of the oxygen core and any matter beyond it, but does not disrupt the entire core. These models then settle back into hydrostatic equilibrium until the instability is triggered again, leading to another, perhaps more energetic ($\gtrsim$1 B) dynamical episode. In most models this repetitive process ends in core collapse; in others a genuine thermonuclear PISN explosion disrupts the remaining core.

The ejection of multiple shells by repeated dynamical, but not catastrophic, pair instability can lead to collisions of subsequent shells with prior shells. If a collision occurs at a radius at which energy can be efficiently radiated (diffusion time comparable to dynamical time; Sect. 5.3), the result can be a very bright optical display comparable to or brighter than a supernova (Whalen et al. 2014a; Chen et al. 2014b). The succession of shell ejections may also be followed by a full-fledged PISN, the ejecta of which collides with the previously ejected shells. This repetitive shell-collision process, with or without a final PISN, is called a pulsational pair-instability supernova (PPISN; Woosley et al. 2007).

The range of stellar masses that produce PPISN is again a function of the metallicity and rotation of the progenitor. The oxygen-core mass is in the range 40–60 $M_{\odot}$, rather independent of details. For nonrotating models of low metallicity, this corresponds to a ZAMS mass of about 90–130 $M_{\odot}$, but this range can be substantially reduced to around 50–70 $M_{\odot}$ with rapid rotation and associated mixing that lead to higher core mass for a given ZAMS mass (Chatzopoulos and Wheeler 2012a; Yoon et al. 2012a). For rapidly rotating models that mix, the ejected shells can be hydrogen deficient but rich in helium and oxygen (Chatzopoulos and Wheeler 2012b). The kinetic energy of the ejected shells tends to decrease with greater rotation. Hydrogen-deficient circumstellar shells may be relevant to some superluminous supernovae (Sect. 18.3).

10.8 Summary

PISN are predicted to occur under rather generic conditions for oxygen cores in the range from about 60 to 130 $M_{\odot}$. The corresponding ZAMS masses depend on metallicity and rotation. Off-center instability leads to compression, ignition, and explosion of the oxygen core with no remnant. At the upper end of the oxygen-core mass range, large kinetic energy, $\sim$100 B, and tens of solar masses of ^{56}Ni are predicted. The resulting explosion can be very bright if not tamped by a massive hydrogen envelope. In the absence of extensive presupernova mass loss, PISN models show a range of light curves and spectral evolution depending on the ZAMS mass and metallicity. During the presupernova RSG phase, however, a PISN progenitor might develop pulsational instabilities in the envelope that could remove the hydrogen-rich layers. If so, interaction between the PISN ejecta and dense, hydrogen-rich CSM could strongly affect the observational properties.

For oxygen cores in the range 40–60 $M_{\odot}$, models predict the repetitive ejection of several shells that may collide to produce luminous displays (PPISN), prior to core collapse or a final disruptive PISN explosion.

Because cosmological models raise the possibility that the first stars born at nearly zero metallicity may be especially massive, there is some expectation that some of the first stars may explode as PISN and PPISN and be observable with *JWST*.

Most published work on PISN and PPISN is based on spherically symmetric models. In spherical models, energy released in connection with the pair instability can only go one way: outward. All this needs to be checked with multidimensional models of the progenitor evolution and explosion.

Chapter 11
Supernova 1987A

11.1 Introduction

Supernova 1987A (Figs. 11.1 and 11.2) in the Large Magellanic Cloud (LMC; Fig. 11.3), the first naked-eye supernova in the era of modern astronomy, was observed throughout the electromagnetic spectrum and was the first detected astronomical source of neutrinos other than the Sun. Numerous reviews of SN 1987A have been published (Arnett et al. 1989; Hillebrandt and Höflich 1989; Imshennik and Nadyozhin 1989; McCray 1993). SN 1987A was a supernova of Type II that presented many surprises. It continues to be a challenge, with much yet to be learned about the evolution of its progenitor, the mechanism and dynamics of the explosion, the compact remnant, and its evolution to become a young SNR.

11.2 The Progenitor Star and the Triple-Ring System

The progenitor of SN 1987A was identified in archival images as a star known as Sk-69° 202 (Sanduleak 1969; White and Malin 1987; Walborn et al. 1989). The star showed no sign that it was about to explode, a fact that has been little remarked and virtually unstudied.

One surprise was that the progenitor was a blue supergiant, of spectral type B3 I. Given the distance to the LMC of 50 kpc (170,000 light years), the bolometric luminosity of the progenitor was $\sim 1.3 \times 10^5\, L_\odot$. According to standard single-star evolutionary models, this luminosity implies a helium-core mass of about 5–6 $M_\odot$ and a ZAMS mass of about 18 $M_\odot$. Barring an exotic binary evolution, the mass of the immediate presupernova would have been lower than the ZAMS mass, owing to mass loss by winds during the progenitor's evolution. Analysis of the light curve and spectra suggest a hydrogen envelope of about 8 $M_\odot$, for a total mass of about 13 $M_\odot$ at the time of explosion. There is evidence that the envelope was enriched

D. Branch, J.C. Wheeler, *Supernova Explosions*, Astronomy and Astrophysics Library, DOI 10.1007/978-3-662-55054-0_11

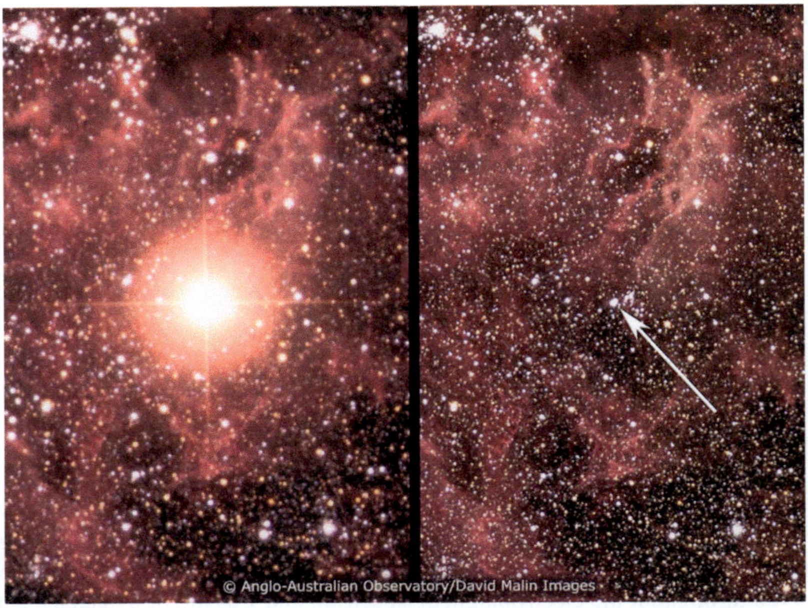

Fig. 11.1 SN 1987A, after and before. The *arrow* in the right panel points to the progenitor star, Sk-69° 202. Credit: Hubble Heritage, Anglo-Australian Observatory, David Malin

in helium, with mass fraction $Y \simeq 0.4$. The expectation was that such a moderately massive star should explode as an RSG. The blue supergiant nature of the progenitor of SN 1987A accounted for the unorthodox light curve, but the physical reasons for its explosion in this state remain uncertain. Much theoretical effort has gone into exploring why the progenitor was so compact, with a radius of about 3.5×10^{12} cm (45 $R_{\odot}$). A low opacity in the envelope, owing to the helium enrichment as well as to the low metallicity of the LMC (30% of solar metallicity), may have been a factor, but this does not seem to be the whole story. With judicious choices of convective overshooting and mass-loss history, stellar evolutionary models can be constructed such that core collapse occurs during a blue supergiant phase, but a consistent explanation must also account quantitatively for the populations of red and blue supergiants in the LMC, as well as the properties of and abundances in the CSM associated with Sk-69° 202.

The highly structured nature of the CSM was another surprise (interaction with the CSM is discussed in more detail in Sect. 11.6). The most dramatic circumstellar feature was the triple-ring system (Fig. 11.4). The inner (or "equatorial," by assumption) ring is circular, with angular radius $\lesssim$1 arcsecond, linear radius 6.2×10^{17} cm (0.7 light years), and a thickness of 15% of its radius. The ring appears as an ellipse on the sky because it is tilted about 45° with respect to the line of sight, with the northern edge toward Earth. Above and below this inner ring, with axes nearly aligned with it, are two fainter, outer (or "polar") rings, both roughly three

Fig. 11.2 SN 1987A in the vicinity of the 30 Doradus nebula. Credit: Hubble Heritage; Anglo-Australian Observatory, David Malin

Fig. 11.3 The Large Magellanic Cloud, an irregular galaxy that was host to SN 1987A. The bright 30 Doradus nebula is to the left of center. Credit: Bert Van Dankelar

Fig. 11.4 The triple-ring system of SN 1987A. Note also the distinctly elongated shape of the ejecta in this projected image. The actual shape of the ejecta is complex (Sect. 11.4.4). Credit: ESA,STScI, HST, NASA

times larger than the inner one and fainter in surface brightness by a factor about 10. The overall geometry of the triple-ring system is roughly bipolar, reminiscent of an hourglass, with the inner ring as the narrow neck and the outer rings as the upper and lower rims. It may be, however, that the three rings are all there is: rings, without hourglass "walls" (Burrows et al. 1995; Morris and Podsiadlowski 2009). Spectroscopic analysis revealed that, in all three rings, helium is enriched and nitrogen is enhanced at the expense of carbon and oxygen (Lundqvist and Fransson 1996; Maran et al. 2000), indicating that the matter in the rings had been processed by the CNO cycle before being ejected from the progenitor when it was in a post-helium-core-burning phase (Podsiadlowski 1992). The inner ring expands at about $10\,\mathrm{km\,s^{-1}}$ and the outer ones a few times faster (Crotts and Heathcote 2000), thus the sizes and velocities of the rings are consistent with the three being ejected coevally, or nearly so, about 20,000 years before explosion; a short time compared to the 10^7 year lifetime of the progenitor. The three rings are not exactly co-aligned, suggesting some dynamics of the progenitor system that has not been fully elucidated.

Beyond the ring system and roughly aligned with its axis is another structure, called the Napolean's Hat nebula because of its apparent morphology of a central peak and two opposing upswept rims. This feature presumably also formed by mass loss from the progenitor star.

One interacting-winds model posits that the inner ring formed from a slow, dense, equatorially-concentrated RSG wind that was later compressed by a faster, lower density, blue supergiant wind (Lundqvist and Fransson 1991; Luo and McCray 1991; Blondin and Lundqvist 1993; Burrows et al. 1995; Chevalier and Dwarkadas 1995). The equatorial density enhancement could have been due to the presence of a binary companion (Wang and Mazalli 1992; Blondin and Lundqvist 1993), although any surviving companion must be dimmer than an F6 main-sequence star (Graves et al. 2005). It may be that the existence and nature of the inner ring can be understood in this way, but accounting for the outer rings is more difficult.

Another possibility is that the progenitor was the product of a merger of two stars. One model proposed a merger of stars that were initially 15 and 5 $M_\odot$ and in a wide orbit, $P \simeq 10$ years (Morris and Podsiadlowski 2007, 2009). In this model, the primary filled its Roche lobe after helium-core burning; its convective envelope led to dynamically unstable mass transfer, a common envelope, and eventual merger of the 5 $M_\odot$ star with the core of the primary. During the early inspiral of the secondary within the common envelope, mass ejection peaked at mid-latitudes (about 45°) owing to the rotational distortion of the common envelope. As the immediate postmerger RSG contracted, it lost angular momentum via a rotationally-enforced equatorial outflow of a few solar masses, some of which became the inner ring. The merger product soon contracted to become a blue supergiant, and a fast wind swept up the mid-latitude material to form the outer rings. Hydrodynamical simulations can be tuned to produce a morphology that is consistent with the circumstellar environment as well as the properties of the progenitor, including helium enhancement and other abundance anomalies (Morris and Podsiadlowski 2009). If there had been a merger, then attempts to deduce the ZAMS mass from the luminosity and presumed helium core mass may be quantitatively incorrect. Other suggestions to account for the rings require neither an RSG phase nor a binary companion (Sect. 11.7).

11.3 Neutrinos and the Compact Remnant

The most fundamental prediction of core-collapse theory is neutrino emission (Chap. 9), and the most important observation of SN 1987A was the detection on February 23, 1987, of neutrinos (actually anti-electron neutrinos) in three experiments across the globe (Bionta et al. 1987; Hirata et al. 1987; Alexeyev et al. 1988). The detection of two dozen neutrinos, among the $\sim 10^{57}$ that were emitted, confirmed the basic process of core collapse. The distribution of neutrino arrival times and energies, as well as the inferred total energy emitted in neutrinos ($\sim$100 B), are generally regarded to have been in satisfactory agreement with standard core-collapse theory (Pagliaroli et al. 2009; Vissani 2015). The information content was not sufficient, however, to usefully constrain the explosion mechanism.[1]

The realization that neutrinos had been detected came days after the optical discovery, but reconstruction of the time history showed that the neutrinos preceded the first detected photons to Earth by 16 h (Hirata et al. 1987; Arnett et al. 1989). Traveling at essentially the speed of light, the neutrinos arrived at the surface of the star about 100 s after core collapse. The shock wave from the collapse, propagating closer to the speed of sound, took about 2 h (a somewhat model-dependent result; Ensman and Burrows 1992) to reach the surface and initiate the photon emission. A simple estimate of this timescale follows by assuming that the shock obeyed a constant-density Sedov similarity solution in the hydrogen envelope so that, for a progenitor radius of 3.5×10^{12} cm, a hydrogen-envelope mass of 10 $M_\odot$, and an explosion energy in units of 1 B, the breakout time would be $t_b \simeq 2.2\text{h}\ (M_{10}/E)^{1/2}$ (Arnett et al. 1989).

The fact that the photons maintained that lag of only a few hours during the 170,000 years of travel time to Earth allowed a number of interesting constraints to be placed on the properties of neutrinos: e.g., upper limits to the mass of 5.8 eV (Pagliaroli et al. 2010); to the electric charge, in units of the electron charge, of 3×10^{-17} (Arnett et al. 1989); and to the magnetic moment, $\lesssim 10^{-13}\ \mu_B$ (Goyal et al. 1995). The speed of the neutrinos was the same as that of the photons, the speed of light, to within one part in 10^8, comparable to the measurement errors of the speed of light (Longo 1987; Stodolsky 1988). Some of the other physics implications of neutrino observations of SN 1987A and future core-collapse events are discussed in Sect. 25.5.1.

The neutrino signal lasted for 10 s, which means that a neutron star existed for at least that long. The finite time also shows that the neutrinos were trapped and subject to diffusion (Sect. 9.4.1). Three years after explosion, the photon luminosity dropped below that of the pulsar-driven Crab Nebula (Fig. 11.5). In spite of intense searches, no credible evidence for pulsed radiation has been found in any waveband (Ng et al. 2011), although tentative evidence for synchrotron emission from a pulsar wind nebula has been seen (Zanardo et al. 2014). Even if pulsar beams were not directed towards Earth, energy input from a pulsar would energize the ejecta and raise its luminosity. If a pulsar is present, it must be less energetic than the known pulsars in young SNRs. The lack of a luminous pulsar could be the consequence of an initially low rotation rate and/or weak magnetic field. Alternatively, if the initial magnetic field were very strong ($\sim 10^{15}$ G, that of a magnetar; Sects. 5.4.4 and 24.2.1), the neutron star may have spun down rapidly (Ögelman and Alpar 2004). At present, a pulsar with a period greater than 100 ms and $B < 10^{12}$ G would not be detectable, and many such pulsars, although not young ones, are known to exist in our Galaxy.

[1] Motivated mainly by a possible detection of a burst of five neutrinos almost 5 h prior to the main burst (Aglietta et al. 1987), a minority school of thought maintains that the neutrinos came in several bursts, possibly owing to a two-stage collapse of the core (de Rújula 1987; Imshennik and Ryazhskaya 2004).

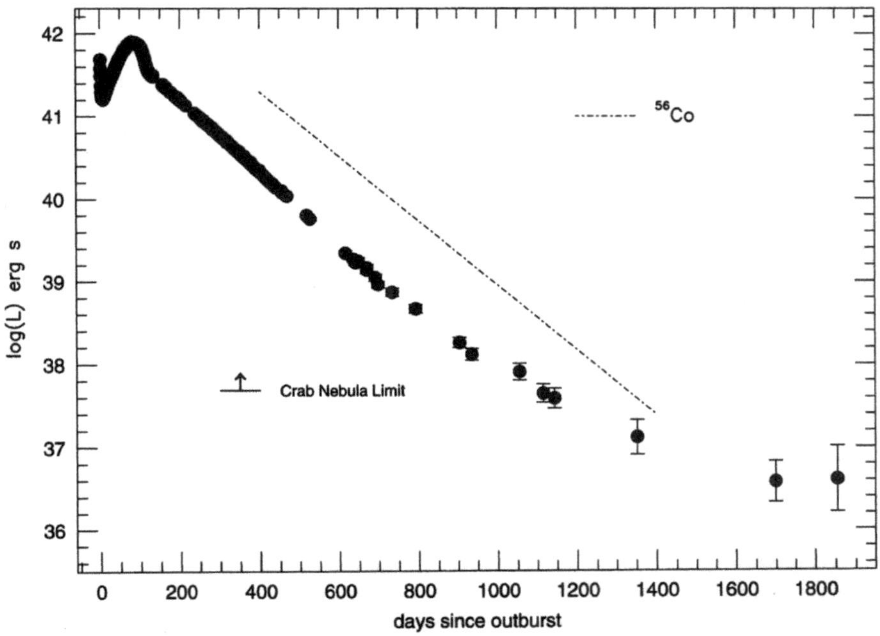

Fig. 11.5 The bolometric light curve of SN 1987A for the first 5 years. The *dashed line* shows the slope appropriate to the decay of ^{56}Co. The luminosity of the pulsar in the Crab Nebula is an underestimate, the current value being 4.6×10^{38} erg s^{-1} (Abdo et al. 2010a). Credit: N. Suntzeff

Upper limits to steady thermal radiation from the surface of a neutron star are rather stringent: $\lesssim 10^{34}$ erg s^{-1} in the optical and X-rays (Park et al. 2004; Graves et al. 2005; Shtykovskiy et al. 2005; Ng et al. 2009).

In the optical and UV, a neutron star could be hidden in several ways, e.g., by a photon-driven wind from the surface of the neutron star that prevented accretion of fallback material (Fryer et al. 1999), or by absorption by dust in the ejecta (Wesson et al. 2015). Another possibility is that SN 1987A produced the same sort of *central compact object* as Cas A (Sect. 7.4.1). At the distance of SN 1987A, such an object would not be detectable with current techniques. Further possibilities are that owing to mass fallback or some change to the equation of state, the compact remnant is now a black hole, accreting at a low rate (Fryer et al. 1999), that the neutron star made a transition to a quark ("strange") star that has been cooled by means of brief, intense pulses of neutrino emission (Chan et al. 2009), or that the transition was to a rapidly cooling, solid quark-cluster star (Liu et al. 2012b). Improving the optical and UV limits on a point source in SN 1987A will be difficult, but it may prove possible to establish tighter limits in the IR.

11.4 Ejecta

Extensive UVOIR photometry and spectroscopy (Menzies et al. 1987; Blanco et al. 1987; Hanuschik and Dachs 1988; Catchpole et al. 1989; Whitelock et al. 1989; Hamuy et al. 1988; Suntzeff et al. 1988; Suntzeff and Bouchet 1990; Bouchet et al. 1991) were obtained from southern hemisphere observatories, and UV emission was observed from space with the *IUE* (Sonneborn et al. 1987; Kirshner et al. 1987). The ejected matter can be only partially resolved spatially (with *HST*, the *Atacama Large Millimeter/Submillimeter Array* (*ALMA*), and ground-based adaptive optics), but it is well resolved in velocity space, giving some 3D information.

11.4.1 Breakout and Fireball

The shock wave generated in the core of the blue supergiant progenitor took 2 h to reach the surface (Sect. 11.3). Shock breakout was not observed directly, because the breakout timescale was short compared to the time of discovery a day after the explosion. The structure at breakout is expected to be complex, with the luminosity varying throughout the photospheric region, the gas temperature being out of equilibrium with the radiation temperature, and hence the color temperature departing from the effective temperature (Ensman and Burrows 1992). Detailed models of breakout from a blue supergiant like Sk-69° 202 produce a peak luminosity of $\sim 10^{45}$ erg s^{-1} and a total radiated energy of $\sim 10^{47}$ ergs. The peak effective temperature of models is about 6×10^5 K and the peak color temperature a factor of several higher. The high radiation temperature implies a flash of hard UV and soft X-ray emission. Most models examining this phase are spherically symmetric. In that case the light crossing time of the progenitor's photosphere was only 100 s. Although the breakout shock was almost surely asymmetric and controlled by the lateral hydrodynamic timescale, the breakout phase was undoubtedly short.

Inferences about the breakout flash of SN 1987A can be drawn from its ionizing effects on the inner circumstellar ring; these are consistent with the expected properties of the breakout flash (Sect. 5.4.1). Scattering of UV breakout photons by interstellar dust produced an echo that was observed more than a year after discovery and provided a glimpse of the UV spectrum that was emitted prior to discovery. The spectrum of the echo revealed a hot UV continuum and a C IV λ1550 absorption that apparently extended to 40,000 km s^{-1} (Gilmozzi and Panagia 1999), the highest ejecta expansion velocity detected for SN 1987A.

The first known photo by Rob McNaught and the first naked eye observations by Oscar Duhalde and Albert Jones (Kunkel et al. 1987) were during a phase of rising optical luminosity, within a day of shock breakout (Arnett et al. 1989). Although the visual luminosity rose rapidly at first, this was due not to breakout but to the cooling fireball as the flux shifted from hard UV into the optical band. The UV and bolometric emissions already were declining at the epoch of first optical detection. The UVOIR bolometric luminosity beginning a few days after the explosion declined for a week as the fireball expanded and cooled (Fig. 11.5). During this phase, the optical emission rose rapidly for a few days and then less steeply. UV spectra showed evidence of P Cygni absorption in the Mg II $\lambda\lambda$2796,2803 lines up to at least 30,000 km s^{-1}. Optical spectra revealed that by the end of the fireball phase the velocity at the photosphere had declined to 4500 km s^{-1}. Detailed radiative-transfer calculations, especially those including NLTE effects, for presupernova models evolved from an 18 M$_\odot$ main-sequence model produced reasonable agreement with the observed light curve and spectra during the fireball phase (Blinnikov et al. 2000; Dessart and Hillier 2010). The early spectroscopic evolution was caused by the recession of the photosphere and by ionization and excitation changes within the constant composition of the outermost solar mass of ejecta. During this phase, agreement between models and observations was obtained without invoking ejecta clumping or any effects of radioactive decay.

11.4.2 Radioactivity

The visual and bolometric light curves of SN 1987A reached a minimum at 1 week and then began a smooth rise toward a second maximum 3 months after explosion (Fig. 11.5). During this phase, the optical spectra were dominated by strong P Cygni lines of the Balmer series (Fig. 11.6). Early portions of this rise may be attributed to the behavior of the atmosphere, a slow increase in the radius of the photosphere at essentially constant T_{eff} (Dessart and Hillier 2010), but the major effect was energy deposited by the radioactive decay of ^{56}Ni and ^{56}Co. In models, the precise timing of the light-curve minimum depends on the amount and degree of mixing of radioactive matter and NLTE effects. Energy input from the decay of ^{56}Co became dominant beginning 3 weeks after explosion. The relatively massive hydrogen envelope caused a long diffusion time, so the luminosity at the second peak was not a direct measure of the rate of radioactive decay at that time (Sect. 5.4.2). At the secondary peak, some of the energy deposited earlier by the ^{56}Ni decay contributed along with the later energy deposited by the ^{56}Co decay. By the time of the second maximum, the photosphere had receded to 2500 km s^{-1} and reached its maximum radius of $\sim 10^{15}$ cm (Jeffery and Branch 1990).

After the second peak, the bolometric light curve declined and then flattened into a well-defined exponential tail (Fig. 11.5). Four months after explosion, a month after the second maximum, the ejecta became transparent in the optical continuum and the spectrum became nebular (Fig. 11.6), dominated by emission lines having widths corresponding to $\lesssim$2500 km s^{-1}. Because the timescale for radiation to escape from the envelope became much shorter than the age of the supernova, the luminosity (including the rising X-ray and γ-ray power) tracked the radioactive energy source and declined at the ^{56}Co decay rate (0.0098 mag d^{-1}) for almost 1000 days. From the luminosity during this phase an accurate measure of the initial ^{56}Ni mass, 0.069 ± 0.003M$_\odot$, was inferred (Bouchet et al. 1991). This implied (in a

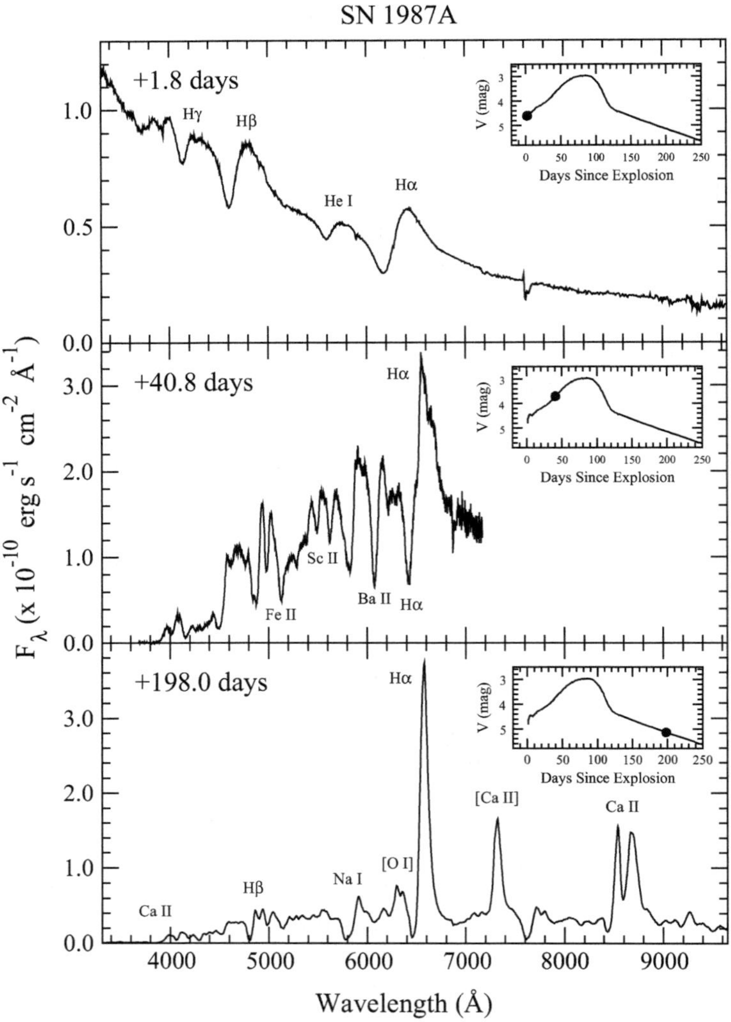

Fig. 11.6 Early (*top*), on the rise to maximum (*middle*), and nebular-phase spectra of SN 1987A. Courtesy of Mark Phillips

spherical model) that the mass cut between the collapsing core and the ejecta was near 1.6 $M_\odot$. A neutron star of baryon mass 1.6 $M_\odot$ has gravitational mass 1.45 $M_\odot$.

Prior to 1987, the role of ^{56}Ni decay in supernovae had been reasonably well established, but SN 1987A provided firm proof of the theoretical expectation of ^{56}Ni production. In addition to the strong evidence from the light-curve slope, direct evidence was provided by the detection and flux evolution of IR ([Co II] 1.547 and 10.52 μm) and γ-ray (847 and 1238 keV) lines of ^{56}Co (McCray 1993; Fig. 11.7).

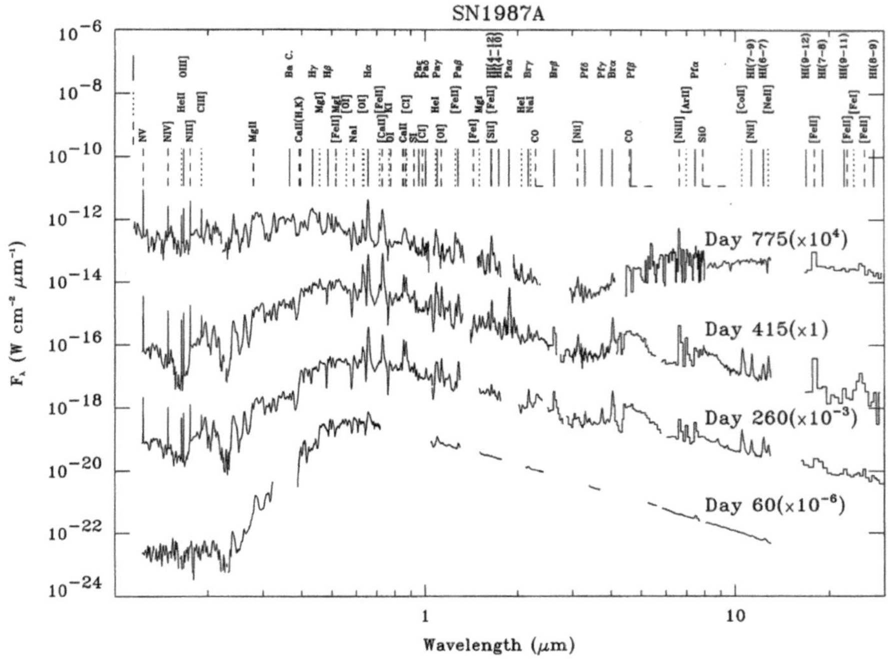

Fig. 11.7 Spectral evolution of SN 1987A to 2 years, from UV to mid-IR. From "Supernova 1987A revisited" (McCray 1993), reproduced with permission of Annual Reviews of Astronomy and Astrophysics, Volume 31, © Annual Reviews, http://www.annualreviews.org

Hydrodynamical and radiative-transfer modeling produced light curves, expansion velocities, and spectra in general agreement with observations (Utrobin 2007b; Li et al. 2012). To obtain a good fit to the observed light curve, the ^{56}Ni and ^{56}Co could not be confined to the center of the explosion and the hydrogen could not be confined to the outer envelope. In quasi-spherical models, the radioactive species had to be artificially "mixed" outward and the hydrogen inward. Some of the ^{56}Ni resided in material moving at least as fast as 3000 km s^{-1}. Current 3D models can account for this mixing (Utrobin et al. 2015). While these mixed models do not provide perfect representations of the rather smoothly varying light curve, they suggest that better results occur when the outer hydrogen envelope is more compact, with a smaller radius than basic single-star progenitor models would suggest. As remarked in Sect. 11.2, differences in the progenitor from single star models would be expected if there had been a merger. Whether a merger would fully account for the light curve behavior is not clear. The issue of mixing is discussed further in Sect. 11.4.4.

At 2.7 years, the rate of decay of the light curve decreased (Fig. 11.5). Two factors were at play: (1) the effects of ionization freezeout when timescales for recombination and cooling became comparable to the supernova age (Sect. 4.5); and

(2) the increasing importance of the decay of ^{57}Co (half-life 270 days), the daughter of ^{57}Ni. The 14 and 122 keV γ-ray lines of ^{57}Co were directly detected at 4 years.

SN 1987A provides an opportunity to study the evolution of a supernova to very late phases. After 4 years, the UVOIR light curve became essentially flat as the decay of ^{44}Ti became the dominant energy source (Figs. 11.5 and 11.12). About 10^{-4} $M_\odot$ of ^{44}Ti was ejected in the explosion (Sect. 11.5). Because of the long ^{44}Ti half-life of 60 years and the corresponding low rate of energy input, the subsequent emission was from cool ($\sim$100 K) heavy-element-rich radioactively heated gas (most of this emission, in the far-IR, went unobserved), plus some freezeout emission from the hydrogen envelope. The spectrum of the ejecta changed only very slowly. The light curve of the ejecta faded in accord with ^{44}Ti decay for 14 years. Then, from 2004 to 2009, the flux rose by a factor of two as a consequence of heat deposited by X-rays coming from CSI (Sect. 11.6). As the density of the ejecta continues to fall owing to expansion, the X-rays penetrate to deeper layers, allowing the composition of these layers to be probed for a second time (Larsson et al. 2011; Fransson et al. 2013).

11.4.3 *Molecules and Dust*

About 100 days after explosion, IR spectra showed the presence of molecules, especially CO, an efficient coolant (Catchpole et al. 1988), and SiO, a precursor to silicate dust (Aitken et al. 1988; Roche et al. 1991), as illustrated in Fig. 11.7. By 500 days, when the temperature had fallen below 1000 K, signatures of dust formation in the ejecta appeared. The UVOIR light curve fell faster than the ^{56}Co rate (Whitelock et al. 1989) as mid-IR emission rose to compensate (Moseley et al. 1989; Suntzeff and Bouchet 1990; Wooden et al. 1993). Peaks of nebular emission lines developed blueshifts of 600 km s^{-1} (Lucy et al. 1989), showing that dust had formed deep in the ejecta, $\lesssim$2500 km s^{-1}, and was blocking redshifted emission from the far side. The dust appears to have had two components: small grains that mainly affected short wavelengths, and very optically thick (black) clumps of grains that blocked all wavelengths equally (but with covering factor less than unity, so that the supernova was not extinguished completely). Dust formation appeared to cease around 700 days, by which time the inferred amount of warm ($\lesssim$200 K) dust was at least 10^{-4} $M_\odot$ (McCray 1993).

Much later, in 2010 and 2012, far-IR and submillimeter emission observed with *Herschel* was interpreted as evidence for a large amount (0.5–0.8 $M_\odot$) of cold (20 K) dust in the ejecta (Matsuura et al. 2011, 2015; Indebetouw et al. 2014; Fig. 11.8). Almost all of the refractory elements in the ejecta had condensed into dust. Analysis of far-IR and submillimeter observations extending from 2 to 25 years after explosion indicated that most of this dust formed at late times, $\gtrsim$10 years, probably by accretion onto and coagulation of dust grains that formed at earlier epochs (Wesson et al. 2015).

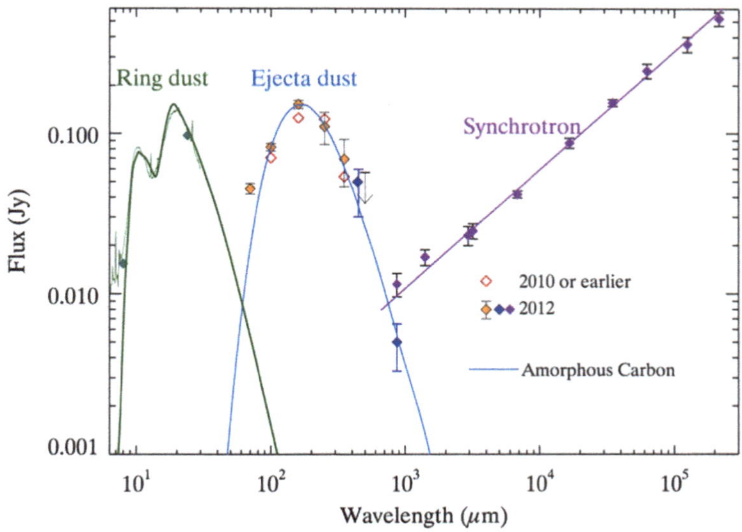

Fig. 11.8 Emissivity of SN 1987A showing the presence of a large amount of cold dust indicating that a substantial fraction of all the ejected matter that could form dust did so. From "A Stubbornly Large Mass of Cold Dust in the Ejecta of Supernova 1987A" (Matsuura et al. 2015). © AAS. Reproduced with permission

High-resolution radio observations with *ALMA* confirmed the presence of CO and SiO molecules (Kamenetzky et al. 2013) and provided evidence for other molecules, CS, SO, SO_2, and HCO^+, deep in the ejecta (Matsuura et al. 2017; Fig. 11.9). The CO lines are sharply peaked, indicating that it fills the center of the ejecta. The SiO is double-peaked, showing that it has low density at low velocities or manifests other aspects of asymmetry. The complex of molecules indicates that at least some microscopic mixing between element layers occurred in the progenitor star or in the explosion. Other observations (abundances, line ratios) limit the extent of microscopic mixing (Fransson and Chevalier 1989; Lepp et al. 1990). Molecular hydrogen, H_2, was predicted (Culhane and McCray 1995) and was detected beginning about 18 years after the explosion (Fransson et al. 2016).

11.4.4 Asymmetries and Mixing

To an unanticipated degree, SN 1987A revealed numerous signs of nonradial flows, asymmetries on both global and small scales, and radial composition mixing, including ^{56}Ni out to $\gtrsim 3000\ \mathrm{km\,s^{-1}}$ and hydrogen down to $\lesssim 700\ \mathrm{km\,s^{-1}}$. In this context, microscopic mixing must be distinguished from macroscopic mixing by clumps or plumes. Microscopic mixing implies the ability of various atomic species to directly interact, for instance by charge exchange, whereas macroscopic mixing, in principle,

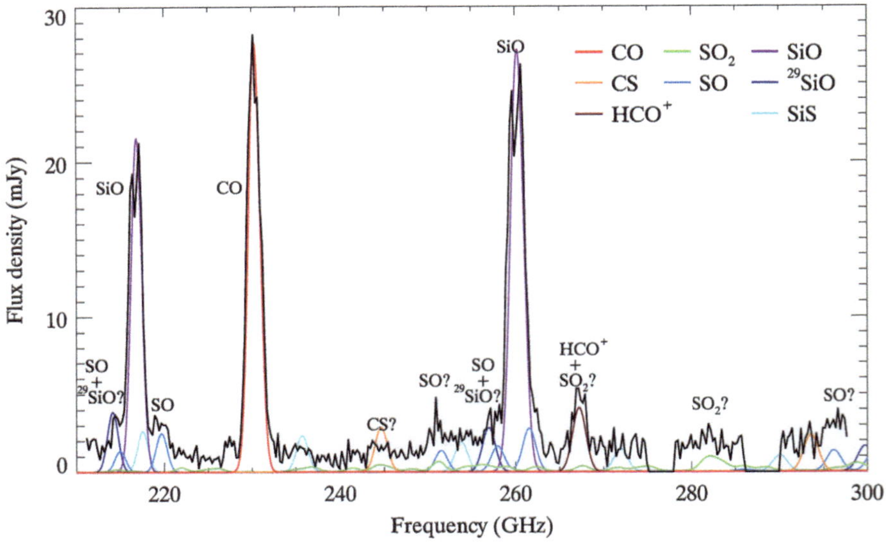

Fig. 11.9 *ALMA* spectrum of the ejecta of SN 1987A (*black*). The *colored curves* are model molecular spectra. From "ALMA spectral survey of supernova 1987A—molecular inventory, chemistry, dynamics and explosive nucleosynthesis" (Matsuura et al. 2017). © AAS. Reproduced with permission

maintains some chemical separation in angle even if various compositions co-exist at a given radius. Since the turbulent cascade is limited by expansion, and the timescale to mix by atomic diffusion is long, there is an expectation that microscopic mixing was limited in SN 1987A. From the observations, much of the mixing must have been macroscopic because lines of, e.g., calcium (a temperature proxy for primordial hydrogen), oxygen, and iron, with similar line profiles and hence radial distributions, had different temperatures (McCray 1993). It is likely that helium was not substantially microscopically mixed with carbon and oxygen, or molecular CO could not have formed because of competing charge-exchange reactions on helium ions (Lepp et al. 1990). There is ample evidence that the ^{56}Ni distribution was not spherically symmetric, but that some of the ^{56}Ni was concentrated in "clumps," "fingers," "plumes," or "jets."

In models, the seeds of asymmetries were planted very early, by instabilities of the standing shock (Sect. 9.4). Density gradients at composition interfaces (Si/O, [C+O]/He, He/H) lead to lower-density matter decelerating higher-density matter in the ejecta, the classic environment for the Rayleigh–Taylor instability. Shocks hitting steep density gradients also can generate Richtmyer–Meshkov instabilities. Both of these instabilities lead to plumes, shear, and Kelvin–Helmholtz instabilities. The structures induced by these various instabilities tend to freeze out as the expansion becomes homologous and pressure forces disappear. This also tends to halt the cascade of turbulence to smaller scales.

Whether the observed mixing was due to Rayleigh–Taylor and other instabilities expected at composition interfaces, or to some global dynamics associated with the asymmetry of the explosion remains unclear. Detailed hydrodynamic simulations in 2D have difficulty explaining the deduced distribution of ^{56}Ni with Rayleigh–Taylor mixing alone, but in 3D, drag forces on clumps are weaker and the mixing is more efficient (Hammer et al. 2010; Wongwathanarat et al. 2013). It may be that future, fully-resolved 3D calculations will show that Rayleigh–Taylor and other fluid instabilities are able to account for much of the outward mixing of ^{56}Ni and the deep inward mixing of hydrogen, but the clear evidence of macroscopic clumps or plumes suggest that some of the mixing arises in more global processes. Many models have invoked mixing in spherically-symmetric geometries by spreading various elements "by hand." These models can give some insight into the mixing processes, but are clearly not to be taken literally. Current nonrotating, nonmagnetic 3D models can provide reasonable estimates of plume-like mixing of the hydrogen and ^{56}Ni, but still struggle to reproduce the observed bolometric light curve that requires smoothing in 1D models (Utrobin et al. 2015).

SN 1987A was the first supernova for which a reasonably well-sampled set of spectropolarimetric data was obtained (Schwarz 1987; Bailey 1988; Mendez et al. 1988; Cropper et al. 1988; Jeffery 1991). The amplitude of the V-band linear polarization was small at first, but increased to more than 1% by 200 days, implying, perhaps, that the asymmetry increased with depth. A nearly constant position angle suggested the presence of a single well-defined axis of symmetry, nearly parallel to the axis of the inner ring (offset by 15°; Wang et al. 2002). The spectropolarimetry also showed definite signs of departures from strict axial symmetry, e.g., in terms of loops in the Q/U plane (Sect. 4.6) associated with the Hα line (Fig. 11.10).

Broadband and spectropolarimetric data on SN 1987A are worth revisiting in the context of recent developments and deeper understanding than were available at the time. Relevant issues might include the interstellar polarization, the accuracy of the broadband data, the nature of asymmetries suggested by the loop structure, and theoretical suggestions that the polarization signal might be low in some circumstances despite the presence of significant morphological asymmetries (Dessart and Hillier 2011b).

Speckle observations showed that the ejecta were globally asymmetric (Papaliolios et al. 1989). Additional possible evidence of asymmetry was the aptly named *mystery spot*, which appeared in speckle observations 5–7 weeks after explosion (Nisenson et al. 1987; Meikle et al. 1987). The spot was offset from the supernova by 0.06 arcseconds and at its brightest emitted 3×10^{40} erg s^{-1}, only three magnitudes fainter than the supernova. The spot fell on the common geometrical and polarization axes on the sky. A second spot may have appeared on the opposite side of the supernova, suggestive of the emergence of relativistic jets (Nisenson and Papaliolios 1999). No models based on jets, reflection, or fluorescence (Felten et al. 1989) have gained general acceptance. The mystery spot remains a mystery.

Early evidence for outward mixing of ^{56}Ni was provided by the *Bochum event*: the appearance from 19 to 70 days of bumps on the red sides of the Hα, Hβ, Na I, and near-IR hydrogen line profiles, and a blue bump in Hα that was not

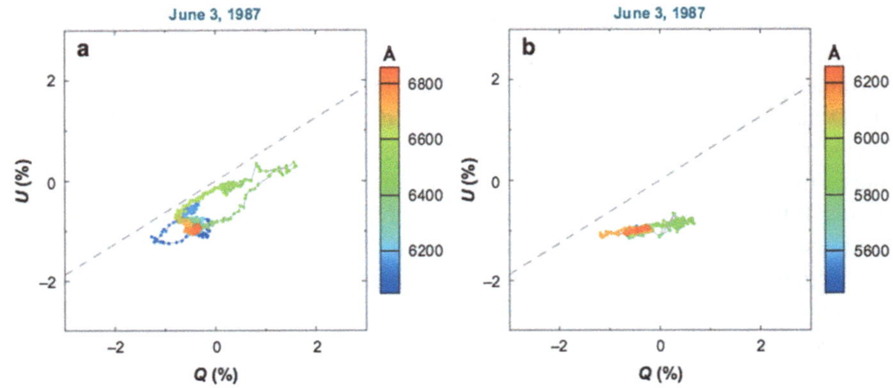

Fig. 11.10 Polarization of Hα and adjacent continuum (*left*, **a**) and of He I λ5876 and adjacent continuum (*right*, **b**) in the Q/U plane for SN 1987A on June 3, 1987, as the optical light curve began to fade from maximum. The loop structure of Hα is prominent. The He I line shows a large excursion to $Q \simeq 0.7\%$ at 5800 Å at absorption minimum. The data at $Q \simeq 1.2\%$ at 6100 Å correspond to the absorption minimum of the adjacent Ba II line. All these features fall closely along the same locus. The *dashed line* corresponds to the speckle angle, $\theta \simeq 16°$. From "Spectropolarimetry of Supernovae" (Wang and Wheeler 2008), adapted from Cropper et al. (1988), reproduced with permission of Annual Reviews of Astronomy and Astrophysics, Volume 46, © Annual Reviews, http://www.annualreviews.org

conspicuous in the IR lines of hydrogen (Hanuschik and Thimm 1990; Phillips and Heathcote 1989; Fig. 11.11). The red bumps tracked the photospheric velocity and probably signaled the emergence of a hot spot of hydrogen heated by a single, rather well-defined clump of $\sim 10^{-3}\,M_\odot$ of ^{56}Ni moving at 4700 km s^{-1} at about 45° to the line of sight, roughly along the ring axis (Utrobin et al. 1995). The blue feature was interpreted as resulting from the nonmonotonic radial dependence of the Sobolev optical depth (Utrobin et al. 1995). The shapes of the light curve and spectral lines also displayed clear evidence for mixing of hydrogen and helium at the core/envelope interface. The lack of distinct flat-topped hydrogen line profiles showed that some hydrogen was mixed down to velocities at least as low as 700 km s^{-1} (Kozma and Fransson 1998), almost to the center of the ejecta.

One factor in mixing is the likelihood of driving by a "nickel bubble." Once a supernova is in homologous expansion with no pressure gradient, even a small injection of energy can create pressure gradients and move material around. The energy input from the radioactive decay of ^{56}Ni and ^{56}Co can provide such an input. In the case of SN 1987A, deposition of this energy near the middle of the ejecta could have led to nearly half the decay energy going into the kinetic energy of the bubble rather than into emitted radiation; the resulting expanding bubble could reach over 1000 km s^{-1} at late times (Arnett et al. 1989). A single, expanding low-density bubble should be Rayleigh–Taylor unstable, perhaps producing fingers of radioactive matter that extend into the surrounding ejecta. Analysis of iron, nickel, and cobalt lines in SN 1987A gave evidence that there was not a single nickel

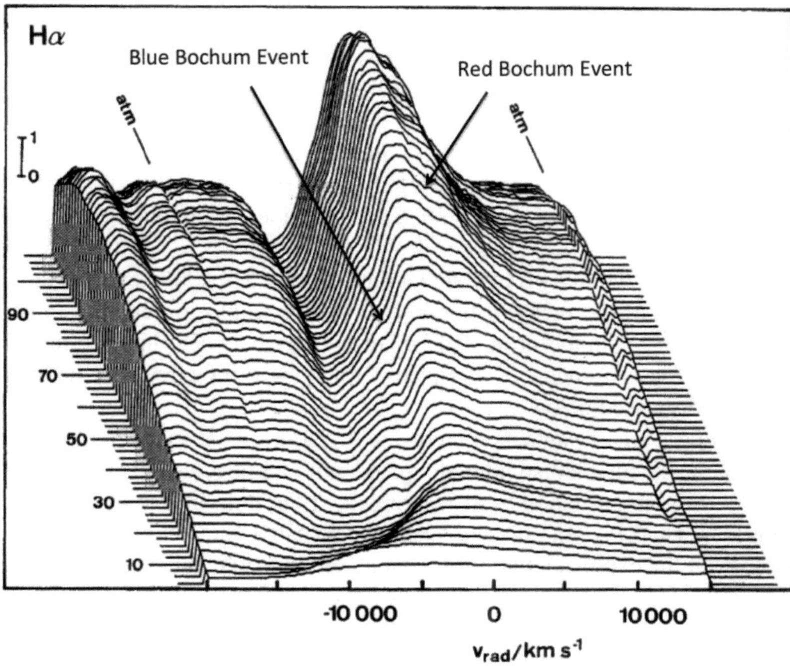

Fig. 11.11 The evolution of the Hα feature in SN 1987A from a few days (near side of the plot) to about 100 days after explosion (far side of the plot). The "Bochum event" was the appearance of satellite emission features on the blue and red sides of the line as indicated by *arrows*. The red feature appeared in other lines, but the blue feature did not, suggesting a different physical origin for the two features (see text). Adapted from "Spectroscopic fine-structure in Supernova 1987A" (Hanuschik and Thimm 1990) reproduced with permission, © ESO

bubble, but of order 100 clumps that expanded within the substrate of the core ejecta and gave the core structure the frothy morphology of toast (Li et al. 1993). For comparison, the Cas A SNR, of a Type IIb supernova, has provided evidence for an intermediate result of several bubbles (Sect. 7.4.1). Some nickel-bubble structure was probably generated in SN 1987A, but the total energy liberated by the ^{56}Ni, 4×10^{48} ergs, was substantially less than the kinetic energy, $E \simeq 0.3$ B, of the mass moving with velocity $\lesssim 3000\,\mathrm{km\,s^{-1}}$, and was not sufficient to accelerate enough nickel to the velocities deduced from the early emergence of X-rays and γ-rays. The nickel-bubble process is also unlikely to account for the particular geometrical prominence of the Bochum event.

Other evidence for mixing came from the detection of hard (16–28 keV) X- and γ-rays (Dotani et al. 1987; Sunyaev et al. 1987, 1989; Tueller et al. 1990; Inoue et al. 1991) earlier than had been predicted by nonmixed hydrodynamical explosion models. Hard X-ray emission, generated by Compton down-scattering of γ-rays, was observed beginning about 130 days after explosion and reached a peak around day 320. The flux in ^{56}Co γ-ray lines at 0.847 and 1.238 MeV peaked at about the same time. A significant fraction of the ^{56}Ni, ~10% (~10 times more than in the Bochum clump), must have been transported in clumps out to velocities of 3000 km s^{-1} or more (Kumagai et al. 1989; Pinto and Woosley 1988; Arnett and Fu 1989; Shigeyama and Nomoto 1990). The shape of the bolometric light curve (Woosley 1988; Utrobin 2004) and the widths of nebular-phase emission lines in the IR (Moseley et al. 1989; Spyromilio et al. 1990; Haas et al. 1990; Li et al. 1993) also required outwards mixing of ^{56}Ni. Further evidence of mixing is discussed in Sect. 11.5.

Spectra of light scattered off interstellar dust, obtained in 2006, revealed the time-integrated spectra of the first few hundred days after explosion emitted in several different directions (Sinnott et al. 2013). Differences between the spectra showed clear evidence for asymmetry, with an axis of symmetry in agreement with earlier inferences as discussed below.

As the supernova expanded, imaging with the *HST* eventually also showed directly that the ejecta are distinctly elongated, with a projected axis that falls close to, if not exactly along, the axes of the rings and the polarimetry (Jakobsen et al. 1993; Figs. 11.4 and 11.13). This and other evidence motivated models for the global morphology of the explosion that invoked a jet of ^{56}Ni-rich matter ejected along the common axis of the rings, roughly north on the sky and tilted 45° away from the observer (Wang et al. 2002). Contrary to that expectation, the hydrogen and other lines are predominantly blueshifted in the north and redshifted in the south. The *HST* images illustrate the transition from a roughly elliptical shape prior to 5000 days to an irregular, edge-brightened morphology as the dominant power source for the ejecta switched from radioactive decay to X-rays from CSI (Sect. 11.6). The *HST* images show a "hole" that implies that most of the X-rays are absorbed outside the inner core where the density gradient is steep and the hydrogen is abundant (Larsson et al. 2013).

Adaptive optics integral-field spectroscopic observations in the J, H, and K bands at 18 years partially resolved the inner ejecta ($\lesssim$3000 km s^{-1}, somewhat less than the hydrogen line widths at a similar epoch) in a North–South direction (Kjær et al. 2010). The ejecta structure was based primarily on the 1.644 μm emission line, thought to be mainly [Si I] blended with [Fe II]. As for the hydrogen lines, the 1.644 μm line was predominantly blueshifted in the north and redshifted in the south, indicating that the bulk of the elongated ejecta that contribute to this feature lie not along the axis of the rings, but more nearly in the plane of the inner ring. The observations of the 1.644 μm line show that the hole in the *HST* images is not due to dust absorption. The images in the 1.644 μm line do not show a hole, establishing that this line is still driven by radioactive decay, not the external X-ray illumination (Larsson et al. 2013). The emission in the 1.644 μm line comes primarily from two clumps spanning 800–3500 km s^{-1}. The northern clump is closer than the southern clump to the ring plane. There are no obvious emission structures on the redshifted side of the ring plane.

Twenty-seven years after explosion, the 68 and 78 keV lines of ^{44}Ti decay were detected by *NuSTAR*. These lines were redshifted by about 700 km s^{-1} (Boggs et al. 2015). This, together with other evidence, could be indicative of a very asymmetric "single-lobe" explosion (Hungerford et al. 2005).

These observations show that there is a rich 3D structure to the observed ejecta, and that the simplest bipolar jet-induced model with the iron-group ejecta along the axes of the rings does not apply to SN 1987A. There is no obvious symmetry with respect to the ring plane or axis; there could be a one-sided and slightly tilted geometry reminiscent of the Bochum event and the 15° shift between the polarization axis and the ring axis. The current geometry of the inner ejecta emission of SN 1987A may reflect some innate symmetry and may be affected by dust, column depth, and projection effects, but it certainly does not conform to any simple geometrical model.

Nevertheless, by various measures—the polarization, the early speckle images, the mystery spot, and the circumstellar rings—SN 1987A "pointed" in a certain direction, a position angle of $\theta \simeq 15°$ and about 45° out of the plane of the sky, and maintained that orientation throughout its development. The persistent polarization position angle across a broad wavelength range and the lack of significant evolution of the position angle strongly argue that the geometry of the photosphere at early and late epochs share the same geometric structure. Although there were significant and interesting departures, the data also show that the overall structure of SN 1987A was substantially axially symmetric from deep inside out to the hydrogen envelope and on to the circumstellar rings. This requires a largescale, systematic, directed asymmetry that cannot be accounted for by smallscale convection or turbulence. The totality of the data suggests that the angular momentum of the progenitor system played a crucial role in the explosion of SN 1987A. There is much more to do to fathom the import of all the asymmetries displayed by SN 1987A and their significance for the general population of core-collapse supernovae.

11.5 Nucleosynthesis

Prior to explosion, beneath the hydrogen envelope the progenitor of SN 1987A is presumed to have had a layered composition, ranging from a helium zone to a central iron core. After the iron core collapsed, the supernova shock propagated outwards through the overlying layers of heavy elements, driving explosive nucleosynthesis. The extent of nuclear transmutation depends on the peak temperature behind the shock. A simple estimate for that temperature comes from equating the energy within the volume of the shock in a radiation-dominated gas, $\frac{4}{3}\pi r^3 \times \mathrm{a}T^4$ (where a is the Stefan–Boltzmann constant), to the energy of the explosion, E, so that $T \simeq (\frac{3E}{4\pi r^3 \mathrm{a}})^{1/4}$, which implies that the postshock temperature declines as $r^{-3/4}$ (Arnett et al. 1989). In spherical models, each layer tends to be burned to the composition of the next heavier group: silicon to iron peak, oxygen to silicon, carbon to oxygen, neon, and magnesium, etc. Elements heavier than carbon are predicted to be heated sufficiently to undergo nuclear burning, thus the bulk of any such elements that were ejected was synthesized during the explosion. Most of the ejected oxygen and essentially all of the carbon and lighter elements probably were present in the progenitor before the explosion.

Care must be taken with elements that could be either primordial or freshly synthesized. In SN 1987A, the Ca II near-IR triplet was formed by primordial calcium within the hydrogen envelope, but the [Ca II] $\lambda\lambda$7291,7323 emission was formed by a mix of primordial and freshly synthesized calcium (Kozma and Fransson 1998; Fig. 11.7). At very late phases, the [Ca II] emission was mainly from primordial calcium because the freshly synthesized calcium did not have sufficient capture cross section for γ-rays. Similar phenomenology is expected in SN IIP. The predictions of detailed explosive nucleosynthesis calculations (Hashimoto et al. 1989; Thielemann et al. 1990) were generally in good agreement with abundances inferred from spectroscopic analysis of nebular-phase IR emission lines (Kozma and Fransson 1998).

Calculations of nucleosynthesis in massive stars predict enhancements of some s-process nuclides in the helium-burning shells (the *weak s-process*; Sect. 24.5). If this material is mixed to the surface during normal post-main-sequence burning, there could be enrichments in scandium, strontium, and barium by factors of several to 10. Observations of SN 1987A during the photospheric phase seem to show enhancements by about that amount. The Ba/Sr ratio exceeded expectations of the basic s-process models, but this may have been affected by the ionization state of the envelope. Consideration of time-dependent ionization of hydrogen gives more ionized hydrogen than models that assume ionization equilibrium and that, in turn, affects the Ba/Sr ratio (Utrobin and Chugai 2005; Dessart and Hillier 2008). There remains the issue of why SN 1987A displayed stronger Ba II lines than other SN II. The strength of the Ba II lines may also depend on the mass of ^{56}Ni ejected (Pastorello et al. 2005). If the progenitor of SN 1987A went through some unusual and exotic process of binary evolution and coalescence, the degree to which one can invoke standard s-process calculations is not clear.

Nucleosynthesis in SN 1987A is expected to produce other radioactive species beside ^{56}Ni and ^{56}Co (Fig. 11.12). The nuclide ^{57}Ni is produced along with ^{56}Ni in nuclear statistical equilibrium; it decays into ^{57}Co, the half-life of which is 270 days. The ratio of ^{57}Co to ^{56}Co produced in the explosion, manifested as the ratio of ^{57}Fe to ^{56}Fe in the ejecta at late times, is expected to be roughly the solar ratio. Allowing for nonionization equilibrium at late times, the rate of decay of the bolometric light curve and directly measured line ratios of [Co II] 1.547 μm to [Fe II] 1.533 μm were consistent with a ratio of the final stable decay products, ^{57}Fe/^{56}Fe, of about twice the solar value, providing a useful constraint on nucleosynthesis during the explosion (Fransson and Kozma 1993; McCray 1993; Seitenzahl et al. 2014).

The ^{44}Ti expected to be produced in the explosion has a half-life of 60 years. Its decay releases hard X-rays and γ-rays that escape and positrons that are trapped by a tangled magnetic field (but not necessarily by a radially-combed field), leading to

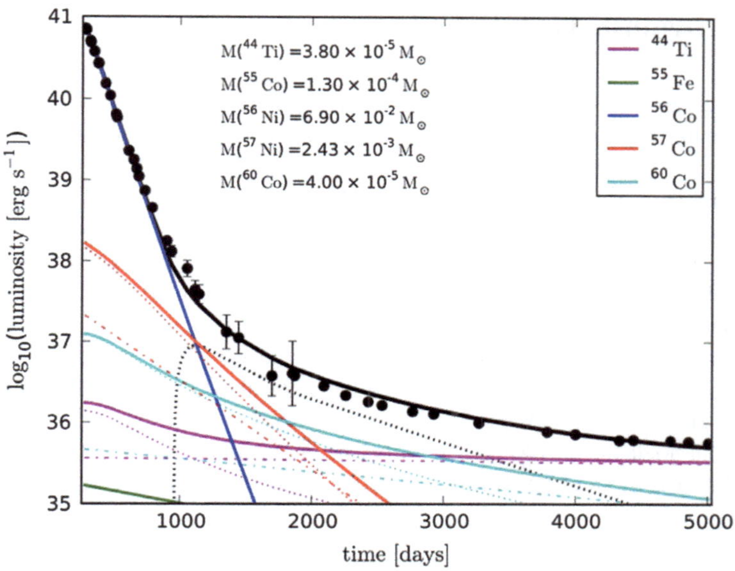

Fig. 11.12 Late-time light curve of SN 1987A illustrating the contributions from various slowly-decaying radioactive species. The *dotted lines* give the γ-ray contributions; the *dot-dash lines* give the electron and positron contributions. The *solid black line* is a model bolometric UVOIR luminosity accounting for delayed recombination. From "The Light Curve of SN 1987A Revisited: Constraining Production Masses of Radioactive Nuclides" (Seitenzahl et al. 2014). © AAS. Reproduced with permission

the longterm decay input mentioned above (Sect. 11.4.2). The production of ^{44}Ti is rather model dependent, so its detection constrains models. Analysis of late nebular-phase emission lines (Jerkstrand et al. 2011), light-curve fitting (Seitenzahl et al. 2014), and detection of direct-escape lines of ^{44}Ti at 68 and 78 keV (Boggs et al. 2015) indicated that $\sim 10^{-4}$ M$_\odot$ of ^{44}Ti was ejected in the explosion, consistent with most models (Seitenzahl et al. 2014).

Both ^{55}Co and ^{55}Fe are produced in nuclear statistical equilibrium and in incomplete explosive silicon burning. The amounts produced depend (at least) on core mass and metallicity. The ^{55}Co decays in a few days, so the ejecta of SN 1987A are expected to contain $\sim 10^{-3}$ M$_\odot$ of ^{55}Fe, which has a half-life of 2.7 years for decay by inner shell electron capture to ^{55}Mn. While attenuation in the ejecta may be an issue, this amount of ^{55}Fe is predicted to be detectable a couple of decades after explosion, though not yet reported.

The ^{57}Ni/^{56}Ni ratio can, in principle be, observed by direct detection of γ-rays from the decay of ^{56}Co and ^{57}Co or by evolving spectral features. Knowing this ratio can give constraints on Y_e in the inner ejecta and hence on the position of the mass cut and the time interval between collapse and explosion (Thielemann et al. 1996).

11.6 Circumstellar Interaction

As outlined above, SN 1987A was surrounded by a complex circumstellar environment. The luminosity of the supernova and the expansion of the ejecta led to various radiative and hydrodynamical interactions with the CSM (Fig. 11.13).

The first sign of CSI was nonthermal radio emission (Turtle et al. 1987). Radio flux was detected on day 2, reached a peak flux at 2 GHz of 130 mJy on day 4, and remained detectable for a few hundred days (Ball et al. 1995). The emission was compatible with interaction of the outermost ejecta with the wind of the progenitor (Chevalier and Dwarkadas 1995). Owing to the low density of the fast blue-supergiant wind, the radio emission was weak and it peaked early; this emission could not have been observed if SN 1987A had been at the typical distance (10 Mpc) of other detected radio supernovae.

At about 80 days, narrow emission lines began to appear in the optical (Wampler and Richichi 1989) and UV (Fransson et al. 1989). These were recognized to come

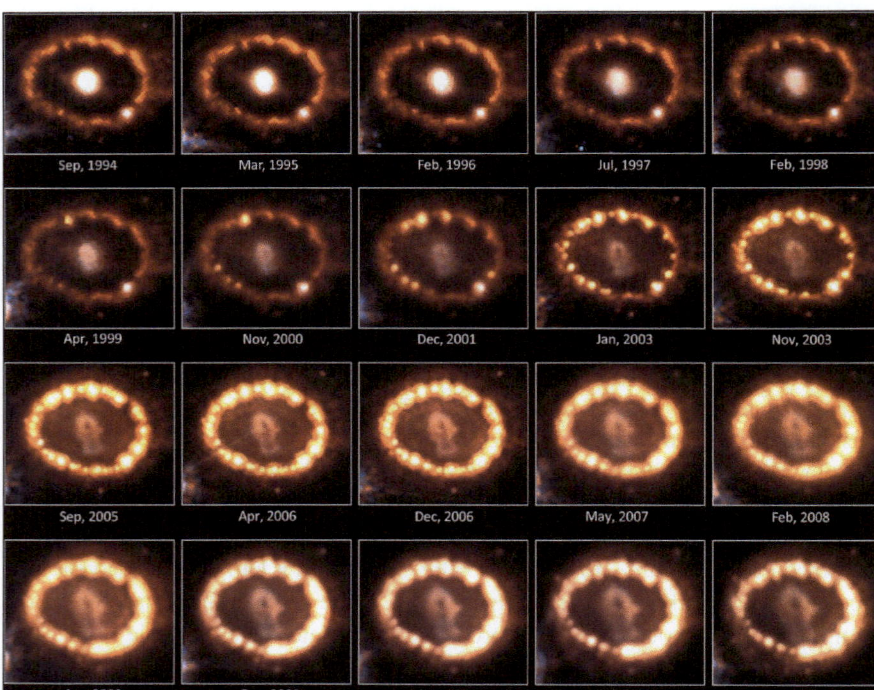

Fig. 11.13 The evolution of the ring and ejecta structure of SN 1987A illustrating the collision of the ejecta with the inner ring and the illumination of the ejecta by X-rays from the reverse shock. The bright spot on the ring at about 5 o'clock is an unrelated star in the LMC that is projected on the ring. The bright spot at about 11 o'clock in the February, 1998, and subsequent images is the first sign of the collision of the ejecta with fingers protruding inward from the ring. From "The Destruction of the Circumstellar Ring of SN 1987A" (Fransson et al. 2015). © AAS. Reproduced with permission

from CSM that had been heated and ionized by radiation from shock breakout. Analysis revealed high N/C and N/O ratios (Fransson et al. 1989; Mattila et al. 2010), indicating that the emitting material had undergone processing by the CNO cycle and therefore must have come from the progenitor. The emitting material later proved to be $\lesssim 0.1\,M_\odot$ of ionized matter in the inner ring. Subsequent observations revealed that the two outer rings and the Napoleon's Hat nebula responded similarly, although diluted by greater distance from the supernova.

Direct imaging of the ring (Wampler and Richichi 1989; Crotts et al. 1989; Wampler et al. 1990; Jakobsen et al. 1991; Plait et al. 1995) provided its angular size. Its absolute size was inferred from the time elapsed between the onset of the emission lines, when photons from the near side arrived, and the time of maximum flux, when photons from the far side arrived. The angular and absolute sizes gave a direct determination of the ring distance, 51.4 ± 1.2 kpc (Panagia 2005).

At about 3 years, radio emission was detected again (Gaensler et al. 1997), and soft X-rays became detectable a few years later (Hasinger et al. 1996). This emission may have been from an H II region between the progenitor and the inner ring that formed in an interaction between the RSG wind and the faster wind of the B3 Ia progenitor star and was then ionized by the blue supergiant. The density associated with this H II region led to the correct prediction that the ejecta would collide with the inner ring around 2003 (Chevalier and Dwarkadas 1995).

The radio and soft X-ray emission increased linearly until 10 years, then both began to increase exponentially (Park et al. 2005). At about the same time, the first of several bright, optical hot spots in the inner ring was detected (Fig. 11.13). The forward shock was beginning to interact with the ring protrusions (Pun et al. 2002), and for the first time the emission from CSI exceeded that of the ejecta. By 15 years, several hot spots were seen on the ring (Burrows et al. 2000; Lawrence et al. 2000; Bouchet et al. 2006). *HST* observations traced the expansion of the forward shock and its collision with the inner ring (Michael et al. 2000). Cooling of the shocked gas gives rise to the optical emission from the hot spots. The hot spots have some radial motion and are still spatially unresolved. The protrusions have been slow to merge into a continuous ring, indicating that they were long compared with their widths. The issue of what determined the particular number of spots with their rather regular spacing eludes satisfactory explanation.

At 17 years, the X-ray and radio emissions increased yet more steeply as the shock began to interact with the main body of the ring (Park et al. 2006; Zanardo et al. 2010). This also brought a softening of the X-ray spectrum, X-ray brightening of the hot spots on the inner ring followed by a filling in of the whole ring in X-ray images, and a decrease in the expansion velocity of the forward shock as it encountered increased density (Racusin et al. 2009). The X-ray flux continued to rise (Maggi et al. 2012; Helder et al. 2013a).

X-ray spectra of the inner ring show that oxygen, neon, magnesium, and iron are underabundant, while silicon and sulfur are overabundant compared to LMC values. This may put constraints on merger models in which the ring generated after the merger is predicted to be iron-rich (Heng et al. 2008). VLBI observations in 2008 suggested that at that time the shock had almost completed its transit of the ring (Potter et al. 2009). The mass of ionized gas in the ring was estimated to be 6×10^{-2} $M_{\odot}$ (Mattila et al. 2010). Observations in 2013 and 2014 indicated that interaction with the ejecta was gradually destroying the ring (Fransson et al. 2015).

The inner ring also displayed evidence of emission by pre-existing warm ($\lesssim$200 K) silicate dust in mid-IR spectra (Bouchet et al. 2006; Dwek et al. 2008, 2010). The dust mass was estimated to be $\sim 10^{-6}$ $M_{\odot}$. This dust presumably was collisionally heated by the hot gas responsible for the X-ray emission. The ratio of mid-IR to X-ray flux was nearly constant in time, suggesting that gas-grain collisions were primarily responsible for cooling of the shocked gas.

HST and *Chandra* observations also showed the presence of a reverse shock, reflected from the inner ring (Michael et al. 2003; Heng et al. 2006; Zhekov et al. 2010; France et al. 2010, 2011; Dewey et al. 2012; Fransson et al. 2013). The emission lines gave evidence for the penetration of the reverse shock by neutral hydrogen in the ejecta that only became ionized after passage through the reverse shock. Lyman-α emission generated by the reverse shock was observed far from the equatorial plane and at a larger radius than that of the equatorial ring. The Lyman-α emission, as well as radio emission (Zanardo et al. 2013), also revealed substantial East/West asymmetry, again showing that SN 1987A, while having a substantial preference to align along a single axis, cannot be represented by a simple bipolar geometry. SN 1987A provides unprecedented diagnostics of the shock structure and dynamics involved in this interaction process, including insight into isotropization by turbulent, magnetic plasma.

11.7 Related Progenitor Stars and Supernovae

Several Galactic blue supergiants are observed to have circumstellar environments closely resembling that of the progenitor of SN 1987A (Smith 2007). The B-type supergiant SBW1 has a triple-ring system that is a near twin of that of SN 1987A. The rings of SBW1 may be the result of a presupernova mass-ejection episode in the equatorial plane, followed by interaction between matter photoevaporated from the equatorial ring with the stellar wind to form the polar rings (Smith et al. 2013a). This raises the possibility of the same origin for the SN 1987A rings, so neither a presupernova RSG phase nor a binary companion is necessarily required for their explanation. A single-star, interacting-winds model was proposed for another SN 1987A-like triple-ring system, the blue supergiant Sher 25 (Chita et al. 2008).

More than a dozen supernovae have shown some resemblance to SN 1987A and are thought to have been explosions of blue supergiants. Most, but not all, of these events occurred in mildly low-metallicity (LMC-like) environments (Taddia et al. 2013). The fraction of all core-collapse supernovae that have blue-supergiant progenitors has been estimated to be 1–3% (Pastorello et al. 2012). Some such events (SN 2006V and SN 2006au) were somewhat brighter and had higher kinetic energies than SN 1987A (Taddia et al. 2012b), while SN 2009E was fainter and less energetic (Pastorello et al. 2012). From hydrodynamical modeling of SN 2000cb,

Utrobin and Chugai (2011) estimated a progenitor radius of $R \simeq 35\ R_{\odot}$ and explosion parameters of $E \simeq 4$ B, $M_{ej} \simeq 22\ M_{\odot}$, and $M_{Ni} \simeq 0.08\ M_{\odot}$.

11.8 Summary

SN 1987A has educated and confounded us and after 30 years shows no signs of ceasing either. An immense amount has been learned from and about this explosion, but some of the most fundamental issues remain: what was the nature of the progenitor evolution that led it to be blue and surrounded by its fascinating CSM? How did the explosion develop a globally-directed flow? What sort of compact remnant was left behind?

While the breakout phase in SN 1987A has been studied, this is a topic of growing interest for a variety of supernovae. This topic is probably worth revisiting in the context of the various direct and indirect observations of breakout and its immediate aftermath in SN 1987A.

The propagation of the forward shock into the equatorial ring and more distant CSM as SN 1987A becomes a young SNR will teach us about shock physics and perhaps reveal whether the CSM is an hourglass or three isolated rings (plus Napolean's Hat). The collision of the forward shock with the outer rings should happen in another 20 years or so. The fact that there are other ring-encircled stars and supernovae that resemble SN 1987A encourages the idea that SN 1987A, while very special, was not unique, so there are more general lessons to be learned. The notion that SN 1987A resulted from a merger in a binary system to yield a single, rotating, star is a fascinating one that deserves to be revisited with new simulations.

In the future, the reverse shock and X-rays from the forward shock will further illuminate the structure and composition of the ejecta. The portion of the ejecta still illuminated from within by radioactive decay shows a complex structure that does not easily conform to any extant explosion model, but hints of rudimentary symmetries that may yet map into evidence from the speckle imaging, the mystery spot, the polarization, and the origin of the Bochum event.

The neutrinos told us that a compact object formed; it has remained frustratingly elusive. Observations peering into the continuum between emission lines may be pushed deeper, but there are limits to what direct observations can do. If the remnant in SN 1987A is a central compact object of the luminosity of that in Cas A, it will remain very hard to detect in any waveband; but what a prize if it can be determined what it is, how it formed, and how it evolves!

The general issue of dust formation in supernovae and in SN 1987A in particular is currently a hot topic. Little evidence for dust has been seen in mid-IR observations. The issue is whether there are substantial amounts, approaching 1 $M_{\odot}$, of cold dust in SN 1987A that can be detected only with appropriate far-IR and submillimeter observations. This issue is important to understand the nature of SN 1987A, but it also means that SN 1987A can be a dusty Rosetta Stone with cosmological implications.

Chapter 12
Type IIP Supernovae

12.1 Introduction

By traditional definition, SN IIP show Balmer lines and an extended light-curve plateau phase of slowly decreasing luminosity that typically lasts 100 days.[1] The optical spectral evolution of an SN IIP was shown in Fig. 1.7, and UVOIR light curves of four SN IIP were shown in Fig. 5.4.

In a volume-limited sample of the contemporary Universe, SN IIP constitute about half of all core-collapse supernovae. The environments of SN IIP show that they are produced by young populations, and direct imaging of some of their progenitors confirms the theoretical expectation that they are produced by RSGs (Sect. 3.2). The basic physics of their light curves is reasonably well understood (Sect. 5.5.4). In this chapter we focus on inferences from observations of SN IIP, most of which have been obtained during the plateau and radioactive tail phases. The plateau is sensitive mainly to progenitor radius, ejected mass, and explosion energy, while the tail depends mainly on M_{Ni}.

At maximum light, SN IIP have a mean $M_B \simeq -16.7$, and they typically are only a few tenths of a magnitude dimmer at the middle of the plateau. SN IIP are important nucleosynthetic events, they may have been important sources of dust production in the early Universe (Sect. 24.6), and they may prove useful as distance indicators for cosmology, independent of SN Ia (Sect. 25.3).

[1]In view of the lack of a clear statistical division between the light curves of SN IIP and SN IIL, some authors consider that they are two parts of a continuum. Without taking a position on this issue, we will follow tradition and discuss SN IIP and SN IIL (Chap. 13) separately.

D. Branch, J.C. Wheeler, *Supernova Explosions*, Astronomy and Astrophysics Library, DOI 10.1007/978-3-662-55054-0_12

12.2 Case Study: SN 1999em

SN 1999em was discovered (Li 1999) only a few days after its explosion in the nearby (12 Mpc) galaxy NGC 1637. The lack of detection of a progenitor in archival images indicated that the ZAMS mass of the star that exploded was $\lesssim 15\,M_\odot$ (Smartt et al. 2009). Extensive optical and some near-IR photometry and spectroscopy showed SN 1999em to be a typical SN IIP (Hamuy et al. 2001; Leonard et al. 2002b; Elmhamdi et al. 2003).

The discovery occurred near the time of V-band and UVOIR maximum, when shorter-wavelength bands were fading and longer-wavelength bands were still brightening (Fig. 12.1). The UVOIR light curve (Fig. 12.2) declined for 20 days before entering a slowly-declining plateau phase (0.003 mag d^{-1} in V) that lasted about 90 days. After an abrupt plunge of two magnitudes in 25 days, a plateau tail phase (Sect. 5.5.4) lasted 25 days before the light curve began to decline at the rate of ^{56}Co decay.

During the photospheric phases, the spectral development was largely as expected for an expanding, cooling, hydrogen-rich envelope (Fig. 12.3). The early optical spectra showed P Cygni Balmer lines and He I λ5876 at characteristic expansion velocities of 10,000 km s^{-1}, and probably several lines of N II (Dessart and Hillier 2005), suggesting the presence of CNO-cycled material at the photosphere. Near-IR spectra showed Paschen lines and He I 1.0830 μm. An early *HST* spectrum revealed that blanketing in the UV, by lines of Fe II, Ni II, and other ions, was very strong from the beginning (Baron et al. 2000; Fig. 12.4). As the envelope cooled, lines of Ca II, Fe II, Na I, and eventually Sc II and Ba II developed. During the plateau phase, as the photosphere receded deeper into the envelope of the ejecta, the expansion velocity decreased from 4000 to 2000 km s^{-1}.

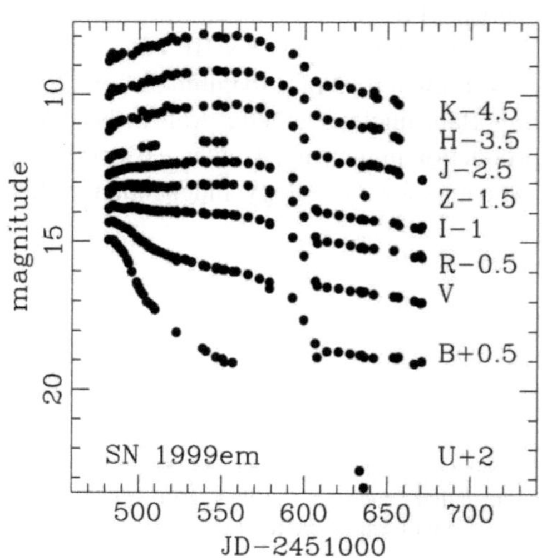

Fig. 12.1 Light curves of SN 1999em. From "The Distance to SN 1999em from the Expanding Photosphere Method" (Hamuy et al. 2001). © AAS. Reproduced with permission

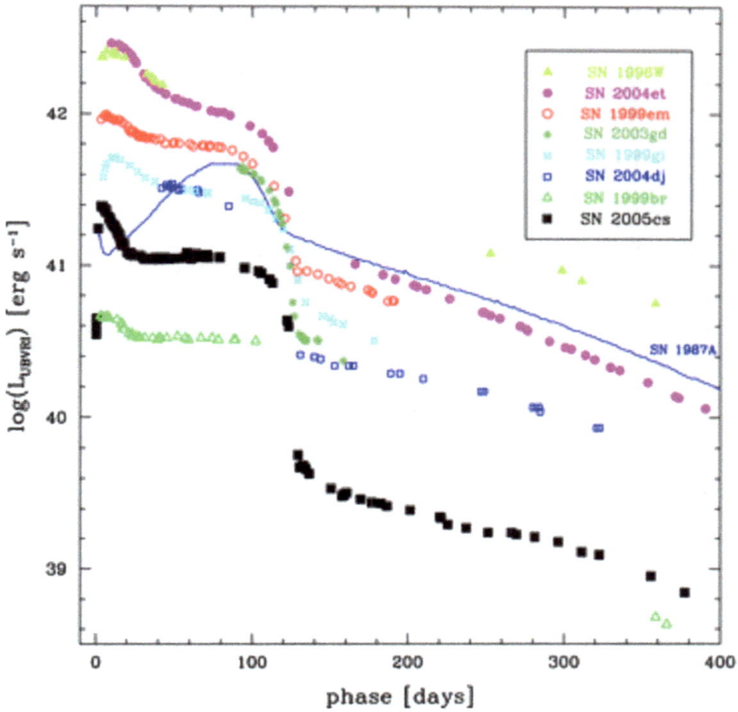

Fig. 12.2 Comparison of the UVOIR light curves of SN 1999em, several other SN IIP including the "weak" SN 2005cs (Sect. 12.5.2), and SN 1987A. Figure 9 of "SN 2005cs in M51—II. Complete evolution in the optical and the near-infrared" from Pastorello et al. (2009), by permission of Oxford University Press on behalf of the Royal Astronomical Society

In the nebular phase, emission lines of Hα, [Ca II] $\lambda\lambda$7291,7323, the Ca II near-IR triplet, [O I] $\lambda\lambda$6300,6364, the Na I D lines, and [Fe II] λ7155 were prominent (Fig. 12.5).

At early times the continuum polarization was weak ($\sim$0.2%), but line polarization was significant. By the end of the plateau phase, the continuum polarization had roughly doubled (Leonard et al. 2001), possibly but not necessarily (Sect. 12.9) revealing increased asymmetry in the deeper layers of the ejecta.

Early radio and X-ray observations were consistent with a modest presupernova mass-loss rate of 2×10^{-6} M$_\odot$ y^{-1} for an assumed wind velocity of 10 km s^{-1} (Pooley et al. 2002). Other indications of later CSI included possible high-velocity absorptions in Hα and Hβ (Faran et al. 2014a).

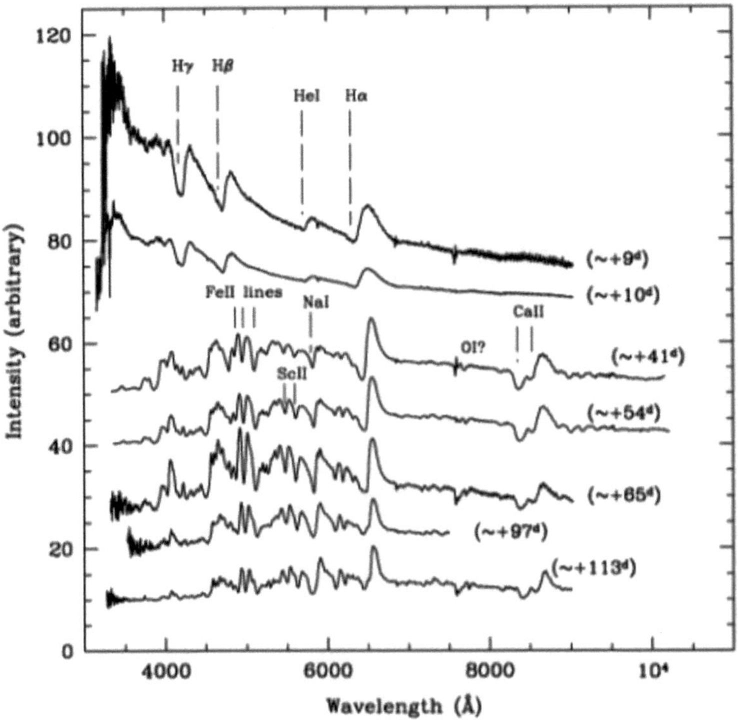

Fig. 12.3 Spectral evolution of SN 1999em in the photospheric phase. Spectra are displaced vertically for clarity. Epochs are days after explosion. Figure 6 "Photometry and spectroscopy of the Type IIP SN 1999em from outburst to dust formation" from Elmhamdi et al. (2003), by permission of Oxford University Press on behalf of the Royal Astronomical Society

As for SN IIP in general, owing to radiative-transfer effects (Dessart and Hillier 2005) emission-line peaks were blueshifted; these blueshifts vanish as the nebular phase begins (Anderson et al. 2014a). Later, between days 465 and 510, the profiles of the [O I] and Hα lines again became skewed towards the blue, this time indicating the formation of dust deep in the ejecta, interior to 800 km s^{-1}. The development of a deficit in the optical luminosity, compared to that expected from ^{56}Co decay, reinforced the dust-formation interpretation (Elmhamdi et al. 2003).

Discussion of hydrodynamic modeling of SN 1999em, aimed at deriving the progenitor radius, ejected mass, kinetic energy, and M_{Ni}, is deferred to Sect. 12.5.1.

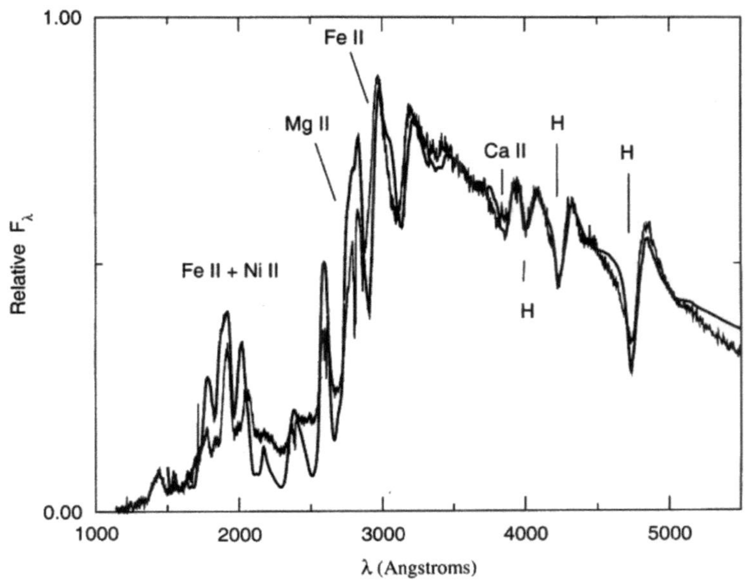

Fig. 12.4 A SYNOW fit to an *HST* spectrum of SN 1999em obtained 11 days after explosion and an optical spectrum obtained 1 day later. From "Preliminary Spectral Analysis of the Type II Supernova 1999em" (Baron et al. 2000). © AAS. Reproduced with permission

12.3 Expected Progenitor Structures

A nonrotating, massive, single star of solar metallicity with ZAMS mass about 8–30 $M_\odot$ is expected to be a red giant or an RSG when its core collapses. After the main-sequence phase, the core contracts and the envelope expands. The core then evolves towards collapse with little effect on the surface properties of the star. Because the extended envelope masks the evolution of the core, we cannot be sure if a particular star is destined to explode by ignition and explosion of a degenerate C/O core, by collapse of an ONeMg core, or by collapse of an iron core (Sect. 8.5). All three could show the characteristic light-curve plateau and hydrogen-rich spectra of SN IIP .

Stellar evolutionary calculations predict density and composition structures of the RSG progenitors of SN IIP just prior to core collapse (Woosley et al. 2002; Dessart et al. 2010a). In the nonrotating models of Woosley et al. for ZAMS masses from 11 to 30 $M_\odot$, presupernova masses interior to the outer boundary of the oxygen-rich layers (commonly referred to as oxygen core masses) increase from 1.7 to 7.7 $M_\odot$, while masses interior to the outer boundary of the helium-rich layers (helium core masses) increase from 1.8 to 9.2 $M_\odot$. Masses of the hydrogen-rich envelopes range from 8.6 to 2.2 $M_\odot$, but nonmonotonically and with the lowest value for the 30 $M_\odot$ model owing to higher mass-loss rates of more massive stars. Thus the presupernova model for a ZAMS mass of 11 $M_\odot$ has a helium core that is

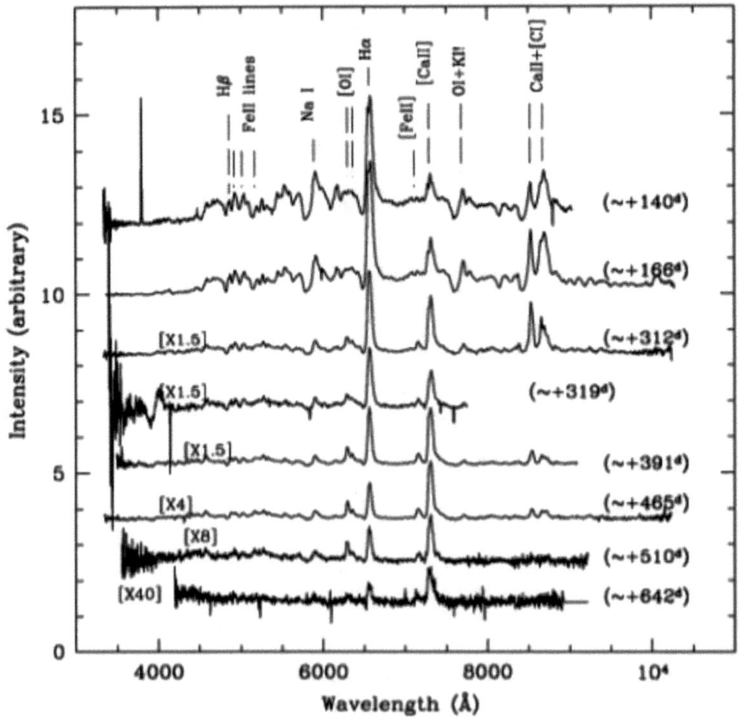

Fig. 12.5 Like Fig. 12.3, but for the nebular phase of SN 1999em. Figure 6 "Photometry and spectroscopy of the Type IIP SN 1999em from outburst to dust formation" from Elmhamdi et al. (2003), by permission of Oxford University Press on behalf of the Royal Astronomical Society

more than four times less massive than the hydrogen-rich envelope, while this mass ratio is reversed in the 30 $M_{\odot}$ model. In these models, the total presupernova mass varies (nonmonotonically) only from 10.6 to 14.7 $M_{\odot}$ and the radii vary only from $\gtrsim 4\times10^{13}$ to $\lesssim 10^{14}$ cm. The presupernova consists of a dense, nondegenerate, helium core that occupies only $\sim 10^{-6}$ of the star's volume, surrounded by the loosely bound hydrogen envelope. The density gradient between the helium core and the hydrogen envelope is steeper in stars of lower ZAMS mass. A steep gradient is favorable for ejection of the envelope when the core collapses.

Rotation leads to more massive cores, longer RSG phases, larger presupernova radii, and enhanced mass loss. For a typical ZAMS equatorial rotation velocity of 200 km s^{-1}, the maximum ZAMS mass of model SN IIP progenitors is reduced from about 30 to 20 $M_{\odot}$ (Heger et al. 2000). For both nonrotating and rotating models, the ratio of the mass of the oxygen core to the mass of the hydrogen envelope in the presupernova model increases steeply with ZAMS mass (Dessart et al. 2010b). We will return to this in Sect. 12.6.

Dust-driven mass loss during the RSG phase may facilitate a transition to the Wolf–Rayet phase and reduce the maximum ZAMS mass of an SN IIP progenitor (Chieffi and Limongi 2013), which could turn out be as low as 15 $M_\odot$.

12.4 Breakout, Fireball, and Rise-Time

The electromagnetic display begins when the shock breaks out of the progenitor's photosphere. Multigroup radiation-hydrodynamics calculations of spherical breakout for SN IIP (Ensman and Burrows 1992; Tominaga et al. 2009, 2011) predict that the luminosity typically peaks at $\gtrsim 10^{44}$ erg s^{-1}, at which time the color temperature is $\sim 3 \times 10^5$ K and the SED peaks at ~ 100 Å. The total energy radiated within one magnitude of peak luminosity is $\sim 10^{48}$ ergs. With increasing progenitor radius, which increases mildly with progenitor ZAMS mass and metallicity, the duration and total energy radiated increase and the color temperature decreases, while the peak luminosity remains about the same. With increasing explosion energy, the peak luminosity, color temperature, and total energy radiated increase, while the duration remains nearly constant. Analytic treatments of breakout (Matzner and McKee 1999; Katz et al. 2012) are in rough agreement with results of detailed calculations. Because it is so brief and at short wavelengths, breakout emission is difficult to observe. An observed UV burst from SN PS1-13arp of about 1 day duration may have been emission from a prolonged breakout in a relatively dense CSM (Gezari et al. 2015). Future detections of breakout will provide a sensitive probe of pre-explosion conditions.

Early fireball radiation, within hours or days of breakout, has been observed from several SN IIP. SN 2006bp was observed in optical, UV, and X-rays (Immler et al. 2007; Quimby et al. 2007c; Dessart et al. 2008). Narrow high-ionization lines in the optical (e.g., He II, C IV) were observed for a few days and then faded. If the emitting region were overrun by the supernova shock in just a few days, then these lines were emitted by nearby CSM that had been ejected from the progenitor no earlier than decades before explosion (Quimby et al. 2007c). Subsequently, broad high-ionization lines that formed in the outer layers of the ejecta were observable for a few days (Dessart et al. 2008).

SNLS-04D2dc (Schawinski et al. 2008; Gezari et al. 2008b) and SNLS-06D1jd (Gezari et al. 2008b) were observed in the UV. Observations of the fireball phase of SNLS-04D2dc were consistent with predictions for a progenitor having a ZAMS mass of 20 $M_\odot$, an explosion energy of 1.2 B, and a radius of 800 $R_\odot$ (Tominaga et al. 2009). The observations also constrained any asphericity of breakout to be small (Couch et al. 2009), consistent with the low observed polarization of SN IIP at early times (Wang and Wheeler 2008). SN 2010aq was well-observed in the UV and optical at early times (Gezari et al. 2010). A brief UV peak was not from breakout itself, but due to the peak of the SED shifting through the UV during the early fireball phase. The observations permitted an estimate of the radius of the progenitor: $700 \pm 200\ R_{\odot}$. A UV excess 2 weeks after breakout may have been due to low metallicity or to CSI.

Future observations of fireball emission from SN IIP will yield information about progenitor radii, surface compositions, explosion energy per unit ejected mass, and line of sight extinction (Ganot et al. 2016).

Observed rise-times to optical maximum of SN IIP are typically about 1 week, somewhat shorter than standard predictions. The radii of the RSG progenitors may be smaller than expected, or the rise-times could be caused by prolonged breakouts in CSM (González-Gaitán et al. 2015).

12.5 The Plateau Phase

Several groups have calculated detailed synthetic spectra for the plateau phase (Baron et al. 2007; Kasen and Woosley 2009; Dessart et al. 2010b; Inserra et al. 2012a). The synthetic spectra fit observed spectra well and provide the ability to estimate the time dependence of the photospheric velocity, reddening by dust, the metallicity of the hydrogen envelope (Dessart et al. 2014b), and distances to SN IIP.

For an adopted presupernova model (1D density and composition structures), modelers artificially deposit energy at the outer edge of the iron core, and the ensuing hydrodynamics, light curve, and spectra are calculated at various levels of approximation. The difference between the amount of energy deposited and the binding energy of the envelope is essentially equal to the final kinetic energy, known as the explosion energy. Numerical presupernova evolutionary models have steep density and composition gradients at composition interfaces; these have been found to lead to bumps in calculated light curves that are not observed in real SN IIP (Utrobin and Chugai 2008). Even if the steep gradients in the presupernova models are correct (but see Hirschi et al. 2004; Heger et al. 2005; Arnett and Meakin 2011a), there is expectation (Kifonidis et al. 2006; Wongwathanarat et al. 2015) and evidence (Blinnikov et al. 2000) that during shock propagation, 3D effects such as Rayleigh–Taylor instabilities act to reduce the gradients. Therefore, instead of carrying out hydrodynamic modeling based on evolutionary models, some modelers prefer to use "non-evolutionary" models constructed by artificially smoothing the density and composition gradients of the evolutionary models (Utrobin and Chugai 2008), or even to use two-component polytropic density structures with parameterized radial compositions (Bersten et al. 2011). The need to smooth 1D models could reflect, in addition to Rayleigh–Taylor instabilities, larger-scale asymmetries of real 3D SN IIP.

The properties of plateaus produced by light-curve calculations depend primarily on progenitor radius R, explosion energy E, ejected mass M_{ej} (Nagy et al. 2014), and, if ^{56}Ni is sufficiently mixed towards the surface, on M_{Ni} and the radial extent of the mixing (Kasen and Woosley 2009; Dessart et al. 2010b; Nakar et al. 2016). For increasing R, the brightness and duration of the plateau increase and the velocity at the photosphere at a given time on the plateau, v_{pl}, decreases. For increasing E, the brightness and v_{pl} increase and the duration decreases. For increasing M_{ej}, the duration increases and the brightness and v_{pl} decrease. Radioactive heating by ^{56}Ni mixed towards the surface increases the duration of the plateau and the brightness of the late plateau (Kasen and Woosley 2009; Bersten et al. 2011).

When all of the hydrogen in the expanding envelope has recombined, the plateau phase ends and the luminosity plunges, typically a few magnitudes in days or weeks. Following a plateau tail phase that may be unnoticeable or extended depending on whether M_{Ni} is high or low, the luminosity declines to the value that corresponds to the instantaneous ^{56}Co radioactivity luminosity. Then, if products of ^{56}Co decay are completely trapped in the massive ejecta, the luminosity depends only on M_{Ni}. With adopted values of distance and extinction, M_{Ni} can be inferred from the luminosity, which must include mid-IR emission when there is significant absorption of UVOIR radiation and reradiation in the mid-IR by dust (Sect. 12.8).

Before being compared to calculated light curves, UVOIR light curves must be corrected for reddening and extinction by dust. Given the relative simplicity of spectrum formation in a spherical hydrogen-rich envelope, the reddening can be inferred by detailed spectral fitting (Dessart and Hillier 2006; Baron et al. 2007). For example, if the adopted reddening correction is too high, the corrected continuum will be too blue, and the model temperature required to fit that continuum will be too high to account for the strength of the temperature-sensitive Ca II lines in the observed spectrum. Procedures for estimating reddening that do not require detailed spectral fitting, all subject to systematic errors, were discussed by Faran et al. (2014a).

When calculated bolometric light curves are compared to UVOIR observations, a bolometric correction should be applied to the observations. In several well-observed cases, early-time UV and later-time near-IR observations have shown that bolometric corrections are tightly correlated with UVOIR colors (Bersten and Hamuy 2009; Maguire et al. 2010b; Lyman et al. 2014a; Pejcha and Prieto 2015). Adopting such corrections for less well observed SN IIP is preferable to simple procedures such as adopting a constant bolometric correction for all epochs, or extrapolating to the UV and near-IR on the assumption that the photosphere radiates like a blackbody.

12.5.1 *Typical SN IIP*

In addition to SN 1999em (Sect. 12.2), well-observed, typical SN IIP include SN 1999gi (Leonard et al. 2002c), SN 2004et (Maguire et al. 2010a), SN 2012A (Tomasella et al. 2013), SN 2012aw (Bose et al. 2013), SN 2012ec (Barbarino et al. 2015), and SN 2013ab (Bose et al. 2015).

Hydrodynamic modeling of SN 1999em (Sect. 12.2) was carried out by three groups, each seeking a model that accounted for the light curve(s) as well as the photospheric velocity, v_{pl}, at some chosen epoch on the plateau, as inferred from blueshifts of absorption components of lines of moderate strength (e.g., Hβ, Fe II

λ5169, Sc II λ6246). Baklanov et al. (2005) used the multigroup STELLA code to calculate light curves for comparison with multiband photometry and derived $R \simeq 1000\ R_\odot$, $E \simeq 1$ B, $M_{ej} \simeq 18\ M_\odot$, and $M_{Ni} \simeq 0.06\ M_\odot$. Utrobin (2007a) carried out single-group calculations for comparison with the bolometric light curve and derived, for the same quantities, 500 $R_\odot$, 1.3 B, 20.6 $M_\odot$, and 0.036 $M_\odot$. From single-group calculations and the bolometric light curve, Bersten et al. (2011) derived 800 $R_\odot$, 1.25 B, 19 $M_\odot$, and 0.056 $M_\odot$. Considering that the hydrodynamical modeling was based on somewhat different techniques and assumptions, and that the reasons for some of the discrepancies are apparent (e.g., the lower value of M_{Ni} inferred by Utrobin is due to the use of a smaller bolometric correction during the tail phase), the agreement of these parameters is good, although the inferred values of R spanned a factor of two. The extent to which ^{56}Ni was mixed towards the surface in these non-evolutionary models differed substantially: uniform mixing throughout the innermost 15 $M_\odot$ by Baklanov et al. and Bersten et al., as opposed to only throughout the innermost 2.5 $M_\odot$ by Utrobin. Bersten et al. emphasized that extensive ^{56}Ni mixing improved their fit to the plateau. This "mixing" in the context of spherical models is ad hoc. While rather small-scale Rayleigh–Taylor instabilities might be captured to some approximation in spherical calculations, the actual mixing may also be in the form of distinct "plumes" that require multidimensional simulations (Sect. 9.4.6).

Hydrodynamical modeling of a somewhat stronger SN IIP, SN 2004et (Maguire et al. 2010a; Sahu et al. 2006), yielded $R \simeq 1500\ R_\odot$, $E = 2.3$ B, $M_{ej} \simeq 24.5\ M_\odot$, and $M_{Ni} \simeq 0.068\ M_\odot$ (Utrobin and Chugai 2009). Values estimated for SN 2012A, SN 2012aw, and SN 2012ec using the radiation-hydrodynamics code of Pumo and Zampieri (2011) were 260 $R_\odot$, 0.5 B, 12.5 $M_\odot$, and 0.01 $M_\odot$ (Tomasella et al. 2013), 430 $R_\odot$, 1.5 B, 20 $M_\odot$, and 0.06 $M_\odot$ (Dall'Ora et al. 2014), and 230 $R_\odot$, 1.2 B, 12.6 $M_\odot$, and 0.04 $M_\odot$ (Barbarino et al. 2015), respectively. A larger value of $R \simeq 700\ R_\odot$ was estimated for SN 2012A by Utrobin and Chugai (2015).

12.5.2 Weak SN IIP

Compared to SN IIP to which we have referred as typical, some SN IIP are notably weak (Pastorello et al. 2004). They are less luminous, their expansion velocities are lower (Fig. 12.6), they eject less ^{56}Ni, and hydrodynamical modeling indicates that their explosion energies are low. For the well-observed SN 2005cs (Pastorello et al. 2006, 2009) the characteristic ejection velocity inferred from spectra near the middle of the plateau was only 1500 km s^{-1}. Early UV and later near-IR observations allowed construction of the bolometric light curve (Fig. 12.2). Maximum light occurred only a few days after explosion. The plateau was not as subluminous as in some more extreme cases (e.g., SN 1999br; Pastorello et al. 2004), but the plateau was followed by a rapid, unusually large plunge of more than 3 magnitudes. Even after this steep decline, the radioactive tail was not yet reached; instead, a plateau tail phase persisted until 300 days after explosion. Subsequently, the decline was

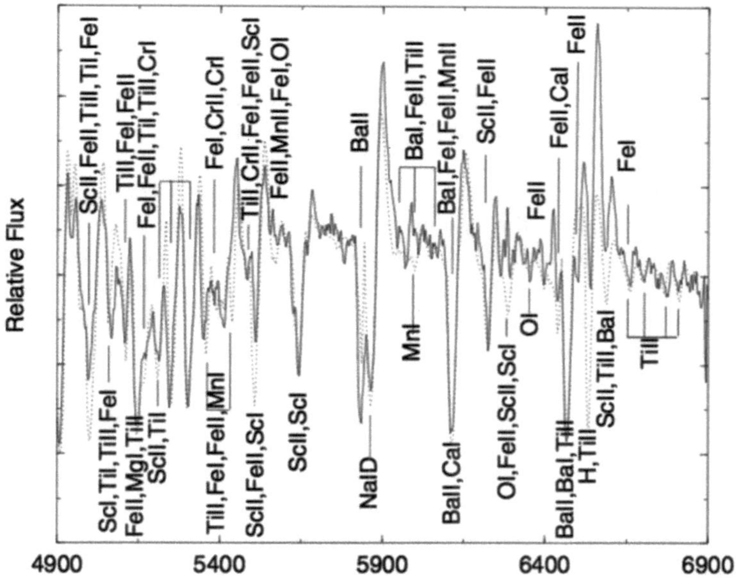

Fig. 12.6 A SYNOW fit to a section of a spectrum of SN 1999br obtained 100 days after explosion. The narrow lines (absorption blueshifts of $\lesssim$1000 km s^{-1}) permitted detailed line identifications. Figure 25 "Low-luminosity Type II supernovae: spectroscopic and photometric evolution" from Pastorello et al. (2004), by permission of Oxford University Press on behalf of the Royal Astronomical Society

consistent with ^{56}Co decay. From hydrodynamical modeling, Utrobin and Chugai (2008) estimated $R \simeq 570$ R$_{\odot}$, $E \simeq 0.5$ B, $M_{ej} \simeq 14$ M$_{\odot}$, and M_{Ni}=0.0015 M$_{\odot}$.

Two possibilities for the progenitors of these weak events have been suggested: those of sufficient ZAMS mass ($\gtrsim$20 M$_{\odot}$) for black-hole formation and extensive fallback, resulting in low M_{Ni} (Turatto et al. 1998; Zampieri et al. 1998), or those near the low end of the core-collapse mass range ($\sim$8 M$_{\odot}$), where little ^{56}Ni ejection is expected owing to the steep density gradient at the outer edge of the collapsing core (Chugai and Utrobin 2000). For SN 2005cs, evidence supports the latter suggestion; the hydrodynamical estimate of the ejected mass was more consistent with a low ZAMS mass (Pastorello et al. 2009), as were estimates based on archival imaging of the progenitor (6–13 M$_{\odot}$; Maund et al. 2005; Li et al. 2006; Takáts and Vinkó 2006; Eldridge et al. 2007). Spectra in the nebular phase showed the usual forbidden lines of typical SN IIP, but the lines were narrower, and a high ratio of [Ca II] to [O I] emission-line strengths (Sect. 12.6) also suggested a progenitor of modest mass. Several other weak SN IIP also have been inferred from archival imaging to have low ZAMS masses (Fraser et al. 2011). Still, most weak SN IIP have been too distant and faint to have useful limits of progenitor masses from archival imaging, so the possibility that some of these events were produced by more massive black-hole-forming progenitors has not been excluded.

12.5.3 Continuity and Correlations

Following the recognition of the weak events discussed above, some other SN IIP such as SN 2008in (Roy et al. 2011), SN 2009js (Gandhi et al. 2013), SN 2009N (Takáts et al. 2014), and SN 2013am (Zhang et al. 2014a) were found to be of intermediate strength, so "typical" and "weak" may be two portions of a continuum of SN IIP rather than distinctly bimodal. When SN IIP in nearby galaxies are classified in the traditional way (excluding SN IIL) the plateau lengths cluster around 100 days and the light-curve characteristics are qualitatively similar, but quantitatively the observational parameters, including plateau lengths (Takáts et al. 2015), cover wide ranges. (The parameters of our case study, SN 1999em, tend to be not far from the middle of these ranges.) Absolute luminosities both at peak and at the middle of the plateau vary by a factor of $\sim$100. Photospheric velocities decline roughly as $t^{1/2}$ during the plateau and at mid-plateau they vary by a factor of five. The rate of luminosity decline during the plateau also varies significantly. The value of M_{Ni}, derived from post-plateau observations, varies by a large factor. Photospheric velocity and luminosity in the middle of the plateau, the rate of luminosity decline during the plateau, and M_{Ni} all are positively correlated. Various analyses appealing to semi-analytic and numerical modeling indicate that these quantities, as well as ejected mass and kinetic energy, increase with ZAMS mass (Hamuy 2003; Maguire et al. 2010a; Poznanski 2013; Spiro et al. 2014; Chugai and Utrobin 2014; Fig. 12.7), although with considerable scatter (Pejcha and Prieto 2015). This behavior is not necessarily predicted by theory (Chap. 9). Correlations of distance-dependent luminosity on the plateau with other observables that are distance-independent is at the root of using SN IIP as extragalactic distance indicators (Sect. 25.3).

On the grounds that there is no clear bimodality between SN IIP and SN IIL, some authors (Anderson et al. 2014b; Sanders et al. 2015; Holoien et al. 2016) have favored considering them as a single continuum of events. In this case, the correlations between observables mentioned above become extended to higher luminosities and more rapid light-curve decline rates.

12.5.4 Estimating Masses and the Red Supergiant Problem

The expectation from standard stellar evolution theory is that SN IIP come from stars of ZAMS masses of 8 up to about 30 $M_\odot$. The so-called "red supergiant problem" (Sect. 3.2) is that ZAMS masses and upper limits derived for progenitors of SN IIP from archival presupernova images, combined with stellar evolution theory, tend to be low, $\gtrsim 8\,M_\odot$, leaving a dearth of progenitors of $\gtrsim 16\,M_\odot$. In contrast, SN IIP ZAMS masses inferred from hydrodynamical modeling tend to be high. For example, if the ejected mass of SN 1999em really was 19 $M_\odot$ (Sect. 12.5.1), then allowing 1 $M_\odot$ for presupernova mass loss and 1.5 $M_\odot$ for

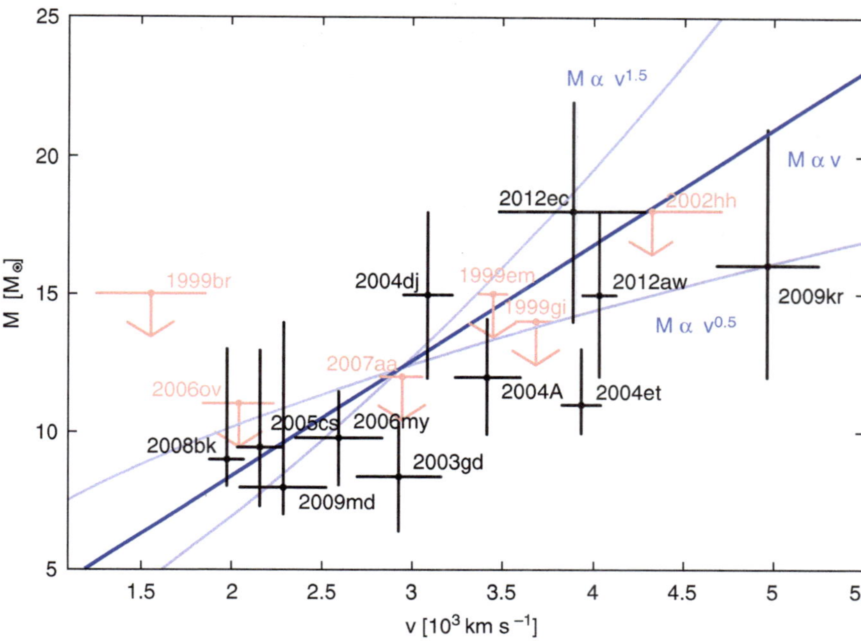

Fig. 12.7 Possible correlation between photospheric velocity 50 days after explosion with progenitor ZAMS masses as inferred from pre-explosion images. The *dark line* is the best-fitting power law. *Pink arrows* denote upper limits to ZAMS masses. Figure 1 "An emerging coherent picture of red-supergiant supernova explosions" from Poznanski (2013), by permission of Oxford University Press on behalf of the Royal Astronomical Society

a neutron star leads to an estimated ZAMS mass of 21.5 $M_\odot$, compared to an estimate of $\lesssim 15\,M_\odot$ from imaging of the progenitor (Smartt et al. 2009). Similar reasoning leads to 27 $M_\odot$ for SN 2004et (Sect. 12.5.1) compared to an upper limit of 13 $M_\odot$ from imaging (Crockett et al. 2011). Discrepancies between archival and hydrodynamical masses are not always so extreme (Tomasella et al. 2013; Dall'Ora et al. 2014), but hydrodynamical masses often seem to be too high, even though they measure only the hydrogen-rich ejecta, not the helium-rich ejecta (Dessart et al. 2013a). For a Salpeter initial-mass function, the median ZAMS mass for the range 8–30 $M_\odot$ is 13 $M_\odot$, lower than most of the values inferred from hydrodynamical modeling. Hydrodynamical masses may be too high because 3D effects during explosion are not adequately mimicked by artificial smoothing of density and composition gradients (Utrobin and Chugai 2009; Dessart et al. 2010b) or by the neglect of clumping in the ejecta (Utrobin and Chugai 2015). On the other hand, in some cases it may be difficult to reconcile low progenitor masses inferred from archival imaging with observed relatively high expansion velocities and plateau durations of ~100 days (Maguire et al. 2010a). The issue of which progenitor stars explode as SN IIP and which do not may be affected by "islands of explodability" (Sect. 9.4.5).

Some of the lowest ZAMS masses inferred from archival imaging are for weak events. These masses are in the range where white dwarfs, C/O cores, and ONeMg cores are expected to form (Sect. 8.5). This might be evidence that the weakest SN IIP are produced by collapsing ONeMg cores, which are expected in some models to produce modest energy and little ^{56}Ni. The progenitors of these events appear to have been too dim ($\log[L/L_\odot] \simeq 4.3$), however, to correspond to AGB or SAGB stars, which are expected to have $\log(L/L_\odot) \simeq 5.0$ (Smartt et al. 2009; Fraser et al. 2011). The implication is that these progenitors did not experience the second dredge-up (Sect. 8.5) and the associated formation of a degenerate core of any composition. They seem to have somehow avoided the second dredge-up and proceeded directly to the iron-core phase (perhaps after off-center neon burning), although this is not predicted by current stellar evolution models.

12.6 The Nebular Phase

Beginning about 100 days after explosion, the light curves of most SN IIP undergo a tail phase of decline at the ^{56}Co decay rate (0.0098 mag d^{-1}). For these, the value of M_{Ni} can be estimated. A minority of SN IIP decay faster than the ^{56}Co rate from the beginning of the tail (Anderson et al. 2014b), so that only a lower limit on M_{Ni} is obtained.

The late-time nebular spectra of SN IIP follow a pattern that is reflected in many other core-collapse events (and helped to cement the connection of SN Ib/c with core collapse). The spectra show prominent forbidden emission lines, although some especially strong permitted lines may continue to contribute. The late-time optical spectra are characterized, from blue to red, by Mg I] λ4571, Hβ, Na I D, [O I] $\lambda\lambda$6300,6364, Hα, [Fe II] $\lambda\lambda$7155,7172, [Ca II] $\lambda\lambda$7291,7323, and the Ca II near-IR triplet. Among the best observed SN IIP at late times are SN 1990E (Benetti et al. 1994a), SN 1999em (Leonard et al. 2002b; Elmhamdi et al. 2003), SN 2004et (Sahu et al. 2006; Maguire et al. 2010a), SN 2007it (Andrews et al. 2011a,b), and SN 2007od (Inserra et al. 2011). In the absence of mixing in SN IIP, emission lines of hydrogen and helium would be broad and have flat tops, while those of heavy elements would be narrower. Observationally, none of the lines have flat tops, and the widths of lines of different elements tend to be similar, which indicates extensive mixing in velocity space (Maguire et al. 2012a).

Among the abundance constraints obtained from analysis of nebular spectra, one of the most striking is a high ratio of stable ^{58}Ni to ^{56}Fe (produced as ^{56}Ni), about 3.5 times the solar value, in SN 2012ec (Jerkstrand et al. 2015b). Such a high value indicates that at least part of the deep silicon shell of the progenitor star was burned and ejected (Jerkstrand et al. 2015c).

Another way to estimate the ZAMS mass of a progenitor is based on the widths and strengths of the [O I] $\lambda\lambda$6300,6364 emission lines in nebular-phase spectra (Fransson and Chevalier 1989; Dessart et al. 2010b). According to stellar evolutionary models, the mass of oxygen in the presupernova increases strongly with ZAMS mass (Sect. 12.5.3). Furthermore, because the amount of presupernova mass loss also increases with ZAMS mass, the outer boundary of the oxygen-rich region is closer to the surface, in mass coordinate, in more massive stars. Therefore, for a given explosion energy, the widths and strengths of the [O I] lines increase with ZAMS mass. Thus the width, measured directly, and/or the strength, often normalized as the ratio of the [O I]/[Ca II] fluxes, provides an estimate of the velocity at the outer edge of the oxygen core, which in turn gives an estimate of the mass of the helium core and the ZAMS mass. This rather model-dependent approach, which may be complicated by the empirical correlation of energy with ZAMS mass, tends to give ZAMS masses between the low masses inferred from archival imaging and the high masses inferred from hydrodynamical modeling (Dessart et al. 2010b; Pastorello et al. 2009; Tomasella et al. 2013; Bose et al. 2013).

Based on detailed modeling of spectra of SN 2004et (Fig. 12.8) and the inferred nucleosynthetic output, Jerkstrand et al. (2012) estimated a ZAMS mass of the progenitor of 15 $M_\odot$, in agreement within uncertainties with the mass of 13 $M_\odot$ from archival imaging, but significantly smaller than the 27 $M_\odot$ inferred from hydrodynamical modeling (Sect. 12.5.1). A ZAMS mass of about 15 $M_\odot$ was inferred for SN 2012aw (Jerkstrand et al. 2014), consistent with some estimates from archival imaging (Kochanek et al. 2012a), but lower than 20 $M_\odot$ from hydrodynamical modeling.

The kinetic energy of the ejecta has been inferred from the observed profiles of the [O I] $\lambda\lambda$6300,6364 emission. The intrinsic ratio of the strengths of the bluer and redder transitions is 3; owing to greater saturation of the stronger component,

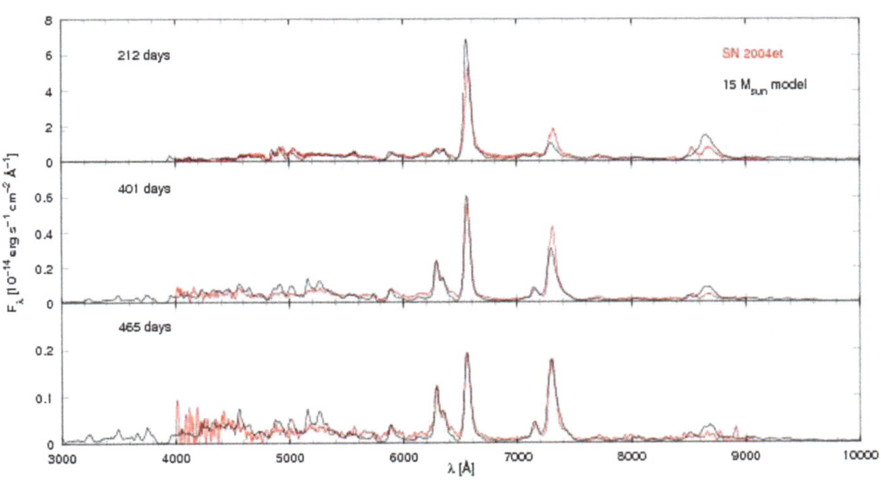

Fig. 12.8 Comparison of observed and model nebular spectra of SN 2004et. The strongest line, Hα, is flanked by lines of [O I] and [Ca II]. From "The progenitor mass of the Type IIP supernova SN 2004et from late-time spectral modeling" (Jerkstrand et al. 2012) reproduced with permission, © ESO

the profile of the blended doublet reveals the line optical depths (provided that they exceed unity) and hence an estimate of the density in the oxygen-rich zone. The density leads in turn to an estimate of the kinetic energy. Such an analysis is one indication that kinetic energies of SN IIP increase strongly with ZAMS masses of the progenitors (Chugai and Utrobin 2014).

Theoretical analysis of nebular-phase spectra needs to be done carefully, distinguishing contributions to the emission by freshly synthesized elements from elements that were present in the presupernova. These contributions can be overlapping in velocity space due to mixing between core and envelope material. The bulk of the hydrogen envelope can trap radioactive decay products that may heat and excite other elements, even if those scarce elements by themselves may not have the stopping power. A proper analysis must determine the major contribution to a given line at a given epoch using spectral synthesis techniques. This is especially important for [O I] $\lambda\lambda$6300,6364, discussed above, and for the Ca II near-IR triplet and [Ca II] $\lambda\lambda$7291,7323 (Sect. 11.5). The mass of freshly-synthesized elements can then, in principle, be determined by isolating their contribution to the line profiles from the envelope contributions. This process may be more complex if the mixing is in the form of plumes.

12.7 Circumstellar Interaction

At the time of core collapse, RSGs are expected to be losing mass at modest rates, $\lesssim 10^{-6}$ to $\lesssim 10^{-4}$ $M_\odot$ y^{-1} due to winds, rates which increase with ZAMS mass. For expected wind velocities $\gtrsim 10$ km s^{-1}, the circumstellar densities are high enough to produce currently detectable radio and X-ray emission only for SN IIP in relatively nearby galaxies, and radio and X-ray observations are generally in accord with expectation (Chevalier et al. 2006). From a lack of observed X-ray bright SN IIP, the maximum presupernova mass-loss rate and an upper limit to the ZAMS masses of SN IIP progenitors were estimated to be $\sim 10^{-5}$ $M_\odot$ y^{-1} and 19 $M_\odot$, respectively (Dwarkadas 2014).

Typical circumstellar densities are not high enough to influence UVOIR photometry, but there may be spectroscopic signatures in the Balmer lines and He I 1.0830 μm (Chugai et al. 2007; Inserra et al. 2012b; Takáts et al. 2014; Gutiérrez et al. 2014; Faran et al. 2014a) as well as in Ca II H&K and Na I (Chugai and Utrobin 2008).

In exceptional cases, higher mass-loss rates and stronger CSI may affect even UVOIR photometry. SN 2009kf (Botticella et al. 2010) was an unusually strong SN IIP. The luminosity was high, especially in the UV at early times, as were expansion velocities, e.g., 9000 km s^{-1} at 60 days after discovery. Hydrodynamical modeling based on the assumption that CSI did not affect the UVOIR light curve resulted in high values of $R = 2000$ $R_\odot$, $M_{ej} = 28$ $M_\odot$, and $E = 22$ B, prompting the suggestion that SN 2005kf was powered by black hole formation (Utrobin et al. 2010). On the other hand, hydrodynamical modeling that allowed for interaction of the ejecta with an unusually dense wind ($\dot{M}_w \gtrsim 10^{-4}\,M_\odot\ y^{-1}$) could account for the bright early UV emission as well as the bright optical light curve (Moriya et al. 2011). Thus, when CSI is taken into account, less extreme values for the explosion parameters may be allowed. Some mechanism such as pulsational instability (Yoon and Cantiello 2010) may cause extensive mass loss in the most massive RSGs within the century before explosion. SN 2009kf was not observed spectroscopically until 2 months after discovery; at earlier times its spectra may have been those of an SN IIn.

Late-time CSI is common among SN IIP. In the optical, there are several indications. CSI emission may cause the optical/near-IR light curve to flatten compared to the ^{56}Co decay rate or even rebrighten (Andrews et al. 2010). Emission lines may develop "boxy" components that originate in a spherical shell that is excited by CSI (Kotak et al. 2009; Inserra et al. 2011). Narrow Hα emission from heated but unshocked CSM, related to phenomena seen in SN IIn, may become detectable (Andrews et al. 2010; Inserra et al. 2012b). Additional CSI is associated with dust.

12.8 Dust

The amount of dust associated with SN IIP, whether pre-existing circumstellar dust or new dust formed in the ejecta, is an important quantity, since SN IIP are the most common supernova type. Furthermore, because the massive ejected envelope causes the deeper, metal-rich layers to expand at relatively low velocity and therefore high density conducive to dust formation, SN IIP are expected to be more efficient producers of dust than SN IIL, SN IIb, and SN Ib/c (Kozasa et al. 2009). Observations, primarily the development of a mid-IR excess, confirm that dust is associated with SN IIP. Both pre-existing and new ejecta dust are implicated, although in some cases it is difficult to distinguish between the two.

A long-wavelength ($\gtrsim$10 μm) cool (several hundred degrees) mid-IR component has been detected in several SN IIP (Kotak et al. 2009; Meikle et al. 2011). This emission is probably due to absorption of the early supernova optical and UV emission and reradiation by interstellar dust. Optical/UV echoes due to scattering of the supernova light by interstellar dust can flatten the light curve and produce a broadening of the Hα emission line (Andrews et al. 2010, 2015). Echoes constrain previously unobserved emission from shock breakout and the fireball phase (Van Dyk et al. 2015).

Pre-existing circumstellar dust may have formed in mass lost by the progenitor. Any such dust within $\sim 10^{17}$ cm of the progenitor is likely to be vaporized by breakout radiation, but for a typical RSG wind speed of 10 km s^{-1} this distance corresponds to only 3000 years, so most pre-existing dust probably survives the supernova. This warm ($\lesssim 10^3$ K) dust, heated by the breakout radiation, then radiates in the mid-IR. For example, a mid-IR excess of SN 2007it was attributed to 10^{-4} to $10^{-3}\ M_\odot$ of pre-existing circumstellar dust (Andrews et al. 2011).

Beginning as early as months after explosion, dust may also form in the CDS produced by CSI. The dust is radioactively heated and radiates warmly in the mid-IR. Signals of CDS dust formation include the decline of the optical light curve to below the level expected on the basis of ^{56}Co radioactivity and the development of a mid-IR excess. The radius of the CDS, as inferred from blackbody modeling of this excess, is much smaller ($\sim 10^{16}$ cm) than in the case of emission from pre-existing dust. Dust formation in a CDS has been inferred for several SN IIP, e.g., SN 2004dj (Meikle et al. 2011); SN 2004et (Kotak et al. 2009; Fabbri et al. 2011), SN 2007it (Andrews et al. 2011), and SN 2007od (Andrews et al. 2010; Inserra et al. 2011).

Finally, after a year or so, dust may form deep in the metal-rich layers of the ejecta, where it is heated by radioactive decay and/or by the reverse shock. A unique signal of dust formation deep in the ejecta is attenuation of the red sides and blueshifting of the peaks of optical and IR emission lines from the ejecta. Deep ejecta dust formation has been inferred in, e.g., SN 1999em (Elmhamdi et al. 2003), SN 2004dj (Szalai et al. 2011; Meikle et al. 2011), SN 2004et (Kotak et al. 2009; Fabbri et al. 2011), SN 2007it (Andrews et al. 2011), and SN 2007od (Andrews et al. 2010; Inserra et al. 2011).

Mid-IR spectroscopy can reveal the presence of molecules that are precursors to dust formation and that provide information on the nature of the dust itself. The fundamental and first overtone bands of CO are at 4.6 and 2.3 μm, respectively, while SiO has features in the range 7.5–9.3 μm. Silicate and graphite dust have features in the range 8–14 μm and near 30 μm, respectively. Molecules of CO were detected in SN 1999em (Spyromilio et al. 2001) and silicate dust and SiO were detected in SN 2004et (Kotak et al. 2009). When dust is known to be present, but spectroscopic signatures of SiO, silicate dust, and graphite dust are absent, the dust is presumed to be amorphous carbon, which has no distinct features.

Dust formation in both a CDS and deeper in the ejecta appears to be common in SN IIP, but estimating the *mass* of ejecta dust involves substantial uncertainties. Observational estimates of the total dust mass associated with an SN IIP tend to be at the level of 10^{-4} to 10^{-3} $M_\odot$, although these may be underestimates if some of the dust is in clumps of high density. Submillimeter observations with Herschel revealed the existence of a large amount, $\sim$0.1–1 $M_\odot$, of cold ($\sim$20 K) dust in SN 1987A (Sect. 11.4.3). If such a cold dust component is common in SN IIP, including those at high redshift, then SN IIP could be major contributors to the large amounts of dust inferred to be present in the early Universe (Sect. 24.6). Theoretically, however, an SN IIP is expected to produce $\lesssim 0.2\,M_\odot$ of dust (Sarangi and Cherchneff 2015).

12.9 Polarization and Asymmetry

A general discussion of asymmetries in supernovae was given in Sect. 1.7 and of spectropolarimetry in Sect. 4.6. The polarization properties of SN 1999em were briefly described in Sect. 12.2. Early spectra of SN IIP generally show low

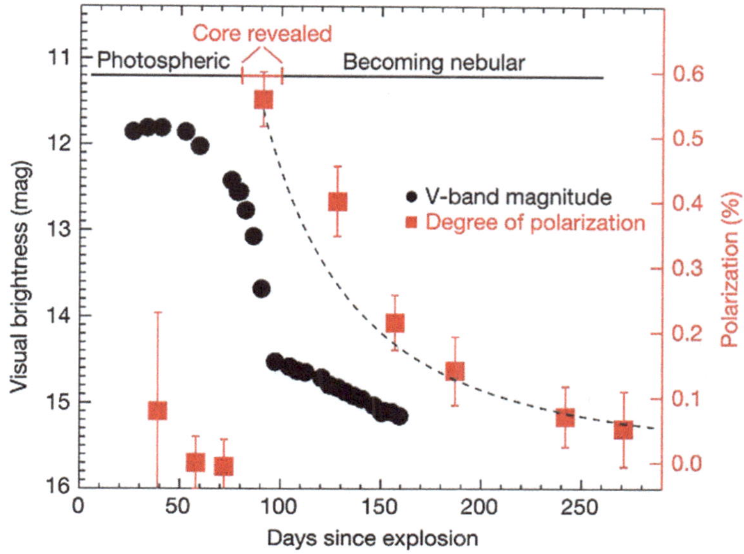

Fig. 12.9 V-band data for SN 2004dj (*black points*) compared to the degree of continuum polarization (*red points*). Polarization was negligible on the plateau, but jumped to $\simeq$0.6% at the end of the plateau, suggesting that the inner core was substantially more asymmetric than the envelope. Reprinted by permission from Macmillan Publishers Ltd.: NATURE, Leonard et al. (2006)

polarization. At later times the polarization is higher (Wang and Wheeler 2008), including continuum polarization of $\sim$1% after the end of the plateau (Leonard et al. 2006; Chornock et al. 2010). Figure 12.9 illustrates the example of SN 2004dj, for which the polarization rose significantly as the light curve plummeted at the end of the plateau. In some cases, the polarization rises to a high level well before the end of the plateau (Leonard et al. 2012).

The time evolution of continuum and line polarization due to electron scattering in asymmetric models for SN IIP can be calculated using Monte Carlo radiative-transfer techniques. Even in models with asymmetric structures in the outer envelopes the continuum polarization can be low, with complicated wavelength-dependent, aspect-angle effects during the plateau phase. The calculated polarization on the plateau is generally low and evolves in a nonmonotonic way, then rises as the ejecta become optically thin before vanishing as the optical depth becomes low. Thus the observed behavior of low continuum polarization on the plateau with rising polarization after the plateau does not necessarily imply a change in the asymmetry of the ejecta (Dessart and Hillier 2011a). These models should be checked with additional careful measurements of the polarization during SN IIP plateaus.

Models of asymmetric SN IIP suggest that there could be complex, multipeaked emission lines. Fluid instabilities can produce clumps of helium within the hydrogen envelope. If plumes of ^{56}Ni are generated, then asymmetric excitation is likely.

Nickel plumes can produce asymmetries that account for the observed structure of the Hα and He I 1.0830 μm lines. The relativistic jets of gamma-ray bursts penetrate the host star; this appears not to be the case for SN IIP because it would cause large asymmetries and therefore time-extended breakout behavior that would have been observed in some cases (Couch et al. 2009).

12.10 Summary

SN IIP are the most common supernovae in the contemporary Universe. They occur in young populations and their progenitors are RSGs. Stars having ZAMS masses in the range of about 10–30 $M_\odot$ are expected to be RSGs when their cores collapse, but archival imaging suggests a dearth of SN IIP progenitors above about 16 $M_\odot$, hence the red-supergiant problem. Hydrodynamical modeling and light-curve fitting tends to suggest the opposite, an excess of high-mass progenitors. Other evidence hints that the truth may be between these two extremes. It is not yet clear whether SN IIP are all due to collapse of iron cores or that some are due to electron capture in the lower end of the core-collapse mass range.

Although shock breakout from an SN IIP has not yet been unambiguously observed, the early post-breakout fireball phase has been. The prospects are good that breakout will be observed at both low and high redshift, providing special constraints on progenitor radii and explosion energies as well as an opportunity to study nonequilibrium shock physics.

When core collapse occurs in an RSG, the massive hydrogen envelope yields the characteristic plateau in the light curve as a recombination front sweeps inward in mass due to the expansion of the ejecta. Models of light curves and spectra depend on input parameters such as the progenitor radius, the explosion energy, and the ejected mass, as well as ^{56}Ni and its mixing. Thus observations can constrain these parameters.

Most calculated progenitor models are nonrotating and nonmagnetic, only approximations to reality. Light-curve and spectral-synthesis modeling require such models to be mixed by hand. In reality, the mixing may be on small scales, as expected from Rayleigh–Taylor and other hydrodynamic instabilities, but also by larger-scale plumes or jets. Ad hoc mixing applied to calculated presupernova models may affect the light-curve fitting results, including ejected mass. Evidence for associated asymmetries can be sought from spectropolarimetry and line profiles.

Some SN IIP are weak. They are dimmer than typical SN IIP, with lower ejection velocities and with less ejected ^{56}Ni. The origin of these events remains uncertain. Early suggestions that some of them represented the formation of black holes have not been confirmed. Their collapse may be initiated by electron-capture in degenerate ONeMg cores, but other evidence suggests that no SN IIP arises from a degenerate core.

As the ejecta expand, the envelope recombines and becomes optically thin. When the recombination wave reaches the base of the envelope, the luminosity declines steeply by several magnitudes before eventually beginning a slower decline powered, in most cases, by the radioactive decay of ^{56}Co. Nebular-phase spectra yield information on the structure and composition of the inner core. Optical and IR spectra can be used to probe the masses of individual ejected elements, which together with evolutionary models lead to an estimate of the ZAMS mass of the progenitor.

SN IIP show various signs of CSI, although not as strongly as SN IIn. There also is ample evidence for dust associated with SN IIP, some pre-existing the supernova and some formed after the supernova in a CDS and/or deeper in the ejecta.

Chapter 13
Type IIL Supernovae

13.1 Introduction

The category of Type IIL supernovae was introduced by Barbon et al. (1979). Candidate SN IIL from SN 1926A through SN 1987B were listed by Patat et al. (1993) and an early compilation of B and V light curves was provided by Doggett and Branch (1985). SN IIL have hydrogen lines in their spectra, making them Type II. They do not display the distinct light-curve plateau characteristic of SN IIP, but rather a nearly linear (in magnitudes) decline in visual luminosity from peak. A linear decline corresponds to an exponential decay of luminosity. The difference between SN IIP and SN IIL is thus primarily based on photometric, not spectroscopic, behavior, although there are some differences in the spectra as well. In a volume-limited sample, SN IIL account for 6% of all core-collapse supernovae. An important issue is that SN IIL often show narrow emission lines in their spectra, and SN IIn light curves often show phases of linear decline. Thus it is not clear that SN IIL can be rigorously distinguished from SN IIn (Chap. 14). It also is not clear that SN IIL light curves can be rigorously distinguished from those of SN IIP (Sect. 12.5.3).

The prototypes of SN IIL are SN 1979C and SN 1980K, both of which were somewhat more luminous at peak, $M_B \lesssim -19$, than the typical $M_B \simeq -18$ for SN IIL. The light curves of some superluminous supernovae ($M_V < -21$; Chap. 18), also have phases of linear decline. The limited available evidence from archival images and supernova rates suggests that the ZAMS masses of SN IIL progenitors tend to be about 20 $M_{\odot}$, somewhat higher than those of SN IIP (Sect. 3.2).

D. Branch, J.C. Wheeler, *Supernova Explosions*, Astronomy and Astrophysics Library, DOI 10.1007/978-3-662-55054-0_13

13.2 Case Studies

13.2.1 SN 1979C

SN 1979C in M100, discovered by amateur astronomer Gus Johnson (Mattei et al. 1979) was at its time one of the intrinsically brightest SN II known, with $M_B \simeq -19.5$. It showed an early, nearly linear decline, but with a hint of a plateau in the U, B, and V bands about 30 days after discovery (de Vaucouleurs et al. 1981; Fig. 13.1). The overall decline in V in the first 100 days was at a rate of 0.04 mag d^{-1}, but the V-band light curve in this early phase could alternatively be described as an early fireball phase with a decline of nearly a magnitude during 10 days, a truncated plateau lasting another 20 days, then a decline of 1 magnitude in 30 days, followed by a flattening to a linear decline at a rate of 0.024 mag d^{-1} for the next 50 days. Four months later, after solar hiatus, the slope had decreased to 0.015 mag d^{-1} (Barbon et al. 1982). The U and B bands showed qualitatively similar behavior, but with steeper slopes. The $B - V$ color reddened steadily until the end of the plateau, then remained roughly constant for the next 80 days.

The early optical spectra of SN 1979C displayed a blue continuum with weak, narrow emission lines (Branch et al. 1981). Later spectra showed strong Hα emission, but with weak or no blueshifted Hα absorption (Panagia et al. 1980; Branch et al. 1981; Barbon et al. 1982; Schlegel 1996), consistent with ongoing CSI (Fig. 13.2). From 10 to 30 years after explosion, the flux of Hα decreased by one-third while blueshifted emission of forbidden oxygen lines increased to dominate the emission (Milisavljevic et al. 2009; Fig. 13.3). The velocity width of the Hα line was essentially constant near 6700 km s^{-1} during the 20-year span, but the double-peaked emissions of [O I]$\lambda\lambda$6300,6364 and [O II] $\lambda\lambda$7319,7330 narrowed substantially. While the Hα emission was from CSI, the narrowing and the strength of the oxygen lines suggested that this emission was from ejecta with enhanced oxygen abundance, perhaps heated by the reverse shock. The late spectra showed no sign of Mg I] λ4571, a typical feature of core-collapse nebular spectra that appears in some, but not all, SN IIL. For a distance of 20 Mpc, the luminosity in Hα at 5000 days was $\sim 10^{38}$ erg s^{-1}, nearly an order of magnitude brighter than SN 1980K (Sect. 13.2.2) at comparable epochs.

Early spectra gave evidence that the progenitor had a dense stellar wind that had been radiatively accelerated such that CSM close to the supernova had a velocity proportional to r^{-2}, rather than constant as in the simplest case (Falk and Arnett 1977; Klein and Chevalier 1978; Fransson 1982). In this picture, photospheric photons that were Compton scattered by the hot postshock gas, together with free–free emission, produced a UV excess (Fransson et al. 1984). The UV emission lines had asymmetric profiles suggesting that they formed in a narrow shell close to the photosphere (Fig. 13.4). The shell presumably formed by CSI, with the radiation from the forward shock exciting the UV lines by photoionization. An enhanced ratio of N/C $\simeq$8 showed that the envelope of the progenitor had been stripped down to CNO-processed material. Hydrodynamical modeling suggested that at the time

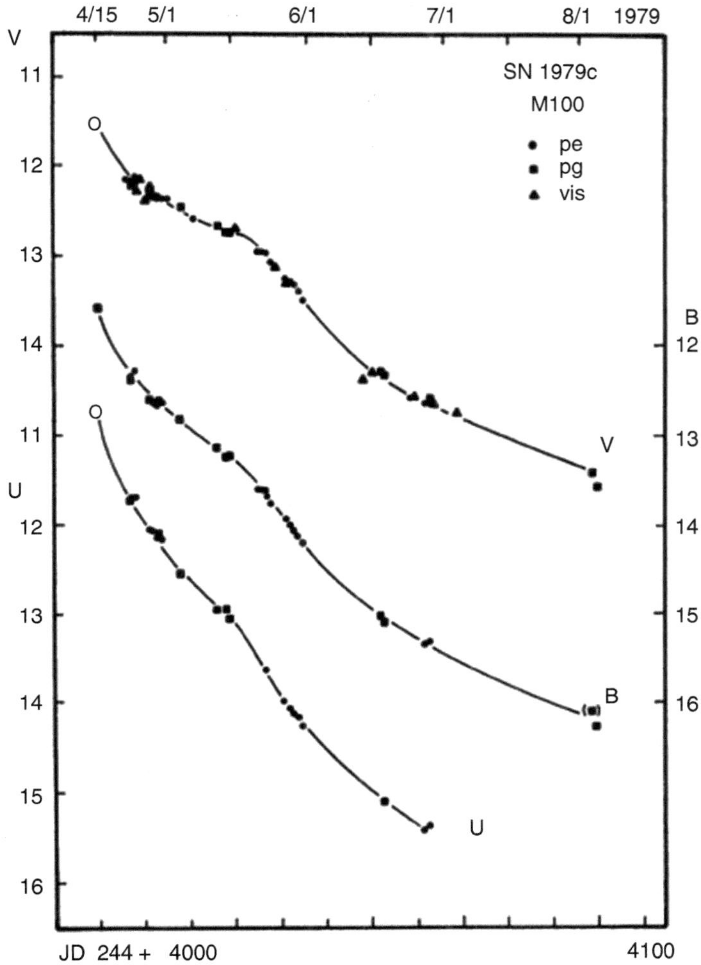

Fig. 13.1 U, B, V light curves of SN 1979C. Figure 5 "The bright supernova 1979C in M100," from de Vaucouleurs et al. (1981) © The Astronomical Society of the Pacific. Reproduced by Permission. All Rights Reserved

of explosion the mass of the hydrogen envelope was rather low, 1–2 $M_{\odot}$ (Blinnikov and Bartunov 1993).

SN 1979C displayed strong radio emission (Weiler et al. 1986, 1991) consistent with a model in which the emission arose from the forward shock colliding with a wind of approximately constant velocity (more distant CSM that was less subject to radiative acceleration at breakout) and a mass-loss rate of $\sim 10^{-4}\ M_{\odot}\ y^{-1}$, rather high for a red supergiant. The radio emission showed periodic variations in time (Weiler et al. 1992; Fig. 13.5), suggesting that the mass loss had been modulated, perhaps by a binary companion. Radio VLBI observations showed that the ejecta

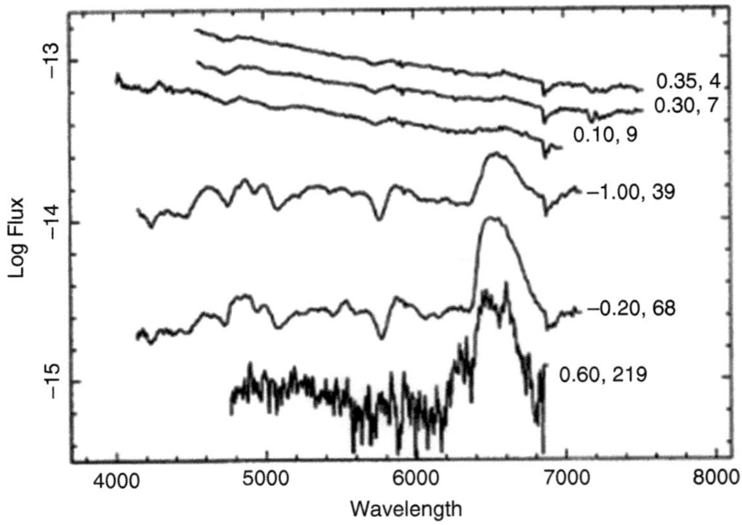

Fig. 13.2 Evolution of optical spectra of SN 1979C. From "On the Early Spectroscopic Distinction of Type II Supernovae" (Schlegel 1996), based on data from Branch et al. (1981). © AAS. Reproduced with permission

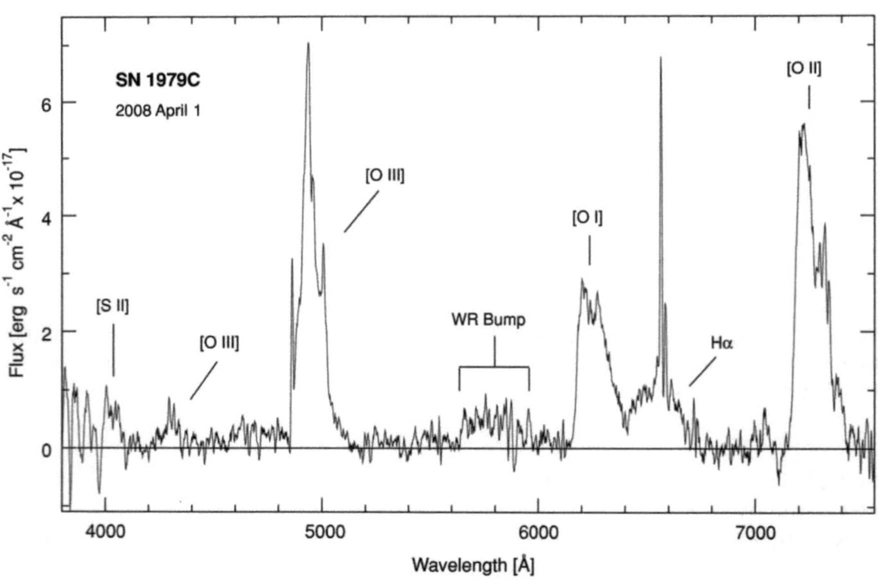

Fig. 13.3 A spectrum of SN 1979C taken in 2008, 29 years after the explosion. A blue continuum was subtracted from the observed spectrum. From "The Evolution of Late-Time Optical Emission from SN 1979C" (Milisavljevic et al. 2009). © AAS. Reproduced with permission

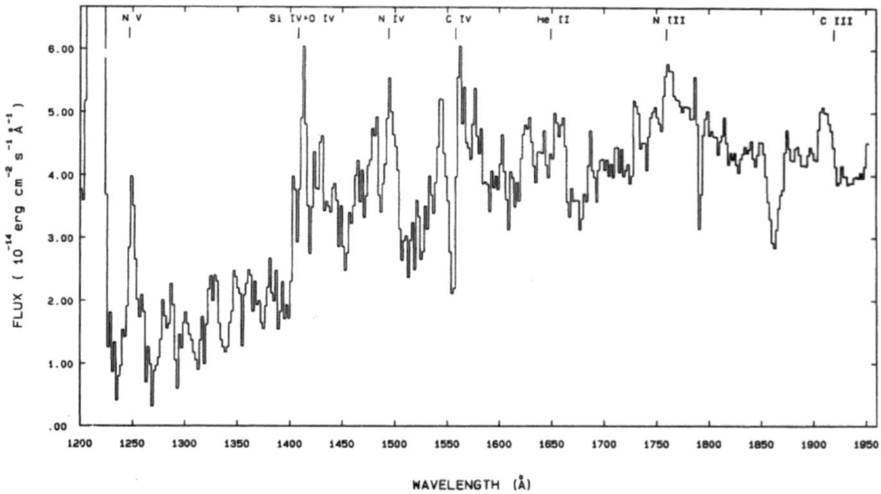

Fig. 13.4 The early UV spectrum of SN 1979C. From "Physical conditions in the UV line emitting region of supernova 1979C in NGC 4321" (Fransson et al. 1984) reproduced with permission, © ESO

were in nearly free expansion for at least 25 years (Bartel and Bietenholz 2008; Marcaide et al. 2009b).

The X-rays from the forward shock (Panagia et al. 1980) were susceptible to absorption in the CDS that is expected to have formed by radiative cooling at the contact discontinuity between ejecta and CSM. Reprocessed radiation from the CDS could have given rise to the low-ionization optical spectra in the ejecta and in the shocked shell, with high-ionization lines being produced in the freely expanding ejecta interior to the reverse shock (Chevalier and Fransson 1994). Strong late-time line emission in the optical and UV was interpreted in terms of interaction between the forward shock and CSM (Fesen and Matonick 1993; Fesen et al. 1999a; Milisavljevic et al. 2009).

Analysis of archival data of SN 1979C showed that the X-ray flux was essentially constant at 6.5×10^{38} erg s^{-1} from 1995 to 2007, consistent with accretion from a fallback disk or a binary companion onto a black hole (Patnaude et al. 2011). On the other hand, an X-ray brightening of SN 1970G, a probable Type IIL, 40 years after explosion may have signalled the development of a pulsar wind nebula and hence the presence of a neutron star (Dittmann et al. 2014). A pulsar wind nebula in SN 1979C cannot be ruled out.

We still have no basic understanding of why SN 1979C had the linear light curve it did, or what sort of star exploded to create it. The data are consistent with the progenitor being a red supergiant, but there are no firm constraints on its radius. The enhanced oxygen lines at very late times suggests the presence of oxygen-rich material, but that alone does not constrain the progenitor in any substantial way. The inferred masses of the hydrogen envelope and the CSM were each around

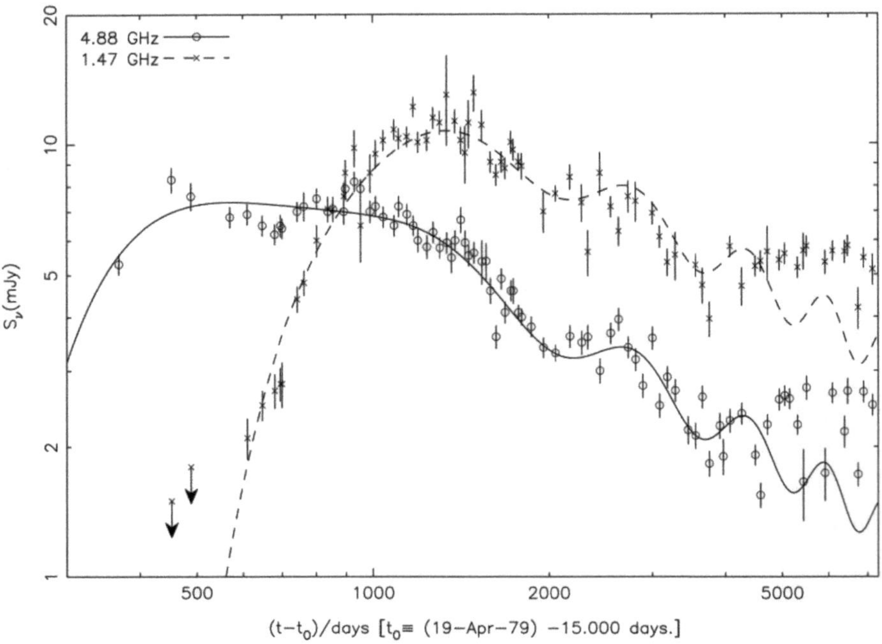

Fig. 13.5 Radio light curves of SN 1979C for 18 years at 20 cm (*crosses*) and 6 cm (*circles*). *Solid and dashed lines* are best-fit (to day 4300) model light curves that include a sinusoidal term. From "Radio Observations of SN 1979C: Evidence for Rapid Presupernova Evolution" (Montes et al. 2000). © AAS. Reproduced with permission

1 $M_\odot$. Pre-explosion images suggest that the ZAMS mass of the progenitor was about 20 $M_\odot$, in which case much more expelled matter should be at larger radii.

13.2.2 SN 1980K

After its discovery (Wild and Barbon 1980) just past maximum light, SN 1980K in NGC 6946 displayed the classic linear decline at a rate of 0.04 mag d^{-1} in the V band. The V light curve flattened to a more shallow decline at 150 days after maximum (somewhat sooner in B; Fig. 13.6). At maximum, SN 1980K had $M_B \simeq -19$. Spectra 60 days after explosion showed distinct Hα emission with only a weak blueshifted absorption (Barbieri et al. 1982), again consistent with ongoing CSI. It is not clear in the low-resolution spectra of the era whether SN 1980K had a narrow emission-line component.

Two years after explosion, the Hα flux was observed to be declining nearly exponentially, consistent with the decay of 0.1 $M_\odot$ of ^{56}Ni (Uomoto and Kirshner 1986). Given the evidence for extended CSI, this conclusion perhaps warrants some

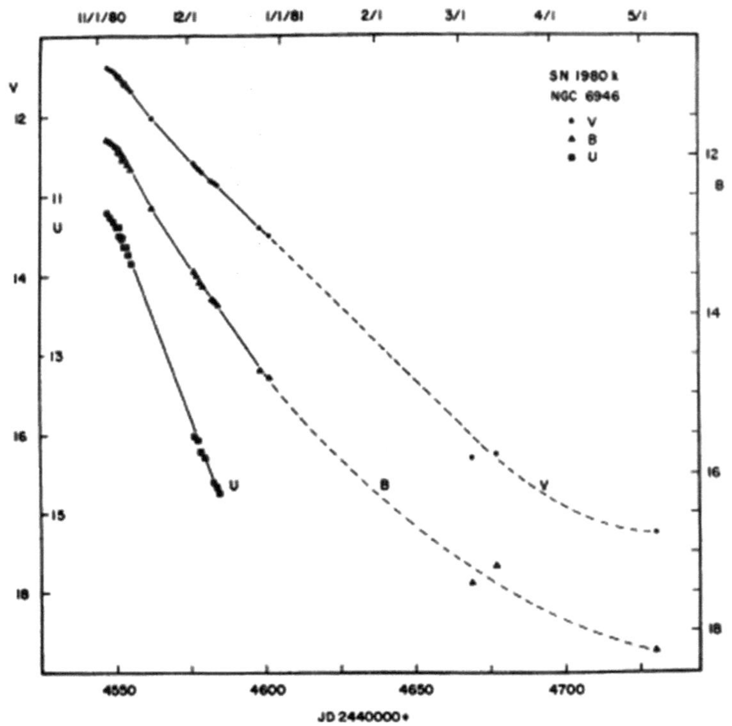

Fig. 13.6 The U, B, and V light curves of SN 1980K. Figure 4 "Photometric observations of the bright Type II supernova 1980K in NGC 6946" from Buta (1982) © The Astronomical Society of the Pacific. Reproduced by Permission. All Rights Reserved

caution. Spectra of SN 1980K acquired 7–14 years after explosion showed that the Hα flux was roughly constant, although unlike SN 1979C the line became narrower at the later epochs (Fesen and Becker 1990; Uomoto 1991; Fesen et al. 1995). Broad emission lines of width 5500 km s^{-1} of Hα, [O I] $\lambda\lambda$6300,6364, and [O II] $\lambda\lambda$7319,7330 were detected 15 and 17 years after explosion (Fesen et al. 1999a). By 30 years, the Hα flux had declined by only 25% (Sugerman et al. 2012; Milisavljevic et al. 2012; Fig. 13.7). The slow rate of change may be related to scattering and thermal light echoes from extended CSM. The flux in [O I] became comparable to that in Hα at late times, suggesting emission from inner ejecta. No signature of Mg I] λ4571 was evident. The evolution of the Hα and [O I] line widths may be due to shocking of high-velocity metal-rich clumps of ejecta (Sugerman et al. 2012) rather than to the standard CSI picture of ejecta colliding with slowly-moving CSM.

SN 1980K was detected in the radio (Weiler et al. 1992) and X-rays (Canizares et al. 1982; Schlegel 1994). In the radio, the decay was steady for a decade, consistent with shock propagation into a steady-state wind. Observations 14–16 years after explosion showed that the radio emission had declined significantly, a factor of two below the previous trend line (Montes et al. 1998). The data are

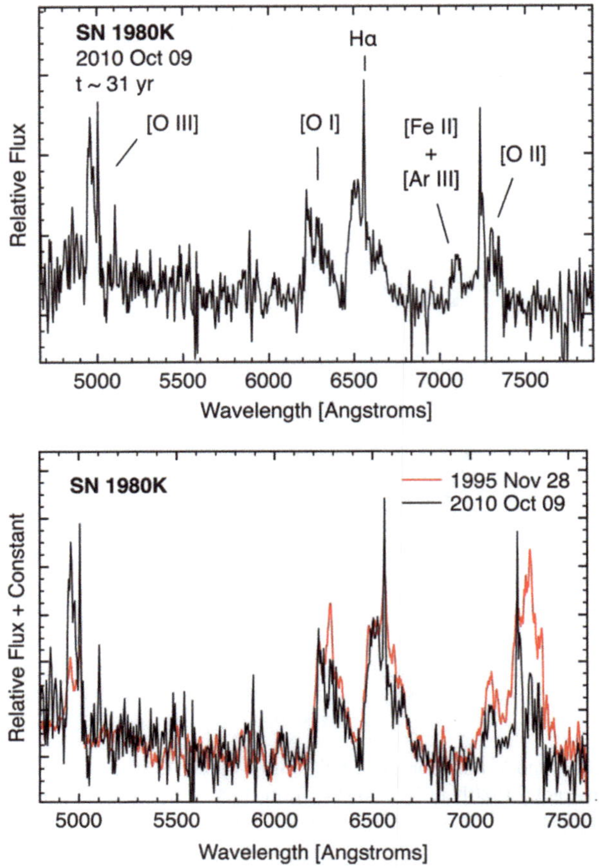

Fig. 13.7 Spectrum of SN 1980K 30 years after explosion (*top*) and the same spectrum compared to one at 15 years after explosion (*bottom*). From "Late-time Optical Emission from Core-collapse Supernovae" (Milisavljevic et al. 2012). © AAS. Reproduced with permission

consistent with an increase in the mass-loss rate 10,000 years prior to explosion that lasted for about 4000 years.

The radio observations do not establish an absolute scale for the wind density or mass-loss rate. If the CSM probed by the early radio were the result of a steady-state wind with a constant velocity, then the transition 10^4 years before the supernova corresponds to a distance from the progenitor of $\sim 4 \times 10^{17}$ cm, for a wind velocity of 10 km s^{-1}. If the mass-loss rate had been constant at 10^{-4} M$_\odot$ y^{-1}, then the mass expelled in 10^4 years would have been ~ 1 M$_\odot$. The mass remaining in the outer, lower-density CSM remains unconstrained.

As for SN 1979C, decades of study of SN 1980K have revealed many interesting aspects, but no deep understanding of the reason that it had a linear light curve, nor what sort of star exploded to create it. Speculation that the progenitor of SN 1980K may have undergone a *blue loop* might suggest that the progenitor had a rather modest mass, 8–10 $M_{\odot}$, but this remains uncertain. Pre-explosion images put a loose limit on the ZAMS mass of $\lesssim 20\,M_{\odot}$.There might have been 1 $M_{\odot}$ in the CSM in the immediate vicinity of the progenitor. The enhanced oxygen lines at very late times suggest the presence of oxygen-rich material, but, as for SN 1979C, this does not substantially constrain the progenitor. There is no hint of a surviving binary companion, nor of a compact remnant.

13.3 Synthesis of SN IIL Characteristics

Apart from SN 1979C and SN 1980K (Sect. 13.2), reasonably well-studied SN IIL include SN 1990K (Cappellaro et al. 1995), SN 2009kr (Fraser et al. 2010), and SN 2013by (Valenti et al. 2015).

Most SN IIL are less luminous at discovery than SN 1979C and SN 1980K, but brighter than SN IIP and with longer rise times to maximum light (Gall et al. 2015). A sample of V-band light curves of SN IIL is shown in Fig. 13.8. The light curves of SN IIP have several distinct phases, including fireball, plateau, and tail. Analogous phases exist for some SN IIL. SN 1979C had a brief rapid decline shortly after discovery that might correspond to a fireball phase, but this phase has not often been seen in SN IIL. Some SN IIL show a short, truncated plateau, as did SN 1979C. The slope of the linear decay in the V band for SN IIL varies from event to event,

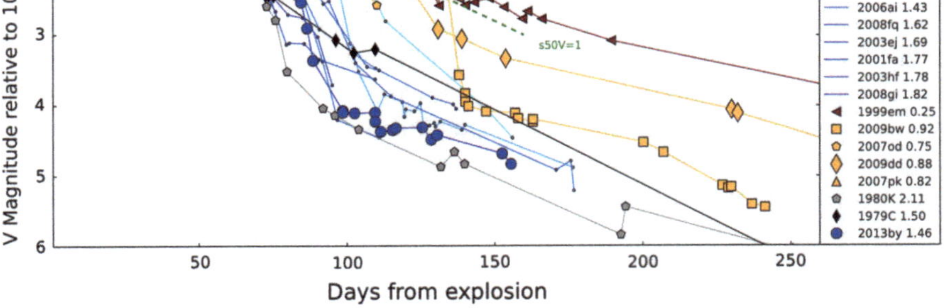

Fig. 13.8 A collection of candidate SN IIL V-band light curves plus the prototypical Type IIP SN 1999em for contrast. The numbers in the legend at right are the declines in V magnitude during the 50 days after peak brightness. Figure 3 "Supernova 2013by: a Type IIL supernova with a IIP-like light-curve drop" from Valenti et al. (2015), by permission of Oxford University Press on behalf of the Royal Astronomical Society

but typical rates of decline are 0.04 mag d^{-1}, steeper than a plateau or ^{56}Co decay, but less steep than the post-plateau plunge phase of SN IIP. SN IIL often show a transition to a less steep late-time decay that may be consistent with that expected for ^{56}Co decay, although in some cases (SN 1996L, SN 1990K) the late decline rate was steeper and in others it was shallower (SN 1994aj). Quasi-bolometric light curves of SN IIL are rare in the literature. The physical nature of the linear phase thus remains ill defined.

SN IIP redden during the plateau, somewhat more rapidly during the plunge, and then become bluer on the late tail. SN IIL tend to evolve to redder $B - V$ during the linear phase, but at slightly bluer colors than SN IIP at comparable epochs. SN IIL do not show any pronounced tendency to become bluer after the end of the linear phase. The end of the SN IIP plateau is red whereas early colors of SN IIL are blue, so the linear phase of SN IIL is not directly related to the plunge of SN IIP, although these phases may have some physics in common. The shape of the plunge in SN IIP is expected to be affected by deposition of radioactive-decay products (Kasen and Woosley 2009). The linear phase of SN IIL may be powered in part by radioactive decay coupled with the effects of a finite diffusion time and the presence of a CDS in an envelope of modest mass.

Early spectra of SN IIL often show a blue continuum and relatively weak, narrow emission lines; in this sense they resemble SN IIn. Later they develop strong Hα emission, typically with little or no broad, blueshifted absorption (Fig. 13.9). The lack of broad absorption makes the spectra of some, but not all, SN IIL different than those of SN IIP, presumably because the line source function falls off less steeply with radius in SN IIL than in SN IIP. This would be consistent with a substantial CSM surrounding SN IIL that is less dense or absent in SN IIP.

The nebular spectra of SN IIL show that hydrogen emission persists for decades. The spectra of some SN IIL (e.g. SN 1986E, SN 1990K) contained the standard emission lines of core-collapse supernovae, including Mg I] λ4571, [O I] $\lambda\lambda$6300,6364, and [Ca II] $\lambda\lambda$7291,7323, but others (SN 1979C; SN 1994aj) do not. In the latter case, these forbidden lines that arise in the ejecta may be suppressed by high densities or obscured by the CSM that produces the extended Hα emission. The evolution of Hα can be indicative of asymmetries in the interaction of ejecta with CSM (Benetti et al. 1998). Similarities of emission-line profiles with those of Cas A suggest that observed emission-line asymmetries may have been associated with dust in the ejecta, and that substructure often interpreted as clumps of ejecta may be due instead to large-scale rings of supernova debris, such as seen in Cas A (Milisavljevic et al. 2012). Thermal IR excesses in SN 1979C and SN 1980K 7–9 months after explosion might have been due to a light echo as the supernova emission was absorbed and re-emitted by dust in the CSM (Dwek et al. 1983; Chugai 1992a).

Some SN IIL show distinct narrow P Cygni features in the Balmer lines. These features are usually seen, if at all, at early phases, but in some cases they are persistent. Events displaying these features include SN 1984E (Dopita et al. 1984; Henry and Branch 1987; Fig. 13.10), SN 1979C (Branch et al. 1981), SN 1994aj

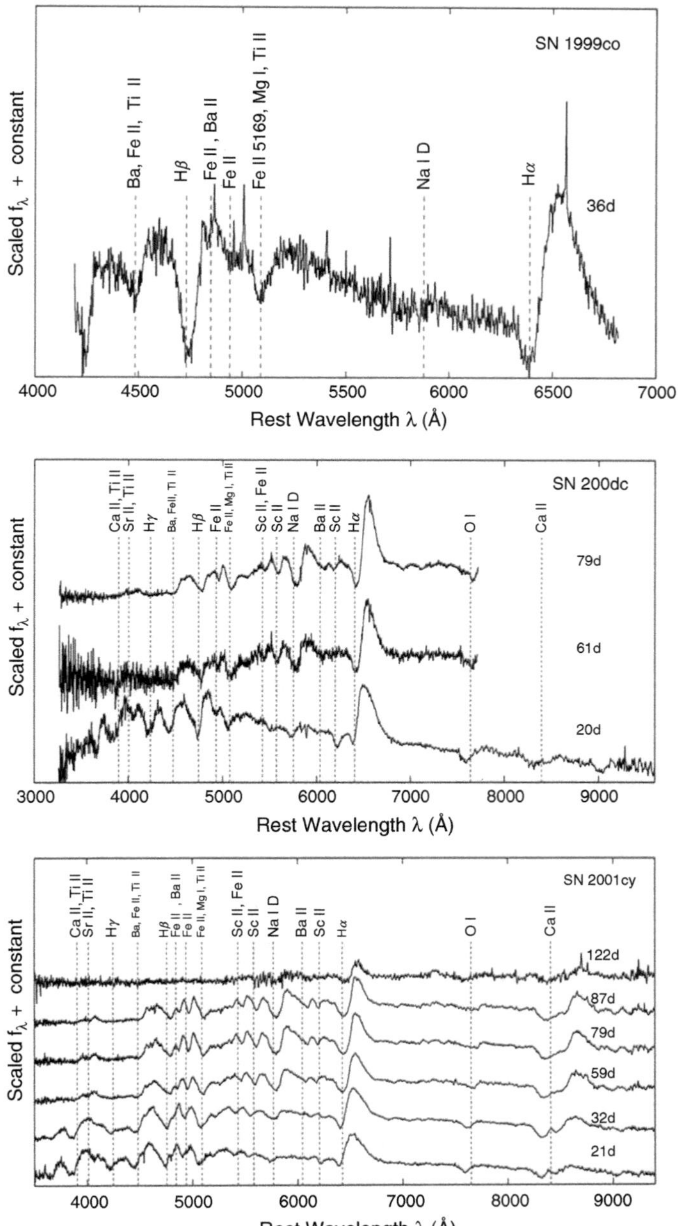

Fig. 13.9 Spectra of a sample of SN IIL (SN 1999co, SN 2000dc, and SN 2001cy). The spectral features are identified by their corresponding blue-shifted absorption. The features tend to be less distinct in early spectra than those of SN IIP. Features become more distinct about a month after explosion, but the absorption of Hα remains muted compared to SN IIP at comparable phases. The absorption profiles of O I and the blend of NaD/He I λ5876 tend to be stronger relative to Hα than they are in SN IIP. Figure 9 "A sample of Type IIL supernovae" from Faran et al. (2014b), by permission of Oxford University Press on behalf of the Royal Astronomical Society

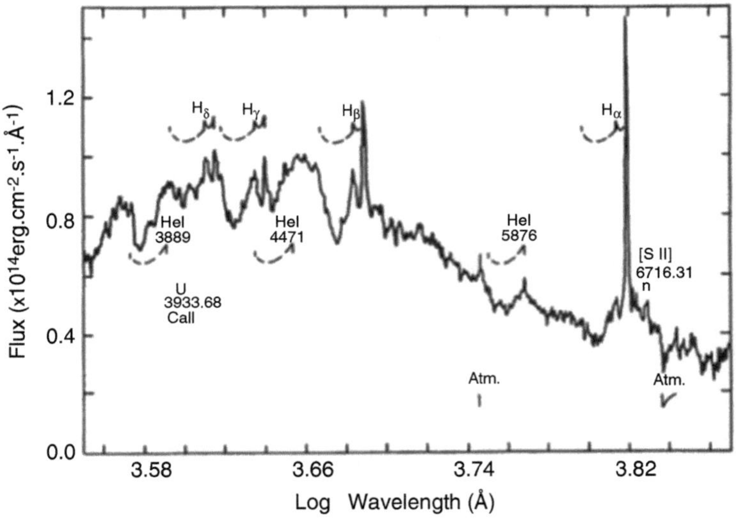

Fig. 13.10 Narrow Balmer lines in a spectrum of the Type IIL SN 1984E near maximum light. From "The 1984 supernova in NGC 3169—Evidence for a superwind" (Dopita et al. 1984). © AAS. Reproduced with permission

(Benetti et al. 1998), and SN 1996L (Benetti et al. 1999). The narrow features are thought to arise by scattering in a slowly-moving wind, as would be characteristic of a red supergiant, that was ejected decades before explosion. When the narrow features disappeared quickly (SN 1984E), the CSM was presumed to have been near the progenitor and quickly swept up by the shock. For events showing the narrow features, emission from the collision between the forward shock and the CSM that illuminates the low-velocity wind material is typically observed a few years after explosion, presumably when the ejecta reach the wind material. A reasonable hypothesis is that the low-mass envelope of the progenitor that characterizes SN IIL is a consequence of extensive presupernova mass loss.

The dilute CSM of a typical SN IIP may correspond to a steady-state wind with a density profile of $\rho \propto r^{-2}$ that extends outward from near the photosphere of the progenitor, whereas the CSM of a typical SN IIL may include in addition a detached shell. The shell would give SN IIL common properties with SN IIn. The presence of a detached shell might be revealed by early observations of the luminosity, line flux, and ionization state of the CSM. The late-time flattening of the light curve that is common to many SN IIL could then plausibly represent the phase when radioactively-powered decay gives way to CSI, but such a transition in light-curve slope could represent instead the transition of the shock between portions of the CSM of different density profiles.

Both SN 1979C and SN 1980K were detected in the radio, and SN 1979C also was detected in X-rays. This is not common for other SN IIL, at least in part because they are more distant.

The luminosity in Hα and a measured shock velocity inferred from the width of Hα allows an estimate of the wind parameter, $w = \dot{M}/v_w$, a measure of the density of the shell, modulo an uncertain factor of the efficiency of converting kinetic energy into Hα flux (Chugai 1990). Interpretation of the wind parameter in the context of a steady-state wind may not be appropriate to time-dependent shell ejection, but a quasi-steady-state condition is nevertheless often assumed. Given the forward shock velocity, the time of onset of the flattened light curve gives an indication of the location of the inner edge of the shell, and any subsequent strong decrease in Hα luminosity indicates the outer boundary of the shell. These tend to be $\sim 10^{16}$ cm and $\gtrsim 10^{17}$ cm, respectively. The assumption of a wind velocity, or better, the measurement of the expansion velocity of the CSM from narrow absorption or emission lines, then allows an estimate of the times of onset and halting of the enhanced mass-loss period that created the shell. For SN IIL, estimates of the onset range from tens to thousands of years prior to explosion, while estimates of the halt are only years or decades before explosion. Estimates of the wind parameter and the wind velocity then also allow an estimate of the mass-loss rate. Typical mass-loss rates are $\sim 10^{-3}\,M_{\odot}\,y^{-1}$. The mass in the CSM is thus, very roughly, 1 $M_{\odot}$. As for SN 1979C, this probably does not account for all of the mass of the CSM, since according to models the progenitor has only a low-mass envelope at the time of explosion. If the ZAMS masses of the progenitors are about 20 $M_{\odot}$, then more mass remains to be discovered.

Where SN IIL fit in with other subclasses of SN II, especially SN IIP and SN IIn, is not clear. SN IIL may not be a distinct class, but represent a portion of a continuum from those with distinct plateaus to those with abbreviated plateaus to those with no discernable plateau (Patat et al. 1994; Faran et al. 2014b). All SN II may be produced by core collapse, as commonly assumed, with the wide diversity of observed properties being due to variations in the mass and composition of the envelope, the presupernova radius, and the mass and distribution of the CSM (Blinnikov and Bartunov 1993; Chatzopoulos et al. 2012). In some cases, complex effects of binary evolution, including common envelopes and mergers (Nomoto et al. 1995), may be involved.

13.4 Summary

SN IIL comprise an interesting subclass of SN II, but the lines between SN IIP, SN IIn, and SN IIL are blurred. SN IIL may represent the low end of the distribution of presupernova envelope mass and plateau behavior rather than a distinctly separate class of explosions from SN IIP, but the envelope mass cannot be so low that the result is an SN IIb. There is ample evidence in the narrow emission lines and the late-time flattening of the light curves that CSI is important, and hence that it is difficult to cleanly discriminate SN IIL from SN IIn. There is a strong suggestion that the modest presupernova envelope mass and the CSI effects are related; the paucity of envelope mass plausibly accounts for the presence of the CSM as matter ejected prior to explosion. SN IIL can show substantial luminosity a decade or more after explosion, evidence of continued CSI. On the other hand, it is not clear that this bookkeeping accounts for all the mass of the original main-sequence progenitor, nor how this qualitative aspect is different from that for SN IIn. The outstanding issues are then which stars eject sufficient mass to produce SN IIL rather than SN IIP, and the mechanism of that mass ejection. Complementary issues are what distinguishes the CSM in SN IIL from that of SN IIn. Some SN IIn do show phases of linear decline (Sect. 14.3.1) and in some others the greater flux from the CSI may mask early rapid declines characteristic of SN IIL.

Some SN IIL show spectral evolution rather similar to SN IIP, but others have differences. Some show narrow P Cygni profiles indicative of a slowly-expanding CSM. Some SN IIL have typical nebular spectra of core-collapse events, but others lack the characteristic Mg I], [O I], and [Ca II] lines.

The notion that SN IIL represent a distinct range of progenitor masses has been suggested, but remains unclear. Although the mass of the hydrogen envelope appears to be low in SN IIL, there is evidence that the progenitors are red or perhaps yellow supergiants. The ZAMS masses of SN IIL progenitors may be around 20 $M_\odot$, but the data are sparse.

Future observations of SN IIL should seek to determine the ZAMS mass from archival observations of the progenitor and perhaps from measurements of the ejected oxygen abundance. Premaximum data, including photometry on the rise, would be valuable. Early spectra to characterize the narrow emission lines and their evolution would help to establish whether the CSM extends down to the photosphere of the progenitor and/or is in a shell. Spectra in both the photospheric and nebular phases should be examined for evidence of asymmetry. There is no published spectropolarimetry on SN IIL. Evidence for shock breakout and for compact remnants is not yet available. Since the implied loss of envelope from the progenitor suggests that the remaining star is more exposed than for SN IIP, near-IR spectra might provide evidence of enrichment in helium and other elements.

Chapter 14
Type IIn Supernovae

14.1 Introduction

Type IIn supernovae ("n" for narrow; Schlegel 1990) are characterized by narrow (tens to hundreds of km s^{-1}) and/or intermediate-width (a few thousands of km s^{-1}) emission lines of hydrogen in optical spectra. Such lines are attributed to CSI. An issue that complicates both categorization and physical understanding is that some supernovae of other types also show evidence in optical spectra of CSI. About 70 SN IIn have been identified. Although most of them have not been studied in detail, these events seem to span a large range of underlying types and physical circumstances. Superluminous supernovae (SLSN; $M_V \lesssim -21$) some of which resemble SN IIn spectroscopically, are discussed in Chap. 18.

At maximum light, SN IIn have a mean $M_B \simeq -18.7$. The rise time often is longer than 20 days (Kiewe et al. 2012). Some SN IIn display a long, slow decay in their optical light curves, while others have more rapid linear declines, reminiscent of SN IIL, preceded or followed by flatter declines. The rapid decline rates can range from 0.03 to 0.07 mag d^{-1}, with the latter approaching the high rate of decline of SN IIP during the plunge from the plateau.

In a volume-limited sample, SN IIn constitute about 10% of all core-collapse events. The rate may be higher at lower metallicity (Smith 2014). SN IIn occur in spiral galaxies, but they are neither tightly confined to spiral arms nor closely correlated with star formation or H II regions (Habergham et al. 2014). They are associated with Hα emission in much the same way as the general population of SN II (Anderson and James 2008). Since current evidence suggests that most SN II arise from stars near the lower end of the core-collapse ZAMS mass range, the progenitors of most SN IIn probably are in a similar mass range; some, however, may be from stars of much higher mass. The progenitor of SN 2005gl was an LBV (Gal-Yam et al. 2007; Gal-Yam and Leonard 2009). *HST* observations revealed a possible blue, and hence possibly massive, progenitor of SN 2010jl (Sects. 3.2 and 14.2.3).

D. Branch, J.C. Wheeler, *Supernova Explosions*, Astronomy and Astrophysics Library, DOI 10.1007/978-3-662-55054-0_14

SN IIn tend to have a blue optical continuum and a high UV flux, which is favorable for their detectability at high redshifts (Cooke 2008). SN IIn up to $z = 2.4$ were detected in archival data (Cooke et al. 2009).

14.2 Case Studies

14.2.1 SN 1988Z

SN 1988Z (Pollas et al. 1988) was one of the early prototypes of SN IIn (Stathakis and Sadler 1991; Turatto et al. 1993), with multiwavelength observations spanning more than a decade. It appeared in a low-surface brightness galaxy at $cz = 6700\,\mathrm{km\,s^{-1}}$. The metallicity of its environment may have been lower than that of most SN IIP and of more steeply-declining SN IIn such as SN 1998S (Taddia et al. 2015a).

The rise to maximum was not observed. When first detected, SN 1988Z was somewhat luminous for a core-collapse event, $M_R \simeq -18.7$. The optical light curve declined at a rate of $0.008\,\mathrm{mag\,d^{-1}}$ for 2 months (Fig. 14.1). After steepening briefly, it subsequently declined with an ever more gradual slope, reaching $M_R \simeq -13$ at 8 years (Aretxaga et al. 1999). The power radiated in Hα over a long time scale is given in Fig. 14.2.

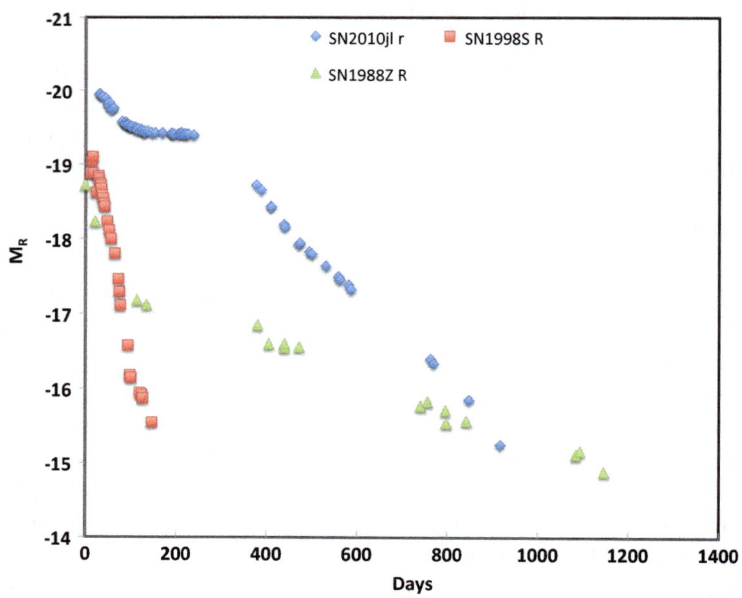

Fig. 14.1 R-band light curves of Type IIn SN 1988Z, SN 1998S, and SN 2010jl. Data are from Turatto (1993), Fassia et al. (2001), and Zhang et al. (2012)

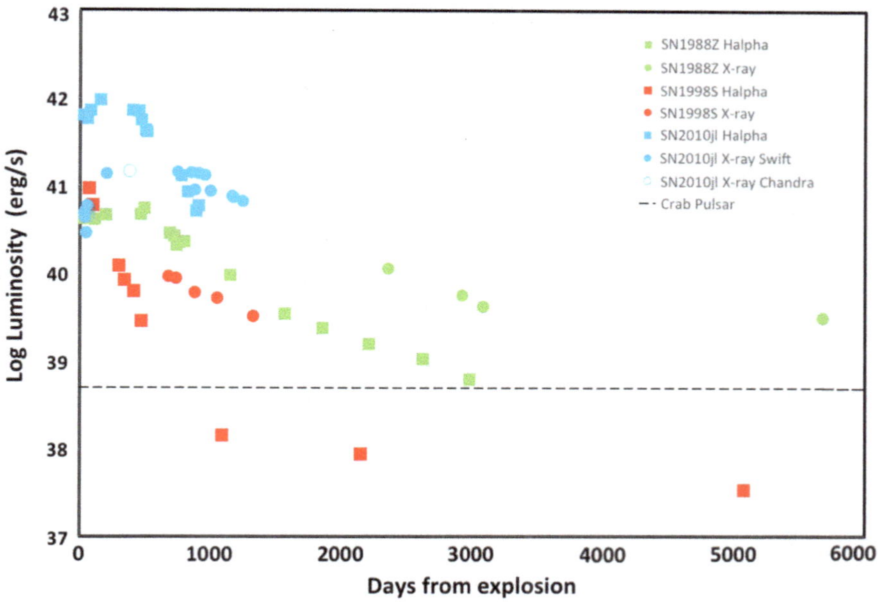

Fig. 14.2 Light curves for the Hα and X-ray luminosity of the Type IIn SN 1988Z, SN 1998S, and SN 2010jl. Data are from Aretxaga et al. (1999), Mauerhan and Smith (2012), Pooley et al. (2002), Fransson et al. (2014), and Ofek et al. (2014b), the latter updated by personal communication with unpublished data. The spin-down luminosity of the Crab pulsar is also shown (Abdo et al. 2010a). Error bars have been suppressed (see original literature), but some measure of the uncertainty in the *Swift* X-ray data can be seen in the scatter around day 700

The initial decline in the light curve was somewhat slower than the ^{56}Co rate of 0.0098 mag d^{-1}, while the later phases were much slower, consistent with ongoing CSI. The light curve at $\gtrsim$1 year might be roughly comparable to the $L \propto t^{-2}$ input from magnetic dipole radiation of a pulsar, but such input would be far too bright at earlier epochs and probably can be ruled out. The late-time luminosity is also far more than could be attributed to any plausible long-lived radioactive nuclide such as ^{44}Ti. SN 1988Z is often taken to be a canonical example of SN IIn, but, like many "prototypes," it may not be representative; many SN IIn do not show a long slow optical decline.

The strong narrow emission lines (FWHM $\simeq$ 200 km s^{-1}) were immediately attributed to dense CSM (Stathakis and Sadler 1991; Fig. 14.3). Broad emission features ($\sim$10,000 km s^{-1}) from the supernova ejecta persisted for over a year. Nearly a year after explosion, Hα and He I emission lines had intermediate-width components of 1200–1800 km s^{-1}; the line peaks were blueshifted with respect to the narrow CSM lines. These intermediate-width lines may have formed in CSM that had been accelerated by the forward shock, although electron scattering may have contributed to the line width at some epochs. The lines of intermediate width broadened with time, which is not easy to understand in either a shock-acceleration

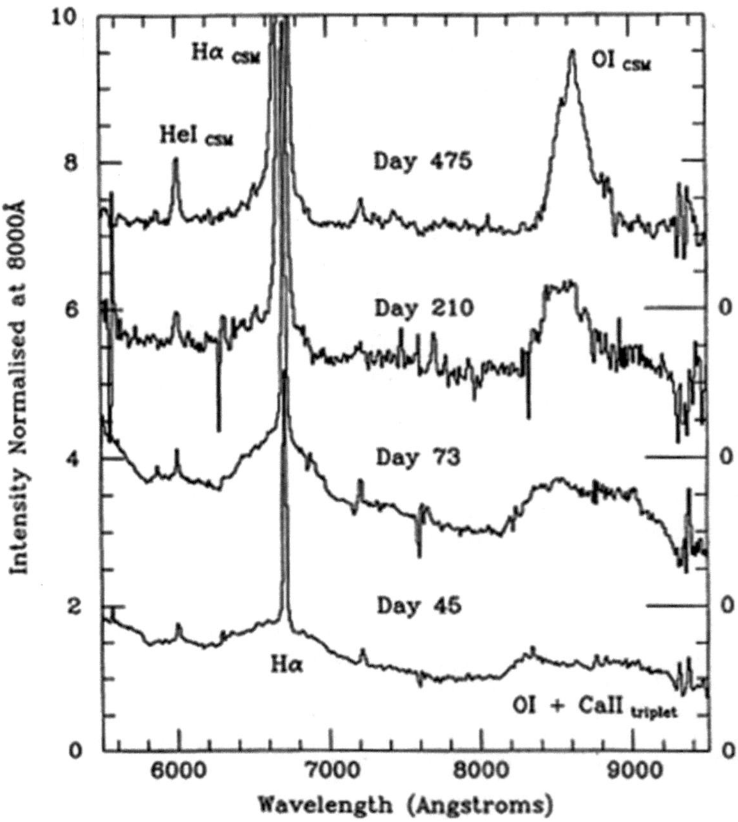

Fig. 14.3 Spectra of SN 1988Z. Figure 5 "What was supernova 1988Z?" from Stathakis and Sadler (1991), by permission of Oxford University Press on behalf of the Royal Astronomical Society

picture in which the shock should decelerate or in the electron-scattering picture where the electron density should decrease. A scattered-light echo might account for the increasing width by mixing in broad Hα emission from earlier epochs. Typical nebular-phase lines of [O I] and [Ca II] did not appear, suggesting a dense environment and slow evolution of the physical conditions. The presence of narrow, unresolved forbidden emission lines of [O III], [N II], [Ne III], [Fe II], [Fe III], [Fe VII], and [Fe X] also was attributed to CSI (Turatto et al. 1993).

The CSM may have contained filaments or clumps, complicating the interpretation. Some models of SN 1988Z invoke interaction of the ejecta with CSM containing both rarified and dense components to account for the broad and intermediate-width lines (Chugai and Danziger 1994).

SN 1988Z was a strong source of radio (Van Dyk et al. 1993; Williams et al. 2002) and X-ray (Fabian and Terlevich 1996; Aretxaga et al. 1999; Schlegel and Petre 2006; Fig. 14.2). The radio light curve could be modeled in terms of a highly clumpy or filamentary CSM. About 5 years after explosion, the radio luminosity began to decline more rapidly than predicted by an extrapolation of the previous behavior, which suggested a lower mass-loss rate earlier than 10^4 years before the explosion. A high presupernova mass-loss rate of $\sim 10^{-3}$ $M_\odot$ y^{-1} was inferred from the X-ray observations.

SN 1988Z radiated of order 0.1 B in the optical bands and a total energy in excess of 1 B (Aretxaga et al. 1999), suggesting that most of the kinetic energy of the ejecta was thermalized by CSI. Models and extrapolations are consistent with substantially more total energy being radiated in the X-ray than the optical bands. The high radiated energy is reminiscent of superluminous supernovae (Chap. 18), the principle difference being the timescale required to thermalize the kinetic energy: a decade for SN 1988Z and only months for superluminous events. The large radiated energy might demand an appreciable effect on the kinematics of the shock, slowing it considerably over the period of observations.

14.2.2 SN 1998S

The well-studied SN 1998S in NGC 3877, discovered (Li et al. 1998) only a few days after explosion, also provided multiwavelength data for an extended time, at least 15 years. At maximum light, SN 1998S had $M_R \simeq -18.5$ (Fassia et al. 2001). After maximum, the R-band light curve declined at a rate of about 0.02 mag d^{-1} for 60 days, then declined rapidly by 0.06 mag d^{-1} for 25 days before beginning a long decline that was consistent with that expected from the decay of 0.15 $M_\odot$ of ^{56}Co (Fassia et al. 2000; Fig. 14.1); alternatively, the latter decline could have been due at least in part to collision with CSM of appropriate density gradient. Blackbody fits to the UVOIR SED give a total radiated energy of $\sim 2 \times 10^{50}$ ergs. SN 1998S may have exploded in a relatively high-metallicity environment similar to SN IIP (Taddia et al. 2015a).

The Hα flux also declined at a rate comparable to the decay of ^{56}Co for about 3 years, although with a luminosity about one tenth of that of SN 1987A at comparable epochs. Around 3 years, the Hα flux began a much slower decline over the next 11 years (Mauerhan and Smith 2012; Fig. 14.2). This phase may have been powered by CSI, but input from a pulsar is possible. The late-time luminosity in Hα was $\lesssim 10^{38}$ erg s^{-1}, five to ten times the luminosity of SN 1987A at similar epochs when SN 1987A was powered by the decay of ^{44}Ti. The late-time bolometric input from magnetic-dipole radiation, declining as t^{-2}, would have been somewhat steeper than the observed behavior of the Hα flux. The evolution of the light curve and spectra was rather similar to that of the Type IIL SN 1980K (Sect. 13.2.2). The Hα luminosity of SN 1980K declined somewhat more steeply than that of SN 1998S for 3 years and then leveled off to $\sim 2 \times 10^{37}$ erg s^{-1} by 11 years. Was SN 1980K, which was not observed spectroscopically at early times, a Type IIn? Was SN 1998S, deep down, a Type IIL?

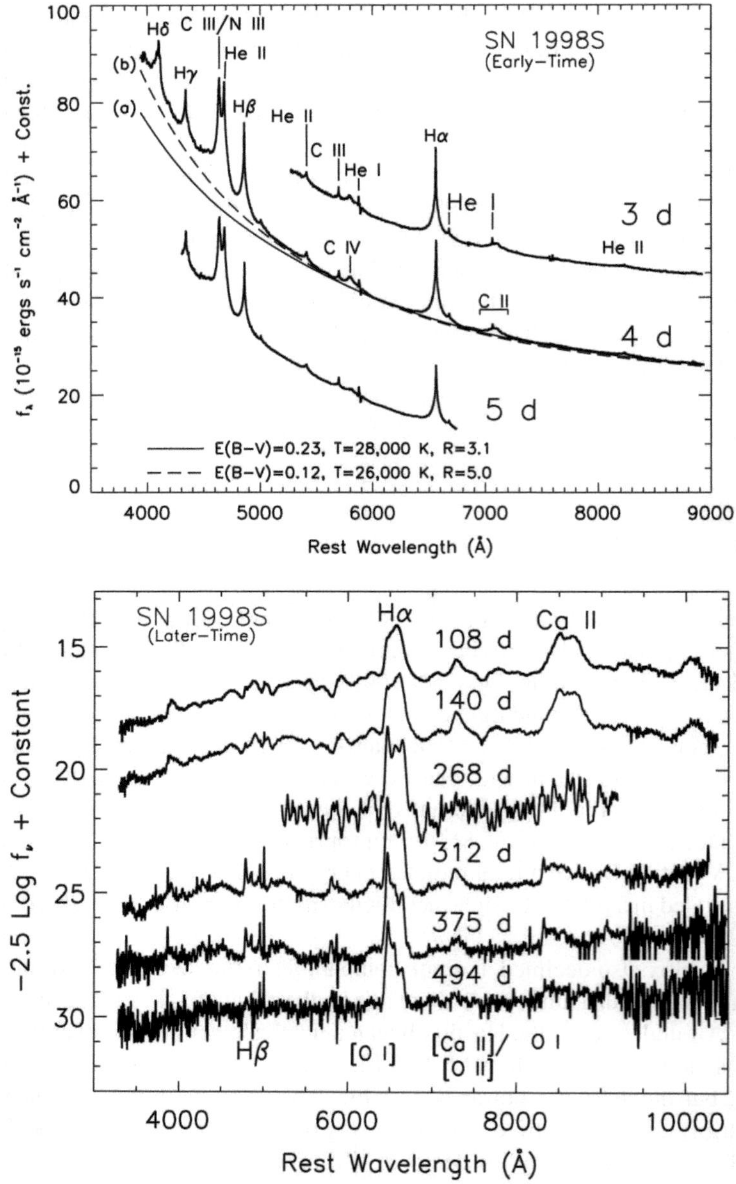

Fig. 14.4 Early (*top*) and late (*bottom*) optical spectra of SN 1998S. From "Evidence for Asphericity in the Type IIn Supernova SN 1998S" (Leonard et al. 2000). © AAS. Reproduced with permission

The spectral evolution of SN 1998S is given in Fig. 14.4. An early blue continuum of SN 1998S was not consistent with a single-temperature blackbody; the color temperature increased at shorter wavelengths (Fassia et al. 2000). The continuum may have arisen in the CDS (Sect. 6.2) produced by the CSI. The optical spectral evolution was followed in some detail (Leonard et al. 2000; Bowen et al. 2000; Fassia et al. 2001; Fransson et al. 2005; Mauerhan and Smith 2012). A high-resolution optical spectrum obtained within a few days of breakout showed lines of hydrogen, helium, carbon, oxygen, and neon with characteristic line widths (FHWM) of only 30–100 $\mathrm{km\,s^{-1}}$, consistent with an RSG progenitor, but not, for instance, a Wolf–Rayet star (Shivvers et al. 2015). During the first week after discovery, the spectra developed strong, broad, symmetric H, He I, C III and N III emission lines with narrow ($<300\,\mathrm{km\,s^{-1}}$) peaks. The smooth continuum and nearly Lorentzian shape of the broad base to the emission lines suggested that the CSM was rather optically thick and hence that electron scattering dominated the line width for the first 10 days. The intermediate component at later times may have been due to shocked clumps of matter that were more optically thin (Chugai 2001).

At 2 weeks, the broad, symmetric, emission components of the H, He I, C III, and N III lines disappeared, while narrow H and He I absorptions superimposed on shallower, broader absorptions appeared. Narrow emission lines of [N II], [O III], [Ne III], and [Fe III] appeared at about the same time. The early disappearance of the scattering wings suggests a relatively smaller mass-loss rate for SN 1998S compared to other SN IIn, e.g. SN 1988Z (Sect. 14.2.1) and SN 2010jl (Sect. 14.2.3). Line identifications in an *HST* UV spectrum obtained at 20 days were provided by Lentz et al. (2001). After a month, the optical spectrum consisted of broad, blueshifted absorption features corresponding to H, He I, Si II, Fe II, and Sc II. Broad P Cygni profiles showed an expansion velocity of $\sim 7000\,\mathrm{km\,s^{-1}}$. Unlike SN IIP or SN Ib/c, in which the UV flux declined rapidly at late times, SN 1998S had substantial UV flux at later phases.

After 6 weeks, the broad emission components appeared again in the lines of H and He I. These lasted for over a year and became increasingly asymmetric with double or perhaps even triple peaks in the H and [O I] lines. The line profiles suggested that the CSM may have been in a disk, but more complex structures are possible. (Spectropolarimetry of SN 1998S is discussed in Sect. 14.3.5.) After 2 months, optical spectra showed flat-topped profiles characteristic of emission from a circumstellar shell, with Hα the most prominent line. After a year, the red wings of both Lyman-α and Hα were suppressed, consistent with dust formation in the ejecta, perhaps behind the reverse shock (Leonard et al. 2000; Pozzo et al. 2004; Fransson et al. 2005).

The late-time spectra of SN 1998S, unlike those of SN 1988Z (Sect. 14.2.1) or SN 2010jl (Sect. 14.2.3), showed the classic core-collapse broad nebular emission lines of [O I] and [Ca II]. Spectra at 14 years showed strong, broad emission features of [O I], [O II], and Hα as well as weaker features of [O III], Hβ, and [Fe II] (Mauerhan and Smith 2012; Fig. 14.5). The oxygen lines had strengthened compared to the H lines, indicating that the reverse shock had entered oxygen-rich matter. The emission lines, especially [O III] λ5007, were suppressed in the red wings, giving more evidence that dust had formed in the ejecta. The spectra of SN 1998S at 14 years, including details of asymmetric line profiles, again closely resembled very late-time spectra of the Type IIL SN 1980K (Sect. 13.2.2).

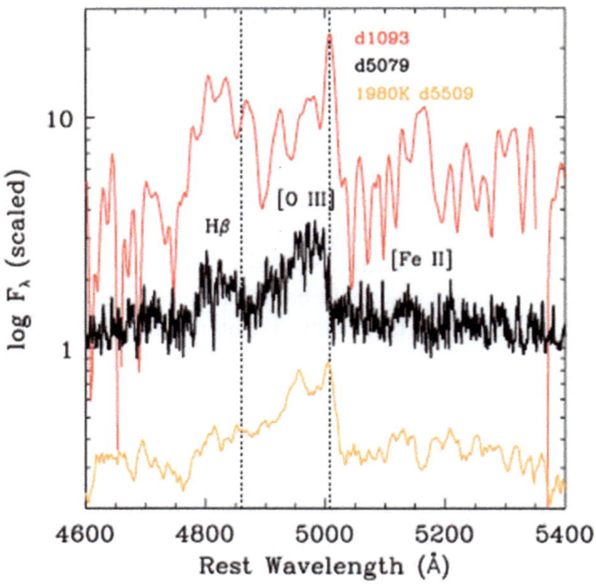

Fig. 14.5 Spectra of SN 1998S at 3 and 14 years after explosion in comparison with SN 1980K at 15 years. Figure 3 "Supernova 1998S at 14 years postmortem: continuing circumstellar interaction and dust formation" from Mauerhan and Smith (2012), by permission of Oxford University Press on behalf of the Royal Astronomical Society

The spectra of SN 1998S suggested at least two components to the CSM (Leonard et al. 2000; Chugai 2001). An outer component having a velocity of $\lesssim$100 km s^{-1} extended from 3×10^{15} cm to beyond 10^{17} cm (Fassia et al. 2001; Shivvers et al. 2015). This material, perhaps from an RSG wind, was subject to ionization by the shock-breakout radiation. Assuming constant velocity, this component was lost from the progenitor between about 20 and 200 years before the explosion and contained about 0.003 $M_{\odot}$. The mass-loss rate during this phase was about 2×10^{-5} $M_{\odot}$ yr^{-1}, low for a typical SN IIn, but consistent with an RSG progenitor. An inner component of the CSM expanded at 350 km s^{-1} and extended only to about 10^{15} cm. This material came from a phase of higher mass-loss rate during the final 10 years before explosion; its ionization was maintained by radiation from the forward shock. The late-time Hα luminosity indicated that mass also was lost at a high rate at least 1000 years before the explosion.

SN 1998S was also monitored at longer wavelengths. Strong emission appeared in the near-IR by day 44. A large excess in the K band 130 days after maximum light is likely to have been thermal emission from pre-existing dust in the CSM, heated by the breakout flash or by X-rays from shocks (Fassia et al. 2000). Observations revealed first-overtone emission in carbon monoxide that disappeared after a year (Gerardy et al. 2000; Fassia et al. 2001). The data suggested a CO mass of $\sim 10^{-3}$ $M_{\odot}$ with a velocity of about 200 km s^{-1}. Multipeaked H and He I lines in the near-IR (Gerardy et al. 2000; Pozzo et al. 2004) may have resulted from interaction with a circumstellar disk. Decreasing flux on the red side of line profiles again gave evidence that dust had formed deep in the ejecta.

SN 1998S was a bright source of radio and X-rays (Pooley et al. 2002; Fig. 14.2). X-ray spectra showed that IMEs and iron-group elements were overabundant with respect to solar values and suggested (with caveats) a progenitor mass of about 20 $M_\odot$. The X-ray data indicated that at times mass-loss rates may have been as high as 10^{-3} $M_\odot$ y^{-1}.

14.2.3 SN 2010jl

SN 2010jl in UGC 5189 (Benetti et al. 2010) was observed for more than 1000 days by a wide array of techniques. The host galaxy was of low metallicity (Stoll et al. 2011). A possible progenitor appeared in archival *HST* images (Smith et al. 2011b), but it is difficult to distinguish a single star from a young stellar cluster.

The R-band light curve of SN 2010jl peaked at $M_R \simeq -20$ (Stoll et al. 2011) and declined for 3 years in a manner similar in slope to SN 1988Z, but about 1.5–2.0 magnitudes brighter intrinsically (Zhang et al. 2012; Fig. 14.1). The light curve behavior was different in each band, but the optical bands declined by about 0.8 mag from discovery to roughly 175 days, and then settled onto a plateau lasting about 300 days before starting a long, steep, exponential decline. The R-band, affected by Hα emission, was especially flat on the plateau. The long-term evolution of the Hα luminosity is given in Fig. 14.2.

The J, H, and K-band flux reached a minimum and began to increase at about 300 days as the optical began its steep decline. The long-wavelength bands peaked about 500 days after discovery. The longer-wavelength components dominated the bolometric luminosity in the later phases. All the bands from U to K declined less steeply than ^{56}Co on the final tail. The host galaxy contribution became appreciable after 350 days.

At maximum, the bolometric luminosity was $\sim 3 \times 10^{43}$ erg s^{-1}. The luminosity had a relatively steep early decline and then also reflected the plateau until about 300 days, at which time it had a relatively gentle break in slope and began a long shallow decline at the rate of $\sim$0.003 mag d^{-1}. As late as 900 days, the luminosity was comparable to normal SN IIP on the plateau, $\sim 10^{42}$ erg s^{-1}. Estimates of the total radiated energy depend on models of the long-wavelength contribution, but were comparable to the total kinetic energy of a typical core-collapse supernova.

As noted above, SN 2010jl was brighter at all observed epochs in the optical band than SN 1988Z. This may have been an artifact of the distribution of the radiated energy, with SN 1988Z dominating throughout bolometrically. SN 1988Z had greater energy emitted in the X-ray band than in the optical, but this was not the case for SN 2010jl, for which the integrated UVOIR energy was considerably larger than the X-ray energy for the first 3 years (Ofek et al. 2014b). A collisionless shock may radiate mostly in the hard bands, but if the optical depth to X-rays in the matter beyond the forward shock is substantial, much of the hard flux may be converted to optical bands; this presumably was the case for SN 2010jl for the epochs observed and not so for SN 1988Z. The declining optical flux of SN 2010jl became comparable to the X-ray luminosity about 4 years after the explosion. This may indicate that at that time the forward shock in SN 2010jl was proceeding into material of smaller optical depth to X-rays. As for SN 1988Z, the large loss of energy to radiation in SN 2010jl may have had a significant effect on the shock kinematics.

The overall spectrum of SN 2010jl changed very little over 800 days aside from the reddening of the continuum. The spectrum was dominated by the continuum at early epochs but by line emission at later times (Fig. 14.6). Like SN 1998S, SN 2010jl had substantial UV flux at later phases in contrast to SN IIP or SN Ib/c, for which the UV flux declines substantially at late times.

The strong line of Hα displayed characteristics typical of SN IIn, but also some novel aspects (Fransson et al. 2014). The Hα flux rose monotonically to a peak at about 300 days, and then declined steadily. Throughout its evolution, the spectrum of Hα comprised at least two components. One was a narrow P Cygni component

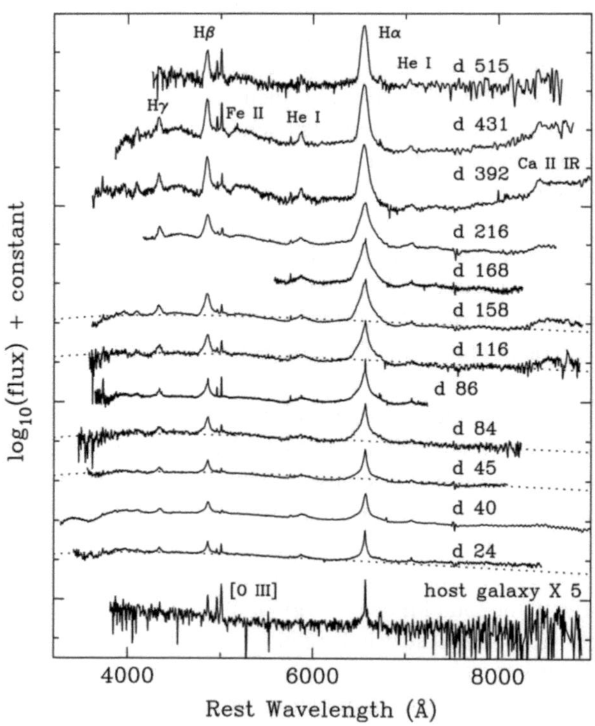

Fig. 14.6 Spectral evolution of SN 2010jl. From "Type IIn Supernova SN 2010jl: Optical Observations for over 500 Days after Explosion" (Zhang et al. 2012). © AAS. Reproduced with permission

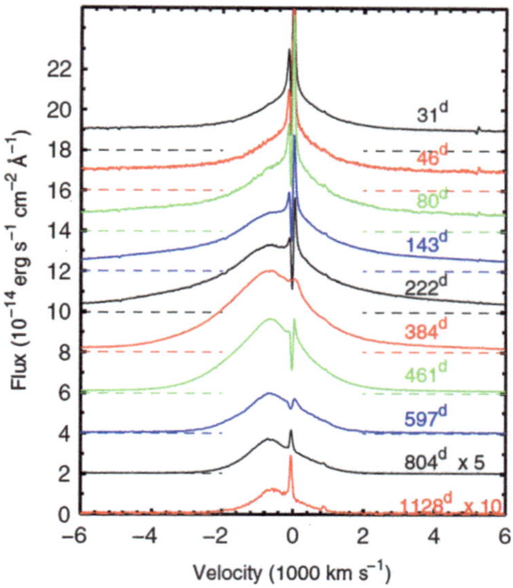

Fig. 14.7 Evolution of the narrow and intermediate components of Hα in SN 2010jl. From "High-density Circumstellar Interaction in the Luminous Type IIn SN 2010jl: The First 1100 Days" (Fransson et al. 2014). © AAS. Reproduced with permission

of $\sim$100 km s^{-1} determined from the blue absorption feature (Fig. 14.7). In the early phases, the narrow component generated in the CSM dominated the Hα line center. The significant depth of the absorption feature suggested that the material in which the P Cygni feature formed had a large covering factor with respect to the underlying continuum emission. The velocity of the P Cygni component shifted very little for 600 days, after which this component faded. This implies that the effective photosphere at the base of the P Cygni line-forming region moved at nearly constant velocity during this extended interval. This photosphere might have formed in the CDS at the contact discontinuity that in turn might be ionized by irradiation from the reverse shock. A second component of Hα was approximately symmetric and of intermediate velocity width, FWHM of about 2000 km s^{-1}. This component was probably due to multiple electron scattering. A multiple-scattering base also was observed for other lines. There may also have been a third, broader component of Hα with FWHM about 6000 km s^{-1} that faded with time (Borish et al. 2015; Jencson et al. 2016).

The novel aspect of the evolution of Hα in SN 2010jl was that the peak of the intermediate-width component shifted blueward by about 1000 km s^{-1} at about 100 days, roughly at the onset of the plateau phase, but remained roughly symmetric about this shifted centroid (Fig. 14.7). The peak of the intermediate-width component shifted to lower velocities beginning about the time of the break from the plateau to the final tail at about 300 days and was at about 500 km s^{-1} at 1000 days. This shift in line centroid while maintaining approximate line symmetry is not well understood. One proposal argues for a role for dust formed in the CDS (Smith et al. 2012b; Gall et al. 2014). Another attributes the shift to acceleration by the supernova radiation rather than to dust absorption (Fransson et al. 2014).

In this interpretation, the blueshift of the peak of the symmetric emission lines may require a planar flow of the radiation-accelerated material that subtends a rather small solid angle. Structure reminiscent of these requirements is observed in η Car. In this model, the IR flux that dominated the late-time luminosity was due to a light echo from moderately distant CSM, $\sim 10^{17}$ cm. Yet another proposal is that dust in the nearby CSM was evaporated in the explosion, but then reformed at about 300 days, leading to the end of the plateau in the optical and the nearly simultaneous increase in the IR flux (Jencson et al. 2016).

During the observed epochs, there was no optical evidence for the standard broad permitted or forbidden lines that characterize the photospheric and nebular phases of normal core-collapse supernovae. This suggests that radiation from any fast-moving ejecta was blocked by the CSM, including, perhaps, the CDS. By the final epoch of observation, about 1000 days, the reverse shock and subsequent hydrogen recombination front may have propagated to the base of the hydrogen envelope so that, even if the CSM became optically thin at that epoch, there would be no strong, broad hydrogen lines. There was no evidence for a substantial production of ^{56}Co (for reasons that require deeper understanding), so any internal power from ^{56}Co at late times would have been weak. This might account for the lack of traditional broad nebular lines. Near-IR observations of the He I 1.0830 μm feature did show evidence for motion at speeds of $\sim$5500 km s^{-1} 100–200 days after the explosion (Borish et al. 2015). Shocks of similar velocities are required to produce the observed X-ray flux. Some material with typical supernova velocities thus did exist. Reconciling evidence for high-velocity material with the lack of normal photospheric and nebular spectra should be explored with self-consistent models.

The nature of the spectra and the relatively shallow late-time decay both indicate that SN 2010jl was dominated by CSI at all observed epochs. The bolometric light curve is consistent with the propagation of a radiative shock into a steady-state wind that represented a mass-loss rate of $\sim$0.1 $M_\odot$ y^{-1}, but it is not clear that such a high mass-loss rate is consistent with a steady-state process. The total mass in the CSM was $\sim$3 $M_\odot$.

The UV showed lines of highly-ionized elements, like C III, C IV, N III, N V, O III, O IV, Si II, and Si IV along with lower-ionization lines of C II, O I, and Mg II. The UV lines showed that the material lost from the progenitor had undergone CNO processing. Narrow optical coronal lines are most likely to have been excited by X-rays produced in the forward shock. X-ray observations by *Chandra, XMM, Swift* and *NuSTAR* (Immler et al. 2010; Chandra et al. 2012b; Ofek et al. 2014b) showed a hard spectrum with kT $\simeq$12 keV at 59 and 373 days after detection and column density, probably intrinsic to the CSM, of $\sim 10^{24}$ and $\sim 3 \times 10^{23}$ cm^{-2}, respectively. There was no detected radio flux.

Evidence for a massive CSM, expansion velocities of the CSM too high for RSG winds, and CNO enhancement invites comparison to giant LBV eruptions. The observed CSM velocity, $\sim$100 km s^{-1}, is on the low side of velocities attributed to LBVs. The optical spectra, spectropolarimetry (Sect. 14.3.5), X-ray flux, and dust all imply significant large-scale asymmetries in the CSM, although the narrow lines suggest that the more distant CSM was more symmetric. A clumpy but otherwise spherical CSM would be less likely to account for the evidence of large-scale asymmetries. A disk would have problems accounting for the shift in the peak of the Hα and other lines while maintaining approximate line symmetry.

The observed properties might be consistent with the supernova ejecta expanding into a CSM characteristic of a giant LBV eruption with a bipolar morphology such as η Car. This morphology might explain why there are high shock velocities in some directions, along the equator, that give rise to hard X-rays but also slower shocks in other directions, the polar directions, where the CSM is still optically thick. In such a picture, the shock would emerge from the optically-thick regions at different times at different latitudes, thus giving a smoother evolution of the light curve than would be the case for a spherical shell with a single sharp break in density at the outer edge. Evidence for an asymmetric CSM calls into question assumptions of a steady-state wind and suggests caution in interpreting the observed velocities, given the strong velocity anisotropies observed in η Car.

14.3 Synthesis of SN IIn Characteristics

Other well-observed SN IIn are SN 1994W (Chugai et al. 2004), SN 1995N (Fransson et al. 2002), SN 2005gl (Gal-Yam et al. 2007), SN 2005ip (Stritzinger et al. 2012), SN 2006jd (Stritzinger et al. 2012), SN 2009kn (Kankare et al. 2012), SN 2011ht (Mauerhan et al. 2013b), and SN PTF11iqb (Smith et al. 2015).

14.3.1 Light-Curve Subtypes

SN IIn display a wide variety of optical light curves and spectra, some of which are probably affected by dust and light echoes. Much of the diversity undoubtedly depends on the vagaries of the timing and extent of possibly repetitive phases of extensive mass loss. It may be that massive stars can explode as normal SN IIP long after episodes of bursting mass loss and as SN IIn if the explosion occurs during or shortly after such an episode (Di Carlo et al. 2002). The photometric and spectroscopic properties of SN IIn do not seem to be well correlated, and it is not clear which of these properties should be the principle organizing component. Because some patterns in the photometric properties seem to emerge, we opt to organize by light-curve subtypes (see also Taddia et al. 2015a). Figure 14.8 gives the light curves of a number of SN IIn discussed below. Note that some SN IIn (SN 2010jl, SN 2006tf) have peak luminosities corresponding to $M_R \lesssim -20$. These events have been referred to as "gap-fillers" in some quarters because they tend to fill the luminosity gap between most supernovae, including most SN IIn, and the superluminous events (SN 2006gy in Fig. 14.8) discussed in Chap. 18. Table 9 of Kiewe et al. (2012) gives a summary of observed and derived properties for many SN IIn events.

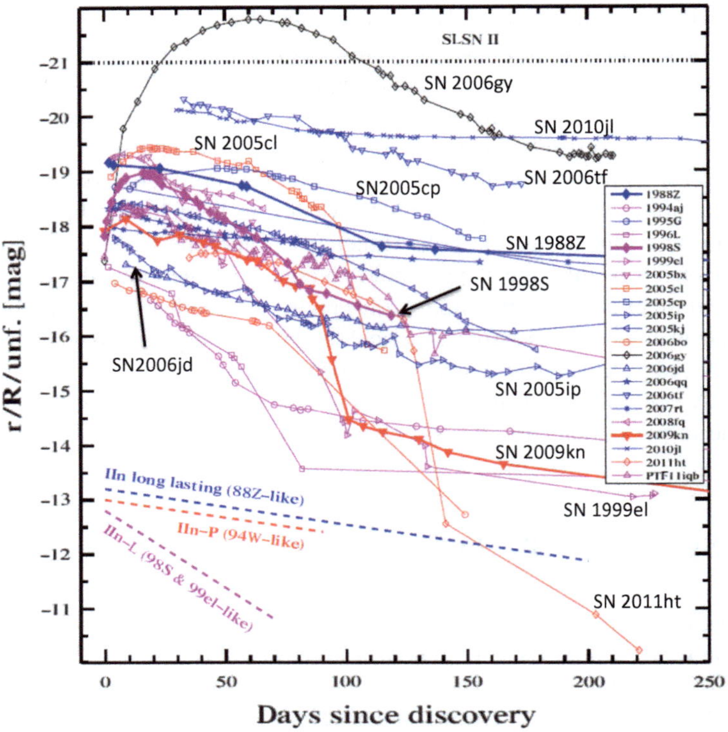

Fig. 14.8 A sample of absolute-magnitude R-band, r-band, and unfiltered CCD light curves of SN IIn and a few other events. Examples of SN IInE (SN 1988Z-like), SN IInP, and SN IInL are given (see text). The *dashed lines* give characteristic early-time slopes of those three categories. From "Early-time light curves of Type Ib/c supernovae from the SDSS-II Supernova Survey" (Taddia et al. 2015a) reproduced with permission, © ESO

14.3.1.1 Type IInE

As mentioned in Sect. 14.2.1, SN 1988Z (Figs. 14.1 and 14.8) has been regarded as a prototype SN IIn with its bright, long, slow decline. As we will see below, many SN IIn display rather different light curves, but there is a subset with extended light curves that do resemble SN 1988Z; we refer to these as SN IInE.

SN 2010jl had a bolometric light curve that was roughly similar to that of SN 1988Z (Figs. 14.1 and 14.8), with an overall shallow decline lasting for at least 1000 days. The light-curve type to which SN 2010jl is assigned depends on the interpretation of the infrared component (Sect. 14.2.3). If the long-wavelength component were due to a light echo, then it was reprocessed radiation and not directly due to an ongoing CSI. Neglecting this component, the light curve of SN 2010jl had a plateau followed by a steep decline after 300 days. If the continued, even growing, infrared component is regarded as output of radiation due to contemporaneous CSI, then the light curve was clearly extended. Based on the observed behavior of the total light output, we classify SN 2010jl as SN IInE.

SN 1995G had $M_R \simeq -18.5$ when discovered (Fig. 14.8). In the optical, it initially declined at a slow rate of about 0.008 mag d^{-1} for 2 years (Pastorello et al. 2002), then began to decline at a faster rate, suggesting that the forward shock had emerged from high-density CSM, although an interpretation involving the supernova envelope becoming optically thin may be possible.

SN 1995N was discovered at $M_R \simeq -17$ about a year after it exploded (Fox et al. 2000). The R-band light curve was virtually flat for 200 days. Few, if any, other SN IIn show this extremely flat phase. Note, however, that the lack of early observations of SN 1995N means that its history is quite uncertain and, despite the flat phase, it could belong to another light-curve category. After the flat phase, SN 1995N began a long slow decline with a decreasing slope that was comparable to, but 1.5 magnitudes dimmer than, that of SN 1988Z. Over the next 8 years, SN 1995N declined at an average rate of only 0.0013 mag d^{-1}. As for SN 1988Z, this late decline was too bright to be plausibly related to radioactive decay, and dipole radiation from a pulsar, when extrapolated to earlier times, would have been too bright. SN 1995N was detectable in the radio for at least 11 years (Chandra et al. 2009b).

SN 2007rt also was discovered some time after maximum light (Trundle et al. 2009; Fig. 14.8). Within the uncertainties, the absolute R-band light curve was virtually identical to that of SN 1988Z from 50 to 450 days after discovery, before which time SN 1988Z had declined by somewhat more than one magnitude. After 450 days the R-band light curve of SN 2007rt declined somewhat more rapidly than that of SN 1988Z.

The events described here have some common light-curve behavior, but also noticeable differences (including spectroscopic), so it is not clear that they belong together as a physical class.

14.3.1.2 Type IInP

After maximum light, some SN IIn show plateaus of gentle slope followed by steep declines; we refer to these as SN IInP (as have others, Mauerhan et al. 2013b; still others have used SN IIPn). Clear examples of this behavior are SN 1994W, SN 2008kn, SN 2011ht, and SN PTF11iqb (Sollerman et al. 1998; Chugai et al. 2004; Kankare et al. 2012; Mauerhan et al. 2013b; Smith et al. 2015; the latter three are represented in Fig. 14.8). SN 1994W and SN 2008kn rose to maximum in about 30 days with peak $M_R \simeq -19$ and -18, respectively, then faded at a rate of about 0.02 mag d^{-1} until about 100 days after explosion, after which they declined at steeper rates. The decline of SN 1994W was especially steep, about 0.3 mag d^{-1} for 12 days. Other candidates for this category include SN 2005cl, SN 2005cp, and SN 2005db (Kiewe et al. 2012; Kankare et al. 2012; Fig. 14.8). SN IInP radiate a few times 10^{49} ergs on the plateau.

SN 1998S might be placed in this light-curve category, although its early plateau with a decline of 0.02 mag d^{-1} lasted only about 50 days and the subsequent drop was only about 1.5 magnitudes before it flattened to a decline roughly consistent with ^{56}Co decay (Sect. 14.2.2). SN 1998S was distinguished from other SN IInP by its spectral characteristics. Although it displayed a distinct, long, plateau, SN PTF11iqb was also distinguished by its spectral evolution.

Despite the similarity of the late-time decline of some SN IInP to that of ^{56}Co, it is possible that this phase was driven by CSI. Setting aside the uncertain evidence for radioactive decay, there is no firm evidence that these events resulted from core-collapse explosions; they could, in principle, have resulted from collisions of multiple shells ejected by the progenitor (Dessart et al. 2009), and some might be the result of real-time variable outflow from a star (Humphreys et al. 2012). Some SN IInP may be electron-capture supernovae interacting with their SAGB-star winds (Moriya et al. 2014).

Although there is some variation in luminosity on the plateau and on the tail, a subset of the SN IInP are remarkably similar, especially when their spectral evolution is considered. They have some range in plateau luminosity, but break sharply from the plateau at about 120 days, whereas other SN IInP tend to end the plateau more gently. All are very dim in the last phases suggesting that they cannot produce a great deal of ^{56}Co. Clear members of this subset are SN 1994W, SN 2009kn, and SN 2011ht. After its steep postplateau decline, SN 1994W continued to decline at a rapid rate of about 0.07 mag d^{-1} for at least 100 days, until it reached $M_R \simeq -11$. In contrast, SN 2009kn declined in this later phase by about 0.01 mag d^{-1}, consistent with the decay of 0.02 M$_\odot$ of ^{56}Co. The steep decline of SN 1994W suggests that it ejected substantially less ^{56}Co. We revisit this subset in Sect. 14.5.

14.3.1.3 Type IInL

Were it not for their spectroscopic classification, some SN IIn could be characterized as SN IIL by the behavior of their light curves. As illustrated in Fig. 14.8, some, e.g., SN 1999el (Di Carlo et al. 2002) and SN 2005ip (Smith et al. 2009b), began a long ($\gtrsim$100 days) approximately linear decline right after maximum light. SN 2005ip was unremarkable in terms of its peak luminosity, $\sim 4 \times 10^{42}$ erg s^{-1} (Smith et al. 2009b; Stritzinger et al. 2012). At later phases, the light curve was dominated by near-IR flux. The total luminosity decayed at roughly the rate of ^{56}Co. The total radiated energy was $\sim 4 \times 10^{49}$ ergs from the photospheric emission and about $\sim 7 \times 10^{49}$ ergs from the dust component.

Other candidate SN IInL, e.g., SN 1987F (Cappellaro et al. 1990), began to decline nearly linearly somewhat after maximum. The plateau phase of the events referred to as SN IInP (Sect. 14.3.1.2) could be characterized as a rather linear decline. The late decline of SN 1994W could also be characterized as a steeply-declining linear light curve. SN 1998S fits into the SN IIL light-curve category as well as into any other (Taddia et al. 2015a). All of this adds to the ambiguity of the SN IIL/SN IIn classification.

14.3.1.4 Miscellaneous Light-Curve Behavior

SN 1986J was discovered as a strong radio source (Rupen et al. 1987). At that time, it had an absolute magnitude of $M_R \simeq -11$. Both the continuum and line fluxes faded slowly over time (Milisavljevic et al. 2008). The very dim, slow decline has no obvious counterpart among other supernovae. The optical flux was about 1% of the X-ray flux. The light curve could be consistent with a number of sources of power; ^{44}Ti decay, a pulsar, or CSI. The strong radio emission suggested that CSI was a significant, if not dominant, contributor.

For an SN IIn, SN 2005gl (Gal-Yam et al. 2007; Gal-Yam and Leonard 2009) was rather dim at maximum light, $M_R \simeq -17$. The light curve might have had a brief plateau decline at a rate of 0.02 mag d^{-1} that lasted for 30 days before declining linearly at a fast rate of 0.07 mag d^{-1} for the next 60 days. SN 2005gl might then belong in either the SN IInP or SN IInL categories, but the overall relatively dim, narrow light curve stands out in comparison to most other SN IIn.

In the V, R, and I bands, SN 2006jd (Fig. 14.8) declined roughly linearly at $\sim$0.01 mag d^{-1} for about 150 days after discovery (Stritzinger et al. 2012). The V and I bands then showed a plateau phase that lasted for about 500 days, followed by another approximately linear decline of $\sim$0.001 mag d^{-1}. In contrast, after the linear decline from discovery at $M_R \simeq -17.6$, the R band had a distinctly nonmonotonic postmaximum light curve. After solar hiatus, at about 600 days, it had brightened by about one magnitude and reached a secondary peak of $M_R \simeq -17.4$. The subsequent sparsely-sampled R-band decline was somewhat slower than the rise to the secondary peak, with a decay rate of about 0.001 mag d^{-1}, closely parallel to the V and I bands. The behavior of the R band probably reflected an increase and decrease in the Hα flux that was attributed to interaction with a circumstellar shell that did not otherwise affect the other bands or the bolometric luminosity. SN 2006jd emitted about 10^{50} ergs in photospheric emission and $\sim 2 \times 10^{50}$ ergs in a dust component (Fransson et al. 2014).

SN 2008iy had an especially long rise, about 400 days, to a maximum of $M_R \simeq -19.1$, and then a comparable, nearly symmetric, slow decline (Miller et al. 2010a). The decline in the R band was nearly linear at a rate of 0.004 mag d^{-1} for at least 300 days, during which time the flux in the K-band rose. The R-band light curve of SN 2008iy is reminiscent of the secondary peak in SN 2006jd discussed above. The total energy radiated by SN 2008iy was $\sim 2 \times 10^{50}$ ergs. The light curve might be explained by an increasing density of optically-thick clumps within a less dense CSM (Miller et al. 2010a), with the intermediate-width spectral component caused by the forward shock propagating into the clumps; alternatively, the light curve can be reproduced by an appropriate choice of the optically-thick CSM density profile with the intermediate-width spectral component broadened by electron scattering (Chatzopoulos et al. 2013b).

SN 2009uj was detected serendipitously on the rise by GALEX prior to optical discovery (Ofek et al. 2010). The UV flux rose rapidly on a timescale of a few days, suggestive of shock breakout from a dense CSM. The decline in the optical was rather rapid compared to other SN IIn. The progenitor might have undergone pre-explosion mass loss at a rate of $\sim 0.1\ M_\odot\ y^{-1}$.

14.3.2 Spectra

Early spectra of SN IIn usually show a blue continuum, and some are nearly featureless, but eventually Hα becomes the dominant optical feature (Fig. 14.6). The Hα feature frequently displays narrow and/or intermediate-width lines associated with the CSM, sometimes on a broad-based emission from the ejecta. The Hα and other emission lines frequently are complex, indicating substantial asymmetries in the CSM.

The narrow emission lines, sometimes accompanied by narrow blueshifted absorptions—e.g., in SN 1999el (IInL; Di Carlo et al. 2002), SN 1994W (IInP; Dessart et al. 2009), and SN 2009kn (IInP; Kankare et al. 2012)—are excited in the unshocked CSM by hard radiation from shock breakout or from the forward shock propagating into the CSM. Narrow coronal emission lines such as [Fe VII], [Ar XIV], and [Ni II] from the unshocked CSM also are seen in some events, e.g., SN 1998S (IInP? IInL?; Fig. 14.4), SN 1997eg (uncertain light curve subtype; Hoffman et al. 2008), and SN 2005ip (IInL; Smith et al. 2009b). These lines are also excited by X-rays from the forward shock or breakout. SN 2006jd (SN IInP?) displayed coronal lines at 1542 days (Stritzinger et al. 2012). Analysis of narrow lines in UV spectra of carbon, nitrogen, and oxygen in SN 1998S (Fransson et al. 2005), SN 1995N (IInE; Fransson et al. 2002), and SN 2010jl (IInE; Fransson et al. 2014) revealed nitrogen overabundances relative to carbon and oxygen, indicating that the CSM had been processed by the CNO cycle while in the progenitor star. Circumstellar lines may serve as a diagnostic of a significant X-ray flux even in the absence of direct X-ray detection.

Emission lines of intermediate width, FWHM $\sim$2000 km s^{-1}, with Lorentzian profiles frequently appear as broader bases to the narrow emission lines (Figs. 14.4 and 14.6). This component to the line profile is due to broadening by multiple scattering by circumstellar electrons (Sect. 4.4.3). For winds or other CSM with a decreasing density profile, the Hα flux arises predominantly from the region close to the shock. This also tends to be the region with the largest optical depth to electron scattering. Some SN IInE and SN IInP have emission with symmetric intermediate-width line profiles and narrow P Cygni profiles; others seem never to display the narrow P Cygni features. Spectra in the special case of SN 1994W and related events are discussed in Sect. 14.5.

At late times, the optical spectra of SN IIn are usually dominated by strong Hα emission, often with multiple components indicating asymmetry or obscuration by dust. The [O I] $\lambda\lambda$6300,6364 and [Ca II] $\lambda\lambda$7291,7323 emissions that are characteristic of core-collapse events in the nebular phase during the first years after explosion are seldom observed in SN IIn, with SN 1998S (Sect. 14.2.2) being an exception (Fig. 14.4). After about 8 years, SN 1995N developed lines of [O I], [O II], and [O III] (Fransson et al. 2002); these lines are observed in decades-old core-collapse events of various spectroscopic types (Milisavljevic et al. 2012).

The Hα line is an especially important spectral feature in SN IIn because it dominates the optical spectra, reveals multiple components—narrow, intermediate, and broad—in both absorption and emission, and evolves in time. Whereas SN IIP show distinct P Cygni profiles in Hα, and SN IIL tend to lack the blue-shifted absorption, the Hα profiles of SN IIn can show both these behaviors and more. The ratio of absorption to emission components may be a principle component relating to diversity in other aspects of the explosions (Gutiérrez et al. 2014), presumably because the Hα morphology itself is sensitive to the mass and density distributions of the supernovae and the CSM. At high densities ($\gtrsim 10^{11}$ cm^{-3}), the gas behind the shock is nearly thermalized and in LTE. In this situation, the efficiency of conversion of shock energy to line emission is low. The efficiency of conversion of X-ray flux to Hα line flux can be as much as 5–7% under conditions that the column density for the X-ray attenuation is appreciable and the ambient density $\lesssim 10^{10}$ cm^{-3}. In general, emission from preshock and postshock gas will give asymmetric profiles. Line profiles sometimes are flat-topped, suggestive of formation in the CDS.

The lines of SN IIn, especially Hα, often show depression of their red wings at later times, indicative of dust formation (Fig. 14.5). The red wing of the Hα line in SN 1995N was consistent with dust formation at $\sim$1000 days (Fransson et al. 2002), and SN 2006jd showed a similar effect strongly by 124 days (Stritzinger et al. 2012). Dust might form in the CDS or in inner ejecta, with different implications for the line profiles. Infrared spectra also frequently show evidence for light echoes from pre-existing dust that must be taken into account in the analysis of new dust formation. In some cases, a blueshift may arise not from dust suppression of the red wing, but due to acceleration of the scattering region by the supernova radiation (Sect. 14.2.3).

One of the difficult aspects of the study of SN IIn is that the dense CSM often blocks any view of the underlying supernova ejecta that might allow a comparison with other common supernova spectral properties. There are exceptions where broad lines characteristic of supernova ejecta are seen. Some events that showed broad ($>$ 5000 km s^{-1}) emission or P Cygni profiles formed in the supernova ejecta are SN 1995N (Fransson et al. 2002), SN 1998S (Sect. 14.2.2), SN 2005ip (Smith et al. 2009b), and SN 2006jd (Stritzinger et al. 2012). SN 2005ip and SN 2006jd showed maximum velocities of over 15,000 km s^{-1} in early spectra. At 22 days after discovery, SN 2006jd showed a broad base on Hα with a somewhat flat top that sloped to the red. By the end of the plateau phase, this broad feature peaked in the blue. A high-velocity wing could still be discerned at 1500 days. The reasons for this spectral evolution and the possible connection to the supernova ejecta are not yet clear.

The broad Hα lines of some SN IIn could form in the CDS or in the hydrogen envelope of the ejecta. These two cases should, in principle, show different velocities, line profiles, and temporal behavior. In general, lines formed in the ejecta should be broader, consistent with more rapid expansion; be sharply peaked if formed in a classic photosphere; and become narrower with time as the photosphere recedes in mass. Lines formed in the CDS might be less broad, consistent with deceleration of swept up matter, perhaps flat-topped if formed in a shell, and perhaps of nearly constant velocity if the shell coasts with nearly constant speed during the epoch of observations. The picture may not be so clear in the face of asymmetries and scattering regions beyond the photosphere, wherever the line forms.

SN PTF11iqb was an interesting case in point. It revealed multiple phases that seemed to correspond to interaction with CSM with associated narrow lines, then to show a plateau with typical P Cygni profiles, and finally to develop indications of renewed CSI (Smith et al. 2015). This event did not show classic nebular forbidden lines, so the CDS presumably remained optically thick. Although altered by dust and perhaps asymmetry, the Hα profile was roughly consistent with a double peak that could have formed in a ring moving with about constant velocity.

There is evidence for helium-rich ejecta in some events, e.g., SN 1997eg (Hoffman et al. 2008), SN 2007rt (SN IInE; Trundle et al. 2009), and SN 2010jl (Borish et al. 2015). In the case of SN 2007rt, the CSM also appeared to be helium-rich, indicating that the mass loss extended into the helium-rich layers of the progenitor. These events may be related to the SN Ibn events discussed in Sect. 16.5. Transition types between SN IIn and SN Ibn have been suggested (Pastorello et al. 2015a). SN 1995N presented evidence for oxygen-rich ejecta (Fransson et al. 2002).

Given the intrinsically complex configuration, even in rough spherical symmetry, the formation of the spectra of SN IIn is also complex. Perhaps the only totally clear conclusion is that the narrow lines that define the class arise through excitation of slowly-moving matter expelled from the star before the supernova explosion. Broader emission or P Cygni profiles can form in the outer CSM shocked by the forward shock if it is dense enough to be optically-thick, in the CDS, or in the ejecta. A significant feature of line formation in this environment is that multiple electron scattering can lead to intermediate-width features that are a measure of the electron temperature, not the bulk Doppler motion, with roughly symmetric wings. These multiple-scattering wings might form in optically-thick CSM beyond the photosphere (Chugai et al. 2004), but may also form within the line-forming photospheric region (Dessart et al. 2009). In the context of the Sobolev approximation, the scattering wings form and broaden as the ejecta velocity decreases (Dessart et al. 2009). In models, line profiles smoothly transition from classic P Cygni profiles to lines with broad wings and narrow P Cygni components as the ejecta speed monotonically decreases (Sect. 4.4.3).

14.3.3 Mass Loss and Circumstellar Medium

The mass loss that creates the CSM of SN IIn probably comes from both quasi-steady-state winds and impulsive ejections (Smith 2014). High mass-loss rates ($\sim 10^{-3}$ to $\sim 10^{-2}\,M_\odot\ y^{-1}$) sustained for at least decades before explosion are common (Moriya et al. 2014b; Moriya and Maeda 2014). Impulsive precursor ejections also are common; on average, about one during the year prior to explosion (Ofek et al. 2014a). To some, impulsive ejections suggest giant-eruption LBV progenitors (Hoffman et al. 2008; Kiewe et al. 2012), and in pre-explosion images an LBV-like star has been identified as the progenitor of SN 2005gl (Gal-Yam and Leonard 2009). The mass loss is usually associated with late phases in the evolution of the progenitor; a time of decades or less before explosion corresponds to late central carbon burning or beyond. There are suggestions that much of the mass loss associated with SN IIn is disk-like, although η Car also has prominent axial lobes.

The optical evidence of CSI raises the issue of radio or X-ray emission. Most SN IIn do not show strong radio emission (Van Dyk et al. 1996), perhaps due to absorption by CSM exterior to the interaction region. Those that are intrinsically strong radio sources, such as SN 1988Z (Sect. 14.2.1), SN 1998S (Sect. 14.2.2), and SN 1995N (Chandra et al. 2009b) appear to be relatively rare. SN IIn that are intrinsically strong X-ray sources, such as SN 1988Z and SN 1998S (Fig. 14.2), as well as SN 1995N (Fox et al. 2000), SN 2005ip (Katsuda et al. 2014), SN 2006jd (Chandra et al. 2012a), and SN 2010jl (Chandra et al. 2012b), are not uncommon. High presupernova mass-loss rates are inferred, ranging from $\sim 10^{-4}$ to several $M_\odot$ y^{-1}, with total mass in the CSM ranging from ~ 0.1 to tens of $M_\odot$.

While LBV giant eruptions are frequently invoked to account for the CSM associated with SN IIn, mass ejection from RSGs is not ruled out. In models, a single moderately rapidly-rotating star with mass in the range 20–25 $M_\odot$ forms a somewhat more massive helium core than a nonrotating model. Such a model can have a phase of substantial mass loss in the RSG phase and ends its evolution as a hot supergiant surrounded by dense CSM (Groh et al. 2013). Radiative-transfer models of this configuration that incorporate NLTE yield spectra with a significant number of emission lines that resemble those of LBVs. In this rotating environment, the mass loss is likely to be anisotropic. During and after core neon burning, internal gravity waves excited by core convection may transport energy to the stellar envelope, ejecting some of it (Shiode and Quataert 2014). This mechanism may also work in single stars. There are suggestions that in some cases the CSM could be a relic disk from the protostar phase (Metzger 2010).

Other mechanisms to provoke substantial mass loss involve binary systems of which η Car is an example. Sudden mass loss also may attend a common-envelope phase or a stellar merger. A merger might involve a compact companion, a neutron star or a black hole (Chevalier 2012). In some cases, the envelope may instead expand to be partially accreted by a binary companion, producing a presupernova brightening (Mcley and Soker 2014).

High mass-loss rates are sometimes deduced by making assumptions regarding the wind parameters that are contradicted by the results themselves; the implied mass loss must be dynamic, not the result of a steady-state wind. In addition, a high density need not necessarily imply a high wind mass-loss rate. Even with a low mass-loss rate, a fast wind sweeping up the surrounding medium may produce a high-density shell. High densities may also occur in clumps, or be due to an LBV eruption before the supernova explodes (Dwarkadas 2011). In the presence of external radiation from nearby hot stars, a photoionization-confined, nearly static dense shell may be formed around RSGs (Mackey et al. 2014).

The possibility that the CSM is clumpy brings other possible interpretations of the data. Some models propose that the intermediate-width optical lines result from the inward propagation of shocks into dense clumps that have been overrun by the forward shock (Chugai and Danziger 1994). In this model, the forward shock sweeps past clumps in the progenitor wind, but the high pressure of the shocked low-density CSM drives a shock into the dense clumps. If the clump shocks are radiative, substantial optical luminosity may result. The lines may be broadened by Doppler motion or perhaps multiple scattering. In other contexts, the shocks propagating into the clumps may not be radiative, and the result would be a substantial X-ray flux (Chugai 1993). A possible issue with this clumpy model is that the bulk of the CSM is proposed to be of low density and perhaps optically thin. In that case, one might expect to see the inner ejecta, which for SN IIn frequently is not the case.

14.3.4 Infrared Observations and Dust

Among core-collapse supernovae, SN IIn have special properties in the IR, many showing strong IR excesses after several months (Gerardy et al. 2002; Fig. 14.9). The typical near-IR luminosity is $\sim 10^{41}$ to 10^{42} erg s^{-1}, often emitted over several years. The IR excesses are attributed to pre-existing dust in the CSM and new dust formed in a CDS and/or deeper in the ejecta. The CDS is not expected to form dust until after it has radiated substantial optical and UV flux and cooled to less than $\sim$2000 K. In many cases, it is difficult to determine whether the emitting dust was pre-existing or new and how the dust was heated.

The presence of pre-existing dust in the CSM was inferred from near- and mid-IR excesses in, e.g., SN 1998S (Fassia et al. 2000), SN 1995N (Van Dyk 2013), SN 2005ip (Fox et al. 2010), and SN 2010jl (Fransson et al. 2014). Potential sources of heating of this dust to $\sim 10^3$ K are supernova radiation and X-ray and UV flux generated by the forward shock in a dense CSM. In a sample of 68 SN IIn observed with *Spitzer* at late times (>100 days after explosion), 10 were detected (Fox et al. 2011). In most cases, the mid-IR emission was consistent with that from pre-existing dust. This emission should truncate when the forward shock runs beyond the dust shell. Typical estimates of the mass of pre-existing dust are $\sim 10^{-2}\,M_\odot$.

The formation of new dust was inferred from suppression of the red sides of emission lines from the ejecta, e.g., in SN 1998S (Mauerhan and Smith 2012) and SN 2007rt (Trundle et al. 2009). SN 2005ip (Fox et al. 2010) and SN 2009kn (Kankare et al. 2012) showed late-time near-IR emission that probably was emitted by newly-formed dust. Typical estimates of the mass of newly-formed dust are $\sim 10^{-4}\,M_\odot$. No observations of SN IIn have been capable of detecting cold dust like that which was detected in SN 1987A with *Herschel.*

14.3.5 Polarization and Asymmetry

Polarization observations of SN IIn are relatively rare, but suggest that the CSM and/or the ejecta are asymmetric. SN 1994Y (possibly spectropolarimetric type D0; Sect. 4.6) was detected in broadband polarimetry (Wang et al. 1996). SN 1998S (Sect. 14.2.2; spectropolarimetric type D1; Fig. 14.10) was observed to be significantly polarized about 2 weeks before maximum light (Leonard et al. 2000) and

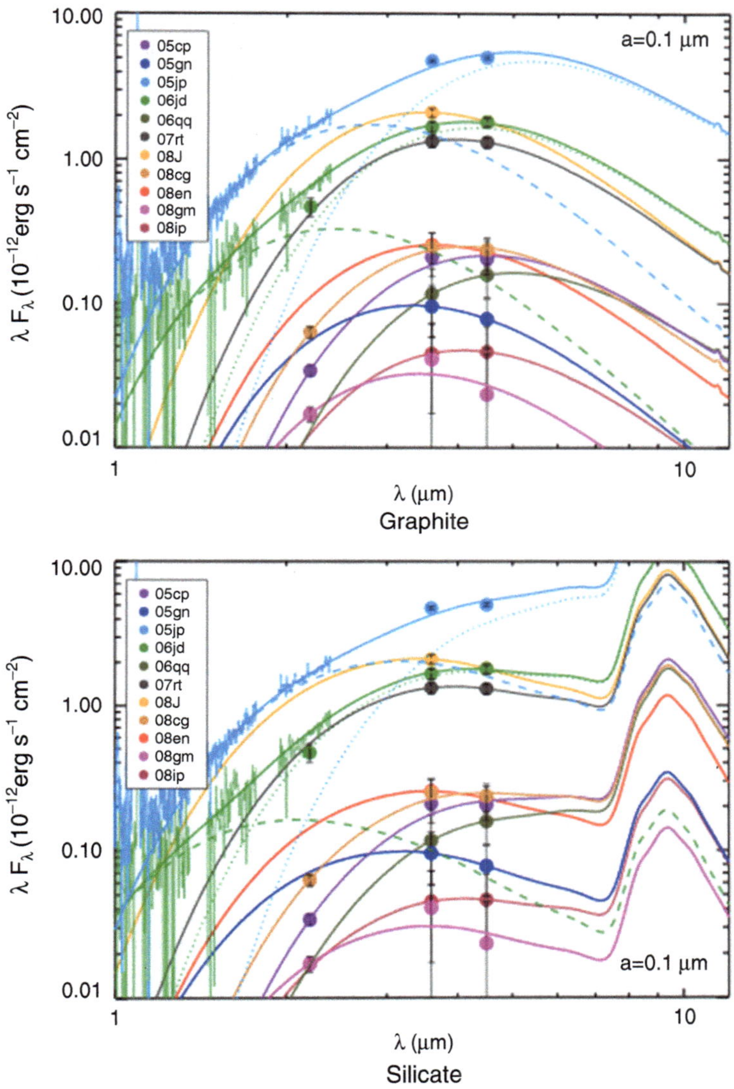

Fig. 14.9 Infrared spectral energy distributions for selected SN IIn. Also shown are best-fit models incorporating graphite or silicon grains. Warm dust with temperature in the range 400–800 K agrees with most of the data. From "A Spitzer Survey for Dust in Type IIn Supernovae" (Fox et al. 2011). © AAS. Reproduced with permission

about 10 and 40 days postmaximum (Wang et al. 2001). SN 1998S displayed narrow emission lines, but also broad lines characteristic of Wolf–Rayet stars, suggesting that it might not be generally representative of all SN IIn. (Leonard et al. 2000) suggested that the broad lines were unpolarized and the narrow lines were polarized. Based on two more epochs of data, Wang et al. (2001) estimated a significantly

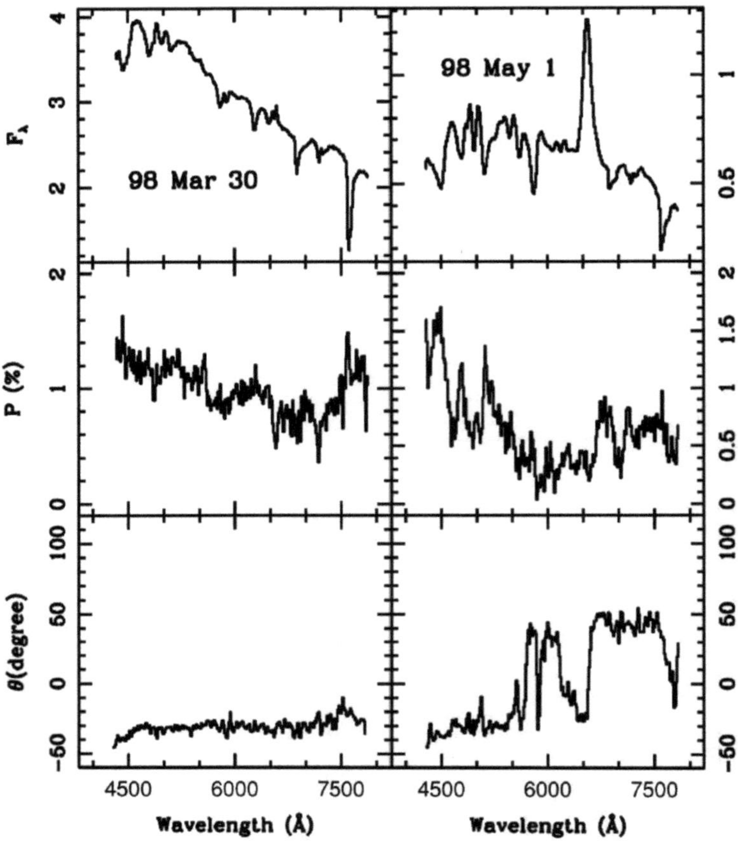

Fig. 14.10 Total flux spectra (*top*), polarized flux (*middle*), and polarization angle (*bottom*) for SN 1998S about 10 days (*left*) and about 40 days (*right*) after maximum light. From "Bipolar Supernova Explosions" (Wang et al. 2001). © AAS. Reproduced with permission

different interstellar polarization (ISP) that implied that both the broad and narrow lines were polarized in the premaximum spectrum, with the continuum displaying an essentially constant position angle and polarization of $\sim$1.6%. Ten days after maximum, the continuum showed a somewhat smaller polarization, with a rotation of $\sim$34° compared to the premaximum data. The polarization rose to a high level of about 3% in the latest data, at 40 days past maximum, while maintaining the same position angle. This exercise illustrates the importance of the difficult task of evaluating the ISP.

SN 1997eg (spectropolarimetry type D1) showed narrow, intermediate, and broad emission-line components, with no obvious evidence of P Cygni absorptions. The polarization across the broad Hα, Hβ, and He I λ5876 lines indicated a departure from axial symmetry (Hoffman et al. 2008). Hoffman et al. argued for a geometry in which aspherical ejecta were misaligned with respect to and occulted by a disk of CSM. In this model, the axis of polarization of the narrow H lines was tilted by 12°

compared to that of the elongated ejecta and may have indicated the orientation of the CSM. Hoffman et al. suggested that the model is consistent with the notion that the progenitors of some SN IIn are LBVs.

Two weeks after discovery, SN 2010jl displayed a high level of continuum polarization, about 2%, that required an axis ratio of about 0.7 (Patat et al. 2011b; Fig. 14.11). Line depolarization at Hα and Hβ suggested that the line-forming region was well above the photosphere. The polarization properties of SN 2010jl resembled those of SN 1997eg and SN 1998S.

The multiple-outburst source SN 2009ip (Sect. 14.4) was polarized at a level of about 1% (Mauerhan et al. 2014; Reilly 2015). The amplitude and angle of the continuum polarization changed between the first and the second, brighter, outburst in 2012. This result suggested a second morphological component to the geometry of SN 2009ip that was roughly orthogonal to the geometry of the previous epoch. In the second outburst, spectral features of Hα, Hβ, He I λ5876, and the Ca II near-IR triplet were prominent. The polarization data showed a complex structure and evolution, with significant shifts in the polarization angle of the continuum and lines, and in the amplitude and velocity of the polarized spectral features (Fig. 14.12).

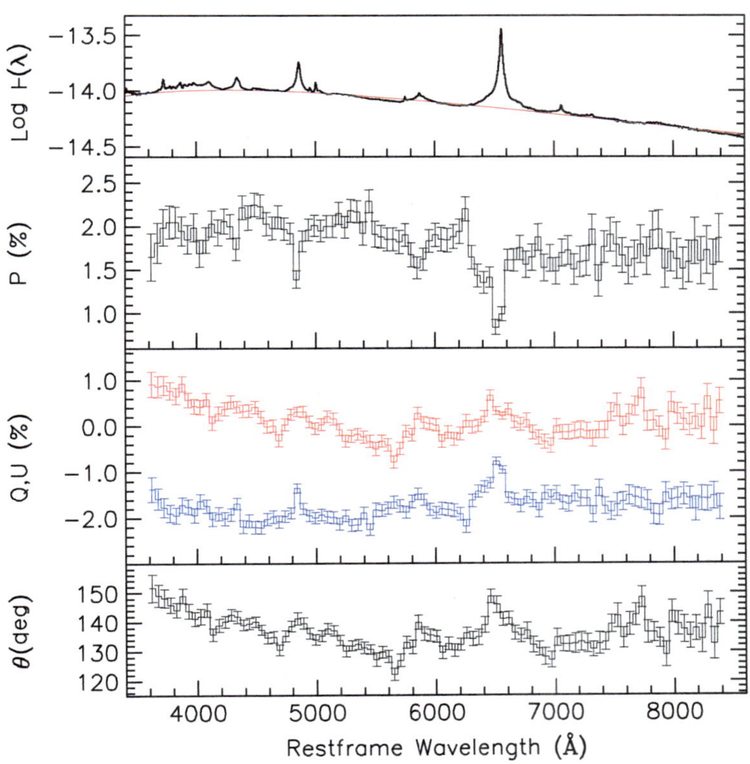

Fig. 14.11 Total flux spectra (*top*), polarized flux (*second*), Stokes parameters Q and U (*third*), and polarization angle (*bottom*) for SN 2010jl about 14 days after maximum. From "Asymmetries in the type IIn SN 2010jl" (Patat et al. 2011b) reproduced with permission, © ESO

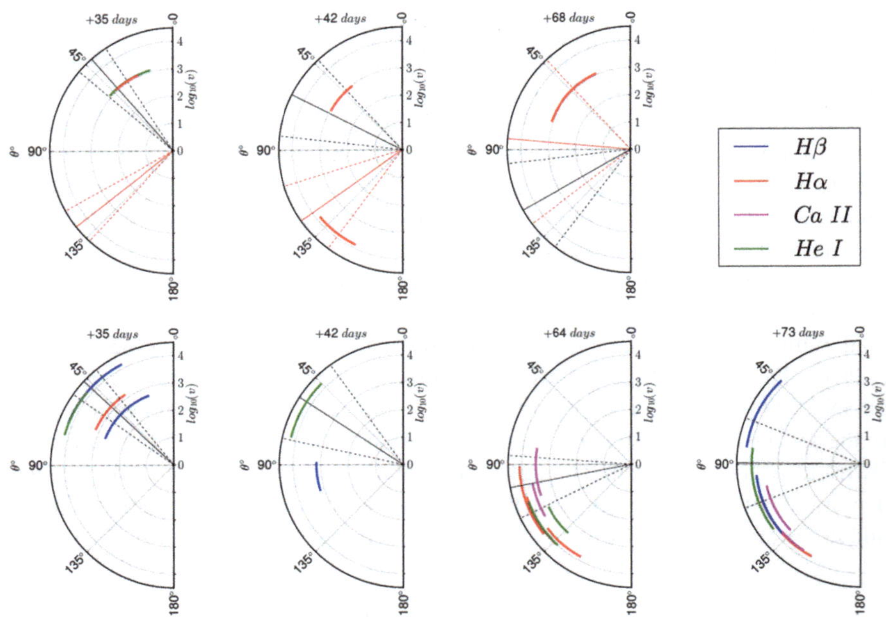

Fig. 14.12 Polar plots of the polarization angle versus radial velocity for the evolution of various spectral features in SN 2009ip. Higher resolution (R = 1200; *top*) is presented for Hα and He I λ5876 and lower resolution (R = 300; *bottom*) for Hα, Hβ, He I λ5876 lines and the Ca II near-IR triplet. The velocity is given in radial units of the log of v in km s^{-1}. The continuum is indicated with the *black lines* bounded by 1σ uncertainty as the *dashed lines*. From Reilly (2015), by permission

A bump in the light curve 30 days after the second outburst was accompanied by a shift in polarization amplitude and rotation in angle. At this epoch, photospheric lines were observed that showed significant variations in the polarization across the lines. Significant polarization was seen in Hα and the He I/NaD blend at ~10,000 km s^{-1}. The polarization signal was similar for the two lines, suggesting that the polarized H and He I lines arose in a common geometry (lower left panel of Fig. 14.12, day +35). The Hα line also showed a series of small "notches" at velocities of 6000 to 12,000 km s^{-1} that had polarization aligned with the broader Hα, but distinctly different from the narrow P Cygni component and the continuum. Line polarization of the narrow P Cygni absorption of Hα suggested a disk-like geometry. By 60 days after the second outburst, the continuum polarization had declined to only 0.2%. The results were consistent with a bipolar explosion in the first outburst colliding with a disk-like or toroidal geometry at the second peak. The asymmetric geometry may imply that the CSI only occurred in restricted solid angle and hence spherical models may underestimate the total energy involved in the explosion.

For SN 2009ip, the broad emission wings and the core of the broad component showed similar polarization. This is in contrast to the case of SN 2010jl (Fig. 14.11) in which the polarization was enhanced in the wings, a result attributed to upscattering of photons in a dense, fully-ionized plasma (Patat et al. 2011b). The lack of higher polarization in the wings of SN 2009ip suggests that there was little reprocessing, and that the line-forming region was in the rapidly-moving ejecta, as deduced for the regions that formed the broad emission lines in SN 1997eg and SN 1998S.

14.3.6 Statistics

An important issue with SN IIn is whether they are merely an extension of the well-known category of SN IIP or represent a separate category in an important physical sense. There may be a continuum in some parameter, mass-loss rate being an obvious candidate, with SN IIP at one extreme with low mass-loss rates, perhaps SN IIL in the middle, and SN IIn with extreme mass loss (Schlegel 1990). SN PTF11iqb is a good example of an event showing both SN IIn and SN IIP characteristics (Sect. 14.3.1.2). One possibility is that about half of the stars above 23 $M_{\odot}$ lose their hydrogen envelopes in binary mass transfer to become SN Ib/c or SN IIb, while the other half retain their envelopes and become SN IIn (Smith et al. 2011a). The association of some SN IIn with candidate LBV stars also tends to promote the notion that some arise in progenitors of especially high mass. The evidence that the environments of SN IIn and SN IIP in their host galaxies are similar (Sect. 14.1) indicates, however, that most SN IIn are not from very massive stars. As remarked in Sects. 14.2.1 and 14.2.2, there is evidence that SN IInL such as SN 1998S arise in regions of metallicity similar to that of SN IIP, but that SN IInE such as SN 1988Z arise in lower-metallicity environments.

14.4 Impostors

Some stars brighten dramatically, eject massive shells, and display spectra with narrow emission lines that resemble spectra of SN IIn, but these events, even if they are given supernova designations, are not necessarily supernovae; they may be *supernova impostors.* A supernova impostor leaves the underlying star fundamentally intact; a supernova obliterates the star or signals a radical transformation, e.g. from star to neutron star.

The original impostor candidate was SN 1961V, which Zwicky (1964) referred to as a Type V supernova. The early optical spectra were dominated by narrow emission lines of H, He I, and Fe II, with a maximum expansion velocity of only 2000 km s^{-1} (Branch and Greenstein 1971). SN 1961V appeared to be associated with a strong nonthermal radio source (Branch and Cowan 1985) that was detectable for at least 15 years and assumed to represent a radio-bright supernova (Stockdale et al. 2001), but that association was questioned (Van Dyk and Matheson 2012). Some subsequent studies led to suggestions that SN 1961V was not a supernova but more like the giant eruption of η Car in 1850 (Goodrich et al. 1989; Filippenko et al. 1995a). The true nature of SN 1961V continues to be debated (Kochanek et al. 2011, 2012b; Van Dyk and Matheson 2012), but other events that are quite likely to have been impostors were identified (Van Dyk et al. 2000). The occurrence rate of impostors is a substantial fraction of the core-collapse supernova rate (Thompson et al. 2009).

There are at least two kinds of impostors (or supernovae posing as impostors), with distinctly different degrees of association with ongoing star formation as judged by Hα emission. One kind, exemplified by "SN 2008S," is associated with recent but not ongoing star formation, suggestive of moderate ZAMS masses of about 10 $M_{\odot}$, while another kind, e.g., "SN 1997bs," is associated with ongoing star formation, implicating higher ZAMS masses (Habergham et al. 2014).

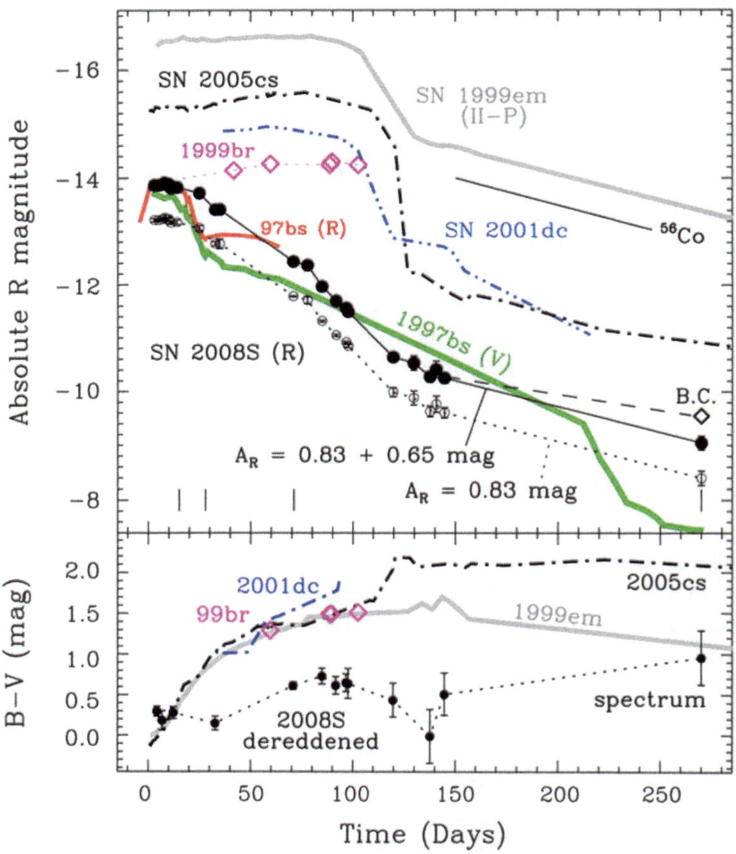

Fig. 14.13 Absolute R-magnitude light curve of SN 2008S with two corrections for extinction in comparison with other, possibly related, events (*top*) and $B - V$ color curves (*bottom*). From "SN 2008S: A Cool Super-Eddington Wind in a Supernova Impostor" (Smith et al. 2009a). © AAS. Reproduced with permission

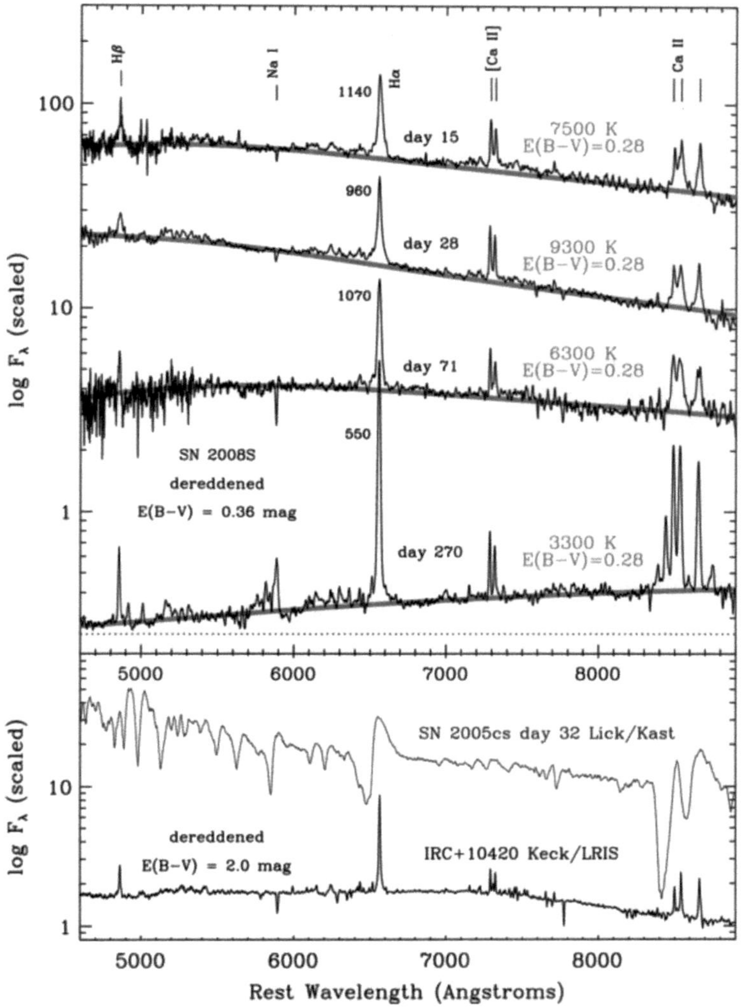

Fig. 14.14 Spectral evolution of SN 2008S and other, possibly related, events. From "SN 2008S: A Cool Super-Eddington Wind in a Supernova Impostor" (Smith et al. 2009a). © AAS. Reproduced with permission

The SN 2008S-like events are associated with large amounts of dust. The light curve of SN 2008S and other events is given in Fig. 14.13 and the spectral evolution in comparison with other events in Fig. 14.14. During its outburst, SN 2008S had narrow emission lines suggesting classification as an SN IIn, but it may have been something else entirely. SN 2008S was not observed on the rise. After discovery, it had a brief flat peak for about 20 days at $M_R = -13.2$ and then declined nearly linearly by 3 magnitudes in about 100 days, after which it fell more slowly at roughly the rate of ^{56}Co decay (Smith et al. 2009a). The spectrum showed strong emission lines of H and Ca II on a smooth continuum. The lines showed velocity widths declining from about 1000 to 600 km s^{-1} with no sign of blueshifted absorptions; the profiles were probably due to multiple electron scattering. The spectra showed no dramatic change over the course of nearly a year.

The progenitor of SN 2008S was detected in archival *Spitzer* mid-IR images up to 4 years prior to the explosion (Prieto et al. 2008), but could not be found in deep optical images, indicating that the progenitor had become strongly obscured by circumstellar dust. Other candidates in this category include SN 1999bw, SN 2001ac, SN 2002bu, SN 2008-OT (also known as NGC 300-OT, where OT refers to "optical transient"), and SN 2010dn. A rather high rate of SN 2008S-like outbursts together with few progenitors detected in the mid-IR suggests that the phase of high pre-outburst obscuration is short-lived, lasting perhaps only 10^4 years. The pre-existing dust is destroyed by the outburst and then reforms as the transient fades (Kochanek et al. 2012b).

The luminosity of the progenitor of SN 2008S was consistent with a progenitor ZAMS mass of about 10 $M_\odot$, suggesting that it was an AGB or SAGB star. This ZAMS mass falls in a range for which collapse might be due to electron capture rather than iron-core collapse, but this is far from certain. The ejected mass of ^{56}Ni must have been small.

The impostors that are associated with ongoing star formation, the SN 1997bs-likes, are a diverse group (Kochanek et al. 2012b). SN 1997bs had a flat peak at $M_R \simeq M_V \simeq -14$ for 10–20 days, a steep drop by about 1 mag in 10 days, then

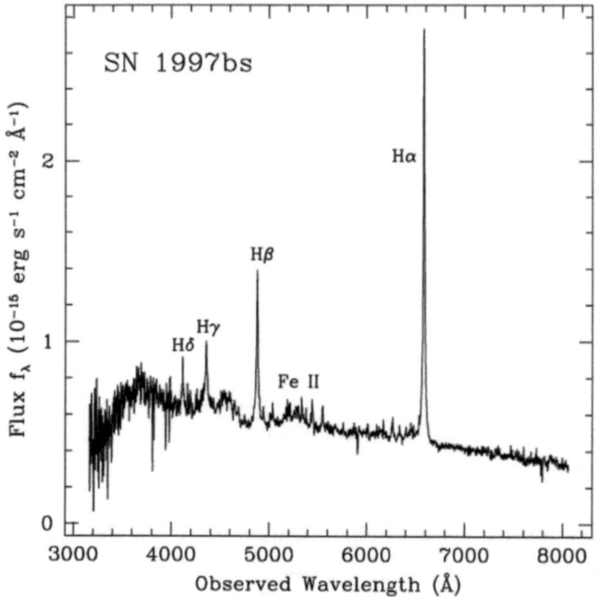

Fig. 14.15 Spectrum of SN 1997bs 1 day after discovery. Figure 2 "SN 1997bs in M66: Another Extragalactic η Carinae Analog?" from Van Dyk et al. (2000) © The Astronomical Society of the Pacific. Reproduced by Permission. All Rights Reserved

a much slower decline for about 200 days, followed by yet another even steeper decline (Fig. 14.13). The spectrum showed strong narrow Hα (Fig. 14.15). By these characteristics, SN 1997bs was not that dissimilar to SN 2008S. The category includes SN 1954J that may have been a weaker analog of η Car, LBV outbursts such as SN 2002kg, non-LBV events such as SN 2000ch that are variable stars with little or no dust formation, and the apparently massive-star event SN Hunt 248 (Mauerhan et al. 2015a), the light curve of which showed triple peaks (Kankare et al. 2015). The archetype of this group, SN 1997bs, may even have been a subluminous supernova (Adams and Kochanek 2015). The difficulty of deciding whether SN 2011A was a supernova or an impostor was highlighted by de Jaeger et al. (2015a). Much more work in the optical, near-IR, and mid-IR is required to sort out the physics and phenomenology of these events.

SN 2009ip was observed to undergo multiple outbursts between 2009 and July, 2012 (Smith et al. 2010b; Foley et al. 2011; Fig. 14.16) before the brightest one ($M_V \simeq -18$) in September, 2012 (Mauerhan et al. 2013a). The outburst of July, 2012, which reached $M_V \simeq -17$, may have been a bright impostor, or it may have been a supernova, with the September brightening resulting from collision of the supernova ejecta with a circumstellar shell (Mauerhan et al. 2014). Alternatively,

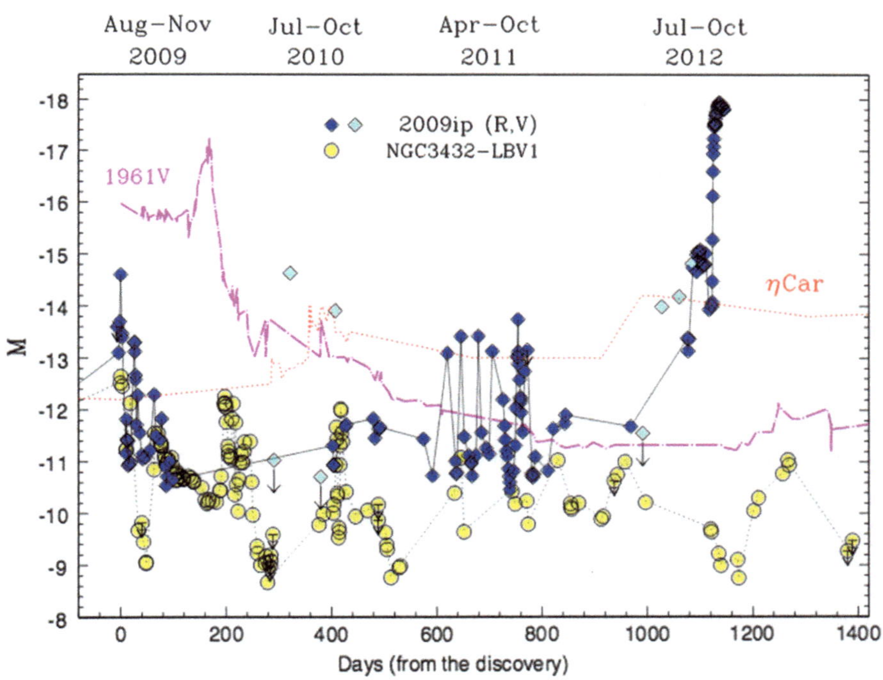

Fig. 14.16 Absolute-magnitude light curve of SN 2009ip in the R and V bands. Also shown are light curves of η Car, SN 1961V, and the variable in NGC 3432. From "Interacting Supernovae and Supernova Impostors: SN 2009ip, is this the End?" (Pastorello et al. 2013). © AAS. Reproduced with permission

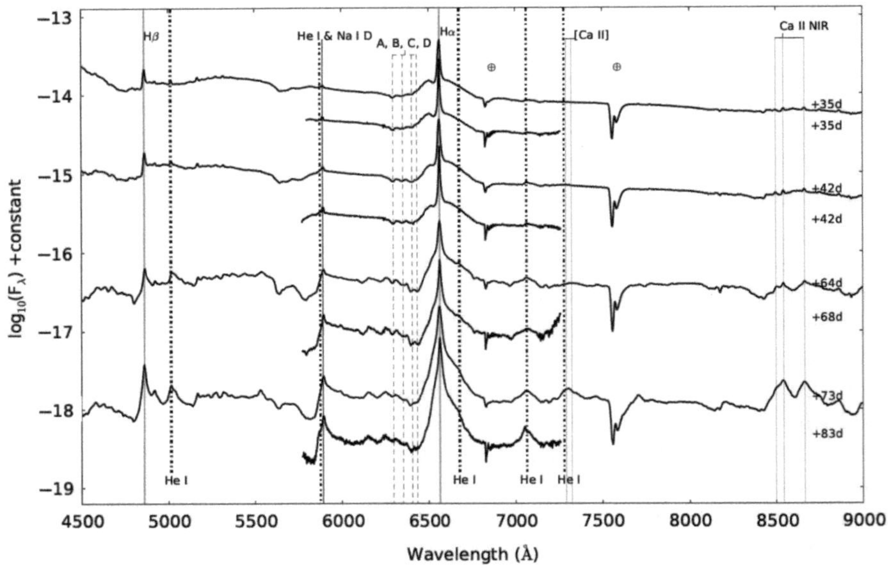

Fig. 14.17 Spectral evolution of SN 2009ip. High-velocity narrow absorption "notches" of Hα are indicated as A, B, C, and D. From Reilly (2015), by permission

the September outburst may have been a supernova. The total energy radiated was $\sim 2 \times 10^{49}$ ergs (Fransson et al. 2014). Whether SN 2009ip was a supernova at all is a subject of controversy (Pastorello et al. 2013; Margutti et al. 2014a; Smith et al. 2014a; Graham et al. 2014; Mauerhan et al. 2014).

SN 2009ip was dominated by strong, narrow ($\lesssim 2000\,\mathrm{km\,s^{-1}}$) emission lines of hydrogen, helium, calcium, and iron for 2 years (Fig. 14.17). Evidence for [Fe II] λ7155 and [O I] λλ6300,6364 lines appeared in late 2013, but there remained little evidence for strong, broad nebular emission lines characteristic of core collapse (Fraser et al. 2015). The emission lines were approximately symmetric. A red bump at $550\,\mathrm{km\,s^{-1}}$ developed on the wing of Hα in 2014. The lines showed no overall blueshift as might be expected if dust formed, and there was no evidence for near-IR or mid-IR flux. Spectropolarimetry of SN 2009ip (summarized in Sect. 14.3.5) revealed a complex geometry. Velocities as large as $12{,}500\,\mathrm{km\,s^{-1}}$ were observed in the minor outbursts prior to the brightest outburst in 2012 (Pastorello et al. 2013). Velocities of this order do not, by themselves, establish that a supernova has occurred. Some spherical models are consistent with a small mass of CSM and modest energy that do not demand a supernova (Moriya 2015), but if the CSI is asymmetric, so that there is less interaction in some directions, a larger energy might be required. By 800 days, there was no unambiguous evidence that either of the 2012 outbursts was a core-collapse event, although that demise cannot yet be ruled out.

14.5 SN1994W-Likes

Among the bewildering complexity of SN IIn phenomena, a small subset has remarkably similar, if ill-understood, properties. These are the subset of SN IInP of which SN 1994W (Sollerman et al. 1998), SN 2009kn (Kankare et al. 2012), and SN 2011ht (Mauerhan et al. 2013b) are the best studied. SN 1994W-likes are defined in part by their light curve behavior (Sect. 14.3.1.2; Fig. 14.18), but even more so by their unique spectral behavior (Fig. 14.19). These events display characteristics seen in other SN IIn, even other SN IInP, but in a very homogeneous fashion. On the plateau, the Hα line shows approximately symmetric scattering wings with FWHM of about 2000 km s^{-1} interrupted by a blue P Cygni absorption at a velocity of $\sim$600 km s^{-1}. At the end of the plateau in each case, the intermediate-width symmetric wings disappear. A narrower emission core and the P Cygni absorption at the same velocity remain. Neither SN 1994W nor SN 2009kn showed the classic nebular emission lines of core collapse, but there is some evidence for weak emission features of [O I] $\lambda\lambda$6300,6364 and [Ca II] $\lambda\lambda$7291,7323 in SN 2011ht, diluted by CSI. Taken at face value, the weak [O I] $\lambda\lambda$6300,6364 might be consistent with a relatively low-mass progenitor (Dessart et al. 2010b).

The end of the plateau is certainly not due to dust formation because there is no associated sudden reddening at that phase. Colors shift to the red after the plateau,

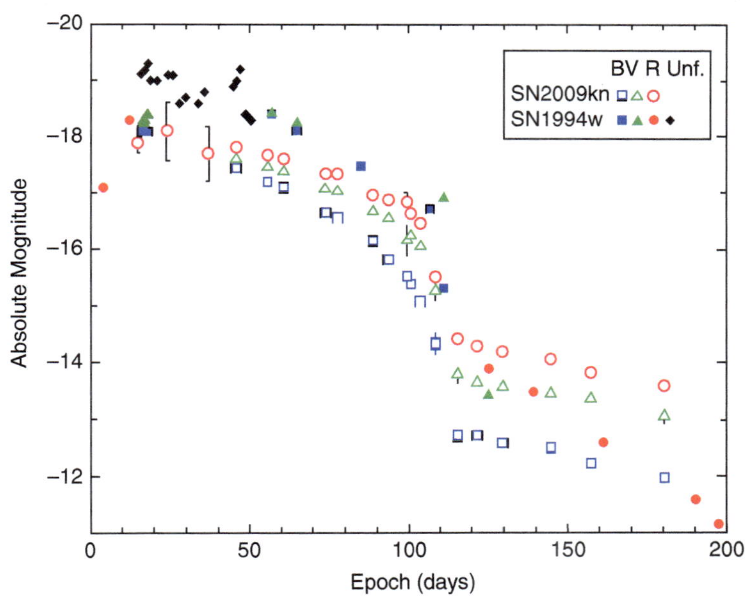

Fig. 14.18 Absolute-magnitude light curves of two SN 1994W-like SN IInP, the prototype SN 1994W and SN 2009kn. Figure 3 "SN 2009kn - the twin of the Type IIn supernova 1994W" from Kankare et al. (2012), by permission of Oxford University Press on behalf of the Royal Astronomical Society

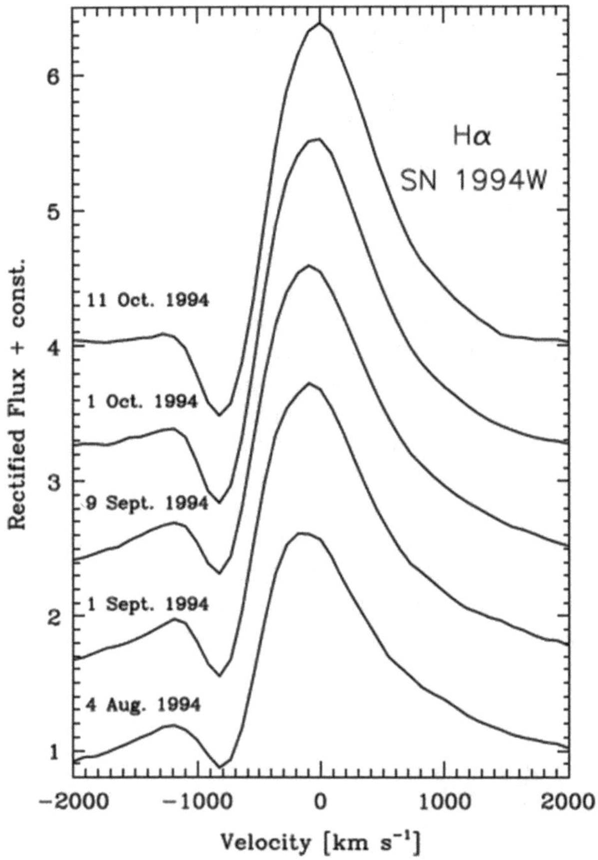

Fig. 14.19 Evolution of the Hα feature of SN 1994W (spectra shifted vertically for visibility) illustrating the peculiar characteristics of this sub-category of SN IInP. The basic profile is an intermediate-width emission line with a narrow P Cygni absorption. The strength of the blue wing decreases with time and that of the red wing increases. The velocity of the P Cygni absorption is remarkably constant. The emission peak shows a slight blueshift at early times. Figure 10 "SN 1994W: an interacting supernova or two interacting shells?" from Dessart et al. (2009), by permission of Oxford University Press on behalf of the Royal Astronomical Society

but it is difficult to attribute this definitively to dust formation. Later dust formation is difficult to rule out, but to account for the reddening of the colors and optical decline in the nebular phase dust would have to form rather rapidly.

The Hα line profile and its evolution is not in itself so special, but the symmetric line width and the velocity of the blue absorption are basically constant on the plateau in each event and are virtually identical among the three well-observed events. Whatever SN 1994W-likes do, they know how to do it very well. The nature of these events and the basic issue of whether they are even supernovae remains a source of controversy.

If the observed line width were from Doppler expansion, then the velocity might be indicative of a supernova (Chugai et al. 2004), but the intermediate-width symmetric profile is almost surely due to multiple scattering, not bulk motion. The velocity of the blue absorption, $\sim$600 km s^{-1}, is far below a supernova ejecta velocity and the velocity of a Wolf–Rayet star wind, perhaps comparable to LBV giant eruptions, and far above the wind velocity of an RSG. The constant velocity of this absorption might be consistent with formation in a CDS moving at constant speed, but that alone does not prove a supernova origin. The observed velocity is small for that expected for a CDS formed in CSM of less mass than the ejecta mass.

Any physical picture must also account for the plateau nature of the light curve. For SN IIP, the plateau represents the recession of the recombination front through the ejected hydrogen envelope. The plateau in SN 1994W-likes is about 120 days long, somewhat longer than typical plateaus of SN IIP; however, if the plateau is due to recombination of the ejected envelope, the velocity of the absorption is expected to evolve to lower values, and one would expect to see normal nebular lines after the end of the plateau. An alternative is that the plateau represents the recession of a recombination front through the CDS. The problem then again arises that at the end of the plateau one might expect to see the inner ejecta and normal core-collapse nebular lines. One possibility is that the entire inner ejecta has recombined and is cool and in a state of low excitation because so little ^{56}Co was produced in the explosion. This demands a rather special set of circumstances but is consistent with the dim light-curve tails (Sect. 14.3.1.2) that allow little ^{56}Co. The CDS picture does not give a natural explanation for the length of the plateau being comparable to normal SN IIP. The observed Hα line is not flat-topped as one might expect if the line formed in a shell. There are also quantitative questions as to whether an optically-thick CDS can form in the conditions that are otherwise consistent with the observations of this class of objects.

Another possible explanation for SN 1994W-likes is collisions among multiple CSM shells with no real supernova explosion (Dessart et al. 2009). The peak visual magnitude, about -17, and the total energy radiated, $\sim 3 \times 10^{49}$ ergs, are characteristic of supernovae, but might be accommodated by sufficiently violent shell collisions. While the late-time decays of SN 2009kn and SN 2011ht were roughly consistent with the decay of ^{56}Co, the decay of SN 1994W was decidedly steeper in the R-band. A rigorous argument for ^{56}Co decay is thus difficult to make, leaving room for some decaying shell-collision emission. Prediscovery observations of SN 2011ht show that the progenitor could not have been as bright as η Car (Roming et al. 2012), but a somewhat dimmer giant-eruption progenitor cannot be ruled out. Models in which the continuum and lines form in a single region with substantial multiple scattering give a good representation of the Hα line profile and its evolution, including a trend for the peak to shift slightly to the blue and the blue wing to decay somewhat more rapidly than the red wing while maintaining the blue absorption at about 600 km s^{-1} (Dessart et al. 2009). The model line peaks are not boxy. These models require a slow expansion of the radiating matter to produce the strong scattering wings rather than the dominant P Cygni profile characteristic of SN IIP. In that sense, they may be more pertinent to shell collisions than to supernova/CSM collisions, but might also apply to the latter in some circumstances.

This class of models calls for a single photosphere at all epochs and hence a dense, optically-thick region.

A third proposed explanation for these events is that they are not some form of explosion for which we are witnessing the propagation of a forward shock through a CSM of varying properties, but rather the temporal behavior of a single intact star (Humphreys et al. 2012). The proposition is that we are observing mass ejection in a dense, optically-thick wind with a typical velocity of $\sim$600 km s^{-1} and then a transition to a less dense, optically-thin wind. The first phase gives the effective plateau, the multiple-scattering wings, and the P Cygni feature. The second phase accounts for the disappearance of the scattering wings. Ordinary broad photospheric and nebular lines are not observed because nothing has exploded. One drawback of this model is that the early high luminosities would demand a luminosity that exceeds the Eddington limit by about a factor of 10^3 for a static star of about 10 $M_\odot$. Whether such a condition would correspond to a wind, as formally proposed, or would itself call for explosive conditions, is debated (Mauerhan et al. 2013b).

SN PTF11iqb showed early CSI, then a basically spectroscopically normal SN IIP plateau, followed by a return of CSI (Sects. 14.3.1.2 and 14.3.6). SN PTF11iqb may be an intermediate event between the SN 1994W-like events and normal SN IIP.

14.6 Summary

SN IIn can show conspicuous emission lines of narrow, intermediate, and broad width as well as blueshifted absorptions of various widths that are indications of interaction with a substantial CSM. The narrowest lines form in CSM that lies outside the forward shock; the broadest lines represent shocked matter, including that in the supernova ejecta proper. The intermediate-width lines may be the result of multiple electron scattering rather than bulk Doppler expansion.

Some SN IIn are quite bright. The most luminous SN IIn with absolute magnitude brighter than $M_V = -20$ probably are cases in which the CSM remains at $\sim 10^{15}$ cm, where production of radiant energy from kinetic energy is most efficient. The fact that SN 2010jl was nearly as bright as the SLSN-II (Sect. 18.2) may mean that there is a continuous distribution in luminosity, and that the apparent gap between the brightest SN IIn and the dimmest SLSN-II is one of poor statistics. If the CSM shell has expanded to larger distance, it will tend to have lower optical depth and the conversion of kinetic to radiant energy will be less efficient, resulting in a generally dimmer event corresponding to more normal SN IIn. These events with thinner CSM may then be more prolific producers of radio and X-ray flux compared to the brighter events.

The light curves tend to decline slowly, at least in certain phases. They also can have phases, both early and late, where the decline is precipitous. The slow decline might represent some other energy source, such as a pulsar, but the bulk of the photometric and spectroscopic data implicate strong CSI. X-rays and radio emission associated with CSI are seen in some cases, but strong radio emission is relatively rare. In some cases, the X-ray flux dominates the optical flux.

SN IIn display many common features in their spectral evolution, mainly the character of their emission lines, but, as for their light curve behavior, there are many differences in detail, e.g., whether certain features appear and if they do, at which phase. Given the likelihood that many, if not most, SN IIn have undergone sporadic mass loss rather than mass loss by steady-state winds, some of the variation may simply arise from timing. Other aspects may point to true physical variation that harbors clues to the progenitor evolution and explosion mechanism. Most of the observations have been in the optical, but given the evidence for dust formation in the CDS and within the ejecta, probing SN IIn with longer wavelength observations should be fruitful.

It is manifestly clear that SN IIn involve CSI. This constraint leaves much leeway in understanding the phenomena involved. Although asymmetries are expected and yield complications, the basic picture is that a forward shock propagates into the CSM, a CDS forms at the contact discontinuity, and a reverse shock propagates back into the inner ejecta. The radiant energy is usually modest (but see Chap. 18), so the kinetic energy involved also can be modest, even if only a fraction is converted to radiant energy. This means that the phenomenology of SN IIn can involve full-fledged supernovae colliding with CSM, but may in some circumstances involve multiple ejections and collisions of shells. If the CSM has substantial mass, the reverse shock may slow the ejecta to the point that the inner broad forbidden lines that characterize core-collapse supernovae are never seen. In most cases, the CSM, while substantial, is deduced to be considerably less than the ejecta of a typical supernova, so much of the inner ejecta should be present but perhaps hidden behind optically-thick CSM or dust.

The SN IInE are likely to involve full-fledged supernovae. The earliest spectra of SN 1988Z showed velocities of $\sim$10,000 km s^{-1}. The total energy radiated by SN 1988Z and SN 2010jl was a significant fraction of 1 B. The underlying kinetic energy must have been at least this much, and perhaps more if the conversion of kinetic energy to radiant energy were at all inefficient. The CSM must be extensive in radius to maintain the interaction, and perhaps also in mass. The fact that characteristic forbidden lines of the nebular spectra of core-collapse are not seen in many cases is presumably because the CDS or the outer CSM or both remain optically-thick for a long time. The locus of the reverse shock and the state of the ejecta thus remain obscure. Definitive statements about the radiant energy cannot be made about the other SN IInE described in Sect. 14.3.1.1 since the events often were discovered well after explosion and often were dimmer than SN 1988Z at roughly comparable phases.

For SN IIn P, the precipitous decline after the end of the plateau strongly suggests that the matter, the outer CSM, the CDS, and the inner ejecta have become optically thin. The frequent failure to see the characteristic nebular lines is then another puzzle. SN 1998S is the exception that proves the rule, having displayed [O I] $\lambda\lambda$6300,6364 and [Ca II] $\lambda\lambda$7291,7323 for an interval of time before apparently colliding with another concentration of CSM that once again obscured the inner ejecta. One possibility is that the events that do not display the characteristic core nebular lines represent multiple-shell collisions, not supernovae, so the star is still there, and the core collapse, if destined to be, has not yet happened. If there has been a core collapse, then it might be that the reverse shock has slowed the ejecta so it does not radiate in the same way, but there is little suggestion that the CSM for these events outweighs the ejecta.

Substantial evidence indicates that the CSM is often formed not from a steady-state wind but rather by transient episodes of very high mass-loss rates, perhaps up to $10^{-2}\,M_\odot\,y^{-1}$. Estimates of mass-loss rates span a large range, but they are subject to error by assuming spherical symmetry and neglecting clumping and asymmetries. Parallels have been drawn between the sorts of transient mass loss attributed to SN IIn and giant-eruption LBVs, but in most cases there is no direct evidence for that specific, phenomenologically-defined, process. The mass loss might also be connected with some binary-star evolution process such as common-envelope evolution or merger, although there is no direct evidence for binary companions. The sparse polarization measurements of SN IIn suggest that the CSM and/or the ejecta are asymmetric. The substantial CSM may identify SN IIn as requiring a separate evolutionary process, but the issue of their being distinctly different from, or just a variant of, SN IIP and SN IIL remains unresolved.

SN IIn often show IR excesses. Some of the data are consistent with a contribution from newly-formed dust, but other cases seem to involve an IR echo from pre-existing dust that formed in material ejected by the progenitor. Care must be taken to separate these two contributions. Pre-existing dust tends to be identified by mid-IR excesses and newly-formed dust is most easily detected by the suppression of the red sides of emission lines that form in the ejecta.

The CSI tends to produce a substantial amount of UV flux. At high redshift, this UV flux is shifted into the optical bands. This implies that SN IIn may be especially detectable at high redshifts. Redshifted Lyman-α flux might be detectable for about 2 years with 8-m class telescopes.

SN IIn have a complex array of properties, despite the common physical manifestation of narrow emission lines. We have attempted to organize these events in light-curve categories, but there should be a concerted attempt to search for correlations among the wide variety of properties they display. Among these are the characteristics of the spectra—narrow lines by definition, featureless blue continuum, narrow P Cygni lines, intermediate width, 20–40 Å, 1000–2000 km s^{-1}, lines, and their evolution in width, intermediate-width P Cygni absorption, broad emission, broad P Cygni absorption, photospheric absorption in late spectra, whether Hα or [O III] dominate standard core-collapse nebular lines or not, spectra change at break in light-curve decay, broad or narrow lines in late spectra—and other features such as signs of line asymmetry, multiple components, signs of dust, X-rays of what epoch and luminosity and radio of what epoch and luminosity.

Issues to be resolved include the nature of the mass loss, when line widths represent multiple scattering rather than bulk motion, the degree to which the CSM is clumpy rather than smooth, whether the CSM is asymmetric, and if so, how and why?

Chapter 15
Type IIb Supernovae

15.1 Introduction

SN 1987K was the first supernova observed to make the transition from spectral Type II to spectral Type Ib (Filippenko 1988) and to be classified Type IIb. The defining property of SN IIb is the emergence of strong P Cygni lines of He I after a previously H-dominated optical spectrum. The nebular spectra are very similar to those of SN Ib/c (and SN IIP apart from the H lines), with strong emission lines of Mg I] λ4571, [O I] $\lambda\lambda$6300,6364, and [Ca II] $\lambda\lambda$7291,7323. The progenitors of SN IIb are nearly, but not completely, stripped of hydrogen by winds or mass transfer so that the photosphere rapidly recedes through the small mass of hydrogen-rich ejecta into the underlying helium-rich layers. From then on, SN IIb resemble SN Ib. They are transition events. The nomenclature "Type IIt," for transitional, did not catch on; Type IIb (Woosley et al. 1988, 1994a) did. The "II" denotes the presence of H lines in the spectrum and the "b" in the context of SN Ib has come to denote the presence of He I lines. SN IIb are one component of the broad category of *stripped envelope* supernovae along with SN Ib and SN Ic. The preponderance of the evidence suggests that most if not all SN IIb arise in binary systems, and they may be a separate physical category of progenitor evolution rather than just a portion of a continuum. Liu et al. (2015b) found that the strength of Hα can be used quantitatively to differentiate between SN IIb and SN Ib at all epochs; however, they did not exclude a continuum of observational properties of these two types.

At maximum light, SN IIb have a mean $M_B \simeq -17$. In a volume-limited sample, SN IIb constitute about 10% of all core-collapse events. The environments in their host galaxies tend to be bluer and of lower surface brightness than those of SN Ib, SN Ic, and SN IIP (Chap. 3).

D. Branch, J.C. Wheeler, *Supernova Explosions*, Astronomy and Astrophysics Library, DOI 10.1007/978-3-662-55054-0_15

15.2 Case Study: SN 1993J

SN 1993J was detected in the nearby (3.6 Mpc) galaxy M81 on March 28.9, 1993 (Ripero et al. 1993). It has been especially well studied and hence serves as a prototypical event, although SN IIb display a range of features not all represented by SN 1993J. The time of the unobserved shock breakout was estimated to have been March 28.0 $\pm$ 0.1 (Wheeler et al. 1993). SN 1993J was quickly recognized as being something new and probably related to a massive star that had been stripped of most, but not all, of its hydrogen envelope (Podsiadlowski et al. 1993; Schmidt et al. 1993; Wheeler et al. 1993; Swartz et al. 1993a; Filippenko et al. 1993; Hoflich et al. 1993; Nomoto et al. 1993; Utrobin 1994; Woosley et al. 1994a).

SN 1993J was observed across the electromagnetic spectrum and with optical spectropolarimetry. The progenitor system was identified and characterized: the star that exploded was a yellow supergiant of type K0 Ia (Aldering et al. 1994) and its probable companion was a hot B2 supergiant. SN 1993J interacted with CSM, generating detectable radio and X-ray emission. SN 1993J triggered two technically separate but interrelated efforts to identify the nature of the progenitor system: one to directly search for the system in pre- and postexplosion images (Sect. 3.2), and the other to model the light curve (Sect. 15.2.1) and spectral (Sect. 15.2.7) evolution. The photometric and spectral evolution were consistent with a star of ZAMS mass about 15 $M_\odot$ that lost mass to a binary companion. This left a progenitor with an inner helium core of about 4 $M_\odot$ and an extended, helium-rich, hydrogen-contaminated envelope of $\lesssim$0.5 $M_\odot$, of which $\sim$0.05 $M_\odot$ is hydrogen, as detailed in subsequent sections. A compendium of early observations and modeling of SN 1993J was presented by Wheeler and Filippenko (1996).

15.2.1 Light Curve

The light curve showed an early rapid rise to a first peak, followed by a decline and then a slower rise to a second maximum (Richmond et al. 1994), reminiscent of SN 1987A. The V-band light curve first peaked 2.1 days after shock breakout at $V = 10.7$. The observed color at that time was $B - V \lesssim 0.1$, with the intrinsic color slightly negative. The UBVRI light curves all showed a minimum, a second maximum, and a fairly rapid subsequent decline, followed by a more gradually declining tail (Fig. 15.1). The minimum occurred 8.3 days after shock breakout. The secondary V-band maximum was attained on day 21. The second peak had $M_R = -17.5$ to -18.1, depending on estimates of interstellar extinction (Richmond et al. 1994). On the initial light-curve tail, the slope in V, R, and I was roughly 0.02 mag d^{-1}; in B the decline was slower at 0.01 mag d^{-1}. After 400 days, the light curve flattened to about 0.0005 mag d^{-1}, owing to emission from CSI (Zhang et al. 2004; Fig. 15.2).

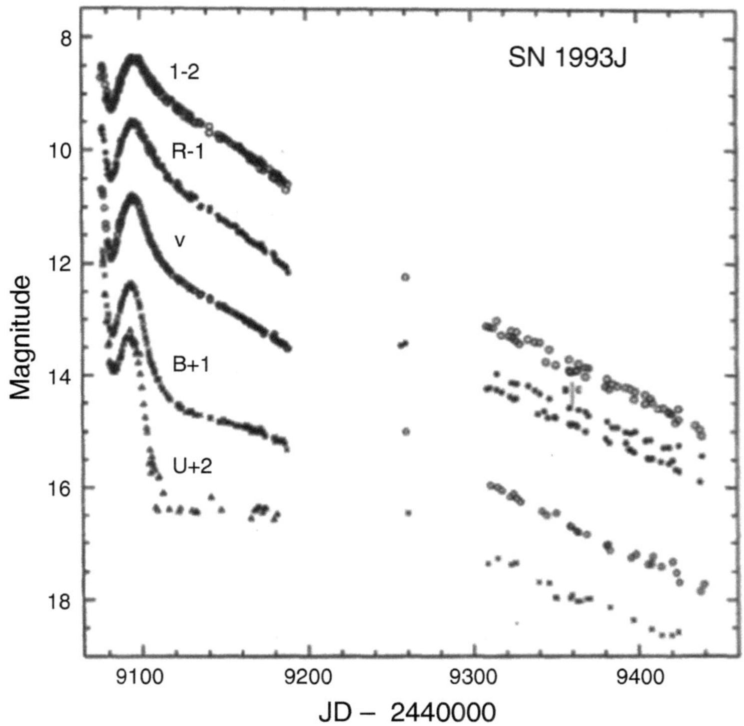

Fig. 15.1 Light curves of SN 1993J to 1 year. From "UBVRI Photometry of SN 1993J in M81: Days 3 to 365" (Richmond et al. 1996a). © AAS. Reproduced with permission

In models (Sect. 15.2.6), the first peak in the light curve was primarily due to the hot, expanding and cooling fireball phase that follows shock breakout. The bolometric luminosity declined, but at first the optical flux increased due to the changing SED. In this phase, the inner layers expanded adiabatically and the outer layers radiated by diffusion (Sect. 5.4.1). The early luminosity may also have been affected by CSI. Given evidence for asymmetry from polarization (Sect. 15.3.5), it is possible that a nonspherical breakout shock affected the details of the fireball phase. As for SN 1987A, the secondary peak in SN 1993J was caused by delayed diffusive release of energy of radioactive decay from about 0.08 $M_\odot$ of ^{56}Ni. To fit the bolometric and V-band light curves, the outer layer of about 0.5 $M_\odot$ was required to be helium-rich with a hydrogen mass fraction near 0.1. Light-curve models required that there be no pure helium mantle in contrast to predictions of most evolutionary models (see Sect. 15.2.5).

For a presupernova separation between the progenitor and its B-type supergiant companion of about 10 AU and a characteristic supernova ejecta velocity of 10,000 km s^{-1}, the companion would have been engulfed in less than 2 days, near the first optical peak of the supernova. It is unlikely that this resulted in an

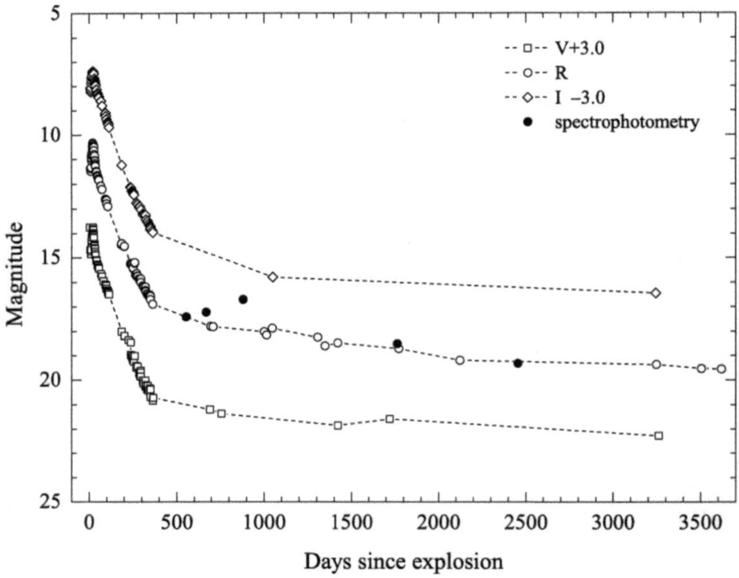

Fig. 15.2 Light curves of SN 1993J to 10 years. From "Optical Photometry of SN 1993J: 1995 to 2003" (Zhang et al. 2004). © AAS. Reproduced with permission

observable optical display, since the kinetic energy of the fraction of the envelope impacting the surface of the companion during a few days was only $\sim 10^{39}$ ergs.

15.2.2 Spectra

The earliest optical spectra were nearly featureless, with several broad peaks and valleys (Fig. 15.3). The early continuum corresponded to a temperature of 15,000 K. By the time of the first optical minimum, P Cygni profiles of the Balmer lines were conspicuous and indicated velocities as high as 15,000 km s^{-1}. Hα emission became strong, with a flat-topped profile. At 18 days, the UV spectra were relatively featureless, indicating veiling of the photospheric spectrum by emission from CSI (Jeffery et al. 1994; Fig. 15.4).

The optical spectra changed dramatically 25 days after outburst. The most distinct new feature was the double-peaked nature of the Hα line. Filippenko et al. (1993) recognized that strong P Cygni lines of He I were developing and that the dip in the Hα emission was absorption by He I λ6678 (Figs. 15.3 and 15.5). The optical He I lines arise from metastable levels near 20 eV and thus require nonthermal excitation from products of the ^{56}Ni decay chain (Sect. 16.4.3). Analysis of the long-term spectral evolution of SN 1993J was presented by Matheson et al. (2000a, 2000c; see also Lewis et al. 1994).

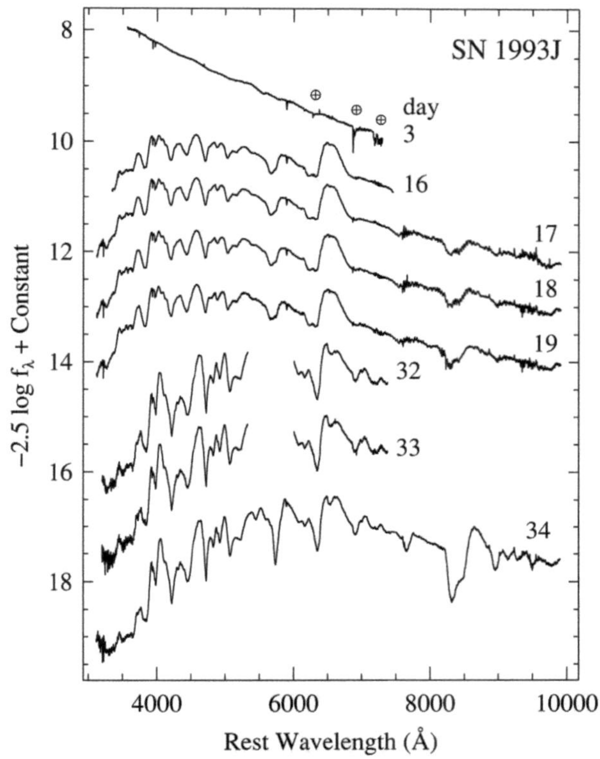

Fig. 15.3 Early spectra of SN 1993J. Epochs are days after explosion. The spectra are shifted vertically for clarity. From "Optical Spectroscopy of Supernova 1993J During Its First 2500 Days" (Matheson et al. 2000a). © AAS. Reproduced with permission

In the late nebular phase, lines in the UV were broad, with box-like shapes (Fig. 15.6). They were deduced to arise in the CDS formed at the contact discontinuity between the forward and reverse shocks (Fransson et al. 2005). The shapes of the lines implied inner and outer velocities of the CDS of about 7000 and 10,000 km s^{-1}. The relative line strengths indicated a strong nitrogen enrichment of N/C $\simeq$ 12.4.

15.2.3 Polarization and Asymmetry

SN 1993J was the first SN IIb to be studied with spectropolarimetry (Trammell et al. 1993; Doroshenko et al. 1995; Höflich et al. 1996; Tran et al. 1997). These works inferred different levels of polarization and its evolution. The conclusions depended on the correction for interstellar polarization (ISP) and probably were influenced by contamination of the Hα line with He I λ6678. Trammell et al. found interstellar polarization $P_{ISP} = 1.6\%$ at rotation angle $\chi = 49°$, Doroshenko et al.

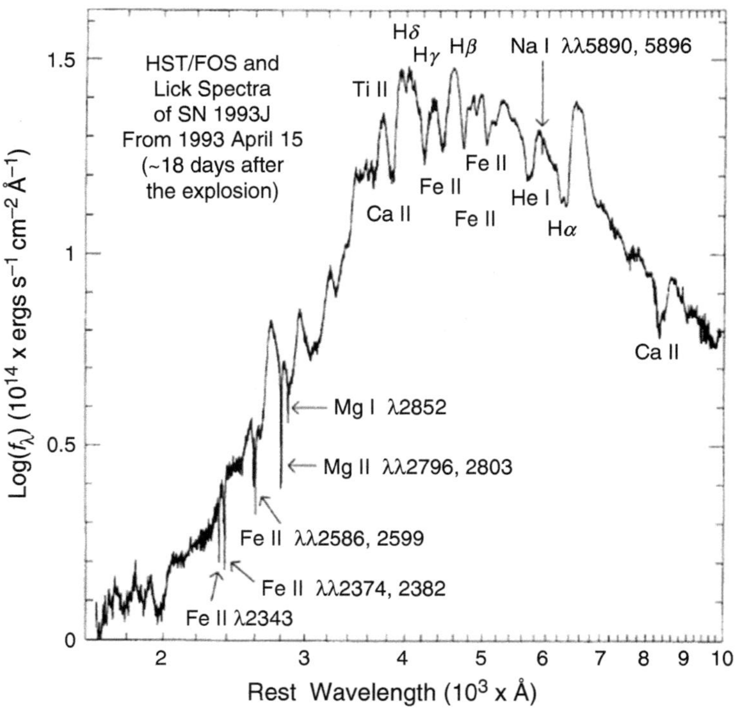

Fig. 15.4 UV and optical spectrum of SN 1993J 18 days after explosion. From "A Hubble Space Telescope ultraviolet spectrum of SN 1993J" (Jeffery et al. 1994). © AAS. Reproduced with permission

found $P_{ISP} = 0.2\%$ at $\chi = 0°$, and Tran et al. found $P_{ISP} = 0.63\%$ at $\chi = 171°$. Trammell et al. and Tran et al. assumed that Hα was intrinsically unpolarized, but later work on SN 1991ig (Maund et al. 2007a,b,c) showed that while Hα itself may be unpolarized, the He I λ6678 line can be polarized, so the assumption that the observed, blended feature is unpolarized in SN 1993J is likely to be incorrect. A technique to estimate ISP that was suggested later in the context of SN Ia (Howell et al. 2001) was to examine the level of polarization in the blue, where, for SN Ia, the continuum is a quasi-continuum formed by many blended, depolarizing lines of Fe II. Applying this technique to the data on SN 1993J gives $P_{ISP} \simeq 0.6$, consistent with Tran et al, but inconsistent with Trammell et al. and Doroshenko et al. On the other hand, the same technique suggests that the polarization angle of the ISP was near 20°, in disagreement with that of Tran et al. If the angle was 20°, then the interpretation of the polarization and its evolution in SN 1993J is drastically altered. A caveat is that the notion of depolarizing in the blue has not been rigorously tested, and it may depend on which portion of the blue spectrum is used to determine the ISP (Sect. 15.3.5).

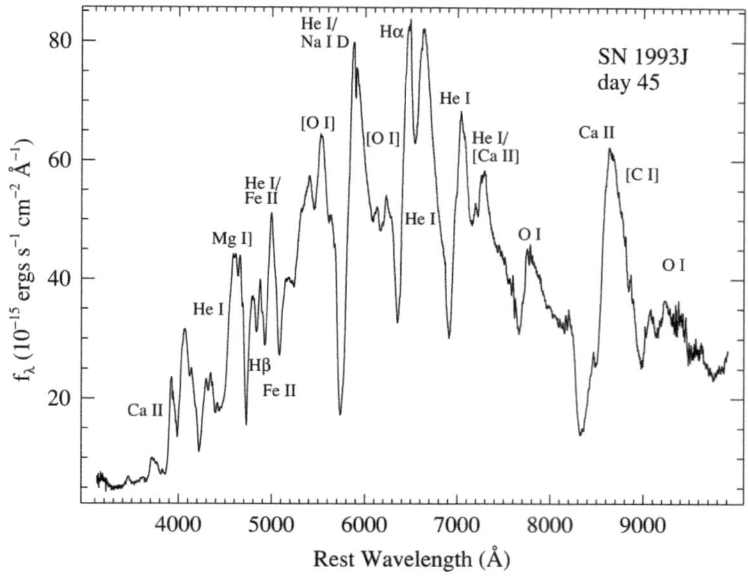

Fig. 15.5 Spectrum of SN 1993J 45 days after explosion, showing strong P Cygni He I lines. From "Optical Spectroscopy of Supernova 1993J During Its First 2500 Days" (Matheson et al. 2000a). © AAS. Reproduced with permission

If the ISP was about $P_{ISP} = 0.6\%$ at $\chi = 20°$, the polarization evolved differently than has been discussed in the literature. With the revised ISP suggested here, the early data were polarized with $P \sim 0.5\%$ and there was a large change in polarization angle between the time of the light-curve minimum and 23 days after explosion (about 2 days after primary V-band maximum), as well as between 23 days and 26 days after explosion, when the photosphere receded into the helium. The continuum polarization at the later epoch, 49 days after explosion, would be rather low. Tran et al. argued that the rotation of the position angle across line features in their latest data showed that the emission lines of He I, Fe II, [O I], and H were all intrinsically polarized with position angles that differed from that of the continuum and from each other (although note that a forbidden line is unlikely to be polarized). That different species can have different position angles has been noted for other supernovae.

The progenitor may have filled its Roche lobe at the time of the explosion and thus the hydrogen envelope may have been deformed with axis ratio about 3:2 (Woosley et al. 1994a). This would have implications for the polarization, as would asymmetry in the underlying explosion.

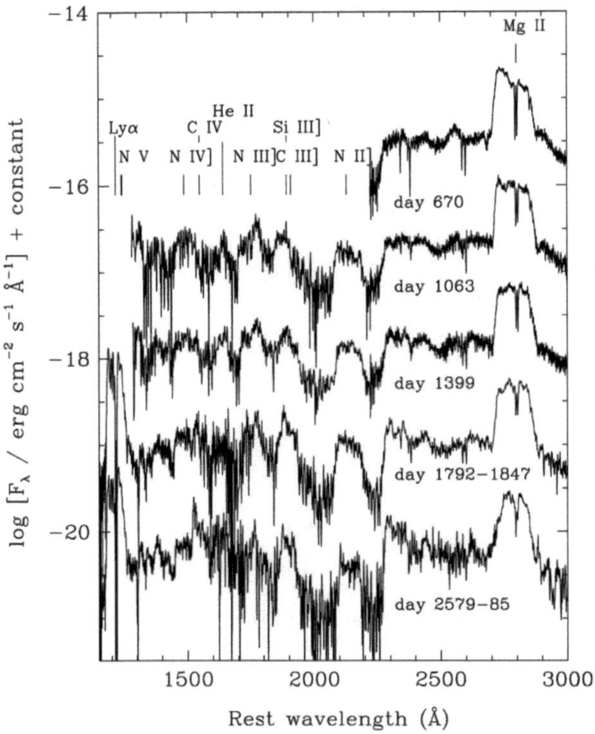

Fig. 15.6 Late UV spectra of SN 1993J. The spectra are shifted vertically for clarity. From "Hubble Space Telescope and Ground-based Observations of SN 1993J and SN 1998S: CNO Processing in the Progenitors" (Fransson et al. 2005). © AAS. Reproduced with permission

15.2.4 Circumstellar Interaction

Three days after explosion, high-resolution spectra showed narrow emissions of Hα, He II λ4686, and high-ionization lines such as [Fe X] λ6374 and [Fe XIV] λ5303; these lines disappeared within a week (Benetti et al. 1994b). The narrow width, high flux, prompt appearance, and rapid disappearance showed that the lines arose in slowly-moving, dense material close to the progenitor, perhaps a stellar wind characteristic of a late-type supergiant. The velocity widths of the narrow H and He II lines were about 170 km s^{-1}, somewhat high for an RSG wind but perhaps affected by the hotter companion star. The presence of the high-ionization lines showed that some CSM was close enough to the exploding star to be ionized and heated by the UV and X-rays from shock breakout. Evidence of CSI continued for well over a decade (Zhang et al. 2004; Fransson et al. 2005).

SN 1993J showed strong radio emission within days of outburst (Fig. 6.2). VLBI observations resolved the radio emission, a campaign that continues to this day (Bartel et al. 1994, 2007; van Dyk et al. 1994; Marcaide et al. 2009a; Martí-Vidal et al. 2011b). Despite the clear evidence for asymmetry of the ejecta from spectropolarimetry (Sect. 15.2.3), the radio image was and remains quite circular. This must reflect the distribution of the CSM.

Analysis of the radio observations found that a fit with a synchrotron spectrum modulated by external free–free absorption and synchrotron self-absorption was preferred to free–free absorption alone. A CSM with $\rho \propto r^{-2}$ was found to be adequate to account for the radio and perhaps the X-ray emission (Fransson and Björnsson 1998). With that assumption, the presupernova mass-loss rate was $\lesssim 10^{-5}\,M_\odot\,y^{-1}$ (Björnsson and Lundqvist 2014). There have been suggestions of an evolution in the radio opacity of the ejecta material, a radial decline in the magnetic field within the radiating region, and a changing radial density profile of the CSM.

In the first few weeks, *ROSAT*, *ASCA*, and *OSSE* observed a hard free–free X-ray spectrum with $kT \simeq 100$ keV (Zimmermann et al. 1994; Kohmura et al. 1994). Six months later, *ASCA* observations showed that the X-ray spectrum had softened (Uno et al. 2002), consistent with a decrease in the optical depth in the expanding cool shell, which allowed more of the soft X-rays to escape. The observations were roughly consistent with the forward shock encountering a previously ejected RSG envelope, or perhaps a common envelope, not necessarily a simple $\rho \propto r^{-2}$ wind. According to models, during the first few months most of the observed X-ray emission arose from the Rayleigh–Taylor unstable region associated with the contact discontinuity (Björnsson 2015). This result is consistent with the circular shape of the image revealed by VLBI, but does not obviously follow from a complex binary evolution of the progenitor system (Sect. 15.2.5).

The radio and X-ray light curves both suggest a sudden increase in the progenitor's mass-loss rate ~8000 years before the explosion (Weiler et al. 2007; Martí-Vidal et al. 2011b). Emission in Hα was comparable to the X-ray emission, perhaps due to clumps of neutral hydrogen in the hot X-ray emitting gas (Chandra et al. 2009a).

15.2.5 Evolutionary Models

Depending on the efficiency of mass transfer and other aspects of stellar evolution, the evidence suggests that the members of the presupernova binary system had similar ZAMS masses of about 15 and 14 $M_\odot$ and an initial orbital period of 6 years (Maund et al. 2004; Stancliffe and Eldridge 2009; Claeys et al. 2011). Evolutionary models suggest that at the time of explosion the supernova progenitor had a mass of 5.4 $M_\odot$ with a post-helium-burning core of 5.1 $M_\odot$. These masses are somewhat higher than some of the earlier evolutionary models predicted. At the time of explosion, the mass of the secondary was predicted to be about 22 $M_{\odot}$ and the orbital period to be about 25 years. The surviving companion should be helium-rich from the mass transfer.

There is a concern (Hoflich et al. 1993) that a low-mass progenitor ($M_{ZAMS} \lesssim 20\ M_{\odot}$), which needs binary mass transfer in order to lose a large fraction of its envelope, may not be able to maintain a sufficiently strong wind during its final evolutionary phase to account for the radio and X-ray observations (Sect. 15.2.4). Binary progenitor models that match the properties of the K-supergiant progenitor of SN 1993J have difficulty accounting for the B-type companion (Stancliffe and Eldridge 2009). We return to these issues in Sect. 15.4.1.

15.2.6 Model Light Curves

Information on the progenitor of SN 1993J from prediscovery images was complemented by extensive observations and modeling of the light-curve and spectral evolution. The light curve depends on the mass and composition of the ejecta, the opacity, and the mass and distribution of ^{56}Ni. In general, as the ejecta turn optically thin the light curve falls and joins the decay powered by radioactivity. Although the amount of ^{56}Ni (as well as a number of other parameters) affects the shape and time of maxima, the principal constraint on M_{Ni} is the amplitude of the second peak of the bolometric light curve (Fig. 15.7).

Fitting the bolometric and color curves depends on the treatment of the atmosphere. Inward mixing of hydrogen enhances the interior opacity and maintains the photosphere at larger radii and cooler temperatures for a longer time. Outward mixing of ^{56}Ni causes the light curve to decline more steeply than the ^{56}Co decay rate after the peak, because γ-ray transparency sets in earlier. With ^{56}Ni mixing, the time of the light-curve minimum is reduced, since heating due to radioactivity starts earlier. The degree of mixing of hydrogen, ^{56}Ni, and other elements is thus constrained by the photometric and spectral evolution.

The "mixing" of ^{56}Ni in published models is done in an ad hoc manner, with the presumption that it mimics the effects of Rayleigh–Taylor mixing. Given the evidence for bulk asymmetries in SN 1993J and other core-collapse supernovae, the mixing of the ^{56}Ni might be in "fingers" or "plumes," with quantitative if not qualitative effects on the interpretation of the data. In any case, care must be taken to avoid invoking a "missing opacity," because mixing of H and ^{56}Ni will act to increase the local opacity.

Light-curve models suggested a progenitor that had lost almost all of its hydrogen envelope to a close binary companion (Podsiadlowski et al. 1993; Nomoto et al. 1993; Utrobin 1994; Woosley et al. 1994a). Early conclusions that SN 1993J occurred in a binary system were confirmed, as outlined in Sect. 15.2.5. The mass of ejected ^{56}Ni was comparable to that of SN 1987A, and the compact remnant was likely to be a neutron star. The lack of detection of an X-ray synchrotron nebula suggested the absence of a bright pulsar (Ramirez-Ruiz and Serenelli 2004).

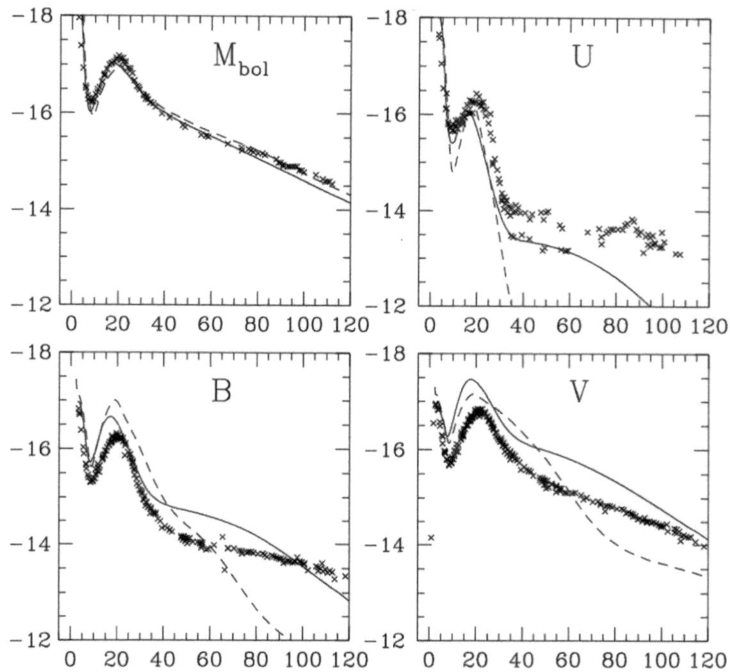

Fig. 15.7 Light curves of SN 1993J compared to model calculations with two different codes, illustrating that light curves in individual bands are more difficult to fit than bolometric light curves. From "A Comparative Modeling of Supernova 1993J" (Blinnikov et al. 1998). © AAS. Reproduced with permission

Within this framework, most light curve models converged to a progenitor with an initial radius of several hundred $R_\odot$, an ejecta mass of 2–3 $M_\odot$, and an outer envelope of $\lesssim$0.5 $M_\odot$ with a hydrogen mass fraction less than 0.5. The kinetic energy was assumed or derived to be $\gtrsim$1 B. Allowing for a neutron star of 1.5 $M_\odot$, the progenitor mass was deduced to be about 4 $M_\odot$, somewhat less than deduced by evolution models. The mass of the helium core, about 3.5 $M_\odot$, implied a ZAMS mass of 15 $M_\odot$. The latter is quite close to that deduced by Maund et al. (2004) from archival observations, but the core mass is somewhat lower. More massive light-curve models were considered (Hoflich et al. 1993), but ruled out because they required large reddening that was not observed.

15.2.7 Models of Photospheric Spectra

Supernova atmosphere models of this era, having been stimulated by SN 1987A, were becoming more sophisticated. Several groups applied NLTE models (Baron et al. 1993, 1994, 1995; Utrobin 1996).

The evolution of the location of the photosphere and the resultant implications for the predicted photometric and spectral evolution depend on details of the model. Atmosphere models assuming homologous expansion and an ejecta density profile following either a power law or an exponential decrease implied that the relatively narrow lines were not due to low photospheric velocity, but instead to forming in an atmosphere of steep density profile (Baron et al. 1995). Given the evidence for CSI, it is not clear that homology is a good approximation for the early spectra.

As the ejecta expanded and cooled below 6000 K, hydrogen recombined, decreasing the opacity and allowing the photosphere to recede. For a massive hydrogen envelope, this phase corresponds to the plateau of constant luminosity in SN IIP. In some of the low-mass models for SN 1993J, this phase is completely incorporated in the first decline. In other models, the recombination phase is postponed because of heating by radioactive decay of ^{56}Ni and ^{56}Co. In all published models, the light curve turns up because of heating by radioactive decay.

In models with little or no outward mixing of ^{56}Ni, the photosphere begins to recede shortly after shock breakout and the first minimum occurs when the photosphere reaches the base of the hydrogen envelope (Wheeler et al. 1993; Woosley et al. 1994a). Without a secondary energy source, the light curve would continue to plummet. With little or no mixing of ^{56}Ni, the diffusion wave from the radioactive heating arrives at about the time the photosphere recedes through the outer hydrogen envelope and the light curve rises to the second maximum. The underlying helium layer is already rather cool and the opacity is low, so the photosphere recedes rapidly through the helium and into the heavy-element core (Wheeler et al. 1993). With ad hoc mixing of ^{56}Ni into the helium layers, the helium is heated and ionized, which retards the recession of the photosphere through the helium until the second peak. This latter behavior may be more consistent with the color evolution and the onset of strong He I lines after the second peak, but the evolution of the He I lines constrains excessive ^{56}Ni mixing. If the ^{56}Ni were mixed throughout the helium layer, then even hydrogen recombination would be delayed due to heating until near the second peak. This delayed recession of the hydrogen recombination probably would not be consistent with the spectral evolution (Shigeyama et al. 1994).

The evolution of the spectra during the first decline represents a challenge. The hydrogen abundance was low, with mass fraction near 0.1. The constraints on the reddening, continuum shape, and colors all indicated that the continuum was that of a well-defined, cooling blackbody during the decline. For too low a temperature, the H lines were predicted by atmosphere models to be too strong. Shortly after the first minimum and after the development of strong Hα emission, fits to the spectra indicated that a temperature of 8000 K was allowable, but that 7000 K was too low (Baron et al. 1993). Contemporaneous photometry, however, yielded a temperature of 6500 ± 500 K (Ray et al. 1993). The very early spectra need more consideration.

The photospheric velocity remained nearly constant for a week after the optical light-curve minimum. This might be consistent with low-mass models for which the photosphere was in the helium core during this time, although the observed value of the photospheric velocity, 11,000 km s^{-1}, was rather high for such a picture. In

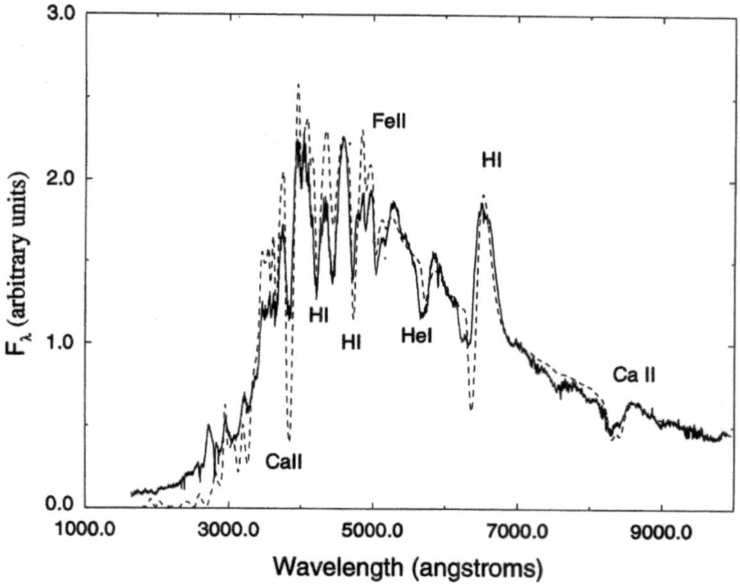

Fig. 15.8 The 18-day UV and optical spectrum of SN 1993J compared to a PHOENIX synthetic spectrum. From "Non-LTE spectral analysis and model constraints on SN 1993J" (Baron et al. 1995). © AAS. Reproduced with permission

addition, it is not clear that having a photosphere in the helium core was consistent with the spectra in this phase; it was difficult to produce the strong net emission in Hα (Baron et al. 1993; Fig. 15.8). It may be that the Hα profile was due to altered level populations produced by radiation from CSI or by a flatter density profile external to the steeply falling profile near the photosphere.

Most model light curves were computed with flux-limited diffusion. These models tended to turn optically thin shortly after the second peak, and the UVOIR luminosity dropped too rapidly. Atmosphere models (Woosley et al. 1994a) smoothed out the sudden decrease in luminosity that plagued low-mass models when the photosphere reached the bottom of the metal core.

Atmosphere models incorporating nonlocal deposition of γ-rays and associated NLTE excitation of helium appropriate to a phase of about 30–40 days after shock breakout were able to account for the development of strong He I lines in the optical and near-IR (Swartz et al. 1993a). In these models, He I lines become visible after (1) the outer hydrogen-helium layer has recombined to expose the helium layer; (2) the onset of substantial γ-ray deposition in the helium layer; and (3) the density of free electrons in the layer has declined sufficiently due to expansion and cooling to eliminate veiling of the He I lines by electron scattering.

In the NLTE models of Swartz et al. (1993a) with a nominal explosion energy of 1 B, an ejecta mass of 4.5 $M_\odot$ proved too large and that of 1.8 $M_\odot$ too small to match the expansion velocities implied by the observed spectra. An ejecta mass of 2.5 $M_\odot$

worked well. Of this mass, an envelope with 0.4 $M_\odot$ of helium contaminated with about 10% hydrogen by mass gave a good spectral fit. These atmosphere models showed that if hydrogen were homogeneously mixed with all of the helium, the He I lines would be veiled and absent from the spectrum. The portion of the ejecta from the core that formed the P Cygni He I lines must therefore have been substantially more hydrogen deficient than the 10% by mass in the outer layers of the model. These atmosphere models constrained the mass of ^{56}Ni to be $\lesssim$0.1 $M_\odot$, so as not to completely ionize hydrogen at 40 days past explosion. The ^{56}Ni could not be mixed too far out into the ejecta or hydrogen would again be excessively ionized. Homogeneous mixing of ^{56}Ni through the helium core seems to be excessive by this criterion. The models predicted that H and He I lines should have been visible for ~100 days, but should have disappeared by ~200 days (Swartz et al. 1993a), as confirmed by observations. Later observations showed broad Hα, but that feature resulted from CSI.

The NLTE spectral models of SN 1993J showed that even a small amount of hydrogen, ~0.1 $M_\odot$, would show up in the spectrum. While not designed to apply directly to the conditions of SN Ib and SN Ic, these models are relevant to the interpretation of such events when H and He I lines are not prominently seen (Sects. 16.4.3 and 17.5.3).

15.2.8 Models of Nebular Spectra

A detailed NLTE study of nebular-phase spectra favored a model with a helium core of 3.2 $M_\odot$ and 0.08 $M_\odot$ of ^{56}Ni (Houck and Fransson 1996). This helium-core mass is somewhat lower than the 4 $M_\odot$ favored by Swartz et al. (1993a) and derived from models of the evolution (Sect. 15.2.5) and light curves (Sect. 15.2.1). Agreement was improved when some iron-rich material was macroscopically mixed outward to a velocity of at least 3000 km s^{-1} and some of the oxygen-rich material mixed inward to about 1000 km s^{-1}. Irregularities in spectral features were due to line blending rather than asymmetries, e.g., the profile of [O I] $\lambda\lambda$6300,6364 was affected by scattering by Hα in the hydrogen envelope. The mixing invoked by Houck and Fransson was done in the context of spherical models and probably needs to be revisited with multidimensional models, given the evidence for polarization of SN 1993J and similar supernovae. Houck and Fransson also concluded that the outer envelope had a mass of 0.2–0.4 $M_\odot$, that the bulk of this mass was ejected with velocity between 8500 and 10,000 km s^{-1}, and that this mass was enriched in helium. These models were consistent with a mass fraction of hydrogen of $\lesssim$0.4, and a total hydrogen mass of $\lesssim$0.2 $M_\odot$. After a year, an additional power source was necessary to reproduce the observed Hα luminosity, most likely X-rays from CSI.

15.3 Synthesis of SN IIb Characteristics

In addition to SN 1993J, well-observed SN IIb include SN 1996cb (Qiu et al. 1999), SN 2008ax (Taubenberger et al. 2011a,b), SN 2010as (Folatelli et al. 2014b), SN 2011dh (Sahu et al. 2013), SN 2011fu (Morales-Garoffolo et al. 2015), SN 2011hs (Bufano et al. 2014), and SN 2013df (Morales-Garoffolo et al. 2014).

Cas A (Sect. 7.4.1), known to be the remnant of an SN IIb, shows intriguing complexity. No evidence has been found for a surviving companion star in Cas A. One possibility is that a binary progenitor system experienced a common-envelope phase and a merger prior to explosion.

15.3.1 SN IIb Subtypes

There may be two distinct categories of SN IIb progenitors: extended and compact. SN 1993J, SN 2011dh, SN 2011hs, and SN 2013df are examples of the former; SN 1996cb, SN 2001ig, SN 2003bg, SN 2008ax, and SN 2008bo may be examples of the latter. The physical dividing line may be the mass of the helium-rich, hydrogen-contaminated envelope that can support an extended structure—a few tenths of a solar mass. The compact category may be characterized by a low-density, high-velocity wind, weak optical emission associated with shock heating, weak Hα emission in the nebular phase, rapidly evolving radio emission, and X-rays dominated by nonthermal processes (Chevalier and Soderberg 2010). Chevalier and Soderberg proposed calling the compact SN IIb events Type cIIb and the extended ones Type eIIb (see also Qiu et al. 1999). An important step would be to compute spectral models to see if the low-mass compact envelopes give the observed spectral evolution. A broader issue is whether there are two distinct categories of SN IIb or a continuum of properties.

15.3.2 Light Curves

Very few SN IIb have been discovered early enough to probe the fireball phase, not to mention breakout. The shapes of the light curves around maximum light are fairly homogeneous (Fig. 15.9). There is a paucity of late-time light curves where an interesting diversity may be revealed. At several hundred days after explosion, SN 1993J declined more rapidly than the ^{56}Co decay rate, indicating that neither γ-rays nor positrons were being completely trapped. Some other SN IIb decay in a manner similar to that of SN 1993J. SN 1996cb followed SN 1993J fairly closely for a month or so after maximum light, but may then have flattened to a slope more consistent with ^{56}Co decay. SN 2002jd resembled SN 1993J near the peak, but then declined more rapidly a month or so after maximum. The similarity of the light curves near maximum and the dispersion in the late-time tails suggests a breakdown

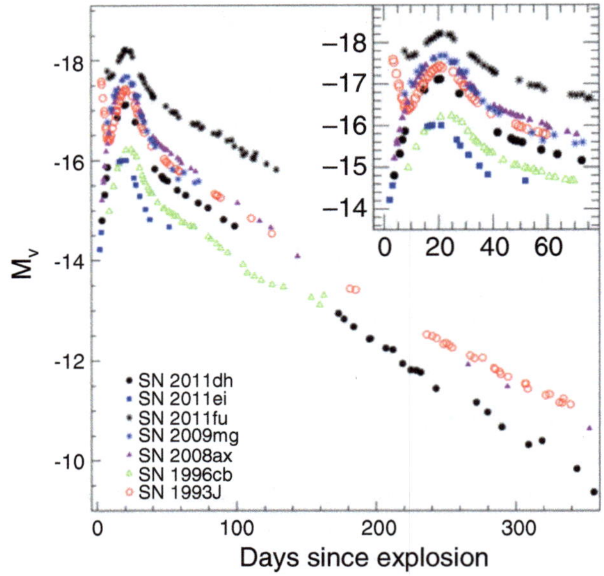

Fig. 15.9 Comparison of light curves of SN IIb. Figure 6 "One year of monitoring of the Type IIb supernova SN 2011dh" from Sahu et al. (2013), by permission of Oxford University Press on behalf of the Royal Astronomical Society

in simple constant-opacity diffusion and γ-ray deposition models, perhaps due to the effects of recombination (Nagy and Vinkó 2016; Fig 15.13) or to asymmetries in the explosion. This notion is explored more thoroughly in Sect. 17.5.2. Especially for the extended subtypes, recombination may affect the peak amplitude and shape.

A prediscovery limit to the brightness of SN 2008ax constrained the time of shock breakout to within a few hours (Pastorello et al. 2008a). The light curve showed only a single peak in the optical, but signs of the fireball were seen in the UV, where the light curve went through a sharp minimum 4 days after explosion (Roming et al. 2009). This suppression of the fireball phase is consistent with a compact progenitor. SN 2011dh evolved rapidly photometrically. There was some evidence of the fireball in optical bands, and a minimum in the U band about a day after discovery showed the tail end of that phase (Tsvetkov et al. 2012). Near-IR emission contributed about 30% of the UVOIR flux and the UV contribution was negligible (Marion et al. 2014).

15.3.3 Spectra

Hints of He I lines in the optical spectra of SN 2008ax could be seen as early as 11 days after explosion, when He I lines already were prominent in near-IR spectra

(Pastorello et al. 2008a; Taubenberger et al. 2011). In the early phases, a high-velocity Ca II feature was observed (Taubenberger et al. 2011; also in SN 2007Y), as in some SN Ia and some SN Ib/c. Late-time near-IR spectra showed double-peaked He I lines that suggested some asymmetry.

SN 2011dh was a classic SN IIb (Sahu et al. 2013; Ergon et al. 2014), although the UV flux was low compared to other SN IIb due to line blanketing by Ti II and Co II (Marion et al. 2014). The H lines formed in a region that expanded about $4000\,\mathrm{km\,s^{-1}}$ more rapidly than the layers in which the He I lines formed. SN IIb have traditionally been identified by the observability of He I in optical spectra. Near-IR spectra revealed the presence of He I in SN 2011dh 3 days before the B-band light curve reached maximum, and prior to evidence for He I in optical spectra (Marion et al. 2014; Fig. 15.10). Observations in the near-IR may be a more powerful means of revealing He I.

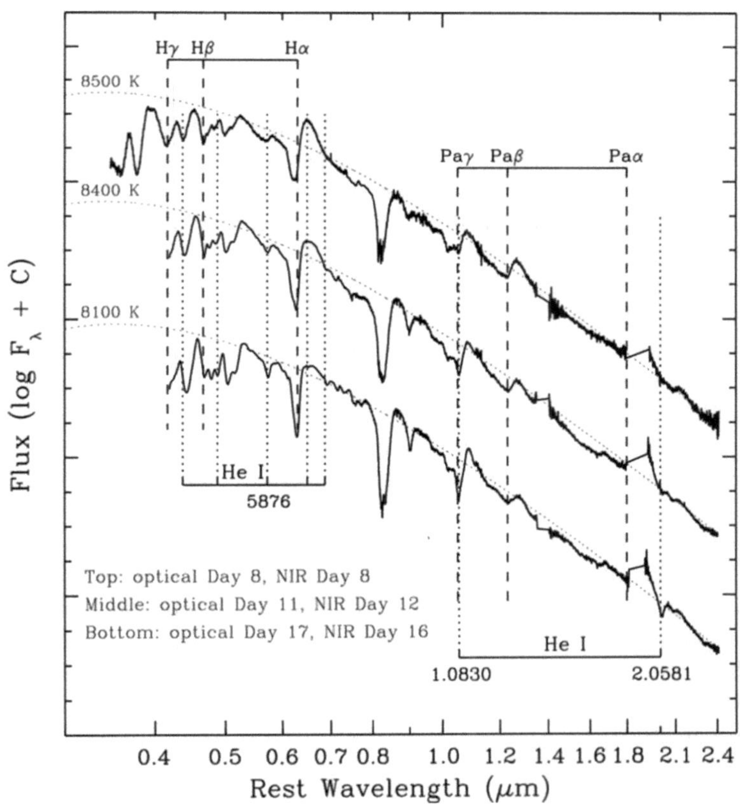

Fig. 15.10 Early optical and near-IR spectra of the Type IIb SN 2011dh. In the near-IR, He I lines were present at day 16 and perhaps at day 12. From "Type IIb Supernova SN 2011dh: Spectra and Photometry from the Ultraviolet to the Near-infrared" (Marion et al. 2014). © AAS. Reproduced with permission

SN 2010as, regarded as a Type IIb by Folatelli et al. (2014b), as well as several other stripped-envelope events, showed low expansion velocity, $\lesssim 8000\,\mathrm{km\,s^{-1}}$ at maximum light and even before. A flat velocity evolution during the weeks after maximum suggested the presence of a dense shell in the ejecta. It is not clear that these SN 2010as-like events should be classified Type IIb rather than Type Ib. As in some SN Ib at early times, Hα appeared in absorption, but Balmer lines did not dominate the early spectra as they do in typical SN IIb.

Spectra of SN 2013cu obtained only 15 h after explosion showed Wolf–Rayet-like features, prompting the suggestion that the progenitor was a WN(h) star (Gal-Yam et al. 2014). The line widths, however, were $\lesssim 100\,\mathrm{km\,s^{-1}}$, much lower than a typical velocity of a Wolf–Rayet wind and more characteristic of a supergiant progenitor (Groh 2014; Shivvers et al. 2015; Smith et al. 2015).

15.3.4 Circumstellar Interaction

In addition to SN 1993J, many SN IIb show signs of CSI. SN 2003bg was a bright radio source. The radio light curve was similar to that of SN 2001ig, suggesting a common progenitor evolution with quasi-periodic mass-loss episodes shortly before explosion (Soderberg et al. 2006a). A fading X-ray source and a rather weak radio source associated with SN 2008ax suggested interaction with a dilute CSM (Roming et al. 2009). From radio and X-ray emission of SN 2013df, the presupernova mass-loss rate (for a wind velocity of $10\,\mathrm{km\,s^{-1}}$) was estimated to be $\lesssim 10^{-4}\,M_\odot\,\mathrm{y}^{-1}$ (Kamble et al. 2016). A similar rate was inferred from late optical spectra of SN 2013df, which developed broad, flat-topped Hα emission and strongly resembled the late spectra of SN 1993J (Maeda et al. 2015). UV spectra of a small sample of SN IIb showed strong diversity in both continuum levels and line features, attributed to differences in the strengths of CSI (Ben-Ami et al. 2015).

SN 2011dh was the youngest radio-loud supernova yet detected, only 14 days after discovery (Martí-Vidal et al. 2011c; Krauss et al. 2012; Bietenholz et al. 2012), and was also an X-ray source (Soderberg et al. 2012; Campana and Immler 2012). From thermal X-ray emission 500 days after explosion, the presupernova mass-loss rate was estimated to be $\sim 3 \times 10^{-6}\,M_\odot\,\mathrm{y}^{-1}$ during the final 1300 years before explosion, consistent with the yellow supergiant progenitor. Such mass loss is insufficient to have expelled most of the hydrogen envelope; that apparently occurred by binary mass transfer some time earlier than 1300 years before explosion (Maeda et al. 2014a).

15.3.5 Polarization and Asymmetry

The complex and still uncertain nature of the polarization of SN 1993J was discussed in Sect. 15.2.3. The spectroscopic and spectropolarimetric behavior of

SN 1996cb was similar to that of SN 1993J. Both had sharp polarization changes across Hα and He I λ5876 suggesting an intrinsic polarization of $\sim$1.5% (Wang et al. 2001). Wang et al. argued that the polarization for SN 1996cb increased with time, suggesting that the asymmetries were stronger deeper within the ejecta. The similar polarization properties of SN 1993J and SN 1996cb suggested that the two events shared a similar geometry, including orientation to the observer. Wang et al. suggested that SN IIb might require a special viewing angle, and could appear as a different type, perhaps SN Ib or SN Ic, if viewed from a different angle. A possible circumstance might be if the hydrogen in a disk could only be seen in absorption if the disk is edge-on.

The Type IIb SN 2001ig was observed at multiple epochs, tightly constraining the ISP (Maund et al. 2007a).There was strong depolarization by blended Fe II lines in the blue that gave a consistent ISP, but only in the restricted wavelength range 4600–4800 Å. The intrinsic polarization was low, $\sim$0.2%, 13 days after explosion, and there was a sharp increase in polarization by 31 days when the helium-rich, hydrogen-contaminated envelope became optically thin. The polarization angle rotated by 40° between these two epochs. The most highly polarized lines showed wavelength-dependent loop structures suggesting nonaxisymmetric geometry (Sect. 4.6). The polarization properties of SN 2001ig were similar to those of SN 1993J and SN 1996cb, although not identical. This suggested that all three events occurred in similar binary systems, but that they may not require identical observer angle, as suggested by Wang et al. (2001).

Spectropolarimetry was obtained 6 and 9 days after the explosion of SN 2008ax and again late in the nebular phase (Chornock et al. 2011). The early data showed that the Hα line was polarized at more than 3% and the continuum at about 0.6%. The lines of Hα, He I, and O I shared a common polarization angle with the continuum, while the Ca II and Fe II lines had different angles. Polarization in the nebular phase suggested deep asymmetries that were also indicated by the shapes of nebular lines in the flux spectra.

Seven epochs of spectropolarimetry of SN 2011dh were obtained from 9 days to about 3 months after the explosion (Mauerhan et al. 2015b; Fig. 15.11). The first epoch corresponded to the phase when the photosphere was still in the outer helium/hydrogen envelope. The continuum polarization was $P \simeq 0.5\%$ for the first 2 weeks after explosion and then decreased, contrary to the usual trend in which the continuum polarization increases with the depth of formation in core-collapse events. The amplitude and position angle varied across P Cygni features. The Hα and He I λ5876 features were most prominent about 30 days after explosion. Their position angles were approximately aligned with one another and with the earlier continuum. The features of O I and the Ca II near-IR triplet showed a different orientation and hence geometry.

To some extent, SN IIb may generally resemble the highly asymmetric SNR Cas A, the only SN IIb that we can directly image.

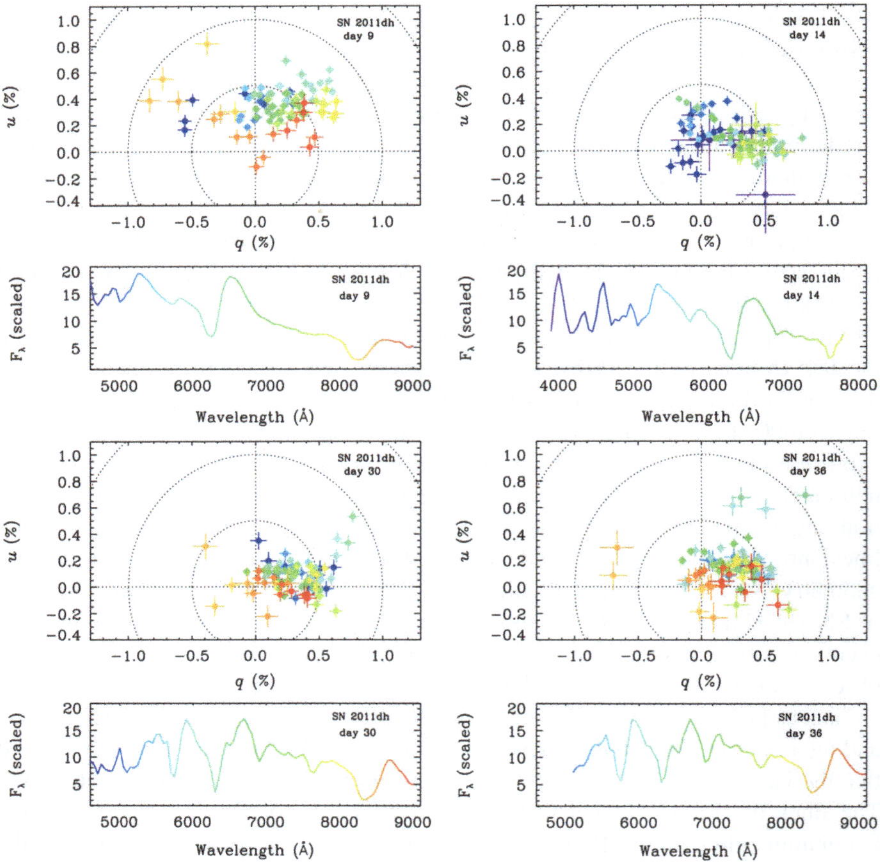

Fig. 15.11 Evolution in the Q/U plane of the polarization spectrum of SN 2011dh 9, 14, 30, and 36 days after explosion. *Colors* refer to the total flux spectra, as shown in the lower panels. Figure 3 "Spectropolarimetry of SN 2011dh in M51: geometric insights on a Type IIb supernova progenitor and explosion" from Mauerhan et al. (2015b), by permission of Oxford University Press on behalf of the Royal Astronomical Society

15.4 Models of SN IIb

15.4.1 Evolutionary Models

Along with observations that suggest a range of envelope masses and radii for SN IIb progenitors come a variety of constraints on the progenitor evolution. Current models of single stars with solar metallicity cannot produce the majority of stripped-envelope supernovae (See also Sects. 16.4.1 and 17.5.1). According to models, the final masses of wind-stripped cores are more than $\sim$10 $M_\odot$ (Eldridge and Tout 2004; Yoon et al. 2010; Georgy et al. 2009, 2012). This mass corresponds to observed

Wolf–Rayet stars and may account for some uncommon stripped-envelope events (Claeys et al. 2011; Groh et al. 2013), but probably not typical ones.

In addition, single massive stars are statistically improbable. More than 70% of all massive stars are in binary systems that are expected to undergo mass transfer (Sana et al. 2012). One third of these interactions may lead to merger of the two stars. It is thus likely a priori that binary evolution dominates the evolution of massive stars. Population synthesis studies have shown that stars evolving in close binary systems can produce a sufficient number of stripped-envelope progenitors (Podsiadlowski et al. 1992; De Donder and Vanbeveren 1998; Yoon et al. 2010; Eldridge et al. 2013). The issue is what sort of binary evolution produces the progenitor systems needed to account for SN IIb, SN Ib, SN Ic, and variations thereupon.

Models that focus on the primary star and do not consider common-envelope evolution, and merger models that retain substantial amounts of helium, more than 0.2 $M_\odot$, are unlikely to account for typical SN Ic (Sect. 17.5), but such models may give insight into the origin of SN IIb and SN Ib (Yoon et al. 2010; Eldridge et al. 2013). For stripped cores of less than 3 $M_\odot$, models predict the helium envelope to expand substantially during C/O core contraction or during core carbon burning (Case ABB or Case BB; Sect. 8.8). After mass loss, these low-mass models are nearly devoid of hydrogen and hence are candidates to be progenitors of SN Ib. For more massive cores and lower metallicity, models retain some hydrogen, $\lesssim 0.1$ $M_\odot$, after the first stage of mass transfer, making them candidates for SN IIb.

Binary evolution of moderately massive stars proceeds until the Roche lobe of the mass-losing star reaches the inner, helium-rich layers left behind by the retreating, convective, hydrogen-burning core. At this point, the envelope radius drops as further mass is lost, the envelope retreats inside the Roche lobe, and transfer stops. Podsiadlowski et al. (1993) argued that such transfer will result not in a typical M-type supergiant, but a K-type of about the right color and luminosity to account for the pre-explosion observations of SN 1993J.

In binary scenarios, the originally more massive primary star will have evolved off the main sequence, and the condition of the secondary at the time of primary explosion depends on the initial mass ratio and separation (Yoon et al. 2010; Claeys et al. 2011). Claeys et al. distinguished three different binary scenarios (Fig. 15.12) that may contribute to the formation of SN IIb. These scenarios depend on the initial mass ratio being ever closer to unity so that the two stars have ever more similar evolutionary timescales: (1) accretion that begins while the companion is still on the main sequence; (2) accretion that begins after the companion has left the main sequence, but before it has evolved up the giant branch (such a star is said to be in the Hertzsprung gap in the H–R diagram, a phase of rapid evolution so that few stars reside in this region of luminosity and effective temperature); and (3) accretion that begins when the companion is already itself an RSG. The last case requires an initial mass ratio very close to unity for the two stars to evolve on nearly the same timescale.

In the first case, the main-sequence accretor can adjust its internal structure so that it evolves as if it were a younger single star of increased mass. This leads most

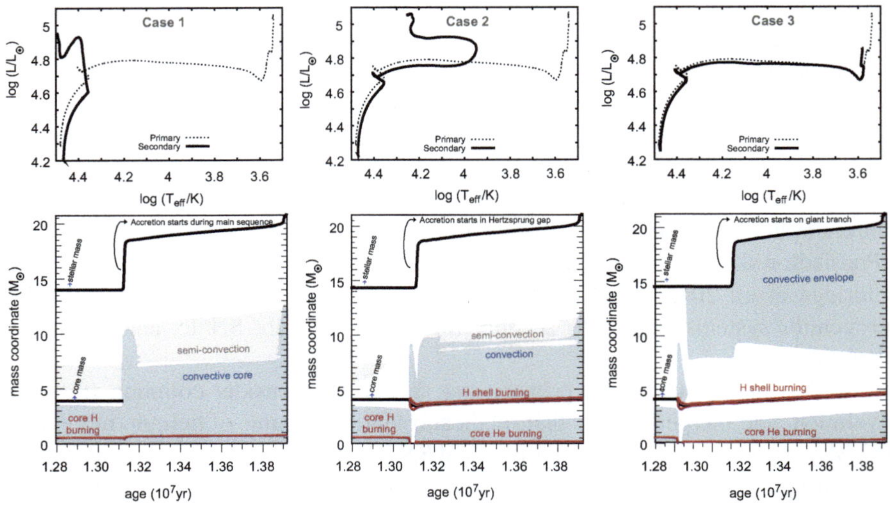

Fig. 15.12 Three binary evolution scenarios as discussed in the text. The *top row* shows evolutionary tracks in the HRD and the *bottom row* shows the evolution of the internal structure of the accreting star around the time that accretion begins. From "Binary progenitor models of type IIb supernovae" (Claeys et al. 2011) reproduced with permission, © ESO

naturally to an O-type companion. If the transfer begins when the companion has recently left the main sequence, the companion can become an overluminous B-type supergiant, as was the case for SN 1993J. If the transfer occurs when the companion is also a supergiant, the secondary will usually be an RSG when the primary explodes, although this circumstance may also lead to common-envelope evolution and rather different outcomes. Details depend on winds and the efficiency of accretion. With standard assumptions, Claeys et al. (2011) concluded that SN IIb should be only about 0.6% of all core-collapse events, a factor of at least five lower than inferred from observations.

The majority of SN IIb resulting from stable mass transfer are predicted to have O-type companions. Only a few percent of all SN IIb are predicted to have a B-type supergiant companion as did SN 1993J. As for many other "prototypes," SN 1993J may be atypical of the breed in some respects. Detecting O-type companions is more difficult than detecting B or K-type supergiant companions because of increased extinction at shorter wavelengths. If the initial mass ratio is sufficiently near unity (an infrequent but possible circumstance), the secondary can gain sufficient mass to explode first as a SN IIP, with the primary exploding later as a SN IIb. This possibility was raised for Cas A by Claeys et al. (2011). The possibility of common-envelope evolution, also suggested for Cas A, represents another complication. It is not clear that the environment of Cas A gives evidence for either an initial primary star or a common envelope.

A yellow supergiant was observed at the location of SN 2011dh. This star is anomalous for its surroundings if it evolved as a single star, but it may be consistent with models of binary-star evolution. Benvenuto et al. (2013) favored a model in which stars of ZAMS masses 16 and 10 $M_{\odot}$ evolved in a binary system. The primary star ended as a yellow supergiant and the secondary remained near the ZAMS at its increased mass. At the time of its explosion, the model primary contained only $\sim 5 \times 10^{-3}$ $M_{\odot}$ of hydrogen; it is not clear that this small mass would be consistent with the SN IIb spectral evolution.

15.4.2 Model Light Curves

Much effort has gone into modeling the light curves of SN 1993J (Sect. 15.2.6) and other SN IIb. For the few events that have early data, the light curves give information on the structure of the outer envelope, including the thin helium-rich, hydrogen-contaminated blanket that characterizes the type. The early light curve depends on the radius of the progenitor, but also on other aspects such as the extent of the outward transport of ^{56}Ni.

These considerations are illustrated by various models of SN 2011dh. From the early color evolution and the rather cool temperature indicated by the spectrum, Arcavi et al. (2011) deduced that the progenitor of SN 2011dh was relatively compact with a radius $\gtrsim R_{\odot}$. Soderberg et al. (2012) reached similar conclusions based on the nature of the CSM deduced from radio and X-ray observations and modeling of the fireball phase. From modeling the early light curve, however, Bersten et al. (2012) concluded that a progenitor with an extended radius of $\sim$100 $R_{\odot}$ was required, compatible with the actual yellow-supergiant progenitor. SN 2011dh is thus generally considered to fall in the category of extended-envelope SN IIb. Ergon et al. (2014) estimated a mass in the outer helium/hydrogen envelope of only 0.01–0.04 $M_{\odot}$, rather low compared to estimates for other SN IIb. From modeling the light curve observed to 2 years, the helium-core and ZAMS masses of SN 2011dh were inferred by Ergon et al. (2015) to be about 3 and 12 $M_{\odot}$. Modeling of SN 2011hs suggested an extended progenitor radius of 500–600 $R_{\odot}$, again putting it in the category of extended-envelope events, and helium core and ZAMS masses of 3–4 $M_{\odot}$ and 12–15 $M_{\odot}$, respectively (Bufano et al. 2014).

Analytic light-curve modeling of a sample of stripped-envelope events indicated that SN IIb have somewhat lower values of ejected mass, M_{Ni}, and kinetic energy than SN Ib/c, and that the ejected mass of nearly all stripped-envelope supernovae is too low for the single-star progenitor channel but consistent with the binary channel (Lyman et al. 2016). These models may not adequately account for recombination and asymmetry.

The light curves, spectra, and evolutionary models of SN IIb all demand that they retain helium/hydrogen envelopes of small mass, a few tenths of $M_{\odot}$, that may be extended or compact in radius. To produce an effective fit to the light curve, a proper treatment of the recombination effects that dominate SN IIP light curves may be necessary. Figure 15.13 shows a model fit to the bolometric light curve of SN 2013df, with and without the effects of recombination of hydrogen and

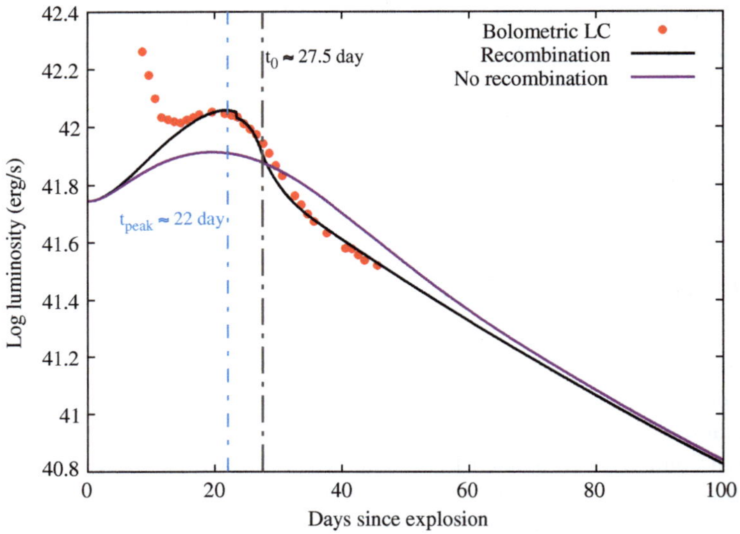

Fig. 15.13 Simple model of the bolometric light curve of the Type IIb SN 2013df incorporating the effects of hydrogen recombination in the manner of Arnett and Fu (1989). The parameter t_o is a diffusion time that depends on the adopted opacity in the ionized matter. Figure 10 "The continuing story of SN IIb 2013df: new optical and IR observations and analysis" adapted from Szalai et al. (2016), courtesy A. Nagy, by permission of Oxford University Press on behalf of the Royal Astronomical Society

helium taken into account. It is difficult to fit the peak/tail contrast in the absence of recombination. Recombination yields a smaller average effective opacity for a given total ejecta mass. A decent fit can be obtained if recombination effects are included near the peak and the effects of both γ-ray and positron deposition are included on the tail (Nagy and Vinkó 2016). Some general issues associated with modeling light curves of stripped-envelope supernovae are discussed in Sect. 17.5.2.

15.4.3 Models of Photospheric Spectra

The photospheric spectra of SN IIb other than SN 1993J have not been modeled extensively, but see Kumar et al. (2013) and Marion et al. (2014) for studies of line identifications in SN 2011fu and SN 2011dh, respectively. Hachinger et al. (2012a) compared models of spectra of the Type IIb SN 2008ax with those of the Type Ic SN 1994I and confirmed the results of Swartz et al. (1993b) that only about 0.1 $M_{\odot}$ of helium, and even less hydrogen, will produce recognizable features in model spectra. Models for SN Ib that also pertain to the early phases of SN IIb are discussed in Sects. 16.4.2 and 16.4.3.

15.4.4 Models of Nebular Spectra

Much of the modeling of the nebular-phase spectra of SN IIb has been done in the context of SN 1993J (Sect. 15.2.2), but see Jerkstrand et al. (2015a) on SN 2008ax and SN 2011dh in addition to SN 1993J. These authors concluded that the progenitors of all three SN IIb had ZAMS masses in the range 12–16 $M_\odot$ and are likely to have been in binaries. They also concluded that except in SN IIb such as SN 1993J that undergo strong late-time CSI, after 150 days any emission feature near the wavelength of Hα is dominated by [N II] $\lambda\lambda$6548,6583.

The absence of H or He I features in nebular spectra of SN Ib and SN Ic implies that SN Ib contain substantially less hydrogen than SN IIb and that SN Ic contain substantially less helium than either.

15.5 Summary

The prototypical, well-observed Type IIb event was SN 1993J. It's optical light curve exhibited a rise to a first peak in several days, a minimum a week later, and a rise to a second peak 3 weeks after explosion. The first peak was driven by the fireball phase that follows shock breakout and the second was powered by the ^{56}Ni decay chain. Optical spectra were nearly featureless for a few days, but within a week developed conspicuous H lines. A few weeks later, strong He I lines developed, with λ6678 absorption cutting into the peak of the fading Hα emission. Spectropolarimetry provided evidence of asymmetry. Evidence of CSI was provided by the late-time light curve and spectra as well as radio and X-ray emission. For a steady-state wind, the presupernova mass-loss rate was $\sim10^{-5}\,M_\odot\,y^{-1}$. Modeling of the optical observations suggested a progenitor radius of several hundred $R_\odot$, an ejected mass of 2–3 $M_\odot$, a kinetic energy of $\sim$1 B, and $M_{Ni} \simeq 0.08\,M_\odot$. The mass of ejected hydrogen was $\lesssim$0.1 $M_\odot$. The progenitor was seen from archival observations to be a type K0 Ia supergiant, and postexplosion images revealed a former binary companion, a B-type supergiant. Binary-evolution modeling indicated that the members of the progenitor binary system had similar ZAMS masses of about 15 and 14 $M_\odot$ and pre-explosion masses of about 4 and 22 $M_\odot$. The former companion should be helium-rich from the mass transfer; confirmation of this prediction remains for the future.

Observationally, SN IIb are reasonably homogeneous. A minority may have ejected masses high enough to be consistent with core-collapse explosions of Wolf–Rayet stars, which have ZAMS masses $\gtrsim$20 $M_\odot$ and pre-collapse masses of $\gtrsim$10 $M_\odot$. Typical SN IIb, however, have lower ejected masses and are core-collapse explosions of stars the envelopes of which were stripped during binary evolution. The ZAMS masses of the progenitors of SN IIb may be in the range of about 12–16 $M_\odot$, and their pre-explosion masses are about 3–5 $M_\odot$.

Open issues concern whether all SN IIb arise in binary systems; the mode of mass loss that leaves some hydrogen behind, thus distinguishing SN IIb from SN Ib (which may have wisps of hydrogen); the mass range or ranges that give rise to SN IIb; and a variety of interesting issues regarding interaction with CSM arising from the progenitor or its companion.

Much more work must be done on the early light curve and spectral modeling of SN IIb to incorporate the effects of CSI, recombination, and asymmetry. Forward and reverse shocks and the hard radiation from them affect the density and temperature distribution of the outer ejecta. The evolution near the first peak in SN 1993J and related events is unlikely to be homologous. The models with envelopes of $\lesssim 0.2\ M_\odot$ become homologous in a day or so, but increasing the mass of the envelope or including CSI delays the approach to homologous expansion. The weak lines in very early spectra of SN 1993J need more consideration. The spectra on the first decline are a particular challenge. The features tend to be of very low amplitude even though the temperatures are moderately low and simple models predict stronger lines. The expected low hydrogen abundance may play a role and the hydrogen may be excessively ionized by the shock-produced X-rays. The X-rays must not, however, heat the photosphere excessively since the observations show a well-defined Planckian continuum on the first light-curve peak of, for instance, SN 1993J. The H lines and other features begin to become distinct shortly before the first minimum in SN 1993J. This sets important constraints on the distribution of the density and composition, including the mass of ^{56}Ni and its mixing. The latter affects the location of the (possibly asymmetric) photosphere and hence the column depth that contributes to the strength of the lines.

The second peak of SN 1993J and many other SN IIb is similar in width to the peaks of SN Ia and SN Ib/c. Some SN Ib and SN Ic decline at the same rate on the tail as SN 1993J, despite their different spectral types and associated structure and composition. SN 1993J was predicted to decline more slowly on the tail than models of comparable mass with no extended envelopes, because the core/envelope interaction slows the expansion of the core and traps the γ-rays longer; nevertheless, the light curve of SN 1993J was even flatter on the tail than most models predicted, presumably due to CSI.

All the models of dynamics, light curves, and spectral evolution of SN IIb (certainly of SN 1993J) have been spherical with ad hoc mixing. Many of the features revealed by close-up study of Cas A may apply to SN IIb, and even core-collapse events in general. This subject is ripe for revisiting.

It would be useful to better understand any similarities and differences between the host galaxy environments of the SN IIb with extended and compact progenitor envelopes.

Chapter 16
Type Ib Supernovae

16.1 Introduction

What we now refer to as Type Ib and Type Ic supernovae were first noticed as events that were hydrogen deficient, making them Type I by spectral definition, but they did not show a strong Si II $\lambda 6355$ P Cygni feature (Bertola 1964; Bertola et al. 1965) that defined typical Type I events, which later came to be called Type Ia. Bertola referred to these events, SN 1962L and SN 1964L, as peculiar Type I. The spectral data on these early supernovae were photographic and otherwise less than ideal. The light-curve shapes resembled those of SN Ia. The peak luminosity seemed dimmer, although uncertain, given the rudimentary state of distance determinations at the time.

Confirmation of a new category of SN I came with modern CCD data and other multiwavelength data on SN 1983N (Richtler and Sadler 1983) and SN 1984L (Wheeler and Levreault 1985). Members of the small early sample of peculiar SN I were associated with star-forming environments. The width of the light-curve peaks required similar diffusion times and hence ejected mass for peculiar SN I and SN Ia, but the peaks were about two magnitudes dimmer than those of SN Ia, suggesting only $M_{Ni} \simeq 0.1$ for the peculiar events. SN 1983N and SN 1984L were also strong radio emitters, implying substantial CSM. These data suggested that SN 1983N, SN 1984L, and other peculiar SN I resulted from core collapse in massive stars that had lost their outer envelopes rather than from thermonuclear explosion, and thus that peculiar SN I were closely related to SN II rather than to SN Ia (Wheeler and Levreault 1985).

The terminology Type Ia and Type Ib was first employed by Elias et al. (1985) in their study of near-IR spectra. SN Ib showed neither a secondary maximum in near-IR photometric bands nor an absorption at 1.2 μm, both characteristic of SN Ia, and their light curves declined more slowly than those of SN Ia. We now know that the 1.2 μm "absorption" in SN Ia is rather a lack of emission that peaks at somewhat longer wavelength due to blends of iron-group elements, especially ^{56}Co and ^{56}Ni.

D. Branch, J.C. Wheeler, *Supernova Explosions*, Astronomy and Astrophysics Library, DOI 10.1007/978-3-662-55054-0_16

The absence of this feature in SN Ib was a clue to the typically lower ^{56}Ni mass ejected in SN Ib ($\sim$0.1 $M_\odot$) compared to SN Ia ($\sim$0.6 $M_\odot$). It was later recognized that one of the events referred to as Type Ib by Elias et al., SN 1983I, was yet something else again (Type Ic; Chap. 17).

The nebular-phase spectra of SN Ib also provided important evidence. The features were typical of SN II and quite different from those of SN Ia, cementing the notion that SN Ib were related to core collapse, not thermonuclear explosion. Spectra of the helium-rich SN Ib and the helium-poor SN Ic are similar in the late nebular phase, where the differences in helium abundance are not manifest. Even when helium is present, it becomes dilute by expansion, and the deposition from radioactive decay, needed to nonthermally excite the helium, becomes low.

Another important step came from radiative-transfer theory and the recognition that SN 1983N and SN 1984L showed evidence for lines of He I that were far from LTE (Harkness et al. 1987). The absence of hydrogen, the weakness of Si II, and the presence of conspicuous He I lines in optical spectra became the defining characteristics of SN Ib.

At maximum light, SN Ib have a mean $M_B \lesssim -17$. In a volume-limited sample, they constitute $\lesssim$10% of all core-collapse events. An object in archival images of the Type Ib SN iPTF13bvn was likely to be a progenitor of ZAMS mass 10–20 $M_\odot$ in an interacting binary system. The environments in their host galaxies tend to have metallicities between those of SN Ic and SN IIP (Sect. 3.3).

Some comparatively uncommon events called Type Ibn, which show narrow emission lines of He I, are discussed in Sect. 16.5.

16.2 Case Studies

16.2.1 SN 1983N

SN 1983N was discovered in M83 on July 3, 1983, by the redoubtable Reverend Robert Evans (Thompson et al. 1983) and reached maximum light about 2 weeks later (Fig. 16.1). It was one of the first supernovae to trigger a global, multiwavelength response and was studied from UV to radio. SN 1983N also provided some of the first premaximum spectra of any supernova. UV observations from the *International Ultraviolet Explorer* were quickly begun. Near-IR spectra, a category that remains underutilized today, were also obtained. Much of the UV, optical and near-IR photometry and spectroscopy was summarized in an influential, often-cited, preprint by Panagia et al. that was widely circulated, but never managed to reach the stage of submission to a journal. Some, but not nearly all, of these data appeared in ancillary publications.

At maximum light, the colors of SN 1983N were redder than those of SN Ia and the spectra were reminiscent of SN Ia at later epochs, supporting the notion that the new SN Ib represented a separate physical class (Wheeler and Levreault 1985;

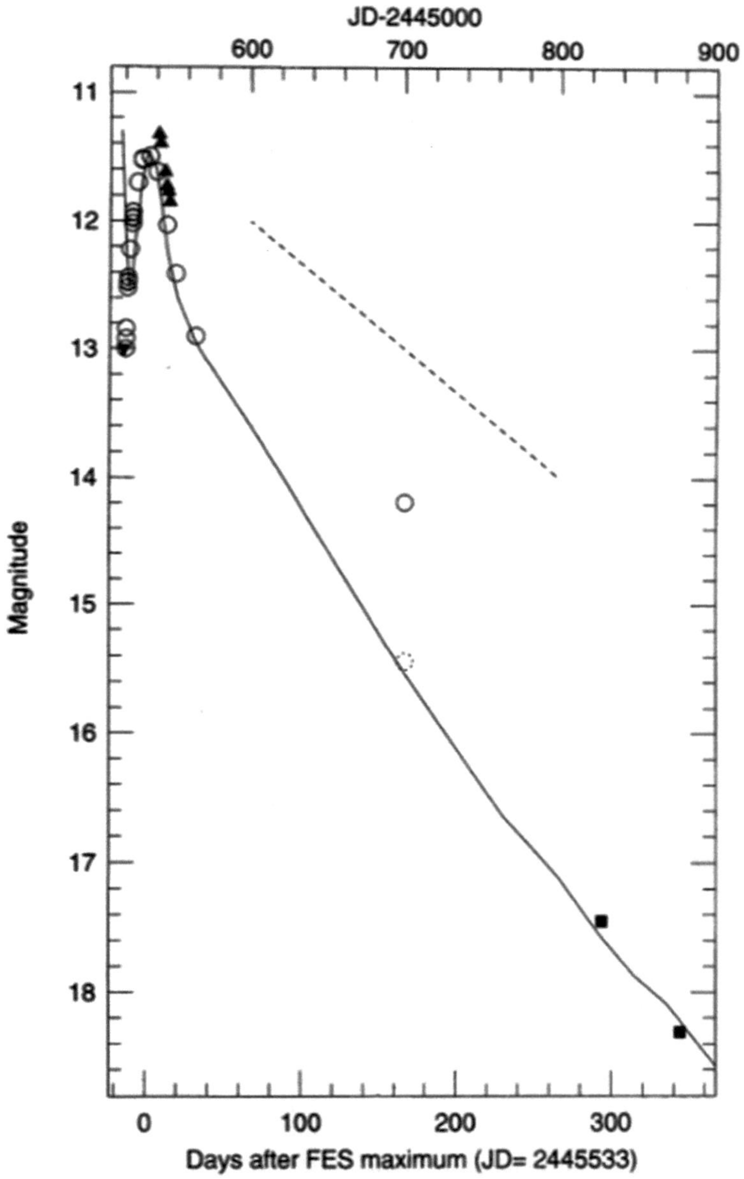

Fig. 16.1 The V-band light curve of SN 1983N. After correction for background light, the discrepant point near 160 days falls on the curve. The *straight dashed line* shows the ^{56}Co decay rate. From "SN 1983N and the Nature of Stripped-Envelope Core-collapse Supernovae" (Clocchiatti et al. 1996a). © AAS. Reproduced with permission

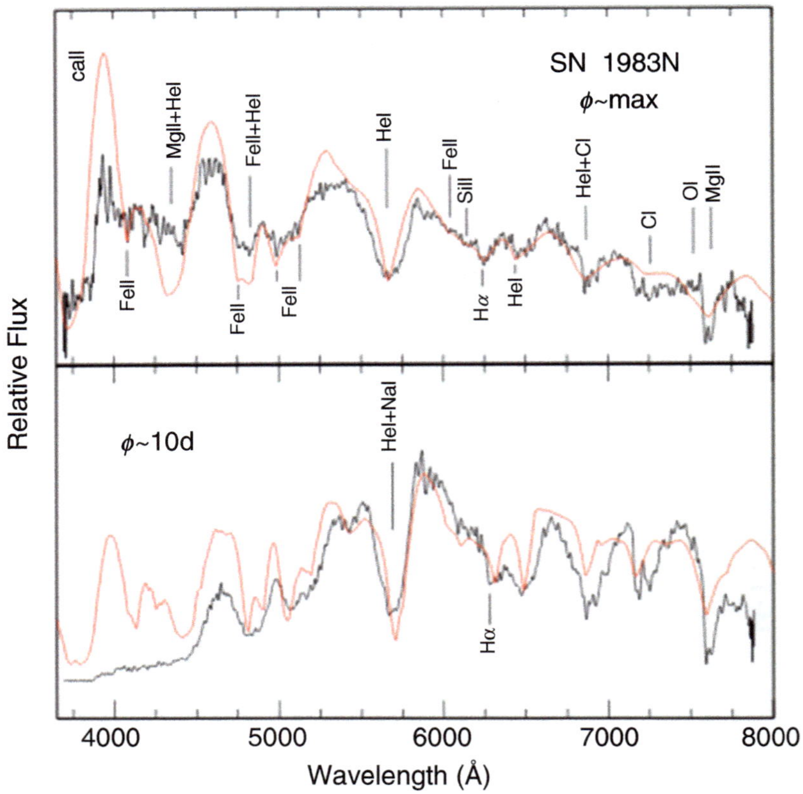

Fig. 16.2 Spectra of SN 1983N at maximum light (*top*) and 10 days later (*bottom*), compared with SYNOW synthetic spectra. From "Hydrogen and helium traces in type Ib/c supernovae" (Elmhamdi et al. 2006) reproduced with permission, © ESO

Uomoto and Kirshner 1985). We now recognize that because of the lower M_{Ni}, SN Ib are cooler at maximum, resulting in both redder colors and in the prominence of Fe II features (Fig. 16.2) that develop only later in SN Ia as they expand and cool.

SN 1983N had a slightly larger contrast between the light-curve peak and tail than did the Type IIb SN 1993J, but a similar late-time decline (Clocchiatti et al. 1996a). The expansion velocities determined from nebular-phase spectra of SN 1983N and SN 1993J also were similar. SN 1983N must somehow have been different from the Type Ib SN 1984L, which had a much slower late-time light-curve decline despite a similar light curve near peak and similar spectra. SN 1983N must somehow have been similar to SN 1993J despite the differences in their early spectra. Models of the light curve of SN 1983N (Ensman and Woosley 1988; Shigeyama et al. 1990) showed that the ejected mass was probably less than about

5 $M_\odot$. Spectra of SN 1983N 8 months after maximum showed broad emission lines of Mg I] λ4571, Na I D, [O I] λλ6300,6364, and [Ca II] λλ7291,7323 (Gaskell et al. 1986). Except for the absence of hydrogen, the features were much like those of SN IIP.

Near-IR spectroscopy was used to search for the [Fe II] 1.644 μm line in SN 1983N about a year after explosion (Graham et al. 1986). These observations seemed consistent with the production of a substantial amount of iron, but it was later realized that [Si I] emission at 1.645 and 1.607 μm was likely to make a substantial or even dominant contribution to the feature (Oliva 1987). In SN 1987A at very late times, the 1.644 μm feature was about 80% [Si I] and 20% [Fe II] (Kjær et al. 2010). Observations of unblended [Si I] at 1.10 μm and [Fe II] at 1.26 μm would help to confirm the nature of the 1.644 μm feature in SN Ib.

Radio emission from SN 1983N was detected just 3 days after optical discovery, representing the first time radio emission was observed from a hydrogen-deficient supernova (Sramek et al. 1984). The radio properties of SN 1983N (Fig. 16.3) were virtually identical in terms of spectral indices, initial thermal absorption, and luminosities to those of SN 1984L, despite the differences in their optical light curves (Sect. 16.3.1).

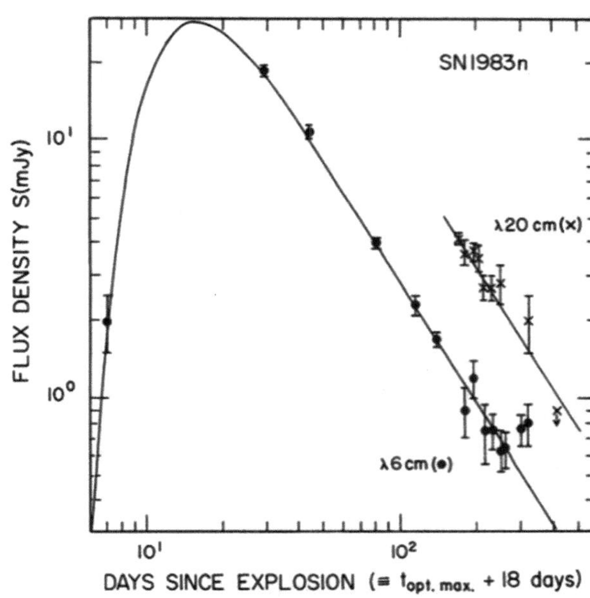

Fig. 16.3 Radio light curves of SN 1983N at 20 and 6 cm, and a model fit. From "Radio supernovae" (Weiler et al. 1986). © AAS. Reproduced with permission

16.2.2 SN 2008D

SN 2008D in NGC 2770 was especially well-studied because it was serendipitously discovered at or near the time of shock breakout (Berger and Soderberg 2008; Soderberg et al. 2008, Fig. 16.4; Mazzali et al. 2008; Modjaz et al. 2009; Malesani et al. 2009). The early light curve gave evidence of $\sim$0.01 $M_\odot$ of ^{56}Ni at 20,000 km s^{-1} (Fig. 16.5), compared to a total ^{56}Ni mass of 0.07 $M_\odot$. The late-time light curve was similar to that of SN 1993J. The spectrum was nearly featureless for the first week, but began to show He I lines in the optical (Fig. 16.6) and near-IR (Fig. 16.7) about 12 days before optical maximum. At 3 and 4 months postmaximum, oxygen emission lines were double-peaked (Fig. 16.8). Estimates of the total ejecta mass and kinetic energy ranged from 3 to 7 $M_\odot$ and 2 to 6 B.

There are a number of open issues regarding the early X-ray emission from SN 2008D (Soderberg et al. 2008; Chevalier and Fransson 2008; Li 2008; Balberg and Loeb 2011; Svirski and Nakar 2014b): whether it was due to thermal emission of shock breakout; whether the emission was nonthermal, a single blackbody, or multiple blackbodies; whether there was a dense wind surrounding the progenitor; whether nonthermal emission arose from a shock-mediated Fermi acceleration or a variation of the forward/reverse shock paradigm of gamma-ray bursts. The common feature of these analyses is that they assumed, or concluded, that the associated processes were spherically symmetric, or nearly so.

There was, however, evidence of asymmetry from line profiles in the nebular phase (Modjaz et al. 2008; Tanaka et al. 2009b). In particular, the spectra revealed Ca II emission lines with single peaks while both forbidden [O I] and permitted O I λ7774 showed double peaks. The implications of this structure are discussed in Sect. 16.3.2. As discussed in Sect. 16.3.3, SN 2008D was also polarized. Given the evidence for asymmetry and the likelihood that the shock erupted from restricted areas of the surface of the progenitor, the nature of the early X-ray emission needs to be reconsidered in the context of asymmetric shock breakout (Couch et al. 2011). In spherical models, dominated by light-travel time, a large radius was needed to account for the 60-s duration of the X-ray flash. This drove the need to adopt a thick wind (Soderberg et al. 2008). In asymmetric models, the peripheral shock propagation time is naturally about 60 s, so a thick wind is not necessary. Asymmetric breakout implies a complex angular and time-dependent photospheric temperature. There is no single blackbody photospheric temperature.

SN 2008D was observed extensively in the radio (van der Horst et al. 2011). VLBI observations ruled out a relativistic jet, even one beamed away from the line-of-sight, and any long-lived relativistic outflow (Bietenholz et al. 2009). The lack of evidence for a contribution to the radio flux by a relativistic jet is consistent with the X-ray burst being produced by shock breakout.

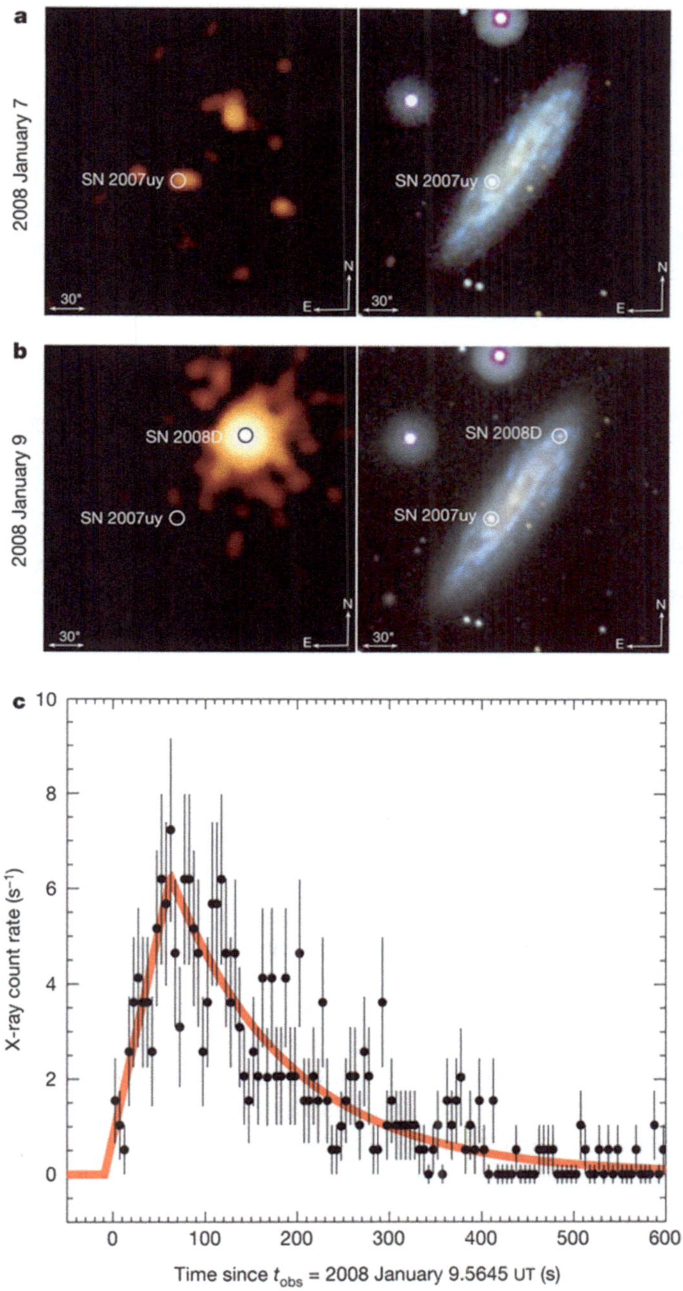

Fig. 16.4 Panel **a**: X-ray (*left*) and UV (*right*) images obtained on 2008 January 7; at this time SN 2008D was not detected. Panel **b**: Discovery images of SN 2008D, obtained 1 day later. Panel **c**: The X-ray light curve, which peaked 63 s after the beginning of the observation. Reprinted by permission from Macmillan Publishers Ltd.: NATURE, Soderberg et al. (2008)

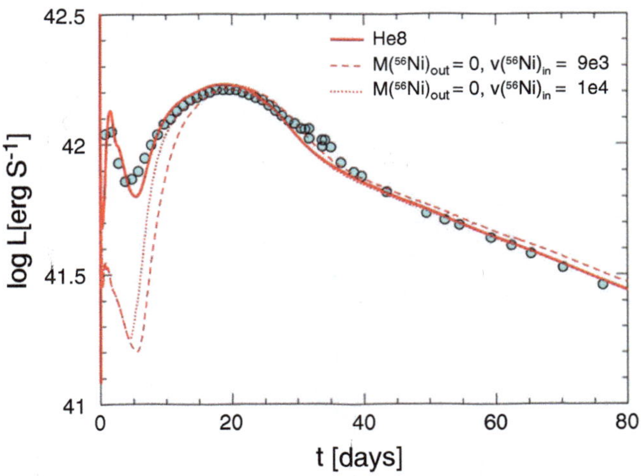

Fig. 16.5 The bolometric light curve of SN 2008D compared to three models. The best-fitting one containing a total ^{56}Ni mass of 0.07 $M_\odot$ with 0.01 $M_\odot$ of ^{56}Ni at 20,000 km s^{-1}. From "Early Ultraviolet/Optical Emission of the Type Ib SN 2008D" (Bersten et al. 2013). © AAS. Reproduced with permission

16.3 Synthesis of SN Ib Characteristics

In addition to SN 1983N and SN 2008D, well-observed SN Ib include SN 1984L (Harkness et al. 1987); SN 1990I (Elmhamdi et al. 2004); SN 1999dn (Benetti et al. 2011); SN 1999ex (Hamuy et al. 2002); SN 2005bf (Folatelli et al. 2006); SN 2007Y (Stritzinger et al. 2009); SN 2007uy (Roy et al. 2013); SN 2009jf (Sahu et al. 2011); and SN iPTF13bvn (Srivastav et al. 2014).

As discussed in Sect. 3.3, much effort has gone into determining whether SN Ib and SN Ic occur in different galactic environments from one another and from SN II. SN Ib occupy the host-galaxy parameter space between SN IIb and SN Ic. SN Ib occur in more metal-rich regions than SN II and SN IIb, and the specific SFR of host galaxies tends to increase from SN II to SN Ib. The central regions of SN Ib host galaxies are less metal-rich and have lower specific SFR than the hosts of SN Ic. Like SN Ic, the distribution of SN Ib is more centrally concentrated than the light of the host galaxy, but, unlike SN Ic, SN Ib do not appear to concentrate in the locally brightest regions of their host. The hosts of both SN Ib and SN Ic are more metal rich than the hosts of SN Ic-bl (Sect. 17.4). NGC 2770, a spiral galaxy of type SASc, was the host of three SN Ib in the span of 10 years, SN 1999eh, SN 2007uy, and SN 2008D (Thöne et al. 2009).

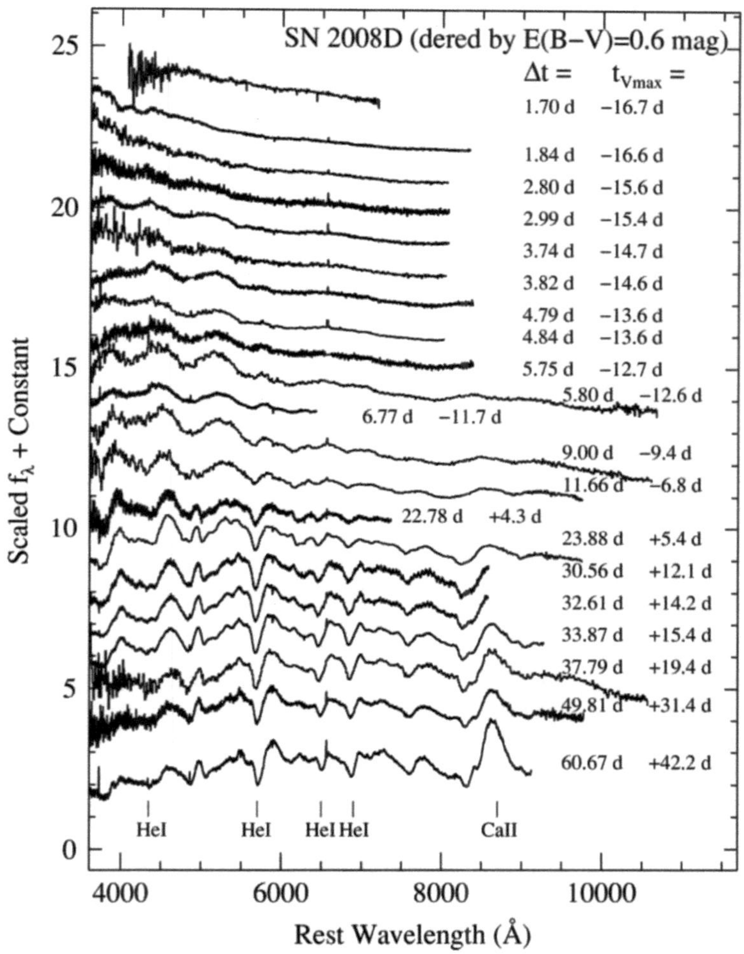

Fig. 16.6 Optical spectra of SN 2008D, showing the development of He I lines. Times (Δt) from shock breakout are given in days and the spectra are shifted vertically for clarity. From "From Shock Breakout to Peak and Beyond: Extensive Panchromatic Observations of the Type Ib Supernova 2008D Associated with Swift X-ray Transient 080109" (Modjaz et al. 2009). © AAS. Reproduced with permission

16.3.1 Light Curves

A central point to emerge from the study of SN Ib/c light curves is that even similar light curves can mask important differences. Neither spectral classification nor photometric classification alone fully captures the nature of SN Ib/c events. This point will be explored below and further in Chap. 17.

The light curves of SN Ib are roughly similar near the peak (Drout et al. 2011; Cano 2013; Lyman et al. 2014a; Taddia et al. 2015b). The bolometric light curves

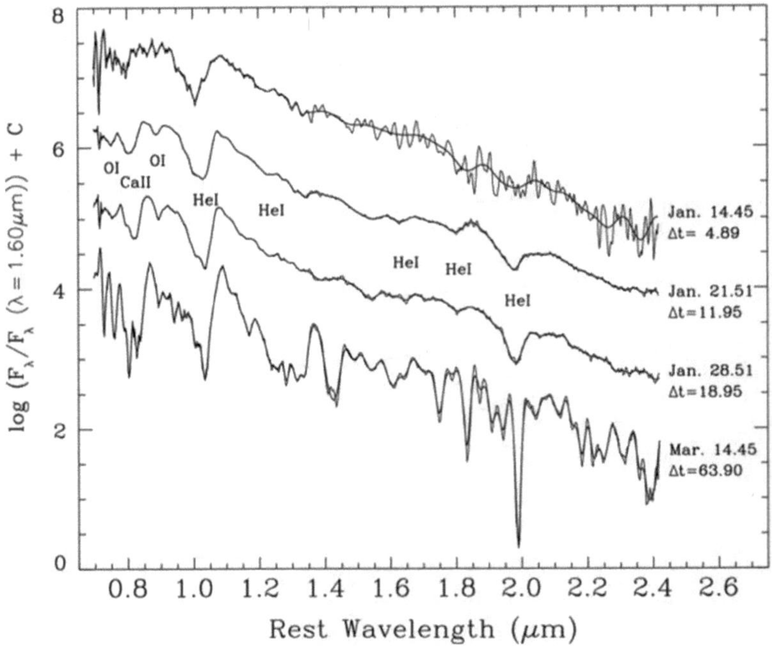

Fig. 16.7 Like Fig. 16.6, but for near-IR spectra. From "From Shock Breakout to Peak and Beyond: Extensive Panchromatic Observations of the Type Ib Supernova 2008D Associated with Swift X-ray Transient 080109" (Modjaz et al. 2009). © AAS. Reproduced with permission

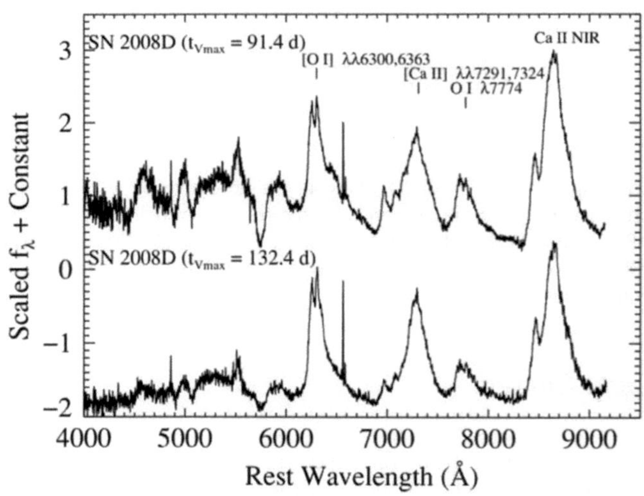

Fig. 16.8 Spectra of SN 2008D at 3 and 4 months past maximum light. The [O I] λλ6300,6364 lines are double-peaked. From "From Shock Breakout to Peak and Beyond: Extensive Panchromatic Observations of the Type Ib Supernova 2008D Associated with Swift X-ray Transient 080109" (Modjaz et al. 2009). © AAS. Reproduced with permission

show a somewhat slower rise to maximum than other stripped-envelope supernovae (Valenti et al. 2011). The postpeak light curve can radiate substantially in the near-IR, a factor that must be taken into account to intercompare UVOIR light curves among SN Ib/c (Benetti et al. 2011).

SN Ib and other stripped-envelope supernovae do not show the secondary peak in the near-IR that characterizes SN Ia (Sects. 20.2 and 20.4). In SN Ia, this behavior is associated with the evolution of the ionization state of the *iron-group elements* (IGEs) and the associated blockage of flux in the UV and relatively low opacity in the near-IR (Sect. 22.3.4). The lack of any secondary near-IR peak in SN Ib and related events may be due to lower mass fractions of IGEs in SN Ib than in SN Ia and to the lower temperature of SN Ib, which affects the ionization structure.

The similarity of light curves often breaks down after a month or two. One of the first well-studied SN Ib, SN 1984L, had a peak that was similar to SN 1983N, but displayed a much flatter late-time V-band light curve (Schlegel and Kirshner 1989), suggesting long-time trapping of ^{56}Co γ-rays and requiring greater than 10 $M_{\odot}$ of ejecta (Swartz and Wheeler 1991). The same may be true of the Type Ib SN 2009jf (Sahu et al. 2011; Fig. 16.9), the Type Ic SN 2011bm (Valenti et al. 2012), and SN 1985F (Filippenko and Sargent 1986); the spectral identity of the latter at maximum light is unknown, but it displayed the nebular-phase spectra of SN Ib/c. Most SN Ib show steeper late-time decline. Some stripped-envelope events display a similar photometric evolution with an intermediate late-time rate of decline like that of the prototypical Type IIb SN 1993J, despite their spectral variety—SN IIb, SN Ib, or SN Ic. Even events with similar peaks and late-time tails can present a conundrum in the sense that the ejecta masses and energies required to fit the peak are incommensurate with the parameters required to fit the tail (Wheeler et al. 2015a; Sect. 17.5.2).

SN 2005bf had an unprecedented, prominent, double-peaked light curve with a first maximum about 16 days after explosion and a second brighter maximum 24 days later (Fig. 16.10). SN 2005bf was extremely faint at late epochs, $\sim$300 days (Fig. 16.11). Early spectra resembled those of a helium-poor SN Ic, but He I features developed later (Tominaga et al. 2005; Folatelli et al. 2006; Parrent et al. 2007). The early spectra appeared to contain weak, high-velocity H lines. Presumably, SN 2005bf at first resembled a SN Ic because the photosphere had not yet receded to where the radioactively-excited He I could be seen. What precisely differentiates SN 2005bf from other SN Ib remains unclear. The odd light curve, the extreme faintness of the late-time photometry, and large blueshifts (1500–2000 km s^{-1}) of the nebular-phase lines remain a challenge. Arguments have been presented in favor of a magnetar, rather than the decay of ^{56}Ni, to power the second peak and a clump of a few tenths of a solar mass to account for the nebular blueshifts (Maeda et al. 2007).

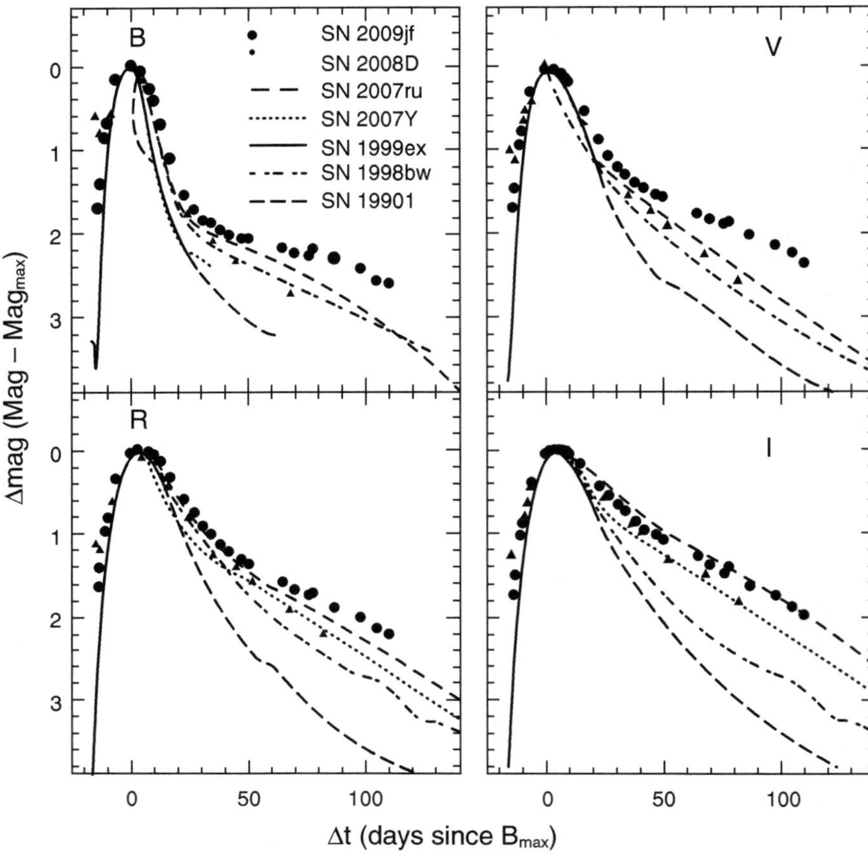

Fig. 16.9 The light curves of the slowly-declining SN 2009jf compared to those of four other SN Ib, as well as the Type Ic-bl SN 2007ru and SN 1998bw. The light curves are normalized at peak to facilitate comparison of their shapes. Figure 3 "Optical studies of SN 2009jf: a Type Ib supernova with an extremely slow decline and aspherical signature" from Sahu et al. (2011), by permission of Oxford University Press on behalf of the Royal Astronomical Society

16.3.2 Spectra

Synthetic spectra of SN Ib computed with the parameterized supernova-spectrum code SYNOW produced a reasonable representation of observed photospheric-phase spectra (Branch et al. 2002). The data were consistent with SN Ib having fairly homogeneous spectral evolution, with exceptions to be noted below. During the month after maximum light, the optical spectra showed lines of He I, O I, Ca II, Sc II, Ti II, and Fe II. The evolution of the photospheric velocity inferred from the Fe II lines, ranging from 17,000 km s^{-1} at the earliest times to 5000 km s^{-1} 1 month postmaximum, suggested that the masses and kinetic energies in the sample

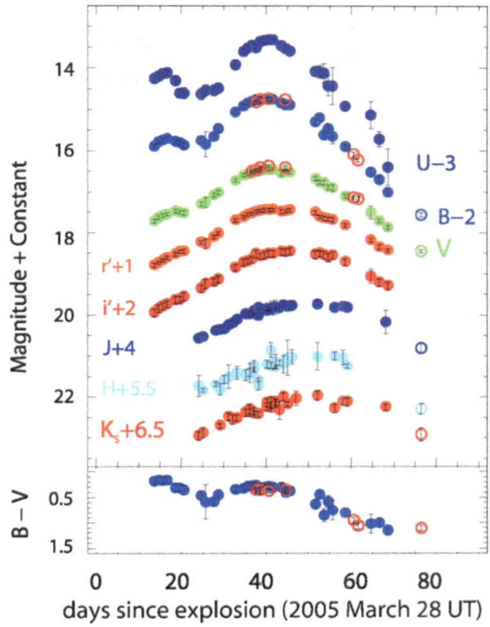

Fig. 16.10 *Top*: Early light curves of SN 2005bf. The first of two peaks can be seen in the U, B, and V bands. *Bottom*: The $B - V$ color evolution. From "The Unique Type Ib Supernova 2005bf: A WN-star Explosion Model for Peculiar Light Curves and Spectra" (Tominaga et al. 2005). © AAS. Reproduced with permission

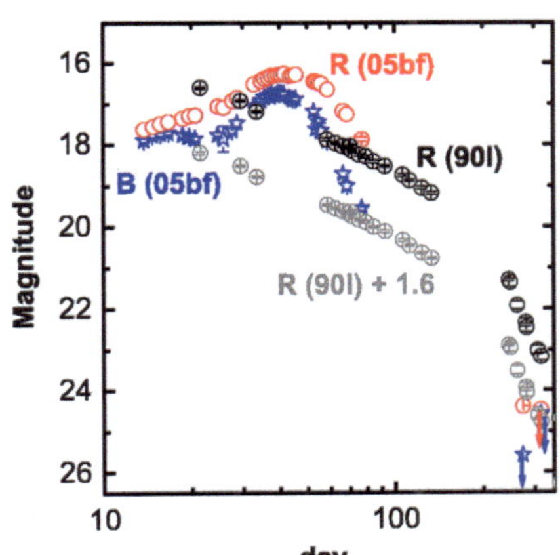

Fig. 16.11 B- and R-band light curves of SN 2005bf, compared to R-band light curves of the Type Ib SN 1990I. To become as dim as the R-band magnitudes of SN 2005bf near 300 days, the SN 1990I light curve is shown shifted downward by 1.6 magnitudes. From "The Unique Type Ib Supernova 2005bf at Nebular Phases: A Possible Birth Event of a Strongly Magnetized Neutron Star" (Maeda et al. 2007). © AAS. Reproduced with permission

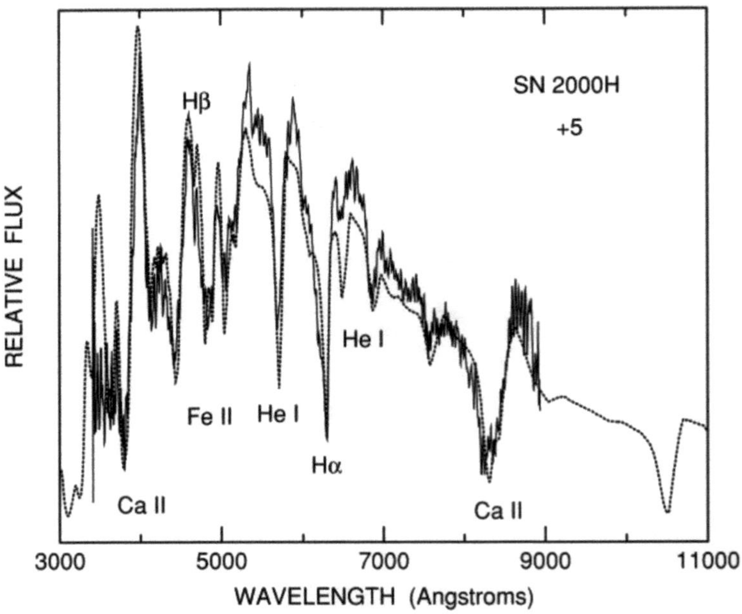

Fig. 16.12 The spectrum of SN 2000H 5 days past maximum light, compared with a SYNOW synthetic spectrum having a velocity at the photosphere of 8000 km s^{-1} and H lines detached at 13,000 km s^{-1}. From "Direct Analysis of Spectra of Type Ib Supernovae" (Branch et al. 2002). © AAS. Reproduced with permission

considered were similar (but see caveats above in Sect. 16.4.2 and further discussion in Sect. 17.5.2). After maximum, the He I features usually showed a velocity greater than the velocity at the photosphere. The minimum velocity of the helium was 7000 km s^{-1}, perhaps lower.

SN Ib often show evidence for small amounts of hydrogen at low optical depth in early spectra (Branch et al. 2002, 2006b, Fig. 16.12; Benetti et al. 2011; Stritzinger et al. 2009; James and Baron 2010; Maurer et al. 2010a). Hydrogen velocities are as high as 10,000–20,000 km s^{-1}. The mass of this hydrogen must be small in order to maintain reasonable energetics. SN Ib usually can be clearly distinguished from SN IIb, which are Balmer-dominated at early times, implying that there may be a distinct difference between the evolution leading to the progenitors of SN IIb and SN Ib (Sects. 15.1 and 16.4.1).

Much effort has gone into determining the amounts of hydrogen and helium in SN Ib/c and the associated physical implications (Sects. 16.4.3 and 17.5.3). Detection of small amounts of either element is a technical challenge because of line blending with other species. Features of Hα can be blended with Si II λ6355 or C II λ6580, and the strongest optical line of He I, λ5876, overlaps with the Na I D lines. The strong near-IR He I line at 1.0830 μm can be confused with C I, Mg II 1.0927 μm, or blends of Si I lines (Millard et al. 1999; Gerardy et al. 2004a; Sauer et al. 2006). There are few clear examples of weak, but distinct He I lines in SN Ib (Matheson et al. 2001; Modjaz et al. 2014; Sect. 17.3.2). The presence of He I lines is nearly always conspicuous (SN Ib) or not (SN Ic).

Near-IR spectra of SN Ib and SN Ic display a strong feature at 1.04 μm, but SN Ib show additional absorption at 1.95–2.0 μm due to blueshifted He I 2.058 μm (Fig. 16.7) while SN Ic do not (Hamuy et al. 2002; Gerardy et al. 2004a; Taubenberger et al. 2006; Valenti et al. 2008b; Stritzinger et al. 2009; Modjaz et al. 2009). For SN Ib, the absorption velocity of the blueshifted He I 2.058 μm line is similar to that of the blueshifted He I 1.0830 μm line, if it is assumed to be the redder absorption trough of a multicomponent feature near 1 μm. The He I 2.058 μm line usually is narrower and more well-defined than the broad blend at 1.04 μm.

Line profiles in nebular-phase spectra may give hints to the structure and asymmetry of the ejecta. While not all SN Ib/c display this behavior (Taubenberger et al. 2009), some stripped-envelope supernovae show double-peaked oxygen emission lines (Mazzali et al. 2005b; Maeda et al. 2007, 2008; Modjaz et al. 2008; Taubenberger et al. 2009; Tanaka et al. 2009b; Valenti et al. 2011; Sahu et al. 2011). A prominent example was SN 2008D (Fig. 16.8), in which Ca II lines had a single peak, but [O I] $\lambda\lambda$6300,6364 and O I λ7774 had double peaks (Sect. 16.2.2). The Ca II emission may have come from outer, more symmetric primordial calcium, while the double peak in the oxygen lines suggests geometric asymmetry in deeper layers. In SN 2008D, the separation between the two peaks was about 2000 km s^{-1} for both the forbidden and the permitted oxygen lines. Optical-depth effects can generate double-peaked [O I] $\lambda\lambda$6300,6364 because of its doublet nature, but that is probably not the cause of the oxygen line profiles in SN 2008D because the observed separation was less than the separation of the doublet (3000 km s^{-1}) and because the permitted line showed the same velocity width between peaks.

When present, these double-peaked features have been interpreted as indicating global asphericities (Sect. 16.3.3), where the emitting oxygen-rich ejecta are in a torus-like structure viewed along the equatorial plane. Maeda et al. (2008) suggested that such global asphericities are common during core collapse, being directly observed in 40% of the sample. Others have argued that it may be incorrect to explain all line profiles by geometric effects alone, especially when the double peaks are close to the natural splitting of the O I doublet (Milisavljevic et al. 2009). When the splitting is different from the natural value, the origin may be global asymmetry of the oxygen distribution. SN 2008D showed that O I λ7774 provides an important check on the nature of the line splitting. When double peaks are seen, the blue peak is often more prominent. SN 2009jf showed double-peaked [O I] $\lambda\lambda$6300,6364 with a stronger blue peak, perhaps due to scattering or dust obscuration (Milisavljevic et al. 2010). The profile of [O I] $\lambda\lambda$6300,6364 can be affected by Hα absorption (Maurer et al. 2010a), but then the Hα must not only be present at late phases but at a particular velocity near 12,000 km s^{-1}. The stronger blue peak might be explained by a sufficiently complex ejecta structure with additional blueshifted emission on top of an otherwise symmetric profile (Taubenberger et al. 2009). Alternatively, the asymmetry may be caused by residual opacity in the central regions blocking the red component. In any case, the asymmetric and multicomponent nebular-phase oxygen line profiles cannot be reproduced in spherical symmetry.

16.3.3 Polarization and Asymmetry

Many SN Ib display ostensibly similar density profiles, ejecta mass, and kinetic energy above the photosphere. This may imply for these events rather little aspect angle effect and hence little global asymmetry (Branch et al. 2002; Benetti et al. 2011; but note the possible unreliability of estimates of mass and kinetic energy from the peak light-curve, Sects. 16.4.2 and 17.5.2). Since the progenitors of SN Ib/c have shed nearly all of their hydrogen layers, they allow a deeper view of the progenitor than do SN II. The trend that stripped-envelope supernovae show more polarization than SN II is an important part of the argument that it is the machinery of the core-collapse explosion that is aspherical. A growing database of detailed spectropolarimetry of SN Ib is revealing more clues to their complex chemo-geometrical structure.

The first event that suggested that SN Ib/c were more polarized than SN II was SN 1997X (Wang et al. 2001; Wang and Wheeler 2008). This event was originally classified as an SN Ic, but was reclassified as an SN Ib based on quantitative analysis of the spectrum using the SNID classification code (Modjaz et al. 2014). SN 1997X showed exceptionally high polarization with a time dependence indicating that there was a substantial intrinsic component that might have been as high as 4%. In the Q/U plane, SN 1997X evolved from spectropolarimetric Type N1 around maximum to Type D1 about a month later. There seems to have been a steep drop in polarization in the first 10 days or so, a factor that still requires explanation.

Spectropolarimetry brings another perspective to the general issue of classification. SN 1997X did not show conspicuous evidence of He I lines in its flux spectrum, although, as mentioned above, careful comparison to samples of SN Ib and SN Ic gave sufficient evidence for He I lines to warrant reclassification. The polarization spectra revealed clear spectral features associated with He I λ5876 and λ6678. This showed that spectropolarimetry is a useful tool to capture the presence and geometry of the helium in the outer layers of the progenitor. The He I absorption features peaked at 15,000 km s^{-1} and the wings extended to 28,000 km s^{-1}. On energetic grounds, the mass of the helium was not substantial. Wang and Wheeler (2008) argued that the spectropolarimetry of SN 1997X supports the notion that there is a continuum between SN Ib and SN Ic, rather than a distinct transition between the two spectral classes; the restriction on the mass of helium seems more consistent with the notion that, while there was helium in SN 1997X and in some SN Ic, there is a qualitative, order of magnitude difference in the amounts of helium in most SN Ib compared to SN Ic and "weak-helium" SN Ib like SN 1997X. From this perspective, SN Ib on the one hand and SN Ic and the "weak-helium" SN Ib on the other may require qualitatively different modes of progenitor evolution (Sects. 16.4.1 and 17.5.1)

The peculiar Type Ib SN 2005bf had substantial spectropolarimetry (Maund et al. 2007c; Tanaka et al. 2009a), obtained 6 days before and 8 days after the second peak (about 34–49 days after explosion). Although interstellar polarization may have been significant, the intrinsic continuum polarization was consistent with global asymmetry of order 10–20%. The data tend to fall along a single locus, suggesting a principal axis, but there was also scatter around that dominant axis. SN 2005bf showed a distinct loop in the He I $\lambda5876$ line, evidence of a departure from axisymmetry. Polarization as high as 4% was observed in the absorption components of Ca II H&K and the Ca II near-IR triplet, rivaling that in the highly-polarized SN 1997X. Lines of Ca II and He I had different polarization distributions in the Q/U plane. The Ca II near-IR triplet showed polarized features at considerably higher velocity ($\sim$13,000 km s^{-1}) than other lines. Lines of Fe II and the prominent O I $\lambda7774$ line were not polarized. SN 2005bf was spectropolarimetric Type D1 overall and Type L in the He I line.

The spectropolarimetry of SN 2005bf can be interpreted in the context of the basic structure of a SN Ib: a progenitor having a C/O core surrounded by a helium mantle (Maund et al. 2007c). In SN 2005bf, the helium mantle may have had its own axis of symmetry, plausibly a rotation axis that differed from the rotation axis of the core. The lack of evidence for excited He I at early times suggested that the photosphere had not yet receded to the jet or plume of radioactive material at the epoch of the first peak in the light curve. The lack of O I polarization about a week before the second peak was consistent with the photosphere remaining in the helium mantle at that epoch, having not yet receded into the C/O core. The lack of Fe II polarization prior to the second light-curve peak was also consistent with any nickel plume or jet having failed to penetrate the photosphere at that epoch.

Spectropolarimetric observations of SN 2008D around maximum and about 2 weeks later showed that the continuum was mildly polarized, implying an asymmetry of the photosphere of less than 10%, and that specific lines were polarized with differences in polarization angles, suggesting nonaxisymmetric structure (Maund et al. 2009). Polarization of SN 2008D was observed in lines of the Ca II near-IR triplet, He I, Mg I $\lambda4571$, Fe II, and possibly O I $\lambda7774$. The spectropolarimetric classification was D1+L for He I and L for Ca II. The polarization of SN 2008D was rather similar to that of SN 2005bf around the time of the second peak of the latter. Maund et al. displayed the polarization data of SN 2005bf and SN 2008D in a manner that allowed the portrayal of chemically distinct line-forming regions in angle and velocity in the ejecta at the expense of representing the polarization amplitude (Fig. 16.13). The polarization angle of the Ca II near-IR triplet was different from that of other species at both epochs, suggesting that high-velocity Ca II formed in a special portion of the ejecta. Maund et al. noted that, like SN Ia, SNI b/c frequently show highly-polarized, high-velocity Ca II features. These may reflect a common aspect of SN Ib/c and SN Ia: origin in a binary system. The polarization of SN 2008D is consistent with a "jet" or "plume" of radioactive material being stalled beneath the photosphere, as opposed to protruding through it.

The relatively slowly-declining SN 2009jf (Fig. 16.9) was observed 9 days after B-band maximum (Tanaka et al. 2012a). It showed prominent polarization in the

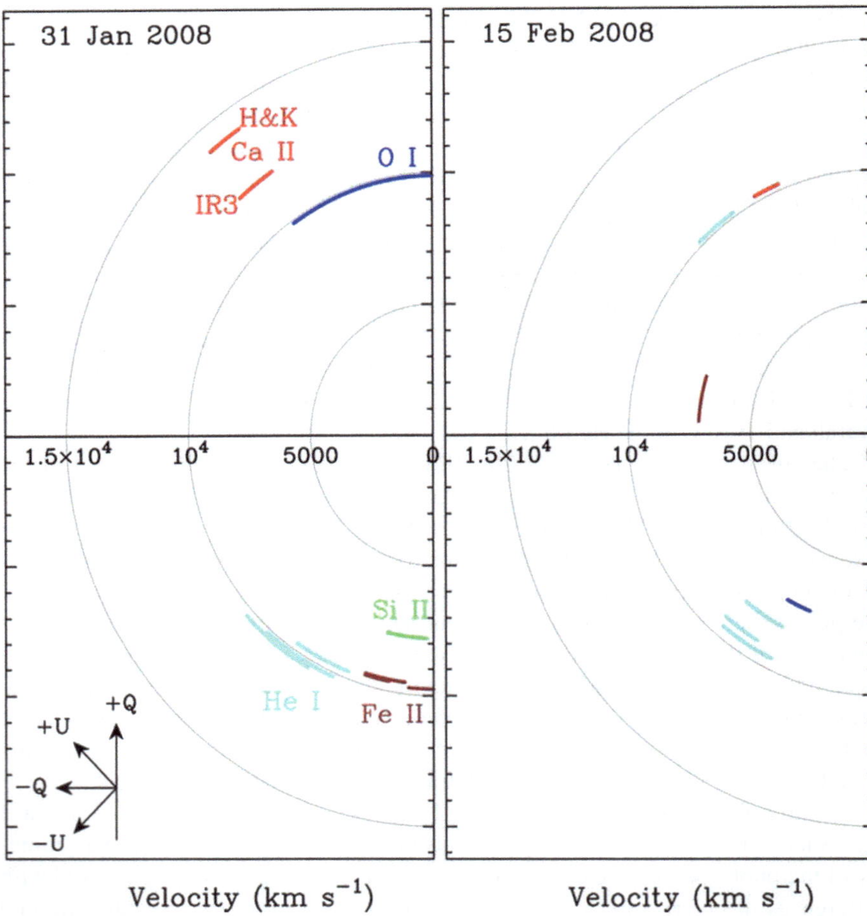

Fig. 16.13 Polar plots presenting the velocity and polarization angle at absorption minimum for Ca II H&K, He I λ5015, λ5876, λ6678, and λ7065, Fe II, Si II λ6355, O I, and the Ca II near-IR triplet for the observations of SN 2008D near maximum and 15 days after maximum, corresponding to 21 and 36 days after the associated X-ray flash (Sect. 16.2.2). The angle of each arc corresponds to the observed polarization angle on the sky, and the radius of the arc corresponds to the velocity. The corresponding orientation of the Stokes parameters is indicated in the *bottom left panel*. From "The Early Asymmetries of Supernova 2008D/XRF 080109" (Maund et al. 2009). © AAS. Reproduced with permission

O I λ7774 line and in the Ca II near-IR triplet (Fig. 16.14). These features both showed loops in the Q/U plane (Sect. 4.6), giving evidence for 3D nonaxisymmetric structure with the oxygen and calcium distributions varying in orientation with depth. Some of the structure of the O I feature falls along a line in the Q/U plane at an angle very roughly consonant with that of the Ca II. This may suggest a preferred axis, albeit with significant departures.

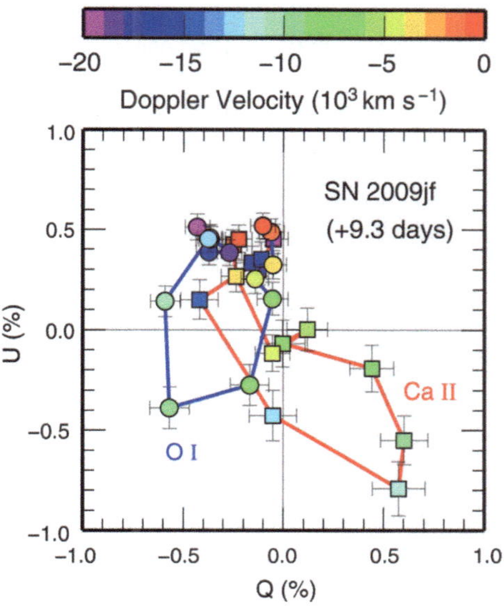

Fig. 16.14 Spectropolarimetry of SN 2009jf 9 days after B-band maximum represented in the Q/U plane. SN 2009jf showed prominent nonaxisymmetric polarization in O I and in the Ca II near-IR triplet. From "Three-dimensional Explosion Geometry of Stripped-envelope Core-collapse Supernovae. I. Spectropolarimetric Observations" (Tanaka et al. 2012a). © AAS. Reproduced with permission

SN iPTF13bvn was the first SN Ib for which a possible progenitor candidate was detected. While a Wolf–Rayet star progenitor was proposed, subsequent work suggested the ejecta mass was too low for that to be the case (Sect. 16.4.1). Evolution models suggested the progenitor was a low-mass helium giant of ZAMS mass $\sim$10 to 20 $M_\odot$ in an interacting binary system. Six epochs of spectropolarimetry observations were obtained for SN iPTF13bvn spanning the phases from 10 days before to 61 days after R-band maximum (Reilly et al. 2016). The continuum was polarized with $p \simeq 0.2$ to 0.4%, implying asphericities of $\sim$10% in the shape of the photosphere. Significant line polarization was associated with He Iλ 5876/Na I D, He I λ6678 and λ7065, Fe II λ4924, O I and the Ca II near-IR triplet. An absorption feature at 6200 Å that is often identified as Si II λ6355 may instead be a high-velocity feature of Hα at 16,400 km s^{-1} (Sect. 16.3.2). At −10 days, Ca II, Fe II and Si II/Hα displayed a common polarization angle across a large velocity range from 6000 to 16,000 km s^{-1} that was slightly rotated from that of the photospheric continuum (Fig. 16.15; left panel). Near peak light, the continuum polarization angle was about 180°, substantially different than either 10 days earlier ($\sim$30°) or 9 days later ($\sim$120°). Near maximum, low-velocity and high-velocity components of features began to emerge. At that epoch, the polarization angle rotated with velocity for both Ca II and He I/Na I D, appearing as counter-clockwise and clockwise spirals, respectively, in a plot of polarization angle versus velocity (Fig. 16.15; middle panel).

By 9 days after maximum, two distinct polarized components, low-velocity and high-velocity, appeared for both He I/Na I D and the Ca II near-IR triplet, indicating the presence of two discrete line-forming regions in the ejecta in both radial-velocity space and in the plane of the sky (Fig. 16.15; right panel). The

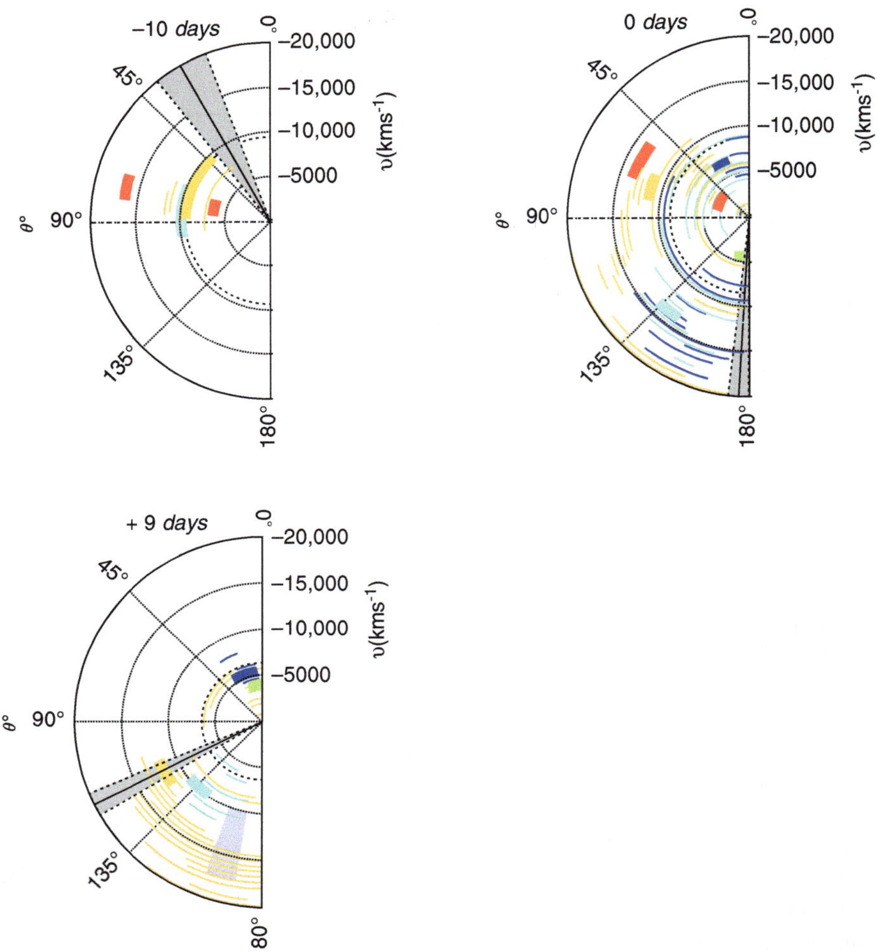

Fig. 16.15 Polar plots presenting the velocity and polarization angle of various components of SN iPTF13bvn at −10 days (*left panel*), peak light, 0 days (*middle panel*), and 9 days after R-band maximum (*right panel*). The *gray areas* represent the range of the continuum polarization angle at each epoch. Features are shown for Hα (*red*), He I/Na I D (*dark blue*), He I λ6678 (*purple*), He I λ7065 (*yellow*), O I (*green*), the Ca II near-IR triplet (*orange*), and Fe II (*light blue*). All three epochs show evidence for low-velocity ($\sim$5000 km s^{-1}), high-velocity (10,000–15,000 km s^{-1}), and perhaps very-high-velocity ($\gtrsim$15,000 km s^{-1}) components that have distinct polarization angles and that vary significantly over time. The polarization angle of the high-velocity He I/Na I D component is marked by the shaded blue region in the right panel. Figure 7 "Spectropolarimetry of the Type Ib Supernova iPTF13bvn: revealing the complex explosion geometry of a stripped-envelope core-collapse supernova" from Reilly et al. (2016), by permission of Oxford University Press on behalf of the Royal Astronomical Society

continuum polarization angle, $\sim 120°$, was intermediate between that at -10 days and that at peak light. From +7 days onwards the continuum polarization angle was approximately aligned with the high-velocity components. Figure 16.15 also shows the possible presence of components of very high velocity, $\gtrsim$15,000 km s^{-1}, that have a polarization angle somewhat different than the high-velocity components arrayed from 10,000 to 15,000 km s^{-1}.

The line of He I λ5876 can be used as a tracer of sources of nonthermal excitation in the ejecta. Reilly et al. (2016) found that the bulk of the radioactive nickel was constrained to be interior to 50–65% of the ejecta radius. The polarization observations favored the explosion of a star with an extended, distorted envelope rather than a compact Wolf–Rayet star.

SN 1997X, SN 2005bf, SN 2008D, and SN iPTF13bvn showed that spectropolarimetry is a valuable probe of the similarities and differences between SN Ib and SN Ic. While the "tilted jet" scenario (Maund et al. 2007c) was designed to qualitatively account for the characteristics of SN 2005bf, its basic features—an asymmetric core that might vary in mass and composition and a ^{56}Ni protuberance at a different angle that carries asymmetric excitation to various depths and at various angles—may help to illuminate other core-collapse events.

16.3.4 Circumstellar Interaction

Several SN Ib, including both of our case studies (Sect. 16.2) as well as SN 1984L (Panagia et al. 1986) and SN 2007uy (van der Horst et al. 2011; Roy et al. 2013), have been detected in the radio. For an assumed wind velocity of 1000 km s^{-1}, presupernova mass-loss rates have been estimated to be in the range 10^{-5} to 10^{-4} $M_\odot$ y^{-1}.

SN Ib have lost nearly all of their hydrogen envelopes. Some of that mass could reside on a binary companion, but some could be ejected by winds or other mass flow (common-envelope evolution, messy mass transfer, mergers) into the circumstellar environment. SN Ib may thus have substantial CSM at some distance from their explosions. Case in point: on a timescale of about 1 year, SN 2014C evolved from a typical SN Ib to a strongly-interacting hydrogen-rich SN IIn. A circumstellar shell, ejected decades to centuries before explosion, was inferred to contain 1 $M_\odot$ of hydrogen at a distance of 6×10^{16} cm (Margutti et al. 2017).

16.4 Models of SN Ib

16.4.1 Evolutionary Models

Gaskell et al. (1986) noted that the features of SN Ib required the progenitors to be "strongly reminiscent" of Wolf–Rayet stars. There remains a tendency to equate hydrogen-poor SN Ib/c supernovae, and models thereof, with Wolf–Rayet star progenitors, but that connection is probably incorrect for typical SN Ib/c.

As discussed in the context of SN IIb (Sect. 15.4.1), single-star models have difficulty shedding sufficient hydrogen to represent SN Ib unless they are quite massive, in which case there are problems with accounting for the SN Ib rate (Sect. 3.6). Most massive stars are in binaries (Sana et al. 2012), and the most successful models for the evolution to produce SN Ib are based on binary evolution.

Evolutionary models for SN Ib/c were reviewed by Yoon (2015). Binary-evolution models that consider the primary stars to be the progenitors of SN Ib favor low-mass stars (Yoon et al. 2010; Eldridge et al. 2013). After mass transfer, stripped cores of less than about 3 $M_\odot$ undergo expansion of the helium envelope during C/O core contraction or core carbon burning. This extended helium envelope is thus also susceptible to loss to the secondary by mass transfer yielding a primary that is stripped of both hydrogen and substantial helium. If the primary retains at least $\sim$0.1 $M_\odot$ of helium it could be a candidate to explode as a SN Ib. If the primary retains substantially less than $\sim$0.1 $M_\odot$ of helium it could represent the progenitor of a SN Ic. More massive cores and those with lower metallicity tend to retain some hydrogen, $\lesssim$0.1 $M_\odot$ after the first stage of mass transfer, making them more likely to represent the progenitors of SN IIb.

The binary-star models of SN Ib have outer layers enhanced with helium and nitrogen and depleted of carbon and oxygen, consistent with being processed by the CNO cycle. These characteristics are reminiscent of Wolf–Rayet stars, but the masses of the models are much too low to correspond to observed Wolf–Rayet stars. Typical masses of Wolf–Rayet stars are 16–18 $M_\odot$, but can range from 5 to 77 $M_\odot$ (WR 22, HD 92740). Masses of the O-type star in Wolf–Rayet star/O-star binary systems range from 14 to 57 $M_\odot$, with a mean of 33 $M_\odot$ (Cherepashchuk and Karetnikov 2003; Crowther et al. 2010). The masses of known Wolf–Rayet stars may be an artifact of the selection criteria to search for them—strong emission lines relative to the continuum—but, as an observed category, they are massive. The low-mass binary models that account for the rate and nature of the SN Ib do not correspond to observed Wolf–Rayet stars, and the observed Wolf–Rayet stars cannot account for the properties of common SN Ib. A possible Wolf–Rayet star progenitor to the SN Ib SN iPTF13bvn was noted (Cao et al. 2013), but the properties of the archival point source (Kim et al. 2015b) and the supernova (Fremling et al. 2014; Srivastav et al. 2014; Bersten et al. 2014; Eldridge et al. 2015; Kuncarayakti et al. 2015) indicated that the mass of the progenitor was too low for a classic Wolf–Rayet star.

At the time of explosion, the rather low-mass SN Ib progenitor models predicted in evolutionary studies could have lower luminosity than the companion and be difficult to observe directly. The Be-star X-ray binaries may be the end product of the general evolutionary scenario that produces low-mass stripped cores (Wellstein et al. 2001; Petrovic et al. 2005). These X-ray systems may be the outcome of SN Ib explosions that produce the neutron-star companion and may be the progenitors of subsequent SN Ib or SN Ic (Sect. 17.5.1).

The Type Ib SN 2010O occurred in the interacting starburst galaxy Arp 299 near to a variable X-ray source and hence may have been consistent with interacting-binary evolution (Nelemans et al. 2010; Kankare et al. 2014). SN 2010O was close to, but not within, a young compact cluster and had no conspicuous optical progenitor star in archival *HST* images. The X-ray variability associated with the active starburst means that a chance alignment is possible. Association of the X-ray source with the supernova might suggest that SN 2010O represented an explosion of the initially less-massive star in an X-ray binary that perhaps resembled Cyg X-3 prior to explosion.

16.4.2 Model Light Curves

Quantitative models of SN Ib light curves are based on helium cores from massive stellar models from which the hydrogen envelope is removed "by hand" or by some prescription of wind or binary mass transfer. Models with an ejecta mass of more than 3 $M_{\odot}$ and energies around 1 B tend to have rise times to the peak of 20–30 days (Dessart et al. 2011). Few if any observed stripped-envelope supernovae have such long rise times (Sect. 16.3.1), so most SN Ib must have lower ejecta masses, higher energy, mass that does not contribute substantially to the diffusion time, or ^{56}Ni closer to the surface.

An enduring problem has been to get the physics of the peak to agree with the physics of the long-term radioactive tail. Rather massive helium-star models retained a substantial fraction of the ejecta mass as helium (Ensman and Woosley 1988). In these models, all the helium and much of the oxygen recombined before maximum. Helium and oxygen recombination affected the rise of the model light curve, adding perhaps half the light, but by the time of maximum light the recombination energy was totally overwhelmed by the energy deposition of radioactive decay of even a modest amount of ^{56}Ni. Nevertheless, in these models all the helium and a substantial portion of the oxygen remained neutral at maximum light. This meant that a significant portion of the ejected mass had low opacity and contributed little to the optical diffusion time that dictated the rise time and the width of the peak. Most of the opacity over the peak was contributed by the relatively low mass of the inner iron and silicon layers. The low effective diffusion mass allowed a fit to the peak of SN 1983N with a rather high ejecta mass, but the whole ejecta mass contributed to the γ-ray trapping, hence yielding excessively flat late-time light curves and less peak-to-tail contrast than the observations. Subsequently, models of stripped-envelope light curves in general and of SN Ib in particular have invoked lower ejecta masses (Shigeyama et al. 1990). This class of models nearly always found the late-time light curve to decline too steeply.

Attempts to correct this peak-tail problem have often addressed adjustments to the structure that would affect the gamma-ray trapping. Suggestions in higher-mass models include outward mixing of ^{56}Ni or clumping (Ensman and Woosley 1988). Schemes to flatten the late-time tail in the lower-mass models included the adoption of two-component structures with a dense inner region, perhaps due to asymmetries, that more effectively trapped γ-rays (Maeda et al. 2003). A more extensive discussion of the problems associated with the light curves of stripped-envelope supernovae is given in Sect. 17.5.2.

In contrast to SN 1983N and the majority of SN Ib, the shallow late-time light curves of SN 1984L and a few other stripped-envelope supernovae such as SN 2009jf (Sect. 16.3.1; Fig. 16.9) are consistent with ejected masses that are high enough to trap γ-rays for an extended time and hence masses that exceed 10 $M_{\odot}$ (Swartz and Wheeler 1991). A caveat is that SN 1984L was a radio source so CSI that mimics ^{56}Co decay cannot be ruled out. Ejected masses exceeding 10 $M_{\odot}$ could be consistent with mass loss from single-star progenitors (Wheeler and Harkness 1990).

Model light curves (and spectra) for stripped-envelope supernovae should account for the effects of time-dependent, NLTE radiative transfer. An interesting feature of such models is a "plateau" associated with the fireball phase, prior to the rise to the peak powered by radioactive decay (Dessart et al. 2011). The computed bolometric light curves typically have a plateau that lasts about 10 days and is about five magnitudes dimmer than the subsequent peak. The Type IIb events SN 1993J (Sect. 15.2) and SN 2008ax (Pastorello et al. 2008a) showed a brief dip between the shock breakout and the subsequent maxima, but these hardly constituted a plateau and were only one magnitude dimmer than the subsequent peak. The Type Ib SN 2008D (Soderberg et al. 2008; Modjaz et al. 2009) showed a brief hiatus on the rise that lasted a few days, but again this was only about one magnitude below peak. Thus the plateau predicted in the models of Dessart et al. seems to be dominated by other effects in observed explosions. Effects of CSI are possible.

16.4.3 Models of Photospheric Spectra

Lucy (1991) showed that the He I lines of SN Ib are plausibly the result of nonthermal excitation by radioactive decay in the hydrogen-stripped core of a massive star with a substantial helium mantle. The degree to which radioactive excitation would reveal H and He I lines in photospheric spectra is also explored in the context of the Type IIb SN 1993J (Sect. 15.2.7) and in the context of SN Ic (Sect. 17.5.3). This work on other spectral types illuminated the conditions relevant to the photospheric spectra of SN Ib. He I lines may not appear in the early photosphere phase, but may appear strongly in the early nebular phase when exposed to the secondary electrons associated with deposition of radioactive decay energy. In models, H and He I lines become prominent when there is substantial γ-ray deposition in layers containing those elements. Observable He I lines also require that the free-electron density be sufficiently low to avoid veiling (Swartz et al. 1993a,b). In the hydrogen-deficient environment of the photospheres of SN Ib, veiling is likely to be less of an issue than in SN 1993J, but charge transfer with helium may play a role (Swartz 1994). The ejecta forming P Cygni He I lines in SN Ib probably contain substantially less than 10% hydrogen by mass (Hachinger et al. 2012a).

Spectral models of the early phases of the explosion can thus elucidate expected visibility of H and He I lines and hence address issues of the classification of stripped-envelope supernovae (James and Baron 2010). Models with only a small outer high-velocity hydrogen mass can produce H lines that fade quickly. Whether such models apply to SN Ib remains under active discussion (Sect. 16.3.2).

Observations of He I lines during the rise to decay-driven maximum can give important clues to the nature and composition of the outer layers, but it remains difficult to determine the mass of helium even with sophisticated NLTE models. Models that include NLTE and nonthermal effects can produce He I lines during the postbreakout fireball phase even without the effects of ^{56}Ni decay (Dessart et al. 2011). In hydrogen-poor but relatively helium-rich models, the levels of He I near 20 eV can be thermally and radiatively excited, especially in progenitor models with extended helium envelopes. In these early phases, the presence of helium affects the radiative transfer by, for instance, controlling the location of the photosphere. Both ionization freezeout and chemical stratification are important effects. SN 2008D may already have provided a test of these thermally-excited models. Its early spectra were nearly featureless whereas the models predict distinct lines of He I and Fe II.

While all stripped-envelope supernovae are thought to be powered by radioactive decay at maximum light, this characteristic alone may not account for the presence near peak of the strong lines of He I that characterize SN Ib. To excite the optical lines of He I, the helium must lie within about one γ-ray mean free path of the source of radioactive decay. Substantial helium might be present in the outer layers of the ejecta, yet not produce conspicuous optical or near-IR lines of He I near peak if the radioactive ^{56}Ni and ^{56}Co remain deep within the ejecta. The observational fact that SN Ib show strong lines of He I may mean that ^{56}Ni is substantially mixed toward the surface in all SN Ib (Dessart et al. 2012a; Cano et al. 2014). Jets or plumes of ^{56}Ni may not be adequate for this aspect, so that some sort of microscopic mixing is required. Such microscopic mixing is not predicted in current models, but all these issues, dynamics and radiative transfer, need to be addressed with fully 3D models. A closely related issue is whether a substantial amount of helium might remain hidden in SN Ic because of a lack of appropriate mixing of ^{56}Ni and excitation of the appropriate levels of He I. This issue will be addressed in Sect. 17.5.3.

16.4.4 Models of Nebular-Phase Spectra

Following the hints that SN Ib/c come from massive stars, Chugai (1986) predicted strong [O I] $\lambda\lambda$6300,6364 in the nebular phase of SN Ib, as was soon confirmed. Gaskell et al. (1986; see also Begelman and Sarazin (1986), who deduced 50 $M_\odot$ of oxygen and a pair-instability progenitor for SN 1985F) argued from the strength of the [O I] $\lambda\lambda$6300,6364 feature in the nebular-phase spectra that there was a major conflict between the ejecta mass derived from the width of the light-curve peak, a few $M_\odot$, and from the [O I] line strengths, tens of $M_\odot$. This was resolved by recognizing that the strength of the [O I] lines is quite sensitive to temperature, that the temperature varies through the oxygen-rich region of the ejecta, and that the important temperature-calibrating line of [O I] λ5577 is often blended and difficult to observe. The strengths of the [O I] lines are affected by mixing and deposition of decay products. These early models were spherically symmetric, and the mixing and deposition were done ad hoc. This needs to be reconsidered in the context of asymmetric models with plumes and jets.

The He I lines that define SN Ib become weaker in late phases when helium is not sufficiently excited by radioactive decay. Models of the Type IIb SN 1993J predict H and He I lines to be visible in the early nebular phase, until $\sim$100 days, but to disappear by $\sim$200 days. The lack of H lines in the early nebular phase of SN Ib complements their lack in the photospheric phase, suggesting that SN Ib contain substantially less than 0.1 $M_\odot$ of hydrogen (Swartz et al. 1993a,b).

In the nebular phase, the flux ratio of [Ca II] $\lambda\lambda$7291,7323 to [O I] $\lambda\lambda$6300,6364 can be a diagnostic of the core mass of a progenitor star, with higher ratios indicative of lower-mass cores (Fransson and Chevalier 1987, 1989) because the mass of explosively synthesized calcium in the ejecta is not sensitive to the ZAMS mass, whereas the ejected mass of oxygen is. The Type Ib SN 1999dn had a rather low flux ratio, [CaII]/[OI] = 0.55, suggesting a progenitor of higher ZAMS mass than those of other stripped-envelope events (Benetti et al. 2011). SN 2009jf may have ejected an especially large amount of oxygen, $\sim$1 $M_\odot$, implying a rather high-mass progenitor (Sahu et al. 2011). Note again that various techniques for estimating mass must be cross-checked for self-consistency. Care must also be taken when the oxygen lines show splitting and the calcium lines do not, since that suggests a different geometry and hence perhaps a different origin for the calcium, e.g., natal rather than freshly synthesized.

16.5 SN Ibn

Supernovae that show little or no evidence for H lines, but do show prominent He I emission lines of intermediate width ($\sim$1000 km s^{-1}), are called SN Ibn (Pastorello et al. 2008b). These events explode inside dense helium-rich CSM, analogous to hydrogen-rich SN IIn. SN Ibn occur in spiral galaxies (Pastorello et al. 2015b;

for an interesting exception, see Sanders et al. 2013) at a rate ~1% of the core-collapse rate. A presupernova outburst of the progenitor of one SN Ibn was observed (Sect. 16.5.1). No quiescent progenitors have been detected in archival images.

16.5.1 Case Study: SN 2006jc

About 2 years prior to explosion, the relatively nearby (25 Mpc) SN 2006jc underwent a luminous outburst of $M_R \simeq -14$ that resembled that of a giant-outburst LBV (Pastorello et al. 2007). This event presumably was the main source of the helium-rich CSM. Implications of the outburst for the progenitor of the supernova are discussed in Sect. 16.5.2.

SN 2006jc was discovered near maximum light and was well observed in the optical (Foley et al. 2007; Pastorello et al. 2007, 2008b; Anupama et al. 2009). The peak absolute magnitude was $M_V \simeq -18$. The postpeak optical light curve (Fig. 16.16) evolved more rapidly than those of typical SN Ib, resembling instead

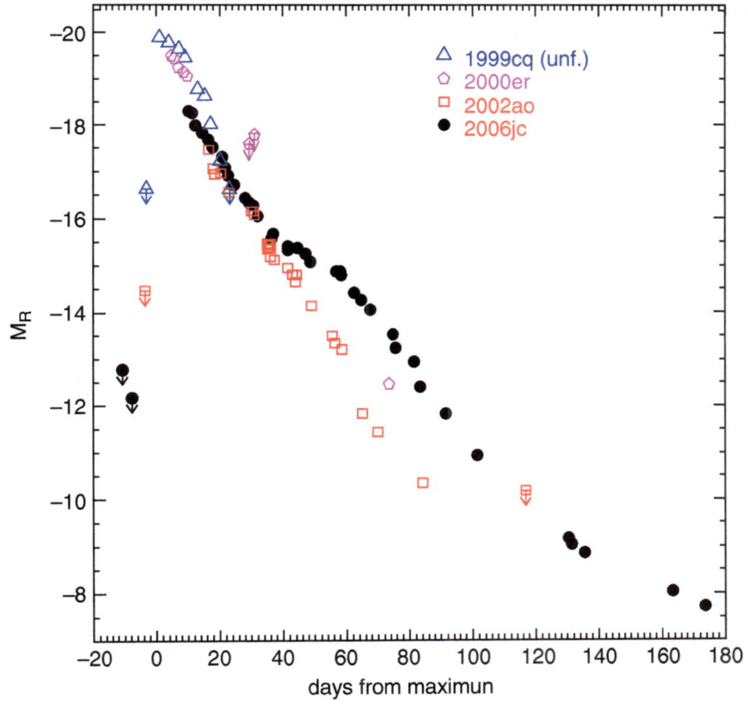

Fig. 16.16 Absolute R-band light curves of SN 2006jc and three other SN Ibn. *Symbols* to the left of the peak denote upper limits. Figure 8 "Massive stars exploding in a He-rich circumstellar medium—I. Type Ibn (SN 2006jc-like) events" from Pastorello et al. (2008b), by permission of Oxford University Press on behalf of the Royal Astronomical Society

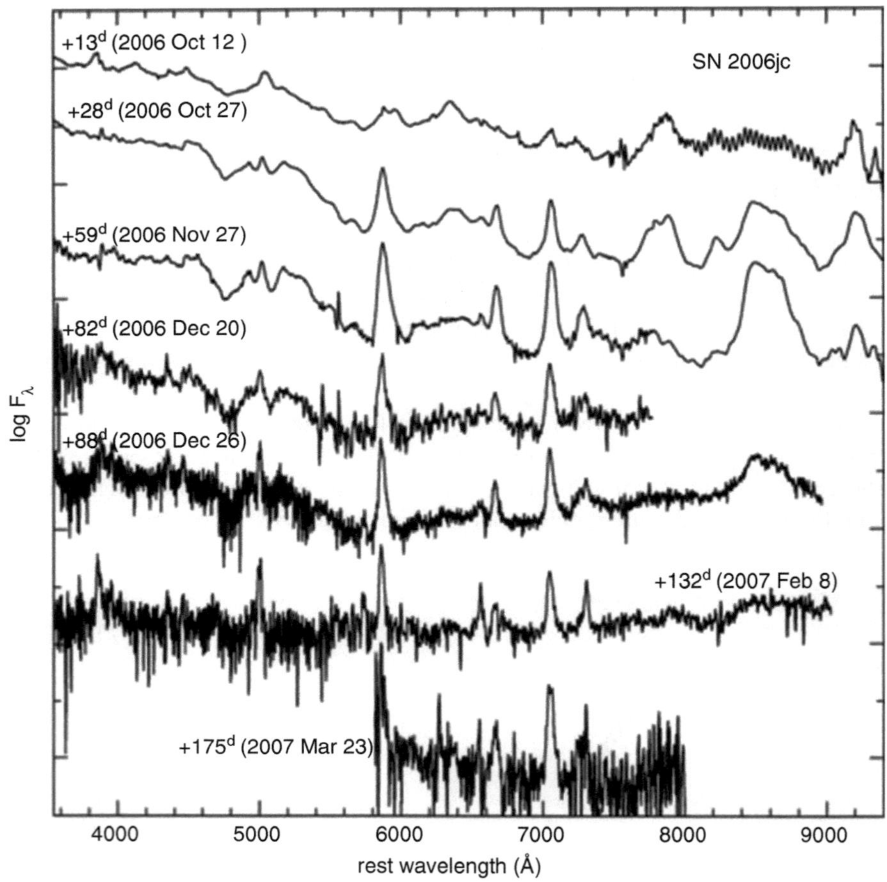

Fig. 16.17 Spectroscopic evolution of SN 2006jc. The most prominent He I emission features are λ5876, λ6678, and λ7065. Figure 5 "Massive stars exploding in a He-rich circumstellar medium—I. Type Ibn (SN 2006jc-like) events" from Pastorello et al. (2008b), by permission of Oxford University Press on behalf of the Royal Astronomical Society

those of the most rapidly-evolving SN Ic (Sect. 17.3.1). Two months after discovery, the SN 2006jc optical light curve began to decline even more rapidly, as the near-IR flux increased.

He I emission lines of Doppler widths near $2500\,\mathrm{km\,s^{-1}}$ (Fig. 16.17) were superimposed on lines of intermediate-mass elements forming in the supernova ejecta with velocities up to $9000\,\mathrm{km\,s^{-1}}$. Narrow P Cygni profiles of He I and O I suggested that the unshocked wind expanded at $500\,\mathrm{km\,s^{-1}}$. In early phases, a steep blue continuum declined to longer wavelengths and joined a flat red continuum at about 6000 Å. The He I emission lines in the blue showed P Cygni profiles with absorption components blueshifted by $1000\,\mathrm{km\,s^{-1}}$ while those in the red did not, and some of the lines in the red had double peaks; these properties suggest an asymmetric CSM. After 3 months, the optical continuum became redder. At later phases, there was no evidence of the broad forbidden emission lines typical of SN Ib/c. Intermediate-width lines of He I continued to dominate the spectrum 6 months after explosion. The broadening of the He I lines at this and earlier phases might have resulted from multiple electron scattering. Estimates of the helium-emitting material ranged from $\lesssim$0.1 to $\sim$1 $M_\odot$. Narrower emission of Hα appeared after 6 months, suggesting an outer CSM more rich in hydrogen.

The rapid decline of the optical light curve at 2 months, subsequent color evolution, and obscuration of the red sides of emission lines were manifestations of the formation of dust (Sakon et al. 2009; Smith et al. 2008; Mattila et al. 2008; Di Carlo et al. 2008; Anupama et al. 2009; Chugai 2009). As the optical light curve declined, the near-IR flux increased such that the quasi-bolometric light curve declined by 0.03 mag d^{-1}, steeper then the ^{56}Co decay rate and the observed decay rate of many typical stripped-envelope supernovae. Near-IR photometry showed clear evidence for $\sim 10^{-4}$ $M_\odot$ of newly-condensed dust, most likely in a CDS produced by CSI. At later epochs, mid-IR flux came from an echo from $\sim 10^{-2}$ $M_\odot$ of pre-existing dust in the wind of the progenitor.

SN 2006jc was not detected in the radio, but X-ray emission from CSI rose by a factor of five over 4 months and then declined steeply, suggesting that at that time the shock had emerged from a circumstellar shell (Immler et al. 2008). The inferred properties of the shell were consistent with ejection during the outburst 2 years before explosion.

On the assumption that the light curve was powered entirely by ^{56}Ni decay, hydrodynamical modeling favored an ejecta mass of $\lesssim$5 $M_\odot$, a high explosion energy of $\sim$10 B, and $M_{Ni} \gtrsim 0.2$ $M_\odot$ (Pastorello et al. 2008b; Tominaga et al. 2008). It is more likely that the light curve was primarily powered by CSI (Chugai 2009).

16.5.2 Synthesis of SN Ibn Characteristics

In addition to SN 2006jc, other SN Ibn include SN 1999cq (Matheson et al. 2000b), SN 2000er (Pastorello et al. 2008b), SN 2002ao (Foley et al. 2007), SN 2010al (Pastorello et al. 2015c), SN PS1-12sk (Sanders et al. 2013), and SN iPTF13beo (Gorbikov et al. 2014). SN 2005la (Pastorello et al. 2008c) and SN 2011hw (Smith et al. 2012a) showed emission lines of both H and He I and appear to be transitional events between SN IIn and SN Ibn.

After maximum light, SN Ibn light curves show a considerable dispersion in properties (Fig. 16.18). Some, like SN 2006jc, evolve more rapidly than typical SN Ib and resemble the most rapidly-evolving light curves of SN Ic such as SN 1994I. Others, like OGLE-2012-006, have light curves (especially in I-band) that are even flatter than ^{56}Co decay. On average, SN Ibn also are about two magnitudes brighter than typical SN Ib/c, owing largely to CSI.

Spectra of SN Ibn often rather closely resemble those of SN 2006jc, the principal characteristic being the intermediate-width emission lines of He I. When narrow

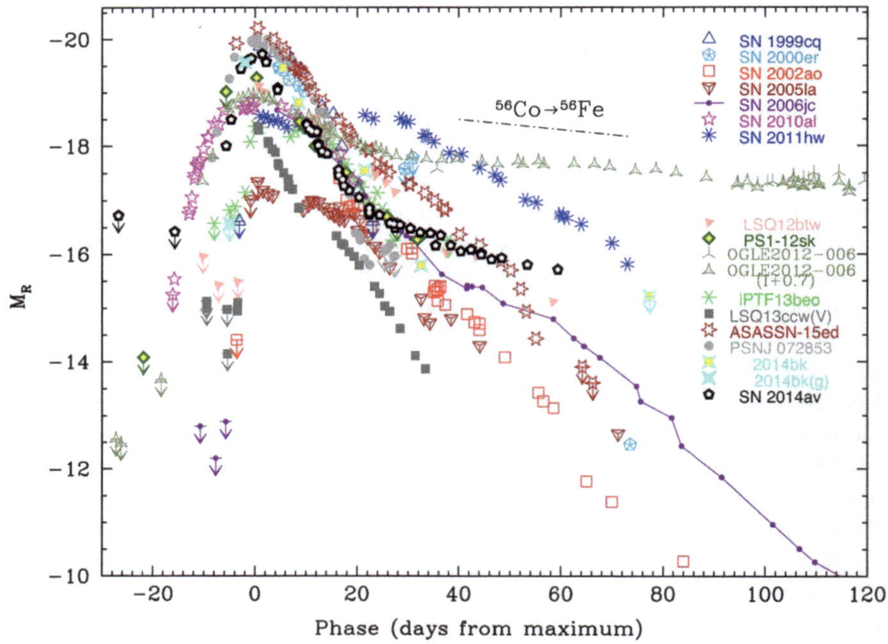

Fig. 16.18 Absolute R-band light curves of SN Ibn. Figure 3 "Massive stars exploding in a He-rich circumstellar medium—IX. SN 2014av, and characterization of Type Ibn SNe" from Pastorello et al. (2016), by permission of Oxford University Press on behalf of the Royal Astronomical Society

absorptions are present, they indicate unshocked wind velocities of $\lesssim$1000 km s^{-1}. Some SN Ibn also show broader He I lines (and in some cases O I and Ca II lines) that form in the ejecta up to $\lesssim$10,000 km s^{-1} (Pastorello et al. 2015d). In these cases, at least, the progenitor star had not been entirely stripped of helium when it exploded. In one event, the spectra evolved from those of an SN Ibn to those of a typical SN Ib (Pastorello et al. 2015c). The broad forbidden emission lines of typical SN Ib in the nebular phase typically are not seen, although [O I] was observed in one case (Pastorello et al. 2015d).

SN Ibn are rare and therefore, apart from SN 2006jc, members of the small observed sample of several dozen events are rather distant and difficult to observe, especially at late times. This may account, at least partially, for the usual lack of detection of the standard late-time forbidden lines as well as the usual lack of IR and X-ray detections.

It is not clear how SN Ibn fit into the overall panoply of stripped-envelope events. As hydrogen-deficient massive stars with winds of ∼1000 km s^{-1}, Wolf–Rayet stars are natural candidate progenitors (Foley et al. 2007; Pastorello et al. 2007, 2008b; Tominaga et al. 2008; Immler et al. 2008). As far as is known, however, Wolf–Rayet stars do not undergo LBV-like outbursts. In addition to the outburst 2 years prior to SN 2006jc (Sect. 16.5.1), the progenitor of SN PTF13beo was inferred, from an early CSI-driven peak in its light curve, to have undergone an unobserved SN 2006jc-like outburst that released $\sim$0.1 $M_{\odot}$ just weeks prior to explosion (Gorbikov et al. 2014). Presumably the helium-rich CSM of all SN Ibn is provided by LBV-like outbursts, not steady-state winds. Thus it has been suggested that the outbursts occur in a Wolf–Rayet star that recently evolved from the LBV phase, or during a transitional phase between an LBV and a Wolf–Rayet star (Smith et al. 2008, 2012a). These ideas conflict with traditional stellar-evolution theory, which predicts that an LBV phase is not soon followed by core collapse, but by $\sim$1 Myr of core helium burning. Nevertheless, observationally it appears that stars can undergo LBV outbursts shortly before SN IIn explosions or even explode as SN IIn (Sect. 14.3.3). In any case, further evidence is needed to firmly associate SN Ibn with Wolf–Rayet stars.

16.6 Summary

The progenitors of most SN Ib as deduced from analysis of light curves and spectra are of low mass compared to observed Wolf–Rayet stars, hence a direct identification may not be appropriate. Careful consideration of time- and spatially-dependent opacities are necessary for more accurate estimates. Current single-star models also produce higher progenitor-star masses than are commensurate with most SN Ib, although such models may be representative of Wolf–Rayet stars. An important task remains to continue to carefully compare the observations of Wolf–Rayet stars, their birth sites, and their distribution in host galaxies with corresponding properties of SN Ib to confirm the trends and as a check on the single-star stellar-evolution models. We must also bear in mind that rotation and magnetic fields may affect both the models and the real supernova progenitors in ways that are not yet fully understood.

Recent results have suggested that most massive stars are born in binary systems and are subject not just to Roche-lobe overflow and common-envelope evolution, but in some cases to stellar mergers. These probabilities, coupled with the observed properties of SN Ib—modest mass and low hydrogen abundance—strongly suggest that most, if not all, SN Ib arise in binary systems. The immediate progenitors may be helium stars orbiting presently bright, more massive secondaries. The rather low-mass progenitors stars may have lower luminosities than their companions and be difficult to detect in archival observations. The outcome of an SN Ib explosion might result in a disrupted binary, but in some cases the result might be something like the Be-star X-ray binaries.

Whether one calls these low-mass helium-star progenitors "Wolf–Rayet" stars may depend on taste and semantics, but we prefer to think of Wolf–Rayet stars as an observed category of stars, not the product of computers. However one comes down on the issue of nomenclature, the identification of the progenitor systems that lead specifically to SN Ib rather than SN IIb or SN Ic remains a high priority.

Invocation of binary systems to account for the origin of SN Ib allows the progenitor stars to have a wide range in initial ZAMS mass, including stars of rather modest mass that lose their hydrogen envelopes through some binary interaction, not by winds alone. This scenario does not, however, provide a simple explanation for why SN Ib are more centrally concentrated in their host galaxies than SN II. This property suggests that the progenitor stars of SN Ib are systematically more massive than those of SN II, the progenitors of which are effectively single stars (including those in wide binaries that never interact).

SN Ib require helium-rich atmospheres to produce the observed strong optical lines of He I shortly after maximum. The strength of He I as a distinguishing feature remains a difficult diagnostic since the helium is strongly out of LTE. Its appearance thus depends not only on abundance, but excitation. While the relevant lines of He I may be produced by thermal and radiative excitation in the extreme conditions shortly after breakout, their appearance at later epochs requires exposure to the fast electrons generated by radioactive decay of ^{56}Ni and ^{56}Co. The implication is that if a substantial abundance of helium is present, the prominent optical He I lines will appear as γ-rays permeate the ejecta. If helium is there to be seen, it will be seen at some phase.

The observational detection of helium also depends on technique. Although He I lines can be seen in the optical, the He I line at 1.0830 μm in the near-IR is a much more sensitive indicator of the presence of helium. That line can be seriously blended and needs to be complemented with observation of the 2.058 μm line. In addition, He I may be identified in spectropolarimetry even when it is not discernible in total-flux spectroscopy.

SN Ib may contain small amounts of hydrogen, but the H lines are weaker than those observed in SN IIb. Thus SN Ib may have a distinctly different progenitor evolution than SN IIb.

Analysis of SN Ib with roughly similar peak light-curve properties implies that there are, nevertheless, significant differences in their physical properties. In particular, SN Ib with similar light curves near peak can show a wide range of behavior at late times.

Calculations of nebular-phase spectra after about 200 days show that SN Ib will not reveal He I in the optical even if helium is present. The spectra at that phase are rather insensitive to variations in the mass and composition, although they may reveal evidence for asymmetry. The similarity of the optical spectra of SN Ib and SN Ic events at late times thus does not mean that their progenitor systems are physically similar.

Spectropolarimetry of a few SN Ib typically shows a mild asymmetry in the continuum with stronger polarization in lines that frequently show different polarization angles from the continuum. The geometry of these explosions is complex. Analysis of the spectropolarimetry suggests that jets or plumes carry radioactive matter that imposes asymmetric exciton of the ejecta. The helium envelope may have a different rotation axis from the inner core. These protrusions of radioactive matter are often constrained to have not penetrated beyond the photosphere near or shortly after maximum light. This property may differentiate typical SN Ib/c and

SN Ic-bl (Sect. 17.4) not associated with gamma-ray bursts from SN Ic-bl that are associated with gamma-ray bursts, where the jet is clearly manifest and escapes into the surrounding CSM. There is a need for a larger sample of spectropolarimetry of SN Ib with a much finer time-sampling than current data provide.

There are still major holes in our understanding of what differentiates SN Ib from SN IIb and SN Ic. These differences probably involve both the progenitor evolution and the explosion mechanism, including the type of compact remnant left behind: black hole or neutron star, and magnetar or normal magnetic field. While SN Ib surely arise in core collapse, there is as yet no direct observational evidence for a compact object.

The defining characteristic of SN Ibn is the presence of intermediate-width He I lines that form in helium-rich CSM. In some cases, broad lines of helium and other elements, forming in the ejecta, also are seen. Near maximum, the light curves of SN Ibn usually evolve more rapidly and are brighter than those of typical SN Ib/c. The light curves appear to be powered primarily by CSI. SN Ibn are core-collapse events that occur in young populations. The prototype, SN 2006jc, was observed to undergo an LBV-like outburst 2 years prior to explosion, and there is further evidence that SN Ibn tend to undergo such pre-explosion outbursts. The progenitors are likely to be stars that are undergoing a transitional phase between an LBV and a Wolf–Rayet star, and/or Wolf–Rayet stars that recently evolved from the LBV phase.

Chapter 17
Type Ic Supernovae

17.1 Introduction

The identification of He I lines in SN Ib led to the recognition that other "peculiar" Type I events did not display He I lines and hence were yet different. SN 1983I and SN 1983V were quite similar to one another and also to SN 1962L and SN 1964L, the peculiar Type I that began this line of inquiry (Sect. 16.1). SN 1983I and SN 1983V had typical peak brightness of $M_V \simeq -17$ to -18, but otherwise their light curves approximately resembled those of the more luminous SN Ia. While SN 1983I and SN 1983V showed some spectral similarity to SN 1983N and SN 1984L, they exhibited little or no evidence of He I lines (Wheeler et al. 1987). These apparently helium-deficient events showed a strong absorption of O I near maximum light that might be adopted as one of the defining characteristics (Wheeler et al. 1994; Matheson et al. 2001; Liu et al. 2015b).

The notion that this helium-rich and helium-poor distinction should be formalized with the labels Type Ib and Type Ic (Wheeler and Harkness 1986) set off a continuing debate as to the proper observational classification of these and related events (Sect. 17.3.2). The burgeoning spectral/light-curve classification scheme (Harkness and Wheeler 1990) has blossomed with details, but still maintains the basic observational and probably physical classification, differentiating SN Ib from SN Ic. The classification undoubtedly depends on the epoch when observations are made. Early spectra of SN Ib often show evidence of outer wisps of hydrogen (Sect. 16.3.2), and small amounts of helium may be identified in early phases of some SN Ic (Sect. 17.3.2). There might be a range in outer helium abundance, so that SN Ib and SN Ic could represent extremes of a continuum, rather than physically distinct categories (Filippenko et al. 1990; Wheeler et al. 1994; Hamuy et al. 2002). It is now common to refer to Type Ib/c to capture this ambiguity. There are, however, lines of evidence suggesting that SN Ic are physically distinct from SN Ib. The preponderance of the still-inadequate data suggests that it remains appropriate to distinguish between SN Ib and SN Ic, as we do here.

D. Branch, J.C. Wheeler, *Supernova Explosions*, Astronomy and Astrophysics Library, DOI 10.1007/978-3-662-55054-0_17

Broad-lined SN Ic-bl are discussed in Sect. 17.4. Hydrogen and helium-deficient events also occur among superluminous supernovae; these are discussed in Chap. 18.

17.2 Case Study: SN 1994I

SN 1994I (Puckett et al. 1994), in the nearby spiral galaxy M51, was well-observed and has been the subject of much modeling. Among stripped-envelope events, its light-curve decline was one of the most rapid on record and its early spectra were especially blue. SN 1994I may be somewhat extreme, but it well represents the challenges to be faced in understanding SN Ic.

Extensive optical observations of SN 1994I were presented by Filippenko et al. (1995b), Richmond et al. (1996b), and Clocchiatti et al. (1996b, 2008) among others. The extinction was significant but rather uncertain, $A_v \simeq 1.4 \pm 0.5$ mag. The V- and R-band light curves rose to maximum in about 10 days and then declined rapidly, more than 0.1 mag d^{-1}. The pseudo-bolometric (BVRI) light curve (Fig. 17.1) showed a nearly constant rate of decline from 50 to 150 days after explosion that was somewhat steeper than the slope expected for decay of ^{56}Co but quite similar to the light curves of the SN Ic-bl events SN 1998bw and SN 2002ap (Sect. 17.4). While SN 1998bw and SN 2002ap continued with that slope for at least 300 days, the final *HST* observation of SN 1994I by Clocchiatti et al. (2008) fell

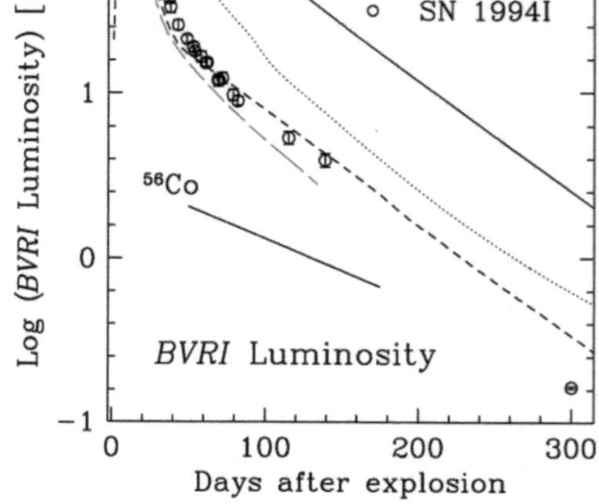

Fig. 17.1 Pseudo-bolometric (BVRI) light curves of SN 1994I as well as the Type Ic SN 1990B, the Type Ic-bl SN 2002ap and SN 1998bw, and the Type Ia SN 1992A. Figure 5 "Late-Time HST Photometry of SN 1994I: Hints of Positron Annihilation Energy Deposition" from Clocchiatti et al. (2008) ©The Astronomical Society of the Pacific. Reproduced by Permission. All Rights Reserved

substantially below the extrapolated curve from 50 to 150 days, by about a factor of 1.6. Light curve models designed to connect the data at about 50 days to that at 300 days were consistent with a freely-expanding envelope in which the γ-ray optical depth decreased as t^{-2}. Those models suggested that deposition of positrons became a significant source of heating as matter became ever more transparent to γ-rays from the decay of ^{56}Co. These models fell below the intermediate data points at about 120 and 150 days, also by factors of order two. The latter might be attributed to observational scatter that was more significant than for the point at 300 days. In that case, the similarity of the light curve from 50 to 150 days to the SN Ic-bl might be spurious, and the steeper late-time light curve of SN 1994I could be attributed to a smaller ejecta mass, for instance. The late-time photometry might then be consistent with much of the decay energy being deposited in the ejecta at 300 days, and that SN 1994I had not yet undergone an infrared catastrophe (Sect. 5.5.1) at 300 days. A general discussion of light curve models for SN Ic is given in Sect. 17.5.2.

SN 1994I was definitely in the helium-poor category (Wheeler et al. 1994; Filippenko et al. 1995b; Clocchiatti et al. 1996b). Spectra near maximum showed strong O I and the Ca II near-IR triplet in absorption (Figs. 17.2 and 17.3). A strong absorption feature at about 1.06 μm was probably a blend, not just He I 1.0830 μm (Sect. 17.3.2).

Observations of radio emission from SN 1994I from 3 days after optical discovery until 8 years later were reported by Weiler et al. (2011; Fig. 17.4). The radio light curves were consistent with shock propagation into CSM from a presupernova steady-state wind of mass-loss rate $\sim 10^{-6}\,M_\odot\,y^{-1}$ for a wind velocity of $100\,km\,s^{-1}$.

SN 1994I was detected in X-rays about 90 days after explosion with a flux (1–2.4 keV) of $\sim 1.6 \times 10^{38}\,erg\,s^{-1}$ and again 7–8 years later with a flux (0.3–2 keV) of $\sim 10^{37}\,erg\,s^{-1}$ (Immler et al. 1998, 2002). The early X-ray observations were consistent with the shocking of $\sim 10^{-3}\,M_\odot$ of matter with a density of $\sim 2 \times 10^5\,cm^{-3}$ within a radius of $\sim 10^{16}\,cm$. The X-ray light curve spanning 8 years was roughly consistent with a steady-state presupernova wind of mass-loss rate of $\sim 10^{-4}\,M_\odot\,y^{-1}$ for a wind velocity of $100\,km\,s^{-1}$, significantly higher than the estimate from radio emission. If clumping in the wind was responsible for the discrepancy (Weiler et al. 2011), the best estimate of the mass-loss rate would be $\sim 10^{-5}\,M_\odot\,y^{-1}$.

The first models for SN 1994I were generated by stripping mass from a helium-star model and then converting some of the remaining C/O to helium in a parameterized way (Iwamoto et al. 1994), or based on helium cores with mass loss by winds (note that Wolf–Rayet mass-loss rates were later significantly revised to allow for the effects of clumping; Sect. 8.4.1) plus some late binary-transfer mass loss (Woosley et al. 1995). The latter models yielded more helium in the outer layers and almost surely had too much mass in helium to correspond to SN Ic (see Sect. 17.5), while the former removed the helium in an ad hoc manner. An ejected mass $\lesssim 1\,M_\odot$ fit the pseudo-bolometric light curves of SN 1994I reasonably well for the first 60 days. The models required substantial reddening, an issue of uncertainty for the light curves of many stripped-envelope supernovae. Iwamoto et al. concluded

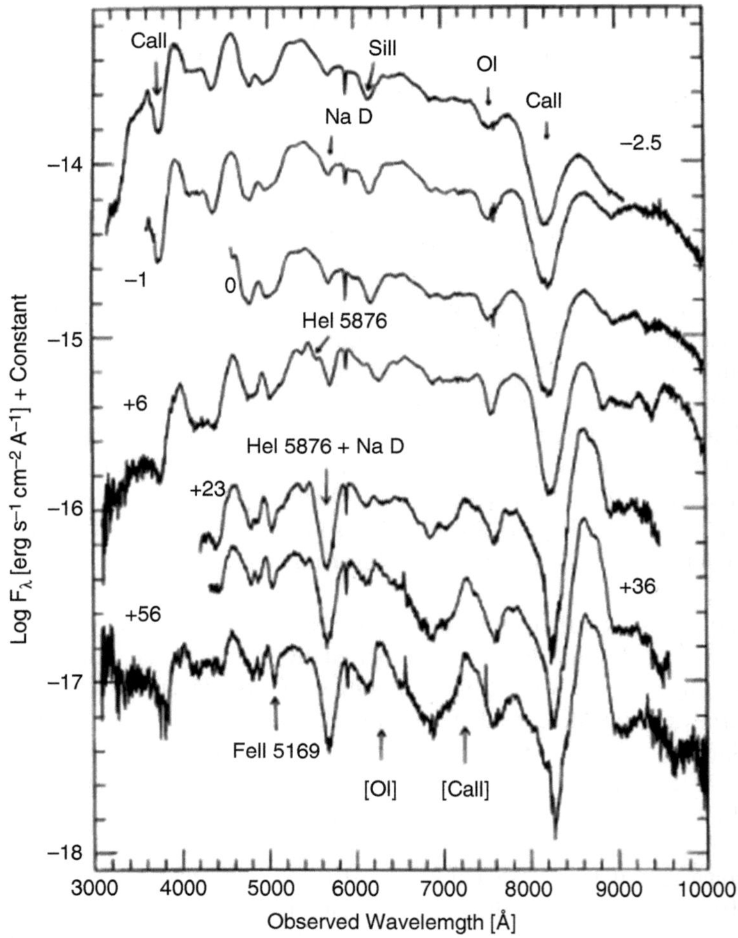

Fig. 17.2 Spectra of SN 1994I. Epochs are relative to maximum light and the spectra are displaced vertically for clarity. From "SN 1994I: Disentangling He I Lines in Type Ic Supernovae" (Clocchiatti et al. 1996b). © AAS. Reproduced with permission

that the progenitor of SN 1994I could not correspond to any classical category of observed Wolf–Rayet stars. Kleiser and Kasen (2014) argued that the effects of oxygen recombination can reduce the effective opacity, narrow the light curve peak for a given ejecta mass, and produce an "oxygen-plateau" light curve in SN 1994I (Sect. 17.5.2).

Atmosphere models of near-maximum spectra of SN 1994I incorporating schematic NLTE effects (LTE abundances scaled with an arbitrary departure coefficient) showed that models with power-law density structure consisting of oxygen with a solar abundance of metals gave a reasonable representation (Wheeler et al. 1994). Important line identifications depended on plausible NLTE

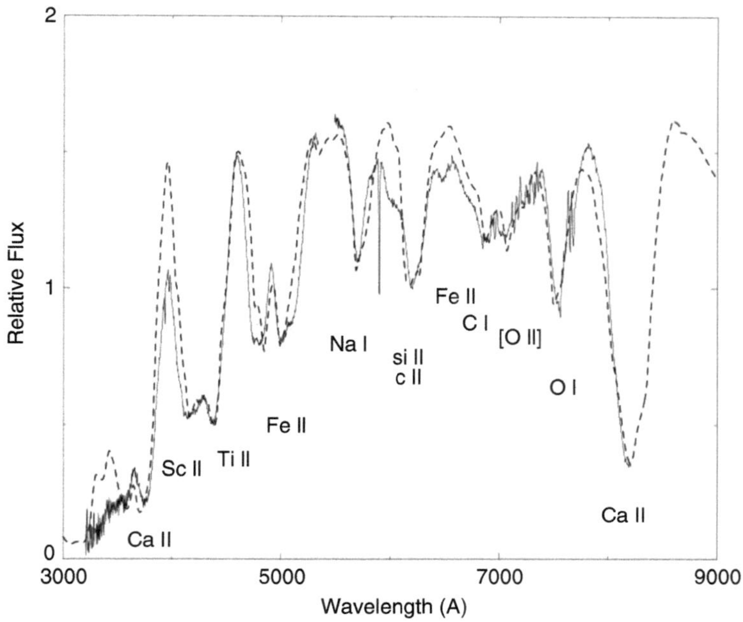

Fig. 17.3 The spectrum of SN 1994I 4 days after maximum light is compared to a SYNOW synthetic spectrum having a photospheric velocity of 11,000 km s^{-1}. The flux is in F_{ν} units. From "Direct Analysis of Spectra of the Type Ic Supernova 1994I" (Millard et al. 1999). © AAS. Reproduced with permission

enhancements. An absorption at 6170 Å could be attributed to Si II λ6355, albeit weaker than this feature appears in SN Ia. In SN 1994I, the feature could be produced with only a solar abundance of silicon (Wheeler et al. 1987). Another controversial feature at 6370 Å in SN 1994I was more likely to be C II rather than He I λ6678. The latter identification would require an unphysical NLTE enhancement and would have imprinted a discernible separate He I λ5876 absorption feature just to the blue of the Na I D-line absorption. NLTE atmosphere models were used to limit the amount of helium in the ejecta to $\lesssim$0.1 $M_{\odot}$ (Hachinger et al. 2012a). The issue of identifying He I in the atmospheres of SN Ic is discussed further in Sects. 17.3.2 and 17.5.3.

The nebular phase began about 2 months after explosion and showed characteristic emission features of [O I] $\lambda\lambda$6300,6364, [Ca II] $\lambda\lambda$7291,7323, and the Ca II near-IR triplet, as well as especially strong [C I] λ8700, which is not typical of core-collapse events (Sauer et al. 2006; Fig. 17.5). Models of nebular spectra suggested a slightly higher mass than that of the 2.1 $M_{\odot}$ model of Iwamoto et al. (1994).

SN 1994I was almost surely the result of core-collapse, but there has been no reported evidence for a compact remnant.

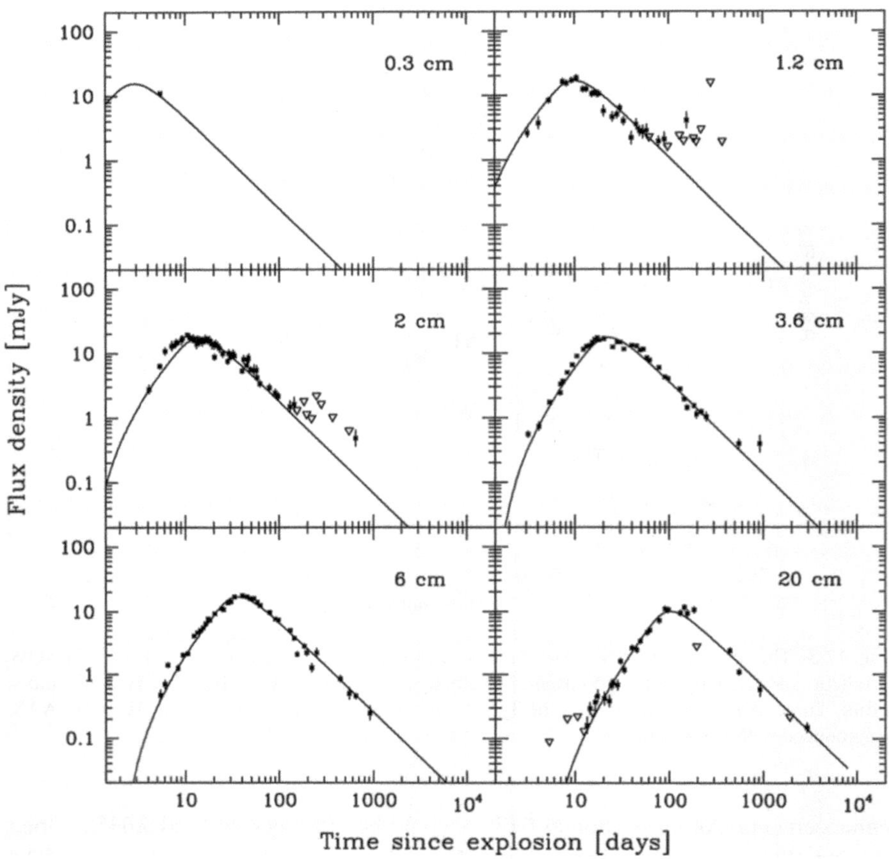

Fig. 17.4 Radio light curves of SN 1994I compared to a best-fit model. From "Radio Emission from SN 1994I in NGC 5194 (M51): The Best-studied Type Ib/c Radio Supernova" (Weiler et al. 2011). © AAS. Reproduced with permission

While SN 1994I may have been somewhat extreme in its properties, any consistent theory of SN Ic must encompass its characteristics: very little, if any helium ($<<0.1\ M_\odot$); a low progenitor mass near $2\ M_\odot$, implying that the progenitor was not a classic Wolf–Rayet star with mass $\gtrsim 10\ M_\odot$ (but see the discussion of uncertainties in Sect. 17.5.2); and CSM roughly consistent with a steady-state wind of moderate mass-loss rate.

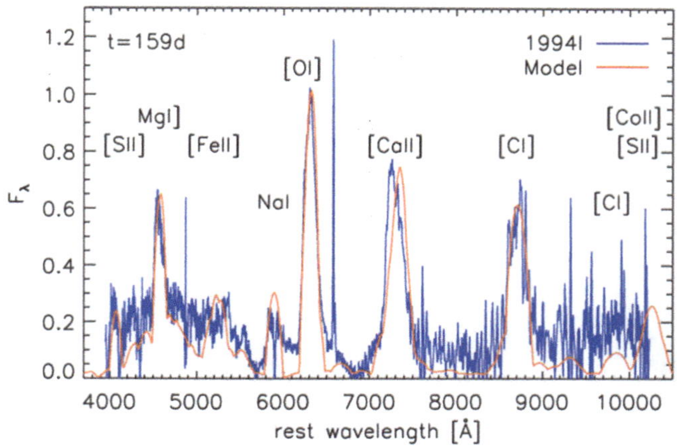

Fig. 17.5 The spectrum of SN 1994I 159 days after explosion is compared to a model nebular spectrum. Figure 2 "The properties of the "standard" Type Ic supernova 1994I from spectral models" from Sauer et al. (2006), by permission of Oxford University Press on behalf of the Royal Astronomical Society

17.3 Synthesis of SN Ic Characteristics

In addition to SN 1994I (Sect. 17.2), well-observed typical SN Ic include SN 2007gr (Chen et al. 2014b) and SN 2004aw (Taubenberger et al. 2006). Proper study of SN Ic requires multi-epoch, multiwavelength observations, both photometric and spectroscopic, and, ideally, with spectropolarimetry. It remains difficult to date the spectral phase of SN Ib/c with "snapshot" data.

At maximum light, SN Ic have a mean $M_B \lesssim -17$. In a volume-limited sample, SN Ic constitute about 15% of all core-collapse events. No progenitors have been detected in archival images. Their environments in their host galaxies tend to be more metal-rich and to have higher star-formation rates than those of SN Ib and SN II (Sect. 3.3). Most SN Ic occur in spiral galaxies, but they may occasionally appear in elliptical or S0 galaxies that have pockets of ongoing star formation (Suh et al. 2011).

17.3.1 Light Curves

Light curves of SN Ib/c from peak to late tail are heterogeneous, but can be put roughly into three categories, slow, intermediate, and rapid decline (Clocchiatti and Wheeler 1997). Near the peak, light curves of SN Ic, including those of SN Ic-bl (Sect. 17.4), are roughly similar in shape (Fig. 17.6). SN Ic and SN Ic-bl tend to have shorter rise times than SN IIb and SN Ib. The immediate decline from maximum tends to be slower than the rise time, but similar for all categories of

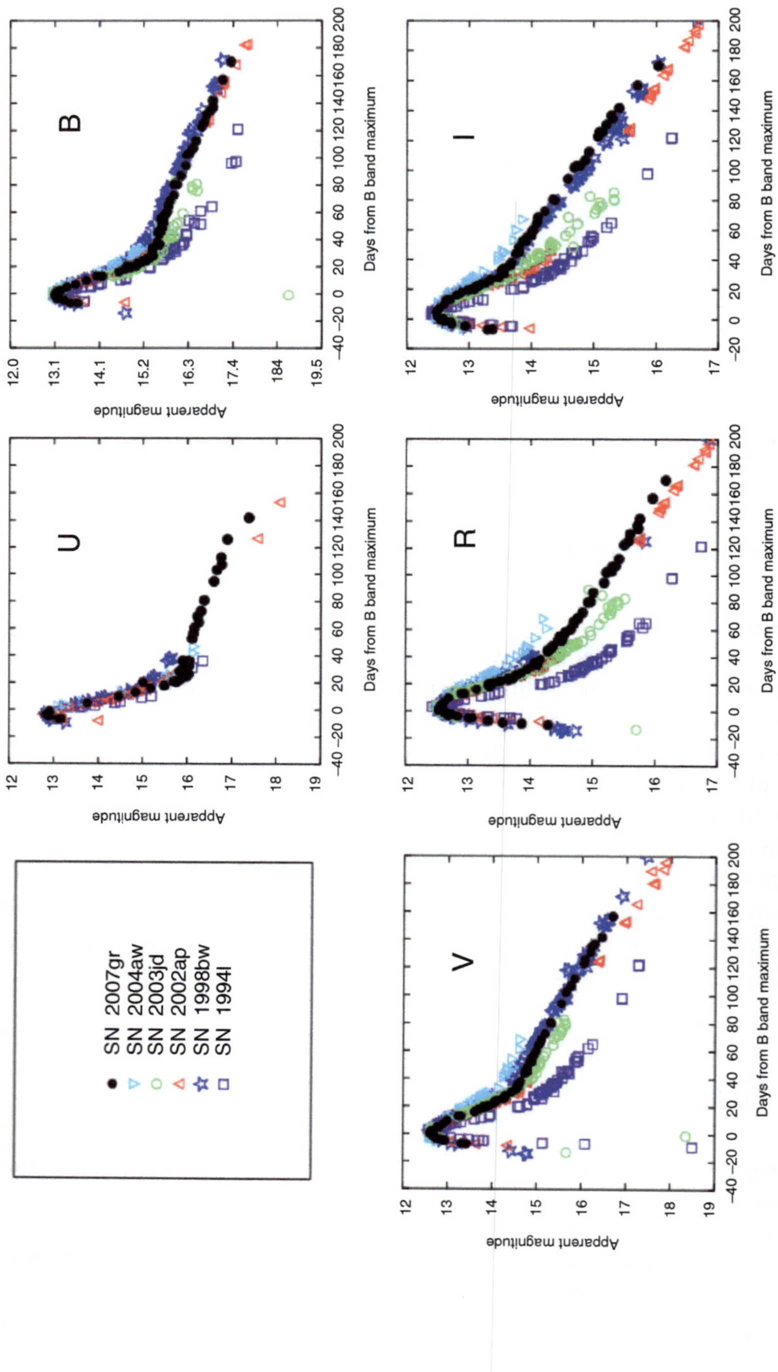

Fig. 17.6 Light curves of the Type Ic SN 1994I, SN 2003jd, SN 2004aw, and SN 2007gr as well as the Type Ic-bl SN 1998bw and SN 2002ap. The light curves are normalized at peak to facilitate comparison of their shapes. From "Extensive optical and near-infrared observations of the nearby, narrow-lined type Ic SN 2007gr: days 5 to 415" (Hunter et al. 2009) reproduced with permission, ESO

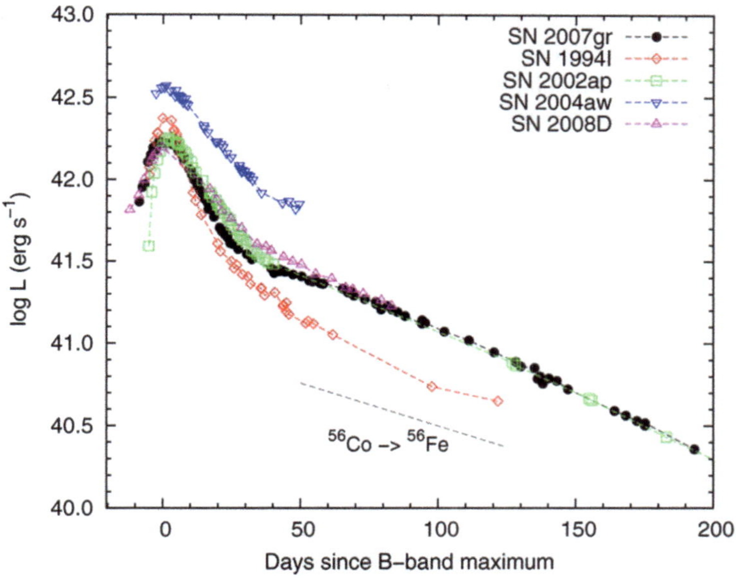

Fig. 17.7 Pseudo-bolometric (UVOIR) light curves of the Type Ic SN 1994I, SN 2004aw, and SN 2007gr as well as the Type Ib SN 2008D and the Type Ic-bl SN 2002ap. From "Optical Observations of the Type Ic Supernova 2007gr in NGC 1058" (Chen et al. 2014b). © AAS. Reproduced with permission

stripped-envelope supernovae (Drout et al. 2011; Taddia et al. 2015b; Bianco et al. 2014). There are exceptions to these general trends: SN 2011bm was slower than most on the rise and the decline from maximum (Valenti et al. 2011), and our case study SN 1994I was significantly faster on the rise and early decline than the broader sample (Figs. 17.6 and 17.7), as were SN 1987M (Filippenko et al. 1990) and SN 2009bb (Pignata et al. 2011). SN Ic do not display significant correlation between the brightness at peak and the immediate postpeak decline rate. Means to estimate bolometric light curves were given by Cano (2013) and Lyman et al. (2014a).

The late-time light curves of SN Ic often follow a common locus with a well-defined exponential decline for several hundred days. In these cases, the later decline is about 0.016 mag d^{-1}, for both normal SN Ic and SN Ic-bl (see SN 2007gr, SN 2002ap, and SN 1998bw in Fig. 17.6), significantly more rapid than ^{56}Co decay. This common decay rate is also seen in other stripped-envelope supernovae (see SN 2008D in Fig. 17.7). Again, there are exceptions to this behavior: SN 2004aw declined more slowly (Tomita et al. 2006) and SN 1990B (Clocchiatti et al. 2001) more rapidly.

As for other stripped-envelope supernovae, SN Ic do not show the secondary peak in the near-IR that characterizes SN Ia (Sect. 20.4). This may be due to the relatively smaller proportion of IGEs in the ejecta that serves to move flux from the UV to the near-IR in SN Ia.

Some supernovae that have spectra resembling SN Ic have light curves with very rapid declines from maximum. A prime example of this behavior is SN 2005ek that peaked at $M_R = -17.3$ and declined by three magnitudes in the next 2 weeks (Drout et al. 2013). Other rapidly declining transients that might be related to SN 2005ek include SN 2002bj and SN 2010X. Whether any of these events prove to be core-collapse supernovae rather than explosions of or on white dwarfs (Sect. 23.4) remains to be established.

17.3.2 Spectra

SN Ic are defined as having little conspicuous evidence for H or He I lines, weak Si II, and strong O I λ7774 absorption near peak light (Fig. 17.8). In the photospheric phase, the O I absorption is typically stronger in SN Ic than in SN Ib, consistent with directly observing the ejecta of a nondegenerate C/O core. Ubiquitous lines of Fe II are also present, the result of relatively cool atmospheres. The nebular spectra show prominent emission lines of Mg I] λ4571, the Na I D-lines, [O I] $\lambda\lambda$6300,6364, and [Ca II] $\lambda\lambda$7291,7323, very similar to those of SN Ib, SN IIb, and SN IIP, setting aside the H lines of the latter. At late times, SN Ic display a range of emission-line widths that is not obviously different from that of SN Ib (Modjaz et al. 2008; Milisavljevic et al. 2010). With the exception of the rather strong O I absorption, early optical spectra of SN Ic are notable for a relative lack of distinct identifying features and the nebular spectra give few if any unique clues to spectral type.

Some SN Ic may show evidence for weak He I absorption, but most do not (Matheson et al. 2001; Liu et al. 2015b). The lack of conspicuous evidence for helium in SN Ic also raises a raft of issues relevant both to identifying the spectral class and constraining the nature of the progenitor evolution. There is evidence in some SN Ic for transient optical He I lines at high velocities (Filippenko et al. 1995b; Clocchiatti et al. 1996b), although the issue of misidentified lines remains (Matheson et al. 2001). A notch corresponding to He I λ5876 in the blue wing of the Na I D feature may be consistent with a corresponding feature of He I 1.0830 μm, but that feature may itself be blended (Baron et al. 1999). Near-IR spectra can reveal evidence for He I, especially the relatively unblended line at 2.058 μm. While both SN Ib and SN Ic show a feature at about 1 μm, events identified by their optical spectra as SN Ib tend to also show the 2.058 μm feature whereas SN Ic do not (Hamuy et al. 2002; Gerardy et al. 2004a; Taubenberger et al. 2006; Valenti et al. 2008a; Stritzinger et al. 2009; Modjaz et al. 2009).

The velocity of any detected helium is an important discriminant. In some cases (Filippenko et al. 1995b; Clocchiatti et al. 1996b), typical helium velocities are near 17,000 km s^{-1}. This implies that the helium is confined to the outermost layers. There must be much less than 0.1 $M_\odot$ of helium to avoid a kinetic energy exceeding

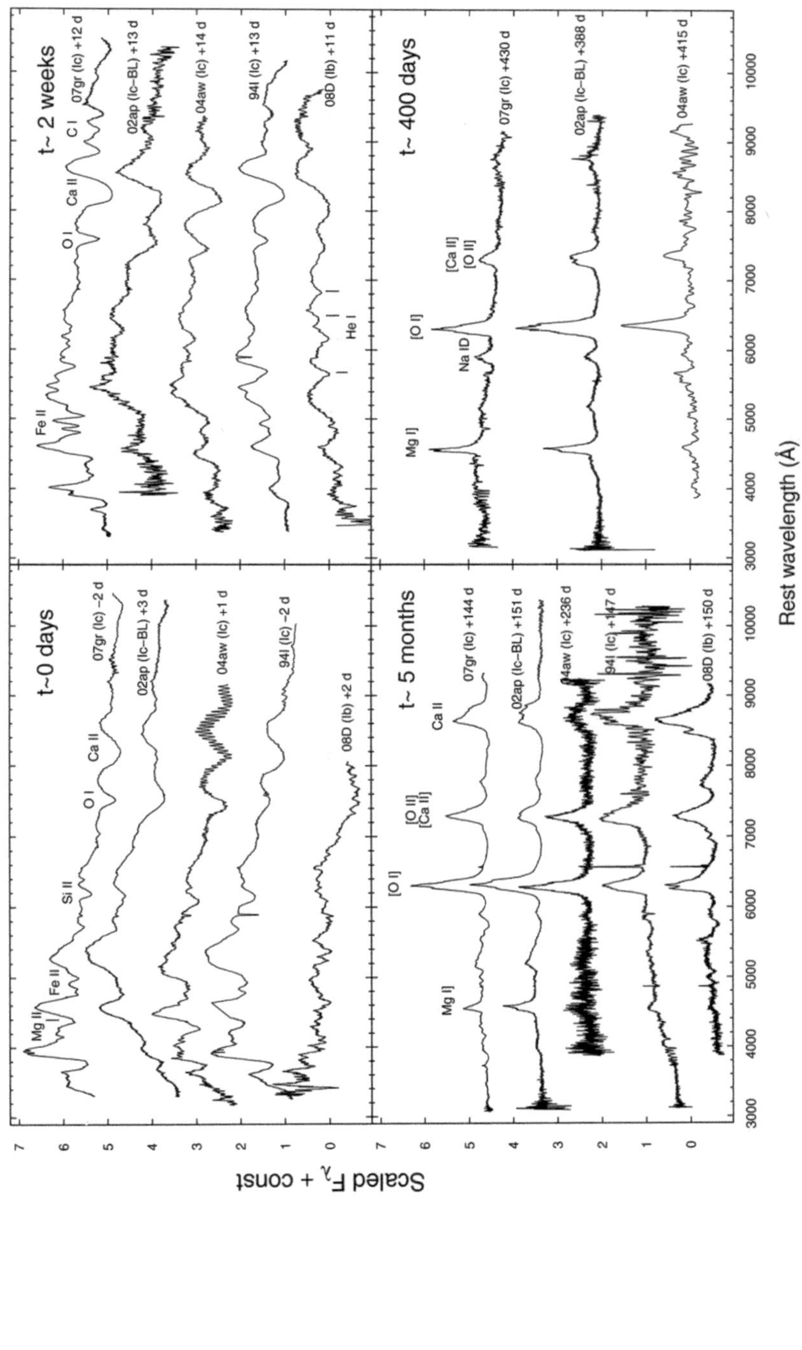

Fig. 17.8 Spectral evolution of the Type Ic SN 1994I, SN 2004aw, and SN 2007gr as well as the Type Ib SN 2008D and the Type Ic-bl SN 2002ap. From "Optical Observations of the Type Ic Supernova 2007gr in NGC 1058" (Chen et al. 2014b). © AAS. Reproduced with permission

1 B in the helium-containing layers alone, and hence substantially less than the several tenths of $M_\odot$ of helium in typical SN Ib. Examples of putative SN Ic with helium at lower velocities have been suggested. This might allow a higher abundance of helium on purely energetic grounds, but as for any identification of weak lines, care must be taken. Fe II lines are weak and ubiquitous. Early spectra of SN 1987M showed evidence for a high-velocity feature of the Ca II near-IR triplet (Filippenko 1992). The spectra of SN Ic may frequently involve coexisting photospheric-velocity and high-velocity features (Parrent et al. 2007). Helium may co-exist in this high-velocity matter but be difficult to establish due to contamination by other lines. The phase of observation is also important. Substantial helium might be present yet unexcited at some phases, but its presence may be revealed at other phases when it is exposed to radioactive decay products (Sect. 17.5.3 and discussion of SN 2005bf in Sect. 16.3.2).

Helium revealed in the near-IR would certainly be important, but if the amount is small compared to that in SN Ib a significant difference in the progenitor evolution might still be implied. Similar issues arise in spectropolarimetry, which can reveal evidence for helium that is obscure in total-flux spectra. A collection of optical spectra of stripped-envelope supernovae typed using the SNID code suggested some revisions of spectral types compared to older literature (Modjaz et al. 2014). The data employed in this study did not include the O I λ7774 and the Ca II near-IR triplet features. A broader wavelength coverage might lead to yet more revision.

The question of whether some SN Ic show evidence for Hα remains uncertain (Parrent et al. 2016). As for SN Ib, some SN Ic show an absorption minimum near 6200 Å around maximum light that may be a combination of photospheric Si II λ6355, C II λ6580, and high-velocity, detached Hα moving in excess of 15,000 km s^{-1} (Branch et al. 2002; Elmhamdi et al. 2006; Folatelli et al. 2006; Branch et al. 2006b). A small abundance of hydrogen might explain the 6300 Å feature observed in some SN Ic without excessive veiling of the other features by electron-scattering (Sect. 15.2.8). As for similar arguments for limits on helium in SN Ic, the high velocity of putative Hα lines implies that the mass of any hydrogen must be much less than 0.1 $M_\odot$, so that its kinetic energy is not extreme and unphysical. While evidence for the existence and kinematics of hydrogen in SN Ic is important to constrain the progenitor system, it is clear that SN Ic have substantially less hydrogen than SN IIb and hence require a substantially different progenitor evolution in some regard. It remains difficult to make such a comparison with SN Ib, since the abundance of hydrogen in both SN Ib and SN Ic is difficult to constrain. For further discussion, see Sect. 17.5.

Events that might represent transitions between SN Ib that have conspicuous evidence for helium and SN Ic that do not include SN 1990W (Wheeler et al. 1994), SN 1990U (Gomez and Lopez 1994), SN 1997X (Munari et al. 1998), and SN 1999ex (Hamuy et al. 2002). See also Valenti et al. (2011) and Modjaz et al. (2014).

Spectra of SN Ic-bl are discussed in Sect. 17.4.2.

17.3.3 Polarization and Asymmetry

With less mass in the outer layers of SN Ic, there is less tendency to generate small-scale Rayleigh–Taylor instabilities at composition boundaries that could lead to signatures of asymmetry (Iwamoto et al. 2000). There may be strong intrinsic asymmetries in the explosion mechanism that drive some sort of mixing, plumes, or jets (Hammer et al. 2010).

SN 2007gr was a well-studied, typical SN Ic that apparently had a low-mass, oxygen-rich core (Valenti et al. 2008b). Spectropolarimetric observations (Fig. 17.9) of SN 2007gr were obtained 21 days after light-curve peak, about 37 days after

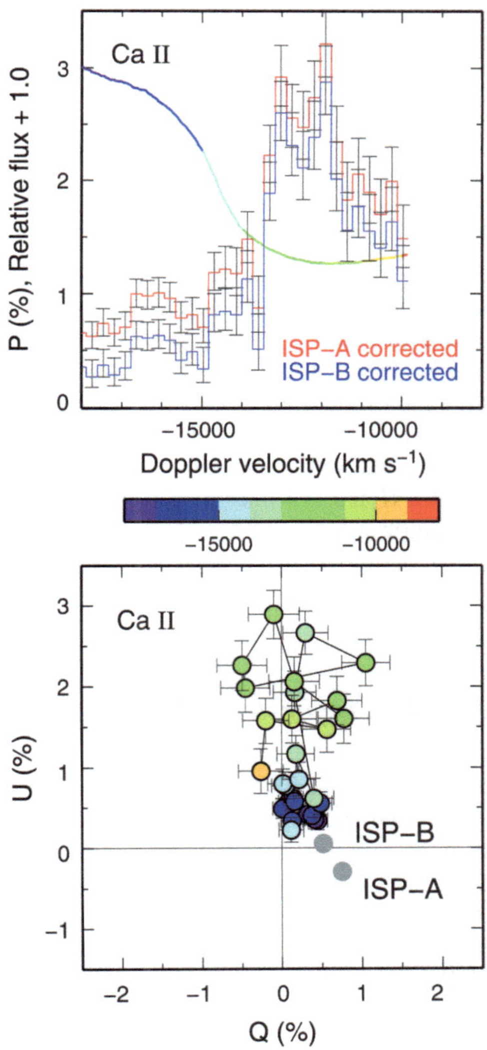

Fig. 17.9 Spectropolarimetric observations of the Ca II near-IR triplet in SN 2007gr about 37 days after explosion at the transition to the nebular phase. The *upper panel* shows the total flux spectrum and the corresponding polarization. The line minimum is substantially polarized, illustrating the inverse P Cygni profile. The *lower panel* presents the same wavelength range in the Q/U plane. The polarization angle of the Ca II near-IR triplet (*green points*) is similar to that of the less polarized continuum (*blue points*). The loop of *green points* represents departures from axisymmetry. From "Optical Spectropolarimetry and Asphericity of the Type Ic SN 2007gr" (Tanaka et al. 2008). © AAS. Reproduced with permission

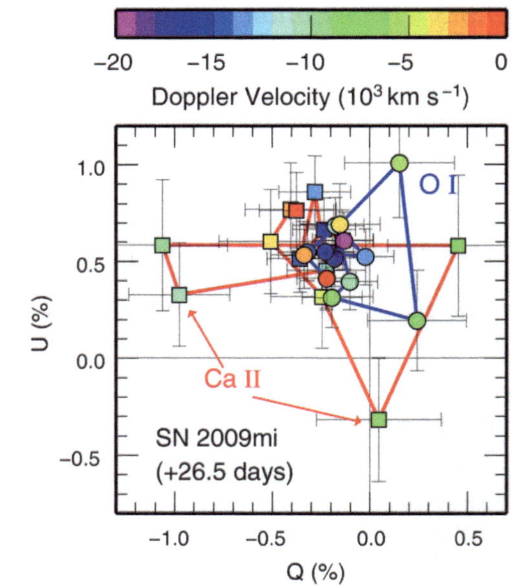

Fig. 17.10 Spectropolarimetry of SN 2009mi represented in the Q/U plane 27 days after B-band maximum, corresponding to the time of the transition to the exponential tail. SN 2009mi showed prominent nonaxisymmetric polarization in the lines of O I and the Ca II near-IR triplet. From "Three-dimensional Explosion Geometry of Stripped-envelope Core-collapse Supernovae. I. Spectropolarimetric Observations" (Tanaka et al. 2012a). © AAS. Reproduced with permission

explosion, approximately at the transition to the radioactive tail and the nebular phase (Tanaka et al. 2008). The Ca II near-IR triplet was significantly polarized at an angle roughly aligned with the milder continuum polarization.

SN 2009mi was observed 27 days after B-band maximum (Fig. 17.10), near the time of the transition to the exponential tail (Tanaka et al. 2012a). It showed prominent polarization in the O I line and in the Ca II near-IR triplet with complex loops in the Q/U plane (Sect. 4.6), evidence for 3D nonaxisymmetric structure. The Ca II and O I lines displayed very different angles and hence very different geometric structure.

Observations of the Type Ic-bl SN 2002ap revealed a complex, composition-dependent geometry as described in Sect. 17.4.1.

17.3.4 Circumstellar Interaction

SN 1994I showed clear evidence of CSI (Sect. 17.2). Late-time radio (Stockdale et al. 2004; Schinzel et al. 2009) and X-ray (Pooley and Lewin 2004) emission from the (probable) Type Ic SN 2001em was stronger than expected for a typical SN Ib/c. A late-time optical spectrum showed strong Hα emission (Soderberg et al. 2004). A plausible interpretation is that the supernova ejecta had encountered the hydrogen envelope that was lost from the progenitor star (Chugai and Chevalier 2006). SN 2012ca may also have been a SN Ic interacting with a hydrogen-rich CSM (Inserra et al. 2014).

The light curve of the atypical Type Ic SN 2010mb had a broad peak, and from about 100 to 400 days after peak it declined more slowly than expected for ^{56}Co decay. Subsequent upper limits demanded a decline more steep than ^{56}Co decay. This behavior is consistent with interaction of the ejecta with a hydrogen- and helium-deficient CSM of several $M_{\odot}$ and finite radial extent (Ben-Ami et al. 2014).

17.4 SN Ic-bl

At the turn of the century, observations revealed a new supernova subtype, *broad-lined* SN Ic, now known as Type Ic-bl. The especially interesting aspect of these events is that some are observationally associated with long, soft gamma-ray bursts (Sect. 17.4.3), and all supernovae associated with such bursts have proved to be SN Ic-bl. The broad lines require high photospheric velocities and perhaps high kinetic energy of the ejecta. Consequently, SN Ic-bl are sometimes referred to as hypernovae. We prefer to make the distinction between the observed SN Ic-bl and *hypernova models*. Such models typically have kinetic energies of $\sim$10 B.

17.4.1 Case Study: SN 2002ap

The Type Ic-bl SN 2002ap in M74 was discovered within a few days of explosion and rose quickly from the first detection by about 1.5 mag to reach $V = 12.3$ about 10 days later. With nominal reddening, at maximum light M_V was about -17.4, not especially bright for a SN Ic. SN 2002ap peaked earlier than some other SN Ic-bl (e.g., SN 1998bw; Sect. 17.4.3), but later than the Type Ic SN 1994I (Sect. 17.2). The optical and near-IR light curves were followed for about 1.5 years (Tomita et al. 2006; Fig. 17.11). The late-time light curve declined at about 0.016 mag d^{-1}, steeper than the ^{56}Co decay rate and similar to that of SN 2007gr (Fig. 17.7), typical SN Ic, the Type IIb SN 1993J, and many other stripped-envelope events of a variety of spectra types (Sect. 17.5.2).

The first spectra were obtained on the rise about a week before maximum (Mazzali et al. 2002). The Si II λ6355 feature was quite broad and the lines of O I λ7774 and the Ca II near-IR triplet were blended together. The breadth of the lines indicated velocities of 30,000 km s^{-1} or more. Shortly after maximum, the Si II line was discernible but relatively weak, as typical of SN Ic, and the O I and Ca II features were clearly separated (Fig. 17.12).

Spectra of SN 2002ap in the nebular phase (Fig. 17.12) showed a weak continuum and emission lines that were not significantly broader than those of typical SN Ic (Foley et al. 2003; Mazzali et al. 2007). Some of the emissions may have had an unresolved narrow component redshifted by 600 km s^{-1} with respect to the host galaxy. The strength of the lines suggested an oxygen mass of near 1 $M_{\odot}$.

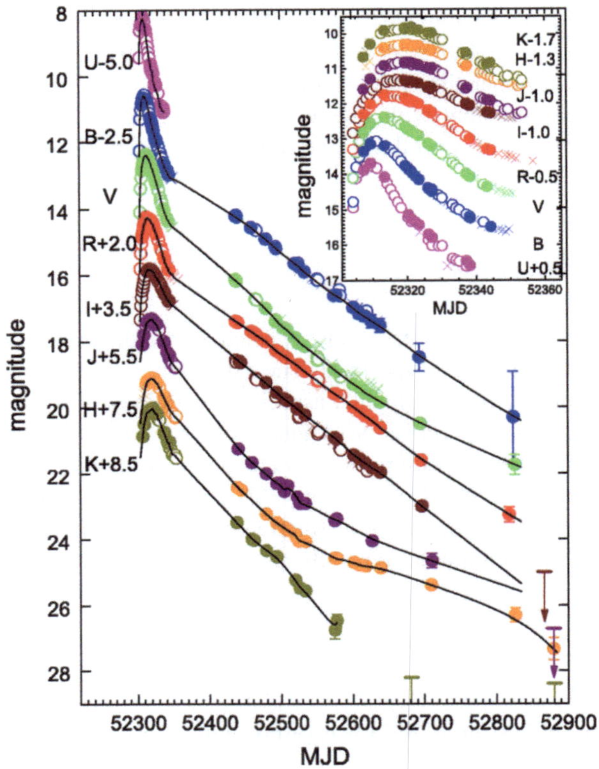

Fig. 17.11 Light curves of the Type Ic-bl SN 2002ap. From "The Optical/Near-Infrared Light Curves of SN 2002ap for the First 1.5 Years after Discovery" (Tomita et al. 2006). © AAS. Reproduced with permission

The explosion was consistent with the ejection of 0.1 $M_{\odot}$ of ^{56}Ni and a total mass of 2–3 $M_{\odot}$, from a core of predominantly carbon and oxygen of 3–4 $M_{\odot}$ (Mazzali et al. 2002; Vinkó et al. 2004). Estimates of the kinetic energy ranged from 4 to 10 B. As for other stripped-envelope supernovae, simple models had difficulty matching the relatively narrow light-curve peak and the subsequent rate of decline on the tail (Sect. 17.3.1). One way to achieve a rapid rise is to have ^{56}Ni mixed outward, but this exaggerates the excessively rapid decline of the model tail. Fits to the decline from maximum and the late tail may require low mean optical opacity (Sect. 17.5.2). Attempts to fit the light-curve and velocity evolution require a flat distribution of mass in the outer ejecta that is not produced by current evolutionary and explosion models.

Spectropolarimetry of SN 2002ap covered about 6 days before to 100 days after maximum light (Kawabata et al. 2002; Leonard et al. 2002a; Wang et al. 2003). The data were complex (Fig. 17.13), showing a shift in the polarization angle of the continuum and different orientations of position angles of the continuum and the

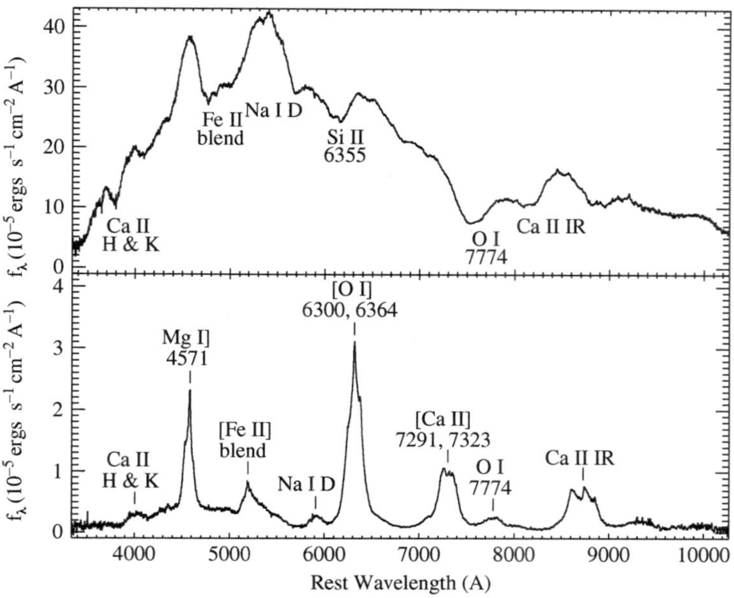

Fig. 17.12 Spectra of the Type Ic-bl SN 2002ap 5 and 156 days after maximum light. Figure 12 "Optical Photometry and Spectroscopy of the SN 1998bw-like Type Ic Supernova 2002ap" from Foley et al. (2003) ©The Astronomical Society of the Pacific. Reproduced by Permission. All Rights Reserved

O I, Ca II, and Fe II lines, all strong evidence for nonaxisymmetry. SN 2002ap was spectropolarimetric Type D0 in the continuum and Type L in strong lines.

Polarization of the Ca II near-IR triplet feature appeared later than that of the O I line. This behavior might be explained by the calcium being primarily natal calcium resident in the photosphere where it was excited to display an asymmetric geometry only as the ejecta thinned out and the outer regions were exposed to asymmetric excitation. Such asymmetric excitation could have been produced by a plume of ^{56}Ni that arose from Rayleigh–Taylor instabilities in the ejecta (Hammer et al. 2010) or by a ^{56}Ni-rich jet of matter produced near the neutron star. An asymmetric ^{56}Ni structure might still have been buried in the oxygen mantle (Wang et al. 2003). A buried plume or jet might thus account for the observations of SN 2002ap with the calcium partaking partly of the photospheric geometry and partly of the asymmetric excitation. Similar properties may pertain to typical SN Ic (Sect. 17.3.3).

Radio emission of SN 2002ap was detected from 4 to 50 days after the presumed date of explosion (Berger et al. 2002). The radio emission was roughly consistent with the interaction of high-velocity ejecta, of order 0.3c, with a progenitor wind. There was no indication of relativistic bulk flow as characteristic of gamma-ray bursts (Sect. 17.4.3). SN 2002ap may belong to a class of stripped-envelope supernovae that rise rapidly to a radio peak in about 10 days, but remain an order of magnitude dimmer than many otherwise similar events with the same spectral

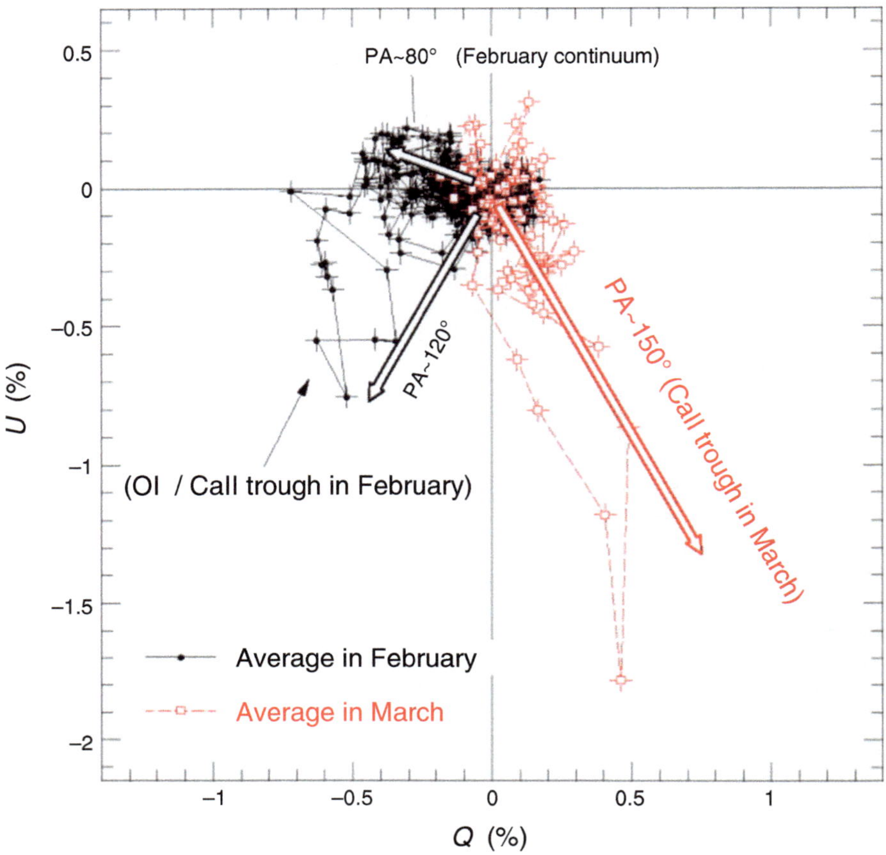

Fig. 17.13 Spectropolarimetry of SN 2002ap represented in the Q/U plane. The data show a shift in the polarization angle of the continuum (from *black points* to *red points*) and different orientations of the continuum and the Ca II near-IR triplet and loops in the Ca II line, all evidence for nonaxisymmetry and chemical/geometric structure that varies in time and hence depth in the ejecta. From "Optical Spectropolarimetry of SN 2002ap: A High-Velocity Asymmetric Explosion" (Kawabata et al. 2002). © AAS. Reproduced with permission

index, and then decline more slowly. This behavior may indicate expansion into a relatively low-density CSM, so that the forward shock suffers little deceleration (Maeda 2013a). Radio observations might then provide more information about the progenitor than would be the case for interaction with a more dense CSM.

Archival pre-explosion images from the *Canada-France-Hawaii Telescope* and *HST* allowed tight limits to be placed on the progenitor of SN 2002ap (Crockett et al. 2007). The absolute-magnitude limit, $M_R \gtrsim -5.1 \pm 0.5$, excluded a star with ZAMS mass $\gtrsim 7\ M_\odot$ that had not lost its envelope. Single-star evolution models with a range of mass-loss rates were incompatible with the observed limit. This suggests that the progenitor evolved in an interacting binary system. The archival observations also constrained any companion to the progenitor. A main-sequence companion would be limited to $\lesssim 20\ M_{\odot}$. The companion could have originally been the more massive star and hence could currently be a neutron star or a black hole. A white dwarf or a helium star would be less likely from an evolutionary point of view. The combination of pre-explosion limits, estimates of the ejected mass, and constraints from X-ray and radio observations suggest that the binary interaction probably was Case B mass transfer (Sect. 8.8), perhaps with a subsequent common-envelope evolution phase (Sect. 17.5.1).

17.4.2 Synthesis of Properties of SN Ic-bl

In addition to SN 2002ap (Sect. 17.4.1), well-observed typical SN Ic-bl unassociated with gamma-ray bursts include SN 2007ru (Sahu et al. 2009), SN 2009bb (Pignata et al. 2011), SN 2010ah (Corsi et al. 2011), and SN PTF10qts (Walker et al. 2014). SN Ic-bl associated with gamma-ray bursts are discussed in Sect. 17.4.3.

SN Ic-bl range in peak magnitude from $M_V \simeq -17$ to brighter than $M_V \simeq -19$ and in luminosity from $\sim 10^{42}$ to $\sim 10^{43}$ erg s^{-1} (Fig. 17.14). This range is presumably related to the amount of ^{56}Ni produced in the explosion. Statistically, SN Ic-bl are brighter at peak than typical SN Ib/c (Drout et al. 2011). The peak absolute magnitudes of SN Ic-bl associated and unassociated with relativistic outflow (either a GRB or an X-ray flash; Sect. 17.4.3) may be drawn from the same

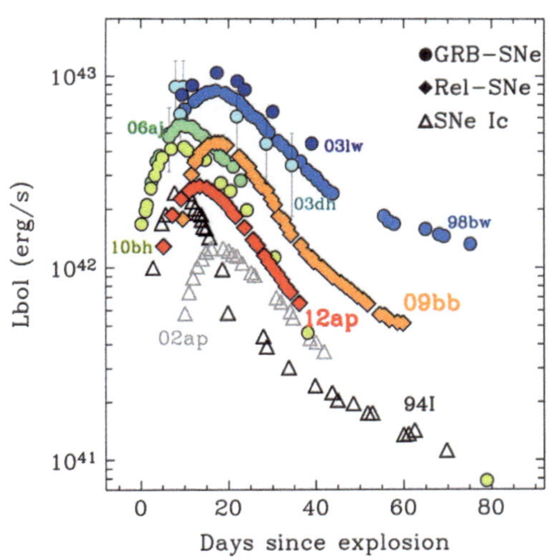

Fig. 17.14 Bolometric light curves of a sample of SN Ic-bl, both with (Sect. 17.4.3) and without (Rel-SNe) associated gamma-ray bursts, as well as the Type Ic SN 1994I. (SN 2002ap, designated in the figure as an SN Ic, is now recognized to be an SN Ic-bl.) From "The Broad-lined Type Ic SN 2012ap and the Nature of Relativistic Supernovae Lacking a Gamma-Ray Burst Detection" (Milisavljevic et al. 2015). © AAS. Reproduced with permission

distribution. The late-time tails of SN Ic-bl often decay at a rate near 0.016 mag d^{-1}, similar to that of SN 2002ap and other stripped-envelope events.

SN Ic-bl represent about 4% of all core-collapse supernovae in large galaxies, but dominate the SN Ic population in low-metallicity dwarf host galaxies (Arcavi et al. 2010). The environments of SN Ic-bl and SN IIb are substantially bluer on average than those of SN Ib, typical SN Ic, and SN IIP (Kelly and Kirshner 2012). Unlike hosts of core-collapse events with more typical ejecta velocities, the hosts of SN Ic-bl tend to have high stellar-mass densities, high densities of star formation, and high gas velocity dispersion for their mass (Kelly et al. 2014a). The evidence that dwarf hosts produce a relative excess of SN IIb over SN Ic may indicate lower mass-loss rates in lower-metallicity stars of a given mass. In this circumstance, stars that would otherwise have produced typical SN Ic may retain some hydrogen and helium and appear as a SN IIb. This line of reasoning does not easily account for the relative lack of SN Ib in dwarf hosts, nor the excess of SN Ic-bl over typical SN Ic compared to the distribution in giant hosts. The preference for SN Ic-bl in dwarf hosts with high stellar-mass density and high star-formation density suggests that some factors are at work in the progenitor evolution other than low metallicity. One possible explanation is that the progenitors of SN Ic-bl are born in regions that are especially efficient in producing massive-star binary systems. Further discussion of progenitor evolution is given in Sect. 17.5.

In early spectra of SN Ic-bl, the broad features are more blended than they are in SN Ic (Figs. 17.15 and 17.16), but the identities of the features appear to be similar.

SN Ic-bl show typical nebular spectra with forbidden emission lines of [O I] $\lambda\lambda$6300,6364 and [Ca II] $\lambda\lambda$7291,7323. There are large variations of line widths and hence velocities of the core ejecta in the nebular phase, with some SN Ic-bl having much higher velocities than typical core-collapse events and some having similar velocities (Maurer et al. 2010b). The high velocities may indicate asphericity of the inner regions.

As for some other stripped-envelope events, some SN Ic-bl show evidence for double-peaked profiles in the nebular emission (Taubenberger et al. 2006; Modjaz et al. 2008, 2014; Maeda et al. 2008; Milisavljevic et al. 2010). The significance of this feature was originally raised in the context of the SN Ic-bl SN 2003jd (Mazzali et al. 2005b), but the double-peaked [O I] feature is not just an aspect of SN Ic-bl. The spacing between the peaks in SN 2003jd was 120 Å. The issue was whether this double-peaked feature was a special characteristic of a supernova associated with a gamma-ray burst or an especially asymmetric explosion with a torus-like structure seen edge-on. In some other cases, the peak-to-peak separation is near 64 Å, the wavelength spacing of the doublet. In general, care must be taken in interpreting the two peaks in [O I] profiles as blueshifted and redshifted components of a torus of ejecta rather than the natural separation of the two components.

Ejecta masses are often derived, with less or more sophisticated techniques, from the width of the light-curve peak, which is a measure of the diffusion timescale (Sects. 5.4.2 and 17.3.1). The ejected masses of SN Ic-bl unassociated with gamma-ray bursts are typically found to be several $M_\odot$. The constraints on the ejected

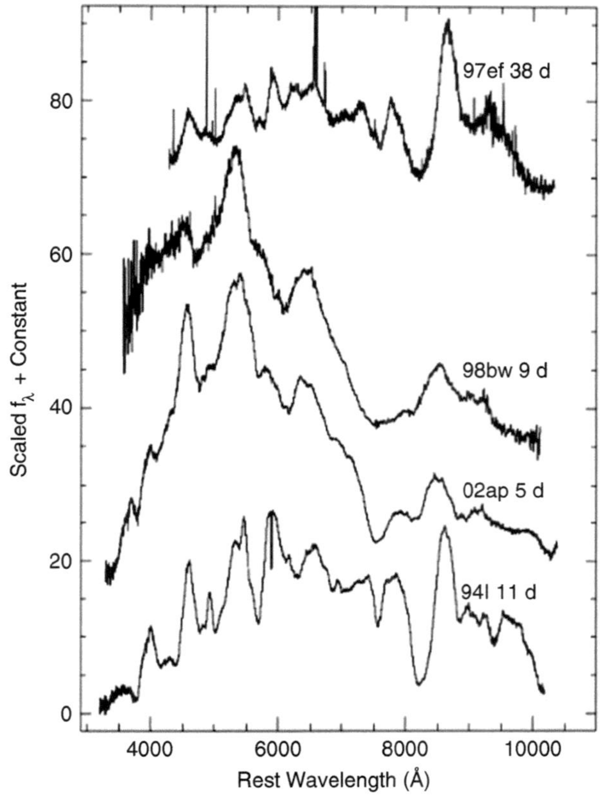

Fig. 17.15 Spectra of three SN Ic-bl and the Type Ic SN 1994I. Note the severe blending in the spectrum of SN 1998bw. Figure 13 "Optical Photometry and Spectroscopy of the SN 1998bw-like Type Ic Supernova 2002ap" from Foley et al. (2003) ©The Astronomical Society of the Pacific. Reproduced by Permission. All Rights Reserved

mass from the width of the light-curve peak coupled with the high ejecta velocities suggested $E \simeq 10$ B, an order of magnitude higher than had long been associated with typical supernovae (but note caveats from Sect. 17.3.1 regarding constraints from the late-time tail). The question of energetics is still somewhat bedeviled with issues of asymmetry and the fact that some SN Ic-bl with very broad lines reveal no connection to gamma-ray bursts.

17.4.3 The Connection Between SN Ic-bl and Gamma-Ray Bursts

As techniques were developed to localize gamma-ray bursts, first circumstantial and then specific evidence emerged to associate at least some long, soft gamma-ray

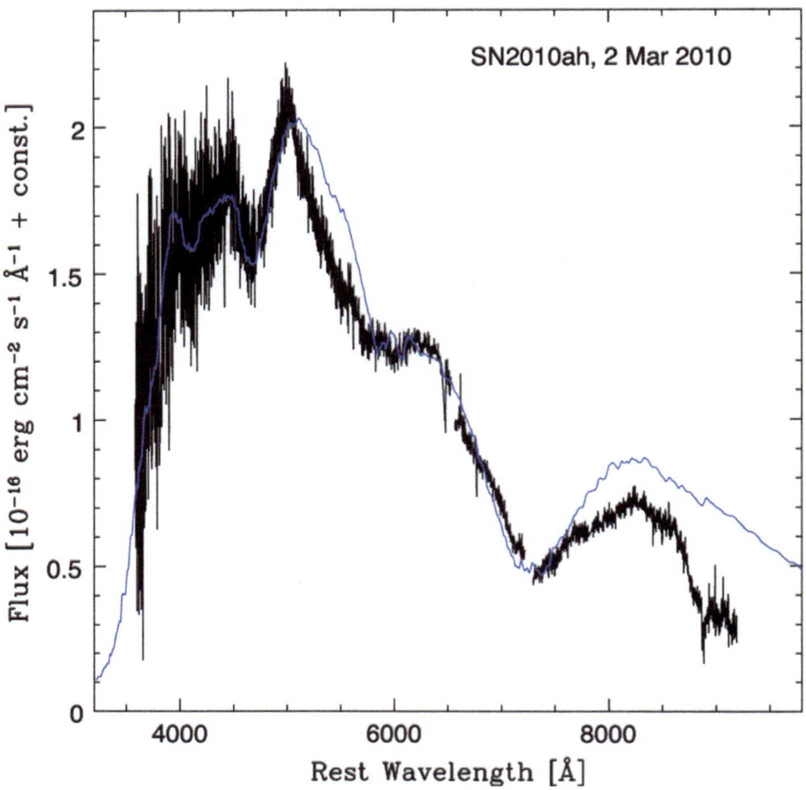

Fig. 17.16 The spectrum of the Type Ic-bl SN 2010ah about 9 days after explosion compared to a synthetic spectrum. In the synthetic spectrum, the flux minima near 4700, 6000, and 7300 Å are primarily due to Fe II, Si II, and a blend of O I and Ca II lines, respectively. Figure 5 "The very energetic, broad-lined Type Ic supernova 2010ah (PTF10bzf) in the context of GRB/SNe" from Mazzali et al. (2013), by permission of Oxford University Press on behalf of the Royal Astronomical Society

bursts with supernovae (Woosley and Bloom 2006). Not every SN Ic-bl is associated with a gamma-ray burst, but every spectroscopically-classifiable supernova directly associated with a gamma-ray burst has proven to be a SN Ic-bl. Whether they have revealed associated supernovae or not, long-soft gamma-ray bursts tend to show the same preference for low-metallicity, dwarf host galaxies as do SN Ic-bl.

Well-studied examples of SN Ic-bl associated with gamma-ray bursts or other evidence for relativistic flow are SN 1998bw (Patat et al. 2001), SN 2003dh (Matheson et al. 2003), SN 2006aj (Sollerman et al. 2006), SN 2010bh (Bufano et al. 2012), and SN 2012ap (Milisavljevic et al. 2014).

The first well-studied supernova associated with a gamma-ray burst was SN 1998bw, associated with GRB 980425. SN 1998bw was especially bright for a stripped-envelope supernova, $M_V = -19.2$, but at its moderate distance, $z = 0.0087$ or 40 Mpc, GRB 980425 was a subluminous gamma-ray burst, so the nature of this event remains somewhat equivocal. What was not equivocal was that SN 1998bw was of a recognizable type, with important variations. It showed no sign of H or He I lines, making it a SN Ic. It did, however, display unprecedentedly broad P Cygni features indicating especially high ejecta velocities, making it a SN Ic-bl and suggesting especially high kinetic energy. The light curve was observed for 4 years (Clocchiatti et al. 2011). SN 2003dh was the first supernova to be associated with a typical, although still relatively nearby ($z = 0.1865$ or 1 Gpc) GRB 030329. Once again, SN 2003dh proved to be a SN Ic-bl.

The gamma-ray burst revolution proceeded in parallel with developments in the polarization of supernovae. Since gamma-ray bursts are the result of strongly-collimated flows, spectropolarimetry may also shed light on the supernova/gamma-ray burst connection. Spectropolarimetry of SN 1998bw suggested a moderate intrinsic polarization that might have increased to the red in the earliest data (Patat et al. 2001) suggesting a dominance of Rayleigh scattering in the atmosphere. SN 2003dh was possibly polarized at 7000–8000 Å (a region contaminated by sky lines; Kawabata et al. 2003). Since the line of sight to a relativistically-beamed gamma-ray burst must be closely along the jet, and hence presumably a global symmetry axis, any polarization thus represents nonaxisymmetry.

SN 1998bw and SN 2002ap (Sect. 17.4.1) and other SN Ic-bl were asymmetric; the unresolved issue is what effect asymmetries had on the spectral and photometric properties. There is a concern that asymmetric explosions could mimic some of the effects of "broad-lined" activity that have been interpreted in terms of high energy with spherical models (Höflich et al. 1999). Others have argued that high energy and ^{56}Ni mass are needed even if the explosions are asymmetric (Maeda et al. 2002).

Some SN Ic-bl are associated with flashes of X-rays that are softer than the classical high-energy output of a gamma-ray burst. These, together with unusually dim gamma-ray bursts such as the one associated with SN 1998bw are sometimes termed low-luminosity gamma-ray bursts. They may occur in progenitors having extended envelopes that partially smother the relativistic jets and prevent the generation of a large γ-ray luminosity (Nakar 2015). A well-studied example was SN 2006aj. A single epoch of spectropolarimetry was obtained for SN 2006aj 10 days after the X-ray flash, near the estimated maximum of the light curve of the supernova (Maund et al. 2007b). The polarization was nearly 4% in the blue, making SN 2006aj another rival, like SN 2005bf, to SN 1997X (Sect. 17.3.3). This raises the issue of whether SN 1997X was associated with an X-ray flash. The polarization spectrum of SN 2006aj was consistent with photometric polarimetry (Gorosabel et al. 2006) and with the spectropolarimetry of SN 2003dh (Kawabata et al. 2003). In particular, there was high polarization of oxygen and calcium. Once again, SN 2006aj is likely to have been observed down the jet axis, and the polarization is presumed to be due to a breakdown in axisymmetry. SN 2006aj was spectropolarimetric Type L (Sect. 4.6).

The Type Ic-bl SN 1997ef and SN 1997dq exploded before the association of SN 1998bw with a gamma-ray burst, but there were suggestions at the time of discovery that SN 1997ef might have some positional coincidence with a gamma-ray burst. After the fact, the broad-line characteristic of these events was recognized and explored more deeply (Iwamoto et al. 2000; Mazzali et al. 2000, 2004). The two supernovae had similar light curves and spectral evolution. Both of these events displayed a long, slow, transition to the nebular phase, in contrast to SN 1998bw. In this transition phase, SN 1997dq displayed both nebular emission in lines of [O I] and [Ca II] and a photospheric spectrum characterized by very low velocities, again suggesting a breakdown of spherical symmetry.

There is a strong presumption that SN Ic-bl are associated with black-hole formation, but magnetars are also considered as a possible energy source. The supernovae associated with gamma-ray bursts are referred to as "engine-driven" supernovae in some quarters. Soderberg et al. (2006b) invoked the phrase to refer to events with an associated relativistic outflow, but defined it to mean those with accretion-fed, rapidly-rotating, compact sources. While there are grounds to suspect rapidly-rotating compact objects in SN Ic-bl, that is conjecture; the presumption that the object is accretion fed, even more so. "Engine-driven" might also imply that energy input continues for a time long compared to a dynamical time. The time to get a relativistic jet out of the progenitor star is typically 10 s for a bare C/O core of radius 10^{10} cm. All core-collapse supernovae are driven by some kind of "engine" and the driving mechanism may last for seconds even for normal supernovae, so this is not a clear distinction. The preferred distinction would seem to be the empirical one, whether or not there is direct evidence for relativistic flow.

Long (typically 30 s), soft gamma-ray bursts are traditionally associated with core collapse, as we have done here. Another family of gamma-ray bursts, the short (typically $\lesssim 2$ s), hard bursts, may be produced by mergers of binary neutron stars. The distinction is blurred by the existence of short bursts with extended emission tails and long bursts that are relatively nearby and hence easy to observe, but have no discernible supernova. Both of these kinds of bursts show an initial hard pulse, characteristic of the classical soft gamma-ray bursts followed by a soft tail of extended duration. Observations hint that this subset of long gamma-ray bursts may be related to mergers rather than core collapse (van Putten et al. 2014).

17.5 Models of SN Ic and SN Ic-bl

The majority of SN Ic do not show any clear evidence of helium. In most of those that do, the helium is at high velocities and must be of mass substantially less than 0.1 $M_\odot$. If there is helium present in SN Ic, it must be in amounts that are qualitatively smaller than in SN Ib. While there are caveats that we will explore below, this analysis suggests that there are distinct differences between SN Ib and SN Ic that must be taken into account in models. A model with substantially more than 0.1 $M_\odot$ of helium is unlikely to represent a typical SN Ic.

17.5.1 Evolutionary Models

As discussed in Sect. 16.4.1, single-star models produce more than 0.3 $M_\odot$ of helium independent of complications of metallicity, rotation, and mass loss (Eldridge and Tout 2004; Georgy et al. 2009, 2012). This amount of helium is likely to appear at some phase when it is exposed to the excitation from radioactive decay and would tend to have too low an ejection velocity to correspond to the small amount of helium that may be present in a minority of SN Ic. Both of these constraints are worthy of being rendered more quantitative with dynamical and radiative-transfer simulations (Sect. 17.5.2). Population synthesis studies have shown that the rate of explosion of stripped-envelope supernovae can be approximately reproduced by invoking evolution in close binary systems of stripped-envelope progenitors (Podsiadlowski et al. 1992; De Donder and Vanbeveren 1998; Eldridge et al. 2008). As for the other stripped-envelope types, most SN Ic probably arise in binary systems.

Even many binary-evolution models have an excess of helium in the primary when it explodes, more than about 0.2 $M_\odot$; this is likely to disqualify them as models for SN Ic (Yoon et al. 2010; Dessart et al. 2011). Producing a nearly bare C/O core may require two stages of common-envelope evolution, the first when the hydrogen envelope expands to engulf the companion and the second when the helium envelope does the same. The latter stage requires either the helium star to be rather low in mass, $\lesssim 3$ $M_\odot$, so that it can expand to red-giant-like dimensions, or a very compact system following the first stage of mass transfer, since a higher-mass helium star can only expand to a few solar radii. The higher the mass and the larger the radius of the secondary, the more difficult it is to accommodate an appropriate post-mass-loss primary. To produce a SN Ic, the binary system also must avoid merger during the first stage of hydrogen mass transfer. All these restrictions mitigate against the primary star in a binary producing most typical SN Ic.

The possibility remains that SN Ic arise when the original secondary explodes. Sufficient stripping of helium might occur through common-envelope evolution if the secondary becomes a helium giant and engulfs a neutron star produced by the primary (Nomoto et al. 1994). A helium star in a very short-period (about 1 day) binary with a neutron-star companion can undergo mass transfer when it expands to only a few solar radii and the mass ratio would be naturally low, thus encouraging rapid, unstable mass transfer and the loss of the entire helium envelope. Such a system can undergo sufficient mass transfer from the helium star to yield a helium-deficient carbon star (Dewi et al. 2002; Ivanova et al. 2003). The remaining helium in helium cores of initial mass 1.5–3 $M_\odot$ is predicted to be $\sim$0.01 $M_\odot$, sufficiently low to qualify for SN Ic. Models for helium cores of 2.4–6.6 $M_\odot$ yield final helium masses closer to 0.1 $M_\odot$. It is less likely that these models could be progenitors of the majority of typical SN Ic, but they might account for some of the events with subtle evidence for helium (Sect. 17.3.2). Mass transfer from a helium star to a neutron star might in rare instances yield an *ultra-stripped* progenitor with only slightly more than the Chandrasekhar mass and hence a very small ejected mass, $\sim$0.1 $M_\odot$ (Tauris et al. 2013a; Suwa et al. 2015). Such events may have been observed (Sect. 17.3.1).

While population synthesis suggests that the overall rate of stripped-envelope supernovae might be reproduced with binary-evolution models, it is less clear that there are sufficient low-mass binary systems of the right properties to account for the rate of SN Ic. It also is not clear that the class of models in which an SN Ic is the explosion of a C/O core orbiting a neutron star produced by the primary accounts for evidence that SN Ic tend to be more closely associated with sites of young star formation than SN Ib or SN II (Sects. 17.4.2 and 3.3). In this class of models, the progenitor of an SN Ic would be, in some phases, a Be-star/X-ray binary. It is possible that association with a strong X-ray source or the fossil remnant H II region of one could affect the ionization of the nearby ISM, giving false evidence of the rate of local star formation.

SN Ic-bl represent yet another challenge to evolutionary theory. Evidence points to their origin in binary systems, but it is not clear that current evolutionary models account for their presence in regions of especially high stellar-mass density and star-formation rate (Sect. 17.4.2). Evolution in dense star clusters might selectively yield conditions especially conducive to the production of SN Ic and SN Ic-bl.

Whatever the driving mechanism of the explosion is, SN Ic-bl that are associated with gamma-ray bursts probably require high angular momentum (Yoon et al. 2010), hence these events are likely to undergo some evolution that is substantially different from typical SN Ic. Examples of such evolution are the mixed and hence chemically-homogeneous evolution of a metal-poor star (Yoon and Langer 2005b; Woosley and Heger 2006; Cantiello et al. 2007), spin-up of the progenitor by a neutron-star or black-hole companion in a very close binary system (Izzard et al. 2004; van den Heuvel and Yoon 2007; but see Detmers et al. 2008, who question this scenario), or specific instances of Case C mass transfer (Brown et al. 2000; Podsiadlowski et al. 2010; Sect. 8.8). If some SN Ic-bl do arise from very massive stars that shed their helium in winds or by mass transfer, the progenitors might be bright bolometrically, but relatively dim in the optical (Yoon et al. 2012b).

17.5.2 Models of Light Curves

Light curve models have failed to explain in a fundamental way why some SN Ic like SN 2011bm decay at the ^{56}Co rate, most decay in a manner parallel to a classic typical event like SN 2007gr and the broad-line event SN 2002ap (Sect. 17.4.1), and some such as SN 1994I (Sect. 17.2) decay rapidly from peak but somewhat more slowly later.

There are four dilemmas associated with the light curves of stripped-envelope supernovae, including SN Ic and SN Ic-bl. (1) Despite relatively similar peak properties, the late-time tails often reveal a spread in properties from rather flat to rather steep declines. (2) For many SN Ic, the properties derived from the peak seem to be incommensurate with the behavior of the tail by predicting too steep a decline.

(3) Events with different spectral types, SN IIb, SN Ib, SN Ic, can, nevertheless, have rather similar late-time decays. (4) Typical SN Ic and SN Ic-bl can have similar late-time decay despite the implication that the broad lines of the latter signify substantial differences in kinetic energy.

As for many SN IIb and SN Ib, simple spherical models that fit the peak of SN Ic often predict a more rapid decline than observed in the late-time tail, the essence of the second dilemma. This is especially true of models that adopt a constant effective optical opacity near maximum, but is also an issue with some models that treat the temporal and spatial variation in the opacity more rigorously. The simple models tend to require low diffusion mass to account for the observed light-curve rise time and photospheric velocity. The low diffusion mass yields a prediction of greater leakage of γ-rays on the tail that is inconsistent with observations of the tail. Other models seek agreement with both the light-curve peak and the tail by adding extra components to the fit, for instance altering the density distribution by adding flat outer layers (Mazzali et al. 2000), by adding a denser core to more efficiently trap γ-rays at later times (Maeda et al. 2003), or by adding an outer region that is not heated by γ-rays (Vinkó et al. 2004). There are no satisfactory light curve models of SN Ic or SN Ic-bl from first principles of evolution and explosion.

Insight into the various dilemmas outlined above, especially the second, may involve consideration of the fraction of the total ejected mass that is involved in diffusion near peak light. That fraction can be formally expressed in terms of the observed rise time, photospheric velocity near peak, and the late-time decay time as [Sect. 5.4.2, Eq. (5.31)]

$$\frac{M_{diff}}{M_{ej}} = \frac{5}{3}\beta C \frac{c}{v_{ph}} \frac{\kappa_\gamma}{\kappa} \left(\frac{t_{max}}{T_\gamma}\right)^2, \tag{17.1}$$

where $\beta \simeq 13.8$ is the scaling parameter for the light-curve peak, C is a dimensionless structure constant dependent on the slope of the density profile and is typically $C \simeq 0.05$, c is the speed of light, v_{ph} is the photospheric velocity near peak, $\kappa \simeq 0.1\,\mathrm{cm^2\,g^{-1}}$ is the optical opacity for ionized matter, $\kappa_\gamma \simeq 0.03\,\mathrm{cm^2\,g^{-1}}$ is the effective opacity to γ-rays, t_{max} is the time from explosion to the time of maximum light, and T_γ is the timescale that characterizes the trapping of the γ-rays on the late-time tail. Note that to obtain an accurate measure of the timescale T_γ, allowance for the contribution of positrons is required (Sect. 5.4.2). For typical parameters, Eq. (17.1) can be expressed as

$$\frac{M_{diff}}{M_{ej}} = 0.1 \frac{(\kappa_\gamma/0.03\,\mathrm{cm^2\,g^{-1}})}{(\kappa/0.1\,\mathrm{cm^2\,g^{-1}})} v_{ph,4}^{-1} \left(\frac{t_{max,10}}{T_{\gamma,100}}\right)^2, \tag{17.2}$$

where $t_{max,10}$ is the rise time in units of 10 days, $T_{\gamma,100}$ is the tail timescale in units of 100 days, and $v_{ph,4}$ is the photospheric velocity in units of $10^4\,\mathrm{km\,s^{-1}}$. This expression for the fraction of the mass involved in diffusion near peak is basically determined by the observed parameters, v_{ph}, t_{max}, and T_γ. The mass ratio derived in this way tends to be ~0.1 for events with typical peak properties and a late-time decline of $T_\gamma \simeq 160$ days or $0.016\,\mathrm{mag\,d^{-1}}$ (neglecting positron deposition) that characterizes many stripped-envelope events. This is in contrast with a common assumption that the entire ejecta mass contributes with essentially equal weight to the width of the peak. Evaluation of the diffusion mass in this way formally resolves the second dilemma defined above. The properties of the peak can be rendered consistent with the properties of the tail by appropriate choice of the fraction of the ejecta that contributes to diffusion over the peak.

A possible factor that could yield a narrower peak for a given ejected mass and energy is recombination that would give a time-dependent discontinuity in the opacity, akin to that in hydrogen in SN IIP. Models of SN Ib based on massive helium cores (Sect. 16.4.2) produced greater peak-to-tail contrast than models with essentially constant opacity because all the helium and much of the oxygen recombined before maximum, yielding a significant mass that had low opacity and contributed little to the optical diffusion time. Only the inner iron and silicon layers that were effectively heated by γ-ray deposition contributed to the opacity. The effective diffusion mass was thus substantially less than the total ejecta mass. The total ejecta mass traps γ-rays and positrons and shapes the late-time tail.

Other models of stripped-envelope light curves have invoked lower ejecta mass. Detailed hydrodynamic models that have included complex prescriptions for the opacity have yielded substantial mean opacities, near $0.1\,\mathrm{cm^2\,g^{-1}}$, in large part because a rather high floor was adopted for the opacity (Bersten et al. 2011). In such a case, the opacity in recombined regions may be overestimated, giving a diffusion mass comparable to the ejecta mass and allowing lower-mass models to fit the peak. These models nearly always found the late-time light curve to decline too steeply. Models that fit peaks but faded too quickly on the tail may have had, by assumption or computation, too large an effective opacity over the peak and thus too low an ejected mass and effective γ-ray optical depth.

The UVOIR opacity that governs the shape of the peak is intrinsically complex with dependence on density, temperature, and composition as well as on NLTE, nonthermal, and expansion effects (Hoflich et al. 1993; Pinto and Eastman 2000). Recombination depends on adiabatic expansion and cooling; subsequent reionization and increase in the opacity depends on the deposition of energy from radioactive decay. Recombination can lead to a situation where the opacity is a sensitive function of both space and time, with the net effect being to decrease the effective opacity and thus yield a narrower peak light curve for a given ejected mass (Ensman and Woosley 1988; Kleiser and Kasen 2014; Piro and Morozova 2014).

Asymmetries in the explosions of stripped-envelope supernovae almost surely play a significant role. From spectropolarimetry, we know that stripped-envelope supernovae are asymmetric and that different elements are ejected at different angles (Maund et al. 2009). It is instructive to contemplate the image of the Cas A remnant of an SN IIb, rent by jets and riddled with irregularities due to nickel bubbles (Sect. 7.4.1). This structure may be a realistic representation of any stripped-envelope supernova.

Gamma-ray deposition on the tail is comparatively simple, being relatively insensitive to composition and to density irregularities. The physics of the γ-ray deposition around the peak depends on the distribution of ^{56}Ni and ^{56}Co. Any outward "mixing" of ^{56}Ni would tend to release γ-rays more quickly and hence yield a more rapid rise, but then also lead to a more rapid late-time decay, which is already a problem with many models. There may be some turbulent "microscopic" mixing, but even Rayleigh–Taylor instabilities tend to form distinct macroscopic plumes in the nonlinear limit (Hammer et al. 2010). The early deposition of γ-rays will not be azimuthally symmetric and hence neither will be the opacity, in contrast to the conditions imposed in any spherically-symmetric model.

The mean opacity around peak may thus be lower than frequently assumed or computed. The opacity is likely to be dependent on asymmetries that have not yet been adequately explored. The low ratio of diffusive to total ejecta mass implied by Eq. (17.2) suggests that the deposition and heating are irregular in real 3D explosions and leave substantial cold, low-opacity, mass near peak that, nevertheless, traps γ-rays on the tail. This perspective also may give some insight into the first dilemma, that there may be more variety in tail behavior than would be expected from the peak properties. Perhaps the peak opacity and diffusion time are controlled by the fraction of the ejecta comprising iron- and silicon-peak elements, and the mass of those elements is more uniform than the total ejected mass.

The third dilemma, that stripped-envelope events of different spectral type, and hence different compositions, ejecta masses, and perhaps energies, can have similar late-time light curves remains a challenge. For given ejected mass and energy, there is a degeneracy between κ and the rise time, t_{max}. From Eq. (17.1) with $E_k \simeq 3/10 M_{ej} v_{ph}^2$, one can write

$$\frac{\kappa}{t_{max}^2} = \left(\frac{5}{6}\right)^{1/2} \beta c \left(\frac{C\kappa_\gamma}{T_\gamma^2}\right)^{3/4} E_k^{-1/4} \frac{M_{ej}}{M_{diff}}. \tag{17.3}$$

For events with similar rise times and decay times, this implies that $\kappa \propto E_k^{-1/4} M_{ej}/M_{diff}$, a scaling that is far from obvious. Note that if M_{ej}/M_{diff} is substantially greater than unity, then care must be taken in differentiating the kinetic energy of the total mass from that associated with the smaller mass involved in diffusion near peak light and hence differentiating relevant mean velocities and their relation to the single observable scaling velocity, v_{ph}.

Similar concerns remain for the fourth dilemma, how both typical and broad-line SN Ic can have similar late-time light curves. The implication is that E_k is somehow correlated with ejected mass in an unexpected way. Some insight into the issue can be obtained by examining the basic scaling applicable to the tail. As presented in Sect. 5.4.2 (Eq. (5.27)), the slope of the late-time light curve depends on the trapping efficiency of γ-rays as characterized by the timescale parameter

$$T_\gamma = \left(\frac{C\kappa_\gamma M_{ej}^2}{E_k}\right)^{1/2}. \tag{17.4}$$

The constraint of similar late-time light curves, those with similar values of T_γ, implies, with $E_k \propto M_{ej}v^2$, that $M_{ej} \propto v^2$ and $E_k \propto M_{ej}^2 \propto v^4$, where v is the appropriate mean velocity of the ejecta. No current core-collapse simulations predict $E_k \propto M_{ej}^2$. A more careful analysis of this kind must account for deposition of both γ-rays and positrons. For a given observed late-time slope, the deposition of positrons will yield a smaller value of T_γ and hence a smaller value of $M_{ej}E_k^{-1/2}$ than if positron deposition were neglected.

This analysis raises the caution that results for ejected mass and energy derived from the peak alone may be suspect for SN Ic. The dependence of M_{ej}/M_{diff} on observed characteristics in accord with Eq. (17.2) implies generally higher estimates of ejected mass than many studies based on the peak properties alone (Wheeler et al. 2015a). Taken at face value, these results would require a substantial reconsideration of the progenitor evolution. Low ejecta masses are not compatible with the observed class of Wolf–Rayet stars; higher ejecta masses might be. Consideration of higher-mass progenitors must also confront issues of predicted rates of explosions compared to observations and issues of the environments of the explosions of the various variety of stripped-envelope events (Sects. 17.3, 17.4.2 and 3.3).

Analysis of the light curves of SN Ic, SN Ic-bl, and other stripped-envelope supernovae should be done in the context of well-observed UVOIR light curves from the rise to peak to the late tail. Deriving model parameters from the light curve depends on details of the models. The value of the timescale T_γ depends on the treatment of positron trapping. Simple models of light curves often invoke the photospheric velocity near maximum as a constraining parameter. Some care must be taken in assigning velocities, because as the matter expands the photospheric velocities naturally decrease, sometimes rapidly. The observed photospheric velocity may also be affected by asymmetries, with material ejected at higher velocities in some directions than in others. Models should match both the photometric behavior and the time evolution of the photospheric velocity. Full models should, of course, match the spectroscopic and spectropolarimetric evolution.

If there is a substantial amount of cold, neutral, undetected helium in SN Ic (despite arguments above to the contrary), it would contribute little opacity near the peak, but would contribute to the total mass and hence the trapping of γ-rays and positrons that determines the rate of decline on the tail. The presence of cold helium may affect the spectrum even if the helium is not directly detectable (Sect. 17.5.3).

17.5.3 Models of Photospheric Spectra

Analysis of spectra may help to resolve some of the conundra associated with the light curves of SN Ic. One of the specific challenges to deeper understanding of the nature of SN Ic and SN Ic-bl is to characterize the conspicuous feature of O I λ7774, typically the strongest feature in the optical spectrum after Ca II H&K and the Ca II near-IR triplet, and the absence of conspicuous optical He I lines in the spectra near maximum light. The strong O I line is evidence for a stellar core stripped to its oxygen-rich layers, but it is still necessary to quantitatively reproduce the line strength and profile in models. A strict paucity in the abundance of helium is the most obvious explanation for the lack of conspicuous He I lines, but the observability of helium is complicated by the need for nonthermal processes in order to produce strong lines. Helium could be present in considerable abundance, but not excited and hence not easily visible.

While relatively few works have specifically addressed model photospheric spectra of SN Ic (Sect. 17.3.2), various models have illustrated the degree to which H and He I features will appear in photospheric (and nebular; Sect. 17.5.4) spectra (Swartz et al. 1993a; Wheeler et al. 1994; Hachinger et al. 2012a; Dessart et al. 2012a).

Models of photospheric spectra underline the paucity of hydrogen and helium required to give a spectrum resembling a SN Ic. Even in early work it was clear that there could not be a great deal of hydrogen in either the helium-rich SN Ib or the helium-poor SN Ic, substantially less than 0.1 $M_\odot$, and that any apparent continuity from SN IIb to SN Ib to SN Ic masked appreciable differences in the atmospheric abundances and conditions. Veiling by free electrons might obscure optical He I lines in some circumstances (Sect. 15.4.3), but the lack of observed strong H lines in SN Ic means that He I lines are not muted in SN Ic by free electrons from ionized hydrogen.

If any helium is sufficiently hot to excite conspicuous He I lines from the metastable states near 20 eV, i.e., during shock breakout, it would show up in early spectra (Dessart et al. 2011). If SN Ic are more devoid of helium than models that show early, thermally-excited He I lines, then even the early He I features should be weak.

There also may be nonthermal effects in very early phases, but they are likely to be of secondary importance. Nonthermal effects are especially critical if the helium is mainly neutral and hence not susceptible to thermal excitation. Nonthermal excitation of helium requires that the helium be exposed to the products of radioactive decay, γ-rays, and secondary electrons (Lucy 1991; Swartz 1991; Hachinger et al. 2012a; Dessart et al. 2012a). If the helium is in the outer layers and the radioactive material sufficiently buried in the ejecta, then helium will not be excited and will not be visible in early spectra. Some mixing of radioactive products into the outer layers of the ejecta will enhance the exposure of any helium to the decay products. Another important factor is the phase of observations. As the ejecta expand and become more optically thin to γ-rays, the likelihood that helium will be nonthermally excited increases.

The phase when any He I lines appear and the subsequent strength of these features is a quantitative issue. It is possible that for sufficiently massive ejecta and sufficiently centrally concentrated decay products, evidence for substantial helium might be suppressed (Dessart et al. 2012a). By including a substantial layer of nonexcited helium, however, this class of models tends to produce a O I λ7774 feature that is too shallow, red, and narrow compared to observations. The nature of the oxygen feature may depend on E_k, which might range up to 10 B. The presence or absence of a substantial mass of cold helium might also be revealed by a careful analysis of the relative line strengths of oxygen and metals.

The totality of the observations of SN Ic suggest that they have modest ejected mass and are oxygen-rich, although ambiguities with regard to the opacity near peak may play a role in quantitatively assessing the ejected mass (Sect. 17.5.2). To observe helium near peak light may require substantial outward mixing of ^{56}Ni (Sect. 16.4.3), but such "mixing" may be expected. SN Ic are polarized and hence intrinsically asymmetric. In addition, mixing tends to yield redder colors near maximum that better reproduce observations (Modjaz et al. 2009). There is no reason to suspect that SN Ic are less susceptible to mixing by strong asymmetries associated with the explosion than other stripped-envelope supernovae. The strong implication is that the reason that helium does not produce conspicuous optical lines near maximum is because the helium is not there.

Since magnetic jets or Rayleigh–Taylor plumes might be expected in SN Ic explosions, the associated distribution of ^{56}Ni is probably very asymmetric; exactly how much helium can be excited at a given phase remains an important issue. If microscopic mixing of ^{56}Ni is required to produce SN Ib, then a physical model of that mixing is required and must also be applied to SN Ic. Ad hoc microscopic mixing also tends to produce broader line features, so there are quantitative issues that must be addressed in complete models.

The discussion here may also pertain to SN Ic-bl. While mixing and asymmetries are likely to lead to nonthermal effects and the revealing of any helium present in the ejecta, it is possible that these events are more massive and perhaps suppress nonthermal excitation of helium. That SN Ic and SN Ic-bl share light-curve properties, including the tail behavior (Sect. 17.5.2), suggests that they are not qualitatively different with respect to mixing, despite the apparent difference in E_k.

17.5.4 Models of Nebular Spectra

Models selected to fit the light curves of SN Ic have been used to explore the nature of the nebular spectra. Models from which hydrogen has been stripped by hand show excessively strong nebular He I lines. In models of SN 1987M, based on the 3.3 $M_\odot$ core of a star of ZAMS mass 13 $M_\odot$, the populations of the metastable He I levels near 20 eV were controlled by nonthermal secondary electrons from radioactive decay (Swartz et al. 1993b). The excitation from these levels was due to collisional thermal processes, whereas Lucy (1991) had argued for radiative processes such as UV photoionization. Lucy (1991) included thermal electrons only from helium, not from metals. Charge transfer may be an issue in these exotic atmospheres (Swartz 1994). Reducing the helium abundance by a factor of 10 compared to single-star evolutionary models can suppress the He I lines.

The ratio of [Ca II] $\lambda\lambda$7291,7323 to the Ca II near-IR triplet is a sensitive measure of the temperature and electron density in the nebular phase. The Mg I] λ4571 line arises naturally in models, but tends to be too narrow in 1D models in which there is no outward transport of the magnesium layer (Swartz et al. 1993b).

The fact that the model optical He I lines eventually disappear accounts for the similarity of SN Ib and SN Ic at late times. On the other hand, the He I 1.0830 μm line persists, carrying ∼0.5% of all the radiated flux in a typical model (Swartz et al. 1993b). This supports the notion that near-IR spectra are necessary to quantitatively explore the differences between SN Ib and SN Ic.

17.6 Summary

SN Ic events require a considerable depletion, if not absence, of helium. The photospheric spectra of SN Ic display evidence for enhanced oxygen compared to other stripped-envelope supernovae. If there is substantially more than 0.1 $M_\odot$ of helium in an outer layer, He I lines are likely to be prominent at some phase, as the ejecta go from optically thick to optically thin in γ-rays. Even if helium is present, but resides only in small amounts at high velocities, as it may in a few SN Ic, these events can still be distinguished in their physical properties and in their evolutionary origins from SN Ib that have substantial helium at lower velocities. Very late optical nebular spectra do not alone clearly differentiate SN Ic from SN Ib because any He I lines are expected to fade. Near-IR spectra may be more discriminating in this regard.

Progenitor models with more than 0.1 $M_\odot$ (and probably much less) of hydrogen or helium are unlikely to be appropriate for SN Ic. The required depletion of helium in SN Ic cannot be accomplished by winds in current models of single massive stars. Some form of binary evolution that thoroughly strips both hydrogen and helium from the immediate progenitor star is strongly implied. The categories of SN IIb, SN Ib, and SN Ic speak to important physical differences in the progenitors.

Observations and models of the photospheric and nebular spectra of SN Ic thus suggest that a different evolution of the stellar core is required for SN Ic compared to events that show prominent He I lines, SN IIb and SN Ib. In particular, the bulk of the evidence argues that SN Ic require a physically-distinct mode of evolution from SN Ib. The progenitors of SN Ic must strip their envelopes very efficiently in a manner that can only be done in contemporary models by invoking some form of binary mass transfer. In many models, hydrogen and helium are removed from the model by hand, and the actual physics of the mass loss by winds or binary mass transfer or a combination of both is not clear. It is also not clear whether common-envelope evolution is different with regard to envelope stripping compared to Roche-lobe overflow.

While it is not unusual to refer to the progenitors of SN Ib/c as Wolf–Rayet stars, it is not at all obvious that the observed high-mass, high-luminosity Wolf–Rayet stars are related to SN Ib/c. Most current estimates of ejecta masses of SN Ic are rather low, a few $M_\odot$, and hence do not correspond to the observationally-defined category of Wolf–Rayet stars that have more than ∼10 $M_\odot$. Considerations of recombination of helium and oxygen and associated variations in the time- and space-dependent opacities may allow a total ejecta mass to exceed the active diffusion mass that determines the light curve peak. Larger ejecta masses might be more compatible with Wolf–Rayet stars. On the other hand, allowance for positron trapping in the late-time tail will tend to lead to less trapping of γ-rays and hence lower estimates of the ejecta mass. The peaks and tails need to be reproduced self-consistently to yield accurate estimates of ejecta masses and clues to the progenitors.

An alternative that might be more compatible with the estimated low ejecta masses is that SN Ic may be associated with progenitor systems that lead to binary X-ray sources with Be-star companions. It may be that a SN Ic results from the explosion of the secondary in a massive-star binary system and thus be the second explosion in the system. There may be significant differences between the progenitor masses and the evolution—in terms of single stars or binaries, winds, Roche-lobe overflow, or common-envelope evolution—that are important to elucidate when attempting to distinguish SN Ib from SN Ic and Be-star binaries from Wolf–Rayet stars.

Future models should consider the velocity information in the widths of nebular lines along with the rise and decay times and photospheric velocity in the constraints on ejected mass and energy and should consider issues of ejecta composition asymmetry. Many models are spherically symmetric, and SN Ic are not. The mass, composition structure, ^{56}Ni distribution, and γ-ray deposition are likely to be very different in asymmetric models.

The fact that SN Ic tend to be more centrally concentrated than the light distribution of their host galaxies, and even specifically populate the brightest patches of their hosts, points to their arising from higher-mass progenitors than other supernovae. It is not clear how to rationalize this with evidence for a modest mass and stripped-core of the immediate progenitor star, but simply invoking a more massive progenitor star does not seem to be an adequate explanation.

SN Ic come not only in the typical variety, but also the SN Ic-bl version. There are still uncertainties about what physically differentiates SN Ic from SN Ic-bl and what differentiates SN Ic-bl that are and are not associated with gamma-ray bursts. As for SN Ic versus SN Ib or SN IIb, the differences again probably involve both the progenitor evolution, the explosion mechanism, and the type of compact remnant left behind. Yet another category of events with SN Ic-like spectra and bright, broad light curves (Sect. 18.3) may be related to pair-instability (Chap. 10) rather than to core collapse.

SN Ic are observed to be asymmetric with spectropolarimetry. Such data may help to differentiate typical SN Ic and broad-line events from those that produce gamma-ray bursts, where the jet is clearly manifest and escapes into the surrounding CSM.

Most SN Ic undoubtedly explode by the mechanism of core collapse and must leave behind a compact remnant. No evidence for such a remnant has ever been directly observed.

Chapter 18
Superluminous Supernovae

18.1 Introduction

Superluminous supernovae (SLSN) are somewhat arbitrarily defined to be brighter at peak than absolute magnitude −21 or a luminosity of about $10^{44}\,\mathrm{erg\,s^{-1}}$, considerably brighter than SN Ia and the SN Ic-bl associated with gamma-ray bursts. The first recognized SLSN may have been SN 1920A (Schaefer and Girard 1999), near a spiral arm of the nearby SBc galaxy NGC 2608. This event displayed a slow decline from M_V between −21.3 and −20.7 at maximum.

The broad recognition of SLSN came with the advent of rolling, unbiased supernova searches. Unlike traditional targeted searches, these searches omitted neither galaxies of low luminosity (thought a priori to be low-probability sites) nor galaxy centers. Among the first of these, the *Texas Supernova Search* (*TSS*; Quimby et al. 2005), and its successor, the *ROTSE* Supernova Verification Project (*RSVP*; Yuan et al. 2007), conducted an essentially unbiased optical census of the transient sky at unfiltered $m < 18.5$ with the four *ROTSE* (Akerlof et al. 2003) telescopes, primarily *ROTSE IIIb* at McDonald Observatory. This survey examined an effective volume of about a billion cubic megaparsecs over several years and provided a rather high spectral completeness. Among the ∼100 supernovae discovered by *TSS* and *RSVP* were some of the most luminous events ever identified. The first discoveries in this new class were SN 2005ap, SN 2006gy, SN 2006tf, SN 2008am, and SN 2008es. In hindsight, SN 1999as and SN1999bd, discovered by the *Supernova Cosmology Project* at Lawrence Berkeley Laboratory, may also have been examples of SLSN (Gal-Yam 2012). A minor mystery was why other unbiased supernova searches did not discover SLSN before *ROTSE*. The SDSS-II SN survey was contemporaneous and had deeper limits. The answer is that SDSS-II did discover SLSN, but they were ignored at the time since the principal goal was to discover and characterize typical SN Ia to be used as distance indicators for cosmology. Revisiting the SDSS-II data revealed several SLSN candidates.

D. Branch, J.C. Wheeler, *Supernova Explosions*, Astronomy and Astrophysics Library, DOI 10.1007/978-3-662-55054-0_18

The SLSN rate is only $\sim 10^{-4}$ of the core-collapse rate (Quimby et al. 2011; McCrum et al. 2015). Nevertheless, other wide-field transient search projects (Chap. 2) such as *PTF*, *PanSTARRS*, and *DES* began to find SLSN, and the *LSST* should discover them in great abundance. At this writing, more than 100 SLSN have been observed. Most SLSN are from 10 to 100 times more luminous than typical SN Ia. Thus SLSN will be observed to rather high redshift; they may be used to explore related issues of cosmology and galaxy formation as well as to provide deeper understanding of stellar evolution. Candidate SLSN have been observed up to $z = 3.9$ (Berger et al. 2012; Cooke et al. 2012a).

By analogy to traditional supernova classification, SLSN that show H lines in their early spectra are referred to as SLSN-II (and sometimes as "superluminous SN IIn".) Some of these show strong emission lines of hydrogen near maximum light. In others, hydrogen appears clearly only somewhat later. SLSN that do not show hydrogen, except at late times in a few cases (Yan et al. 2015), are SLSN-I (or SLSN-Ic in view of the lack of H and He I lines in their early spectra, but the relationship between these superluminous events and typical SN Ic awaits elucidation). SLSN usually occur in galaxies of low metallicity and low luminosity. SLSN-I may be more extreme in this respect (Sect. 18.3.2), which would suggest that the presence or absence of a hydrogen envelope is not the only difference between SLSN-II and SLSN-I.

The high luminosity of SLSN-II is very likely due to conversion of kinetic energy to radiation by CSI with matter that was shed by the progenitor star in the years to centuries prior to the explosion (Smith and McCray 2007). While there had been discussion in recent literature of pair-instability supernovae (PISN) among the first stars born at $z \simeq 20$ at low or zero metallicity (Chap. 10), the relatively nearby ($z = 0.018$) SN 2006gy (Sect. 18.2.1) triggered discussions about the possibility of contemporary PISN occurring at higher metallicity (Sect. 18.4). Other mechanisms proposed to account for the high luminosity of SLSN are interaction between expelled shells ejected by the pulsational pair instability, which does not totally disrupt the star (Chap. 10); interaction of a gamma-ray-burst jet with the stellar envelope (Young et al. 2005; Quimby et al. 2007a); a magnetar (Kasen and Bildsten 2010; Woosley 2010; Dessart et al. 2012b; Inserra et al. 2013; Howell et al. 2013; Wang et al. 2015c); an exceptionally energetic core-collapse explosion (Umeda and Nomoto 2008; Moriya et al. 2010b; van den Heuvel and Portegies Zwart 2013), fallback accretion onto a black hole (Dexter and Kasen 2013), and a quark-gluon phase transition (Ouyed et al. 2012, 2015a).

The story of SLSN-I began with SN 2005ap (Quimby et al. 2007a). Its peak unfiltered magnitude was -22.7 and its luminosity exceeded 4×10^{44} erg s^{-1}. At the phases observed, the spectra showed no H or He I lines. The light-curve peak was rather narrow implying a short diffusion time. Radioactive-decay models of the peak required more mass of ^{56}Ni than ejecta mass, which is physically inconsistent. A caveat, as presented in Chaps. 16 and 17, is that the width of the peak together with simple diffusion arguments based on constant opacity may not give an accurate measure of the ejecta mass of stripped-envelope supernovae; the same might be true of SLSN.

SN 2005ap ($z = 0.28$) was placed in a broader context by Quimby et al. (2011) with the discovery of SN PTF09atu, SN PTF09cnd, SN 2009jh (also known as SN PTF09cwl), and SN 2010gx (also known as SN PTF10cwr). The spectra of these events showed broad features in the blue and a mostly smooth continuum at longer wavelengths, similar to that of SN 2005ap. Identification of the Mg II $\lambda\lambda 2796{,}2803$ doublet, formed in the ISM of the host galaxies, revealed the redshifts. Detection of the same feature in co-added spectra of the luminous, enigmatic SN SCP06F6 (Barbary et al. 2009) gave $z = 1.19$. SN PTF09atu ($z = 0.50$), SN PTF09cnd ($z = 0.26$), SN 2009jh ($z = 0.35$), and SN 2010gx ($z = 0.23$) showed broad features in the rest-frame B band that were similar to those in SN 2005ap. These events put SN 2005ap and SN SCP06F6 in perspective by showing that they belonged in the same category, with the four events filling the redshift gap that had obscured this connection. With this synthesis, Quimby et al. used the SYNOW code to identify five blue features in early spectra of SN 2005ap and several of the other events with O II lines—lines that have not been identified in spectra of other supernova types (Fig. 18.1). These features also are seen in other SLSN-I (Fig. 18.2).

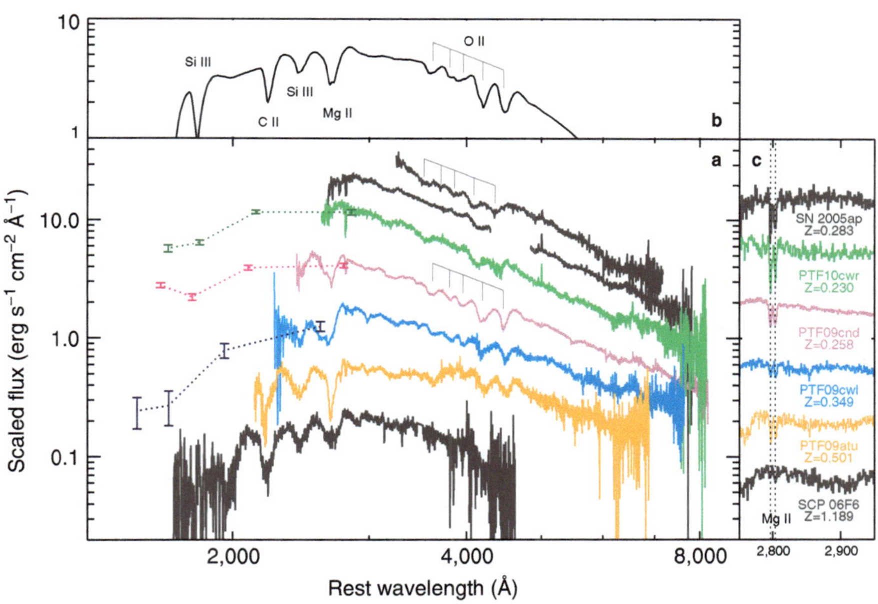

Fig. 18.1 *Bottom*: Spectra of six SLSN-I including SN 2005ap (the top two spectra, obtained 9 days apart). *Top*: A SYNOW synthetic spectrum. *Right*: The narrow Mg II doublets used to determine the redshifts. Reprinted by permission from Macmillan Publishers Ltd.: NATURE, Quimby et al. (2011)

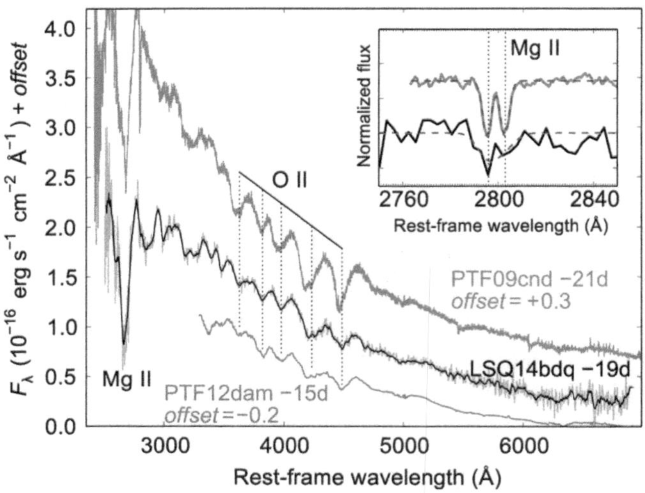

Fig. 18.2 Spectra of three SLSN-I events showing conspicuous O II features. *Inset*: determination of the redshift of SN PTF09cnd by matching double-Gaussian profiles to the observed Mg II doublet. From "LSQ14bdq: A Type Ic Super-luminous Supernova with a Double-peaked Light Curve" (Nicholl et al. 2015a). © AAS. Reproduced with permission

18.2 SLSN-II

18.2.1 Case Study: SN 2006gy

The well-observed SN 2006gy, discovered by Quimby et al. (2006), emerged as a prototype of SLSN-II. At peak, it had $M_V \simeq -22$ and the integrated radiated energy was ~1 B, comparable to the kinetic energy of typical supernovae. The effective temperature near maximum light was near 12,000 K. Unlike many SLSN that occur in dwarf galaxies of low luminosity and low metallicity, SN 2006gy exploded in NGC 1260, a relatively massive host with about solar metallicity (Neill et al. 2011). It appeared close to the active nucleus of the galaxy (Smith et al. 2007), perhaps a region of special conditions within the host.

The spectral evolution is shown in Fig. 18.3. The evolution of the Hα profile delineated three phases (Smith et al. 2010a). The first occurred until 90 days after explosion and included maximum light. In this phase, Hα showed a nearly Lorentzian-shaped symmetric emission profile of FWHM 1350 km s^{-1}. The shock wave was deep inside a dense, opaque circumstellar shell while the photosphere was in the outer, unshocked CSM. The Hα line is presumed to have been excited by photoionization and to have been intrinsically narrow, but broadened by multiple scattering on free electrons produced in the photoionized regions beyond the shock (Chugai 2001; Chevalier and Irwin 2012).

In the second spectral phase, 90–150 days after explosion, the Hα line broadened somewhat and developed a narrow P Cygni feature of width about 260 km s^{-1} that

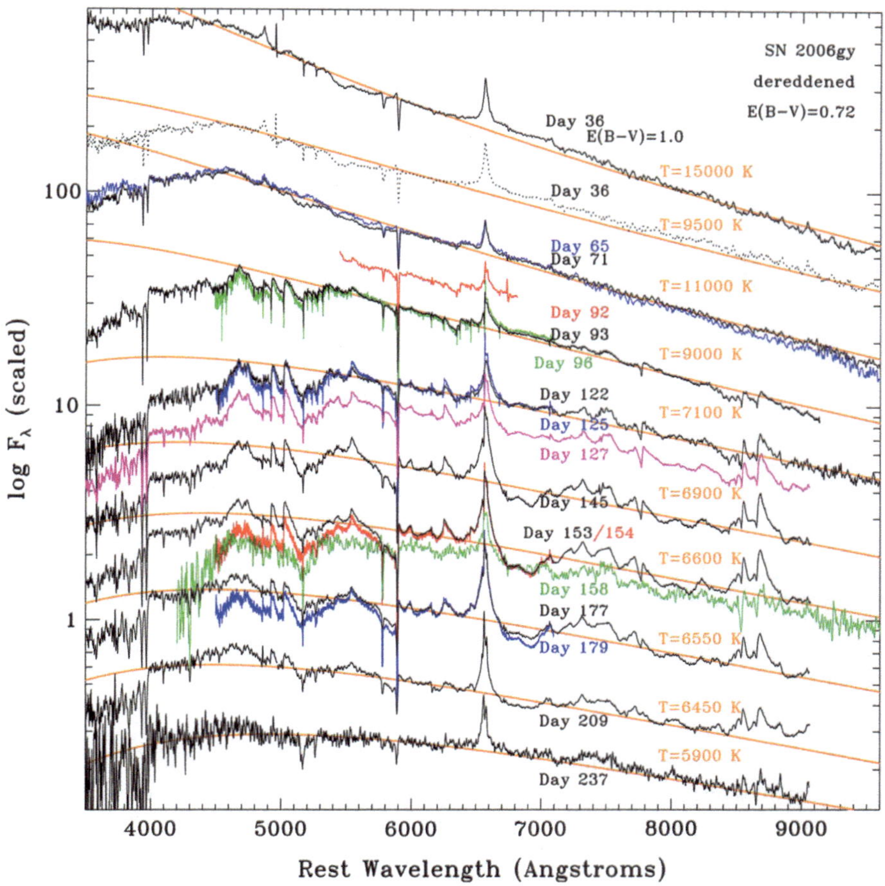

Fig. 18.3 Spectra of SN 2006gy, shifted vertically for clarity, and blackbody curves. Epochs are with respect to the estimated date of explosion. The day 36 spectrum was obtained before the light curve peaked. From "Spectral Evolution of the Extraordinary Type IIn Supernova 2006gy" (Smith et al. 2010a). © AAS. Reproduced with permission

probably represented the expansion speed of the unshocked CSM. In this phase, the forward shock had overtaken the photosphere and was outside it, a situation more closely resembling line formation in a typical supernova. An absorption in the blue wing of Hα , forming in the shocked matter, extended to 4000 km s^{-1}. The red wing remained nearly constant with an intermediate width of about 1800 km s^{-1}, plausibly determined by the Doppler shift of an expanding CDS that formed between the forward and reverse shocks (Sect. 6.8).

In the third spectral phase, 150–240 days after explosion, the Hα line narrowed. The CSI was less strong, the line-emitting region became optically thin, and broadening by electron scattering was reduced.

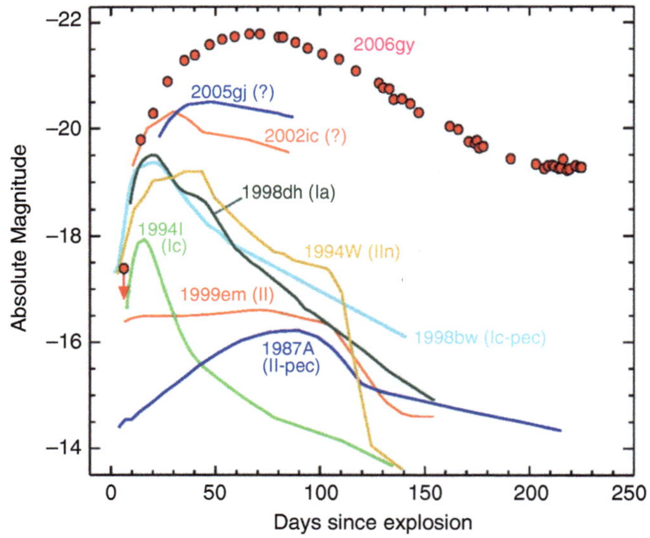

Fig. 18.4 The absolute R-band light curve of SN 2006gy is compared to light curves of supernovae of other types. SN 2005gj and SN 2002ic were Type Ia-CSM events (Sect. 23.2), SN 1998bw was Type Ic-bl, and SN 1999em was Type IIP. From "SN 2006gy: Discovery of the Most Luminous Supernova Ever Recorded, Powered by the Death of an Extremely Massive Star like η Carinae" (Smith et al. 2007). © AAS. Reproduced with permission

The brightness and width of the SN 2006gy light curve (Fig. 18.4) were inconsistent with simple radioactive-decay models. Both the light curve and spectral evolution were characteristic of CSI. The high optical luminosity almost surely derived from the collision of the ejecta with a dense shell of matter previously ejected by the progenitor star. The origin of this CSM remains uncertain. Models require a density structure that is more shell-like than corresponding to a steady-state wind with an r^{-2} density profile (Chatzopoulos et al. 2012, 2013b; Moriya et al. 2013; but see Ginzburg and Balberg 2012). Even at very late phases, there was little evidence of the classic nebular emission lines of core-collapse events, so the CSM and CSI may have continued to shield the inner supernova ejecta. The nature of the underlying explosion thus also remains obscure.

Radio and X-ray observations also constrained the nature of the CSM. SN 2006gy was not detected in the radio, with upper limits making it a much dimmer source than the Type IIn SN 1988Z up to 2 years after outburst (Smith et al. 2008). Weak X-ray emission was detected about 90 days after explosion (Smith et al. 2007). The unabsorbed flux in the 0.5–2 keV band was $\sim 2 \times 10^{39}\,\mathrm{erg\,s^{-1}}$. Six weeks later, only an upper limit at a somewhat lower flux level could be set. This limit on the X-ray flux was about 100 times lower than that of SN 1988Z at a comparable epoch. The low X-ray flux of SN 2006gy was not compatible with the CSM density required to power the optical luminosity by direct postshock emission in a CSI model. The Hα flux also was weaker than simple expectations based on

CSI. One way to reconcile this discrepancy is to have the X-rays be absorbed in an optically-thick shell until the forward shock breaks out of the dense shell that accounted for the high optical luminosity. Plausible shell CSI models are consistent with the occurrence of shock breakout after maximum light (Chatzopoulos et al. 2012; Moriya et al. 2013). In this picture, weak Hα and unabsorbed soft X-rays are emitted as the forward shock propagates into the more dilute CSM beyond the opaque shell even as radiation continues to diffuse from within the dense shell. Since direct postshock emission was insufficient to account for the postpeak optical luminosity, there may be a role for radioactive decay or light-echo effects (Miller et al. 2010b).

18.2.2 Synthesis of SLSN-II Characteristics

In addition to SN 2006gy, well-studied SLSN-II include SN 2006tf (Smith et al. 2010b), SN 2008am (Chatzopoulos et al. 2011), and SN CSS121015 (Benetti et al. 2014).

Nearly all SLSN-II show clear evidence of CSI. The data are roughly consistent with photospheres of radii $\sim 10^{15}$ cm, as is typical for supernovae, and temperatures near 12,000 K, somewhat hotter than those of typical supernovae. The temperature is directly responsible for the extreme luminosity. The underlying reason is that while typical supernovae must expand to reach a photospheric radius of $\sim 10^{15}$ cm, for SLSN-II there is already a substantial amount of matter at that radius, so the matter can radiate efficiently, with optical depth $\tau \sim c/v$, without having to expand and suffer adiabatic losses. Much of the kinetic energy of SLSN-II is converted to radiation, whereas only about 1% of the energy of a typical supernova appears as radiation. A comparison of some SLSN-II light curves is given in Fig. 18.5.

Other SLSN-II showed similar characteristics to SN 2006gy, but also some interesting departures. SN 2006tf peaked at $M_V \simeq -20.8$ (Fig. 18.5) and radiated about 1B. It showed nearly symmetric emission lines of Hα (Fig. 18.6), with the development of narrow P Cygni features superposed on the overall wider Hα profile by 66 days after discovery (Smith et al. 2010b). A year after explosion, the Hα emission line showed a prominent blue plateau extending to about 1000 km s^{-1}. While Hα in SN 2006gy was excited radiatively throughout its evolution, the significant increase in the Balmer decrement a year after the explosion of SN 2006tf suggested that Hα was excited by collisional processes at that time (Smith et al. 2010b). This showed that there was some substantial difference in the environment of the two events. Unlike SN 2006gy, at these later epochs in SN 2006tf there were indications in both the emissions and absorptions for rapidly-moving material, about 7500 km s^{-1}, from the underlying supernova ejecta. This evidence was not sufficient to identify the nature of the explosion in any detail.

SN 2008am was similar to SN 2006gy in many respects (Chatzopoulos et al. 2011), but with an especially rapid rise to maximum in 34 days. SN 2008am peaked at an unfiltered magnitude of about −22 (Fig. 18.5) with a bolometric luminosity

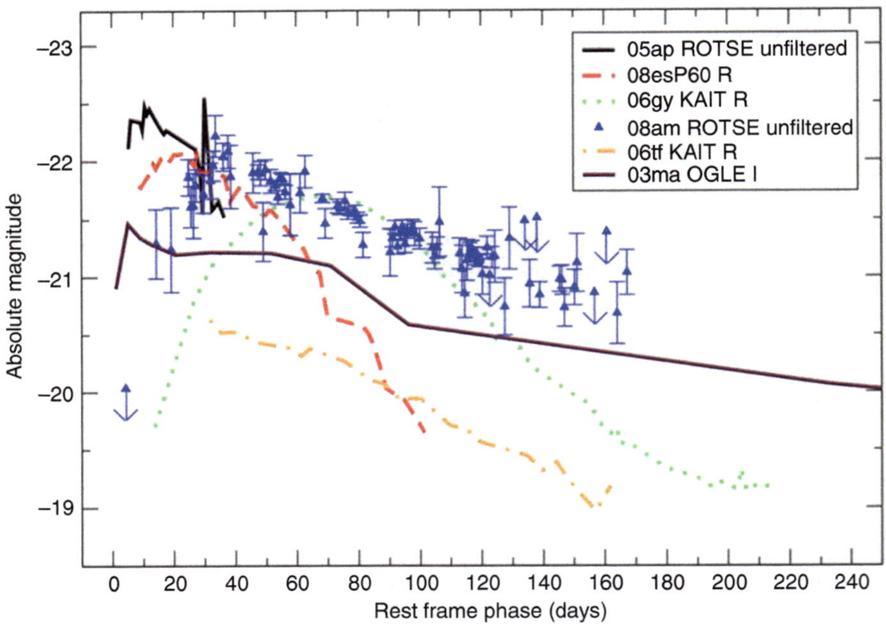

Fig. 18.5 The absolute light curves of five SLSN-II events as well as the SLSN-I SN 2005ap. Epochs are with respect to the estimated date of explosion. From "SN 2008am: A Super-luminous Type IIn Supernova" (Chatzopoulos et al. 2011). © AAS. Reproduced with permission

$\sim 3 \times 10^{44}$ erg s^{-1} and a radiated energy $\sim$2 B. It had symmetric Lorentzian-shaped emission lines near maximum and a narrower Hα line about a year later. The broader and narrow P Cygni features that appeared in SN 2006gy 20–80 days after maximum light were not observed in SN 2008am, but this may be due to the sparse temporal sampling of the spectra. Other SLSN-II of note include SN 2003ma, which declined from maximum by only three magnitudes in 5 years (Fig. 18.5) and radiated $\sim$4 B (Rest et al. 2011a). The shape of the light curve resembled that of the Type IIn SN 1988Z (Sect. 14.2.1), long and flat, but that of SN 2003ma was ten times brighter.

Among the first SLSN, SN 2008es was in a category of its own (Gezari et al. 2009). It peaked at $M_V \simeq -22$ (Fig. 18.5) and a luminosity of $\sim 3 \times 10^{44}$ erg s^{-1}. The radiated energy was $\sim$0.6 B. The spectrum was very blue and nearly featureless in early phases and the light curve showed a nearly linear decline, reminiscent of SN IIL. Hydrogen emission lines developed at a later epoch. The light curve cannot be explained by radioactive decay, but there also was no obvious evidence of it being powered solely by the CSI that characterizes SLSN-II as described above. Some CSI was occurring, so it may be that SN 2008es is a variation on that theme, but powering by a magnetar is possible (Kasen and Bildsten 2010; Woosley 2010).

Studies of SLSN-II have the potential to yield the density distribution in the CSM and hence important clues to the mass-loss history and nature of the progenitor. There are strong indications that a portion of the CSM is optically thick, requiring

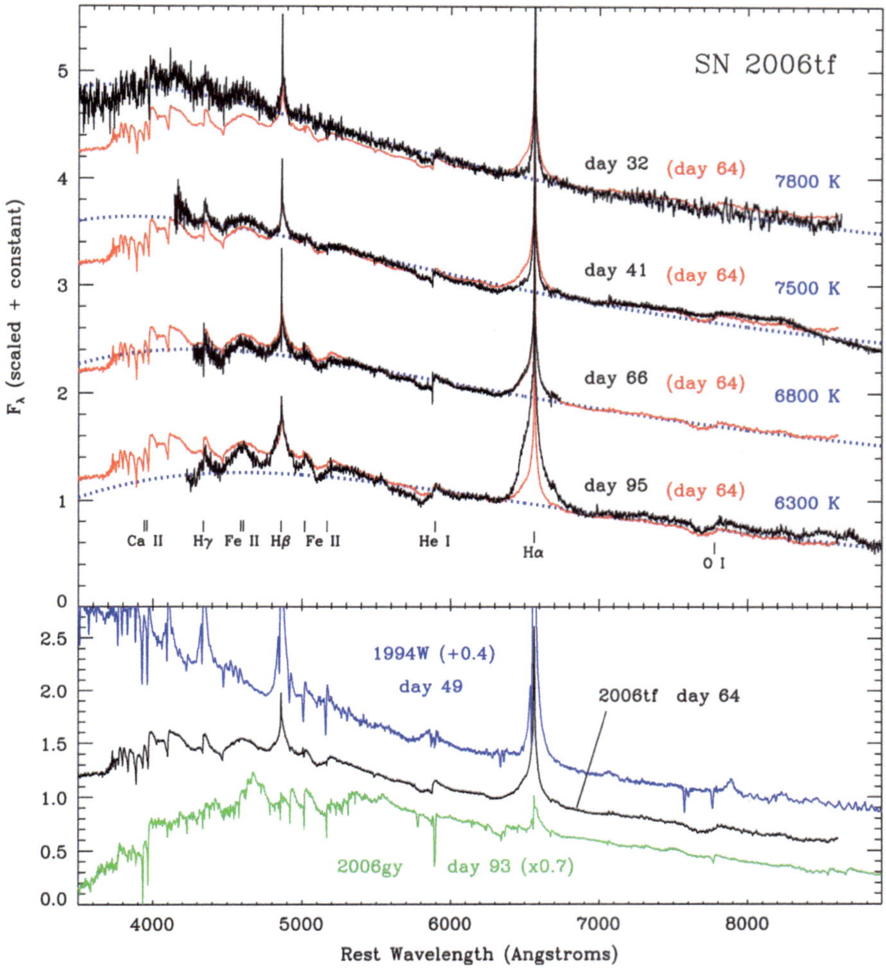

Fig. 18.6 *Top*: Spectra of the SLSN-II SN 2006tf. A day 64 spectrum is plotted in *red* four times, for comparison with the other spectra. Epochs are with respect to the estimated date of explosion and the spectra are shifted vertically for clarity. *Bottom*: A spectrum of SN 2006tf is compared with a spectrum of the SLSN-II SN 2006gy as well as the Type IIn SN 1994W. From "SN 2006tf: Precursor Eruptions and the Optically Thick Regime of Extremely Luminous Type IIn Supernovae" (Smith et al. 2010b). © AAS. Reproduced with permission

high CSM densities. This could result from high mass-loss rates as attributed to giant-outburst LBVs (Humphreys and Davidson 1994), by compression and cooling of a low-velocity wind by a subsequent high-velocity wind (Dwarkadas 2007), or by pulsational instabilities in the envelopes of red-supergiant PISN progenitors (Moriya and Langer 2015; Sect. 18.4).

Despite the strong presumption of CSI, no SLSN-II has been detected in the radio. This paucity is partially due to the great distances of SLSN, but may also be another clue to the density distribution of the CSM. Since the CSM can be the product of winds of various rates or episodic mass loss, it can have a variety of density, velocity, and temperature profiles, including complex morphological structures. This may naturally account for the heterogeneity among SLSN-II, but mapping a given model structure into given observed observational properties remains a challenge awaiting appropriate multidimensional radiative-hydrodynamics simulations. One complication compared to typical supernovae is that SLSN-II driven by CSI depart from homologous expansion, so this simplifying assumption cannot be adopted in addressing the radiative transfer.

Owing to shielding by the optically-thick CSM, few of the standard diagnostics of the ejecta mass, composition, and velocity are available so the nature of the underlying supernova in SLSN-II remains unknown. While "normal" core collapse seems most likely, pair-instability (Chap. 10) cannot be ruled out.

There is an expectation that both the supernova ejecta and the surrounding CSM may be aspherical, but there has been very little spectropolarimetry of SLSN of any variety. One exception is SN 2006tf for which a single epoch of spectropolarimetry was obtained about 2 months after discovery (Smith et al. 2010b). These data showed wavelength variations in the broad, blue components of Hα and O I λ7774 that suggested intrinsic asymmetries, but the data were sparse and little else was learned. Efforts are underway to obtain more spectropolarimetry of SLSN.

18.3 SLSN-I

18.3.1 Case Study: SN 2010gx

SN 2010gx, at $z = 0.23$, was discovered independently by the CRTS team (Mahabal et al. 2010) and the PTF survey (Quimby et al. 2010). The host was a dwarf galaxy with a high specific star-formation rate and a low metallicity (Chen et al. 2013).

The light curves of SN 2010gx (Fig. 18.7) rose relatively quickly to a maximum ($M_B \simeq -21.2$) and then declined relatively slowly. As for SN 2005ap (Sect. 18.1), early spectra showed a blue continuum and broad features including those of O II in the blue (Fig. 18.1, where SN 2010gx is labeled PTF10cwr). Several weeks after discovery, the spectral features began to resemble those of young SN Ic (Fig. 18.8). The spectra of SN 2010gx, like the light curves, evolved much more slowly than those of typical SN Ic.

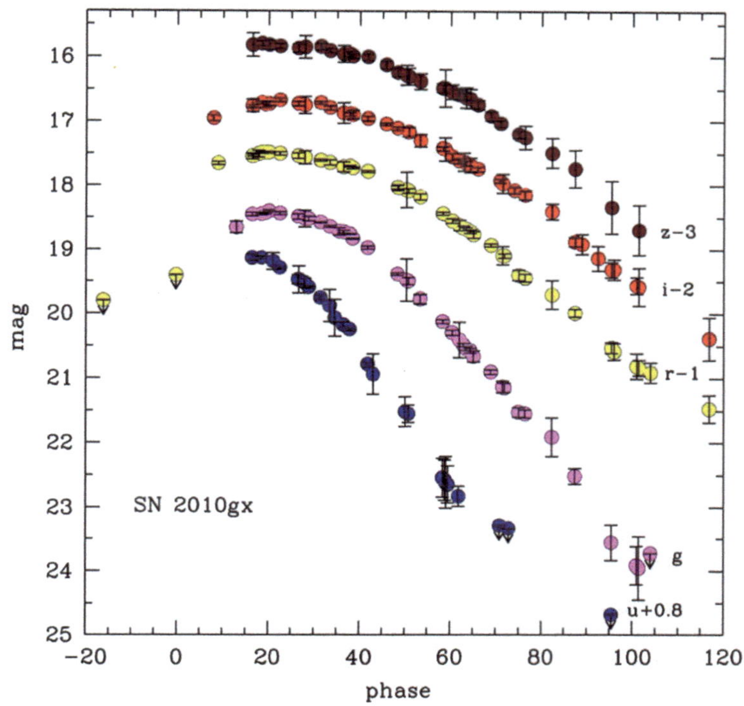

Fig. 18.7 Light curves of SN 2010gx. Phases are relative to the estimated explosion date. From "Ultra-bright Optical Transients are Linked with Type Ic Supernovae" (Pastorello et al. 2010). © AAS. Reproduced with permission

18.3.2 Synthesis of SLSN-I Characteristics

In addition to SN 2010gx, well-studied SLSN-I include SN 2006oz (Leloudas et al. 2012), SN PS1-11ap (McCrum et al. 2014), SN PTF12dam (Chen et al. 2015), and SN iPTF13ajg (Vreeswijk et al. 2014).

SLSN-I tend to occur in metal-poor *extreme emission-line galaxies* (Fig. 18.9) that are of low mass, but characterized by intense star formation and a hard ionizing radiation field (Leloudas et al. 2015; see also Lunnan et al. 2014; Thöne et al. 2015). This suggests an environment even more extreme than the galactic hosts of gamma-ray bursts and that SLSN-I have short-lived progenitors, $\lesssim$5 million years, with ZAMS mass in the range 25–120 $M_{\odot}$. The intense ambient radiation field to which the progenitor stars are exposed may play a role in promoting mass loss and hydrogen deficiency.

The light curves of SLSN-I (Fig. 18.10) tend to evolve rather quickly, with rise times of several weeks to more than a month that are somewhat shorter than those of SLSN-II and declines that are generally more rapid than the rate of ^{56}Co decay. A subset of SLSN-I are dimmer than average by a factor of several, have less

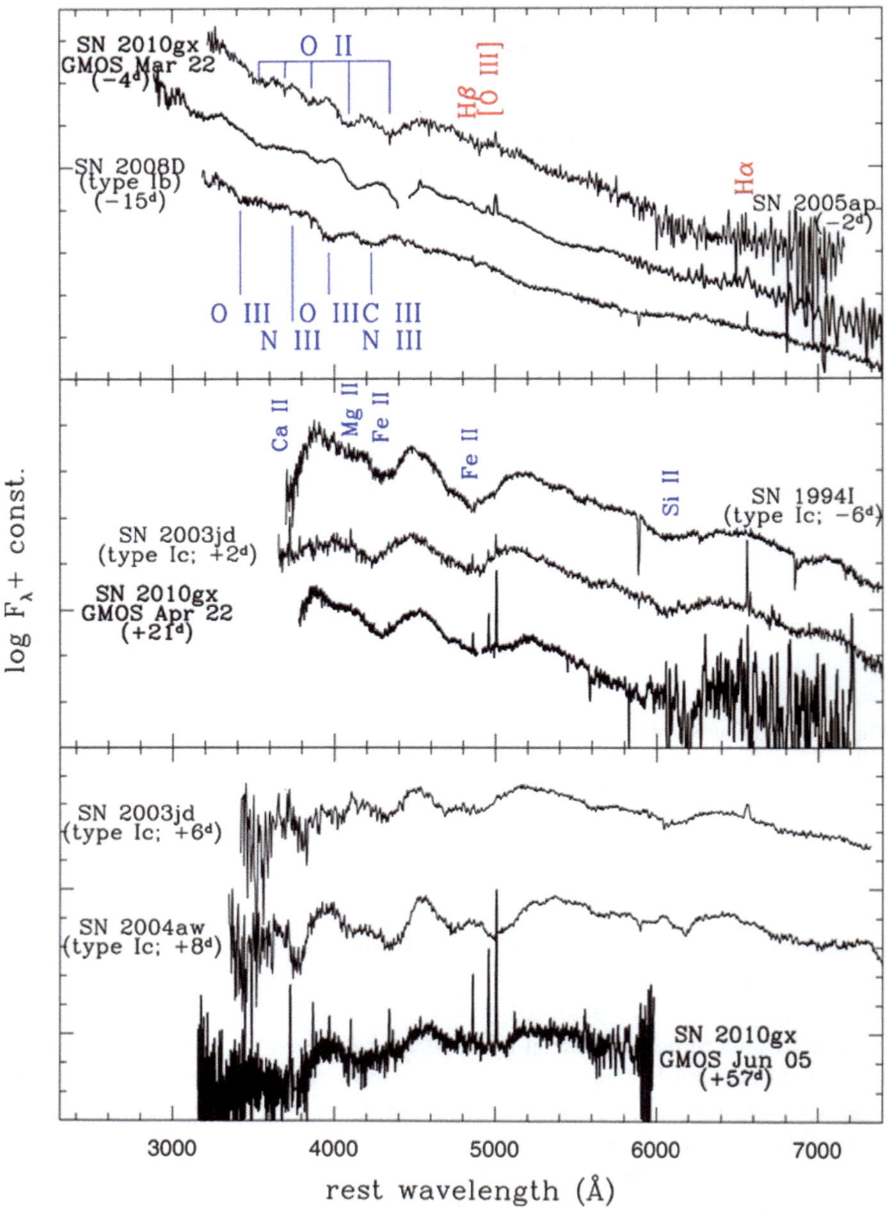

Fig. 18.8 *Top*: premaximum spectra of the SLSN-I SN 2010gx and SN 2005ap as well as a very early spectrum of the Type Ib SN 2008D with line identifications from Modjaz et al. (2009). *Center*: Premaximum and near-maximum spectra of the Type Ic SN 1994I and the Type Ic-bl SN 2002jd, respectively, and a day +21 spectrum of SN 2010gx. *Bottom*: Postmaximum spectra of the SN 2002jd and the Type Ic SN 2004aw and a day +57 spectrum of SN 2010gx. Phases are relative to the time of maximum in the B band. From "Ultra-bright Optical Transients are Linked with Type Ic Supernovae" (Pastorello et al. 2010). © AAS. Reproduced with permission

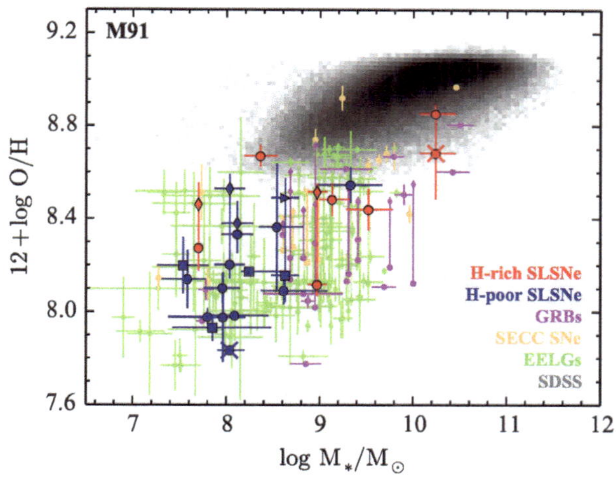

Fig. 18.9 Location of the hosts of SLSN-I with respect to metallicity and host-galaxy mass. The cloud of *black points at the upper right* represents normal galaxies and the *green points* are extreme emission line galaxies (EELG). Also shown are representative data for hosts of SLSN-I (*dark blue*), SLSN-II (*red*), stripped-envelope core-collapse supernovae (SECC SNe; *yellow*), and GRBs (*purple*). The hosts of SLSN-I tend to resemble the EELG. Figure 1 "Spectroscopy of superluminous supernova host galaxies. A preference of hydrogen-poor events for extreme emission line galaxies" from Leloudas et al. (2015), by permission of Oxford University Press on behalf of the Royal Astronomical Society

peak-to-tail contrast, and have late-time tails that decay at about the rate of ^{56}Co decay (Papadopoulos et al. 2015). These events do not seem to be clearly distinguished spectroscopically from more typical SLSN-I. It is thus not clear whether they represent a separate category in some physical sense (Sect. 18.4) or are an extension of a continuous distribution. There are some suggestions of regularities in the light-curve behavior of SLSN-I that might make them standardizable candles of use for cosmology (Inserra and Smartt 2014; Papadopoulos et al. 2015).

The typical spectroscopic evolution of SLSN-I is shown in Fig. 18.11. SLSN-I often have blue continua and a significant UV contribution. As for SN 2010gx (Sect. 18.3.1), spectra of SLSN-I first show a blue continuum and broad O II features and eventually develop features reminiscent of SN Ic. Some SLSN-I showed photospheric velocities of 12,000–19,000 km s^{-1} that did not appear to decrease in three rest-frame weeks around maximum light (Chomiuk et al. 2011). The sparse data on SLSN-I suggest that the photospheric temperature declines from about 20,000 K about 20 days prior to maximum light to 10,000 K at 40 days after maximum. The data suggest, but do not yet establish, that SLSN-I all have about the same temperature near maximum light. The effective temperatures are consistent with the near total ionization of hydrogen if any were present, and the presence of singly- and doubly-ionized helium and singly-ionized oxygen at the photosphere (Hatano et al. 1999b).

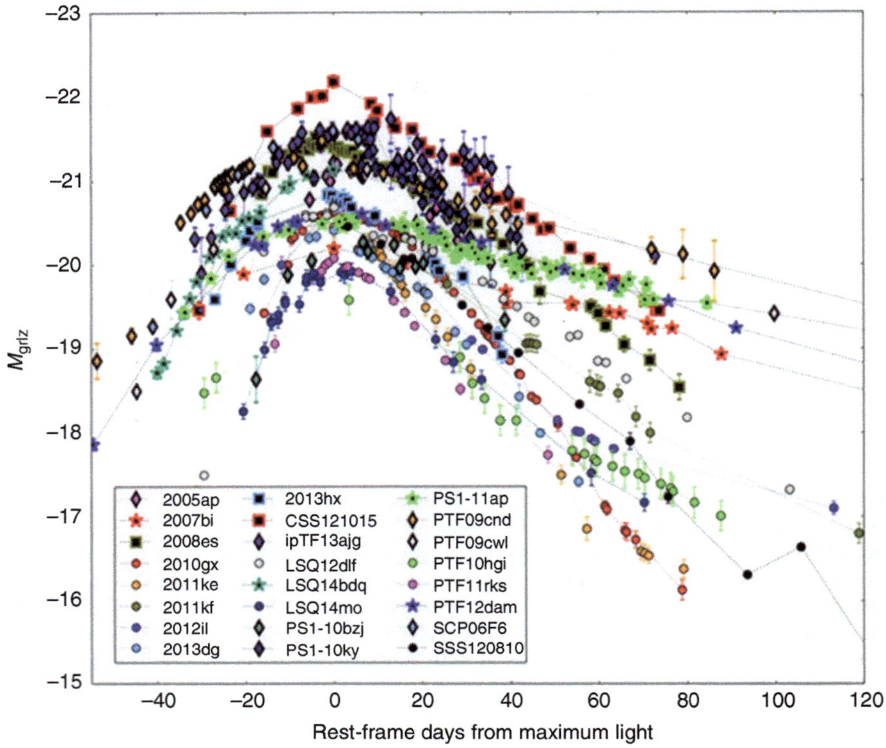

Fig. 18.10 Pseudo-bolometric light curves of SLSN-I. Figure 1 "On the diversity of superluminous supernovae: ejected mass as the dominant factor" from Nicholl et al. (2015b), by permission of Oxford University Press on behalf of the Royal Astronomical Society

Photospheric spectra of SLSN-I are consistent with ejecta rich in carbon and oxygen with a solar abundance of heavier elements (Howell et al. 2013). Nebular spectra are difficult to acquire. At 400–500 days, both SN 2007bi (Sect. 18.4) and SN PTF12dam (Chen et al. 2015) showed nebular emission lines of [O I] $\lambda\lambda$6300,6364, [Ca II] $\lambda\lambda$7291,7323, and O I λ7774 that were reminiscent of the nebular spectra of typical SN Ic and other core-collapse supernovae.

Attempts have been made to fit the light curves of SLSN-I with simple radioactive-decay models. In many cases, the models fail by requiring a greater mass of ^{56}Ni than the total ejecta mass informed from the width of the light curve peak and the implied diffusion mass. As mentioned in Sect. 18.1, there is a possible issue of cold, mostly neutral mass that would contribute little to the opacity and hence to the diffusion mass. In other cases, those with less peak-to-tail contrast and tails that roughly follow ^{56}Co decay, radioactive-decay models are more successful. Even then, there can be issues with spectral constraints (Sect. 18.4). Very late photometric and spectroscopic observations may help to constrain the amount of ejected ^{56}Ni.

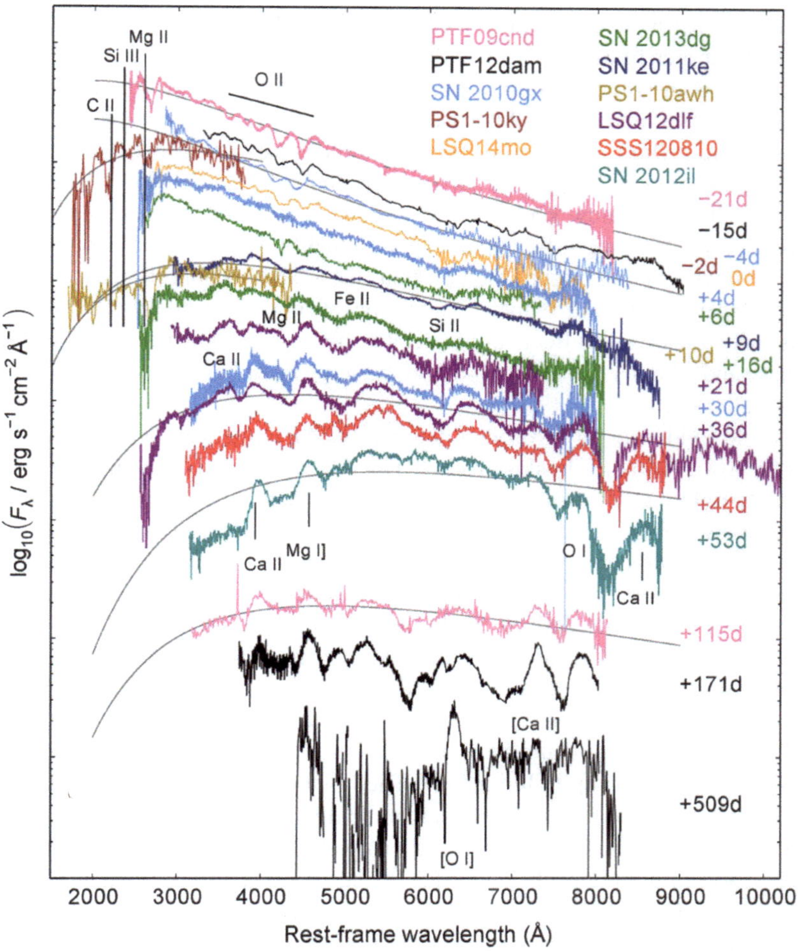

Fig. 18.11 Typical spectroscopic evolution of SLSN-I. Features in postmaximum spectra are like those of SN Ic. Figure 16 "On the diversity of superluminous supernovae: ejected mass as the dominant factor" from Nicholl et al. (2015b), by permission of Oxford University Press on behalf of the Royal Astronomical Society

The lack of strong H and He I lines and any evidence for narrow lines of any species gives pause to a model based on CSI, yet economy of hypotheses suggests that it be considered for SLSN-I, given the abundant evidence for its role in SLSN-II. Given a number of free parameters, it is formally possible to reproduce the light curves of SLSN-I with a CSI model (Chatzopoulos et al. 2013b). In some phases, SLSN-I often show a blue continuum that is characteristic of CSI. This raises the possibility that CSI may contribute to all SLSN, even the SLSN-I that show no obvious spectral evidence for narrow lines that accompany CSI in SN IIn.

Evidence for hydrogen-deficient CSM is given by SN Ibn (Sect. 16.5), which show no hydrogen lines but narrow emission lines of He I. In SLSN-I, there may be configurations in which helium is present in the CSM, but insufficiently excited to produce observable lines. Alternatively, the CSM could be deficient in both hydrogen and helium. For SLSN-I this raises the issue of whether the CSM could be dominated by, e.g., oxygen (Chatzopoulos and Wheeler 2012b; Sorokina et al. 2016). Whether and how such a CSM would be revealed in spectra is an open question. The CSM might be expanding sufficiently rapidly to broaden and mute emission lines from an oxygen-rich plasma. If the shell is sharply truncated on its outer boundary, the forward shock could propagate out of the shell and subsequently produce few observable effects in the nebular phase. The spectra would depend on the phase and whether the shock is still propagating in, or has emerged from, a dense circumstellar shell.

Possible evidence for hydrogen-deficient CSM associated with SLSN-I was given by SN 2006oz (Leloudas et al. 2012). A feature of special interest was the very early light curve that showed a dip in the optical bands and a possible initial cooling phase in the UV before the rise to maximum light. The bolometric light curve showed a precursor plateau with a duration of 6–10 days in the rest frame; this suggested interaction with hydrogen-deficient CSM. The dip and subsequent rise to maximum in SN 2006oz might be understood in the context of CSI if the circumstellar shell was located at some distance from the progenitor star and was originally optically thin before being ionized (Moriya and Maeda 2012). The very early light curve could represent energy released from the supernova ejecta. When the CSI began, the shell could have become ionized, leading to a rapid increase in opacity and hence a drop in luminosity. The breakout of the shock from the CSM might then have led to the eventual peak in the light curve.

Subsequently, Nicholl and Smartt (2016) presented evidence that several other SLSN-I showed double peaks (Fig. 18.12) and argued that such behavior may be ubiquitous in SLSN-I. **?** suggested that the precursor peaks may be the original fireball phase and that the second peak is "engine" driven. Piro (2015b) concluded that breakout even from an RSG will not account for the observations, but that more extended material, CSM, is required.

The nearly constant photospheric velocity seen in some SLSN-I is roughly consistent with expansion of an optically-thick shell of material and hence some version of the CSI that characterizes SLSN-II. An important caveat is that the very large photospheric velocities observed in some cases may be difficult to reconcile with a substantial CSM that should decelerate the ejecta (Howell et al. 2013).

While CSI remains a candidate to account for or at least contribute to SLSN-I, magnetar models also can give good agreement with the observations of some events (Inserra et al. 2013). The second, main peak in SN 2006oz and the events illustrated in Fig. 18.12 could, for instance, be due to input from a magnetar. For SLSN-I, it is easy enough to envisage a magnetar being born in a progenitor star that itself was the core of a star from which hydrogen and helium had been stripped. Such a model can give spectra with blue continuua and spectral features reminiscent of SLSN-I (Dessart et al. 2012b; Howell et al. 2013).

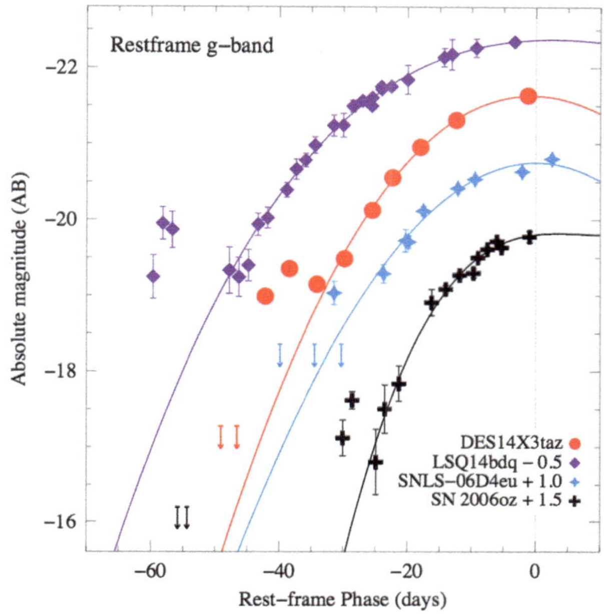

Fig. 18.12 A sample of SLSN-I that show precursor peaks prior to the main peak. The data are shifted vertically for clarity. The *lines* are polynomial fits to the main peak to illustrate the precursor peaks by contrast. Figure 1 "Seeing double: the frequency and detectability of double-peaked superluminous supernova light curves" from Nicholl and Smartt (2016), by permission of Oxford University Press on behalf of the Royal Astronomical Society

Magnetar models also predict a shell of matter that is swept up by the intense pulsar radiation. This shell could account for the nearly constant photospheric velocity in some cases and might also alter the nature of the later-time spectra. Magnetar models predict a hot cavity within the expanding ejecta characterized by a flux of hard X-rays that drive an ionization front out through the ejecta (Metzger et al. 2014; Wang et al. 2015c). Under appropriate conditions, the breakout of this ionization front may lead to detectable UV or X-ray emission, perhaps months after the explosion. Such breakout radiation could be evidence for the role of a magnetar, although other mechanisms may also give rise to X-ray flares (Svirski and Nakar 2014a). The shell associated with magnetar models is likely to be unstable and its morphology to be complex. This could affect the nature and timing of any ionization breakout. A shell might not even form if the magnetar radiation forms a collimated jet (Bucciantini et al. 2009).

The relatively slow, monotonic decline of the effective temperature from pre-maximum to postmaximum could constrain magnetar models. The rapid late-time drop in the luminosity seen in some events may be inconsistent with magnetar models but might be consistent with the post-breakout phase (Nicholl et al. 2014). While there is evidence that some magnetars are born in stars of very high mass

(Muno et al. 2006), most magnetar models require modest ejecta mass that may prove inconsistent with the evidence from host galaxy data that SLSN-I come from especially massive progenitor stars.

SN PS1-10afx, at $z = 1.39$, had a peak absolute optical magnitude of $\simeq -22.3$ and a bolometric luminosity of $\sim 4 \times 10^{44}$ erg s^{-1} (Chornock et al. 2013). The photospheric velocity was about 11,000 km s^{-1}. The light curve rose rather rapidly to maximum in about 12 days and also declined rapidly. The color temperature of about 6800 K was cooler than most other SLSN. Simple radioactive-decay models could not account for the high luminosity and narrow light curve. Simple magnetar models predicted excessive temperature and ejecta velocity. CSI models could account for the light curve, in principle, but did not provide obvious accounts of the spectra and colors. Whether SN PS1-10afx represents a previously unrecognized variant of one of these physical models or requires a new progenitor and explosion mechanism remains to be seen.

An extremely bright ($M_V = -23.5$) transient at $z = 0.23$, designated SN 2015L, may have been a SLSN-I that would challenge all existing models (Dong et al. 2016), or it may have been something else, e.g., a tidal disruption of a star by a black hole.

Any firm conclusion as to the nature of SLSN-I, whether CSI, a magnetar, or something else, is premature in the absence of radiative-transfer calculations for detailed 3D explosion models.

18.4 Superluminous PISN?

There is ongoing discussion of the possibility of PISN (Chap. 10) in the context of the first stars born at zero or very low metallicity. Models of PISN at the upper end of the progenitor mass-range predict that this process could eject large masses of ^{56}Ni with the potential to produce very bright light curves. In models, the peak brightness depends on whether the progenitor retains a hydrogen envelope at the time of explosion and hence on issues such as mass loss and rotation. SLSN tend to occur in low, but not extremely low, metallicity environments. Nevertheless, as mentioned in Sect. 18.1, the issue of whether PISN might contribute to the sample of SLSN arose at the time of the first discoveries of SLSN (Smith et al. 2007). One suggestion is that PISN might arise in contemporary environments by runaway stellar collisions in dense star clusters (Pan et al. 2012d).

Models of PISN show the progenitor must be massive enough to form an oxygen core that exceeds about 60 $M_\odot$. Some SLSN show evidence of rather high ejecta mass, but are unlikely to be PISN because the ejecta mass does not meet the 60 $M_\odot$ criterion or because if their brightness had derived from the radioactive decay of ^{56}Ni, as demanded by the pair-instability model, they would require a greater mass of ^{56}Ni than is nominally allowed for the total mass of the ejecta.

The high radiative output of SLSN-II apparently is the result of CSI with dense, hydrogen-rich matter. In at least some SLSN-II, such matter could be the result of pulsational instabilities during the RSG phases of PISN progenitors (Moriya and Langer 2015).

Several percent of SLSN-I, especially the relative dim variety that have small peak-to-tail contrast and tails roughly consistent with ^{56}Co decay (Fig. 18.13), have some characteristics expected of a PISN. These are sometimes referred to as SLSN-R, for radioactivity. The SLSN-I SN 2007bi (Gal-Yam et al. 2009; Young et al. 2010) was considered a PISN candidate because it showed no evidence for hydrogen or

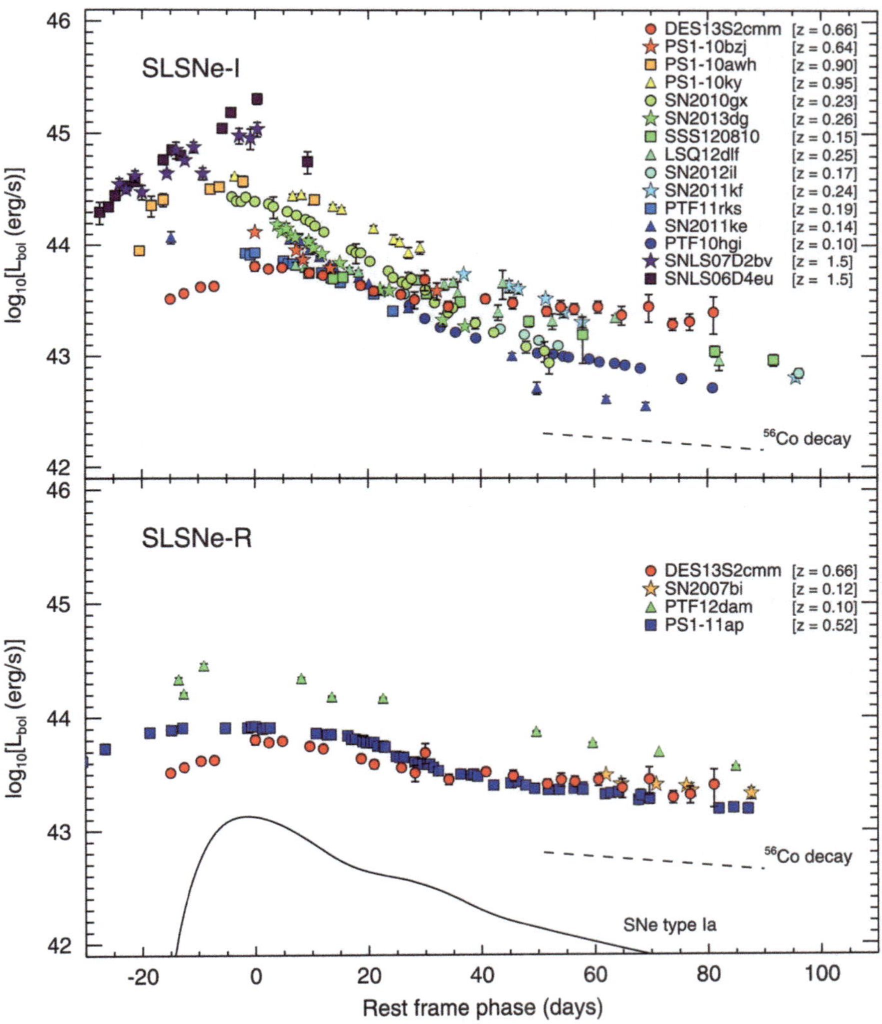

Fig. 18.13 *Top*: A sample of SLSN-I light curves. *Bottom*: A subset of SLSN-I light curves with low peak-to-tail contrast. Figure 5 "DES13S2cmm: the first superluminous supernova from the Dark Energy Survey" from Papadopoulos et al. (2015), by permission of Oxford University Press on behalf of the Royal Astronomical Society

helium or CSI in the form of narrow emission lines even late in the nebular phase. The host galaxy has a metallicity of about 1/3 solar (Young et al. 2010), the predicted maximum metallicity for PISN (Chap. 10). Unlike most SLSN, the peak luminosity of $M_V \simeq -21.3$ and postpeak decline were consistent with certain models of PISN, those that explode devoid of hydrogen envelopes so that they are powered entirely by radioactive decay. The rise time to peak and hence the ejected mass were uncertain. The radiated energy was $\sim$1 B. The peak luminosity and the postmaximum decline of about 0.01 mag d^{-1} were consistent with that of ^{56}Co for 500 days, suggesting that more than 3 M$_\odot$ of ^{56}Ni was ejected. The late-time spectra were consistent with a nebular spectrum powered by several M$_\odot$ of ^{56}Co. With an assumed ejecta mass of $\sim$100 M$_\odot$ and an observed photospheric velocity of 12,000 km s^{-1}, the kinetic energy would have been $\sim$100 B. These are all potential hallmarks of models of pair-instability supernovae.

Despite some suitable properties, however, PISN models have other characteristics that mitigate against their ability to account for SN 2007bi. The most important of these is that models produce a significant amount of iron-peak isotopes, ^{56}Fe, ^{56}Ni, and ^{56}Co. The line blanketing from these elements yields postmaximum spectra that are much redder than the observed spectra of SN 2007bi (the observed colors fall in the range $B-R \simeq 0.2$ to 0.4 while the models predict about 1.5; Dessart et al. 2012b; Chatzopoulos et al. 2015). The nearly-constant observed line widths from about 50 to 400 days after maximum also were inconsistent with predictions of PISN models.

Several other SLSN-I that resembled SN 2007bi at maximum and postmaximum phases were well-observed on the rise to maximum (Fig. 18.13). SN PTF12dam and SN PS1-11ap had rise times that were too short and colors that were too blue to be consistent with PISN models (Nicholl et al. 2013; McCrum et al. 2014). A circumstellar-interaction model with an ejecta kinetic energy of only 4 B provided a good match to the light curve of SN PTF12dam (Baklanov et al. 2015). Given its similarity to these events SN 2007bi, too, is unlikely to have been a PISN. Magnetar models (Nicholl et al. 2013), core collapse in massive stars (Yoshida et al. 2014), or CSI perhaps with contributions from radioactive decay (Chatzopoulos et al. 2013b), are more likely to account for SN 2007bi-like events.

18.5 Summary

The enigmatic and heterogeneous nature of SLSN leaves a host of questions: what shapes the rise and decline of the light curve? What are the effects of the initial radius and density distribution of the supernova progenitor and the location and distribution of CSM? Does the heterogeneity of the current sample of SLSN require very different progenitors and physics, or is there some common theme expressed in different ways?

The spectral evolution of SLSN-II, including the presence of narrow emission lines, strongly argues that, like SN IIn, SLSN-II are powered by CSI. A large variation in the properties of the CSM may naturally produce the heterogeneity among SLSN-II. Understanding the variety of observational properties remains a challenge awaiting appropriate multidimensional radiative-hydrodynamic simulations. A fundamental problem is to understand the episodic mass loss of progenitors of SLSN-II and SN IIn.

SLSN-I show no obvious evidence for CSM and most of them cannot be driven only by radioactive decay. Helium could be present, but not sufficiently excited by ^{56}Ni or other processes. Taken at face value, however, SLSN-I are deficient in both hydrogen and helium, raising the issue of whether the CSM could be dominated by oxygen. The spectral evolution of such a situation is an important open question.

While SLSN-II and most SLSN-I are not powered by radioactive decay alone, the issue of PISN remains of interest. There is an expectation that very massive stars subject to PISN may be selectively produced among the first stars to form in the early Universe. At least some SLSN-I could be circumstellar-interacting PISN. The SLSN-I SN 2007bi had some of the expected characteristics of a PISN, but detailed examination of the expected spectral evolution and comparison to closely related events shows that there are quantitative difficulties in identifying SN 2007bi with a PISN process.

An issue that has not yet been adequately explored is that all models for SLSN based on magnetars assume that at early times radiation from the buried pulsar is thermalized in the ejecta. MHD calculations of this scenario in other contexts suggests that the magnetic flows may form jets that penetrate the ejecta rather than thermalizing.

If there is a single physical mechanism powering the high luminosity of SLSN, it is most likely to be CSI. CSI must be a dominant if not the sole mechanism powering SLSN-II, and there are hints that it is also at work in at least some SLSN-I. Much needs to be done to understand CSI in hydrogen- and helium-poor CSM.

Part III
Type Ia Supernovae

Chapter 19
Degenerate Carbon Burning

19.1 Introduction

The physics of thermonuclear combustion in an electron-degenerate environment shares many elements in common with problems of terrestrial combustion. Important books that address many of the terrestrial issues are Courant and Friedrichs (1948), Landau and Lifshitz (1959), Zel'dovich and Raizer (1967, 1969), Fickett and Davis (1979), Williams (1985), Peters (2000), and Oran and Boris (2005).

19.2 Single Degenerate: The Smoldering Phase

In the single degenerate scenario for Type Ia supernovae (Sect. 21.2), carbon ignition in hydrostatic equilibrium is deemed to begin deep inside a degenerate white dwarf when the carbon-burning rate exceeds the local neutrino loss rate (Arnett 1969; Barkat 1975). The details of the conditions under which carbon ignites depend on the evolution of the star. An accreting white dwarf ignites at somewhat different central density than would the degenerate core of a double-shell burning asymptotic giant branch star (Sect. 8.3.2). In a white dwarf, ignition occurs at a central density of $\rho_c \simeq 3 \times 10^9\,\mathrm{g\,cm^{-3}}$ (but see Sect. 19.3). Immediately after ignition, the carbon-burning rate remains modest. As the nuclear input increases, the temperature gradient becomes superadiabatic, thus leading to convection. If the white dwarf has cooled for a long enough time to crystallize (van Horn 1968; Winget et al. 2009), there could be differences in the systematics, including a phase separation of the carbon and oxygen (Stevenson 1977). Any crystal will melt soon after carbon ignition and is likely to be mixed back to homogeneity by the convection, although this phase has not been investigated in any detail (Bravo et al. 2011). There is a growing understanding that this convective phase contributes to "preconditioning" the white dwarf and hence possibly to the diversity of properties observed in SN Ia

D. Branch, J.C. Wheeler, *Supernova Explosions*, Astronomy and Astrophysics Library, DOI 10.1007/978-3-662-55054-0_19

explosions. The convective preconditioning may itself be affected by the rotational state of the white dwarf.

The phase of convective carbon burning that precedes dynamical carbon burning and explosion has been referred to as the "simmering" or "smoldering" phase.[1] Numerical models of the single-degenerate scenario (Sect. 22.3) show that after about 100–1000 years the nuclear burning becomes more rapid than the convection, leading to dynamical carbon burning and explosion (Sect. 19.4).

The importance of the smoldering phase is that it sets the conditions for the dynamical runaway that ultimately causes the explosion of the white dwarf in this class of models. The convection in this phase may determine whether smoldering leads to single or multiple spots of dynamical burning and how those spots are distributed in radius and time. Typical convective velocities are about $100\,\mathrm{km\,s^{-1}}$, which turns out to be the speed of the initial laminar conductive flame (Sect. 19.5.1) at which the initial dynamical runaway propagates (Sect. 19.6). The velocity field of the smoldering phase can affect the nature of the early runaway phase by advecting the burning regions. Rotational velocities may exceed both the convective velocities and the early flame speeds (Sect. 22.3).

The smoldering phase leads to a gradual conversion of ^{12}C into heavier elements that determines details of the composition structure at the time of explosion. The smoldering phase also allows neutron capture on ^{12}C and other weak interactions that control the neutron abundance at the time of explosion. That, in turn, affects the weak interactions during the explosion and, in particular, the production of iron-group nuclei that are stable to β decay at the expense of radioactive ^{56}Ni that powers the light curve (Brachwitz et al. 2000; Höflich et al. 2004). There is growing evidence that the very centers of some SN Ia explosions are occupied by neutron-rich, stable, iron-group elements (Sect. 22.3.4).

The smoldering phase has been examined with multidimensional models. The problem is a difficult one for a number of reasons, all related to the fact that the stellar configuration is relativistically degenerate and thus has an effective adiabatic index, Γ_1, very close to $4/3$. The adiabatic index affects the convection criterion with the result that the configuration is very close to the boundary of convective instability. Precisely how much of the core participates in the convection is thus a delicate numerical problem. In addition, this value of Γ_1 means the star is very close to the condition of neutral dynamical stability, so the structure is very sensitive to numerical as well as physical perturbations. Sonic noise generated by convection can generate motion in the outer, otherwise convectively stable region. Some of this

[1] At a meeting in Santa Barbara in 2009, Bob Kirshner attempted to eliminate the etymological ambiguity by calling on the assembled community to settle the issue. "Simmering" won by acclamation. Upon further reflection, we are not so sure. Oatmeal simmers on the stove, but there is little likelihood that it will blow up. Smoldering, on the other hand, connotes the inherent potential to erupt in a more rapid burning phase, just as does this class of models for SN Ia. Wikipedia defines smoldering as the slow, low-temperature, flameless form of combustion. We thus think that "smoldering" may be the more apt phrase and will use it here. On the other hand, simmering boils. Neither word quite captures the physics; perhaps "simoldering?"

may be physical, but there are subtle issues of numerical accuracy and boundary conditions that can reflect and concentrate sound waves but are negligible in reality.

Investigations of the smoldering phase with a 2D implicit code found that the composition is not fully homogenized, so that dynamical runaway occurs in a somewhat carbon-depleted environment, that runaway is induced by the compression of regions near the center of the star by the inflow of convective plumes, and that convective flow speeds are likely to exceed the flame speed early in the dynamical runaway phase (Höflich and Stein 2002). Use of a strictly subsonic 3D anelastic code to simulate the convective flow in the final minutes before dynamical runaway revealed a dipolar flow pattern that fluctuated with time and a tendency to ignite the runaway phase in an outflow (Kuhlen et al. 2006; Zingale et al. 2009).

Early 1D explosion models assumed that the explosion, a subsonic deflagration, began in the center of the white dwarf at the end of the smoldering phase (Sect. 22.3). Subsequent 3D, full-star studies have shown that the turbulent ambience in the transition from smoldering to dynamic runaway is likely to be chaotic so that no two models run away with the same conditions and that the distribution of the ignition conditions is also stochastic (Nonaka et al. 2012). The subsequent dynamics (Sect. 19.4) depend on whether the point of runaway is rather near the center or further away (Sect. 19.8). The former may lead to a deflagration driven at the speed of a turbulent flame (Sects. 19.4 and 19.5), the latter to a single buoyant "bubble," with possibly dramatic differences in the outcome.

The simulations show strong outward flow and lower-speed inflow and favor a range of initial offsets from the center of runaway from low values up to $\sim$100 km, with the highest probability in the range of 40–75 km (Nonaka et al. 2012). The *turbulent intensity* (the characteristic velocity on large scales) and the associated *integral length scale* (the largest scale on which turbulence is driven) are smaller than previously estimated, $\sim$16 km s^{-1} and $\sim$200 km, respectively.

The first point to reach a temperature allowing dynamic runaway will tend to run away at an exponential rate from the next hottest spot because of the extreme temperature sensitivity of carbon burning. This behavior favors a single point of runaway as opposed to multiple runaway points (Nonaka et al. 2012). The latter tend to give better agreement with observations of typical SN Ia (Sect. 22.3), but multiple points of runaway are not obviously supported by the simulations of the smoldering phase.

Higher effective Reynolds number, rotation, or magnetic fields may alter the pattern of the smoldering turbulent convection predicted by current simulations. Rotation tends to disrupt the radial convective flow (Zingale et al. 2011) and hence may affect whether runaway occurs in an inflow or outflow and the distribution of runaway points. Rotation may also affect the propagation of flame bubbles from individual ignition points (Sect. 19.8; García-Senz et al. 2016). Convection and shear in a rotating configuration can lead to the generation of magnetic fields (Piro 2008). The latter can affect the turbulent flow pattern and provide a torque that prevents the growth of large shear and may, for instance, limit the ability of accreting white dwarfs to substantially exceed the Chandrasekhar limit, a topic of great current interest (Sects. 22.3.7 and 22.4.1). The convection might generate g-mode oscillations that could accumulate at the surface of the white dwarf, perhaps to ignite a helium layer there (Piro 2011; Sect. 22.4.2).

19.3 Convective Urca Process

An important aspect of the smoldering phase is the associated convective Urca process (Paczyński 1972). Convective circulation driven by carbon burning in degenerate white dwarfs yields first electron capture and then β decay of susceptible nuclei. The Urca cycle yields no net change in composition, but a loss of neutrinos or antineutrinos at each step of the cycle along with their energy. One likely Urca pair is ^{23}Ne/^{23}Na, for which the threshold density for electron capture is 1.7×10^9 g cm^{-3}. Others that may play some role are ^{21}F/^{21}Ne (threshold density 3.5×10^9 g cm^{-3}) and ^{25}Na/^{25}Mg (threshold density 1.3×10^9 g cm^{-3}). Detailed discussion of the properties of susceptible nuclei are given in Ergma and Paczyński (1974) and Iben (1978, 1982). Paczyński argued that the convective Urca cycle of capture and decay would cause an energy loss that would cool the star and postpone dynamical runaway.

Paczyński's argument was contested by Bruenn (1973), who pointed out that, microscopically, each weak interaction added heat to the system despite the loss of the neutrino. As an electron is captured from below the Fermi sea, another electron drops from the Fermi surface to fill the "hole," resulting in heat. On the other half of the cycle, β-decay electrons deposit excess thermal energy above the Fermi sea.

It is important to determine the effect of the convective Urca process because it has direct observational implications for SN Ia. If a C/O white dwarf of the Chandrasekhar mass undergoes thermonuclear runaway at the density at which carbon ignites, then the density is low enough that subsequent electron capture and neutronizing reactions in the explosion are minimized. If, on the other hand, the convective Urca process substantially postpones dynamical runaway to a phase of higher density, neutronization in the thermonuclear combustion can lead to unacceptably high neutron enrichment of the ejecta.

The physics of the convective Urca process was analyzed in detail in classic work by Iben (1978, 1982) and reexamined by Stein et al. (1999), among others. While the neutrinos associated with the convective Urca process carry away energy, the entropy, and the temperature, cannot decline. Rather, the convective Urca neutrino losses slow the convective currents and reduce the rate of increase of entropy associated with the input of nuclear energy. The loss rate by the Urca process cannot exceed the rate of generation of kinetic energy. The best guess then is that the convective Urca process can make a quantitative difference in the rate of increase of nuclear burning, but not a qualitative difference. Dynamical runaway occurs near the density at which carbon first ignites. Nevertheless, rigorous theoretical models should include this basic and inevitable physics, whereas many models continue to ignore it.

19.4 Dynamical Degenerate Carbon Burning

19.4.1 General Considerations

Convection controls degenerate carbon burning shortly after ignition and the onset of the smoldering phase, and would continue to do so if it were not for the ongoing evolution of the star. In the single degenerate scenario (Sect. 21.2), the process that brought the degenerate configuration to carbon ignition, whether binary accretion, shell burning, or a combination of both, continues after carbon ignition. The mass of the core and the central density increase, and the temperature increases by adiabatic compression modified by convective and neutrino losses. As noted in Sect. 19.2, just where and how dynamical burning develops is affected by the smoldering phase, including the convective Urca process (Sect. 19.3). The exact coupling between the convective structure, rotation, magnetic fields, and dynamical runaway has yet to be properly elucidated, but some of the physics of the resulting combustion is independent of these details.

Dynamical combustion involves burning of a given region on a timescale comparable to, or more rapid than, the sound travel crossing time of that region. There are two general modes of dynamical chemical and nuclear combustion. One is a subsonic process called a *deflagration*. Because the burning is subsonic, deflagrations send pressure waves or shocks that lead to dynamical response and expansion of unburned matter in advance of the arrival of the actual burning front. Astrophysical deflagrations can be mediated by conduction or turbulence. The other mode of dynamical burning is a *detonation*. Detonations are shock/reaction complexes in which the shock initiates the burning that in turn powers the shock. For a given fuel, detonations are more violent than deflagrations because of the associated shock. Because they are intrinsically supersonic, they give no warning. Unburned matter is not perturbed in advance of the arrival of the burning front. These differences are of qualitative importance in the combustion of SN Ia and lead to significantly different observational consequences. More detail of the physics of flame fronts, shocks, detonations, and deflagrations is given below.

For dynamic, but subsonic, burning by deflagration, a variety of processes and instabilities come into play. As remarked above, for central carbon ignition in a Chandrasekhar-mass white dwarf, the star is very near the limit of neutral dynamical instability, so it expands when dynamical burning commences. Burning under these conditions is thus very complex, but certain aspects can be elucidated.

Thermonuclear runaway can correspond to the condition that the nuclear burning timescale is substantially shorter than the dynamical timescale, the sound crossing time for the region of interest. In this situation, burning occurs much faster than any possible dynamic response, resulting in initial *isochoric* (ρ = constant) burning (Fig. 19.1). A shock wave can be generated as the burned material expands, and that shock may or may not trigger more burning depending on details such as spherical geometry, central or off-center ignition, and gradients in ρ and T. If insufficient heating and burning is initiated by the shock, the conditions are likely to be strongly

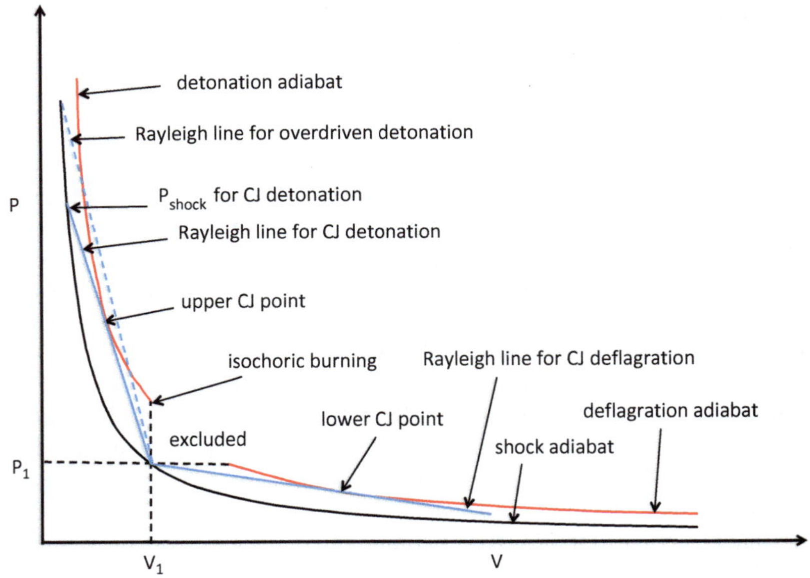

Fig. 19.1 *P–V* diagram illustrating the systematics of deflagrations and detonations. The *black line* represents the shock adiabat that connects the preshock and postshock conditions. The *red line* is the detonation adiabat, the locus after a given energy per gram has been released in the postshock burning. The *blue lines* represent the locus of constant mass flux, or Rayleigh lines, along which the system must evolve after the shock as the burning proceeds. The *solid blue lines* correspond to Chapman–Jouguet (CJ) detonation or deflagration; the *dashed blue line* corresponds to an overdriven detonation. See text for terminology

Rayleigh–Taylor unstable since heated, lower-density matter is pushing on higher density matter (Sect. 19.5.2). The flow, already subject to thermal convection, is apt to become fully turbulent.

In some situations, ignition of burning at faster than the dynamical timescale may lead unambiguously to detonation. The situation is still somewhat ambiguous for central degenerate carbon ignition. There are reasons to think that the physics is not conducive to immediate initiation of detonations because the overpressure from isochoric burning of highly-degenerate matter is small (Mazurek et al. 1977; Khokhlov et al. 1997; Seitenzahl et al. 2009), but there are still open issues. Computations of the smoldering phase with anelastic codes (Nonaka et al. 2012) are limited to relatively large, subsonic hotspots. The critical length scale for the initiation of a detonation at the relevant density is vastly smaller (Sects. 19.7 and 19.8). It is common to switch codes from the smoldering to the dynamic-burning phase and to assume an initial flame and deflagration in the latter, hence skipping a possibly critical step in the physics. There is as yet no firm proof from first principles that dynamic runaway in this context does not directly lead to detonation (Sect. 19.9). In any case, the observed abundances of SN Ia are inconsistent with M_{Ch} central detonation models (Sect. 22.3.1). If central detonation is not immediately initiated, then deflagration burning in one or a number of hot spots in the smoldering phase will propagate. The issue of how the burning from various hot spots merges, the problem of "spotty ignition," is still uncertain. This issue is addressed in Sect. 22.3.4. There remain ambiguities in other contexts of carbon and helium ignition for which there is little direct observational guidance.

If density, temperature, or composition change on scales comparable to the size of the shock and burning region, the conditions may correspond to neither detonation nor deflagration. If the burning and shock pre-accelerate material, a deflagration may develop. If the material is very volatile, with a sufficiently large energy release, continued burning may act like a piston to accelerate material until the shock and burning fronts coalesce, leading to a detonation. The physical processes involved in deflagration-to-detonation transition are not well understood in either the terrestrial or supernova ambiences, although progress is being made in both areas. The mathematical framework for considering detonations and deflagrations is given in Sect. 19.4.2. The problems of initiating burning and the deflagration-to-detonation transition is considered in more detail in Sects. 19.5, 19.9 and 19.11.

19.4.2 The Basics of Supersonic and Subsonic Combustion

A detonation is a shock-reaction complex in which burning is initiated by a shock wave. We will sketch the theory of 1D detonations beginning with the equations of hydrodynamics. From conservation of mass we have the continuity equation

$$\frac{\partial \rho}{\partial t} + \nabla \cdot (\rho \mathbf{v}) = 0, \tag{19.1}$$

and from conservation of momentum we have Euler's equation

$$\frac{\partial \mathbf{v}}{\partial t} + \mathbf{v} \cdot \nabla \mathbf{v} = -\frac{1}{\rho} \nabla P. \tag{19.2}$$

To get an energy equation, we can assume isentropic flow:

$$\frac{ds}{dt} = 0 = \frac{\partial s}{\partial t} + \mathbf{v} \cdot \nabla s. \tag{19.3}$$

It is convenient to introduce the specific enthalpy

$$w = E + PV, \tag{19.4}$$

where $V = 1/\rho$ is the specific volume, so that

$$dw = dE + PdV + VdP = Tds + VdP. \tag{19.5}$$

Hence the energy equation for isentropic flow can be written as

$$dw = \frac{1}{\rho}dP. \tag{19.6}$$

Combining this with Euler's equation, the energy equation can be written as

$$\frac{\partial \mathbf{v}}{\partial t} + \mathbf{v}\cdot\nabla\mathbf{v} = -\nabla w = \frac{\partial \mathbf{v}}{\partial t} + \frac{1}{2}\nabla v^2 - \mathbf{v}\times\nabla\times\mathbf{v},$$

where the last term is zero for 1D and hence vortex-free flow.

In idealized form, a shock wave is pictured as an infinitesimally thin discontinuity. In practice, there will be some thickness, but in the region of the shock a particle rather than a fluid dynamic treatment is needed because the shock is only a few mean free paths thick. This detail can be avoided in common circumstances by using the conservation equations across the discontinuity to join regions where fluid dynamics is valid. The shock is thus assumed to be very small compared to any other relevant length scale in the problem.

Picture, then, a shock wave as a sharp discontinuity moving at velocity S into material at rest. For motion in one direction in steady state $\frac{\partial}{\partial t} \to 0$, hence from Eq. (19.1) we have $\frac{\partial}{\partial x}(\rho v) = 0$, which implies

$$\rho_1 S = \rho_2(S - v_2). \tag{19.7}$$

Using momentum conservation, we have $\rho v\frac{\partial v}{\partial x} = -\frac{\partial P}{\partial x}$, which can be integrated across the front to give

$$\rho_2 v_2^2 - \rho_1 v_1^2 - \int_1^2 v\frac{\partial \rho v}{\partial x}dx = -(P_2 - P_1), \tag{19.8}$$

where the integral is zero by continuity and hence

$$P_1 + \rho_1 S^2 = P_2 + \rho_2(S - v_2)^2. \tag{19.9}$$

Using energy conservation, we have $\frac{\partial}{\partial x}\left(\frac{1}{2}v^2 + w\right) = 0$ in steady state, which yields

$$E_1 + \frac{P_1}{\rho_1} + \frac{1}{2}S^2 = E_2 + \frac{P_2}{\rho_2} + \frac{1}{2}(S - v_2)^2. \tag{19.10}$$

Equations (19.7), (19.9), and (19.10) are variously known as the *shock conditions*, the *jump conditions*, the *Rankin–Hugoniot relations*, or just the *Hugoniot relations*.

For a strong shock, $P_2 \gg P_1$ and $E_2 \gg E_1$, so the jump conditions can be written as

$$\rho_1 S = \rho_2(S - v_2), \tag{19.11}$$

$$\rho_1 S^2 = P_2 + \rho_2 (S - v_2)^2, \tag{19.12}$$

and

$$E_2 + P_2/\rho_2 + \frac{1}{2}(S - v_2)^2 = \frac{1}{2}S^2. \tag{19.13}$$

Combining these equations, we have

$$E_2 = \frac{1}{2}v_2^2. \tag{19.14}$$

Behind a strong shock, the internal energy density equals the kinetic energy density (per gram), independent of the equation of state.

For a strong shock for which P_2 and T_2 are arbitrarily large but the increase in density by compression is not, we can eliminate S and $S - v_2$ to find

$$E_2 = \frac{1}{2}\frac{P_2}{\rho_2}\left(\frac{\rho_2}{\rho_1} - 1\right). \tag{19.15}$$

For an equation of state defined by $E_2 = \frac{1}{\gamma-1}\frac{P_2}{\rho_2}$ we have

$$\frac{\rho_2}{\rho_1} = \frac{2}{\gamma - 1} + 1, \tag{19.16}$$

which yields $\frac{\rho_2}{\rho_1} = 4$ for $\gamma = 5/3$. Thus for a perfect monatomic gas, a common situation in astrophysics, a strong shock increases the density to precisely, but only, a factor of 4.

If we denote the mass flux through the shock by $j = \rho_1 S$ then $S = jV_1$ and $S - v_2 = jV_2$. Substitution into Eq. (19.9) gives

$$P_1 + j^2 V_1 = P_2 + j^2 V_2, \tag{19.17}$$

or

$$j^2 = \frac{P_2 - P_1}{V_1 - V_2}. \tag{19.18}$$

The mass flux is thus given by the slope of the line connecting the preshock and postshock conditions in a P–V diagram. This locus is known as a *Rayleigh line* (blue lines in Fig. 19.1). Now Eq. (19.10) can be written

$$E_1 + P_1 V_1 + \frac{1}{2}j^2 V_1^2 = E_2 + P_2 V_2 + \frac{1}{2}j^2 V_2^2, \tag{19.19}$$

or

$$E_1 - E_2 + \frac{1}{2}(P_1 + P_2)(V_1 - V_2) = 0. \tag{19.20}$$

This equation is known as the *shock adiabat* or *shock Hugoniot* (black solid line in Fig. 19.1). Since $E = E(P, V)$, Eq. (19.20) prescribes the dependence of P_2 on V_2, given P_1, V_1. It is an *adiabat* in the sense that there is no heat flow in or out, but it is not an *isentrope*, since entropy is generated in the discontinuous shock. In many instances, one can equate isentrope and adiabat but not here. The shock adiabat is the locus, for increasing shock strength, of the line connecting the preshock conditions, P_1, V_1, with the postshock conditions, P_2, V_2. The mass flux j, and hence the shock velocity $S = j/\rho_1$, are determined at each point on the shock adiabat by the slope of the chord, the Rayleigh line, joining the preshock and postshock conditions in a P–V diagram (Fig. 19.1).

There are problems for which it is desirable to consider details of the burning region (cf Sect. 19.10), but usually we can assume that the combined width of the shock front and the burning front is much less than any other length scale in the problem. The resulting solution is known as a *Zel'dovich–von Neumann–Doering* or ZND detonation (Sect. 19.9). In this case, the width of the shock-reaction complex separating the burned from the unburned material is infinitesimal. Thus we can write

$$E_2 + P_2V_2 + \frac{1}{2}(S - v_2)^2 = E_1 + P_1V_1 + \frac{1}{2}S^2 + q, \tag{19.21}$$

where q is the specific energy density (typically in erg g^{-1}) released by burning. This expression can be written

$$E_1 + q - E_2 + \frac{1}{2}(P_1 + P_2)(V_1 + V_2) = 0 \tag{19.22}$$

which represents the *detonation adiabat* or *detonation Hugoniot* (red line in Fig. 19.1). The locus described by this equation is an adiabat in the sense that while energy is released in the reactions, there is no heat input from, or loss to, the external environment. Because of the heat release denoted by the term q, all points P_2, V_2, must lie *above* the shock adiabat in a P–V diagram.

The expression for the mass flux, $j^2 = \frac{P_2 - P_1}{V_1 - V_2}$, is still valid for a detonation since it follows from mass and momentum conservation only. The quantity j^2 will be a constant for given initial conditions for all intermediate partially burned states as well as for the final products. The whole process of shock and burning takes place on the chord $P = P_1 + j^2(V_1 - V)$ (Fig. 19.1). The initial shock is given by $P_{shock} = P_1 + j^2(V_1 - V_{shock})$. The pressure in the burned products is given by $P_{burn} = P_1 + j^2(V_1 - V_{burn})$ with $P_{burn} < P_{sh}$ and $\rho_{burn} = \frac{1}{V_{burn}} < \rho_{sh}$. In a P–V diagram, the shock reaction complex is described by the initial shock that instantaneously (in the approximation of infinitesimally-thin shock) raises the pressure from P_1 to P_{shock} with simultaneous compression of the matter from V_1 to V_{shock}. At this point

P_{shock} is less than the pressure on the detonation adiabat corresponding to the same V_{shock}. As the matter burns, the system must evolve along the chord $j^2 = \frac{P_2 - P_1}{V_1 - V_2}$, so the pressure decreases and the specific volume increases. The burning is complete (all of q released) when the chord j^2 = constant intersects the detonation adiabat.

In general, the chord will not be tangent to the detonation adiabat at the point of intersection with it. This is the case for an *overdriven detonation* (dashed blue line in Fig. 19.1) that can arise in a transient state or where an external energy source, such as a piston as well as the release of the energy, q, drives the shock. The lower segment of the chord between the initial conditions and the detonation adiabat cannot be reached by a physical process in which the combustion is caused by the compression and heating of a shock. Any segment of the chord that lies above the detonation adiabat is inaccessible because insufficient energy is available.

For given initial conditions there is a minimum j, and hence front velocity S, that can satisfy the conservation equations. The condition that the slope of the chord representing conservation of mass be an extremum (maximum) is that the chord be tangent to the detonation adiabat, or

$$\frac{dj^2}{dP_2} = 0. \tag{19.23}$$

One can show that $\frac{ds_2}{dP_2} = 0$ at the tangent point. For this condition we have

$$\left.\frac{dP_2}{dV_2}\right|_{tan} = -j^2 = -\rho_2^2 \left.\frac{dP_2}{d\rho_2}\right|_s = -\rho_2^2 c_2^2. \tag{19.24}$$

With

$$j^2 = \rho_1^2 S^2 = \rho_2^2 (S - v_2)^2, \tag{19.25}$$

we find $|S - v_2| = c_2$, where c_2 is the sound speed behind the burning front. The condition that the mass conservation chord be tangent to the detonation adiabat is thus equivalent to the condition that the front propagates at precisely the sound speed with respect to the material behind the front. One can also show that above the tangent point $S - v_2 < c_2$, hence sound waves can catch up to the front and accelerate it. This is the condition for an overdriven detonation. To maintain such a condition, something (a piston) must be supplying extra push (energy) over and above the energy released as q. If there is no such energy, the shock will decay until the tangent condition is reached. In the limit $q = 0$ the shock will decay until $S - v_2 \sim S \sim c_2 \sim c_1$, that is, until it becomes a sound wave.

The condition $S - v_2 > c_2$ corresponds to nonintersection of the detonation adiabat with the chord j^2 = const. For this case, the burning front runs away from the matter behind it, leading to a rarefaction which slows the shock. If the burning releases enough energy, the shock can be forced to accelerate in a transient phase. For ignition of rapidly burnable material, the process of ignition drives a shock. At first the shock may be too slow, but burning can build up pressure and accelerate it. If the shock temporarily propagates too rapidly (as an overdriven detonation), it will outrun the burning, which then decays. A stable equilibrium state is available at the tangent point. A self-consistent, self-propagating, steady-state solution to the jump conditions corresponds to the extremal solution at the tangent point. Such a detonation is called a *Chapman–Jouguet detonation* (CJD; Fig. 19.1).

The tangent to the shock adiabat at point $P_1,\ V_1$ is

$$\left(\frac{\partial P}{\partial V}\right)_1 = -\rho_1^2\left(\frac{\partial P}{\partial \rho}\right)_{s_1} = -\rho_1^2 c_1^2. \tag{19.26}$$

Since the tangent to the detonation adiabat has slope $-j^2 = -\rho_1^2 S$, this condition implies $S > c_1$. A Chapman–Jouguet detonation corresponds to maximum front velocity with respect to the postfront material and minimum velocity with respect to preshock material. The condition $S - v_2 = c_2$ implicitly defines the front speed as a function of initial conditions and the energy release, $S = S(P_1, V_1, q)$.

Based on conservation of mass and momentum we have $\rho_1 S = \rho_2(S - v_2)$ and $P_1 + \rho_1 S^2 = P_2 + \rho_2(S - v_2)^2$, which can be expressed as

$$v_2 S = (P_2 - P_2)V_1. \tag{19.27}$$

For a detonation $P_2 > P_1$, v_2 and S have the same sign, and both are supersonic with respect to the preshock matter. There is also a solution for which $P_2 < P_1$ and for which v_2 and S are oppositely directed. In this case, burned material expands away from the burning front, and the front is subsonic with respect to unburned material. Such a burning front is a deflagration (Fig. 19.1). The slope of the chord connecting P_1, V_1 to P_2, V_2 is greater than that of an isentrope (less negative in the P–V diagram) so that the velocity of a deflagration is subsonic. The deflagration adiabat is an extension of the detonation adiabat to the regime $V_2 > V_1$. The situation for which the Rayleigh line is tangent to the detonation/deflagration adiabat in the deflagration regime is again an extremum, corresponding to a *Chapman–Jouguet deflagration*. The region along the detonation/deflagration adiabat corresponding to $P_2 > P_1$ and $V_2 > V_1$ is unphysical since it would correspond to $j^2 < 0$. The shock-reaction complex of a detonation can be decomposed into the shock plus a subsequent deflagration beginning at the conditions P_{shock}, V_{shock}.

19.5 Astrophysical Deflagration: Subsonic Burning and Combustion

Because the propagation speed of a subsonic burning front, S, is less than the sound speed of the unburned matter, $S < c_{fuel}$, the front sends out pressure waves that influence material ahead of the burning. In common circumstances relevant to thermonuclear combustion in white dwarfs, those pressure waves can coalesce into a shock that precedes the subsonic burning and is separated from the burning by a substantial distance. This situation is transient and does not correspond to a detonation since the preceding shock does not directly initiate the burning. The shock can accelerate the outer portions of the white dwarf well beyond the burning region. In the presence of stellar gravity, a deflagration front may propagate by conduction (Sect. 19.5.1) or turbulence (Sect. 19.5.2), preceded by a transient shock that runs far ahead of the burning front.

19.5.1 Laminar Flames

In the context of the smoldering phase, there is a transition from slow burning controlled by neutrino losses and convection to the formation of a thermonuclear flame. This occurs when the burning becomes more rapid than a convective eddy turnover timescale. Under this condition, the cold fuel can be separated from the hot ash by a thin burning front, the condition to form a well-defined flame front. The flame is subsonic so the pressure is nearly constant across the front, but the density decreases from the fuel to the ash.

A basic component of thermonuclear burning that is not initiated by a shock is a *laminar flame* that is propagated by thermal conduction of degenerate electrons. During deflagration, the actual burning occurs in a thin region. The width of the front is controlled by thermal diffusion propagated by the conductive electrons. The diffusion timescale is

$$\tau_{diff} \approx \frac{\delta_{cond}^2}{D_{cond}} \approx \frac{\delta_{cond}^2}{\lambda c}, \tag{19.28}$$

where δ_{cond} is the width of the conduction front, $D_{cond} = \lambda c$ is the diffusion coefficient, λ is the electron mean free path, and c is the speed of sound. Here

$$D_{cond} \approx \frac{\sigma}{\rho C_p}, \tag{19.29}$$

where the conductivity is

$$\sigma = \frac{4acT^3}{\rho \kappa_{cond}}, \tag{19.30}$$

which defines the conductive opacity, κ_{cond}, and where C_p is the specific heat per unit mass at constant pressure.

The timescale of the conductive flame is also given by the condition

$$\tau_{diff} \sim \tau_{burn} \sim \frac{E}{\dot{\epsilon}_{nuc}}, \tag{19.31}$$

where E is the internal energy per unit mass and $\dot{\epsilon}_{nuc}$ is the rate of production of nuclear energy per unit mass. It follows that

$$\delta_{cond} \approx \left(\lambda c \frac{E}{\dot{\epsilon}_{nuc}}\right)^{1/2}. \tag{19.32}$$

One can also write the velocity of the flame front as

$$v_{cond} \sim \frac{\delta_{cond}}{\tau_{burn}} \approx \frac{\delta \dot{\epsilon}_{nuc}}{E} \approx \left(\frac{\lambda c \dot{\epsilon}_{nuc}}{E}\right)^{1/2}. \tag{19.33}$$

With the assumption that the burning is *isobaric* and subsonic, one can also assume that the pressure equilibrates between the burned and unburned matter. Equation (19.10) can then be written as

$$\frac{dE}{dt} + \frac{P\partial(1/\rho)}{\partial t} = \frac{1}{\rho}\frac{\partial}{\partial x}\left(\sigma \frac{\partial T}{\partial x}\right) + \dot{\epsilon}_{nuc}. \tag{19.34}$$

This equation can be solved by explicit integration. Alternatively, with the steady state condition

$$\frac{d}{dt} \rightarrow v_{cond}\frac{d}{dx} \tag{19.35}$$

Equation (19.10) can be written

$$v_{cond}\left(\frac{dE}{dx} + \frac{Pd(1/\rho)}{dx}\right) = \frac{1}{\rho}\frac{d}{dx}\left(\sigma\frac{dT}{dx}\right) + \dot{\epsilon}_{Enuc}, \tag{19.36}$$

where now v_{cond} has the role of an eigenvalue. An approximate fit to the solution of the eigenvalue problem gives, for the conditions of degenerate carbon burning,

$$v_{cond} = 92\,\mathrm{km\,s^{-1}}\left(\frac{\rho}{2\times 10^9\,\mathrm{g\,cm^{-3}}}\right)^{0.805}\left(\frac{X_{12}}{0.5}\right)^{0.839}, \tag{19.37}$$

where X_{12} is the mass fraction of ^{12}C (Timmes and Woosley 1992; Khokhlov et al. 1997). This expression is accurate to about 10% in the range $0.01 \leq \rho_9 \leq 10$. The dependence of the laminar conductive flame speed and thickness on density are illustrated in Fig. 19.2.

In the conditions characteristic of the center of a white dwarf of M_{Ch}, the laminar flame propagates at a characteristic velocity of v_{cond} ~100 km s^{-1}. This is much smaller than the sound speed, $c_s >$ 1000 km s^{-1}. It is clear that the laminar conductive burning cannot produce a SN Ia. For densities $\gtrsim 10^9$ g cm^{-3}, the velocity of the laminar flame is comparable to the speed of convection in the early phase of carbon burning, but the flame speed slows as the white dwarf expands. For even

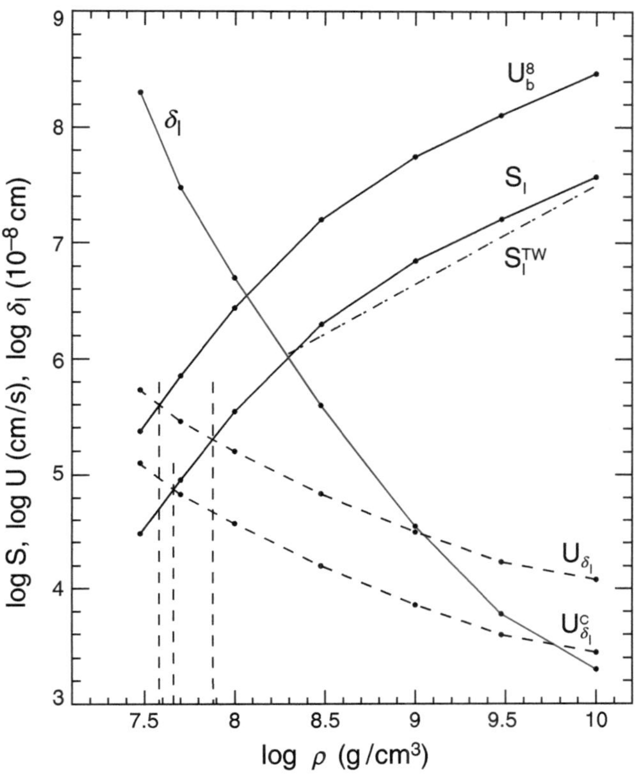

Fig. 19.2 Velocities and flame thickness as a function of density. The parameters $S_l \equiv v_{cond}$ and $\delta_l \equiv \delta_{cond}$ are the laminar flame speed and thickness from Khokhlov et al. (1997); S_l^{TW} is the laminar flame speed according to Timmes and Woosley (1992); U_b^8 is the critical turbulent velocity to overcome the nonlinearly stabilized Landau–Darrieus instability (Sect. 19.6.1) assuming K = 8 for the amplification factor (Sect. 19.11); $U_{\delta_l} = v_{turb}(\delta_{cond})$ (Sect. 19.6.3) is the estimated turbulent velocity at the scale of the flame front during the supernova explosion, neglecting turbulence freezeout; $U_{\delta_l}^c$ assumes turbulence freezes out in the expansion. *Vertical dashed lines* indicate density conditions for which the laminar flame speed is comparable to critical turbulent velocities for various assumptions. From "Deflagration-to-Detonation Transition in Thermonuclear Supernovae" (Khokhlov et al. 1997). © AAS. Reproduced with permission

this laminar burning, the conditions in the burned matter must satisfy $T > T_{crit} \simeq 5 \times 10^9$ K, the temperature at which $\rho\dot{\epsilon}_{nuc}E = \frac{d}{dx}\left(\sigma\frac{dT}{dx}\right)$. A temperature smaller than this cannot produce enough energy to drive the diffusion. For conditions typical of central carbon ignition at M_{Ch}, $\rho \sim 3 \times 10^9\,\mathrm{g\,cm^{-3}}$, the width of the laminar flame front is $\delta_{cond} \sim 10^{-4}$ cm, and the characteristic mass, $\delta m \sim \rho\delta_{cond}^3$, is very small, only 10^{-3} g. Even at a density of $10^7\,\mathrm{g\,cm^{-3}}$ that may characterize the conditions of deflagration-to-detonation transition (Sect. 19.11), $\delta_{cond} \sim 10$ cm, still much less than the radius of the white dwarf. It is thus impractical to resolve the flame front and the gross structure of the star simultaneously in numerical simulations. The conductive flame can be resolved with appropriate direct numerical simulations.

As discussed in Sect. 19.6.1, the laminar flame can be subject to a variety of perturbations. The nature of the thin conductive flame establishes a lower limit on the scale of perturbations that can affect the laminar flame, the *flame-polishing scale*. For sufficiently small perturbations, the flame can burn through them more rapidly than they can be dissipated by pressure gradients or other effects.

19.5.2 Rayleigh–Taylor Instability

As burned matter expands behind the laminar flame front, the density decreases even though P is approximately constant, and the gradient of the density is directed opposite to that of the pressure. Light material is supporting and accelerating dense material, the classic condition for the *Rayleigh–Taylor instability* (Chandrasekhar 1961). This is basically a buoyancy instability—hot stuff floats. The Rayleigh–Taylor instability is especially important in this context, because it can drive turbulent heat exchange.

The characteristic timescale for the Rayleigh–Taylor instability is

$$\tau_{RT} = \frac{2\pi}{\omega_{RT}} = \sqrt{\frac{2\pi\lambda}{g\mathcal{A}}} \tag{19.38}$$

where λ is the length scale of an infinitesimal perturbation of the form $\xi = \xi_0 e^{i\omega_{RT}t}$, g is the local acceleration of gravity, and the *Atwood Number*,

$$\mathcal{A} = \frac{\rho_{fuel} - \rho_{ash}}{\rho_{fuel} + \rho_{ash}}, \tag{19.39}$$

is related to the density contrast across the burning front with density ρ_{fuel} in the unburned material and ρ_{ash} in the burned matter and $\rho_{fuel} > \rho_{ash}$. The characteristic speed of Rayleigh–Taylor overturn is

$$v_{RT} = \lambda\omega_{RT} = 2\pi\frac{\lambda}{\tau_{RT}} = \sqrt{2\pi g\lambda\mathcal{A}}. \tag{19.40}$$

At the center of a spherically symmetric star, $g \to 0$, so the Rayleigh–Taylor velocity goes to zero and the overturn time, τ_{RT}, is infinite. The instability is suppressed in the center, but becomes active once the burning proceeds off-center. For the gravity characteristic of a white dwarf, $g \sim GM_{WD}/R_{WD}^2$, we have

$$v_{RT} \sim 3 \times 10^5\,\mathrm{cm\,s^{-1}} \left(\frac{M_{WD}}{\mathrm{M_\odot}}\right)^{1/2} R_{WD,8}^{-1}\mathcal{A}^{1/2}\lambda^{1/2}, \tag{19.41}$$

where $R_{WD,8}$ is the radius in units of 10^8 cm and λ is in cm. As noted above (Sect. 19.2), the convective velocities in the smoldering phase can be competitive with the Rayleigh–Taylor velocities early in the Rayleigh–Taylor phase. For $\lambda \gtrsim 10^6$ cm, $v_{RT} \gtrsim v_{cond}$ for the conditions of degenerate carbon burning ($\mathcal{A} \simeq 0.1$), where v_{cond} is the velocity of the laminar, conductive flame (Sect. 19.5.1). In this regime, turbulence generated by the Rayleigh–Taylor instability (and remnant convection from the smoldering phase) can severely distort the laminar flame. This characteristic size for λ is attained when the burning has extended through about 10^{28} g or 10^{-5} $M_{\odot}$, a very small portion of the star. Turbulence generated by Rayleigh–Taylor instability begins early in the propagation of the laminar flame, about 0.1 s after dynamic burning begins, still just a fraction of the global dynamic time of the white dwarf, ~1 s (Sect. 19.8). For $\lambda \gtrsim 10^8$ cm, approaching the size of the white dwarf, $v_{RT} \gtrsim 10^8$ cm s^{-1} $\sim c_s$. In principle, such rapid turbulent burning could cause the star to explode, but in practice other physics probably comes into play (Sects. 19.6.4 and 19.11). The physics of the Rayleigh–Taylor instability can be invoked to construct subgrid models of turbulent flame propagation (Sect. 19.6.4).

19.6 Interaction of Buoyancy, Turbulence and Flames

The Rayleigh–Taylor instability naturally leads to turbulence. The important physics, and a major uncertainty of deflagration flames in the context of SN Ia, then becomes the interaction of turbulence and flames. Many insights of turbulent burning in supernovae can be imported from the study of terrestrial turbulent flames as reviewed by Hillebrandt and Niemeyer (2000).

19.6.1 Landau–Darrieus Instability

If the laminar flame can be regarded as a discontinuity on relevant length scales, the flame will be subject to the *Landau–Darrieus instability* (Darrieus 1938; Landau 1944). The physical nature of the instability is related to the curvature of the flame front in response to a perturbation in the flow. Consider a planar flame front that is "wrinkled" by a perturbation with small amplitude, ξ, at length scale, λ, along the flame front (Fig. 19.3). Flow incident on the "lagging" portion of the flame front crosses it at an oblique angle, which generates vorticity behind the flame. The sense of the vorticity is opposite on the two sides of the maximum lag and in the sense that it accelerates the flame. This causes the portions of the flame adjacent to the maximum lag point to propagate toward one another, increasing the angle of the front to the flow and thereby enhancing the creation of vorticity and the strength of the effect. The Landau–Darrieus instability is enhanced by the buoyancy effect that causes the leading portions of the perturbed flame to propagate "upwards" and

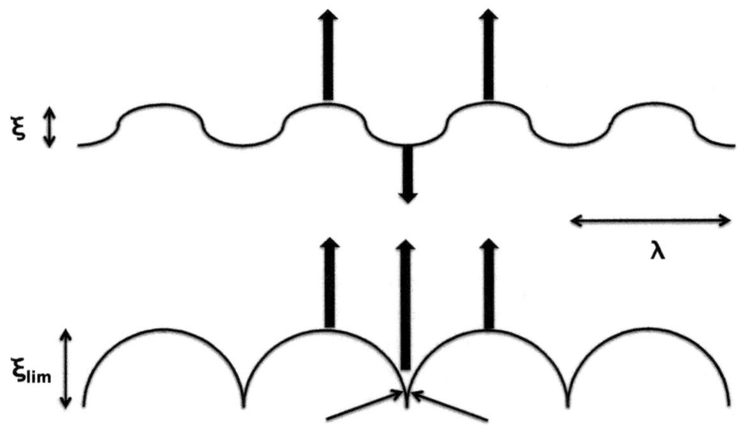

Fig. 19.3 Schematic representation of a laminar conductive flame perturbed with wavelength λ and amplitude ξ. Through the effects of vorticity, the leading edge will propagate more rapidly than the trailing edge, generating an absolute linear instability to a thin flame (*top*). The Landau–Darrieus instability is stabilized in the nonlinear limit by the formation of cusps at the lagging front (*bottom*). The phase velocity of the merging cusps is greater than the flame speed, thus enhancing the speed of the lagging portion of the flame and leading to a limiting amplitude ξ_{lim}. See "Propagation of Turbulent Flames in Supernovae" (Khokhlov 1995)

inhibits the lagging portions from sagging "downwards." This also tends to drive the portions of the flame on the two sides of the maximum lag point toward each other. In the nonlinear limit, the evolution of the flame front is stabilized at amplitude ξ_{lim} by the formation of cusps, the phase velocity of which exceeds the conductive velocity (Zel'Dovich 1966; Khokhlov 1995).

The nonlinear limit of the Landau–Darrieus instability is thus a wrinkled, but still continuous, flame front. The rate of propagation of the flame is proportional to the rate of ingestion of fuel, which is proportional to the effective area of the flame. Because a wrinkled flame has a larger effective area than a planar flame, the wrinkled flame propagates more rapidly with

$$S_{flame} = \left(\frac{A_{flame}}{A_0}\right) v_{cond}, \tag{19.42}$$

where v_{cond} is the velocity of the flame driven by pure conduction, A_0 is the surface area of the unperturbed flame, and A_{flame} is the effective area of the wrinkled surface.

This speedup of the flame propagation by the Landau–Darrieus instability must occur in the supernova ambience as it does for any flame, but the Landau–Darrieus instability does not set the ultimate effective speed of the flame front. Rather, it may provide the seed for the Rayleigh–Taylor instability that comes to dominate the effective rate of propagation of a subsonic flame.

19.6.2 Buoyancy-Driven Flames

If the Rayleigh–Taylor buoyancy instability is strong enough, it can disrupt the nonlinear stabilization of the flame and lead to new physical processes. One can define a critical length scale to be the scale at which the Rayleigh–Taylor velocity is comparable to the conductive flame speed, i.e., from Eq. (19.40),

$$v_{RT}(\lambda_{crit,RT}) = \sqrt{2\pi g \lambda_{crit,RT} \mathcal{A}} = v_{cond}. \quad (19.43)$$

The resulting critical length can also be expressed as

$$\lambda_{crit,RT} = \lambda_0 \left(\frac{v_{cond}}{v_{RT}}\right)^2, \quad (19.44)$$

where λ_0 is a fiducial scale, often taken to be the *integral scale* or *driving scale*, the large scale on which the turbulence is driven. In the supernova case, the integral scale is somewhat smaller than the largest scale in the system, the radius of the white dwarf. For typical conditions, the critical length can be written as

$$\lambda_{crit,RT} \sim 1.2 \times 10^3\,\mathrm{cm} \left(\frac{M_{WD}}{\mathrm{M}_\odot}\right)^{-1} R^2_{WD,8} \mathcal{A}^{-1} v^2_{cond,7}, \quad (19.45)$$

where $v_{cond,7}$ is the conductive speed in units of $10^7\,\mathrm{cm\,s^{-1}}$. Similar arguments can be made comparing v_{RT} to the cusp velocity of the nonlinearly stabilized front (Khokhlov 1995).

For $\lambda > \lambda_{crit,RT}$, the bulk fluid motion associated with buoyancy moves faster than conduction. Note that at the center of the star, $r = 0$, $g = 0$, and hence $v_{crit} \to 0$ and $\lambda_{crit,RT} \to \infty$. Right at the center the conductive flame always wins. The density ratio, ρ_{fuel}/ρ_{ash}, is smallest at the center where the degeneracy is largest, so the Atwood number, $\mathcal{A}$, increases outward, while v_{cond} decreases. The gravity, g, rises as $M_r \to M$, then declines as $r \to R$ for M_r approximately constant. The critical length scale for which the buoyancy overwhelms the intrinsic conductive speed of the laminar flame front is thus a rather complex function of radius that needs to be evaluated for a given numerical model. In practice, buoyancy rapidly becomes the dominant factor. The resulting critical length scale is $\lambda_{crit,RT} \sim 10^5\,\mathrm{cm}$ at a density of $3 \times 10^9\,\mathrm{g\,cm^{-3}}$.

19.6.3 The Gibson Scale

The overturn and eddies associated with Rayleigh–Taylor buoyancy are likely to generate turbulence. Consider the spectrum of turbulent velocities, $v_{turb}(\lambda)$, which is the turbulent velocity on length scale λ, for which the velocity decreases at smaller length scales. Another important critical length scale, λ_G, can be defined as

$$v_{turb(\lambda_G)} \simeq v_{cond}. \tag{19.46}$$

This is known as the *Gibson scale* (Peters 2000). Note that the criterion for this critical scale is similar to that for the scale $\lambda_{crit,RT}$ (Eq. (19.43)). The difference is that the turbulent cascade has a different dependence on scale and v_{turb} generally exceeds v_{RT}.

For $\lambda < \lambda_G$, the turbulent velocity is less than the laminar conductive speed. For $\lambda > \lambda_G$, the bulk flow of the turbulent velocity exceeds the conductive flame speed. As the expansion of a supernova proceeds and the density declines, the conductive length grows and the Gibson scale declines primarily due to the rapid decrease in the conductive velocity (Fig. 19.4). In the ambience of an exploding white dwarf with central density exceeding $10^9\,\mathrm{g\,cm^{-3}}$, $\lambda_G >> \delta_{cond}$. For densities $\lesssim 10^7\,\mathrm{g\,cm^{-3}}$, λ_G for carbon burning becomes substantially less than δ_{cond}, in which case λ_G is not conceptually well defined.

There was long expected to be a qualitative difference in the flame propagation in both terrestrial and astrophysical contexts depending on whether the Gibson scale is greater than or less than the thickness of the laminar flame, δ_{cond}. If $\lambda_G > \delta_{cond}$, the flow at large scales exceeds the laminar flame speed, but the turbulence is slow on the scale of the flame thickness. The result is a flame that is contorted as it is convected around on large scales, but, while perhaps wrinkled, the flame is smooth and continuously connected. The difference was thought to occur for conditions for which $\lambda_G < \delta_{cond}$. For a turbulent cascade that corresponds to *Kolmogorov turbulence*, for which

$$v_{turb}(\lambda) \sim v_{turb}(\lambda_0) \left(\frac{\lambda}{\lambda_0} \right)^{1/3}, \tag{19.47}$$

the Gibson scale can be written as

$$\lambda_G \sim \left(\frac{v_{cond}}{v_{turb}(\lambda_0)} \right)^3 \lambda_0. \tag{19.48}$$

The condition that $\lambda_G < \delta_{cond}$ thus corresponds to the condition that the rate of dissipation of turbulent energy exceeds the rate of specific energy generation in the flame,

$$\dot{\epsilon}_{turb} \sim \frac{v_{turb}(\lambda)^2}{\tau_{turb}} \sim \frac{v_{turb}(\lambda)^3}{\lambda} > \frac{v_{cond}^3}{\delta_{cond}}. \tag{19.49}$$

The argument was made that, in this case, turbulence convects the matter at the flame faster than the flame can propagate, with the implication that the flame is broken and distorted with a complicated topology. In particular, it was hypothesized that the transition to this regime might play a role in the deflagration-to-detonation

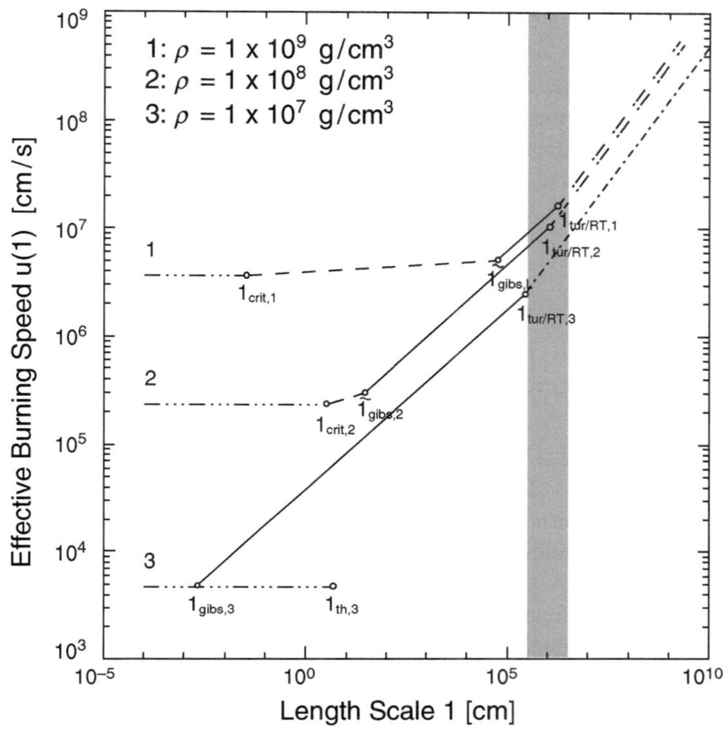

Fig. 19.4 Speed of the deflagration burning front, $u_l \equiv S_{turb}$ (Sect. 19.6.4), as a function of length scale at three densities showing the thickness of the thermal conductive flame, $l_{th} \equiv \delta_{cond}$, the flame thickness associated with the nonlinearly stabilized Landau–Darrieus instability, $l_{crit} \equiv \xi_{crit}$ (Sect. 19.6.1), the Gibson scale, $l_{gibs} \equiv \lambda_G$, and the thickness of a turbulent flame front, $l_{tur/RT}$. For scales greater than the Gibson scale, the propagation of the front is primarily determined by large-scale turbulent phenomena, not the physics of the conductive front. Note the decrease in the Gibson scale with declining density. The Gibson scale becomes comparable to l_{crit} at a density $\sim 10^8\,\mathrm{g\,cm^{-3}}$. The *shaded region* represents the typical resolution of multidimensional simulations, $\sim 10\,\mathrm{km}$. From "The Thermonuclear Explosion of Chandrasekhar-Mass White Dwarfs" (Niemeyer and Woosley 1997). © AAS. Reproduced with permission

transition that occurs in many terrestrial chemical combustion environments and arguably occurs in the thermonuclear combustion of SN Ia. We return to this issue in Sects. 19.6.5 and 19.11.

Although it is common to invoke a Kolmogorov cascade in this and related contexts, it is worth noting that in the context of exothermic processes, the nature of turbulence changes and the cascade may depart substantially from the simple Kolmogorov scaling. Simulations show that turbulence in the presence of burning is a nonequilibrium process with substantial backscatter and high intermittency (Hamlington et al. 2011, 2012).

19.6.4 Turbulent Flames

The viscosity of degenerate matter is low, $\nu \lesssim 0.3\,\mathrm{cm^2\,s^{-1}}$, and the cascade of turbulent velocity fluctuations should proceed to small scales, $\lambda \ll \lambda_G$. At the *dissipation scale* or *Kolmogorov scale* set by the viscosity, the smallest-scale eddies are estimated to have scale length

$$\lambda_{min} \sim \lambda_0 Re^{-3/4} \lesssim \delta_{cond}, \tag{19.50}$$

where *Re* is the Reynolds number of the medium,

$$Re = \frac{D_{turb}}{\nu} \sim \frac{v_{turb}(\lambda_0)\lambda_0}{\nu} \sim 10^{14}, \tag{19.51}$$

where $D_{turb} \sim v_{turb}(\lambda_0)\lambda_0$ is the coefficient of turbulent diffusion, $v_{turb}(\lambda_0) \sim 10^7\,\mathrm{cm\,s^{-1}}$, and $\lambda_0 \sim 10^7\,\mathrm{cm}$. Eddies with $\lambda < \delta_{cond}$ do not affect burning because all the fuel burns before one eddy turnover, so details of the dissipation of turbulence on the smallest scales are probably not significant for the white dwarf combustion problem. A caveat is that there are suggestions that the ratio of the kinetic energy in turbulence to the potential energy released in the turbulent cascade may have a slight dependence on *Re* (Cabot and Cook 2006). Since *Re* is huge in these conditions, $\sim 10^{14}$, this effect could influence flame propagation. No current numerical simulations can approach *Re* this large.

A relation between the propagation of conductive flames and the rate of propagation of the turbulent flame was proposed by Damköhler (1940), who suggested that for small turbulent intensity the flame would propagate at the speed of the laminar flame, v_{cond}. For large turbulent velocities, the turbulence should penetrate the flame and broaden it. The turbulence would also stretch and fold the flame and thereby increase its surface area, A_{turb}, in such a way that the effective speed of the turbulent flame would satisfy the relation

$$S_{turb}/v_{cond} = A_{turb}/A_{flame}, \tag{19.52}$$

where A_{flame} is the area of a portion of the laminar flame (compare to Eq. (19.42)). That is, the flame speed would be increased over that of the laminar flame by the effective area of the distorted flame. This model for the flame may apply for modest levels of turbulence, but it neglects any feedback of the flame on the turbulent flow.

Numerical results and further work suggest that once the buoyancy instability and associated turbulence has set in, the turbulent flame is much faster than the laminar burning front, as hypothesized by Damköhler. The burning front is contorted and overwhelmed by the buoyancy-driven turbulence. The effective speed of the burning front becomes *independent* of the laminar flame speed and of details like the Landau–Darrieus instability and its nonlinear stabilization. The flame speed is rather set by overturn on the largest scales where the turbulent cascade is initiated.

The flame wrinkles on smaller scales in just such a way as to increase its effective area in order to consume matter at the rate dictated by the rate of buoyancy overturn on the largest scales where the turbulent velocity is most rapid (Khokhlov 1995). For the same macroscopic velocity of the burning front, a larger value of v_{cond} gives a less wrinkled flame and a larger value of λ_G, and a smaller value of v_{cond} gives a more wrinkled flame and a smaller value of λ_G. It is the wrinkling of the flame front that adjusts to the change in v_{cond}, not the bulk speed of the flame.

When buoyancy begins to be competitive with conduction, the flame is said to be in the *wrinkled flame regime*. When turbulent velocities greatly exceed the laminar flame conductive velocity, the flame is said to be in the *corrugated flame regime* or the *multiple flame-sheet regime*. In this regime, the flame front is severely wrinkled and perhaps even broken into multiple flame sheets as the flame intersects itself and fragments break off from the flame sheet. On the scale of the flame thickness, δ_{cond}, however, the flame is unperturbed. This corrugated flame regime characterizes common conditions in supernova deflagrations, but is relatively rare in terrestrial combustion where fuels are often so volatile that laminar burning speeds overwhelm the buoyancy characteristic of Earth's gravity. The conditions for a corrugated flame can be provided in terrestrial experiments with slow-burning fuels, for instance dilute methane in sufficiently large tubes. Another option is to increase g by means of a centrifuge.

Besides interesting physics, the conclusion that the flame speed in the corrugated flame regime is independent of details of the laminar flame propagation is important for the numerical simulation of buoyancy-driven flames in supernova models. Models need only resolve burning from the largest scales that establish the gravitational acceleration and the sphericity down to modest scales, $\lambda \gg \lambda_G$ and $\lambda \gg \delta_{cond}$, where the turbulent cascade is determined. This smaller length scale is approximately set by the condition $\tau_{RT}(\lambda) < \tau_{dyn}$ with $\tau_{RT} \propto \sqrt{\lambda}$ (Eq. (19.41)). This typically defines a length scale $\sim 0.01 R_{WD}$. For shorter length scales, less than the grid scale, Eq. (19.52) can be used as a subgrid model to establish the flame propagation speed. The ability of contemporary computers to resolve this limited range of scales, a dynamic range of only ~ 100, opens the way to an accurate computation of the deflagration of a whole 3D white dwarf. It is crucial to do 3D computations of the buoyancy problem. The results of 2D and 3D models are qualitatively different since the resulting turbulence cascades to larger scales in 2D and to smaller scales in 3D (although there can be stochastic backscatter to larger scales even in 3D; Oran and Boris 2005). Two dimensional calculations are at best difficult to interpret and, at worst, wrong.

This discussion was predicated on the existence of a global deflagration front roughly centered on the white dwarf; for an alternative possibility, see Sect. 19.8.

19.6.5 Distributed Flames?

As the turbulent velocity increases, the Gibson length can, in principle, be reduced to less than the flame thickness, δ_{cond}. In this case, the implication is that the turbulence moves the flame before it can burn through its own length. The flame front may be broken into pieces of size $\lesssim \delta_{cond}$ that are carried by the turbulent velocity faster than the flame can propagate. This putative burning regime is known as the *distributed flame regime*. It has long been hypothesized that in this distributed regime turbulence affects both the unburned fuel and the region where burning is active, the transfer of heat from the burning regions to the fuel, and hence the overall speed of propagation of the flame (Peters 2000). In the supernova community, it has been proposed that a detonation cannot occur unless the flame enters the distributed flame regime, with the concomitant speculation that transition to the distributed flame regime with turbulence-enhanced flame speed may be the key to induce the formation of a detonation (Sect. 19.11)

To explore this issue, we need to characterize the structure of turbulent flames in somewhat more depth. The *flame brush* is the region over which the burning proceeds from pure fuel to pure burning product. Studies of chemical reactions often differentiate flames that require a diffusion of fuel reactants from those that are "premixed." The supernova problem constitutes a situation in which the fuel, the degenerate carbon and oxygen, is naturally premixed. Within the flame brush there is a *preheat region* where thermal energy is diffused into the fuel and a *reaction region* where fuel is converted to ash.

Various regimes pertinent to the structure of the flame brush can be characterized by dimensionless numbers. Of basic importance are the Prandtl number

$$Pr = \frac{\nu}{K}, \tag{19.53}$$

where ν is the kinematic viscosity and K is the thermal diffusion coefficient, and the *Lewis number*

$$Le = \frac{K}{D_{specie}}, \tag{19.54}$$

where D_{specie} is the diffusion coefficient for various species. For the white dwarf problem, thermal conduction with diffusion coefficient $K \sim v_{cond}\delta_{cond}$ dominates over either the kinematic viscosity or the particle diffusivity, so $Pr << 1$ and $Le >> 1$. For large Le, flames tend to pulsate (Sect. 19.11.2).

One can also define the Damköhler number

$$Da = \frac{t_{turb}}{t_{burn}} = \frac{\lambda}{v_{turb}(\lambda)} \frac{v_{cond}}{\delta_{cond}}, \tag{19.55}$$

which is the ratio of the characteristic turbulent fluid motion time scale to a characteristic reaction time. The Damköhler number depends on the length scale considered and can be quite complex for chemical burning. Common practice is to invoke the Damköhler number, Da_0, that specifically applies to the largest scale of the turbulence. Large $Da_0 >> 1$ corresponds to very rapid nuclear reactions (for the supernovae case) compared to the characteristic turbulent time at a given length scale. Small $Da_0 << 1$ corresponds to very slow reactions.

Another useful dimensionless number that characterizes turbulent burning is the *Karlovitz number* that specifies the ratio of the burning timescale to the time to dissipate the turbulence at small scales:

$$Ka = \frac{t_{burn}}{t_K}, \tag{19.56}$$

where the Kolmogorov time

$$t_K = \left(\frac{\nu}{\dot{\epsilon}}\right)^{1/2} \tag{19.57}$$

represents the timescale to dissipate the turbulent energy at the dissipation scale and $\dot{\epsilon} \sim v_{turb}(\lambda_0)^3/\lambda_0$ is the rate of energy density cascade, a constant in the Kolmogorov model for turbulence (Eq. (19.49)). With this definition of $\dot{\epsilon}$, we can write for a conductive, laminar flame front

$$t_K = \left(\frac{\nu}{K}\frac{K}{\dot{\epsilon}}\right)^{1/2} = Pr^{1/2}\left(\frac{\delta_{cond}v_{cond}}{v_{turb}(\lambda_0)^3/\lambda_0}\right)^{1/2} \tag{19.58}$$

and

$$Ka = Pr^{-1/2}\left(\frac{\delta_{cond}}{\lambda_0}\right)^{1/2}\frac{v_{turb}(\lambda_0)}{v_{cond}}^{3/2} = Pr^{-1/2}\left(\frac{\delta_{cond}}{\lambda_G}\right)^{1/2}. \tag{19.59}$$

In chemical reactions, it is often the case that $Pr \sim 1/Le \sim 1$, and the dependence on the Prandtl number is suppressed by taking it to be unity. In the supernova case, $Pr << 1$, but the Karlovitz number is sometimes defined in the absence of the Pr term in Eq. (19.59), leading to some ambiguity in the comparison of numerical simulations (Aspden et al. 2010). In a physical situation, Ka is set by the true dissipation at the molecular scale whereas in numerical simulations the dissipation is set in practice by the resolution scale of the simulation.

These various dimensionless variables come into the characterization of the flame brush in the following way. The regime where the turbulent diffusion is less than the thermal diffusion, $v_{turb}(\lambda_0)\lambda_0 < v_{cond}\delta_{cond}$, is defined as the laminar flame regime where the flame burns smoothly and steadily and turbulence can be neglected. This tends to characterize the supernova conditions when the flame ignites at high density and the flame is very thin. This regime is generally characterized by small Da_0 and small Ka. When $v_{turb}(\lambda_0)\lambda_0 > v_{cond}\delta_{cond}$, turbulence affects the structure and propagation of the flame. The wrinkled flame regime is characterized by $v_{turb}(\lambda_0) < v_{cond}$ and the corrugated flame regime by $v_{turb}(\lambda_0) > v_{cond}$. These regimes tend to have $Da_0 > 1$; for instance on the boundary between the wrinkled and corrugated flame regime, $v_{turb}(\lambda_0) \sim v_{cond}$, and so $Da_0 = \lambda_0/\delta_{cond} > 1$. The wrinkled and corrugated flame regimes pertain to the subsonic deflagration in the supernova problem when the Rayleigh–Taylor turbulence folds and bends the flame front, broadening the flame brush, increasing the effective area of the flame, and hence enhancing the flame velocity.

Much of the wrinkled and corrugated flame regimes for both the terrestrial and supernova ambiences occur in conditions for which the burning timescale is less than the timescale for the turbulence to dissipate. As the turbulence gets stronger, the rate of turbulent energy dissipation, $\dot{\epsilon}_{turb}$, exceeds the rate of specific energy liberated in the flame (Eq. (19.49)), and the Gibson scale becomes less than the flame width. More drastic changes to the flame brush may occur, including the transition to the distributed burning regime. In the supernova, the evolution to smaller λ_G and larger turbulent dissipation happens naturally as the burning white dwarf expands and burning proceeds at ever lower densities yielding slower, broader laminar flame fronts (Fig. 19.4). Chemical flames with low turbulent dissipation rate are characterized as being in the *thin reaction zone* regime in which the laminar flame is thin compared to the preheat zone that is, in turn, broadened by the diffusion of composition and heat and the effects of turbulence. This regime may also pertain to supernova flames in certain conditions.

A critical question arises in the regime where the burning timescale is long compared to the large-scale turbulent overturn timescale but short compared to the turbulent dissipation time and when the Gibson scale, λ_G, is less than the flame width. The issue is whether and how there is a natural transition from turbulent, subsonic flames to shock-driven detonations as is routinely observed, but not well understood, in terrestrial flames. To understand the turbulent burning, especially in this strongly turbulent regime, it is important to do 3D simulations that resolve the flame brush. This remains a great computational challenge addressed in Sect. 19.11.

19.7 Ignition Kernels

Flames are considered to start with an *ignition kernel* that begins the process. If the ignition kernel is larger than the *quenching length*, then the flame will propagate. The early evolution is characterized by the kernel growth that is a function of fuel and the physical environment, including factors like the sound speed, temperature gradients, and the level of turbulence. Eventually, a successful flame propagates in a steady-state manner that no longer depends on the specifics of the ignition kernel, subsonically or supersonically as circumstances dictate.

In terrestrial combustion, the ignition kernel is usually some accidental or engineered external source, a spark, a shock, a focused laser beam, or some other means to supply the minimum initial energy. The quenching length that determines whether the flame propagates may be determined by losses through container walls, by conduction, or by radiation into the surroundings.

In the context of the smoldering phase, ignition kernels are generated in the heating of the convective zone. There will be lots of potential ignition kernels. The important issues are when a given kernel gets hot enough to generate a flame, whether there are multiple kernels, the flames from which converge, or a single dominant kernel, and whether the flame remains subsonic or develops a self-propagating shock wave. Laminar flame kernels are invoked in Sect. 19.8 and detonation kernels in Sect. 19.9.

Similar issues arise in the context of degenerate, unstable nuclear burning of hydrogen in classical or recurrent novae (Sect. 21.5.1), of helium burning shells in a variety of supernova models (Sect. 22.4.2), and in merging white dwarfs (Sect. 22.4). In each of these situations, material is heated quasi-statically until it reaches a point of runaway and forms an ignition kernel. Astrophysical ignition kernels can be generated in other contexts by converging shocks (Sects. 22.3.6 and 22.3.8) or from colliding white dwarfs (Sect. 22.4).

Because ignition kernels are often very small on the scales of astrophysical objects, the physics of the ignition kernel is often ignored and burning is invoked in an ad hoc fashion. An artificial ignition kernel is defined and imposed when material is "hot enough" or is simply arbitrarily invoked. The issue of whether a given implied ignition kernel leads to deflagration or detonation is often employed arbitrarily. In each of these contexts there needs to be a careful consideration of the ignition kernel. Proper treatment requires fully 3D consideration of the formation and evolution of the ignition kernels, including all the relevant physics and the effects of laminar flame and detonation instabilities (Sect. 19.10). Direct numerical simulations of ignition kernels in a variety of astrophysical contexts are warranted.

The issue of when a spark in turbulent flow leads to a flame and when it leads to a detonation is not understood in the context of terrestrial flames. We should be wary of assuming analogous processes are understood in supernovae.

19.8 Early Runaway: Hot Spots and Bubbles

The general discussion of turbulent flames presented in Sect. 19.6.4 may apply to the formation and evolution of buoyancy-driven turbulent deflagration flames in SN Ia as pictured for a pure deflagration (Sect. 22.3.3) or a quasi-spherical deflagration/detonation scenario (Sect. 22.3.4), but other models have been discussed in the literature. Among these are models that explore the formation and propagation of a few or many "hot spots" and bubbles after dynamic runaway, as suggested by simulations of the smoldering phase (Sects. 19.2 and 22.3.4). Details of the early dynamical runaway probably depend on the environment as affected by convection in the smoldering phase that precedes runaway. Dynamical runaway could occur at a single point or at a number of points in a stochastic way, determined by the nature of the turbulent convection.

As presented in Sect. 19.2, because of the extreme volatility of degenerate carbon burning, the first point to begin to burn dynamically will tend to run away before other points with slightly less temperature and burning rates can do so. If dynamical ignition and runaway do occur at a single point, then the stochastic location of that point could make a qualitative difference in the outcome. Even in the presence of the convection associated with the smoldering phase, in the absence of a prompt detonation, dynamical burning that begins within about 100 km of the center will be dominated by laminar expansion since the laminar flame speed is high, $\sim$100 km s^{-1}, and the effective gravity and hence buoyancy effects are low. Small perturbations are burned out by flame polishing and, in the absence of rotation and magnetic fields, the bubbles are at first nearly spherical. Flames that ignite very close to the center will burn through the center before buoyancy begins to play a substantial role.

If the ignition point is more distant from the center, then the laminar flame bubble will be prone to float before its lower boundary reaches the center of the white dwarf, and the flame will never reach the far side of the white dwarf. These two cases have been differentiated as *central ignition* and *buoyancy-driven* flames. In either situation, the flame bubbles can reach large sizes before significant buoyant motion or deformation occur. This characteristic size represents a minimum scale for fragmentation of flame bubbles. More mass tends to be burned when the ignition kernel is larger (as a result of larger laminar flame speeds) and closer to the center of the white dwarf (Aspden et al. 2011). The further the ignition point is from the center, the stronger the buoyancy effects and the less the effects of background turbulence, but even for central ignition the laminar flame bubble can be advected offcenter by ambient turbulence. For flame bubbles that start sufficiently offcenter, very little burning occurs until the buoyant bubble reaches the surface of the white dwarf, although the ash left behind may be subject to weak interactions and further nuclear processing. The behavior of single, floating flame bubbles is qualitatively different than for a central deflagration that expands roughly spherically from the center, the basic picture invoked for many models (Sects. 19.6.4 and 22.3.4).

The most highly-resolved nonrotating calculations of the dynamics of a single ignition point done to date began with a spherical bubble of ash with a radius of $\sim$ 2 km surrounded by a laminar flame with a numerical resolution of $\sim$100 m per zone (Malone et al. 2014). These simulations assumed that a single ignition kernel grew to this size by the spreading laminar flame. The initial size and mass of the bubble are vastly greater than the critical sizes associated with the initiation and propagation of a detonation (Sect. 19.9), so the assumption has been made that the ignition kernel did not develop into a detonation.

For a fiducial case, the bubble was placed 41 km from the center of the model in a region with an outflow of 10 km s^{-1} from the preceding smoldering phase computed by Nonaka et al. (2012). The flame bubble evolved through several distinct phases (Fig. 19.5). In the early expansion, the rate of expansion of the bubble at the laminar flame speed, $\sim$100 km s^{-1}, was much more rapid than the

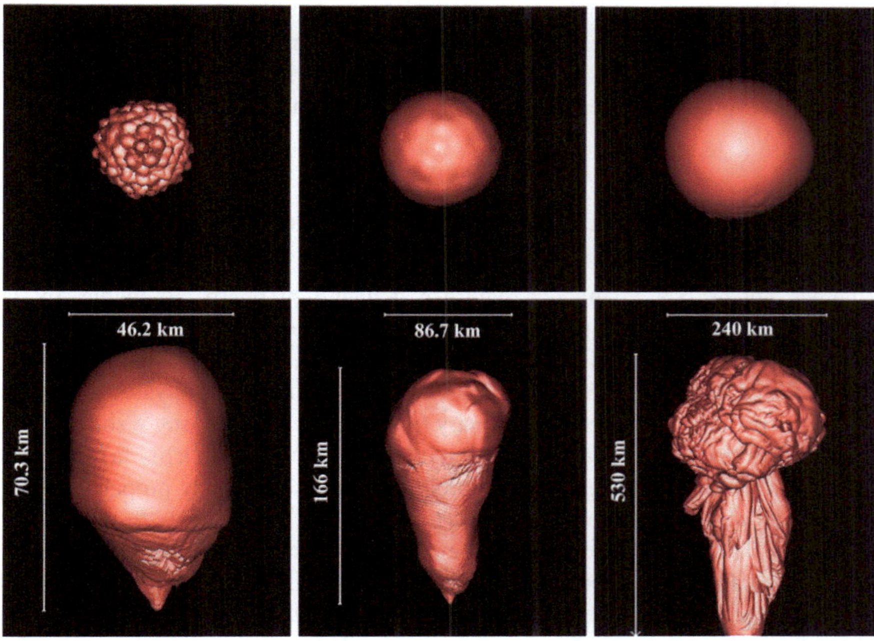

Fig. 19.5 Evolution of a flame bubble after ignition of carbon in a white dwarf of M_{Ch}. Contours of the flame surface correspond to a ^{12}C mass fraction of 0.45. The initial hot spot has a diameter of about 4 km with random perturbations of a few hundred meters. The center of the white dwarf is down. *Top*: t = 0.00, t = 0.0059, and t = 0.0119 s after creation of the perturbed flame bubble. *Bottom*: t = 0.150, 0.265, 0.469 s; note the change of scale. From "The Deflagration Stage of Chandrasekhar-Mass Models for Type Ia Supernovae. I. Early Evolution" (Malone et al. 2014). © AAS. Reproduced with permission

model outflow velocity or ambient transverse velocities from the convection. As a result, the initial perturbed spherical flame bubble burned out the perturbations and expanded nearly spherically. The bubble began to float at about 0.03 s but remained spherical until about 0.06 s when it started to elongate in the radial direction. The elongation occurs because the outward motion of the upper surface is accelerated by buoyancy so the flame propagates more rapidly than the laminar flame speed, while the inward motion of the lower surface is retarded by buoyancy. The latter also tends to retard the propagation of bubbles to encompass the center of the star compared to constant-gravity simulations (Zingale and Dursi 2007). By 0.1 s, when the flame bubble had a size $\sim$ 50 km, buoyancy effects caused noticeable distortion. Buoyancy instabilities began to wrinkle the flame surface, and the upper surface of the bubble developed a classic "mushroom" shape, but turbulence remained mild. Fresh fuel was entrained in the Rayleigh–Taylor roll. Between 0.1 and 0.2 s, a torus structure developed at the top of the bubble. Somewhat later, substantial vorticity developed that broke up the torus shape of the earlier rolling structure. By $\sim$0.6 s, the bubble reached the boundary of the initial convective region, the burning became restricted to the bubble cap rather than by fuel entrained in the bubble interior, reactions were less rapid, and the bubble expanded laterally in reaction to pressure gradients. The bubble reached the surface of the white dwarf at ~0.8 s. The issue of whether the subsequent evolution can reproduce observed SN Ia is addressed in Sect. 22.3.6.

19.9 Astrophysical Detonation: Supersonic Burning

Whereas not all SN Ia models require a deflagration phase, all viable SN Ia models require detonations to produce conditions remotely resembling observations (Chap. 22). The nature of detonation and the transition from subsonic deflagration to supersonic detonation (Sect. 19.11) is a central issue in the physics of SN Ia. The major difference between an ordinary (piston-driven, steady-state) shock and a detonation is that chemical (in a stick of dynamite) or nuclear (in a SN Ia) energy is released in the front. This energy reinforces the shock, and steady propagation can be achieved with no external input (Sect. 19.4.2).

The shock that is the precursor to the combustion front in a detonation wave is taken to be infinitesimal in most applications. The width of the burning region that follows the shock in a detonation, known as the *induction length*, is $L_{in} \sim S_{sh}\tau$, where S_{sh} is the shock speed and τ is the timescale for the reaction, known as the *induction time*. For chemical reactions, the induction time can be a complicated function, but for nuclear reactions where the burning rate is generally monotonic with ρ, T, and fuel composition, the induction time, while very rapid, can be straightforwardly defined as the time to burn some fraction, say 1/2, of the fuel. The distance from the shock to the point where the fuel is essentially entirely consumed is also known as the *Zel'dovich–von Neumann–Doering (ZND) length*, L_{ZND}. A 1D detonation in which the extent of the shock plus burning front L_{ZND}, is negligibly small compared to other length scales is a ZND detonation.

One of the major uncertain areas in the study of detonations in both terrestrial and astrophysical contexts is the issue of initiation. As discussed in Sect. 19.7, the question of when a spark in a turbulent environment yields a subsonic flame or a detonation is a difficult one. Detonations can be initiated by *direct initiation* when a shock generated in some manner propagates into a reactive medium with a velocity exceeding the CJ velocity. This can result in an overdriven shock-reaction complex that can relax into a CJ detonation. Direct ignition might occur if two white dwarfs collide, but even in that case it is not trivial to define the necessary and sufficient conditions for an ignition kernel to lead to a CJ detonation given variations in composition, density, and temperature. In general, if the fuel is sufficiently volatile and releases enough energy, the ignition kernel, a spark, can lead to a sufficiently strong local overpressure even in isochoric burning to produce a strong shock and an initially overdriven detonation that relaxes to the CJ condition. There can be *auto ignition* in a turbulent flow where the detonation kernel is provided by intersecting shocks, but experiments and simulations also show that if the region of shock intersection is too small or moving too rapidly, a detonation may not result. Another alternative, discussed in Sect. 19.11.2, is a spontaneous burning regime that begins with no shock but develops a shock-reaction complex over time.

In the extremely degenerate conditions at the end of the smoldering phase in white dwarfs of M_{Ch}, the overpressure of isochoric burning is rather small, and it seems unlikely that this process alone will trigger a detonation (Seitenzahl et al. 2009). Whether more complex processes will lead to prompt detonation in that situation is not known from first principles (Hillebrandt and Niemeyer 2000; Niemeyer and Woosley 1997). The assumption of prompt detonation leads to a contradiction with observations (Sect. 22.3.1), so it is assumed that prompt detonation does not occur in this particular context. Other aspects of detonation initiation are presented in Sect. 19.11.

In most astrophysical contexts, the shock-reaction complex is far too small to be resolved in full-star simulations. Numerical procedures that artificially spread shocks over several zones still capture the essence of the problem given that the energy released in the detonation front is treated accurately. Such a numerical scheme will propagate a detonation front at the CJ detonation speed.[2] Other numerical procedures, e.g. level-set methods, employ a steady-state solution for the detonation velocity. Either method yields a self-propagating CJ detonation, given that the detonation is artificially initiated, at densities relevant to the smoldering phase. This is why it must be assumed that the physics of detonation initiation is not satisfied in the smoldering environment. These numerical methods will also give a failure of the detonation at sufficiently low density when the burning time gets long compared to the dynamical time. The level-set method may need to be turned off by hand since it prescribes a detonation. Neither method captures the physics of the natural instability of detonations (Sect. 19.10). While standard numerical techniques are apparently adequate to capture detonations for a large range in astrophysical conditions, they are not appropriate to address the questions of detonation kernels and initiation nor the quantitative aspects of detonation death that will depend on the relevant multidimensional instabilities. These require very highly-resolved direct numerical simulations.

19.10 Detonation Instability

If a detonation is initiated, it can propagate self-consistently through most of a C/O white dwarf of M_{Ch} before the star can expand to lower densities. In 1D calculations, a carbon detonation can propagate to $\rho \gtrsim 10^6\,\mathrm{g\,cm^{-3}}$. If promptly initiated at the onset of dynamical burning in a white dwarf of M_{Ch}, virtually the whole white dwarf would burn at relatively high densities and thus be converted to iron group elements. This is the basis of the statement that if central detonation is initiated

[2]Allowance can be made for the variable energy release and hence variable CJ conditions in NSE if the burning leads to that condition (Buchler et al. 1972; Sect. 19.10).

it produces too much iron-group matter. An important but underappreciated fact is that real detonations are *unstable*. They form intersecting shocks that break off pockets of unburned material, which in turn burn behind the primary shock front and create new intersecting shocks. This instability was first shown in terrestrial fuels by Denisov and Troshin (1959) and Voitsekhovsky et al. (1963). One-dimensional detonations are unstable longitudinally. In 3D there are also transverse instabilities that collectively stabilize the full detonation structure. It is likely that the only reason that detonations propagate successfully is because of the 3D structure that accompanies real detonations (He and Lee 1995).

One-dimensional astrophysical degenerate carbon detonations are susceptible to longitudinal instability as the burning gets out of phase with the shock (Khokhlov 1993). In this limited form of the detonation instability, the burning catches up with the leading shock, the peak pressure and density behind the shock reach maximum and the induction length reaches a minimum. The shock then accelerates, the peak pressure and density decline, and the induction length grows. The burning lags, and the shock slows until the burning catches up to begin the cycle again. The period of this longitudinal instability is about 10^{-8} s at a density of 3×10^{7} g cm^{-3}. The proof-of-principle that carbon detonations are susceptible to the full multidimensional transverse as well as longitudinal instability was presented by Boisseau et al. (1996).

To recover the detonation instability manifest in terrestrial experiments, numerical calculations must resolve the burning region, L_{ZND}. The instability in either experiments or computations can be characterized by the locus of points of maximum pressure. The maximum pressure is created at triple points that represent the intersection of forward shocks, transverse shocks, and weaker shocks that connect points of shock interaction. The latter are known as *Mach stems*. The loci of the points (marked by scoured traces in soot on the walls of shock tube experiments; Fickett and Davis 1979, their Fig. 1) define *detonation cells*. Detonation cells represent the dynamic boundary of pockets of incompletely burned fuel that lag behind the main front, belatedly burn, and generate new transverse shocks to regenerate the process. Detonation cells are often diamond shaped, but they can also be irregular. The shape may be an artifact of boundaries, but the size is dictated only by the nature of the fuel and its burning characteristics. The size of the detonation cells results from the complex nonlinear interaction of shocks and cannot be simply predicted from linear theory, but only computed with appropriately resolved numerical simulations.

Two-dimensional calculations of the detonation instability and the resulting cell structure at densities from 1×10^{6} to 3×10^{7} g cm^{-3}, the range over which detonations are expected to be initiated and to propagate in the context of SN Ia, showed the characteristics expected of detonation instabilities: triple points, transverse waves, weak shocks linking approaching points with each other, and Mach stems connecting the points of reflected transverse shocks with each other and connecting the point of intersection with the boundary (Gamezo et al. 1999; Fig. 19.6). The detonation cell structure grows from initial perturbations and does not depend on those perturbations. A weak longitudinal instability is, in general, present, but the

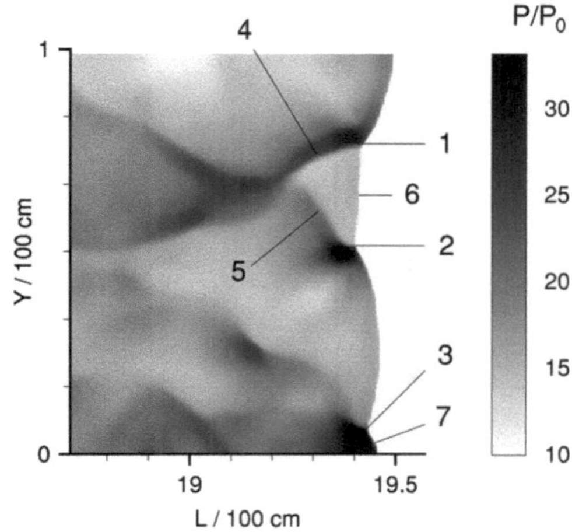

Fig. 19.6 Pressure field behind the cellular detonation front on the ^{12}C burning scale for $\rho_0 = 5 \times 10^6$ g cm^{-3}. Triple points: 1, 2, 3; transverse waves: 4, 5; weak leading shock: 6; Mach stem: 7. From "Multilevel Structure of Cellular Detonations in Type Ia Supernovae" (Gamezo et al. 1999). © AAS. Reproduced with permission

cell structure is driven almost entirely by the transverse instability (Gamezo et al. 2014).

Because the mode of propagation of 3D detonations has more degrees of freedom than a 1D detonation, the average propagation speed is less than the Chapman–Jouguet speed, $\bar{S} < S_{CJD}$. Thus detonations propagate more slowly and die more easily than predicted by 1D models of Chapman–Jouguet or ZND detonations. The existence of this detonation instability may affect the initiation and propagation of detonations, and the DDT. It should be taken into account in any model for SN Ia incorporating detonations—unconfined delayed detonations, pulsating delayed detonations, gravitationally-confined detonation, detonations during double-degenerate evolution, and detonations on the surfaces of white dwarfs and neutron stars.

The presence of the detonation cell structure creates an effective burning zone that is larger than for computations in which the induction length of the burning is resolved, but only in 1D. Detonation cells in terrestrial explosions typically have a length, L_c, that is one to ten times larger than L_{in}, the 1D induction length. The cell size is the length over which a fuel is totally consumed in the presence of a detonation. For a detonation to propagate self-consistently, the scale of a physical system must be larger than a detonation kernel, $L_k \sim 10L_c$, so that a steady-state, but unstable front can develop. A detonation is then not able to propagate unless the physical system (shock tube, star) is larger than 10–100 times L_{in}. The initiation of a detonation typically requires a region of length, L_Z (Sect. 19.11.1), that is 1–4 times L_k, 10–40 L_c, 10–400 times L_{ZND}. These simple estimates should be re-evaluated in the context of the flame-brush instability (Sect. 19.11.3).

The structure of a degenerate carbon detonation can be approximately separated into three consecutive stages of energy release, corresponding to carbon-burning, oxygen-burning, and quasi-equilibrium silicon-burning reactions (Gamezo et al. 1999). The reaction timescales of these stages differ by several orders of magnitude (Fig. 19.7). A further complication is that because of the approach to nuclear statis-

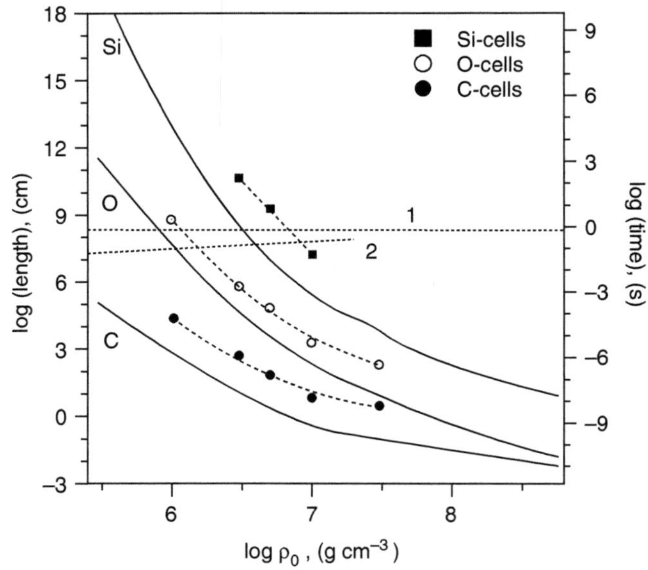

Fig. 19.7 Key length scales and timescales as functions of initial density for self-sustained detonations. The *solid lines* are the half-reaction lengths and times for steady-state 1D detonations. The *points* are the detonation cell sizes for 2D cellular detonations. The *horizontal dotted line 1* shows the typical size of a Chandrasekhar mass white dwarf, $\sim 2 \times 10^8$ cm. The *dotted line 2* gives the scale of the density gradient, i.e., the length over which the density changes by a factor of 2 for a typical preexpanded white dwarf. From "Multilevel Structure of Cellular Detonations in Type Ia Supernovae" (Gamezo et al. 1999). © AAS. Reproduced with permission

tical equilibrium, silicon burning can be endothermic. The result is a *pathological detonation* with front speed $S_p > S_{CJD}$ and $Q_p > Q_{CJD}$, where Q_p and Q_{CJD} are the energy released in the pathological detonation and in the corresponding CJ detonation, respectively. In the pathological detonation, the velocity of the front depends on time-dependent details of the reaction kinetics. The computed cell sizes are about 30, 10, and 100 half-reaction lengths for carbon, oxygen, and silicon, respectively.

The transverse shocks and triple points produce fluctuations in density, temperature, and pressure and generate vorticity. The result is pockets of under-reacted and over-reacted material that lag behind the shock and eventually disappear. This gives a different mean composition than a 1D steady-state detonation. The average 2D half-reaction lengths are larger than those for 1D models by 30–60%. These effects may have observable consequences if detonations are initiated in SN Ia at densities $\rho \lesssim 10^7\,\mathrm{g\,cm^{-3}}$ and propagate to lower densities. Detonation instability is unlikely to affect the process of deflagration-to-detonation transition, but the final composition that determines observed spectra could be affected.

At low densities ($\lesssim 10^6\,\mathrm{g\,cm^{-3}}$) in the outer portions of expanding ejecta, carbon burning is unlikely to proceed beyond the $^{12}C + {}^{12}C$ reaction itself (Domínguez and

Khokhlov 2011). This reduces the energy available in the detonation (compared to the assumptions of Boisseau et al. (1996) and Gamezo et al. (1999)) by about a factor of two. In this circumstance, the characteristic length of the detonation cell structure is greater by a factor of $\lesssim$100 and the detonations may change behavior. The burning may be extinguished only to re-ignite by spontaneous burning (Sect. 19.11). Such an instability may affect the very outer layers of a SN Ia, the initiation and propagation of detonations in gravitationally-confined detonation models (Sect. 22.3.6), and in some double-degenerate models that invoke detonations (Sect. 22.4).

The reason for the restriction of carbon burning at low densities is that the cell sizes for oxygen and silicon get larger than a scale height in the white dwarf, and thus it becomes impossible to sustain a detonation in those fuels. While the carbon cell size tends to remain small, less than 10^5 cm, the oxygen cell size is comparable to the radius of a white dwarf for $\rho \sim 10^6\,\mathrm{g\,cm^{-3}}$ (Fig. 19.7).

The constraints on initiating or maintaining a detonation in silicon are even more stringent. Detonation cells associated with silicon burning to iron-group elements, in particular, grow to become comparable to the size of the white dwarf for $\rho_{det} \lesssim 10^7\,\mathrm{g\,cm^{-3}}$, the density at which detonations are thought to be initiated in SN Ia (Fig. 19.7). For a white dwarf to contain several silicon-burning cell lengths within a density-gradient scale height, the density would need to be substantially greater than $10^7\,\mathrm{g\,cm^{-3}}$. The formation of silicon and nickel is thus delayed compared to 1D detonation models. Part of the reason why there are no iron-group elements in the outer parts of SN Ia may be that under conditions where carbon detonation is initiated, silicon detonations cannot occur. Neglect of the detonation cell structure for silicon may quantitatively affect estimates of the detonation ignition criteria at lower densities in massive white dwarfs (Sect. 22.3) and even at central densities for less massive white dwarfs (Sects. 22.3.8 and 22.4.2).

Another context in which detonation instability may be important is the propagation of a detonation through the structure previously processed by the deflagration. During the deflagration, fingers of unburned carbon and oxygen are carried by large scale Rayleigh–Taylor turbulence to the center of the star. In the subsequent homologous expansion of the explosion, this central carbon and oxygen would be at low velocity, but such low-velocity C/O is not observed. This is one of many arguments that pure deflagration models of do not work for typical SN Ia (Sect. 22.3.3). To match the observations, it is also important that the detonation reach these central regions and scour out most of the inner carbon and oxygen. Röpke and Niemeyer (2007) argued that the detonation cannot penetrate the interwoven fingers of the ash of burned material and hence cannot penetrate these inner regions. This argument ignores the presence of the detonation instability and the large length scales associated with oxygen and especially silicon burning. This large-scale detonation structure can sweep over relatively small-scale ash structure and re-ignite the detonation, including that of carbon, in other layers and fingers (Gamezo and Oran 2007).

The detonation cell structure and its implications for degenerate helium detonations has yet to be studied. This omission should be remedied. The effects of rotation and magnetic fields on detonation cell structure are totally unexplored.

19.11 Deflagration-to-Detonation Transition

The composition structure (abundances versus velocity) of SN Ia inferred from spectroscopy, especially the substantial abundance of intermediate-mass elements at $v \gtrsim 10{,}000\,\mathrm{km\,s^{-1}}$ (Chap. 20), can reasonably be reproduced by a thermonuclear explosion in a C/O white dwarf of M_{Ch} with the condition that the burning begin subsonically and then accelerate substantially after the white dwarf has expanded to a density $\sim 10^7\,\mathrm{g\,cm^{-3}}$ (Sect. 22.3.4). A possibility is for the combustion to begin as a deflagration that makes a transition to a detonation. *Deflagration-to-detonation transition* (DDT) also is an important and ill-understood process in terrestrial combustion for which there are many parallels and some important differences with the astrophysical domain.

An important question is then under what circumstances a deflagration front spontaneously makes a transition to a detonation. Transitions from deflagrations to detonations are routinely seen in laboratory shock-tube experiments. An important difference is that the laboratory context provides a *confined DDT*. The walls and boundaries of the shock tube reflect and concentrate shocks. Numerical studies strongly suggest that confined DDT is produced directly on small scales by turbulence generated on those scales by intersecting shocks, rather than, for instance, a turbulent cascade from larger scales (Khokhlov et al. 1997). In the confined context, detonations ultimately are the result of direct initiation, the analogy of a firing cap in a stick of dynamite.

A SN Ia may require an *unconfined DDT*. There are few, if any, terrestrial experiments that explore this regime since even open-air explosions generally involve shocks reflected from the ground. In a supernova, the spontaneous formation of proper conditions would seem to require that turbulence generated by the deflagration on large scales be strong enough on small scales to pack or disrupt the flame. In the absence of confining walls, the mechanism of colliding shocks generating sufficiently strong turbulence on small scales seems unlikely in the aftermath of the smoldering phase, although there might be reflections of shocks from regions of composition and density gradients or from shear regions set up by accretion onto the white dwarf.

To approach the problem of DDT in supernovae, we need to consider the requirement for detonation. As the white dwarf expands, the laminar flame slows and broadens and the Gibson scale, λ_G, decreases. Eventually, $\lambda_G < \delta_{cond}$, i.e., the turbulent velocity on the scale of the conductive front thickness is comparable to or exceeds the conductive burning speed of the nonlinearly stabilized flame, $v_{turb}(\delta_{cond}) > K v_{cond}$, where K is a constant somewhat exceeding unity (Fig. 19.2). This critical condition can also be expressed in terms of a critical Karlovitz number, *Ka*, as described in Sect. 19.6.5. The interaction of the very rapid turbulence with the flame becomes a key issue. Turbulent flames may spontaneously produce colliding shocks and direct initiation of DDT even in the unconfined case (Sect. 19.11.2).

19.11.1 The Zel'dovich Gradient Mechanism

A particular mechanism for converting a deflagration to a detonation that has been investigated in both the terrestrial and the supernova regimes is the *Zel'dovich gradient mechanism* (Zel'dovich et al. 1970). This mechanism supposes a "prepared" region with oppositely directed gradients in temperature and fuel concentration. The fuel is brought to the condition where it can start to burn on a timescale more rapid than that for dynamical expansion. The initial runaway on dynamic timescales is expected to occur by means of a wave of *spontaneous burning*, where adjacent parcels of matter are hot enough to burn on their own with no interaction between them required. The rate of progress of the spontaneous burning front is dictated solely by the rate of burning in each parcel, which is dictated by the gradients in temperature and fuel abundance; the speed of the burning front is the phase velocity of the spontaneous burning. Under proper circumstances, shock waves generated during the spontaneous burning can accumulate to generate a CJ detonation in the ambience of SN Ia models (Khokhlov et al. 1997; Niemeyer and Woosley 1997; Aspden et al. 2010). In the context of Sect. 19.7, a region that satisfies the Zel'dovich criterion represents a detonation ignition kernel. The Zel'dovich mechanism may play a role in some situations of astrophysical detonation formation, but it may not be the key, or even primary, mechanism in all cases. Much work in the supernova community has focused on the distributed burning regime that has the potential to disrupt the flame, mix fuel and ash to create appropriate gradients in temperature and fuel, and hence to initiate the Zel'dovich gradient mechanism.

There is a minimum length of a gradient region, the *Zel'dovich length*, L_Z, that allows it to undergo spontaneous burning that leads to a detonation. If the gradients are too steep and L_Z is too small, the spontaneous burning will not generate the requisite shock. If the gradients are shallow and hence extend over a sufficiently large length scale, a detonation can, in principle, form. Conditions must be right to initiate the spontaneous burning, nominally by being sufficiently hot for a given density and composition. The overpressure and material velocity increase in the spontaneously-burned material as the phase velocity of the burning front decreases. When the speed of the spontaneous wave approaches the speed of sound, a shock wave is formed, and a shock-reaction complex develops. In a prepared region with opposing temperature and composition gradients, the shock-reaction complex forms somewhere in the middle of the mixed region where there is both enough fuel and a sufficiently high temperature. To form a self-propagating CJ detonation, the shock-reaction complex must survive as it propagates into the unburned material. The shock-reaction complex can survive propagation down the temperature gradient and grow into detonation only if conditions beyond the complex change slowly enough, that is, if L_Z is sufficiently large.

Despite some differences in assumptions (temperature gradients with constant composition versus opposing gradients in temperature and fuel abundance, burning to NSE versus only to silicon) studies give about the same results for L_Z (Khokhlov et al. 1997; Niemeyer and Woosley 1997). At a density of $10^7\,\mathrm{g\,cm^{-3}}$ $L_Z \sim 3\,\mathrm{km}$, at $10^8\,\mathrm{g\,cm^{-3}}$ $L_Z \sim 2\,\mathrm{m}$, and at $10^9\,\mathrm{g\,cm^{-3}}$ $L_Z \sim 1\,\mathrm{mm}$. At the respective densities, L_Z encompasses ever tinier masses, $\sim 6 \times 10^{-10}$ $M_\odot$, $\sim 2 \times 10^{-18}$ $M_\odot$, and $\sim 2 \times 10^{-27}$ $M_\odot$, respectively, for the densities and lengths given. What the final product is, and hence the energy liberated, does not affect L_Z significantly, but the composition of the fuel does. Studies have examined the sensitivity to the fuel abundance, including admixture of helium, and to the form of the temperature gradient in the spontaneous-burning region (Seitenzahl et al. 2009; Shen and Moore 2014).

The scales estimated for L_Z are much larger than L_{ZND}, the corresponding distance to burn fuel to ash in a CJ detonation, the carbon- and oxygen-burning detonation cell sizes (Sect. 19.10), and especially compared to the width of conductive flames at the respective densities. These estimates of L_Z suggest that in a solar mass of white dwarf material there are a huge number of potential Zel'dovich ignition kernels. The difficult problem is ascertaining whether the temperature gradient is sufficiently flat in practice over these length scales. Niemeyer and Woosley (1997) emphasized that if the spontaneous burning were initiated by gradually increasing the temperature (for instance by accretion) to the point at which spontaneous burning begins, the temperature gradient required is very flat, so flat that the turbulent cascade associated with convection might disrupt and eliminate the very shallow gradient. An issue is that the criterion for the maximum allowable gradient invoked by Niemeyer and Woosley is itself a very sensitive function of temperature because of the sensitivity of the burning rates. If the intermittent conditions of turbulence or a shock collision *triple point* (the point of incursion of a third shock along the locus of intersection of two other colliding shocks) raise a region to somewhat above the minimum temperature, the gradient that would allow a CJ detonation to form might be considerably steeper.

The small Zel'dovich lengths and masses, especially at higher densities, might suggest that it is easy to trigger a detonation. This is precluded if sufficiently intense turbulence is required to create the distributed flame regime and produce sufficiently large Zel'dovich kernels. If there are other means to create the tiny Zel'dovich kernels through triple points or shear, detonation might still occur. There is not yet a fundamental understanding of whether detonation is or is not initiated in white dwarfs of given physical conditions. This limitation applies to all the models that are discussed in Sect. 22.3.4. The ability of a CJ detonation to propagate under given conditions of density, temperature and fuel is necessary, but it is not sufficient. Rapid, dynamical burning could lead to deflagration, not detonation.

Some direct numerical simulations have shown a distinct change in the structure of the flame brush for very strong turbulence (Aspden et al. 2008, 2010). Aspden et al. invoked the "small scale" and "large scale" turbulent-burning regimes discussed by Damköhler to define a *well-stirred reactor regime*, where the turbulent motions are more rapid than the laminar flame speed, so the flame is mixed with fresh fuel before it can burn. The local flame actually burns at a speed slower than the laminar speed, but the overall speed is increased because of the growth of the area of the flame. Aspden et al. also described a *stirred-flame regime*, an asymptotic regime where the flame cannot be further broadened by the turbulence, but which is especially characterized by strong fluctuations associated with intermittency (Pan et al. 2008; Schmidt et al. 2010; Hamlington et al. 2012). Aspden et al. argue that this may be the regime where transition to detonation occurs.

19.11.2 Turbulent Flame-Brush Instability

An alternative picture to the Zel'dovich mechanism with surprising implications arises from highly-resolved direct numerical simulations of turbulent hydrogen and methane flames (Poludnenko and Oran 2010, 2011; Poludnenko et al. 2011; Poludnenko 2015). Poludnenko et al. found that for sufficiently strong turbulence, the Damköhler paradigm is not manifest, and the hypothesis that the turbulence enhances the flame speed is not correct. The traditional picture advocated by Damköhler is that turbulence affects both the preheat region and the heat transport from the burning product to the preheat region and hence the turbulent flame speed. The simulations contradict this picture. The structure of the reaction region is basically unperturbed as the turbulent intensity is increased. The traditional picture is based on the behavior of a passive scalar in turbulent, nonreactive flow; however, the response of a flame to turbulence is qualitatively different from that of a passively-advected scalar. With a flame present, turbulent motions do not penetrate to small scales in the same manner. Although the preheat region is broadened by turbulence, turbulence does not determine the local flame speed. The structure in the preheat region is affected by the turbulence, but the structure of the reaction zone in the simulations is essentially unchanged. There is substantially finer wrinkling of the flame surface on the fuel side than on the product side. The result is that the amount of fuel burned and the energy released in the smooth flame regions is the same as in the planar laminar flame.

Turbulence is more efficient at packing than broadening the flame. In detail, the turbulence causes the flame brush to become highly convolved and to develop cusps as portions of the flame impinge on one another from nearly opposite directions (Fig. 19.8). The reaction zone increases in volume as the flames focus on a region of fuel from multiple directions. The burning rate is then related to the phase velocity of the encroaching flame fronts that is more rapid than either the conductive flame or the turbulence on the flame scale. The cusps burn out quickly, resulting in an increased rate of global fuel consumption and a smoothing of the flame surface. The resulting, flatter, slower flame with fewer cusps then becomes susceptible to the turbulent motion and the cycle begins again. The result is that the turbulent energy cascade fails to penetrate the internal structure of the flame even as the speed of the turbulence is increased. The traditional conceptual basis of the Gibson scale exaggerates the effectiveness with which rapid, small-scale, turbulence penetrates the flame and hence the role, significance, and even existence of a distributed flame regime.

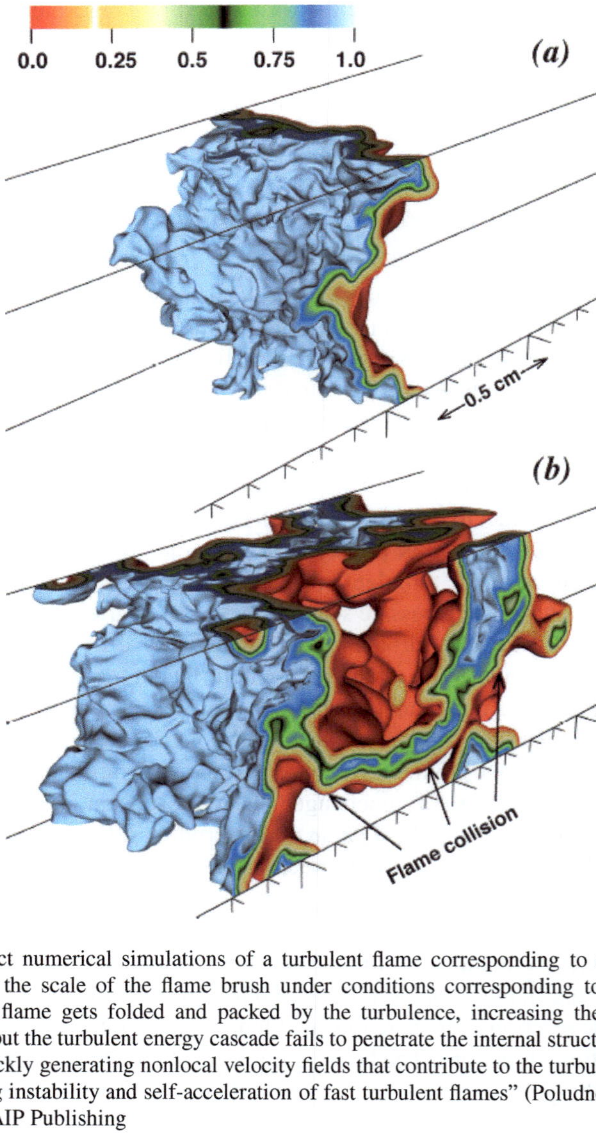

Fig. 19.8 Direct numerical simulations of a turbulent flame corresponding to a terrestrial H_2-air mixture on the scale of the flame brush under conditions corresponding to Lewis number $Le = 1$. The flame gets folded and packed by the turbulence, increasing the volume of the reaction zone, but the turbulent energy cascade fails to penetrate the internal structure of the flame. Cusps burn quickly generating nonlocal velocity fields that contribute to the turbulence. Reprinted from "Pulsating instability and self-acceleration of fast turbulent flames" (Poludnenko 2015) with permission of AIP Publishing

In this context of a severely-folded, contorted flame, the velocity of the turbulent flame front, S_{turb}, once again becomes dependent on the properties of the laminar flame. The flame folding and hence the process of forming flame cusps is affected by the laminar flame thickness. The result is that the propagation of the flame is not a local process. The rate of burning in cusps exceeds that of the plane laminar flame and the formation of cusps is controlled by long-range velocity correlations. The formation and evolution of flame cusps depends somewhat on the stretching of the flame and hence on the ratio of the thermal to species diffusion coefficients, the Lewis number, *Le*. Many of the simulations of terrestrial flames assume the diffusion rates to be similar, $Le \sim 1$, whereas $Le \gtrsim 10^3$ in the degenerate matter relevant to SN Ia.

It is this "packing" of the flame front by turbulence that may trigger detonations. A detonation may thus occur before the conditions are reached that characterize the distributed-flame regime envisaged by Damköhler. If the turbulent intensity is not too high, the creation of flame surface by turbulence can be balanced by the destruction of flame surface in cusps resulting in an equilibrium flame brush structure and propagation of the turbulent flame in nearly a steady state. As the turbulent intensity and the turbulent-flame velocity increase, the steady-state condition must break down in order to accommodate the rate of energy release. Collisions of fast-moving flame surfaces in the hot, shock-compressed fuel can create a high-pressure point that ignites an overdriven detonation that subsequently relaxes to a canonical planar CJD front. The turbulent intensities that drive this transition are not high enough to disrupt the internal flame structure, which remains close to that of a planar laminar flame.

The traditional picture also envisages a turbulent steady state. In practice, the detailed simulations show that as the strength of the turbulence increases, there is an onset of strong time variability in the turbulent flame brush. When the turbulent overturn time becomes comparable to the burning time, the turbulent flames become intrinsically unstable (Poludnenko 2015). The result is a dynamic, unsteady flow in which the rate of consumption of fuel is generally not balanced by the rate of ingestion of fresh fuel into the flame brush. Rather, there are phases when the inflow of fuel dominates; the flame brush thickens and encompasses more fuel. Subsequently, there is a phase when the consumption of fuel is more rapid than the input. The fuel burns quickly, and the flame brush thins. The processes of fuel ingestion and consumption average out, but the imbalance at a given moment and the resulting variability is intrinsic to the evolution of the turbulent flame brush.

In the unstable state, the net effective turbulent flame speed, S_{turb}, develops pulsations, with the speed varying by factors of ten in specific models. The result is termed a *galloping flame* for which the flame speed varies and the average speed can considerably exceed the characteristic turbulent velocity on the scale of the flame brush in the upstream fuel. This unsteady burning results in pressure and density gradients across the flame that produce vorticity and an anisotropic amplification of turbulence within the flame brush that selectively accelerates the flame in the direction of propagation. The variations in the flame speed produce strong pressure pulses or even weak shocks. As the turbulent intensity increases, the area of the flame surface grows exponentially, S_{turb} grows rapidly, and there is a transition to a detonation. This transition may not be abrupt. The flame collisions in cusps are highly intermittent. The magnitude of the pulsations in flame speed depend on the density contrast across the flame, and hence will be different in degenerate matter than in terrestrial matter.

19.11.3 DDT Due To Turbulent Flame-Brush Instability

The condition for creating a detonation within the packed flame brush is that burning liberates an amount of energy comparable to the extant energy in one acoustic time across the flame brush. That requires the flame to be folded to the smallest volume and hence to a critical size with the highest flame front area and net burning speed. This critical size effectively defines the detonation kernel in this context. To achieve this critical size, the turbulence must be such that S_{turb} on the scale δ_{cond} exceeds v_{cond} by a factor of a few.

In a flame, the density, temperature and pressure are not constant, so there is a gradient in the sound speed. The time to burn out the fuel is $\sim\rho_{fuel}/\rho_{ash} \times \delta_{flame}/c_s$ where δ_{flame} is the size of the flame brush. For chemical flames, the density ratio is typically $\sim$10, so that the fuel burns out in several acoustic times and pressure waves can accumulate in the fuel, thus enhancing the tendency to detonate. For astrophysical conditions, the density ratio is close to unity and the fuel burns more quickly. There is less time to accumulate pressure in the fuel, but there will still be shock production and collision. From the packed flame brush, there are many paths to detonation even with relatively mild turbulence.

The critical turbulent flame speed for the onset of the runaway that leads to a detonation corresponds to that of a CJ deflagration (Sect. 19.5), but with an important new feature. For a deflagration, the burning front travels with speed S_{turb} and, by mass conservation, in the frame of the front the velocity of the product is related to the velocity of the front by

$$S_{turb}\rho_{fuel} = v_{ash}\rho_{ash}. \tag{19.60}$$

A traditional Chapman–Jouguet deflagration has the property that $v_{ash} = c_{s,ash}$, the sound speed of the product. This is the maximum speed of a steady-state flame. In the context of the strongly-turbulent flame brush, there is strong intertwining of cold fuel and hot product, and the flame begins to behave like a Chapman–Jouguet deflagration with a build up of pressure in the flame brush when the front propagates at the sound speed in the fuel, $c_{s,fuel}$. When the turbulent flame speed satisfies

$$S_{turb}\rho_{fuel} = c_{s,fuel}\rho_{ash}, \tag{19.61}$$

the energy generated by burning within the flame brush on a sound-crossing time becomes comparable to the extant thermal energy in the flame brush with a resulting increase in pressure. The pressure pulse in the galloping flame can lead to transient conditions for which the flame speed exceeds this Chapman–Jouguet velocity. For terrestrial flames, simulations show that a criterion based on the CJ deflagration speed predicts the onset of DDT for various laminar flame speeds, system sizes, and turbulent intensities.

This condition of Eq. (19.61) thus represents a threshold criterion for the transition to detonation. In the degenerate conditions of a white dwarf, the density ratio is modest, so the criterion is met as the turbulent flame speed approaches the speed of sound in the fuel. When this condition is reached, even a small increase in the turbulent intensity can lead to the transition from the regime of pulsating flames that create shocks to a full self-propagating detonation. Simple estimates based on Kolmogorov scaling suggest that the turbulent speed and size of the integral scale in the context of smoldering, $\sim$100 km s^{-1} and $\sim$10 km, is sufficient to reach the threshold condition for detonation for density of $\sim 2 \times 10^7$ g cm^{-3}. At higher density, the critical size of the flame brush is much less than the integral scale, but the velocity required on the integral scale increases, so initiating a detonation by flame-brush instability becomes less likely. The flame-brush instability leading to DDT can occur in unconfined conditions and in situations where the turbulence in the matter well ahead of the front is distinctly subsonic. There is no need for the special fuel and flame structure and the generation of spontaneous burning waves that characterize the Zel'dovich mechanism, although that mechanism might pertain in other contexts. A caveat is that, with the time dependence of flame pulsations and intrinsic intermittency, there is not a single turbulent flame speed. The problem is characterized by a probability distribution of flame speeds appropriate to a given set of conditions, and hence so is DDT.

While the direct numerical simulations described here open the way to a new paradigm for deflagration-to-detonation transition, there are issues to be resolved. Among these are the amplitude of the flame pressure pulsations and their dependence on the density ratio of fuel to product and on *Re* and hence the scale of the "system," in this case an expanding white dwarf. This DDT mechanism will be a challenge to model in thermonuclear flames because the range of scales that need to be resolved in degenerate matter from the burning scale to the scale of the turbulent flame width is significantly larger than for the reaction models used in terrestrial simulations. In addition, the supernova problem involves a multistage flame, including, for instance, carbon, oxygen, and silicon burning scales (Sect. 19.10). The astrophysical problem may be accessible for certain conditions and the basic mechanism may be amenable to construction of appropriate subgrid models. The current regime of simulations strongly suggests that the turbulent flame will itself intensify the turbulence as the deflagration moves outward in the expanding white dwarf, and that a detonation by means described here is nearly inevitable.

Models of SN Ia that most closely comport with observations require that the transition from subsonic to supersonic burning should occur at $\rho \gtrsim 10^7$ g cm^{-3}. This is approximately the condition for which the laminar flame slows and broadens, and estimated large-scale turbulence driven by Rayleigh–Taylor instability leads to a condition that $\lambda_G \sim \delta_{cond}$. At this point, the structure of the deflagration flame is likely to undergo some qualitative change, which may be close packing of the flame preheat regions, the formation of a galloping flame, shocks, and DDT. Note that while they may suggest similar conditions for DDT in SN Ia, the criterion that $\lambda_G \sim \delta_{cond}$ is qualitatively quite different from the requirement that the turbulent flame speed approach the CJ deflagration speed in the galloping flame. The transition from carbon burning to nuclear statistical equilibrium, producing iron-group elements, to burning only to intermediate-mass elements such as magnesium, silicon, sulphur, and calcium, also occurs at $\rho \gtrsim 10^7$ g cm^{-3}.

A possibility that has been considered is that the deflagration might simply accelerate, but not make the transition to a detonation. From observations alone, it is difficult to distinguish a deflagration that speeds up to $v_{def} \lesssim c_s$ from a DDT that produces $v_{det} \gtrsim c_s$. Both could yield an extended region of partially-burned matter as required by observations. The physics that would cause a deflagration to speed up suddenly is unclear, and it is not clear why acceleration of a deflagration would not proceed to a detonation.

While this chapter is primarily concerned with carbon combustion, it will also be important to address issues of helium detonations from this perspective, both for intrinsic understanding and because in some models of SN Ia carbon detonation specifically is triggered by helium shell detonation (Sect. 22.3.8). Helium is volatile (its nuclear reaction rates are very temperature sensitive) and it has a large specific energy release. Helium burning to NSE releases substantially more energy than does C/O burning, so if helium ignites under conditions of comparable density it might be prone to detonation. In practice, however, the conditions of degenerate helium ignition occur at substantially smaller density, $\sim 10^4$ to 10^5 g cm^{-3}. The density ratio from fuel to ash is still near unity, and the laminar flame speed is lower than for carbon at $\gtrsim 10^7$ g cm^{-3}, so the critical size of a turbulent flame brush in helium will be substantially larger than the helium flame thickness. The nature of the detonation kernel for helium in the context of detonation length scales (Sect. 19.10) merits more attention. Circumstances have been explored for which degenerate helium can undergo deflagration (Timmes 2000; Woosley and Kasen 2011).

19.12 Summary

This chapter has focused on the systematics involved in degenerate thermonuclear burning as it has been explored and elucidated primarily in the context of M_{Ch} models of SN Ia. Thus this material especially applies to carbon ignition in the degenerate cores of AGB stars, if that occurs (Chap. 8), or in white dwarfs in binary systems with nondegenerate companions (Chap. 22). Many of the principles discussed are, however, quite general and may apply to other relevant environments.

In the context of M_{Ch} models, carbon ignition occurs when the core grows to nearly M_{Ch}, the density and temperature climb, and the rate of carbon burning exceeds neutrino losses. Ignition leads to a smoldering phase during which the rate of nuclear input is balanced by neutrino and convective losses. With further compression, this control breaks down and dynamic carbon burning ensues. The convective Urca process may play an important quantitative role in setting the central density at the onset of dynamical burning, but it is unlikely to make a qualitative difference.

The two general modes of thermonuclear combustion are subsonic deflagration and supersonic, shock-driven, detonation. The basic properties of these modes of burning can be described by the equations of fluid dynamics and expressed in steady-state Chapman–Jouguet solutions. Although there is no proof from first principles, dynamical burning must not lead directly to detonation in M_{Ch} models, although direct initiation of detonations could be possible in other contexts.

In M_{Ch} models, the deflagration phase can have a variety of complex aspects. Central or somewhat offcenter ignition of degenerate carbon leads to the formation of a thin, wrinkled flame, the properties of which are shaped by electron conduction. Such a flame will be intrinsically susceptible to the Landau–Darrieus instability, but stabilized in the nonlinear limit. Propagation of the flame will naturally induce buoyancy-driven Rayleigh–Taylor instabilities and associated turbulence. Whether there are a variety of ignition kernels with a merger of flame bubbles, or a single ignition kernel leading to a single buoyant flame bubble, remains a matter of contention.

All models for typical SN Ia invoke detonations. An important but understudied feature of astrophysical detonations is that they are unstable. This aspect may affect the initiation and decay of detonations and the abundances left in their wake. In M_{Ch} models that begin with a phase of deflagrative burning, there must be a transition from deflagration to detonation. This transition is a challenge in both terrestrial and astrophysical contexts. Direct numerical simulations suggest that paradigms based on turbulent transport of passive scalars leading to the suggestion of a key role for distributed flames may not apply to turbulent flames where energy release in the burning is intrinsic to the physics. Rather, the turbulence may affect the preheat region and pack the flame without significantly perturbing the structure of the flame front itself. The formation of packed flame cusps, propagation of pressure waves and shocks, and the resulting nonlocal effects on the velocity field may lead naturally to unconfined DDT when the turbulence is sufficiently strong.

The nature of the deflagration, the DDT, and the detonation may depend on the metallicity, the precise mass of the white dwarf, the ignition density, rotation and even magnetic fields. Many important aspects of the propagation of thermonuclear flames, DDT, and propagation of the naturally unstable detonation remain to be understood theoretically. Observations may help to elucidate key aspects such as the density at which dynamical runaway ensues, the nature of the deflagration phase, the conditions under which DDT occurs, and the nature of the detonation phase. These issues can be raised in the context of M_{Ch} models, sub-Chandra models, double-degenerate models and other proposed models for SN Ia and related events where thermonuclear combustion is involved.

Chapter 20
Observational Properties

20.1 Introduction

SN Ia occur less frequently than core-collapse supernovae, but interest in these events is intense, not only for their intrinsic aspects, but because they contribute about half of the heavy-element abundances in the Universe, and especially because of their utility as distance indicators for cosmology (Sect. 25.2). Review articles on SN Ia include those of Howell (2011), Wang and Han (2012), Hillebrandt et al. (2013), and Maoz et al. (2014).

The observational properties of the majority of SN Ia are strikingly homogeneous. Nature has a way to produce titanic explosions having very similar characteristics. Nevertheless, peculiar SN Ia exist. Even among typical SN Ia, modern high-quality data reveal diversity, as well as correlations among the diverse properties.

This chapter is primarily concerned with the observational properties of the explosions themselves. Issues that bear on the nature of the presupernova binary systems, such as interactions of the supernova with CSM and a donor star, are considered in Chap. 21. Chapter 22 discusses proposed explosion mechanisms and the extent to which they are likely to comport with observation. This chapter, as well as Chaps. 21 and 22, are restricted to supernovae that traditionally would be classified as Type Ia. Events that appear to be SN Ia interacting strongly with CSM, and other events that may be explosions of or on white dwarfs, are considered in Chap. 23.

In addition to our case studies SN 2011fe (Sect. 20.2.1) and SN 2014J (Sect. 20.2.2), the many other well-observed, typical SN Ia include SN 1981B (Branch et al. 1983), SN 1989B (Wells et al. 1994), SN 1994D (Patat et al. 1996), SN 1996X (Salvo et al. 2001), SN 1998bu (Hernandez et al. 2000), SN 2003du (Stanishev et al. 2007), SN 2005cf (Wang et al. 2009c), and SN 2013dy (Pan et al. 2015a).

D. Branch, J.C. Wheeler, *Supernova Explosions*, Astronomy and Astrophysics Library, DOI 10.1007/978-3-662-55054-0_20

20.2 Case Studies

20.2.1 SN 2011fe

SN 2011fe (Fig. 1.1) was discovered by the Palomar Transient Factory in the nearby (about 6.4 Mpc) "Pinwheel" galaxy M101 (Nugent et al. 2011a). Estimates of the explosion time vary from only hours to a few days before the first detection (Nugent et al. 2011b; Piro and Nakar 2013; Mazzali et al. 2014). The early observations put a new focus on the physics of the early light curve, especially the possible *dark time* (Sect. 22.2) caused by drastic adiabatic cooling immediately following explosion, before ^{56}Ni decay reheats the ejecta to produce the luminous output. In addition, the proximity of SN 2011fe and the lack of significant interstellar extinction made it a special opportunity for detailed study of all phases of a typical SN Ia. SN 2011fe was reviewed by Chomiuk (2013) and Kasen and Nugent (2013).

Nondetection of the supernova in archival observations only 7 h before discovery led to estimates of the radius of the star that exploded ranging from $\lesssim$0.02 $R_\odot$ to $\lesssim$0.2 $R_\odot$. Together with the constraint that the mass of the star exceeded 0.5 $M_\odot$ (the M_{Ni} required to power the light curve), these radius constraints implied that the explosion was of a white dwarf or (entirely unlikely) a neutron star (Nugent et al. 2011b; Bloom et al. 2012). The uncertainty in the length of the dark time allows the possibility that the radius of the progenitor could have been somewhat larger than a white dwarf, perhaps the size of a nondegenerate carbon or helium star, but these also are improbable progenitors. Archival nondetections of X-ray, UV, and He II λ4686 emission imposed additional constraints on the white dwarf (Liu et al. 2012a; Graur et al. 2014a).

Multiwavelength nondetections of a companion star, both pre- and postexplosion, excluded the presence of a red giant, a subgiant, or a main-sequence star of more than about 4 $M_\odot$ (Li et al. 2011a). UV limits on reheating the ejecta by collision with a putative companion star constrained such a companion to have been closer than about 10^{11} cm from the explosion, again ruling out a red giant and even constraining a main-sequence star to have $\lesssim$1 $M_\odot$ (Brown et al. 2012a). Nondetections of relatively narrow ($\sim$1000 km s^{-1}) Hα emission in late-time spectra (Shappee et al. 2013b; Lundqvist et al. 2015; Graham et al. 2015a) constrained the mass of any slowly moving hydrogen in the deep layers of the ejecta, stripped from a companion star, to be low, $\lesssim$10^{-3} $M_\odot$, placing a significant constraint on the nature of the presupernova binary system (Sect. 21.6.4).

Limits on radio and X-ray emission from CSI showed that the environment was extremely clean (Horesh et al. 2012; Chomiuk et al. 2012; Margutti et al. 2012), ruling out not only a red giant, which would blow a wind, but also some other scenarios that would demand substantial CSM. Assuming a constant-density CSM, that density was inferred to be less than 6 cm^{-3}. Assuming, instead, CSM from a 100 km s^{-1} steady-state wind, the presupernova mass-loss rate was constrained to be $\lesssim$10^{-9} $M_\odot$ y^{-1}. High-resolution spectroscopy did not detect variability in the narrow Na I D lines, as seen in some SN Ia and attributed to CSM. In SN 2011fe, the narrow Na I lines presumably formed in ISM (Patat et al. 2013). Not all circumstellar environments of SN Ia are so clean (Sects. 21.6.5 and 23.2).

Optical and near-IR spectrophotometry and photometry showed that the photometric evolution was that of a typical SN Ia (Richmond and Smith 2012; Pereira et al. 2013; Tsvetkov et al. 2013; Munari et al. 2013; Fig. 20.1). In the B band, the light curve rose to a maximum brightness of $B = 10$ in 18 days. The light-curve decay parameter Δm_{15}, the decline in magnitudes in the B band during the first 15 days postmaximum, was 1.1, a typical SN Ia value. From 40 to 325 days postmaximum, the rate of decline was 1.43 magnitudes per 100 days. The peak absolute magnitude, for a distance of 6.4 Mpc, was $M_B = -19.10$ and the maximum

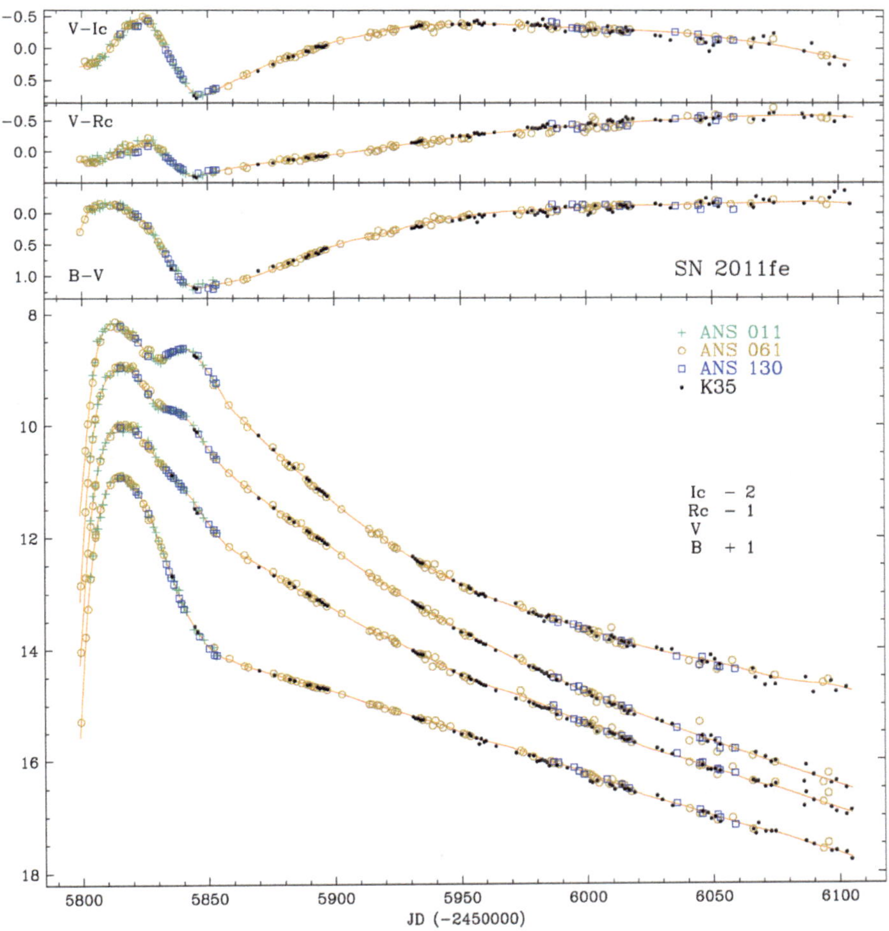

Fig. 20.1 Light curves (*bottom*) and colors (*top*) of SN 2011fe. Reprinted from New Astronomy, Vol. 20, "BVRI Lightcurves of Supernovae SN 2011fe in M101, SN 2012aw in M95, and SN 2012cg in NGC 4424" (Munari et al. 2013) with permission from Elsevier

UVOIR luminosity was 1.2×10^{43} erg s^{-1} (Pereira et al. 2013). Maximum brightness in the V and R bands occurred about 2 days later than in B, while the I band reached maximum 2 days earlier than B. The I-band light curve was double peaked, the second one being 28 days later and 0.45 magnitudes fainter than the first. In the near-IR, the J, H, and K bands also reached maximum a few days before B and were double peaked (Matheson et al. 2012). Ultraviolet photometry from *Swift* showed that emission in the UV was faint compared to the optical and that the UV bands reached maximum a few days before B (Brown et al. 2012a).

The photospheric-phase optical spectra of SN 2011fe also were those of a typical SN Ia (Parrent et al. 2012; Pereira et al. 2013; Figs. 20.2 and 20.3). Early spectra contained the characteristic broad P Cygni features of Si II, Ca II, S II, O I, and Mg II. Blueshifts of the photospheric absorption components decreased with time: that of Si II λ6355 was 14,000 km s^{-1} a few days after explosion and 10,500 km s^{-1} at maximum light.[1] Blends of Fe II lines developed soon after maximum. Postmaximum strengthening of Fe II lines relative to lines of IMEs was a consequence of decreasing temperature as well as of the photosphere receding into deeper, more iron-rich matter. A high-quality series of UV spectra also was obtained (Fig. 20.4). Analysis of *HST* spectra extending from the UV to the near-IR showed that the composition structure was stratified, with the ratio of IMEs to IGEs decreasing with depth (Mazzali et al. 2014; Fig. 20.5).

The earliest optical spectra contained high-velocity absorption features (HVFs) of O I, Si II, Ca II, and Fe II in the range 16,000–27,000 km s^{-1}, detached from the photosphere (Parrent et al. 2012; Pereira et al. 2013). HVFs are seen in many SN Ia (Sect. 20.4).

Until the time of maximum light, optical lines of C II were detected, providing evidence of a small amount of unburned fuel in the velocity range 9000–15,000 km s^{-1}. Near-IR spectra (Hsiao et al. 2013) showed that the C I 1.0693 μm line increased in strength until the time of maximum light before gradually declining. After an early rapid decline, the blueshift of the Mg II 1.0927 μm absorption remained constant at $\gtrsim$11,000 km s^{-1}. Because magnesium is a product of explosive carbon burning, but not explosive oxygen burning, this velocity is thought to be a measure of the inner boundary of the region in which carbon, but not oxygen, was burned (Wheeler et al. 1998; Marion et al. 2006).

Spectropolarimetry obtained prior to and shortly after maximum light revealed low continuum polarization indicative of a slightly asymmetric photosphere. Unlike most typical SN Ia, the continuum polarization rose somewhat toward the red, a behavior more typical of SN 1991bg-like events. SN 2011fe also showed higher, variable polarization in the strong lines, implying a time-dependent large-scale asymmetry in the line-forming layers (Milne et al. 2017).

The nebular phase of SN Ia, in which the optical spectrum is dominated by forbidden emission lines, often is regarded to begin within a month or so of

[1]A SYN++ synthetic-spectrum fit to a maximum-light spectrum of a typical SN Ia, and the contributions of various ions to the synthetic spectrum, is shown in Fig. 4.7.

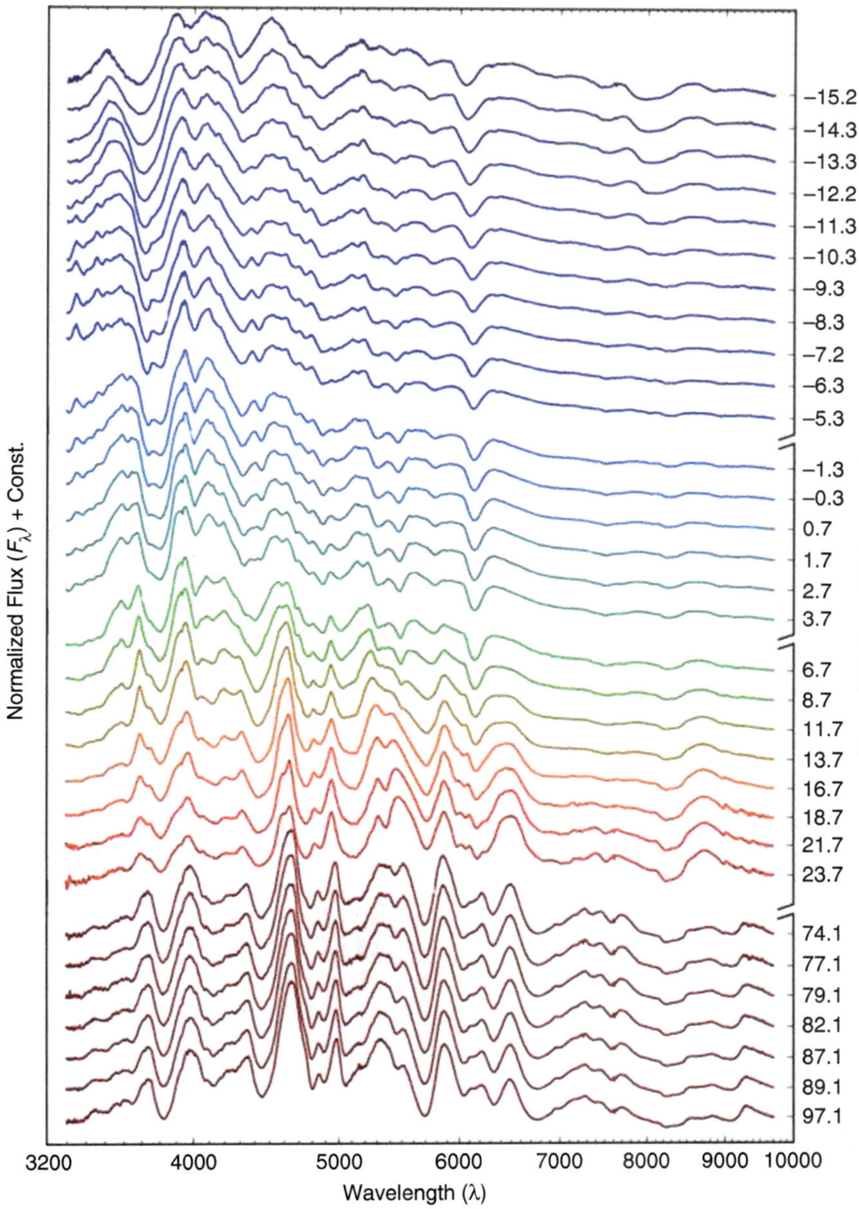

Fig. 20.2 Spectral evolution of SN 2011fe. Spectra are shifted vertically for clarity. For line identifications, see Fig. 20.3. From "Spectrophotometric time series of SN 2011fe from the Nearby Supernova Factory" (Pereira et al. 2013) reproduced with permission, © ESO

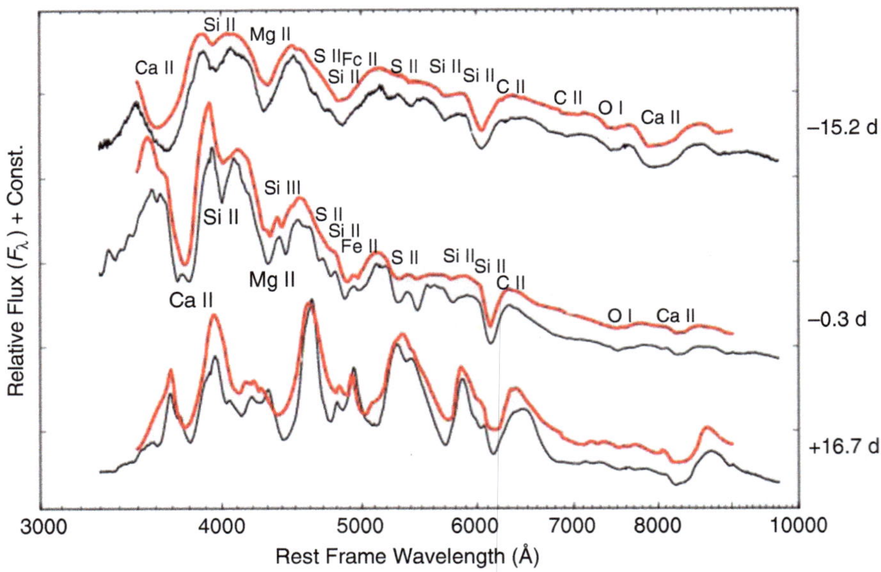

Fig. 20.3 SYNAPPS fits to spectra of SN 2011fe. Epochs are with respect to the time of maximum light in the B band and spectra are shifted vertically for clarity. In the day 16.7 synthetic spectrum, the P Cygni feature to the blue of 6000 Å is due to Na I, and blueward of 5000 Å the spectrum is dominated by overlapping lines of Cr II, Fe II, and Co II. From "Spectrophotometric time series of SN 2011fe from the Nearby Supernova Factory" (Pereira et al. 2013) reproduced with permission, © ESO

maximum light. In SN 2011fe, however, Fe II lines appeared to shape the blue part of the spectrum for at least a year (Black et al. 2016; Fig. 20.6). Thus the transition to the true nebular phase involves a complex mixture of permitted and forbidden lines, and the epoch at which the transition is complete is not clear. During the transition, line identifications, especially in the blue, are uncertain.

From a traditional forbidden-line analysis of optical spectra from 7 to 12 months postmaximum (Fig. 20.7), the inner layers of the ejecta were inferred to consist of 0.23 $M_\odot$ of ^{54}Fe surrounded by 0.47 $M_\odot$ of ^{56}Ni, the bulk of which was contained within 8500 km s^{-1} (Mazzali et al. 2015). Features attributed to forbidden emission lines of singly-ionized IGEs ([Fe II] and [Co II]) were more blue shifted than those attributed to doubly-ionized IGEs ([Fe III] and [Co III]; McClelland et al. 2013; Taubenberger et al. 2015; Lundqvist et al. 2015); an interpretation involving an asymmetric explosion will be discussed in Sect. 22.3.4. Between 7 and 34 months, emission peaks redshifted by about 100 km s^{-1} per month (Taubenberger et al. 2015; Graham et al. 2015a). By 19 months, the previously strong emission feature attributed to [Fe III] λ4701 had disappeared while [Fe II] emissions persisted, reflecting a decreased ionization level. By 33 months, the optical spectrum was dominated by permitted Fe I emission and the UV was still optically thick to permitted lines (Fransson and Jerkstrand 2015). Photometric observations at 30

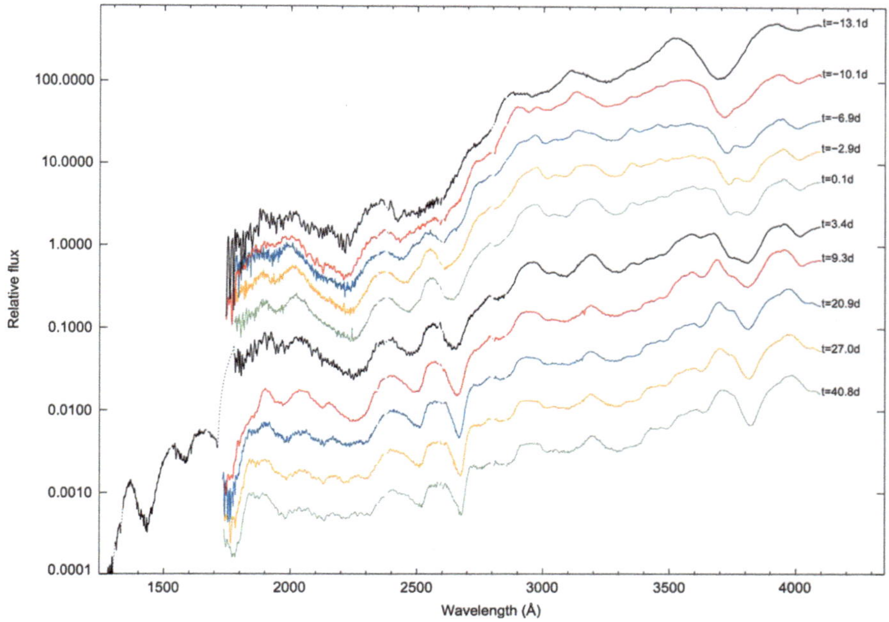

Fig. 20.4 Spectral evolution of SN 2011fe in the UV. Epochs are with respect to the time of maximum light in the B band and spectra are shifted vertically for clarity. The strong absorptions near 2650 and 3800 Å in postmaximum spectra are due to Mg II and Ca II, respectively. Figure 1 "Hubble Space Telescope spectra of the Type Ia supernova SN 2011fe: a tail of low-density, high-velocity material with $Z < Z_{\odot}$" from Mazzali et al. (2014), by permission of Oxford University Press on behalf of the Royal Astronomical Society

months indicated that even at that late date, trapping of positrons from the decay of ^{56}Co was complete (Kerzendorf et al. 2014b), suggesting, perhaps, that an expected infrared catastrophe (Sect. 5.5.1) had not yet occurred. Redistribution of UV emissivity resulting from nonthermal excitation and ionization by multiple scatterings and recombination may, however, have prevented the optical and near-IR emission from dropping catastrophically, even if the ejecta had cooled to ∼100 K (Fransson and Jerkstrand 2015).

20.2.2 SN 2014J

SN 2014J, discovered by Fossey et al. (2014) in the nearby (3.5 Mpc) irregular starburst galaxy M82, was perhaps the nearest known SN Ia since Kepler's SN 1604 in our Galaxy. Discovery occurred 7 days after the time of first light, which was well determined from archival pre- and post-explosion observations (Zheng et al. 2014; Goobar et al. 2015). Extinction was high, $A_V \lesssim 2.0$, all or nearly all attributed

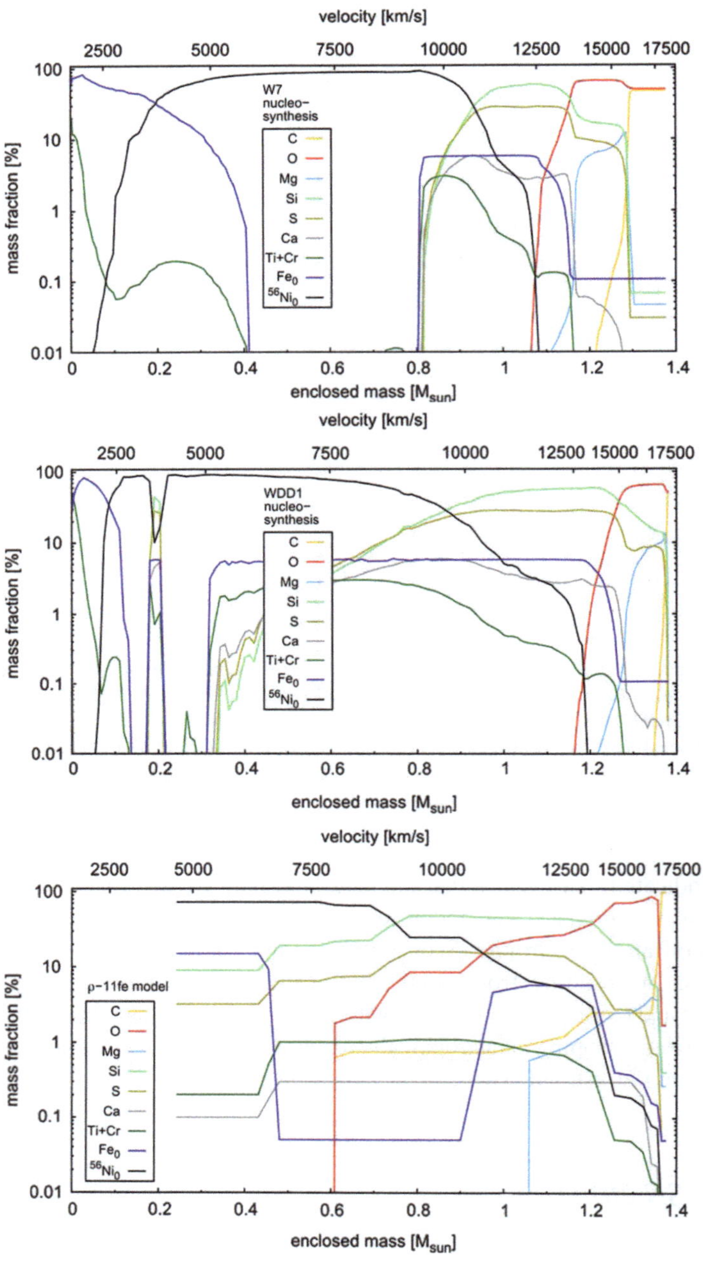

Fig. 20.5 The composition structure of SN 2011fe inferred from spectroscopic analysis (*bottom panel*) and the composition structures of two explosion models as calculated by Iwamoto et al. (1999): model W7 (*top panel*; Sect. 22.3.2) and a delayed-detonation model (*middle panel*; Sect. 22.3.4). Figure 10 "Hubble Space Telescope spectra of the Type Ia supernova SN 2011fe: a tail of low-density, high-velocity material with $Z < Z_\odot$" from Mazzali et al. (2014), by permission of Oxford University Press on behalf of the Royal Astronomical Society

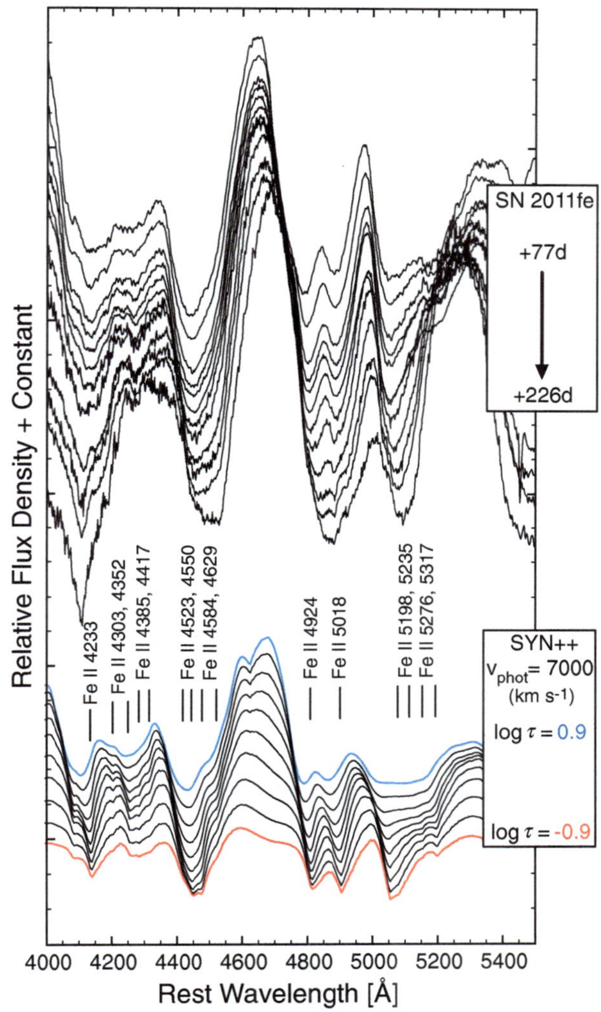

Fig. 20.6 Spectra of SN 2011fe from 77 to 226 days after maximum light and a set of SYN++ spectra with a range of optical depths of the Fe II reference line, $\lambda 5018$. Figure adapted from Black et al. (2016) by J. Parrent

to interstellar dust with a steep extinction curve (Sect. 20.6), $R_V \simeq 1.4$ (Amanullah et al. 2014; Kawabata et al. 2014; Brown et al. 2015b; Patat et al. 2015; Marion et al. 2015). Atomic, molecular, and diffuse-interstellar-band absorptions originating in M82 provided valuable information on its ISM (Sect. 25.4.1)

Archival pre-explosion observations constrained the nature of the progenitor system. Nondetection of X-rays excluded a single-degenerate binary system containing a white dwarf of radius comparable to that of a near-Chandrasekhar-mass white dwarf and a mass accretion rate in the range in which stable nuclear burning

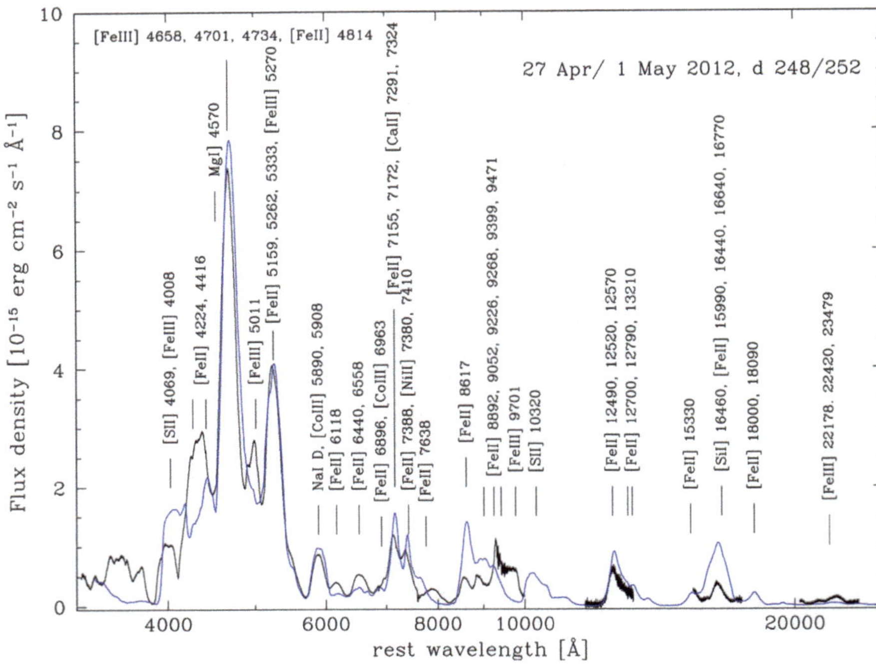

Fig. 20.7 A synthetic spectrum consisting of forbidden-line emission, generated with a nebular-phase code (*blue*), is compared to a spectrum of SN 2011fe obtained about 250 days after explosion (*black*). Note the caveat in the text regarding line identifications in late-time spectra, especially in the *blue*. Figure 5 "Nebular spectra and abundance tomography of the Type Ia supernova SN 2011fe: a normal SN Ia with a stable Fe core" from Mazzali et al. (2015), by permission of Oxford University Press on behalf of the Royal Astronomical Society

can occur (Nielsen et al. 2014a). Nondetection of near-UV to near-IR emission excluded the presence of a bright red giant (Kelly et al. 2014b). As for SN 2011fe, nondetection of postexplosion X-ray (Margutti et al. 2014b) and radio (Pérez-Torres et al. 2014) emission showed that the circumstellar environment was very clean: particle density $\lesssim 1\,\mathrm{cm}^{-3}$ for constant-density, or a pre-explosion mass-loss rate $\lesssim 10^{-9}\,\mathrm{M}_\odot\,\mathrm{y}^{-1}$ for an assumed $100\,\mathrm{km\,s}^{-1}$ steady-state wind. High-resolution spectroscopy did not reveal variability in the narrow Na I D lines (Goobar et al. 2014; Foley et al. 2014a; Jack et al. 2015b). Variability of narrow K I lines was observed and tentatively interpreted in terms of CSM (Graham et al. 2015c; Soker 2015), but see Maeda et al. (2016). Nondetections of narrow Hα emission in late-time spectra (Lundqvist et al. 2015) constrained the mass of any slowly-moving hydrogen to be low, $\lesssim 10^{-2}\,\mathrm{M}_\odot$.

Photometry with the *Kilodegree Extremely Little Telescope (KELT*; Siverd et al. 2015) provided an unprecedented high-cadence, high S/N-ratio light curve, in a band approximating the R band, from before explosion until 80 days later (Fig. 20.8). Early prediscovery data, just a few days after explosion, suggested the presence of

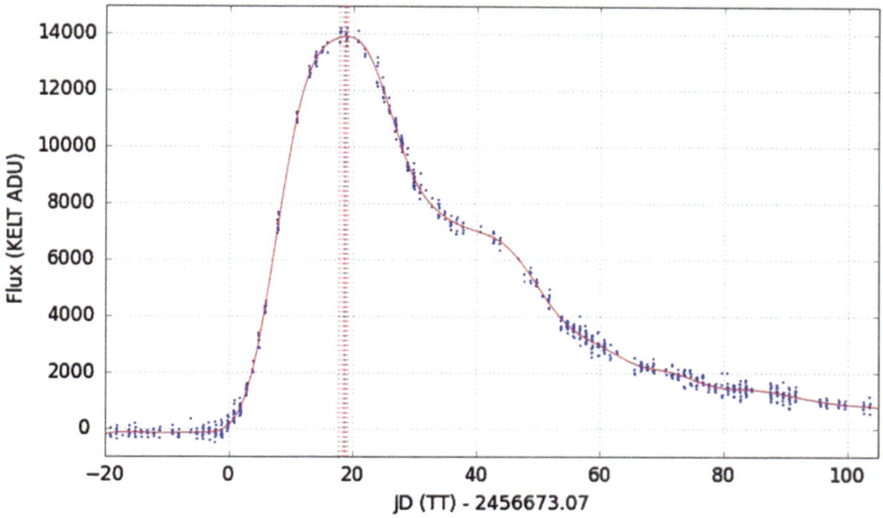

Fig. 20.8 High-cadence light curve of SN 2014J. The flux is on a linear scale. From "Observations of the M82 SN 2014J with the Kilodegree Extremely Little Telescope" (Siverd et al. 2015). © AAS. Reproduced with permission

radioactive material in the outer layers of the ejecta or the cooling of shock-heated material of large dimensions, $\gtrsim 1\ R_\odot$, possibly reflecting interaction between the ejecta and nearby CSM (Goobar et al. 2015).

The early light curves in the UV, optical, near-IR, and mid-IR were those of a typical SN Ia (Foley et al. 2014a; Amanullah et al. 2014; Tsvetkov et al. 2014; Marion et al. 2015; Brown et al. 2015b; Johansson et al. 2017). In the B band, the light curve rose to a maximum brightness of 11.7 in 18 days. The postpeak decline parameter was $\Delta m_{15} = 1.1$. The peak absolute magnitude, for a distance of 3.5 Mpc, was estimated to be $M_B \simeq -19.2$, depending on the adopted extinction.

Photospheric-phase optical (Fig. 20.9), UV, and near-IR spectra (Goobar et al. 2014; Foley et al. 2014a; Marion et al. 2015; Brown et al. 2015b; Ashall et al. 2014; Vacca et al. 2015; Jack et al. 2015b) were similar to those of SN 2011fe, albeit with mildly higher expansion velocities and stronger HVFs. Analysis of optical spectra near maximum light revealed a stratified composition structure (Ashall et al. 2014; Marion et al. 2015), similar to that inferred for SN 2011fe (Sect. 20.2.1).

The presence of a C II $\lambda 6580$ feature was not established, perhaps because of the lack of spectra during the first week after explosion together with the presence of diffuse interstellar bands in the relevant wavelength range. The C I 1.0693 μm line appeared to be present near maximum light (Marion et al. 2015).

Near-IR spectra obtained 1–2 months postmaximum were dominated by permitted lines of Co II and Fe II (Friesen et al. 2014; Vacca et al. 2015; Fig. 20.10). Mid-IR spectra (8–13 μm) observed from 1 to 4 months postmaximum contained a strong line of [Co III] at 11.888 μm (Telesco et al. 2015; Fig. 20.11). In

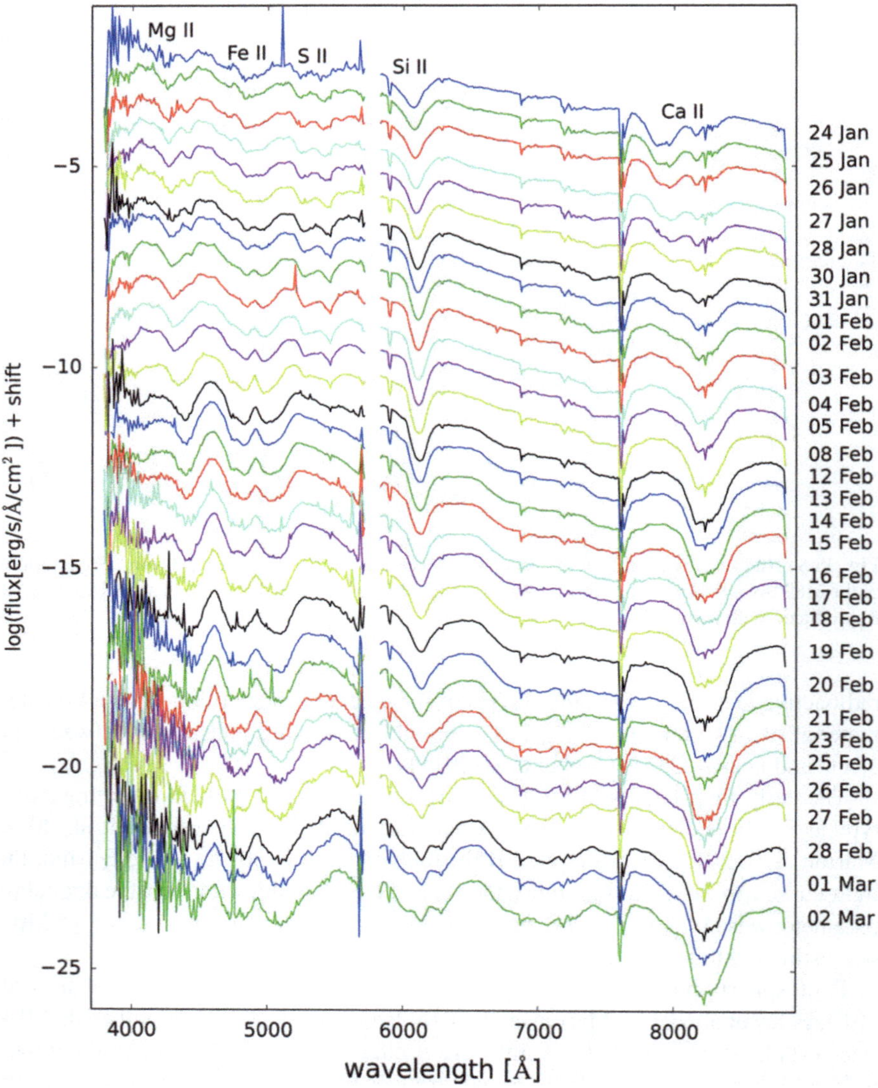

Fig. 20.9 High-cadence, high-resolution spectra of SN 2014J. Figure 1 "Time series of high-resolution spectra of SN 2014J observed with the TIGRE telescope" from Jack et al. (2015b), by permission of Oxford University Press on behalf of the Royal Astronomical Society

10-month spectra, as in SN 2011fe, emissions attributed to singly-ionized IGEs were more highly Doppler-shifted—but in this case to the red—than those attributed to doubly-ionized IGEs (Lundqvist et al. 2015). The near- and mid-IR spectra were consistent with predictions based on Chandrasekhar-mass delayed-detonation models (Sect. 22.3.4).

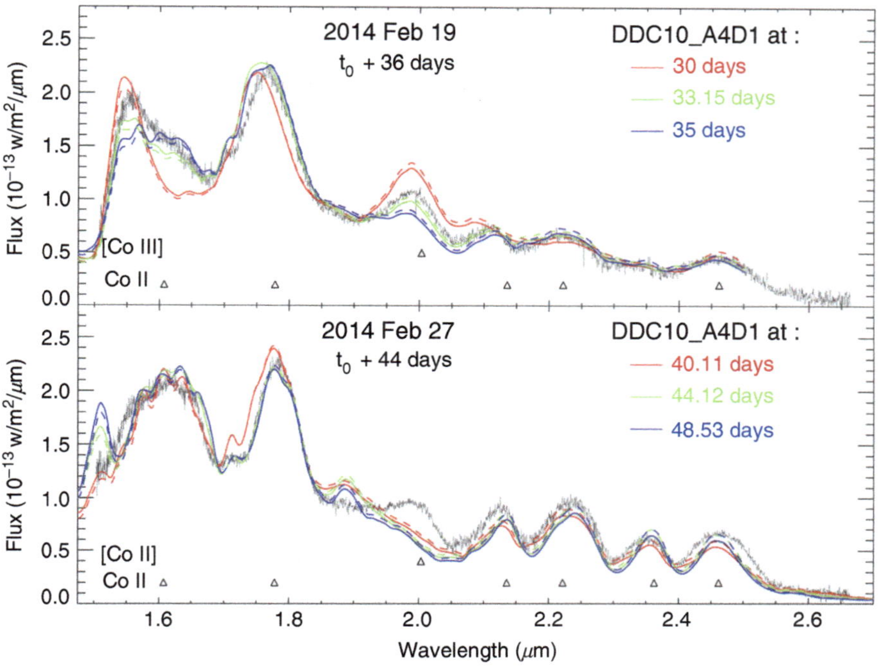

Fig. 20.10 Near-IR spectra of SN 2014J compared to CMFGEN synthetic spectra based on a delayed-detonation explosion model (Sect. 22.3.4), calculated for three different times past explosion. In the synthetic spectra, the features are primarily due to Co II lines. From "Observations of Type Ia Supernova 2014J with FLITECAM/SOFIA" (Vacca et al. 2015). © AAS. Reproduced with permission

SN 2014J was the first, and at this writing the only, SN Ia to be detected in γ-rays, by observations with *INTEGRAL*. Even for this "nearby" event, however, the signal was far below the instrumental background, so the data were limited and noisy. As predicted for plausible SN Ia explosion models that eject about 0.5 $M_{\odot}$ of ^{56}Ni (The and Burrows 2014), the ^{56}Co lines at 847 and 1238 keV rose to a maximum about 80–100 days after explosion (Churazov et al. 2014, 2015; Diehl et al. 2015). An interpretation of the ^{56}Ni line at 158 keV near maximum light invoked a 0.03–0.08 $M_{\odot}$ plume of ^{56}Ni ejected at $\gtrsim 30,000\,\mathrm{km\,s^{-1}}$ (Isern et al. 2016). [For a different interpretation, see Diehl et al. (2014).]

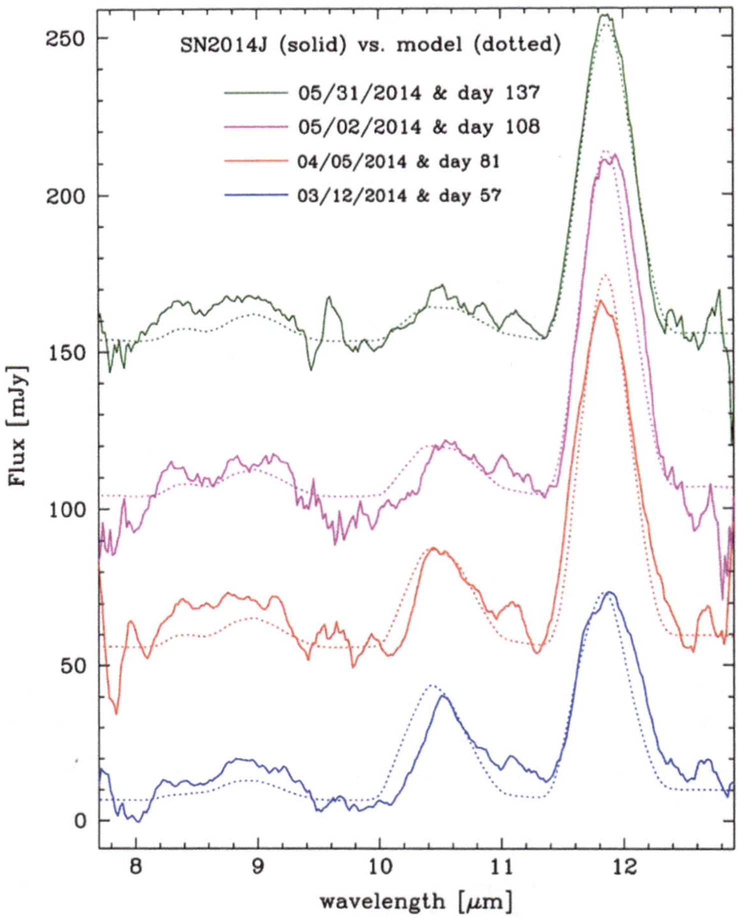

Fig. 20.11 Mid-IR spectra of SN 2014J compared to synthetic spectra based on a delayed-detonation explosion model (Sect. 22.3.4). In the synthetic spectra, the emissions near 10.5 and 11.9 μm are mainly produced by [Co II] 10.523 μm and [Co III] 11.888 μm, respectively. From "Mid-IR Spectra of Type Ia SN 2014J in M82 Spanning the First 4 Months" (Telesco et al. 2015). © AAS. Reproduced with permission

20.3 Homogeneity

The high degree of homogeneity of a substantial fraction of SN Ia deserves emphasis. Examining photospheric-phase spectral line features is a good way to assess homogeneity and diversity, because strengths and blueshifts of lines (in the supernova rest frame) are not affected by distance or extinction. In the optical, typical SN Ia undergo a standard pattern of spectroscopic evolution, with the degree of homogeneity becoming increasingly impressive with time (Fig. 20.12).

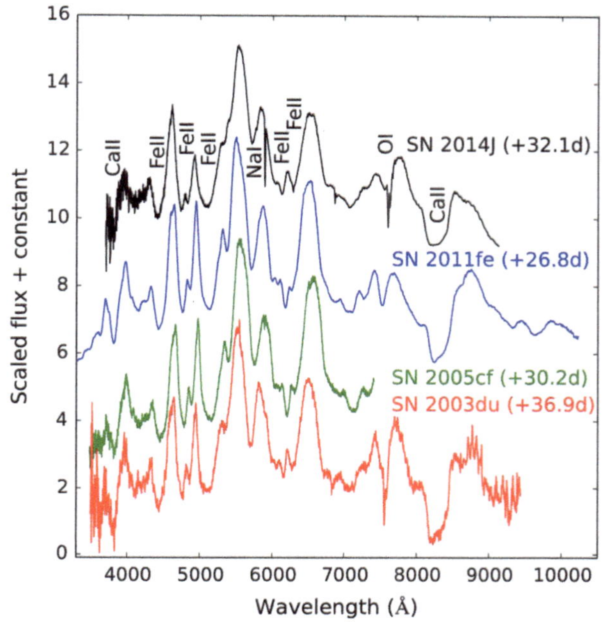

Fig. 20.12 An example of the spectroscopic homogeneity among four typical SN Ia. Epochs are with respect to the time of maximum light. The spectra have been shifted vertically for clarity, and the very narrow absorptions near 7600 Å in SN 2014J and SN 2003du are telluric. Figure 15 "Optical and NIR observations of the nearby type Ia supernova SN 2014J" from Srivastav et al. (2016), by permission of Oxford University Press on behalf of the Royal Astronomical Society

The standard pattern of spectroscopic evolution in the near-IR was shown by Marion et al. (2009), Gall et al. (2012), and Hsiao et al. (2013). The homogeneity of typical SN Ia means that the explosions cannot be very asymmetric, with strong dependence on aspect angle. This is an important constraint on explosion models (Chap. 22). Photospheric-phase SN Ia spectra were reviewed by Parrent et al. (2014).

Late-time optical spectra of typical SN Ia also show a high degree of homogeneity in terms of the relative strengths of the spectral features (Fig. 20.13). As discussed in Sect. 20.2.1, line identifications during the lengthy transition to the true nebular phase are uncertain. After the transition, spectral features are attributed to blends of collisionally-excited or recombination-driven forbidden emission lines of iron-group ions, e.g., [Fe III] λ4701, [Co III] λ5893, [Fe II] λ7155, and [Ni II] λ7378 (Kirshner and Oke 1975; Bowers et al. 1997; Maeda et al. 2010c; Friesen et al. 2017).

The shapes of rest-frame light curves also are independent of distance and (to first order) extinction, and they too are impressively homogeneous (Fig. 20.14). The typical SN Ia shows a rise to maximum luminosity in the B band in about 17 or 18 days (Ganeshalingam et al. 2011), a postpeak decline of 1.1 magnitudes in the first 15 days and 3 magnitudes in the first 40 days, followed by a nearly linear tail that declines by 1.4 magnitudes per 100 days.

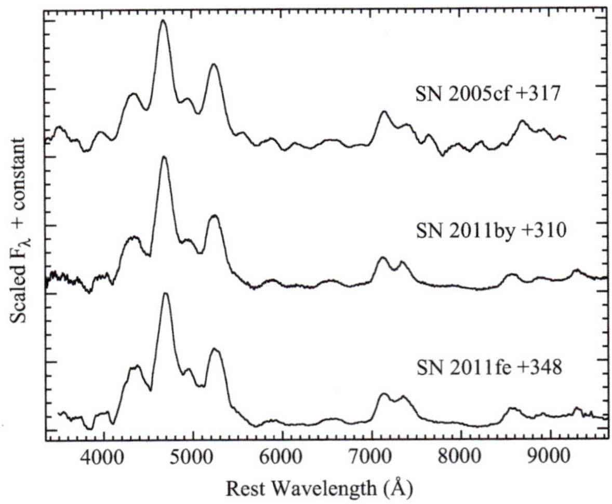

Fig. 20.13 An example of the spectroscopic homogeneity of late-time SN Ia spectra. Figure courtesy of J. Silverman

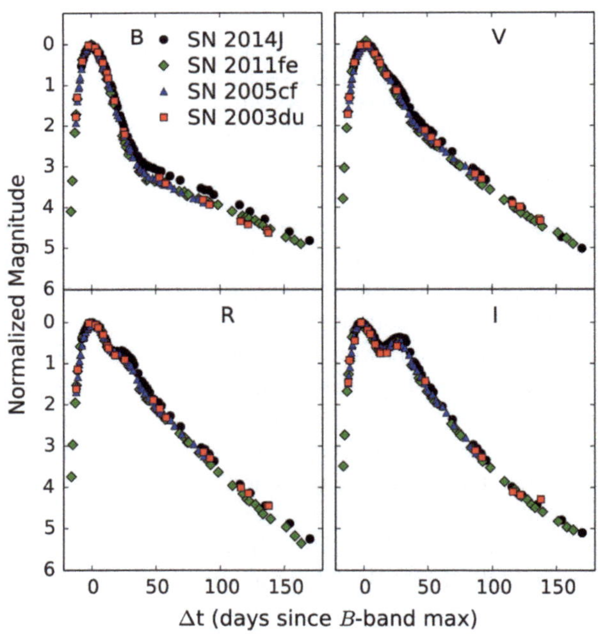

Fig. 20.14 The light-curve-shape homogeneity among the four SN Ia of Fig. 20.12. The light curves are normalized to the peak in each band. Figure 4 "Optical and NIR observations of the nearby type Ia supernova SN 2014J" from Srivastav et al. (2016), by permission of Oxford University Press on behalf of the Royal Astronomical Society

Broad-band colors are subject to reddening by dust in our Galaxy (usually a small effect, and correctable) and in the host galaxy (more of a problem; Sect. 20.6). Until a month postmaximum, colors become redder because decreasing temperature causes the underlying pseudo continuum to cool and the line opacity in the blue, especially from Fe II lines, to increase. A subsequent highly-homogeneous linear decrease of 0.7 magnitudes in $B - V$ from 1 to 3 months after maximum, often referred to as the *Lira relation* (Lira 1995; Lira et al. 1998; Burns et al. 2014; Fig. 20.15), can be used together with $B - V$ observations during this phase to estimate the interstellar extinction of a SN Ia (Phillips et al. 1999; Folatelli et al. 2010).

In order to assess absolute-magnitude distributions, differences in distances must be taken into account. For a sample of 137 SN Ia discovered by the Sloan Digital Sky Survey, with relative distances determined from host-galaxy redshifts, the distribution of M_B, corrected for extinction with a standard Galactic value of $R_B = 4.1$, was found to be approximately Gaussian with FWHM of only 0.24 magnitudes. An even smaller value was obtained by allowing the value of R_B to be lower (Yasuda and Fukugita 2010). The absolute-magnitude scatter can also be seen by plotting SN Ia in a Hubble diagram—SN Ia peak apparent magnitude plotted against distance. Figure 20.16 shows Hubble diagrams in six bands, from B in the optical to H in the near-IR. Ideally, if the SN Ia were perfect *standard candles*, and the plotted data were error-free, there would be no scatter. The data, however, are not

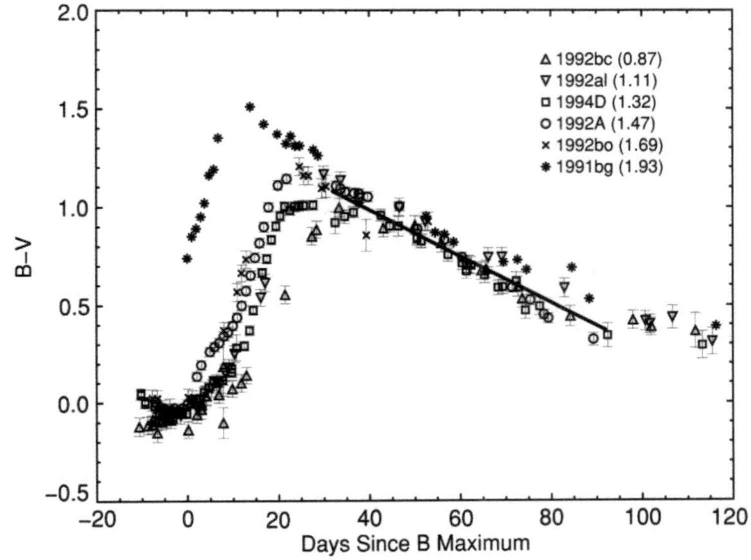

Fig. 20.15 The $B - V$ evolution for six SN Ia with little reddening by dust in their host galaxies. Values of Δm_{15} are given in *parentheses*. The line is the original relation of Lira (1995). From "The Reddening-free Decline-rate Versus Luminosity Relationship for Type Ia Supernovae" (Phillips et al. 1999). © AAS. Reproduced with permission

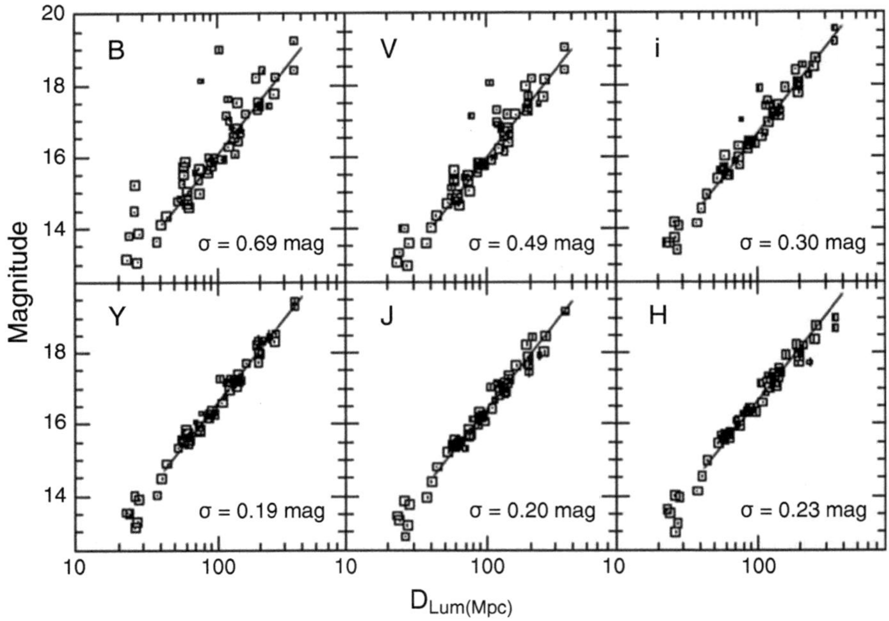

Fig. 20.16 Hubble diagrams for SN Ia. Magnitudes are corrected for extinction in the Galaxy, but not in the host galaxies. From "Near-infrared Properties of Type Ia Supernovae" (Phillips 2012), reproduced with permission

error-free: apparent magnitudes have been corrected for Galactic extinction, but not for extinction in the host galaxy; peculiar velocities of the host galaxies cause errors in distances inferred from redshifts; observational errors in apparent magnitudes are not negligible, especially in the near-IR; and in the B and V bands, the dispersion is inflated by a few dim SN Ia that probably suffer heavy host-galaxy extinction. Thus the standard deviations given in the figure are upper limits to the intrinsic values. In the near-IR, even with the sources of error just mentioned, SN Ia appear to be excellent standard candles (Barone-Nugent et al. 2012; Kattner et al. 2012; Friedman et al. 2015; Dhawan et al. 2015; Stanishev et al. 2015).

A sample of UVOIR light curves is shown in Fig. 20.17. The typical bolometric rise time is 18 days. For a Hubble constant of 70 km s^{-1} Mpc^{-1}, the peak luminosity is $\sim 1.5 \times 10^{43}$ erg s^{-1}, corresponding to an instantaneous radioactivity luminosity of an initial $M_{Ni} \simeq 0.7$ M$_\odot$ (Sect. 5.4.2). The typical time-integrated luminosity is $\gtrsim 2 \times 10^{49}$ ergs, small compared to a typical ejecta kinetic energy of $\sim 10^{51}$ ergs (1 B).

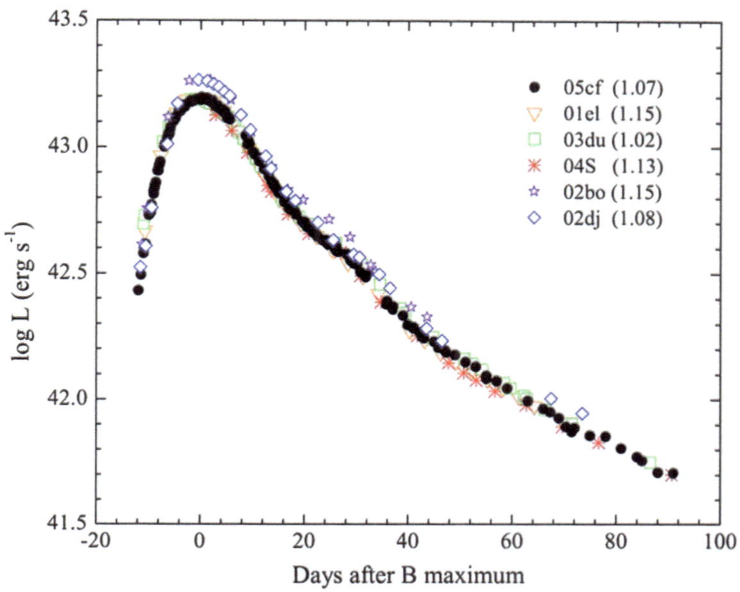

Fig. 20.17 A comparison of the bolometric light curves of six SN Ia. The absolute magnitudes depend on estimates of distance and extinction by dust. From "The Golden Standard Type Ia Supernova 2005cf: Observations from the Ultraviolet to the Near-Infrared Wavebands" (Wang et al. 2009c). © AAS. Reproduced with permission

20.4 Diversity and Correlations

Substantial departures from homogeneity were noted in the cases of SN 1984A (Branch 1987; Branch et al. 1988) and SN 1986G (Phillips et al. 1987) and became widely recognized with the discovery of SN 1991T and SN 1991bg (Fig. 20.18). These two events were strikingly peculiar, but in opposite ways. The former had a broad light-curve peak, overluminosity, and a high temperature that produced conspicuous Fe III features in its near-maximum optical spectra (Filippenko et al. 1992b; Phillips et al. 1992; Sasdelli et al. 2014). The latter had a narrow light-curve peak, underluminosity, and a low temperature that produced conspicuous Ti II features in its spectra (Filippenko et al. 1992a; Leibundgut et al. 1993; Taubenberger et al. 2008).

Because other events were soon found to have properties that fell between these two extremes and the most typical events, SN Ia began to be considered as a single-parameter sequence. Maximum luminosity was shown to correlate with Δm_{15} (the *width-luminosity relation* or *Phillips relation*; Pskovskii 1977; Phillips 1993; Phillips et al. 1999; Fig. 20.19), and with *stretch*, a measure of the width of the light-curve peak, from premaximum to postmaximum (Goldhaber et al. 2001). Values of Δm_{15} and stretch for SN Ia range from about 0.8 to 1.9 and 0.8 to 1.2, respectively, and the two parameters are well correlated (inversely). Maximum luminosity also

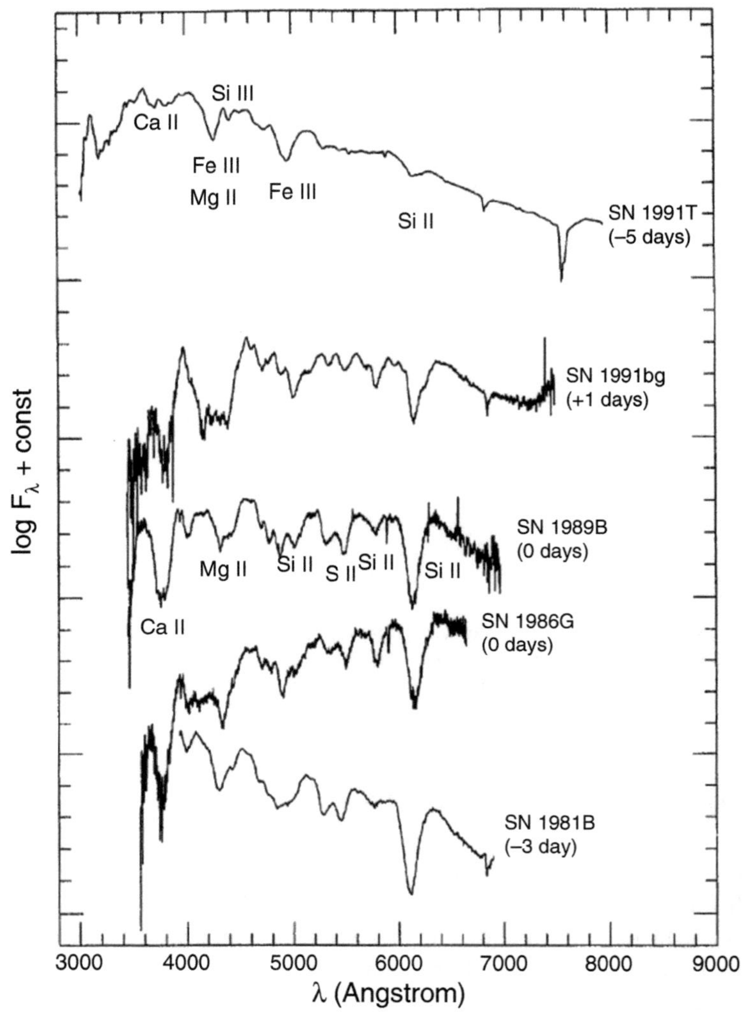

Fig. 20.18 Spectra of the peculiar SN 1991T, SN 1991bg, and SN 1986G (a less extreme version of SN 1991bg) compared to those of typical Type Ia SN 1989B and SN 1981B. The spectra of SN 1989B and SN 1986G are reddened by dust. Epochs are with respect to maximum light in the V band and spectra are shifted vertically for clarity. From "SN 1991bg — A type Ia supernova with a difference" (Leibundgut et al. 1993). © AAS. Reproduced with permission

correlates with colors at maximum (brighter events are bluer; Tripp 1998) as well as with $R(Si\ \mathrm{II})$ (Nugent et al. 1995), the ratio near maximum light of the depths of absorption features near 5750 and 6100 Å, the latter always and the former usually attributed to Si II. The Δm_{15} and $R(Si\ \mathrm{II})$ parameters are correlated. When SN Ia are regarded as a single-parameter sequence, an observed light-curve parameter such as Δm_{15} or stretch usually is used as the empirical ordering parameter. A primary

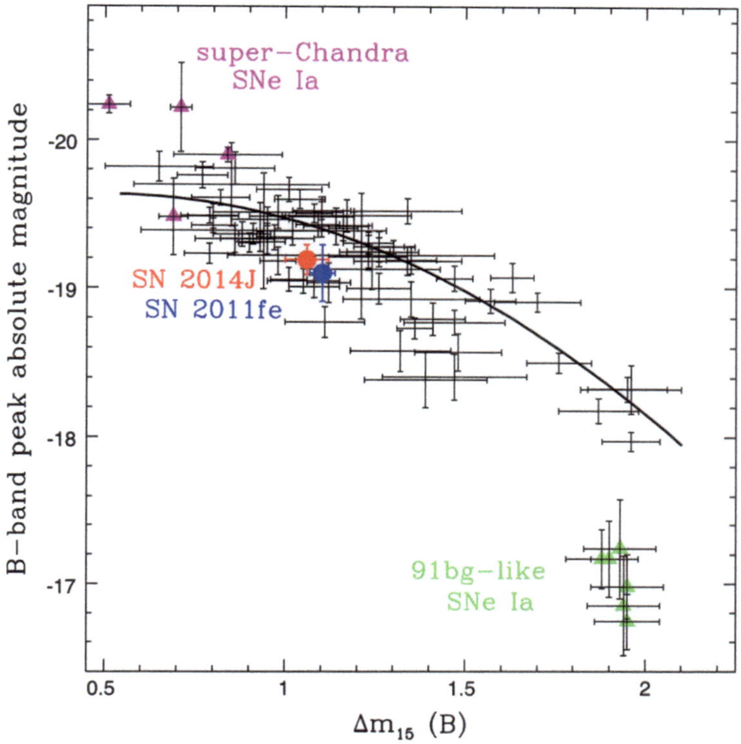

Fig. 20.19 Relation between M_B and Δm_{15}. The line is the relation of Phillips et al. (1999). Events designated super-Chandra are among those that we refer to as substantially super-Chandrasekhar (Sect. 20.7). Figure 6 "SN 2014J at M82 — I. A middle-class Type Ia supernova by all spectroscopic metrics" from Galbany et al. (2016), by permission of Oxford University Press on behalf of the Royal Astronomical Society

physical parameter along the sequence is M_{Ni}, the initial mass of radioactive ^{56}Ni, the decay of which powers the light curve.

SN Ia that undergo more nuclear fusion and make more ^{56}Ni should be more luminous and have more kinetic energy. If all SN Ia ejected 1.4 $M_\odot$ and had the same opacity, more energetic and luminous SN Ia should have shorter diffusion times (Sect. 5.3) and thus narrower light curves, but the opposite is observed. A resolution of this apparent paradox may (among other possibilities; Sect. 22.3.4) be that ejected mass, M_{ej}, is correlated with M_{Ni} (Stritzinger et al. 2006; Scalzo et al. 2014a). With M_{Ni} inferred from maximum brightness and then γ-ray optical depth and hence M_{ej} inferred from the shape of the ^{56}Co-powered tail of the bolometric light curve, the derived distribution of M_{ej} was strongly peaked near 1.4 $M_\odot$, but a low-mass tail of sub-Chandrasekhar events comprising 25–50% of SN Ia extended down to 0.9 $M_\odot$. (In this analysis, about 1% of SN Ia were inferred to be substantially super-Chandrasekhar; Sect. 20.7.) Low M_{ej} would shorten the diffusion time and

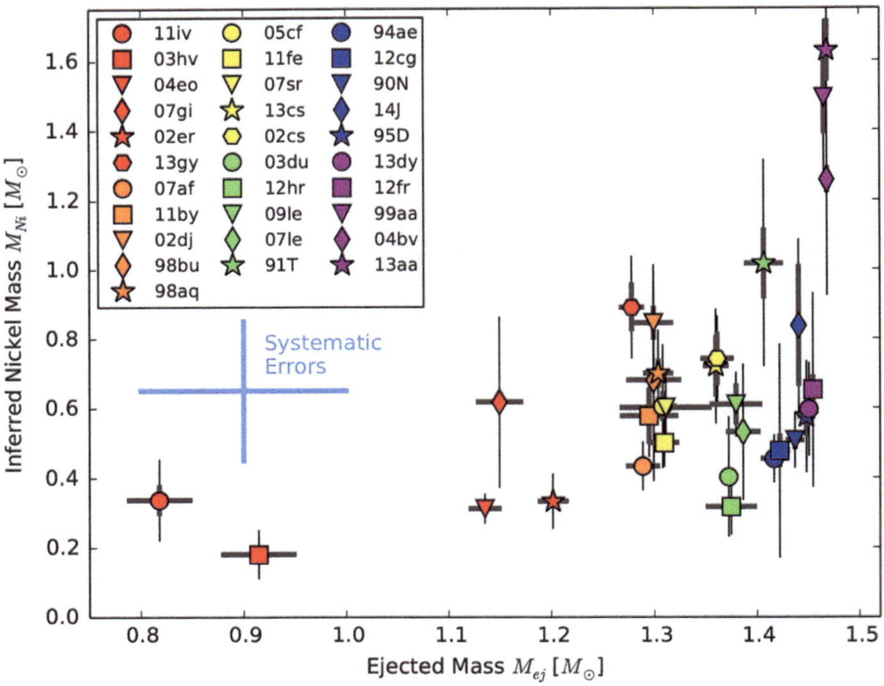

Fig. 20.20 Values of M_{Ni} inferred from [Co III] λ5893 emission in the nebular phase versus values of M_{ej} inferred from the width of light-curve peaks. The high values of M_{Ni} for SN 1999aa and SN 2013aa were attributed by the authors to line blending and distance uncertainty, respectively. Figure 11 "Measuring nickel masses in Type Ia supernovae using cobalt emission in nebular-phase spectra" from Childress et al. (2015), by permission of Oxford University Press on behalf of the Royal Astronomical Society

contribute to the rapid postpeak decline of subluminous, low-M_{Ni} events and help to resolve the paradox (Scalzo et al. 2014c). A correlation between M_{ej} and M_{Ni} as inferred from the strength of the [CO III] λ5893 emission in the nebular phase was obtained by Childress et al. (2015; Fig. 20.20). Mechanisms that could lead to sub-Chandrasekhar mass ejection are discussed in Chap. 22.

The photometric behavior in the double-peaked near-IR bands (Sect. 20.2.1) fits, at least loosely, into a single-parameter description. Although the luminosities and rise times of the first peaks are quite homogeneous, the epochs at which the second peaks occur are correlated with Δm_{15} (Dhawan et al. 2015). The second peaks are the result of evolving line opacity versus wavelength, caused by a decreasing ionization level of iron-group elements, especially iron and cobalt (Sect. 22.3.4). An observed, strengthening emission feature attributed to a blend of high-excitation (11 eV) lines of Fe II near 7500 Å was proposed as the direct cause of the second peak in the I band (Jack et al. 2015a).

Extensive investigations of observational diversity among large samples of SN Ia established additional correlations among observables (Blondin et al. 2012; Silverman et al. 2012c,b; Folatelli et al. 2013; Maguire et al. 2014).

In nebular spectra of typical SN Ia, the emission lines of [Fe II] λ7155 and [Ni II] λ7378 tend to be persistently both blueshifted as in SN 2011fe (Sect. 20.2.1) or both redshifted as in SN 2014J (Sect. 20.2.2). An interpretation in terms of Doppler shifts of these lines is discussed in Sect. 22.3.4. Diversity of nebular spectra is most obvious in SN 1991bg-likes; emission lines are narrower than in typical SN Ia and [Ca II] $\lambda\lambda$7291,7323 is more prominent, perhaps owing to lower temperature (Silverman et al. 2013a).

At early times, continuum polarization of typical SN Ia is lower than that of core-collapse supernovae, consistent with nearly spherical photospheres, and no preferred axis of orientation is readily discerned (Maund et al. 2013b). This is consistent with the overall homogeneity of typical SN Ia. Polarization in the Si II and Ca II features, however, can be substantial. Line polarization seems to correlate positively with Δm_{15}, although there are notable outliers (Wang et al. 2007a; Patat et al. 2012). Postmaximum line polarization generally becomes low, but not always so in the Ca II lines (Patat et al. 2009; Zelaya et al. 2013).

Despite the discussion above, the diversity is more complex than can be described by a single-parameter sequence. Most obviously, some events such as SN 1984A are spectroscopically typical in the sense that the usual features are present, but the features are unusually broad and blueshifted. In addition, correlations between any two observables appear to include intrinsic scatter. More subtle indications of complexity include (1) that the duration of the very early light curve, before the luminosity has reached half its maximum value, does not correlate with the subsequent light-curve shape (Firth et al. 2015), and (2) that the light curves of events having the same values of Δm_{15} can differ at times $\lesssim$5 days before and $\gtrsim$30 days after maximum light (Höflich et al. 2010).

At early times, nearly all typical SN Ia (but not SN 1991bg-likes) observed 5 or more days before maximum have HVFs of Ca II H&K and the Ca II near-IR triplet, usually in the range 30,000–17,000 km s^{-1}, decreasing with time, and detached from the photosphere by about 6000 km s^{-1} (Hatano et al. 1999a; Maguire et al. 2014; Silverman et al. 2015; Fig. 20.21). Less frequently, HVFs of Si II and possibly O I, S II, Si III, and Fe II also have been seen (Branch et al. 2005; Parrent et al. 2012; Marion et al. 2013; Childress et al. 2014a; Silverman et al. 2015). The HVFs tend to be highly polarized and they do not behave in accordance with a single-parameter sequence. These features may have their origin in pre-explosion nuclear burning in accreted layers of the progenitor white dwarf, to composition and/or density features in the ejecta related to burning during the explosion, or to CSI (Thomas et al. 2004; Mazzali et al. 2005a; Piro 2011; Sect. 21.6.5).

If SN Ia are explosions of C/O white dwarfs, spectroscopic identification of carbon constitutes detection of unburned fuel. (This is not necessarily true of oxygen, some of which may be freshly synthesized by carbon burning.) About half of SN Ia observed 10 or more days before maximum light show evidence for carbon (Maguire et al. 2014), primarily in the form of a weak feature due to C II λ6580

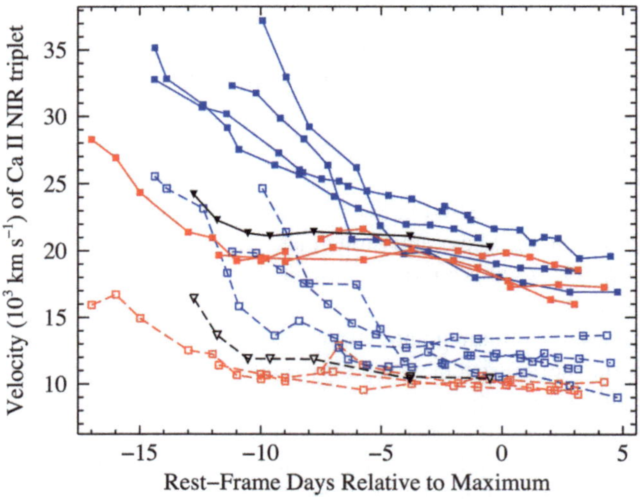

Fig. 20.21 Expansion velocities derived from the absorption minima of the Ca II near-IR triplet photospheric (*open symbols*) and high-velocity (*closed symbols*) features. *Blue and red symbols* denote High-Velocity and Normal-Velocity SN Ia according to Wang et al. (2009b). *Squares and triangles* denote spectroscopically normal and SN 1991T-like events, respectively. Figure 6 "High-velocity features of calcium and silicon in the spectra of Type Ia supernovae" from Silverman et al. (2015), by permission of Oxford University Press on behalf of the Royal Astronomical Society

(Sect. 20.4; Branch et al. 2003; Silverman and Filippenko 2012; Blondin et al. 2012). The mass of carbon is typically estimated to be in the range 10^{-3} to 10^{-2} $M_{\odot}$, a significant constraint on explosion models (Sect. 22.3.3). The C II feature usually forms at velocities coincident with or only mildly faster than the velocities of the photospheric Si II features. Carbon appears to be generally present in the IME line-forming layers rather than confined to the outermost layers or to only a few clumps. The carbon signatures also do not behave in accordance with a single-parameter sequence. The C II feature often (but not always: Cartier et al. 2014) is undetectable after maximum light. The more persistent C I 1.0693 μm line is a useful complement to C II features (Hsiao et al. 2015), although the C I feature is rather weak in most SN Ia (Marion et al. 2006).

Data in the UV are sparse, but it is clear that the UV diversity exceeds that of the optical (Cooke et al. 2011; Walker et al. 2012; Maguire et al. 2012b; Milne et al. 2013; Brown 2014). Owing to high line opacity, the UV spectra form in outer layers of the ejecta where, according to various indications, the diversity in physical conditions is relatively high (Sauer et al. 2008; Hachinger et al. 2013) and not well understood (Brown et al. 2015a).

20.5 Multiparameter Subclassification

In the single-parameter description, SN Ia often are referred to in three ways: as normal, or SN 1991T-like, or SN 1991bg-like,[2] but in view of the richer diversity outlined in Sect. 20.4, more extended subclassification schemes based on photospheric-phase optical spectra have been introduced.

Benetti et al. (2005) assigned SN Ia to three groups (Fig. 20.22) based on five parameters: M_B at maximum light; Δm_{15}; *R*(*Si II*); v_{10}(*Si II*), the blueshift of the 6100 Å absorption 10 days after maximum light (Branch and van den Bergh 1993); and $\dot{v}$, the average rate at which the early postmaximum 6100 Å absorption drifts to the red. Members of the *Low temporal Velocity Gradient* (LVG) group have $\dot{v} \lesssim 70\,\mathrm{km\,s^{-1}\,d^{-1}}$ and $\Delta m_{15} \lesssim 1.5$; this group is itself quite diverse, including some typical SN Ia as well as SN 1991T-likes. Several members of the LVG

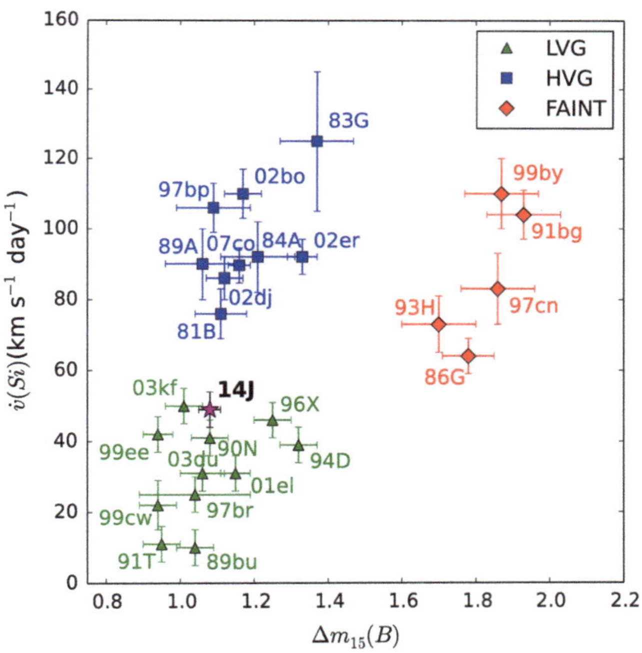

Fig. 20.22 Spectroscopic classification according to Benetti et al. (2005). The position of SN 2014J is denoted by the *star*. The parameters of SN 2011fe (not shown) are $\dot{v} = 60\,\mathrm{km\,s^{-1}\,d^{-1}}$ and $\Delta m_{15} = 1.1$. Figure 20 "Optical and NIR observations of the nearby type Ia supernova SN 2014J" from Srivastav et al. (2016), by permission of Oxford University Press on behalf of the Royal Astronomical Society

[2]The occasionally used term *Branch-normal* simply refers to SN Ia that are neither SN 1991bg-like, SN 1991T-like, nor otherwise obviously peculiar. To avoid confusion with the terms *Core Normal* and *Normal Velocity*, we refer to these as "typical" SN Ia.

group, sometimes referred to as *Si II velocity plateau events*, have exceptionally low postmaximum velocity gradients (Quimby et al. 2007b; Childress et al. 2013b; Zhang et al. 2014b), presumably because silicon is confined to a narrow velocity interval in the ejecta. Members of the *High temporal Velocity Gradient* (HVG) group have $\dot{v} \gtrsim 70\,\mathrm{km\,s^{-1}\,d^{-1}}$ and $\Delta m_{15} \lesssim 1.5$; these include broad-line SN 1984A-likes. Members of the *Faint* group have $\Delta m_{15} \gtrsim 1.5$ and include SN 1991bg-likes.

Branch et al. (2006a, 2009) assigned SN Ia to four roughly equally populated groups (Fig. 20.23), based on the values of $W(5750)$ and $W(6100)$, the pseudo-equivalent widths of the absorptions near 5750 and 6100 Å near maximum light. The *Core Normal* events are highly homogeneous. In a plot of $W(5750)$ against $W(6100)$, the Core-Normal region is densely populated. Three "arms," apparently extending quasi-continuously (although not even the arms are single-parameter sequences) from Core Normals to extreme SN 1991bg-likes, SN 1991T-likes, and SN 1984A-likes, are called *Cool*, *Shallow Silicon*, and *Broad Line*, respectively. Even though the classification criteria of Benetti et al. and Branch et al. do not overlap, the Cools of Branch et al. tend to be the Faints of Benetti et al., the Broad-Line events tend to be the HVGs, and both the Core Normals and the Shallow Silicons are included in the LVGs.

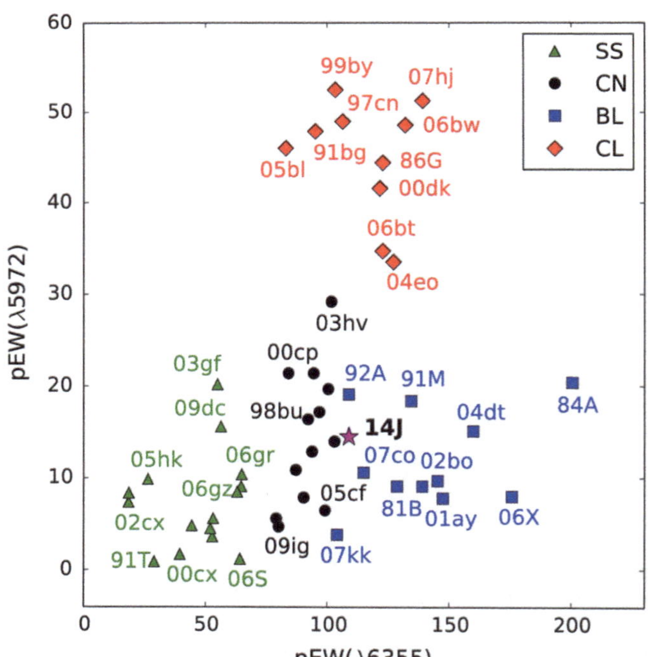

Fig. 20.23 Spectroscopic classification according to Branch et al. (2006a). The position of SN 2014J is denoted by the *star*. The parameters of SN 2011fe (not shown) are $pEW(\lambda5972) = 18\,\text{Å}$ and $pEW(\lambda6355) = 95\,\text{Å}$. Figure 21 "Optical and NIR observations of the nearby type Ia supernova SN 2014J" from Srivastav et al. (2016), by permission of Oxford University Press on behalf of the Royal Astronomical Society

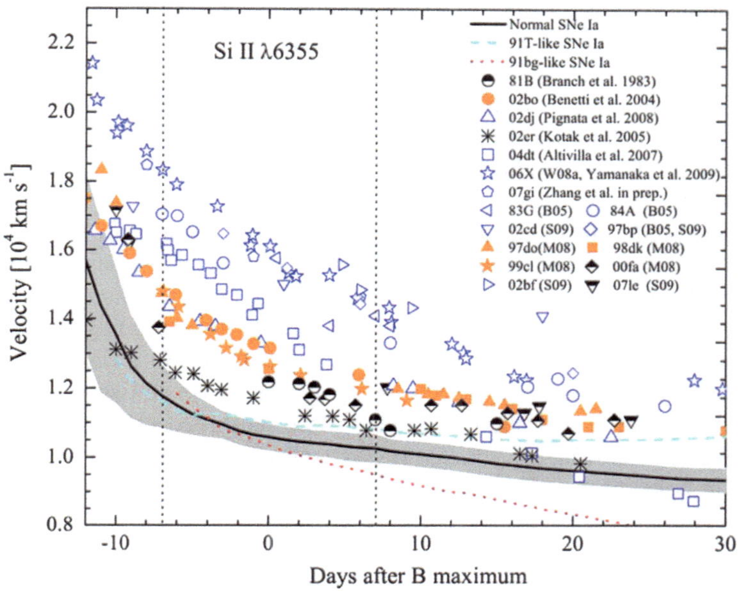

Fig. 20.24 Spectroscopic classification according to Wang et al. (2009b). *Symbols* denote High-Velocity SN Ia, the *solid line* and *gray region* represent the behaviour of Normal-Velocity SN Ia, and the *dashed and dotted lines* illustrate the behavior of SN 1991T-like and SN 1991bg-like events, respectively. From "Improved Distances to Type Ia Supernovae with Two Spectroscopic Subclasses" (Wang et al. 2009b). © AAS. Reproduced with permission

Based on values of $v_{max}(Si\ II)$, the blueshift of the 6100 Å absorption near maximum light, Wang et al. (2009b) divided a sample of SN Ia into two groups (Fig. 20.24), *Normal Velocity* and *High Velocity* (the latter is not to be confused with the HVFs discussed above). Members of the Normal-Velocity group have $v_{max}(Si$ II) in a narrow range centered on 10,600 km s^{-1}, while those of the High-Velocity group have a range of higher values, $\gtrsim$11,800 km s^{-1}. The Normal-Velocity events of Wang et al. are like the Core Normals of Branch et al., although more inclusive, and like some of the LVGs of Benetti et al. Members of the High-Velocity group of Wang et al. are like the Broad-Line events of Branch et al. and the HVGs of Benetti et al. The distribution of $v_{max}(Si$ II) is smooth, thus the division into two groups does not necessarily imply a bimodal distribution.

Our first case study, SN 2011fe, was LVG, Core Normal, and Normal-Velocity. Our other case study, SN 2014J, was near the borderlines between LVG and HVG and between Core Normal and Broad Line, and was a borderline member of the High-Velocity group.

In the scheme of Branch et al., the path from extreme Cools (SN 1991bg-likes) through Core Normals to extreme Shallow Silicons (SN 1991T-likes) may roughly correspond to the single-parameter sequence discussed in Sect. 20.4, with M_{Ni} and perhaps M_{ej} being the main physical variable(s). Similarly, the Faints and LVGs of Benetti et al., taken together, may roughly correspond to the single-parameter sequence. The Broad-Line events of Branch et al., the HVGs of Benetti et al., and the High-Velocity events of Wang et al. indicate, however, that a single-parameter sequence cannot encompass the entire maximum-light spectroscopic diversity.

The above subclassification schemes were discussed on the basis of larger samples of SN Ia by Blondin et al. (2012), Silverman et al. (2012b), and Folatelli et al. (2013).

Doppler shifts of nebular-phase emission lines (assuming that they are unblended) measure the degree to which they form off-center, along the observer's line of sight. Lines of [Fe II] $\lambda 7155$ and [Ni II] $\lambda 7378$ are thought to form in the dense innermost ejecta (Maeda et al. 2010b). An intriguing link between photospheric-phase and nebular-phase spectra is that in SN Ia classified as LVG on the basis of photospheric spectra, these nebular-phase emission lines tend to be blueshifted. In HVGs, these lines tend to be redshifted (Maeda et al. 2010a; Silverman et al. 2013a; Fig. 20.25). HVGs tend to have redder $B - V$ colors at maximum light (Maeda et al. 2011; Blondin et al. 2012), while LVGs are more

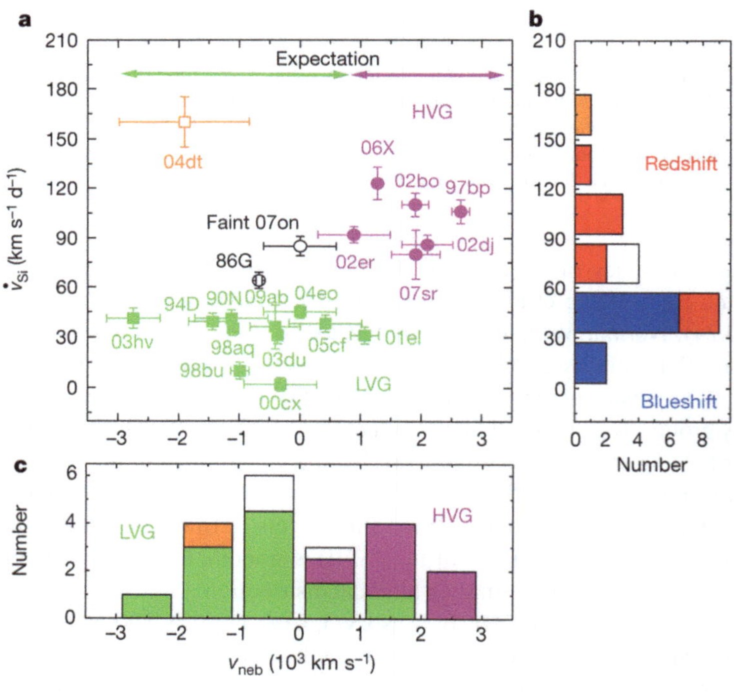

Fig. 20.25 Panel (**a**): The parameter $\dot{v}$ is plotted against v_{neb}, the mean velocity shift of [Fe II] $\lambda 7155$ and [Ni II] $\lambda 7378$ emissions in the nebular phase. *Arrows at the top* of the figure indicate ranges of v_{neb} in which HVG and LVG events are expected, based on a model in which HVGs and LVGs are asymmetric SN Ia viewed from different sides. Panels (**b**) and (**c**) are self-explanatory histograms. Reprinted by permission from Macmillan Publishers Ltd.: NATURE, Maeda et al. (2010a)

likely than HVGs to show C II lines at early times (Parrent et al. 2012; Folatelli et al. 2012). One interpretation of these links between early and late phases is that HVGs and LVGs are otherwise identical asymmetrical explosions viewed from different sides (Sect. 22.3.4).

The issue of whether SN Ia consist of physically distinct subgroups remains open. SN 1991bg-likes often are considered to be at one extreme of a single-parameter sequence, but there are indications that they may be a separate class in some sense. The slope of the width-luminosity relation requires a change in slope to accommodate their low luminosities for their Δm_{15} values (Fig. 20.19). They also show stronger C II features near maximum than typical SN Ia (Höflich et al. 2002). Based on a small sample, unlike typical SN Ia their continuum polarization is significant. It rises toward the red and displays a distinct preferred axis (Howell et al. 2001; Patat et al. 2012). Unlike most typical SN Ia, SN 1991bg-likes have not shown HVFs (Silverman et al. 2015). For a discussion of SN Ia that may be transitional events connecting typical and SN 1991bg-like SN Ia, see Hsiao et al. (2015).

20.6 Colors and Extinction

Galactic extinction by interstellar dust is characterized by $R_V \equiv A_V/E(B-V)$, with an average value of $R_V = 3.1$ (thus $R_B = 4.1$). From SN Ia in external galaxies, various investigations inferred the value of R_V along the line of sight to be significantly lower than 3.1, typically $R_V < 2$ (Branch and Tammann 1992; Tripp 1998; Hicken et al. 2009; Burns et al. 2014; Sect. 20.2.2). A low value of R_V corresponds to a strong variation of extinction with wavelength, implying small grain sizes.

Several effects may contribute to the low value of R_V. Values of R_V often are based on the adoption of a unique relation between Δm_{15} and intrinsic $B-V$ color, but, for example, at a given value of Δm_{15}, the $B-V$ colors of the High-Velocity events of Wang et al. (2009b) are redder than those of the Normal-Velocity events by about 0.06 magnitudes, probably owing to greater line blanketing in the former (Foley and Kasen 2011; Mandel et al. 2014). Analyzing the two groups together on the assumption of a common relation between Δm_{15} and $B-V$ will produce an incorrect value of R_V.

Another consideration is that some SN Ia may be reddened by circumstellar dust. There is a fundamental geometrical difference between extinction by interstellar and circumstellar dust. Interstellar extinction is a single-interaction situation: a photon that is absorbed or scattered out of the observer's line of sight has negligible probability of being returned to the line of sight. Extinction in a circumstellar shell, however, can involve multiple scattering that preferentially attenuates photons of shorter wavelengths, thus steepening the effective extinction law to $R_V < 3.1$ even if the dust itself has the same properties as typical interstellar dust (Wang 2005; Goobar 2008).

Nevertheless, although there is evidence for CSM associated with some SN Ia (Sects. 21.6.5 and 23.2), most of the extinction of SN Ia by dust appears to be caused by interstellar dust (Phillips et al. 2013; Burns et al. 2014; Amanullah et al. 2015; Patat et al. 2015), in which case observations of SN Ia provide a means of studying the properties of "non-standard" interstellar dust in external galaxies. This raises the Copernican issue: what is "standard" interstellar dust? Perhaps not Galactic dust. Note, however, that before SN 2014J occurred, $R_V \gtrsim 3.5$ was measured along the line of sight to that event (Hutton et al. 2015).

20.7 Substantially Super-Chandrasekhar SN Ia

High peak luminosities and broad light curves (Fig. 20.19) of a small minority of SN Ia—SN 2003fg (Howell et al. 2006), SN 2007if (Yuan et al. 2010; Scalzo et al. 2010), SN 2009dc (Yamanaka et al. 2009; Tanaka et al. 2010; Silverman et al. 2011; Taubenberger et al. 2011a), and perhaps several others (Hicken et al. 2007; Maeda et al. 2009; Scalzo et al. 2012, 2014b; Brown et al. 2014)—seem to require high values of M_{Ni} and substantially super-Chandrasekhar mass ejection.[3] Early spectra often show stronger lines of C II and lower expansion velocities than those of typical SN Ia. In nebular spectra, [Fe III] emission is weak or absent, indicating a lower ionization state than typical SN Ia. This presumably is due to the higher density of ejecta that follows from lower expansion velocity (Taubenberger et al. 2013a). SN 2009dc appears to have ejected about 1.8 $M_\odot$ of ^{56}Ni and a total mass of 2.8 $M_\odot$ (Taubenberger et al. 2011a; Fig. 20.26), quantities that challenge all SN Ia models (Hachinger et al. 2012b). Substantially super-Chandrasekhar events are $\sim$1% of all SN Ia and have a strong tendency to occur in metal-poor environments (Khan et al. 2011). Possible progenitors of these events are discussed in Sects. 21.2.5 and 21.3.1.

[3]In some models for typical SN Ia, M_{ej} is mildly super-Chandrasekhar (Chap. 21), therefore we refer to these extreme events as substantially super-Chandrasekhar.

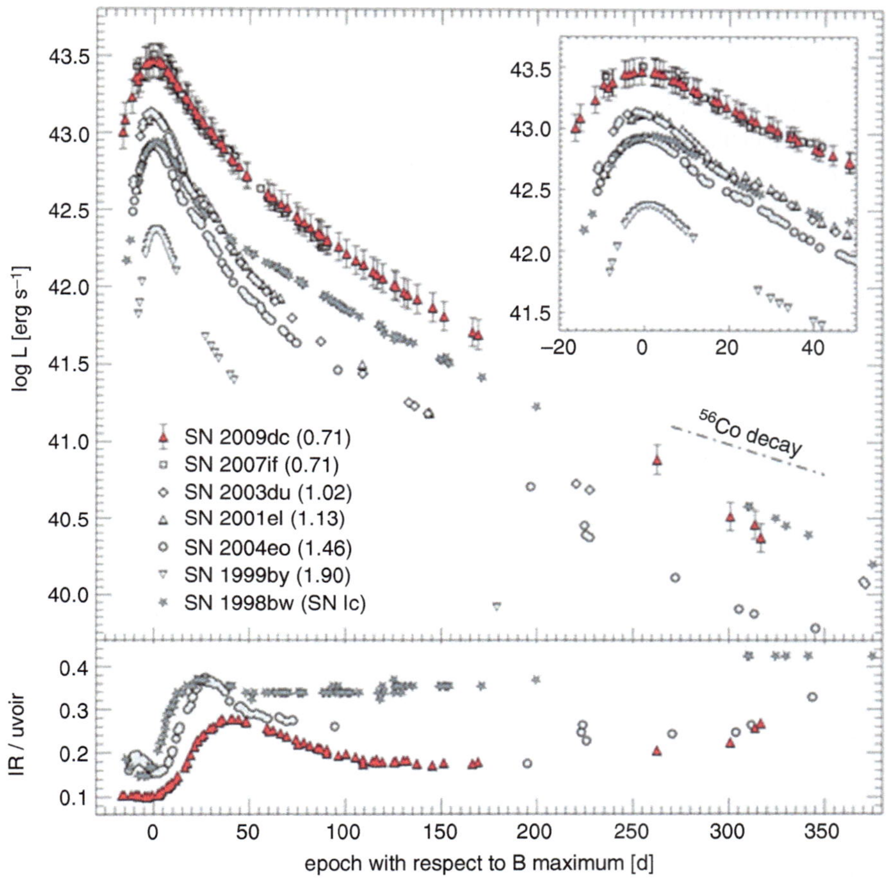

Fig. 20.26 *Top*: UVOIR light curves of the substantially super-Chandrasekhar SN 2009dc and 2007if, the typical Type Ia SN 2003du, SN 2001el, and SN 2004eo, the peculiar SN 1991bg-like SN 1999by, and the Type Ic-bl SN 1998bw. The Δm_{15} values of the SN Ia are given in *parentheses*. *Bottom*: the fractional contribution of the near-IR, for SN 2009dc, SN 2004eo, and SN 1998bw. Figure 7 "High luminosity, slow ejecta and persistent carbon lines: SN 2009dc challenges thermonuclear explosion scenarios" from Taubenberger et al. (2011a), by permission of Oxford University Press on behalf of the Royal Astronomical Society

20.8 Outliers

A small minority of events have displayed strong individual peculiarities that may provide clues to understanding SN Ia in general.

SN 2000cx (not to be confused with the well-known SN 2002cx of Sect. 23.3), a spectroscopic Shallow Silicon, displayed a peculiar photometric evolution including a lopsided light curve with a typical premaximum brightening but a slow post-maximum decline ($\Delta m_{15} = 0.94$; Li et al. 2001). High-velocity ($\gtrsim$20,000 km s^{-1})

features appeared to include those of Ti II (Thomas et al. 2004; Branch et al. 2004b). The properties of another Shallow Silicon, SN 2013bh, were very similar to those of SN 2000cx, including the apparent Ti II HVFs. The environments of SN 2000cx and SN 2013bh suggest that they occurred in relatively old, metal-deficient populations (Silverman et al. 2013d).

The rise to maximum light of SN 2001ay, a spectroscopic Broad Line event, was unusually rapid, but the postmaximum decline was exceptionally slow, with $\Delta m_{15} = 0.68$, and the peak luminosity was typical rather than high (Krisciunas et al. 2011). The peculiarities were suggested to have been due to an unusually high C/O ratio in the progenitor and a consequent high explosion energy (Baron et al. 2012).

SN 2010lp, a spectroscopic Cool event, had a relatively broad light curve for a Cool ($\Delta m_{15} = 1.24$) and narrow (1700 km s^{-1} FWHM) [O I] $\lambda\lambda 6300, 6364$ emission in its nebular spectrum, indicating unburned or partially burned fuel in deep layers of the ejecta (Taubenberger et al. 2013b). The only previous report of [O I] emission in SN Ia was in the typical SN 1937C (Minkowski 1939). SN 2010lp was suggested to be the result of a violent merger of two white dwarfs (Kromer et al. 2013b; Sect. 22.4).

SN 2004dt (Wang et al. 2006; Altavilla et al. 2007), a spectroscopic extreme Broad-Line event, displayed an exceptionally high degree of polarization in the Si II line, approaching 2% at 1 week prior to maximum light, making it an extreme outlier in the correlation between Δm_{15} (1.21 for SN 2004dt) and degree of polarization (Wang et al. 2007a), and requiring a large asymmetry in the Si II line-forming region.

The optical light curves of SN 2006bt ($\Delta m_{15} = 1.09$; Foley et al. 2010a), a spectroscopic Cool event, SN 2006ot ($\Delta m_{15} = 0.85$; Stritzinger et al. 2011), a Broad Line event, and SN PTF10ops ($\Delta m_{15} = 1.12$; Maguire et al. 2011), another Cool, were relatively broad, yet the secondary near-IR maxima were unusually weak. The environments suggest that these SN Ia were produced by very old, metal-poor populations.

20.9 Correlations Between Properties of SN Ia and Host Galaxies

The light-curve widths and peak luminosities of SN Ia correlate with the morphological types and colors of their host galaxies. The younger populations of blue, late-type galaxies tend to produce more slowly-declining, brighter events than the older populations of red, early-type galaxies (Hamuy et al. 1996a; Branch et al. 1996; Howell 2001). On the whole, the widths and peak luminosities also correlate, inversely, with host-galaxy stellar mass (Neill et al. 2009; Sullivan et al. 2010). Among typical SN Ia (setting aside SN 1991bg-likes), this correlation appears to be driven by a subset of those that, as indicated by the strength of local Hα emission, occur in "passive" environments rather than in or near star-forming environments. SN Ia in passive environments (in both spiral and elliptical galaxies) are, on average, less luminous than those in or near star-forming environments (Rigault et al. 2013). Massive galaxies tend to have older stellar populations than less massive galaxies. More massive galaxies also tend to have higher metallicities and, other things being equal, higher metallicity is predicted to lead to lower ^{56}Ni production (Timmes et al. 2003; Howell et al. 2009). Thus metallicity and/or population age may be significant for the presence of low-luminosity SN Ia in massive galaxies.

Some spectroscopic features also correlate with host galaxy properties, e.g., $v_{max}(Si\ II)$ correlates positively, and the strength of the Ca II near-IR-triplet HVF correlates inversely, with galaxy mass (Pan et al. 2015b).

A more tricky proposition is to establish correlations between host-galaxy properties and *Hubble residual.*[4] If SN Ia peak luminosities depended only on light-curve width and color, Hubble residuals would be independent of host-galaxy properties, but this may not be the case (Kelly et al. 2010; Johansson et al. 2013; Childress et al. 2013a; Rigault et al. 2013, 2015; Pan et al. 2014; Jones et al. 2015). If correlations of host galaxy properties and Hubble residual become statistically robust, multiparameter fitting techniques based on SN Ia light-curve widths and color, plus one or more galaxy properties, may result in reduced scatter in SN Ia distance determinations (Sect. 25.2).

An important aspect of correlations between properties of SN Ia and their hosts, or locations within their hosts, is that they restrict the amount of observational diversity that can be attributed to SN Ia asymmetry and observer viewing angle. A sample of identical but asymmetric SN Ia would not exhibit statistically significant correlations of their observational properties with their locations. Compared to the Normal-Velocity SN Ia of Wang et al. (2009b, Sect. 20.5), the members of the High-Velocity group may be more concentrated to inner and brighter regions of their host galaxies and may tend to occur in larger and more luminous hosts, suggesting that they come from younger and more metal-rich populations (Wang et al. 2013b, but see Pan et al. 2015b). If confirmed, this would mean that asymmetry alone cannot be responsible for the differences between members of the Normal-Velocity and High-Velocity groups (and by implication, the differences between members of the LVG and HVG groups; Sect. 20.5), otherwise their locations would be statistically indistinguishable.

---

[4]When using SN Ia as distance indicators for cosmology, the goal is to obtain a redshift-independent *distance* for each event, without necessarily separating the intrinsic colors from the interstellar reddening. Procedures for obtaining such distances are usually referred to as light-curve (actually light-curve and color) fitting techniques. The Hubble residual is the difference between (1) the SN Ia distance modulus obtained by such a technique and (2) the distance modulus that is implied by the host galaxy redshift together with the adopted cosmological parameters. Thus an event having a negative Hubble residual is overluminous *for its light-curve width and color.*

20.10 Variations with Redshift

Overall, SN Ia at high redshift appear to be quite similar to those at low redshift (Meyers et al. 2012; Foley 2012), but since the mean stellar populations and ages of galaxies vary with redshift the mean properties of the SN Ia that they produce are not expected to be exactly the same. Compared to low-redshift SN Ia, high redshift ($z \sim 1$) events tend, on average, to have broader light curves and to be brighter (Howell et al. 2007), to have weaker optical spectral features of IMEs such as the Si II λ4000 absorption (Walker et al. 2011), and to be brighter in the UV (Sullivan et al. 2009). The ratio of "near-UV blue" to "near-UV red" events increases with z (Milne et al. 2015). These tendencies are broadly consistent with expectation based on galaxy demographics: at higher redshift, galaxies were younger and had lower metallicities. To a large extent, the associated variations of peak luminosities with redshift are accounted for by light-curve width and color corrections.

Because of the steepness of the DTD (Sect. 3.8), most SN Ia in star-forming galaxies, at all redshifts, are young, near the peak of the DTD at ~0.3 Gyr. SN Ia in passive galaxies tend to have ages equal to the amount of time elapsed since star formation ceased, and the maximum possible time elapsed decreases with redshift since the host galaxies were younger at higher z. Therefore, because star-forming galaxies tend to be of low mass while passive galaxies tend be of high mass, the ages and other characteristics of SN Ia correlate with galaxy mass in a redshift-dependent way (Childress et al. 2014b). This may be at the root of the negative Hubble residuals of the subset of SN Ia mentioned in Sect. 20.9.

20.11 Summary

SN 2011fe, an example of an extremely well-observed typical SN Ia, illustrated the growing capability to discover supernovae shortly after explosion. Early data established that SN 2011fe surely was the explosion of a white dwarf (or a merger of two), and equally surely it did not have a red-giant companion when the explosion occurred. The detection of γ-rays from SN 2014J confirmed that its light curve was powered by the ^{56}Ni decay chain.

Typical SN Ia display striking homogeneity in their photospheric and nebular-phase spectra, colors, light curves, and absolute magnitudes at maximum light. Mild premaximum diversity gives way to increasing homogeneity at later phases.

High-quality observations reveal ever more precisely characterized diversity. To first order, SN Ia can be regarded as a single-parameter sequence characterized, for example, by light-curve width. The locus from extreme spectroscopic Cools (SN 1991bg-likes) through Core Normals to extreme Shallow Silicons (SN 1991T-likes) may roughly correspond to a sequence of increasing M_{Ni} and perhaps M_{ej}. Nevertheless, various evidence, most obviously the existence of the spectroscopic Broad-Line events, shows that a single-parameter sequence is inadequate to describe the entire diversity. This motivates subclassification of SN Ia based on photospheric-phase data. Asymmetry may play some role in the observational diversity. Typical SN Ia have nearly spherical photospheres but significant line polarization that indicates rich asymmetry in the line-forming layers. Doppler shifts of emission lines in nebular-phase spectra indicate asymmetry in the inner ejecta.

Most or all of the extinction of SN Ia appears to be caused by interstellar rather than circumstellar dust. The extinction curve typically is more steep (lower R_V) than that of Galactic dust.

A small fraction of SN Ia appear to eject large amounts of ^{56}Ni and to have a total ejected mass that is substantially super-Chandrasekhar. Another small fraction of SN Ia display notable individual peculiarities. These extremes show that SN Ia diversity presents much that remains to be explained.

The properties of SN Ia correlate with the properties of their host galaxies in various ways, e.g., SN Ia in the young populations of star-forming environments tend be brighter than those in passive, non-star-forming environments. Such correlations restrict the degree to which SN Ia diversity can be attributed to asymmetric events viewed from various lines of sight, and they have the potential to improve the use of SN Ia as distance-indicators for cosmology. Moderate variations of the mean properties of SN Ia with redshift are broadly consistent with expectations based on galaxy demographics.

Chapter 21
Progenitors

21.1 Introduction

There are several compelling reasons to associate Type Ia supernovae with explosions of C/O white dwarfs. Unlike core-collapse supernovae, SN Ia occur in all types of galaxies, including ellipticals that have little, if any, current star formation (Sect. 3.3). In spiral galaxies that have a mix of young and old stars, SN Ia are not concentrated to the spiral arms. Both of these facts have long been regarded as clues that the progenitors of many SN Ia are long lived and hence of modest ZAMS mass, although the notion that some SN Ia arise in somewhat shorter-lived systems also has a long and respected history (Sect. 3.4). The progenitors are not so old that SN Ia occur preferentially in galactic haloes, or in globular clusters despite their high stellar densities that might enhance binary evolution in some way (Sect. 3.3). The observations show that the progenitors of SN Ia cannot have ZAMS mass exceeding about 10 $M_\odot$. Stars with ZAMS mass less than 10 $M_\odot$ are expected to make white dwarfs (Sects. 8.2–8.5). The star that exploded to make SN 2011fe must have been no larger than a white dwarf (bearing in mind uncertainties about the delay between explosion and early rise of the optical output; Sects. 20.2 and 22.2).

Another reason to associate SN Ia with exploding white dwarfs is that the basic uniformity of explosion properties (Sect. 20.3) suggests that the immediate progenitor is a white dwarf of about the Chandrasekhar mass (M_{Ch}). SN Ia display a fascinating array of properties in detail (Sect. 20.4), but it also is true that typical SN Ia are remarkably homogeneous. Another important part of the argument is that decades of work comparing observations to models have shown that the light curves and spectral evolution of SN Ia are most consistent with classes of models based on the explosion of near-M_{Ch} C/O white dwarfs. Light-curve, spectral, and spectropolarimetric observations of SN Ia constrain the progenitor, the environment, and the level of asymmetry. Dominant fuels other than carbon and oxygen, e.g., helium, fail to produce the observed spectral features of SN Ia. These issues will

D. Branch, J.C. Wheeler, *Supernova Explosions*, Astronomy and Astrophysics Library, DOI 10.1007/978-3-662-55054-0_21

be explored in more detail in Chap. 22. Finally, as mentioned in Sect. 20.2.2, γ-ray observations of SN 2014J were in broad agreement with SN Ia explosion models.

Single stars of ZAMS mass less than about 10 $M_{\odot}$ are not expected to produce supernovae.[1] Following an SAGB phase, stars with ZAMS mass in the range of about 8–10 $M_{\odot}$ are expected to produce ONeMg white dwarfs. After an AGB phase, those between 0.45 and 8 $M_{\odot}$ produce C/O white dwarfs. Stars with ZAMS mass below 0.45 $M_{\odot}$ are fully convective on the main sequence. They skip the red-giant phase and evolve directly from the main sequence to eventually become helium white dwarfs, but this takes more than the current age of the Universe, so these low-mass stars are irrelevant to our discussion.[2] Single white dwarfs produced by stars with ZAMS mass less than 10 $M_{\odot}$ radiate away their internal energy at constant radius and do not explode. Thus SN Ia require special circumstances. Bringing a white dwarf to an explosive endpoint seems to demand binary (or multiple star) evolution.

The search for the progenitors of SN Ia thus turns to white dwarfs that accrete mass from or merge with binary companions. An accreting ONeMg white dwarf that approaches M_{Ch} is expected to collapse to a neutron star, probably with some associated photon display, but it will not be an SN Ia (Sect. 9.4.3). If an accreting helium white dwarf reached about 0.7 $M_{\odot}$, it would explode and it would be bright (Mazurek 1973; Nomoto and Sugimoto 1977), but the event would not look like a SN Ia (or anything else observed so far, which suggests that this "never" occurs). By a process of elimination, and because there have been successes in accounting for observational properties (Chap. 22), attention is focused on accreting and merging C/O white dwarfs as progenitors of SN Ia.

Constraints on the masses of C/O white dwarfs at their formation come from observations and stellar evolution calculations. The C/O white dwarfs with mass $\gtrsim$0.6 $M_{\odot}$ that are of most interest for making SN Ia are descended from stars of ZAMS mass $\gtrsim$2 $M_{\odot}$. The theoretical estimate of the maximum mass at formation of a C/O white dwarf arising from single-star evolution is no more than 1.2 $M_{\odot}$, perhaps no more than 1.07 $M_{\odot}$ (Umeda et al. 1999). White dwarfs that are initially more massive are expected to be composed of ONeMg, although there may be a ZAMS mass range (about 7–9 $M_{\odot}$) that produces hybrid white dwarfs having carbon cores surrounded by shells of oxygen and neon; these white dwarfs could have masses up to 1.3 $M_{\odot}$ (Chen et al. 2014c). Following their formation, white dwarfs can add mass by mass transfer from, or merger with, a companion star.

In broad outline, the two canonical mechanisms to produce SN Ia can be made to sound deceptively simple. In the *single-degenerate* (SD) M_{Ch} channel (Wheeler and Hansen 1971; Whelan and Iben 1973; Nomoto 1982), the more massive star

[1]An SAGB star with a carbon core surrounded by oxygen and neon might produce a single-star explosion (Denissenkov et al. 2013), but such a configuration is unlikely to resemble a typical SN Ia. For novel speculations concerning explosions of single white dwarfs, see Chiosi et al. (2015), Di Stefano et al. (2015), Graham et al. (2015d), and Bramante (2015).

[2]Helium white dwarfs can, however, be produced by binary evolution.

in a binary evolves first, to become a C/O white dwarf. Later, when the companion (donor) star evolves and expands, mass is transferred from the donor to the white dwarf by Roche-lobe overflow (RLOF) or a wind. As the white dwarf increases in mass and decreases in radius ($R \propto M^{-1/3}$), the interior temperature and density rise. As the mass approaches, but does not quite reach, M_{Ch}, temperature and density are $\gtrsim 10^8$ K and $\gtrsim 10^9$ gm cm^{-3}. At this point, carbon burning ignites deep in the interior, releasing nuclear energy in electron-degenerate matter (Sect. 19.2), ultimately leading to a thermonuclear explosion. In the other canonical channel, the *double-degenerate* (DD) super-Chandrasekhar channel (Webbink 1984; Iben and Tutukov 1984), binary evolution involving one or two common-envelope (CE) episodes results in a very close (separation $\lesssim 4$ $R_\odot$, $P_{orb} \lesssim 10$ h) pair of C/O white dwarfs. Loss of angular momentum by emission of gravitational waves from the rapidly orbiting white dwarfs causes gradual inspiral and eventual coalescence (merger). If the combined mass of the white dwarfs exceeds M_{Ch}, an SN Ia may be produced.

The most attractive feature of the canonical SD channel is that if a C/O white dwarf *does* approach M_{Ch} in a reasonably sedate manner, it very likely *will* explode (Chap. 19). In nondegenerate matter, the onset of nuclear burning is accompanied by increasing temperature and pressure, with compensating expansion and adiabatic cooling that stabilizes the burning. In electron-degenerate matter, however, initially there is no significant pressure increase and expansion, so the increasing ion temperature that results from carbon burning increases the burning rate. The ensuing thermonuclear instability eventually elevates the pressure and drives a nuclear combustion front through the entire white dwarf in seconds, unbinding it completely. The kinetic energy of the ejecta is the difference between the nuclear energy release ($\sim$2 B) and the white dwarf binding energy ($\sim$0.5 B). Major issues with the SD channel are whether approach to M_{Ch} occurs often enough to account for a substantial fraction of the observed SN Ia rate, why there has been no spectroscopic detection of hydrogen stripped from the donor star (Sect. 21.6.4), and why there is little direct evidence of the former donor.

An attractive feature of the canonical DD channel is that there would be no hydrogen in the vicinity nor any former donor. Major issues are whether super-Chandrasekhar DD mergers occur often enough to account for the SN Ia rate, whether their explosions would resemble typical SN Ia, and whether mergers result in collapse to neutron stars rather than explosion.

In this chapter we will explore the complexities of the progenitor possibilities and their confrontation with observations in more detail. As we do so, we should keep in mind that in spite of the above discussion of the apparent merits of the canonical SD channel, if the distribution of mass ejection by typical SN Ia includes both a peak near M_{Ch} and a tail extending to lower masses (Sect. 20.4), then more than one kind of progenitor may be contributing. Chapter 22 will discuss how thoughts on progenitors map into explosion processes and match with observational constraints. In this chapter we divide the topic of progenitor evolution into the two broad categories of SD and DD, but we will see in Chap. 22 that while some explosion mechanisms are unique to a single category, others might be common to both. For instance, relatively slow growth of a C/O white dwarf to carbon ignition near M_{Ch} is frequently associated with the SD mechanism, but there are models in which similar basic physics of the explosion arise in a DD context. Likewise, double-detonation models (Sect. 21.2.6) might pertain to either channel. On the other hand, violent mergers with two white dwarfs colliding dynamically are specifically associated with the DD channel. For reviews of the search for the progenitors of SN Ia, see Wang and Han (2012), Maoz et al. (2014), and Ruiz-Lapuente (2014). In later sections (Sect. 21.2.2 and 21.3.2) we will confront progenitor candidates with observations.

One way to assess the viability of a progenitor channel is statistical in nature. One seeks to determine what ZAMS binary systems ultimately can lead to SN Ia. Answering this question for either SD or DD systems is a formidable challenge (Claeys et al. 2014). The problem is often tackled in two parts, the first being to carry out detailed binary evolution calculations, beginning at the onset of mass transfer to a white dwarf. The goal of this part is to determine which combinations of initial white dwarf mass, donor mass, and orbital period lead to the desired result: a plausible SN Ia progenitor system. For this, one must do a few things, including: (1) choose a detailed stellar evolution code (e.g., that of Eggleton (1973), STARTRACK (Belczynski et al. 2008), or MESA (Paxton et al. 2013, 2015)), a ratio of convective mixing length to pressure scale height, a prescription of convective overshoot, and other parameters involved in single-star evolution; (2) adopt prescriptions for mass loss by stellar winds, the efficiency of mass transfer by RLOF (steady and conservative or unstable and nonconservative), and all the factors that affect the retention fraction (Sect. 21.2.1): white dwarf accretion from a wind, common envelope evolution (with, in the usual treatments, a value of $\alpha_{ce}\lambda$ or γ parameter; Sect. 8.8), and possible stripping by momentum transfer and ablation by shock heating of mass from the donor by a fast white dwarf wind (Hachisu et al. 2008; Ablimit et al. 2014; and (3) evolve the two masses, the orbital period, and the transfer rate while taking account of the loss of mass and angular momentum from the system.

The binary evolution beginning at the phase of white dwarf accretion is calculated for thousands of different combinations of initial parameters including, ideally, metallicity. A main-sequence star of high metallicity has a larger radius than one of lower metallicity, which will promote RLOF, so the evolution of the higher metallicity star resembles that of a star of lower metallicity and higher mass. Most work in binary evolution ignores possible magnetic interaction between the white dwarf and the mass-losing star (Sect. 21.2.4), as well as the effects of rotation and magnetic fields internal to the stars. Another complication is that in close binaries, tidal forces may induce mixing that does not occur in the evolution of single stars (Song et al. 2013).

The next part of the procedure is to ask which initial stellar systems defined by two ZAMS masses and the initial orbital period (and hence separation[3]) lead to the desired combinations of white dwarf mass, secondary mass, and orbital period at the onset of mass transfer that were identified in the first part of the procedure. These *binary population synthesis* simulations also constrain the evolutionary state of the donor at the onset of mass transfer. Because of the desire to follow the evolution of numerous (perhaps $\sim 10^8$) initial systems, binary population-synthesis simulations often employ a less detailed but very fast binary-evolution code (e.g., that of Hurley et al. 2002) coupled with a Monte Carlo technique. As initial conditions, one must specify an initial mass function for the primaries (e.g., that of Salpeter 1955; Scalo 1986, or Kroupa et al. 1993), the distribution of the ratios of secondary to primary mass (often taken to be flat)[4], and the distribution of initial separations (often taken to be flat in logarithmic intervals except for a smooth decrease at very small separations, for convergence).

The predictions of SN Ia rates and the ratio of the numbers of SD to DD explosions depend on model assumptions, e.g., the efficiency, α_{ce}, of the common-envelope process. Few binary synthesis studies consider the full range of progenitor models discussed here—core-degenerate models (Sect. 21.4), spin-up/spin-down models (Sect. 21.2.5), triple-star systems (Sect. 21.3.3)—and of explosion models discussed in Chap. 22, e.g., sub-Chandrasekhar detonations and violent mergers. In the following, some of these complicated issues are discussed as they pertain specifically to the SD and DD channels.

21.2 Single Degenerates

In the canonical SD channel, a primary consideration is the extent to which mass transfer to a C/O white dwarf results in an increase of the mass of the white dwarf. This is the problem of the *retention fraction.* Constrained by the retention fraction, the theory of binary evolution is employed to determine the binary systems in which an accreting C/O white dwarf forms and approaches M_{Ch}. This step involves exploring variations in the nature of the other star, the masses of the two stars, and the orbital period. Finally, attention turns to binary population synthesis, beginning with the two ZAMS masses and the initial separation, to estimate the rate (the *realization frequency*) at which stellar populations, both young and old, may be able to produce SN Ia by the SD channel. In the next three sections each of these components are discussed in more detail.

---

[3]The period in days, P, and the separation in units of the solar radius, a, are related by $P^2 = 1.38 \times 10^{-2} a^3/(M_1 + M_2)$, where M_1 and M_2 are the masses of the two stars in solar units.

[4]For a discussion of observational differences between mass-ratio distributions of close and wide companions, see Gullikson et al. (2016). Note also that there are strong selection effects against low secondary masses. Techniques of aperture-masking interferometry coupled with adaptive optics have the power to reveal low-mass companions (Hinkley et al. 2015).

21.2.1 Canonical SD Model: C/O White Dwarfs Accreting to the Chandrasekhar Mass

Numerical simulations of the complex response of C/O white dwarfs to mass transfer necessarily rely on simplifying assumptions. Important parameters include the mass of the white dwarf, the composition of the accreted matter, and the mass-transfer rate, the latter being determined by the time-varying nature of the binary system. Other complications, not treated in many models, are the angular momentum and magnetic field of the white dwarf and of the accreted matter. Basic models treat spherically-symmetric accretion at a fixed rate onto a nonrotating C/O white dwarf. These models give some notion of the problems associated with the retention fraction (Nomoto 1982), but neglect some relevant issues. Here we describe what might be called the canonical approach to retention fractions.

The accreted matter may be hydrogen-rich (Fig. 21.1). For a narrow range of transfer rates between $\lesssim 10^{-7}$ and $\lesssim 10^{-6}$ $M_{\odot}$ y^{-1} that depends on the mass of the white dwarf and the length of time it has cooled since forming, nondegenerate hydrogen burns steadily to helium at $T \sim 10^8$ K, $\rho \lesssim 10^2$ gm cm^{-3}. The accumulated helium also burns steadily at $T \sim 10^8$ K, $\rho \lesssim 10^7$ gm cm^{-3}, and the mass of the white

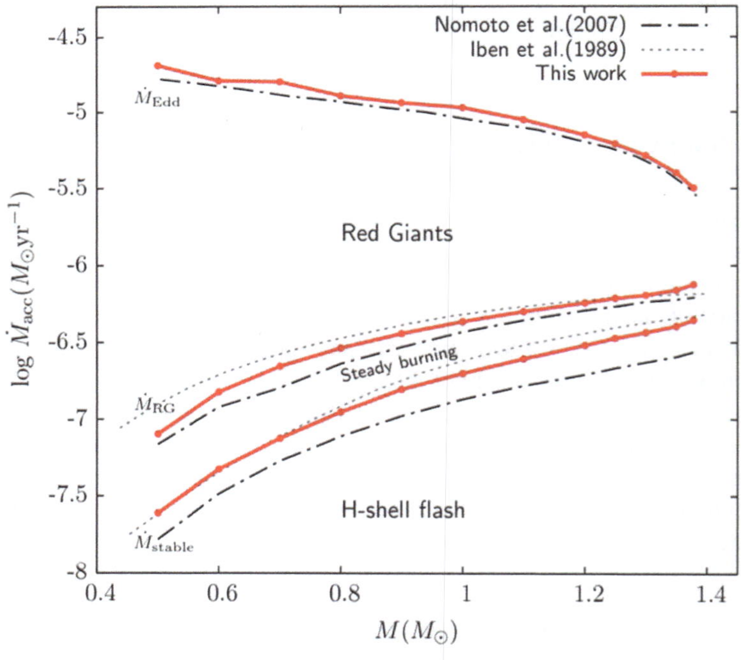

Fig. 21.1 The response of accretion of hydrogen onto a nonrotating C/O white dwarf. From "A Super-Eddington Wind Scenario for the Progenitors of Type Ia Supernovae" (Ma et al. 2013b). © AAS. Reproduced with permission

dwarf increases at the transfer rate. The retention fraction may approach unity, but only for a special set of conditions. These conditions correspond to the observed supersoft X-ray sources (Sect. 21.6.2).

For a transfer rate that is too high for steady burning, but lower than the Eddington accretion-rate limit (Sect. 8.8) of $\lesssim 10^{-5}$ $M_\odot$ y^{-1}, an extended hydrogen envelope develops. A common assumption is that a portion of the accreted hydrogen is burned at the maximum rate for steady burning, thus joining the white dwarf as helium, while the excess hydrogen is carried away from the system by a strong white dwarf wind (Hachisu et al. 1999), with the consequence that the retention fraction is low. This puts greater demands on the donor to provide enough mass to grow the white dwarf to M_{Ch}. For a transfer rate that approaches or exceeds the Eddington accretion limit, the hydrogen envelope may expand to fill its Roche lobe. This would lead to heavy mass loss from the system and a very low retention fraction that prevents a significant increase in the mass of the white dwarf. Alternatively, a super-Eddington wind might carry off the excess hydrogen, prevent expansion from filling the Roche lobe, and allow the white dwarf to accrete at the Eddington rate (Ma et al. 2013b; Wang et al. 2015a).

For a transfer rate that is too low for steady burning, degenerate hydrogen accumulates on the white dwarf and burns in flashes that are qualitatively similar to those in AGB and SAGB stars (Sect. 8.3). The higher the transfer rate, the lower the degree of degeneracy at ignition and the weaker the hydrogen shell flash. A typical assumption is that for a transfer rate within a factor of eight of the steady-burning rate, a weak flash ejects only a fraction of the hydrogen accreted since the previous flash, with the rest being accumulated on the white dwarf as helium. This situation may pertain to recurrent novae (RN), which are less energetic than classical novae and for which the interval between flashes is less than a century. The possible role of RN as progenitors of SN Ia is addressed in Sect. 21.5.1.

For still lower transfer rates, $\lesssim 10^{-8}$ $M_\odot$ y^{-1}, strong hydrogen flashes produce classical nova explosions (Warner 1995). Observations of novae often show an enhancement of carbon and oxygen in the ejecta, suggesting that some of the underlying white dwarf is expelled in the explosion. The spectra of some novae indicate that oxygen, neon, and magnesium are enriched, hinting at an underlying ONeMg white dwarf. There are arguments that the ejected mass may exceed the mass that was accreted since the previous flash, preventing an approach to M_{Ch} (Livio and Truran 1992; Townsley and Bildsten 2005). Conditions that lead to classical novae probably are not conducive to producing SN Ia.

At accretion rates of $\lesssim 10^{-9}$ $M_\odot$ y^{-1}, accretion disk instabilities may lead to brief enhanced episodes of mass loss and the production of *dwarf novae* (Warner 1995). The enhanced accretion rate has been conjectured to cause steady hydrogen burning and an increase in the white-dwarf mass (King et al. 2003; Wang et al. 2010a; Meng and Yang 2010b), but the total amount of mass accumulated during a disk-instability episode is too small to initiate the burning (Lin et al. 1985). Accreted matter accumulates at the low time-averaged accretion rate to eventually produce a strong mass-ejecting flash that prevents an increase of the white dwarf mass (Hachisu et al. 2010).

As just discussed, an important aspect of attempts to estimate the retention fraction in models is the nature of the shell burning of matter that accumulates on the surface of the white dwarf, and especially whether that shell burning is stable. These issues, in turn, depend on matters such as numerical resolution and on the treatment of mixing between the accumulated layers and the underlying white dwarf. Mixing of carbon and oxygen from the white dwarf would, for instance, enhance CNO burning in the hydrogen shell. Helium shell flashes may produce some contamination of the surface layers of the white dwarf with oxygen, neon, and magnesium prior to the explosion of the underlying white dwarf (Shara and Prialnik 1994; Kato and Hachisu 2012). Thus there is ambiguity in the interpretation of oxygen, neon, and magnesium in the ejecta of classical and recurrent novae. Rather than coming from the underlying white dwarf, such material might result from a shell flash and be confined to the very outer layers. Production of IMEs in the outer layers by shell flashes might also affect early observations of the outer layers of SN Ia and play some role in the formation of the observed high-velocity features (Sect. 20.4).

Helium can accumulate on the surface of a C/O white dwarf as a result of accretion and burning of hydrogen or it may be transferred directly from a helium-star companion. As for hydrogen, basic models consider the effects of the transfer of helium at a constant rate to a nonrotating white dwarf (Fig. 21.2).

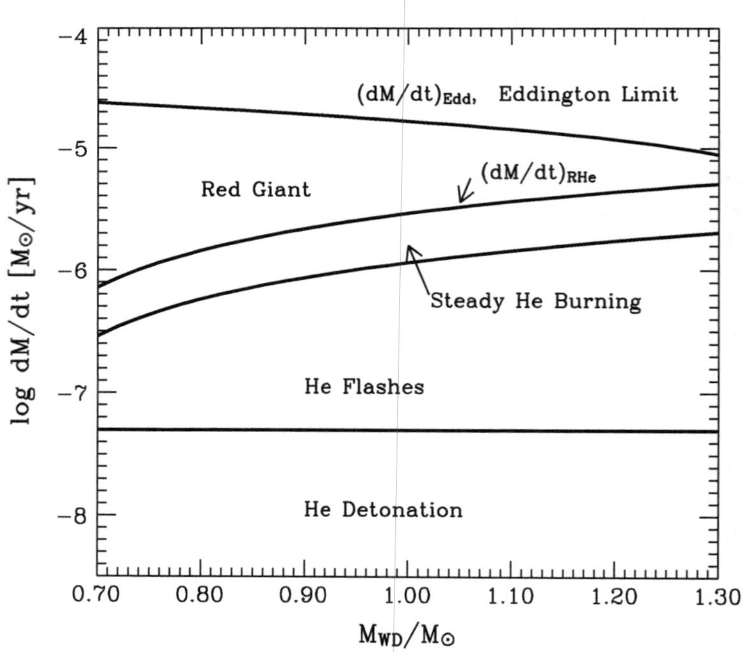

Fig. 21.2 The response of accretion of helium onto a nonrotating C/O white dwarf. From Yoon (2004) with permission

For a narrow range of helium transfer rates near 10^{-6} $M_{\odot}$ y^{-1}, nondegenerate helium burns steadily to C/O, and the white dwarf mass increases at that rate. For a helium transfer rate that is too high for the helium to be consumed by steady burning, but lower than the Eddington limit, the helium envelope surrounding the white dwarf expands. The usual assumption, as for hydrogen transfer, is that helium is burned at the maximum rate for steady helium burning, with the excess being lost from the system in a white dwarf wind, resulting in a low retention fraction. As in the hydrogen case, for a transfer rate that exceeds the Eddington limit, the retention fraction is very low, and the mass of the white dwarf does not increase substantially.

For a transfer rate that is too low for steady helium burning, but greater than about $\sim 5 \times 10^{-8}$ $M_{\odot}$ y^{-1}, a common assumption is that a weak flash ejects only a small fraction of the helium accreted since the previous flash, with the rest accumulating on the white dwarf as C/O (Kato and Hachisu 2004). For a transfer rate lower than $\sim 5 \times 10^{-8}$ $M_{\odot}$ y^{-1}, if ~ 0.1 $M_{\odot}$ of helium is accreted, ignition of degenerate helium leads to a nuclear flash that ejects the entire helium layer and prevents an approach to M_{Ch}. This ignition of helium may lead also to disruption of the C/O core, producing a sub-Chandrasekhar double-detonation supernova (Sects. 21.2.6 and 22.3.8). If the core does not disrupt, the result will be some kind of subluminous transient (Chap. 23).

The goal in the SD channel is to grow the white dwarf to the point of carbon ignition. As the mass of a C/O white dwarf increases, energy is deposited in the outer layers by gravitational energy of accretion, energy released by nuclear burning, and compressional heating. Some energy is radiated away by photons and neutrinos and some is conducted to the interior. These factors govern the thermal evolution of the white dwarf. For example, for an accretion rate of hydrogen $\gtrsim 3 \times 10^{-6}$ $M_{\odot}$ y^{-1} and/or for an initially massive white dwarf there may not be enough time for accretion to heat the deep interior to carbon ignition before ignition occurs far off-center at relatively low density, burning only to oxygen, neon, and magnesium. Then a carbon burning front propagates inward, creating an ONeMg white dwarf that, if it approaches M_{Ch}, collapses (Sects. 8.5.3 and 9.4.2).

Uncertainties abound. It may be that steady burning of hydrogen never occurs (Hillman et al. 2015; Starrfield 2015). In reality, the accretion is not spherical (although in the strong surface gravity the time to distribute accreted material across the surface of the white dwarf is very short). In 1D simulations, ignition of shell burning necessarily occurs simultaneously on a sphere; realism requires 3D. A substantial fraction of the more massive white dwarfs are observed to have strong magnetic fields ($\gtrsim 10^6$ G; Sect. 21.2.4). A prospect that has not been investigated in any detail is that a white dwarf magnetic field of order 10^6 G might alter or suppress the wind that is invoked to carry away extra accreted hydrogen or helium (Wheeler 2012). The efficiency of the strong white dwarf wind at high transfer rates is uncertain even in the absence of magnetic fields. The uncertain $^{12}C(\alpha, \gamma)^{16}O$ rate is important for the outcome of helium burning (Saruwatari and Hashimoto 2010). In the case of hydrogen transfer, interplay between hydrogen- and helium-burning shells is complex (Jose et al. 1993; Starrfield et al. 2013; Wolf et al. 2013; Idan et al. 2013). It will be no surprise if current wisdom about the response of even a nonrotating, nonmagnetic C/O white dwarf to mass transfer turns out to require significant revision.

A key factor that must be considered in any SD model is angular momentum. A white dwarf that accretes mass also accretes angular momentum and spins up to rapid, perhaps differential rotation in the absence of external torques (Sect. 21.2.5). An important issue then becomes how angular momentum is transported to the interior by rotationally-induced instabilities including the dynamical and secular shear, Eddington–Sweet, Goldreich–Schubert–Fricke, and baroclinic instabilities (Yoon and Langer 2004b). The circumstances under which the white dwarf will be limited to solid-body rotation are uncertain. Angular-momentum transport may be too slow to establish rigid-body rotation; the white dwarf would then rotate differentially, with outer layers at higher angular velocity. Differential rotation may play an important role in stabilizing helium shell burning because the centrifugal force and associated composition mixing by rotationally-induced instabilities lead to a thicker, less-degenerate, and more stable helium-burning shell (Yoon and Langer 2004b), thus changing the phasing of hydrogen and helium shell flashes (Yoon et al. 2004). The lifting effect of rotation also is expected to allow the white dwarf to exceed the classical Chandrasekhar mass (Sect. 21.2.5) and play a role in *spin-up/spin-down* models (Sects. 21.2.5, 22.3.7, and 22.4.1).

21.2.2 Canonical SD Model: Binary Evolution and Population Synthesis

A sample of representative binary-evolution calculations for solar metallicity that may lead to SN Ia in the SD channel is shown in Fig. 21.3. These simulations assumed nonrotating white dwarfs, strong white dwarf winds, and the Hachisu et al. (2008) prescription for mass stripping of donors by white dwarf winds. The parameters for which white dwarfs were predicted to approach M_{Ch} are enclosed by solid lines. This parameter space shrinks to zero at a white dwarf mass of $\sim$0.6 $M_{\odot}$ (not shown in Fig. 21.3). The paths from the onset of mass transfer to M_{Ch} are various, and complicated: when M_{Ch} is finally approached, the system may be undergoing (1) an optically thick white dwarf wind phase (filled stars); (2) steady hydrogen burning (filled squares), and (3) a weak hydrogen-flash phase (filled circles). The path denoted by filled triangles involves hydrogen burning during a disk-instability phase and is not expected to be viable (Sect. 21.2.1). Open circles and crosses indicate systems that fail to approach M_{Ch} because of mass ejection by novae and unstable dynamical mass transfer, respectively, situations leading to small retention fractions that leave the donor star unable to transfer a sufficient supply of mass.

At the onset of mass transfer, the donor may be on the main sequence or, more likely, slightly evolved (i.e., really a subgiant, although this situation is commonly referred to as the main-sequence channel); a red giant or supergiant, depending on

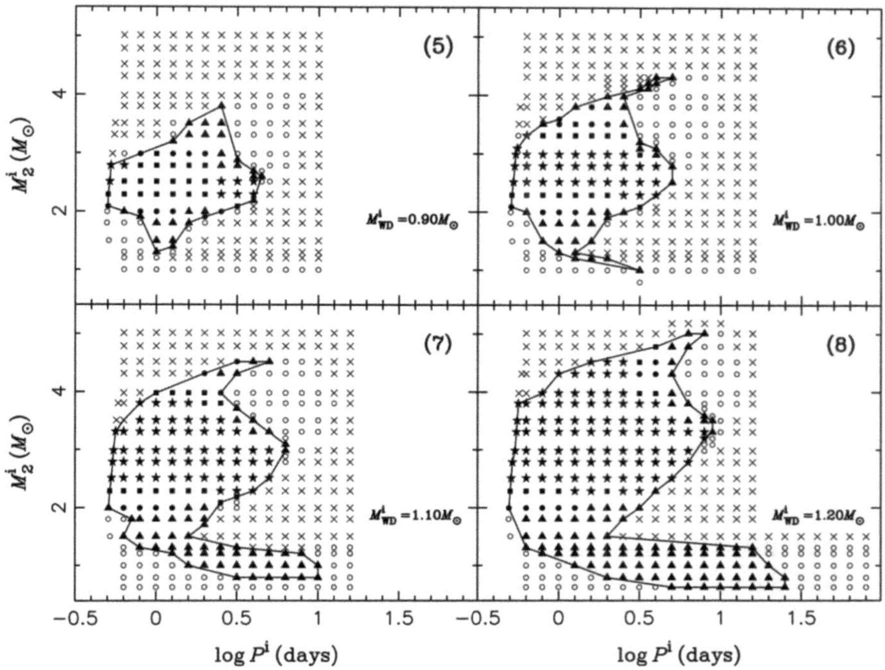

Fig. 21.3 Outcomes of binary-evolution calculations for values of the initial mass of the secondary star, M_2^i, the orbital period, P^i, and the mass of the white dwarf at the time of its formation, M_{WD}^i. For the meaning of the *lines and symbols*, see the text. From "A Comprehensive Progenitor Model for SNe Ia" (Meng and Yang 2010b). © AAS. Reproduced with permission

the donor's initial mass; or a post-AGB helium star. To illustrate the complexity of the possibilities (in spite of the uncertainties), in the following paragraphs binary population-synthesis results obtained by Wang et al. (2010a) and Wang et al. (2009a) are summarized. In this work, white dwarf rotation (Sect. 21.2.5) was neglected. Meng and Yang (2010b) made similar assumptions and found similar results, but the results of some other investigators have been significantly different.

21.2.2.1 From ZAMS Binaries to Hydrogen-Rich Main-Sequence and Subgiant Donors

Wang et al. (2010a) found three channels from ZAMS binaries to hydrogen-rich main-sequence or subgiant donors that might lead to SN Ia. The main difference in the three channels is in the initial separation.

(1) The ZAMS masses of the primary and secondary are in the ranges 4–7 $M_\odot$ and 1–2 $M_\odot$, respectively, and the initial orbital period is in the range 5–30 days. Owing to the small separation, the primary fills its Roche lobe before becoming

a red giant. The low mass ratio causes the mass transfer to be unstable, a CE forms, the orbit decays, and the primary's hydrogen envelope is lost from the system. The resulting helium star later develops a C/O core, expands, fills its Roche lobe again, and this time stably transfers helium to the secondary, producing a C/O white dwarf and a mass-enhanced main-sequence secondary. The secondary later expands to fill its Roche lobe and transfer mass back to the white dwarf. In this channel the matter transferred to the secondary is helium, which presumably mixes into the hydrogen-rich envelope of the secondary, causing a helium enrichment of the matter that is later accreted by the white dwarf. This consequences of this enrichment, which is frequently ignored in treating the response of the white dwarf to accretion, are not clear.

(2) The ZAMS masses are 2.5–6.5 $M_{\odot}$ and 1.5–3 $M_{\odot}$ and the initial period is 200–900 days. Owing to the wider separation, the primary first fills its Roche lobe on the early AGB, with subsequent consequences being similar to those of channel (1): mass transfer to a C/O white dwarf from a helium-enriched main-sequence secondary.

(3) The ZAMS masses are 4.5–6.5 $M_{\odot}$ and 1.5–3.5 $M_{\odot}$ and the initial period is $\gtrsim$1000 days. Owing to the very wide separation, the primary first fills its Roche lobe during the thermal-pulsing AGB phase. After a CE episode, the primary becomes a C/O white dwarf with a main-sequence or subgiant donor. In this case the matter transferred to the white dwarf is not expected to be enhanced in helium.

These evolutionary channels are tightly constrained by the retention issues discussed in Sect. 21.2.1. Other constraints on this class of models come directly from observations. White dwarfs accreting from main-sequence companions are abundantly observed in cataclysmic variable systems. Classical novae are an impediment to retention of mass by the white dwarf, and dwarf novae may undergo classical-nova eruptions at long intervals. A great deal of attention has been placed on recurrent novae (Sect. 21.5.1) since many of them harbor white dwarfs that are more massive than typical field white dwarfs, and these massive white dwarfs are still accreting. Some models with main-sequence or subgiant companions that may qualify as progenitors of SN Ia are in a state where steady hydrogen and helium shell burning is predicted. These systems are predicted to be bright sources of soft X-rays (Sect. 21.6.2).

21.2.2.2 From ZAMS Binaries to Hydrogen-Rich Red-Giant Donors

Binary systems with ZAMS masses of about 5–6.5 $M_{\odot}$ and 1–1.5 $M_{\odot}$ and initial orbital periods $\gtrsim$1500 days may become progenitor candidates with red-giant donors. With allowance for a white dwarf wind that could strip donor mass, these systems might be able to bring the white dwarf to M_{Ch} (Hachisu et al. 2012). Other possibilities for a plausible red-giant donor include accretion from an equatorially-enhanced (Lü et al. 2009) or tidally-enhanced (Chen et al. 2011) wind. Observational constraints on systems with red-giant donors are discussed in Sect. 21.5.2.

21.2.2.3 From ZAMS Binaries to Helium-Rich Donors

Population-synthesis studies have also identified several channels with helium-star donors that may lead to SN Ia (Wang et al. 2009a). The principle difference from the hydrogen-rich donors discussed above is that the helium-rich secondary has a larger ZAMS mass so the initial mass ratio is higher. Basic categories of evolution again depend on the initial separation.

(1) The ZAMS masses of the primary and secondary are 5–8 $M_{\odot}$ and 2–6.5 $M_{\odot}$, respectively, and the initial orbital period is 10–40 days. The primary first fills its Roche lobe as a subgiant or red giant and, owing to a relatively high mass ratio, it stably transfers its hydrogen envelope to the secondary. The resulting helium-star primary eventually develops a C/O core, expands, fills its Roche lobe again, and transfers its helium envelope to the secondary, leaving the bare C/O core. The secondary develops a helium core, expands, and fills its Roche lobe. The ensuing mass transfer is unstable, a CE forms, the orbit decays, and the secondary's envelope is lost from the system. Later, the secondary expands to fill its Roche lobe again as a helium star, finally transferring helium to the C/O white dwarf.
(2) The ZAMS masses are 6–6.5 $M_{\odot}$ and 5.5–6 $M_{\odot}$, the mass ratio is nearly unity, and the period is 300–1000 days. The primary first fills its Roche lobe on the early AGB, mass transfer is unstable, and the hydrogen envelope is lost from the system. Later, the primary becomes a helium red giant, fills its Roche lobe again, and stably transfers its helium envelope to the secondary. The subsequent evolution of the secondary is as in channel (1).
(3) The ZAMS masses are 5.5–6.5 $M_{\odot}$ and 5–6 $M_{\odot}$, similar to those of (2), but the period is $\gtrsim$1000 days. The primary first fills its Roche lobe in the thermal-pulsing AGB phase, the secondary evolves to burn helium in its core, and after ejection of a CE the evolution of the secondary is as in channel (1).

21.2.3 Distribution Predictions

Ultimately, the results of population synthesis are of most value when placed in a broader context. With an adopted star-formation history of a galaxy, population synthesis can be used to predict the time evolution of the SN Ia rates and much more, including the evolution of the distributions of the primary and secondary ZAMS masses and of the initial orbital periods.

For a constant star-formation rate of 5 $M_{\odot}$ yr^{-1}, which serves as a rough approximation for a spiral galaxy such as ours, SN Ia with helium-star donors begin

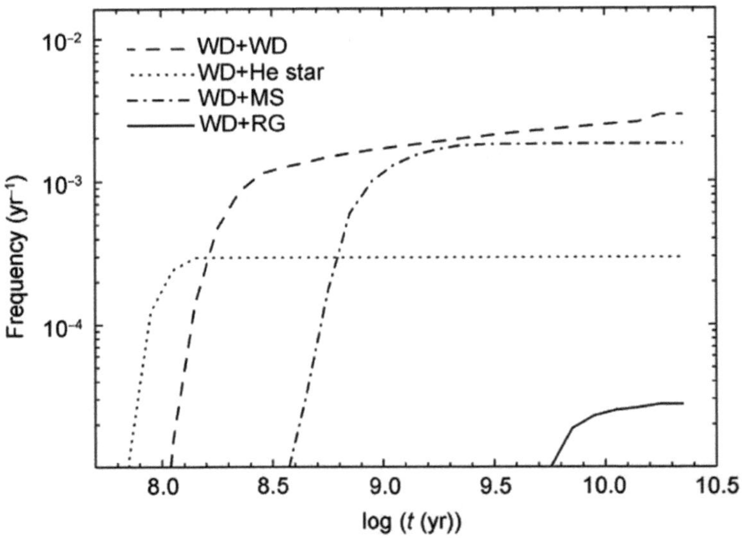

Fig. 21.4 Binary-evolution predictions of the evolution of the SN Ia rate for a constant star-formation rate of 5 $M_\odot$ yr^{-1}. From "Birthrates and delay times of Type Ia supernovae" (Wang et al. 2010b), by permission of Oxford University Press on behalf of the Royal Astronomical Society

to occur $\lesssim 10^8$ years after the beginning of star formation, the lifetime of a star of about 6 $M_\odot$ (Wang et al. 2010b; Fig. 21.4). After $\sim 10^9$ years, most SD SN Ia have main-sequence donors. Eventually, the SD SN Ia rate reaches a steady state of $\gtrsim 10^{-3}$ SN Ia yr^{-1}, compared to the estimated Galactic rate of 5×10^{-3} SN Ia yr^{-1} (Sect. 3.5). Figure 21.4 also presents a result for the DD channel to which we will return below. Other investigators (Mennekens et al. 2010; Chen et al. 2014a), with different assumptions and treatments, found lower rates for the SD channel. Given this dispersion of results, there is currently no robust argument from population synthesis that the SD channel can or cannot provide a sufficient rate of SN Ia.

SN Ia also occur in elliptical galaxies. For an early single 10^{11} $M_\odot$ burst of star formation, which serves as a rough approximation for the history of a massive elliptical galaxy, SN Ia with helium-star donors again begin to occur $\lesssim 10^8$ years after the beginning of star formation (Wang et al. 2010b). After $\sim 10^9$ years, most SD SN Ia are again predicted to have main-sequence or red-giant donors.

The starburst case predicts a delay-time distribution (DTD; Sect. 3.8), the shape of which can be compared to the DTD inferred from observations (approximately proportional to t^{-s}, with $s \gtrsim 1$; Sect. 3.8). Figure 21.5 shows the observational DTD of Totani et al. (2008). The DTD of the SD channel may (Hachisu et al. 2008; Wang et al. 2010b; Meng and Yang 2010b) or may not (Mennekens et al. 2010; Graur et al. 2014a; Chen et al. 2014a) be consistent with observations, depending, for instance, on whether SN Ia can be produced in old populations by systems with main-sequence donors that undergo long-term mass-transfer variability (Toonen et al. 2014) and/or systems with red-giant donors.

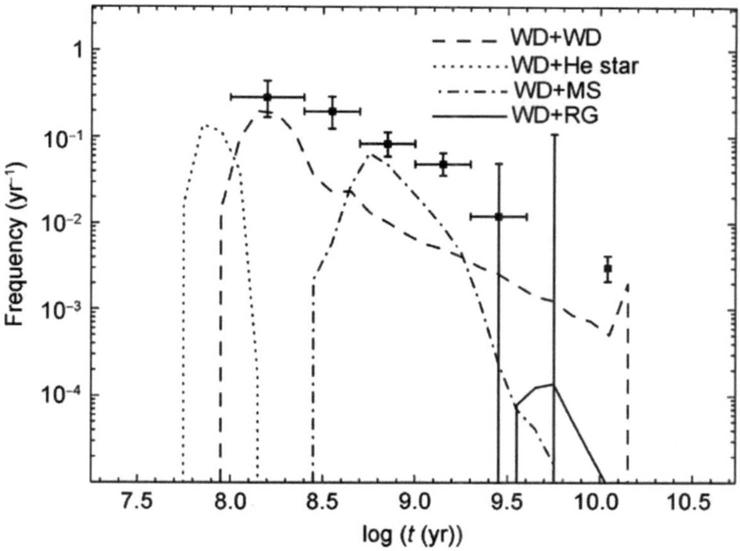

Fig. 21.5 Like Fig. 21.4, but for a single 10^{11} $M_\odot$ starburst. Data points represent the observed DTD. From "Birthrates and delay times of Type Ia supernovae" (Wang et al. 2010b), by permission of Oxford University Press on behalf of the Royal Astronomical Society

The distributions of the properties of the binary systems at the onset of accretion are of interest in connection with attempts to identify individual observed binary systems that may become SN Ia (Sect. 21.5). For example, during the first Gyr after a single starburst the initial white dwarf mass ranges from 0.6 to 1.2 $M_\odot$ and the ZAMS mass of the donors ranges from $\lesssim$2 to $\gtrsim$4 $M_\odot$ (Meng and Yang 2010b).

The distributions of the properties of the donors at the time of explosion also are of interest in connection with their interactions with the supernova ejecta (Sect. 21.6.4) and with attempts to find them in young SNRs (Sect. 21.6.6). Population synthesis can provide predictions, at the moment of explosion, of the distributions of survivor masses, orbital velocities, effective temperatures, surface gravities, and surface abundances, all as functions of cosmic time. For example, main-sequence and subgiant donors are predicted in some models to have typical masses at the time of explosion of $\sim$0.9 $M_\odot$ (Meng and Yang 2010c), before the donors are stripped of $\sim$0.1 to $\sim$0.2 $M_\odot$ by the supernova ejecta (Liu et al. 2012c).

21.2.4 M-Dwarf Donors?

A possibility not generally taken into account in binary population-synthesis studies is that the donors in SD systems may be M-type dwarfs (Wheeler 2012), which have ZAMS mass $\lesssim$0.6 $M_\odot$. This idea is ignored in classic population-synthesis studies because conservative mass transfer from such donors would cause the orbit to widen, terminating the transfer. Consideration of the magnetic fields of the white dwarf and the donor could alter this picture. A significant fraction of white dwarfs have surface magnetic fields of $\sim 10^6$ G, and M dwarfs typically have fields of 10^2 to 10^3 G. Their magnetic fields might lock the two stars together, with the magnetic poles pointing towards one another in the orbital plane, resulting in synchronous rotation and, contrary to the usual situation for an accreting white dwarf, a slowly rotating white dwarf. Mass transfer could be channeled by a "magnetic bottle" to a concentrated area at the magnetic pole of the white dwarf, thus enhancing the effective rate of accretion compared to spherical accretion by a factor of $\sim$100, perhaps more if X-rays from the accretion region enhance the rate of mass transfer from the M dwarf.

It is not clear whether this channel can make a significant contribution to the SN Ia rate. Issues include the likelihood that appropriate rates emerge from population-synthesis studies, that the self-excited mass transfer can be stable, and that an excess of hydrogen-rich CSM may be implied (possibly mitigated by the low-density interior of a wind-blown bubble; Sect. 21.6.5). If these systems are to be viable, they must lose angular momentum to prevent the orbit from widening.

On the positive side, binaries consisting of a white dwarf and an M dwarf are common in the Galaxy. The DTD would be determined by the time for inspiral to RLOF by gravitational-wave radiation following formation of the white dwarf rather than by the evolutionary timescale of the companion as it is in most SD channels. The resulting DTD might be similar in shape to that of the double-degenerate channel (Sect. 21.3.2).

21.2.5 Super-Chandrasekhar Single Degenerates

In the SD channel, accretion of mass and angular momentum from a companion star will produce rapid rotation of the white dwarf. The centrifugal support will allow the white dwarf mass to exceed the classical Chandrasekhar mass at the time of carbon ignition and runaway, even in models intended to account for typical SN Ia.

The maximum mass of a rigidly-rotating, nonmagnetic, electrically-neutral C/O white dwarf is 1.48 $M_\odot$. If a donor exhausts its envelope and stops supplying mass to a rigidly-rotating, mildly super-Chandrasekhar white dwarf, the explosion will be delayed until loss of angular momentum allows the central regions of the white dwarf to spin down and reach a central density of about $3 \times 10^9\,\mathrm{g\,cm^{-3}}$. Estimates of the characteristic timescale are uncertain, but may range from 10^6 to 10^{10} years, depending on the strength and configuration of the magnetic field (Ilkov and Soker 2012). In the meantime, a former red-giant donor may become a white dwarf; a former main-sequence donor may become a white dwarf or a lower-mass main-sequence star. In this *spin-up/spin-down* model (Yoon and Langer 2005a; Di Stefano et al. 2011; Justham 2011; Di Stefano and Kilic 2012; Hachisu et al. 2012), explosions of mildly super-Chandrasekhar white dwarfs could produce typical SN Ia. This variation on the SD channel would be consistent with the lack of a bright companion and detectable CSM around most SN Ia; at the time of explosion the binary would consist of two low-luminosity, non-interacting stars.

The mass of a differentially-rotating white dwarf can be higher. In principle, up to 4 $M_\odot$ is possible, although unlikely to be achieved in nature (Ostriker and Bodenheimer 1968). With realistic constraints of binary evolution taken into account, the maximum achievable mass of an accreting C/O white dwarf is about 2 $M_\odot$ (Chen and Li 2009; Liu et al. 2010; Hachisu et al. 2012). The spindown process is likely to be very complex. Contrary to some assumptions in the literature, there is no minimum accretion-rate to allow differential rotation (Yoon and Langer 2005a). Whether the white dwarf is differentially rotating depends on the effectiveness of internal viscosity. The internal viscosity, however, whether hydrodynamic or magnetohydrodynamic, depends on the distribution of density, temperature, and shear, and the temperature and shear depend on the viscosity. There will be processes of internal redistribution of angular momentum and also possibly loss of mass and angular momentum from the white dwarf. The density at which nuclear runaway occurs will be a function of the mass and the angular momentum of the white dwarf (Yoon and Langer 2004a).

Binary population-synthesis calculations in the manner of Wang et al. (2010a; Sect. 21.2.2), but with an attempt to explore the effects of rotation by allowing white dwarfs to continue to accrete (donor willing) beyond the classical Chandrasekhar mass, were carried out by Wang et al. (2014a). (The quantitative results, based on the assumption that an accretion rate of 3×10^{-7} $M_\odot$ y^{-1} is required to sustain differential rotation, would change somewhat if this assumption were modified.) The majority of the white dwarfs attained final masses that could be supported by rigid rotation, in the range 1.38–1.48 $M_\odot$, but a significant fraction ended in the range 1.48–1.9 $M_\odot$, and a small fraction exceeded 2 $M_\odot$. Many of these white dwarfs would explode after the donor has spun down to become a second white dwarf, but some might explode with considerable hydrogen in the vicinity, possibly producing SN Ia-CSM (Sect. 23.2).

21.2.6 Sub-Chandrasekhar Single Degenerates

If $\sim$0.1 $M_\odot$ of helium accumulates onto the surface of a C/O white dwarf at a rate in the range $\sim 10^{-8}$ to $\sim 10^{-7}$ $M_\odot$ y^{-1}, either from surface burning of accreted hydrogen or direct transfer from a helium star, a helium detonation may form and eject the accumulated layer (Sect. 21.2.1). If only the helium layer is ejected, the result will be a subluminous transient (Chap. 23). If an ingoing detonation forms at the outer edge of the C/O core, however, the core also could explode. If an ingoing detonation does not form, converging pressure waves might compress the C/O core sufficiently to cause it to detonate and explode. Either way, if the core explodes, the result is a sub-Chandrasekhar *double-detonation* supernova (Woosley and Weaver 1994; Shen and Bildsten 2014).

It seems unlikely that double detonations produced by accretion from hydrogen-rich donors occurs at a rate comparable to the SN Ia rate (Yoon and Langer 2004b). Whether double detonations produced by accretion from helium donors can be expected to match the SN Ia rate is not clear (Piersanti et al. 2013; Wang et al. 2013a; Ruiter et al. 2014). The issue of whether the resulting explosion would have observational properties resembling typical SN Ia is discussed in Sect. 22.3.8.

21.3 Double Degenerates

The canonical DD model invokes the merger of two C/O white dwarfs, the combined mass of which at least mildly exceeds M_{Ch}, followed by a slow, viscosity-dependent evolution of the merger remnant to central carbon ignition and explosion (Sect. 21.3.1). Binary population synthesis provides estimates of various stellar populations that may produce SN Ia by the DD channel (Sect. 21.3.2). Noncanonical and sub-Chandrasekhar-mass DD models also are considered (Sect. 21.3.3). The formation and evolution of compact binary stars, including pairs of white dwarfs, was reviewed by Postnov and Yungelson (2014).

21.3.1 Canonical Super-Chandrasekhar DD Model

In the canonical DD model, binary evolution produces a close pair of C/O white dwarfs with a total mass that exceeds M_{Ch} (Iben and Tutukov 1984; Webbink 1984). The white dwarfs inspiral owing to loss of orbital angular momentum by gravitational-wave radiation and interact within a Hubble time if their orbital period when the second white dwarf formed was $\lesssim$10 h. As the separation decreases, tidal interactions alter the white dwarf rotations and transfer a viscosity-dependent fraction of the orbital energy to internal energy of the white dwarfs, heating them and influencing the initial conditions under which the merger occurs (Dan et al. 2011; Fuller and Lai 2012). At the onset of the merger, the white dwarfs are in or near synchronous rotation with the orbit.

Most early studies of the merger of two C/O white dwarfs concluded that the likely outcome was collapse to a neutron star rather than explosion (Saio and Nomoto 1985, 2004). These studies assumed that the less massive white dwarf tidally disrupts into a rotationally-supported thick disk from which the more massive white dwarf accretes near the Eddington rate, $\sim 10^{-5}$ $M_\odot$ yr^{-1}. Under these conditions, the maximum temperature occurs near the surface of the accretor. This temperature depends on uncertain viscosity and angular-momentum transport, but it is likely to be high enough, $\gtrsim 6 \times 10^8$ K, to ignite stable carbon burning. The burning propagates inward by heat conduction and the nuclear energy is carried off by neutrinos. In $\sim 10^4$ years, the C/O white dwarf is converted to an ONeMg white dwarf that collapses by electron capture to form a neutron star (Sect. 9.4.2).

Subsequent work suggested that after the dynamical phase, the merger remnant consists of a cold, slowly-rotating core and a hot, quasi-spherical envelope, surrounded by a centrifugally supported disk. In this case, the rate of accretion from the disk can be sub-Eddington and the temperature at the core-envelope interface might remain low enough, due to local neutrino losses, to avoid carbon ignition there. About 10^5 years after merger, central ignition of carbon may produce a SN Ia (Yoon et al. 2007; Lorén-Aguilar et al. 2009). Other work found that contraction of the hot envelope on a thermal timescale, 10^3 to 10^4 years, compresses the base of the envelope and probably ignites stable carbon burning there after all, leading again to inward-propagating carbon-burning and collapse to a neutron star (Shen et al. 2012). A later suggestion was that rotational instability of the accreting primary may regulate the mean accretion rate to lower values and avoid quiescent carbon burning (Tornambé and Piersanti 2013). The merger remnant may then not explode until sufficient angular momentum is lost, long after accretion ceases. This is the DD version of the *spin-up/spin-down* channel (Sect. 21.2.5), which might allow the merger remnant to ignite central carbon in a manner similar to that of the canonical SD model, possibly leading to an SN Ia. The issue of whether the canonical DD channel can produce SN Ia for at least some combinations of initial white dwarf masses remains open.

21.3.2 Canonical DD Model: Binary Evolution and Population Synthesis

Several binary-evolution channels may lead to super-Chandrasekhar pairs of close C/O white dwarfs that can merge in less than a Hubble time. For example, Mennekens et al. (2010) found that the most frequent path begins with a primary ZAMS mass in the range about 4–7 $M_\odot$, a mass ratio not much less than unity, and an initial orbital period $<$ 100 days (Fig. 21.6; left panel). The primary evolves off the main sequence and transfers its radiative envelope to the secondary by RLOF. After the primary becomes a white dwarf, the now more massive secondary leaves the main sequence and transfers mass to the white dwarf, causing a CE episode with mass loss from the system and inspiral, leaving a close pair of white dwarfs with an orbital period of only hours. The second white dwarf to form often is the more massive one. Depending on initial parameters, the time elapsed between star formation and merger, usually dominated by the time to inspiral due to gravitational-wave radiation, varies from $\sim$0.2 Gyr to more than a Hubble time. Compared to channel (1) for hydrogen transfer in the SD channel (Sect. 21.2.2), the main difference in the initial conditions is that here the ratio of ZAMS masses is closer to unity.

If the initial orbital period is much longer, $>$200 days (Fig. 21.6; right panel), by the time the primary fills its Roche lobe its envelope is convective, causing a CE episode, mass loss from the system, and a drastic decrease in the period. After the

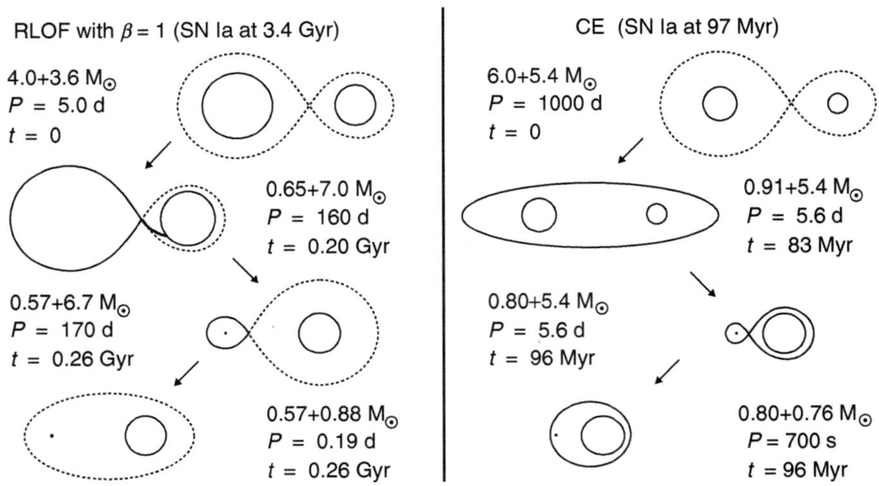

Fig. 21.6 Representation (not to scale) of binary evolution leading to a DD by the RLOF channel (*left panel*) and the double-CE channel (*right panel*). The parameter $\beta = 1$ refers to conservative RLOF. From "The delay-time distribution of Type Ia supernovae: a comparison between theory and observation" (Mennekens et al. 2010) reproduced with permission, © ESO

secondary leaves the main sequence, a second CE episode leads to a very close pair of white dwarfs with an orbital period of a few tens of seconds. The total delay time, in this case dominated by the time to form the second white dwarf, ranges from only 0.05 to 0.2 Gyr (Mennekens et al. 2010).

Regarding the shape of the DTD, super-Chandrasekhar DD models may fare rather well. For most of them, the time from star formation to merger is dominated by the inspiral time of the two white dwarfs, which is proportional to a^4, where a is the separation when the second white dwarf is formed. If the distribution of a is a power law, $n \propto a^{\beta}$, then the DTD will be $\propto t^{(\beta-3)/4}$, and a t^{-1} DTD, as observed, would correspond to $\beta = -1$, which is reasonable. It should be noted that even if the distribution of initial main-sequence separations is a power law as usually assumed, the initial distribution of the white dwarf separations that results from binary evolution is not necessarily also a power law. Nevertheless, some population synthesis calculations (Bogomazov and Tutukov 2009; Mennekens et al. 2010; Yungelson 2010; Toonen et al. 2012) do tend to predict DTD shapes that are consistent with observation, except perhaps for SN Ia occurring within 0.2 Gyr of the starburst (Claeys et al. 2014).

Even if super-Chandrasekhar DD mergers do explode rather than collapse, the predicted occurrence rate falls short of that inferred from observations of typical SN Ia by at least a factor of a few and perhaps as much as a factor of 10 (Toonen et al. 2012; Chen et al. 2012; Meng and Yang 2012; Graur et al. 2014a; Claeys et al. 2014). Observations of binary white dwarfs in the solar neighborhood also indicate that the rate of super-Chandrasekhar DD mergers falls short by a factor of ten or more (Badenes and Maoz 2012).

21.3.3 Noncanonical and Sub-Chandrasekhar Possibilities

In addition to the canonical DD model (Sect. 21.3.1) in which the explosion, if there is one, occurs long after a super-Chandrasekhar merger of two C/O white dwarfs, there are other possibilities. An explosion might occur promptly, at the onset of, during, or shortly after merger. Some of these possibilities may apply to sub- as well as super-Chandrasekhar mergers. The occurrence rate of sub-Chandrasekhar mergers may be comparable to the SN Ia rate (Ruiter et al. 2009; van Kerkwijk et al. 2010; Toonen et al. 2012; Badenes and Maoz 2012), but the traditional view has been that these merger remnants cannot become hot enough to ignite carbon and explode. The models to be mentioned in this section invoke detonations; deflagrations are not involved. As always, care must be taken to discriminate between the conditions for dynamical burning and the more stringent, uncertain, conditions for detonation. As for SD models, if detonation does not occur, then these models will not account for SN Ia. This issue will be discussed in Sect. 22.4.1. The degree to which the possibilities outlined in this section are likely to correspond to the observations of SN Ia will be explored in Sect. 22.4.

In much of the work summarized in Sect. 21.3.1, it was assumed that one of the white dwarfs was significantly more massive than the other. If the white dwarfs have nearly equal masses (high mass ratio, approaching unity), the situation is different. The less massive white dwarf rapidly disrupts in a *violent merger.* If the masses of two merging C/O white dwarfs are at least 0.8 $M_\odot$, an ingoing detonation could be promptly triggered near the surface of the more massive (primary) white dwarf. If the primary explodes while the secondary is largely intact, ^{56}Ni production depends on the central density and thus the mass of the premerger primary. If the secondary has been largely disrupted, the primary is compressed to higher density and additional ^{56}Ni may be produced. The total ejected mass is super-Chandrasekhar, but the result, depending on the precise white dwarf masses, might be the production of a modest amount of ^{56}Ni and an SN 1991bg-like event (Pakmor et al. 2011), or the synthesis of more ^{56}Ni and perhaps a typical SN Ia (Pakmor et al. 2012; Moll et al. 2014; Sato et al. 2015; Sect. 22.4.2), or even the production of as much ^{56}Ni as required by the overluminous events that we refer to as substantially super-Chandrasekhar (Sect. 20.7). Occurrence frequencies conceivably could be comparable to the observed rates of SN 1991bg-likes and substantially super-Chandrasekhar events. Owing to the high masses required, however, the occurrence frequency of violent mergers of two massive C/O white dwarfs is unlikely to be high enough to account for a large fraction of typical SN Ia.

The role of spiral-mode instabilities in accretion disks surrounding either sub- or super-Chandrasekhar mergers of C/O white dwarfs warrants further investigation. For sufficiently high mass ratios, rapid accretion of hot disk material onto the merger remnant may lead to a detonation on a dynamical time scale (Kashyap et al. 2015).

When they form, most C/O white dwarfs have a helium-rich surface layer of $\sim 10^{-3}$ to 10^{-2} $M_\odot$. At the onset of a merger of two C/O white dwarfs, whether sub- or super-Chandrasekhar, the helium layer might detonate and provoke detonation of the underlying C/O of the primary (Pakmor et al. 2013).

If the less massive white dwarf is a helium or He–C/O hybrid white dwarf, instabilities in the helium accretion stream may produce dense knots of material that strike the surface of the primary and raise the temperature high enough to ignite a detonation in the accreted helium layer (Guillochon et al. 2010), possibly causing the C/O white dwarf to detonate centrally in a double-detonation supernova (Sect. 21.2.6). If the accretion stream does not cause a detonation, the subsequently accreted helium could detonate, again perhaps causing a double-detonation event (Dan et al. 2014). A possible problem is that to produce enough ^{56}Ni for typical SN Ia, the mass of the C/O white dwarf should be rather high, around 1.1 $M_{\odot}$ (Piro et al. 2014).

Thermonuclear explosions resulting from collisions of white dwarfs at low impact parameter, perhaps in dense stellar systems such as globular clusters, might produce SN Ia-like signatures (Rosswog et al. 2009; Raskin et al. 2010; Lorén-Aguilar et al. 2010; García-Senz et al. 2013), but probably only at $\sim$1% of the SN Ia rate. In any case, there is little observational evidence to associate SN Ia with globular clusters. An alternative is to consider hierarchical triple-star systems (Kushnir et al. 2013) that are not restricted to dense stellar systems. In a system containing a close pair of white dwarfs, the presence of a third star on a highly inclined orbit may drive the orbit of the close pair to high eccentricity by the Kozai–Lidov mechanism (Kozai 1962; Lidov 1962) and drastically reduce the time until merger, thus raising the merger rate (Thompson 2011). One of the problems with this model is how to avoid a collision of the inner binary when the components are still on the main sequence and their geometrical cross-sections for collision are 100 times greater than when they have become WDs (Hamers et al. 2013)—unless the pair of white dwarfs forms before acquiring the additional companion.

If an explosion does not occur promptly, the differentially-rotating accreted matter is likely to generate a magnetic field and associated viscosity by means of the magnetorotational instability (Schwab et al. 2012). During the 10^4- to 10^8-s viscous phase, a torus of accreted matter may be converted to a nearly spherical envelope. There are several possibilities for explosions, conceivably at a high rate of occurrence, during this phase. Compressional heating of merger remnants of nearly equal-mass C/O white dwarfs, even when their combined mass is sub-Chandrasekhar, may cause central ignition of carbon followed by explosion as an SN Ia (van Kerkwijk et al. 2010; Zhu et al. 2013). In a merger of a C/O white dwarf with a helium white dwarf, viscous heating might produce a detonation in the accreted helium layer and possibly a double-detonation supernova (Schwab et al. 2012). Super-Chandrasekhar merger remnants that explode neither promptly nor in the viscous phase become candidates for explosion much later—in the canonical DD channel (Sect. 21.3.1).

During a merger (Zhu et al. 2015) and the subsequent viscous phase (Ji et al. 2013), the initial white dwarf magnetic fields are strongly amplified, perhaps to $\gtrsim 10^{10}$ G, by the magnetorotational instability. High-resolution 3D MHD simulations will be required to assess the full effects of strong magnetic fields on the fate of the merger remnants.

21.4 Core Degenerates

While the DD channel requires that the two white dwarfs survive the CE phase, binary population-synthesis simulations find that in some cases a white dwarf and a stellar core merge during or shortly after the CE phase (Ruiter et al. 2009; Mennekens et al. 2010). There has been some exploration of whether this merger process could lead to SN Ia. The resulting *core-degenerate* (CD) channel (Kashi and Soker 2011; Ilkov and Soker 2012; see also Livio and Riess 2003) is a relative of the DD channel.

In the CD channel, a primary with ZAMS mass in the range 2–7 $M_\odot$ evolves to become a white dwarf, in the process transferring mass to a secondary of ZAMS mass $\gtrsim 1$ $M_\odot$. The secondary becomes an AGB star and loses mass during a CE episode. Late in the CE phase, or during the planetary nebula phase within $\sim 10^5$ years after the CE phase, while the C/O core of the secondary is still hot, the white dwarf and the hot core merge. Usually the white dwarf is less massive and less dense than the core, so it is disrupted and accreted onto the core. Because the core is hot and therefore oversized, by a factor $\gtrsim 1.2$ in radius, carbon accreted from the white dwarf is in a less deep gravitational potential than in the canonical DD channel and reaches a lower peak temperature. Thus the core is less likely to ignite off-center. This possibility to avoid off-center carbon ignition and subsequent collapse to a neutron star (Sect. 21.3.1) is critical for the CD channel. The merger process and the predicted characteristics of the merger remnant (rather similar to those of a DD merger remnant) were discussed by Aznar-Siguán et al. (2015).

A range of spin-down times is possible in the CD channel, depending on the mass of the rapidly rotating merger remnant and the action of magnetic fields. If the merger product spins down by magnetic-dipole radiation (on a timescale less than a Hubble time for a field substantially exceeding 10^7 G; Ilkov and Soker 2012), carbon may ignite centrally provided that the mass of the merger product exceeds M_{Ch}. More massive merger products are expected to have shorter spin-down times. This property could be related to the tendency for younger populations to produce more luminous SN Ia. In this channel, as in the canonical DD channel, most of the explosions would be only mildly super-Chandrasekhar, but a small minority could be substantially super-Chandrasekhar (Sect. 20.7). The implied evolution proceeds from central carbon ignition to deflagration to detonation as in the canonical SD and DD channels.

A less common situation would be for the white dwarf to be more dense (but not necessarily more massive, since the core is hot) than the core of the secondary (Soker et al. 2013). In this case, the core would be disrupted and accreted onto the cooler white dwarf. A long phase of viscous accretion of the disrupted material might be avoided and the explosion might occur only months or years after the CE phase, with the presence of nearby CSM perhaps accounting for SN Ia that undergo strong interaction with hydrogen CSM (Sect. 23.2). This configuration, however, again could be susceptible to off-center carbon ignition and the production of a neutron star instead of a supernova.

Estimates of the rate at which core degenerates might produce supernovae vary from high, comparable to the SN Ia rate (Ilkov and Soker 2013), to low, a few percent of the SN Ia rate (Zhou et al. 2015).

21.5 Observed Candidate Progenitor Systems

The reality of binary evolution is likely to be far more complex than our current models and understanding. A useful complement to the perilous approach of binary evolution and population synthesis is to consider individual observed binary systems that appear to be SN Ia progenitor candidates. A review of observed binary systems that contain a white dwarf and a nondegenerate companion identified a number of single-degenerate systems that may be SN Ia candidates (Parthasarathy et al. 2007). These systems deserve further observational scrutiny and theoretical modeling.

21.5.1 Recurrent Novae

Between thermonuclear outbursts, RN (Sect. 21.2.1) accrete from years to a century. Theoretically, the short interval between outbursts requires that the accreting white dwarf be near or above M_{Ch} and that the accretion rate be high (Starrfield et al. 1985; Truran and Livio 1986; Yaron et al. 2005; Wolf et al. 2013). Observationally, the white dwarfs of RN do tend to have rather high masses, 1.2 $M_\odot$ or more, suggesting that they were either born as high-mass white dwarfs or are in the process of growing toward M_{Ch}. The high-mass, high-surface-gravity white dwarfs are expected to accrete more mass between outbursts than they eject during outbursts. Thus RN have been a popular candidate for SD progenitors of SN Ia (Hillman et al. 2015). SN PTF11kx and related events (Sect. 23.2) show evidence of CSM resembling that around RN. Nevertheless, RN may not provide a robust solution to the SN Ia progenitor problem, for several reasons. Some RN probably involve ONeMg white dwarfs, which will not produce SN Ia even if they reach M_{Ch}. In addition, the number of RN in the Galaxy and in M31 may be too low for RN to be progenitors of a large fraction of SN Ia (Soraisam and Gilfanov 2015; Shafter et al. 2015). A quarter of Galactic events classified as classical novae could, however, be RN for which a repeated burst did not happen to be observed (Pagnotta and Schaefer 2014). Furthermore, the population of RN in M31 could be much larger than traditionally thought (Williams et al. 2014).

Two examples of well-studied RN, U Sco and RS Oph, are discussed here. Other SD hydrogen-transferring SN Ia candidates include QU Carinae (Kafka et al. 2012), V Sge (Hachisu and Kato 2003), V407 Cyg (Hachisu et al. 2012), T CrB (Kato and Hachisu 2012), CI Aql (Sahman et al. 2013), V617 Sgr (Shi et al. 2014), and T Pyx (Patterson et al. 2014; Godon et al. 2014).

U Sco, a well-known RN with a subgiant donor, erupts once per decade, most recently in 2010. The system is an eclipsing binary, semi-detached, with an orbital period of 1.23 days. From orbital data, the mass of the white dwarf was found to be $1.55 \pm 0.24\ M_\odot$ (Thoroughgood et al. 2001). The mass of the subgiant is $\sim 0.9\ M_\odot$. Additional reasons to think that the white dwarf is a rotationally-supported super-Chandrasekhar object were given by Hachisu et al. (2012). The ejecta of the 2010 outburst may or may not have been enriched in neon (Mason et al. 2012; Mason 2013). Neon enrichment might suggest that the white dwarf is composed of ONeMg rather than C/O, but a caveat is that such an enrichment might result from helium-shell flashes atop a C/O white dwarf (Sect. 21.2.1).

Careful eclipse monitoring provided a measure of the orbital period increase after the 2010 eruption of U Sco and therefore a measure of the mass ejected in the outburst. The amount of mass gained since the previous outburst, based on estimates of the accretion rate during quiescence, may or may not have been less that that ejected in the eruption (Pagnotta et al. 2015).

RS Oph is a well-known RN with a red-giant donor. It erupts about every 20 years, most recently in 2006. It is a symbiotic system with an orbital period of 455 days. The white dwarf accretes from a wind or from RLOF (Wynn 2008). By modeling the 2006 outburst as a thermonuclear runaway, Hachisu et al. (2007) estimated the mass of the white dwarf to be $1.35 \pm 0.01\ M_\odot$ and the accretion rate in quiescence to be $\lesssim 10^{-7}\ M_\odot\ y^{-1}$. The mass of the white dwarf appears to be increasing (Osborne et al. 2011; Skopal 2015), but an orbital period change associated with the eruption has yet to be measured. As with U Sco, an issue is whether RS Oph contains an ONeMg white dwarf.

RS Oph is a representative of the binary-evolution channels in which the donor star is a red giant. There are two serious constraints on this class of models. One is the issue of whether a red-giant donor can drive a C/O white dwarf to M_{Ch}. The other is that observational constraints suggest that few if any typical SN Ia have red-giant companions at the time of the explosion (Sect. 21.5.2). A caveat is that the red giant might become a white dwarf before the first white dwarf spins down and explodes (Sect. 21.2.5).

An RS Oph-like RN in M31 undergoes outbursts on a timescale of only 1 year. The white dwarf in this system is likely to be near-M_{Ch} ($\gtrsim 1.35\ M_\odot$) and accreting between outbursts at a rate $\gtrsim 10^{-7}\ M_\odot\ y^{-1}$ (Tang et al. 2014; Kato et al. 2014; Henze et al. 2015). This RN joins U Sco as a promising single-degenerate candidate (Darnley et al. 2015). Within $\sim 10^6$ years the white dwarf should explode or collapse, depending on its composition, unless it experiences a prolonged spin-down phase.

21.5.2 Other Possibilities

Several interesting helium-donor SN Ia candidates are known. HD 49798, with an associated eclipsing X-ray source RX J0648.0-4418, contains a massive ($1.28 \pm 0.05\ M_\odot$) rapidly rotating (13 s) white dwarf accreting from the wind of a hot, helium-rich subdwarf of mass $1.50 \pm 0.05\ M_\odot$ (Mereghetti et al. 2009). The orbital period is 1.55 days. When the donor expands at the end of its helium-shell burning phase, RLOF may bring the white dwarf to M_{Ch}, or, considering its rapid rotation, beyond M_{Ch}. If the white dwarf is C/O rather than ONeMg, as seems likely (Liu et al. 2015a), this system may be a progenitor of a substantially super-Chandrasekhar SN Ia (Wang and Han 2010; Mereghetti et al. 2013). Another candidate is V445 Puppis (Nova Puppis 2000), which contains a white dwarf of estimated mass $\gtrsim 1.35\ M_\odot$ that accretes from a helium-rich companion (Woudt et al. 2009; Kato and Hachisu 2012).

Searches for close pairs of white dwarfs that would be DD candidates have not identified any that have combined masses exceeding M_{Ch} and that also can merge within a Hubble time (Napiwotzki et al. 2001; Badenes et al. 2009).[5] Some systems that may *evolve* to become good DD candidates have been found. KPD 1930+2752 contains a massive white dwarf and an extreme horizontal-branch sdB star with combined mass 1.36–1.48 $M_\odot$ (Maxted et al. 2000; Geier et al. 2007). The sdB star may become a white dwarf and the two stars may ultimately merge. (Alternatively, if the sdB star fills its Roche lobe before becoming a white dwarf, stable mass transfer may produce a SD SN Ia.) AR Pav is a symbiotic binary of orbital period 605 days containing a $\sim 1\ M_\odot$ white dwarf that is accreting via RLOF from an M6 giant of 2.5 $M_\odot$ (Quiroga et al. 2002). A CE phase may produce a close pair of white dwarfs of total mass $\sim 1.5\ M_\odot$ that could explode as a DD SN Ia, or continued stable RLOF may cause AR Pav to evolve to an SD candidate resembling RS Oph (Mikołajewska 2013).

21.6 Other Evidence Constraining Progenitor Systems

21.6.1 Archival Observations at Supernova Sites

As discussed in Sect. 3.2, detection of SN Ia progenitor systems in archival optical images has not been securely achieved (Maoz and Mannucci 2008; Smartt 2009). Archival optical images at the site of SN 2011fe (Sect. 20.2.1) excluded the presence of red-giant and most helium-star companions, as well as main-sequence stars more massive than 3.5 $M_\odot$ at the time of explosion. Archival *HST* images at the site of SN 2014J (Sect. 20.2.2) also excluded the presence of a red giant.

For SN 2011fe, the limits on the radius of a secondary star that filled its Roche lobe, $\lesssim 0.1\ R_\odot$ constrain even lower-mass main-sequence stars as companions. Transfer by stable RLOF requires a mass of $\gtrsim 1.2\ M_\odot$ (Schaefer and Pagnotta 2012) and thus a radius $\gtrsim 1.2\ R_\odot$. While this constraint applies only to SN 1011fe,

[5] A few systems of unknown orbital inclination may qualify, but the masses of the secondaries are so low that the systems would be expected to evolve to an condition of stable mass transfer and possibly produce underluminous explosions (Brown et al. 2013; Chap. 23).

this event is solidly within the Core-Normal classification. The constraint on main-sequence donors in the progenitor system of SN 2011fe thus has general implications, because many cataclysmic variables that might be generically linked to SN Ia in the SD channel have main-sequence companions.

Accreting white dwarfs emit a high luminosity in supersoft X-rays (20–80 eV). Thus detection in presupernova observations of soft X-rays at the site of an SN Ia would constitute a direct detection of an SD SN Ia progenitor (unless DDs heat to X-ray emitting temperatures well in advance of merger). The only tentative detection so far (Voss and Nelemans 2008) is, however, ambiguous (Roelofs et al. 2008). Nondetection of archival X-ray, UV, and He II λ4686 emission imposed constraints on the progenitor system of SN 2011fe; nondetection of archival X-rays also constrained the progenitor of SN 2014J. Upper limits on X-rays from archival observations at the sites of some other SN Ia are marginally in conflict with expectations of the SD channel (Nielsen et al. 2012), unless the X-rays are reprocessed to longer wavelengths by CSM or the white dwarf was no longer accreting at the time of observation, as in the spin-up/spin-down model (Sect. 21.2.5).

21.6.2 Statistics of Supersoft X-Ray Sources

As outlined in Sect. 21.2.2, one possible class of SN Ia progenitors is the supersoft X-ray sources, which are presumed to be systems in which mass transfer leads to hot, bright, stable, double-shell burning near the surface of the white dwarf. Therefore an alternative to archival X-ray observations at SN Ia sites is to examine the statistics of supersoft sources.

In nearby galaxies, and also within a kiloparsec of the Sun, there are fewer known supersoft sources than expected in the SD channel, by one or two orders of magnitude (Di Stefano 2010a). The DD channel might have a similar problem. Before the CE episode that produces the second white dwarf, the first white dwarf may accrete from a red-giant wind to produce an X-ray emitting symbiotic system (Di Stefano 2010b). The duration of a supersoft phase during the evolution of DD progenitors could, however, be negligible (Nielsen et al. 2014b).

The low integrated flux of X-rays from early-type galaxies also has been interpreted to mean that SN Ia in such galaxies must come from DDs, not SDs (Gilfanov and Bogdán 2010). It has been argued, though, that accreting white dwarfs with red-giant donors (presumably the only SD donors in the old populations of early-type galaxies; Sect. 21.2.2) spend only a small fraction of the time burning steadily (Hachisu et al. 2010; Kato 2013).

If SN Ia do come from canonical SD systems, the X-rays may be reprocessed to longer wavelengths by CSM (Wheeler and Pooley 2013; Nielsen et al. 2013). If X-rays are reprocessed to the UV, the latter might lead by photoionization of the ISM to detectable recombination lines of He II and other emission lines that would not be expected from galaxies with low current rates of star formation (Woods and Gilfanov 2014). An application of the He II test to a sample of galaxies that have had low rates of star formation during the past Gyr indicated that the SD channel can account for no more than 10% of the SN Ia rate in such galaxies (Johansson et al. 2014).

21.6.3 Breakout and Interaction with an Accretion Disk

In the SD channel, the breakout of a detonation at the surface of the white dwarf is expected to produce an X-ray and γ-ray flash lasting a fraction of a second (Höflich and Schaefer 2009; Piro et al. 2010). Interaction of the outermost ejecta with an accretion disk could produce additional high-energy radiation, lasting for seconds (Höflich and Schaefer 2009). No such flashes associated with SN Ia have been found in catalogs of gamma-ray bursts, but the high-energy radiation could be reprocessed by CSM. Breakout should be followed by an optical/UV fireball lasting about an hour (Piro et al. 2010; Nakar and Sari 2010; Rabinak et al. 2012); this remains unconstrained by observations.

In the DD channel, the explosion may (Fryer et al. 2010) or may not (Dan et al. 2012) be sufficiently enshrouded by nearby CSM to cause the breakout flash to be much longer and softer than that from a bare white dwarf. A long, soft flash might conflict with some observations. If the detonation occurs between 1000 s and 1 day after the merger, interaction of the ejecta with matter launched by the accretion disk as a wind may produce UV radiation in conflict with some observations (Levanon et al. 2015).

21.6.4 Interaction with a Donor Star

In the SD channel, a SN Ia may interact with the donor star in several ways. The collision of the ejecta with a donor star was predicted to produce an X-ray flash lasting minutes to hours and having favorable orientation for observation in $\sim$10% of the cases, as well as optical/UV emission that exceeds the radioactivity-powered emission for several days (Kasen 2010). The signatures of interaction were predicted to be somewhat less favorable for observation by Maeda et al. (2014b) and Kutsuna and Shigeyama (2015). Nevertheless, the lack of interaction emission from SN 2011fe excluded the presence of a red-giant companion and even restricted the presence of a main-sequence companion (Sect. 20.2.1). More generally, the lack of such signatures in large samples of SN Ia could limit the fraction that can have red-giant companions at the time of explosion (Hayden et al. 2010a; Bianco et al. 2011; Brown et al. 2012b; Silverman et al. 2012a). Excess UV and blue emission during the first days after the explosion of the typical Type Ia SN 2012cg (Fig. 21.7) was interpreted in terms of interaction with a main-sequence companion of about 6 $M_\odot$ (Marion et al. 2016). Past observations of almost all SN Ia would not have been able to detect such early, faint emission, even if the viewing angle were favorable.

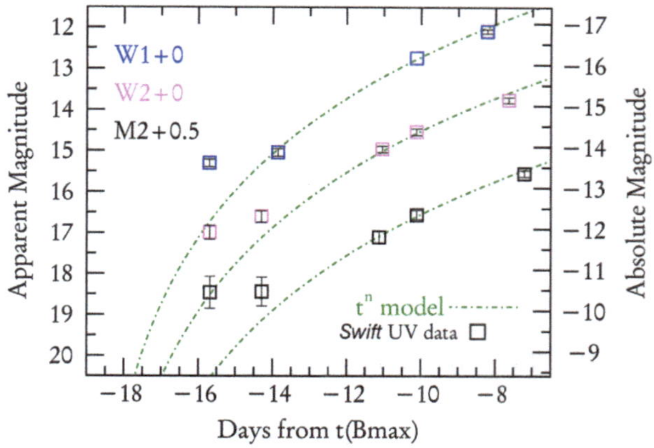

Fig. 21.7 *Swift* photometry for SN 2012cg showing excess flux in the UV that also was seen in blue optical light about 16–14 days before maximum. In each UV band, model light curves that scale as $t^{3.6}$ are plotted for reference (*green dotted-dashed lines*). From "SN 2012cg: Evidence for Interaction Between a Normal Type Ia Supernova and a Nondegenerate Binary Companion" (Marion et al. 2016). © AAS. Reproduced with permission

Most of the matter that is ablated or stripped from the donor star, 0.1–0.2 $M_\odot$ for a main-sequence star and practically the entire envelope of a red giant, is expected to become entrained inside the bulk of the supernova ejecta, expanding at low velocity (Wheeler et al. 1975; Marietta et al. 2000; Pan et al. 2012a; Liu et al. 2012c). Searches for narrow hydrogen or helium lines in nebular-phase spectra of SN Ia have not been successful; in some cases the upper limit to the amount of solar-abundance material has been estimated to be as low as 0.01 $M_\odot$, lower than expected from any donor that filled its Roche lobe at the time of explosion (Leonard 2007; Lundqvist et al. 2013). For SN 2011fe, the upper limit is even more stringent, less than 0.001 $M_\odot$, inconsistent with canonical SD models but not necessarily with spin-up/spin-down models and perhaps M-dwarf donors.

A caveat to the observational estimates of the amount of stripped hydrogen is that they depend on a simple model of the subsequent nebular phase (Mattila et al. 2005). Based on general properties of the models of Marietta et al. (2000), the hydrogen was assumed to be confined to a spherical, constant-density region interior to a velocity of 1000 km s^{-1} within the structure of the W7 explosion model. In practice, the hydrogen distribution will be the result of a complex 3D interaction with the hydrogen initially confined to a rather narrow cone. If this hydrogen is entrained by turbulence in the supernova ejecta, it could be swept out and diluted in a complex distribution of smaller optical depth to γ-rays than assumed by Mattila et al.

Main-sequence donors run away from the explosion site at velocities on the order of their orbital velocities, ranging from 100 to 200 km s^{-1}. They are predicted to be out of thermal equilibrium and overluminous, $L \sim 10$ to 10^3 $L_\odot$, for 10^3 to 10^4 years (Shappee et al. 2013a). Because they had been corotating with their orbits they would be rapid rotators, typically in the range 30–140 km s^{-1}, unless the rotation rate was severely reduced by the interaction with the supernova ejecta (Pan et al. 2012a; Liu et al. 2013).

Helium-star donors are so compact that they are hardly stripped by the explosion. Models suggest that energy deposited by the impact of ejecta causes helium-star donors to expand and evolve to a luminous phase of $L \sim 10^4\ L_\odot$ about 10 years after the explosion; these post-explosion donor stars could be sought at the sites of SN Ia that occurred decades ago (Pan et al. 2013).

21.6.5 Circumstellar Interaction

Upper limits on radio, UV, and X-ray emission (Panagia et al. 2006; Immler et al. 2006; Hancock et al. 2011; Russell and Immler 2012; Brown et al. 2012b) as well as nondetection of narrow lines in optical spectra (Mattila et al. 2005; Lundqvist et al. 2013), can be used to constrain the strength of CSI and the presupernova mass-loss rate of SN Ia progenitors. We defer to Chap. 23 a discussion of a class of supernovae, SN 2002ic-likes, at least some of which appear to be SN Ia so strongly interacting with circumstellar hydrogen that there is potential for confusion with SN IIn. These events are far less numerous than typical SN Ia and they presumably require some special evolutionary sequence, e.g., as discussed in Sect. 21.2.5.

For SN 2011fe (Sect. 20.2.1), nondetections in the radio and X-rays excluded a steady-state wind from a red-giant donor. The limit on the mass-loss rate was $\lesssim 10^{-9}$ $M_\odot\ y^{-1}$ for a wind velocity of $\sim$100 km s^{-1}. The upper limit on a constant-density CSM was $\sim$6 cm^{-3}, higher than the mean ISM density, but low for a substantial CSM. Similarly, for SN 2014J (Sect. 20.2.2), nondetections of radio and X-rays imposed stringent limits: $\lesssim 6 \times 10^{-10}\ M_\odot\ yr^{-1}$ for a wind velocity of $\sim$100 km s^{-1} and $\sim$1 cm^{-3} for constant-density CSM. It is possible that low-density bubbles driven by fast winds or sweeping of CSM into shells by nova outbursts could satisfy these limits (Dimitriadis et al. 2014; Harris et al. 2016; Dragulin and Höflich 2016).

In two particular cases, SN 2006X (Patat et al. 2007; Fig. 21.8) and SN 2007le (Simon et al. 2009; see also SN PTF11kx in Sect. 23.2), time variations in the strength of narrow Na I absorption features in high-resolution spectra were interpreted in terms of CSM. The UV radiation of the supernova partially ionizes circumstellar sodium, which subsequently recombines, causing the Na I absorptions to vary with time in a manner dependent on the distance of the absorbing material from the supernova. In both SN 2006X and SN 2007le, the velocity of the absorbing material, $\lesssim$100 km s^{-1}, suggested red-giant donors. Similar absorptions were observed after the 2006 outburst of the SN Ia progenitor candidate RS Oph (Patat et al. 2011a), an RN that contains a red giant donor (Sect. 21.5.1). The fast ($\sim$1000 km s^{-1}) ejecta of an RN is decelerated to lower velocities by interaction with CSM from the red-giant wind (Moore and Bildsten 2012). Based on the small number of SN Ia that have been observed at high resolution at multiple epochs, the fraction of SN Ia that show variable Na I absorption appears to be $\sim$20% (Maguire et al. 2013; Sternberg et al. 2014). Some other SN Ia may have CSM that is off the

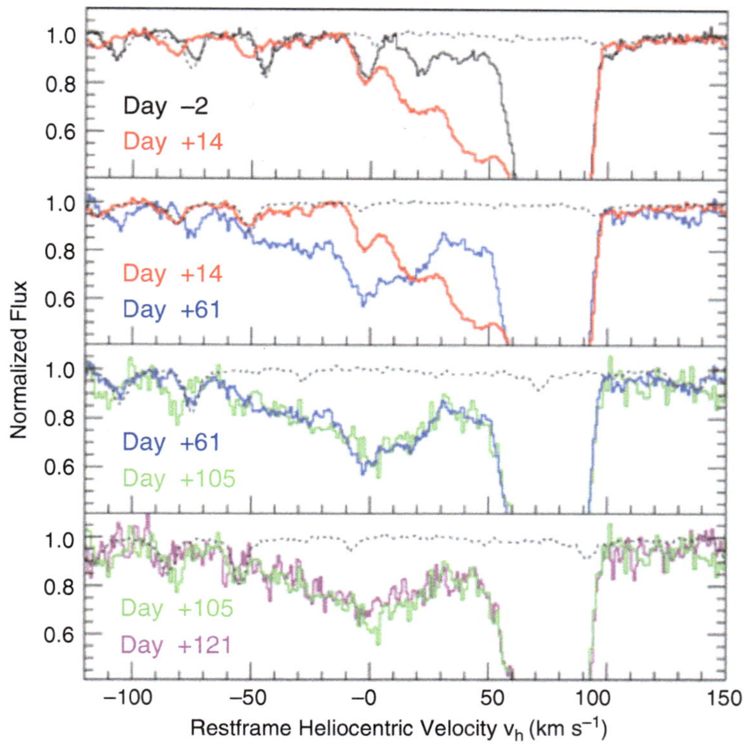

Fig. 21.8 Time evolution of the wavelength region of the Na I D_2 component. Epochs are with respect to the time of maximum light in the B band. The strong absorption centered on 75 km s^{-1} is from ISM in the host galaxy. The time-varying absorptions at lower velocities are interpreted as forming in CSM. The *dotted lines* show the telluric absorption spectrum. From "Detection of Circumstellar Material in a Normal Type Ia Supernova" (Patat et al. 2007), by permission of Oxford University Press on behalf of the Royal Astronomical Society

line of sight of the observer, or too far from the explosion to be ionized. Alternative interpretations to produce variable Na I absorption are tidal tails in the context of white dwarf merger models (Raskin and Kasen 2013) and planetary-nebula shells in the context of the CD model (Tsebrenko and Soker 2015). Variability in narrow lines of K I was observed in SN 2014J.

Narrow interstellar absorption lines that form in the turbulent ISM of host galaxies are expected to display statistically similar distributions of blueshifts and redshifts. In contrast, narrow Na I absorptions in spectra of SN Ia (some known to have variable Na I absorption and some observed at sufficiently high resolution only at a single epoch) tend to be more blueshifted than redshifted (Sternberg et al. 2011), suggesting outflow of CSM. The SN Ia with blueshifted Na I absorptions tend to be spectroscopic Broad-Line events (Foley et al. 2012; Blondin et al. 2012), to have broad light curves (Maguire et al. 2013), and to have high sodium abundances in the CSM (Phillips et al. 2013).

The presence of CSM usually is regarded as support for the SD channel, but four ways in which DD systems also may produce CSM will be mentioned here. First, at formation, a helium white dwarf is expected to have a hydrogen-rich surface layer of $\sim 10^{-3}\,M_\odot$. Hundreds to thousands of years prior to the merger of a helium and a C/O white dwarf, the hydrogen layer could be transferred to the C/O white dwarf and burned as in a classical nova. In response, the white dwarf may expand and overflow its Roche lobe, with the hydrogen being ejected by a common-envelope episode at the orbital velocity of the helium white dwarf, $\sim$1500 km s^{-1} (Shen et al. 2013). In spiral galaxies, blueshifted, narrow Na I absorptions might form in the surrounding swept-up shell provided that enough of the sodium is neutral. No such signatures would be expected to be associated with SN Ia in elliptical galaxies because sodium would be ionized in the low density ISM, and none have been seen. Second, the merger process may be preceded by the ejection of "tidal tails" having velocity of about 2000 km s^{-1} and mass $\sim 10^{-3}\,M_\odot$ that sweep up ISM and eventually produce slow, dense shells in which narrow, blueshifted Na I absorptions could form (Raskin and Kasen 2013). Third, strong amplification of magnetic fields during the viscous phase (Sect. 21.3.3) could lead to magnetized outflows from post-merger accretion disks (Ji et al. 2013). Fourth, the dynamical phase of the merger process might eject $\lesssim 10^{-2}\,M_\odot$ of matter at $\sim$3000 km s^{-1}, leading to shells that could eventually resemble SNRs (Tanikawa et al. 2015).

The early spectra of SN Ia often have high-velocity features (HVFs) that are detached from the photosphere (Sect. 20.4). If HVFs result from sweeping CSM into a shell, the lack of an obvious signature of CSI in the SN Ia light curves requires that at the time of explosion the CSM must be close to the white dwarf, within a distance much less than 10^{15} cm, so that any excess thermal energy from the interaction is rapidly dissipated. The CSM could be $\sim 10^{-2}\,M_\odot$ of material from an accretion disk, RLOF, or a common envelope (Gerardy et al. 2004b; Quimby et al. 2006; Tanaka et al. 2006), or from magnetic bottles associated with M-dwarf donors (Sect. 21.2.4). The possibility that HVFs have causes unrelated to CSM is not excluded (Sect. 20.4).

21.6.6 Supernova Remnants

SNRs that are believed to have resulted from SN Ia are rather symmetric, more so than remnants of core-collapse events (Lopez et al. 2011). This is consistent with the low observed continuum polarization of SN Ia (Sect. 20.4) and in conflict with SN Ia models that would produce highly asymmetric ejecta.

Some young SNRs (Sect. 7.4) can provide clues to the progenitors of the explosions that produced them. Spectra of light echoes showed that Tycho's SN 1572 was a spectroscopic Core-Normal SN Ia (Krause et al. 2008b). A G-type subgiant star near the center of the SNR was suggested to be the surviving donor in a SD progenitor system (Ruiz-Lapuente et al. 2004), but the identification is not conclusive (Kerzendorf et al. 2013; Liu et al. 2013; Bedin et al. 2014). No former companions to SN 1006 (González Hernández et al. 2012; Kerzendorf et al. 2012) or Kepler's SN 1604 (Kerzendorf et al. 2014a) have been found.

Modeling of the X-ray spectra of SN 1572 is consistent with ongoing CSI (Chiotellis et al. 2013). An X-ray emitting arc inside the remnant of SN 1572 might be a result of interaction between the supernova ejecta and material ablated from the donor star in a SD system (Lu et al. 2011).

Fast winds could blow a cavity around an SN progenitor system (Sect. 21.6.5), but such cavities are not compatible with the X-ray emission from the forward shock and from shocked ejecta of some young Galactic SN Ia remnants, particularly SN 1572 (Badenes et al. 2007). On the other hand, the Galactic SNR RCW 86, the putative remnant of SN 185 AD (Zhou et al. 2006), does appear to have been produced by an SN Ia that exploded into a wind-blown bubble, the parameters of which suggest an SD progenitor for this system (Badenes et al. 2007; Williams et al. 2011; Broersen et al. 2014).

In the Kepler SNR, interaction of the ejecta with CSM (Chiotellis et al. 2012; Burkey et al. 2013) and the presence of dust in the CSM (Williams et al. 2012) may suggest an SD progenitor system with an AGB-star donor. This raises the issue of whether SN 1604 could have been a strongly interacting SN 2002ic-like system (Sect. 23.2).

Light-echo spectra of the supernovae that produced SNR 0509-67.5 and SNR 0519-69.0 in the LMC were those of a SN 1991T-like SN Ia and a typical SN Ia, respectively (Rest et al. 2005, 2008). The central regions of SNR 0509-67.5 contain no surviving companion stars brighter than $M_V = 8.4$ (Schaefer and Pagnotta 2012) or $M_V = 8.15$ after correcting for extinction (Di Stefano and Kilic 2012), apparently inconsistent with an SD system containing a helium-star, red-giant, or main-sequence star of more than about 1 $M_\odot$ (Shappee et al. 2013a). The more crowded central regions of SNR 0519-69.0 contain stars that are not excluded as former main-sequence donors in SD systems, but the presence of helium-star and red-giant donors at the time of explosion are excluded (Edwards et al. 2012). Both SNRs are consistent with the DD channel, with lengthy spin-downs in the SD and CD spin-up/spin-down channels, and with M-dwarf donors.

21.6.7 White Dwarf Populations

In the SD channel, all donors eventually become single white dwarfs. The descendants of main-sequence donors might account for an observed population of single white dwarfs that have high space velocities, yet are not particularly old (Hansen 2003). Descendants of red-giant donors might account for an observed population of single, low-velocity, very low-mass (<0.4 $M_\odot$) white dwarfs that is difficult to otherwise explain (Justham et al. 2009). Descendants of helium-star donors would become *hypervelocity* white dwarfs that have space velocities $> 400\,\mathrm{km\,s^{-1}}$ (Wang and Han 2009; Pan et al. 2013; Geier et al. 2013), many of them exceeding the escape velocity of the Galaxy. Most hypervelocity stars are thought to come from disruption of close binaries by the supermassive black hole at the center of the Galaxy, but US 708 (also known as HVS 2), a hypervelocity (1000 km s^{-1}) helium-rich subdwarf (sdO/B) star that originated in the Galactic disk or halo, may have been a former donor in a SD progenitor system (Geier et al. 2015; Brown et al. 2015c).

21.6.8 *Abundance Constraints*

In near-M_{Ch} models, the central density at the time of explosion is high enough, $\sim 10^9$ g cm^{-3}, for electron captures to produce greater concentrations of neutron-rich species such as ^{58}Ni and ^{55}Mn than in most violent mergers and white dwarf collisions, where the maximum densities achieved are only $\gtrsim 10^7$ g cm^{-3}. Measures of density at explosion can then discriminate between these channels. Most abundance constraints favor high density. The strengths of [Ni II] λ7378 (Nomoto et al. 2013a) and the 1.98 μm (Friesen et al. 2014) emission lines in nebular-phase spectra (after ^{56}Ni has decayed) were taken to indicate that ^{58}Ni was produced in high-density explosions. Similar arguments for high-density explosions were based on an analysis of the [Fe II] 1.644 μm emission line in nebular-phase spectra (Diamond et al. 2015), on the solar ratio of Mn/Fe (Seitenzahl et al. 2013a), and on high observed Ni/Fe and Mn/Fe ratios in a SNR thought to be the result of a SN Ia (Yamaguchi et al. 2015).

21.7 Summary

The quest to identify the progenitors of SN Ia is far from complete. Nevertheless, new observations and theoretical developments have refined the constraints to be brought to the problem. While there are two basic progenitor channels, some form of cataclysmic variable in the canonical SD case and merging white dwarfs in the DD case, the options are more varied when complemented with the spin-up/spin-down and related possibilities.

The traditional cartoon-level picture of the SD channel, a red-giant transferring mass to a white dwarf and growing it to M_{Ch}, is under assault. Former red-giant companions or their stripped cores or traces of their stripped envelopes or evidence of ejecta impacting with them are not seen. Evidence for the existence of AGB stars in some cases, due to blue shifted or variable Na I lines, exists, but the preponderance of the evidence is that the majority of typical SN Ia do not arise in systems in accord with the simple view of a red-giant transferring mass up to the time of explosion. Evidence for the role of red giants in some systems should be sought in CSI, perhaps decades after the explosion when the ejecta might catch up with matter ejected long before the explosion or piled up on the boundaries of a wind-blown bubble. X-ray observations of young SNRs imply that the SN Ia did not explode into high-velocity wind-blown shells.

Related, refined arguments based on statistical samples and individual systems also argue against a high fraction of SN Ia arising in systems with subgiant companions or even with solar-mass main sequence companions at the time of the explosion. While variations on cataclysmic variables, especially RN, remain popular candidates, it is not clear that they can do the job. Some probably involve ONeMg white dwarfs, and in some the white dwarf, over time, may be losing mass, not gaining it. Supersoft sources remain candidates. Very low-mass main-sequence stars might survive current observational limits, but making an explosion from them is problematic.

The absence of evidence for a former companion star in young SNRs and in nearby well-studied events like SN 2011fe is consistent with DD models that predict no surviving companion, although it is possible that at the time of explosion the former companion is a white dwarf as in the spin-up/spin-down model. DD models also naturally account for the lack of any observed swept-up hydrogen in the ejecta.

As will be explored in more detail in Chap. 22, explosion models that invoke runaway carbon ignition at near-M_{Ch} are compatible with a range of observations of SN Ia. This statement supports a basic SD model wherein a white dwarf is grown from below to near M_{Ch}. It is also possible, however, that differential rotation can support more than M_{Ch}, so the limit might in some sense be approached from above as a rapidly-spinning white dwarf slows down. Such putative rapidly-rotating progenitors could arise in variations of both the SD and DD channels. Violent mergers in the DD channel might lead to explosions, but whether they robustly account for observations requires further quantitative investigation.

Stellar evolution, binary-star evolution, and population-synthesis calculations are important ingredients in the attempt to determine the progenitors of SN Ia. All are complex with manifest uncertainties.

The variety of detailed properties of SN Ia as presented in Chap. 20 might require a variety of progenitor mechanisms. A key issue is the explanation of the strong majority of typical SN Ia. That would seem to call for a dominant channel of progenitor evolution, but if so, that channel has yet to be firmly identified.

Chapter 22
Explosion Models

22.1 Introduction

All models of SN Ia can be subjected to crude tests, such as whether an appropriate amount of ^{56}Ni and kinetic energy is produced, and whether IMEs are ejected in a suitable velocity range. The more stringent test of a numerical explosion model is when radiative-transfer calculations for the detailed composition, density, and geometrical structure of the model produce light curves and, especially, time series of spectra and spectropolarimetry that can be compared with observations. Among models of SN Ia that can be tested in this way, none succeed entirely, but some fare reasonably well and some fail. Reviews of SN Ia explosion models include those by Hillebrandt and Niemeyer (2000) and Höflich et al. (2013).

A challenge for explosion models of SN Ia is to understand properly both the typical events and the intricate diversity that has been elucidated by ever more thorough observations (Sects. 20.4, 20.5, and 20.8). The impressive overall homogeneity of SN Ia is due in large part to basic physics (Höflich et al. 2013). The structure and binding energy of a near-M_{Ch} white dwarf is primarily dictated by quantum physics. The nuclear energy released is roughly the same for a range of elements from carbon to silicon whether they burn to other IMEs or to IGEs. If a substantial portion of the white dwarf is burned to IGEs, then the amount of ^{56}Ni and the peak luminosity will have relatively little variance. This means that SN Ia arising from near-M_{Ch} C/O white dwarfs have roughly similar properties independent of the details of the progenitor evolution, a sort of "stellar amnesia" with regard to the evolutionary history. Still, accounting for the full diversity of SN Ia requires variations of the explosion process; whether different classes of explosions are required remains to be resolved. Clues to the progenitor evolution (Chap. 21) and explosion physics may have to be sought in more subtle, second-order effects. An implication is that differences in the evolution of SN Ia progenitors may be larger than suggested by the relatively small observational differences among SN Ia. The complementary perspective is that if a model does not fall within

D. Branch, J.C. Wheeler, *Supernova Explosions*, Astronomy and Astrophysics Library, DOI 10.1007/978-3-662-55054-0_22

known observational characteristics, be cautious. Of course if it is your model, then the level of tolerance for discrepancy is greater.

Another important criterion for success of an explosion model is to treat the physics realistically and self-consistently. This is a challenge for models of SN Ia because some of the physics is not sufficiently well understood. Basic issues include angular-momentum transport when white dwarf rotation is considered, turbulent combustion, and the initiation and propagation of detonations capturing the full 3D nature of detonation cell structure (Chap. 19). At the current state of the art, these have to be parameterized at some level. The choices of the parameters are somewhat subjective, but are constrained by observations and physics.

With our current understanding, all reasonably successful models of SN Ia require a detonation. While progress is being made in understanding the turbulent combustion physics that may lead to a transition from deflagration to detonation, models still require assumptions about when a detonation occurs. Some models do this in a more plausible, self-consistent way than others. There is a tendency in some literature to assume that nuclear burning on a shorter timescale than hydrodynamic motion leads to a detonation. This condition is necessary, but it is not sufficient. Dynamical burning could just as well correspond to the onset of subsonic deflagrative burning, as it does in many models; dynamical burning doth not (necessarily) a detonation make. The conditions for which a self-propagating, shock-driven detonation forms, while not totally understood, are surely more stringent. This caveat applies especially to detonations in carbon and oxygen for which dynamical burning under highly degenerate conditions gives relatively mild shocks. Burning of helium releases more energy per gram (1.5×10^{18} erg gm^{-1}) than burning of an equal-parts mixture of C/O (7.9×10^{17} erg gm^{-1}). This makes it more likely to trigger a shock and hence a detonation, but even the initiation of helium detonations lacks a rigorous understanding from first principles. Initiation of a detonation is very likely to involve conditions such as the temperature gradient or the level of turbulence. Virtually no models have incorporated the physics of nonsteady flame brushes (Sect. 19.11.2). Most work that invokes a detonation ignores the possibly important fact that detonations propagate with a cell structure that must be accommodated on relevant length scales for a self-consistent detonation. These cell sizes are different and increasingly large for carbon, oxygen, and silicon burning (Sect. 19.10).

For explosion models in general, the hydrodynamics, coupled with a restricted network of nuclear reactions sufficient to calculate the nuclear energy release, and with allowance for neutrino losses and neutronization, is followed until $\sim 10^2$ s after explosion, by which time the ejecta are nearly in homologous expansion (neglecting any CSI). The time history of the density and temperature having been obtained, the calculation is sometimes "postprocessed" using a much larger network of nuclear reactions involving hundreds of nuclides, to obtain more precise nucleosynthesis predictions. Radiative transfer codes can then be employed to predict the photometric and spectral behavior of the models for direct comparison with observations. These techniques have tended to neglect the phase of dark time (Sect. 22.2), although the technology to do so exists. This lacuna may be filled by the time this book is published.

Explosion models of SN Ia comprise three basic categories: one in which accretion that is slow compared to hydrodynamic timescales brings a white dwarf to the point of dynamical burning and runaway, thus approaching runaway slowly from below in mass; one involving spin-up/spin-down in which rapid rotation suspends dynamical runaway until sufficient angular momentum is lost or redistributed, thus approaching runaway slowly from "above;" and one in which accretion occurs on a rapid timescale (Sect. 21.3.3). The first class typically is associated with the SD scenario. Spin-up/spin-down models might arise in either a single-degenerate or double-degenerate mode. The rapidly-accreting models are uniquely associated with the double-degenerate mode (including white dwarf collisions). In order to comport with the context in which various models are presented in the literature, we will present our discussion in terms of "single-degenerate" and "double-degenerate" scenarios with appropriate subcategories for the various options that are considered.

22.2 Dark Time

A long-standing aspect of models of SN Ia needs to be newly explored in the light of recent data. The explosion time and the time when buried ^{56}Ni begins to produce leaked photons are not exactly the same. In the 1980's, the first rudimentary numerical models of SN Ia that attempted to capture the effects of ^{56}Ni decay showed that the explosion and breakout flash would be followed by a *dark time*, the period between the explosion and the early rise of the light curve. This is because a white dwarf expands from a small radius. After breakout, adiabatic expansion leads to a cooling wave proceeding into the ejecta. In the absence of CSI, the expansion becomes nearly homologous after about a minute, although ^{56}Ni decay introduces subsequent thermal heating that leads to some acceleration and modification of the kinetic energy over the course of about the next day.

The temperature minimum is reached a few minutes to an hour after the explosion. The ejecta typically cool to $\sim$100 K in the absence of any effects of radioactive heating or CSI. The opacity drops to $\sim 10^{-4}$ to $10^{-3}\,\mathrm{cm^2\,g^{-1}}$, much smaller than the Compton scattering opacity, and the photosphere rapidly retreats into the matter. The postbreakout cooling that is adiabatic in the absence of early decay heating could even cool regions containing some ^{56}Ni that have not yet had time for significant decay.

In layers heated by ^{56}Ni, the opacity rises and the retreat of the photosphere is halted. Exactly where the ^{56}Ni is, and how much matter has been heated to "normal" temperatures and opacities by the time the cooling wave reaches the ^{56}Ni, is a quantitative, model-dependent issue. From that point, nonlocal deposition of γ-rays heats the material, raises the opacity, and leads to the outward diffusion of radiation. Direct γ-ray heating of the outer layers can be important a few hours after the explosion, depending on the ^{56}Ni distribution. The reheating will, in turn, lead to the rise of the light curve, a process that needs to be quantified. In models, the photosphere moves outward through the ejecta after the dark time. The net result of the adiabatic cooling and reheating is a delay of a day or so between the breakout flash and *first light* representing the onset of the rise to maximum light. Remarkably, while this qualitative behavior is robust in any numerical model, none of the early papers remarked on it, presumably because the focus was still on the little-understood phase of peak light and because there were no data on the early rise.

Modern observations of SN Ia early on their rise have changed this situation. Now there are data. Elegant but simple one-zone models predict a $L \propto t^2$ rise in the light curve (Sect. 5.4.2). Many light curves observed within 10–15 days before maximum do show this quadratic behavior, and it has been common to extrapolate it to estimate the time of explosion (Hayden et al. 2010b; Nugent et al. 2011b). There are several problems with this procedure. One is that the dark time will enforce some delay between the time of the explosion and the time of first light on the ultimate rise to maximum, and these epochs should be differentiated (Piro and Nakar 2013; Mazzali et al. 2014). Another issue is that some very early observations have shown distinct departures from the simple t^2 rise. The earliest data on SN 2013dy and SN 2014J showed broken power-law rises that were steeper than t^2 (Zheng et al. 2013, 2014).

In addition, detailed considerations do not predict the early light curve to have a t^2 rise. For example, the first light may have distinct differences (Dessart et al. 2014a) depending on whether the model is a standard delayed-detonation (Sect. 22.3.4) or a pulsating delayed-detonation (Sect. 22.3.5). The rise is not straightforward and may require more physics to be properly understood.

A direct way to determine the length of the dark time and constrain models would be to detect the epoch of shock breakout and the delay to the subsequent time of first light (Piro and Nakar 2013). This delay should be a sensitive test of the array of models discussed below, so all models should make quantitative predictions of the duration of dark time for comparison with much data yet to come. If no such delay is seen, that would also provide strong new clues to the nature of the explosion.

Extrapolation of the early light curve back in time is used to predict a putative epoch at which the flux from the supernova was zero. This epoch is unlikely to correspond precisely to the time of explosion. It may or may not prove to be a reasonable estimate of the time of first light of radioactive decay after a dark time, which is currently only a theoretical concept. Care should be taken to differentiate the empirically-derived epoch of zero flux from the time of explosion and the time of first decay light.

The discussion above of the dark time and first light has been framed in the context of the canonical SD M_{Ch} explosions (Sect. 22.3). An explosion triggered by violent merger in the DD context (Sect. 22.4) may not produce a breakout in the usual sense, but there could still be a dark time driven by the subsequent adiabatic expansion until radioactive decay dominates the light output.

22.3 Single-Degenerate Chandrasekhar-Mass Models

The most thoroughly studied models are those for carbon ignition in a nonrotating near-M_{Ch} C/O white dwarf. Prior to explosion, the white dwarf is assumed to be near-M_{Ch} in order that energy production by carbon burning overcomes losses to neutrinos, to be in hydrostatic equilibrium, and to have equal or nearly equal abundances of carbon and oxygen by mass with solar abundances of heavier elements. Some models include an enhanced abundance ($\sim 10^{-2}$ $M_\odot$) of ^{22}Ne to provide a small neutron excess that is important for the detailed isotopic outcome of explosive nucleosynthesis. The ^{22}Ne forms by $^{14}\mathrm{N}(\alpha,\gamma)^{18}\mathrm{F}(\mathrm{e}^+)^{18}\mathrm{O}(\alpha,\gamma)^{22}\mathrm{Ne}$ during core helium burning in the progenitor of the white dwarf.

Ignition of carbon usually is regarded to require a central density of about $3 \times 10^9\,\mathrm{gm\,cm^{-3}}$, but there are potential complications, such as the convective Urca process, that tend to be ignored in most work (Sect. 19.3). In 1D models, the dynamical burning front originates at the center of the white dwarf, while in 2D and 3D models the nuclear runaway following the smoldering phase may take place off-center, at one, several, or many places.

22.3.1 *Central Detonation*

Calculating a detailed model for an explosion that began as a central detonation in a nonrotating near-M_{Ch} C/O white dwarf would be tractable, but nearly the entire white dwarf would burn at sufficiently high densities to synthesize IGEs, mainly ^{56}Ni, and almost no IMEs (Arnett 1969; Hansen and Wheeler 1969). This model cannot account for the observed spectra of SN Ia. Since the explosion would be even more luminous than normal SN Ia, yet no such event has been observed, it seems that central detonation "never" occurs in a nonrotating near-M_{Ch} white dwarf. This strong observational constraint is consistent with current theory. While details are debated, the systematics of the onset of dynamical burning—carbon ignition, smoldering, and runaway—suggest that the dynamical phase will begin with the unstable turbulent bubbles and plumes of subsonic deflagration rather than detonation.

22.3.2 *Model W7*

A well-known SN Ia explosion model, the venerable model W7 (Nomoto et al. 1984), has traditionally been referred to as a 1D deflagration model. The propagation of the burning front was based on time-dependent mixing-length theory, with the ratio of mixing length to pressure scale height chosen to be 0.7 (hence the

7 following the white dwarf W) to obtain IMEs at velocities in accord with observation. The density and composition structure of W7 (Fig. 20.5 of Chap. 20) is such that calculated light curves and optical, near-IR, and UV spectra (Branch et al. 1985; Wheeler and Harkness 1986; Mazzali et al. 1993; Baron et al. 2006; Hsiao et al. 2013) agree, very generally, with observation. Consequently, W7 long served as a first-order indication of the composition and density structures that are required of a successful SN Ia explosion model.

Close inspection revealed that the forced deflagration speed without transition to a detonation was not physically self-consistent. W7 provided no IMEs above $15{,}000\,\mathrm{km\,s^{-1}}$ (possibly due to coarse zoning in the outer layers of the ejecta), so as it stands it cannot account for the higher velocities of silicon and other elements that are observed in many SN Ia, and, of course, no 1D model can account for the observed polarization of SN Ia. An additional problem (shared by the central detonation model discussed above) is that excessive neutronization by electron captures produced a high ratio of ^{54}Fe to ^{56}Fe. This is in conflict with the ratio in the solar system, to which past SN Ia made a strong contribution. A deflagration is an inherently 3D process, and modern 3D deflagration models (Sect. 22.3.3) bear little resemblance to W7, so perhaps W7 should simply be referred to as a parameterized 1D model, rather than as a deflagration model. As discussed below, more physically-consistent 1D models based on the concept of the DDT have become available and are now preferred as first-order test beds with which to compare observations.

22.3.3 3D Deflagrations

Deflagrations, being subsonic, produce pressure waves that proceed at the speed of sound, more rapidly than the burning can follow. In models, this leads to expansion of the outer layers of the white dwarf at slightly faster than the local sound speed. Models that invoke only burning by deflagration must eject a substantial amount of unburned matter.

In 3D, models depend on the assumed number, location, and timing of ignition kernels ("hot spots"), but because an intrinsic characteristic is that a substantial fraction of the white dwarf remains unburned, the ^{56}Ni production and the kinetic energy are too low to account for typical SN Ia. The amount of ejected carbon exceeds that observed in typical SN Ia (Sect. 20.4) and the composition structure is insufficiently stratified, with unburned fuel extending down to the center (Gamezo et al. 2003; García-Senz and Bravo 2005; Röpke et al. 2007; Ma et al. 2013a; Figs. 22.1 and 22.2). The latter leads to predicted lines of [C I] and [O I] in nebular-phase spectra that are not observed (Kozma et al. 2005). Pure deflagration models are of interest in connection with less luminous transients (Chap. 23).

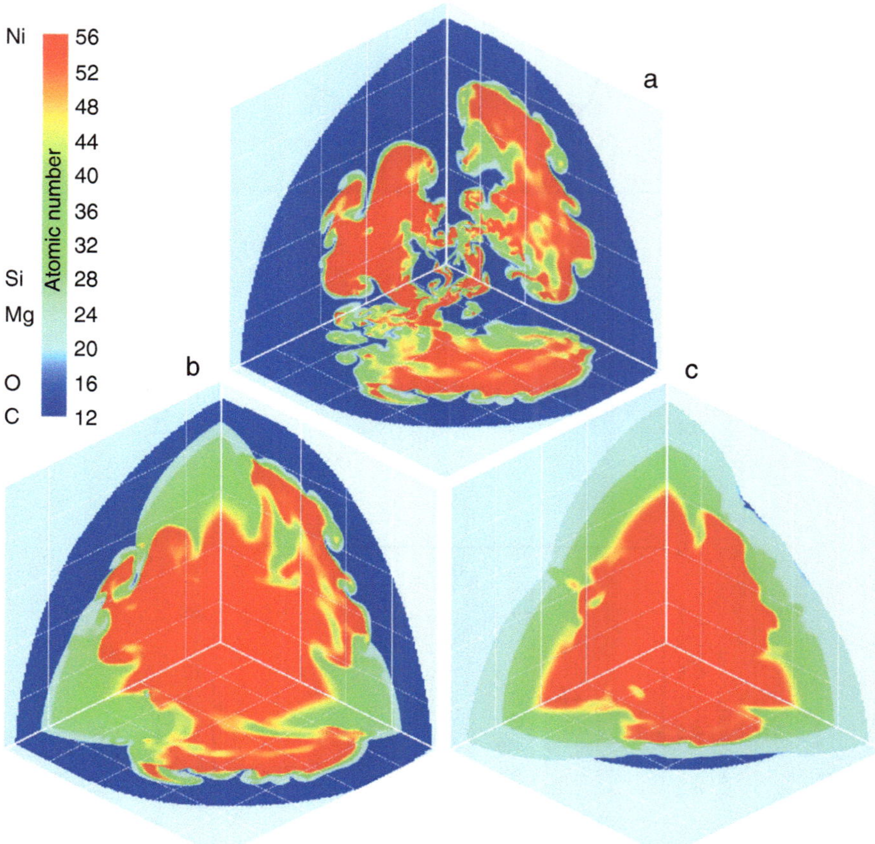

Fig. 22.1 Cut-out views of the composition structures of a 3D deflagration model (panel **a**) and two delayed-detonation models (panels **b** and **c**), about 2 s after explosion. From "Three-dimensional Delayed-detonation Model of Type Ia Supernovae" (Gamezo et al. 2005). © AAS. Reproduced with permission

22.3.4 Delayed Detonations

The most promising SD Chandrasekhar-mass model is perhaps the *delayed-detonation* model (Khokhlov 1991a; Woosley and Weaver 1994). In this model, the explosion begins at the end of the smoldering phase as a deflagration at or near the center of the white dwarf. When the burning front encounters a critical density, $\sim 10^7\ \mathrm{g\,cm^{-3}}$, about a second after the onset of the deflagration, a DDT is assumed to occur. The most critical aspect of this model is that the supersonic detonation catches up with the mildly supersonically-expanding outer layers so that they, too, are substantially burned. The deflagration and the initial detonation phases produce IGEs, but expansion of the white dwarf during the deflagration phase lowers the

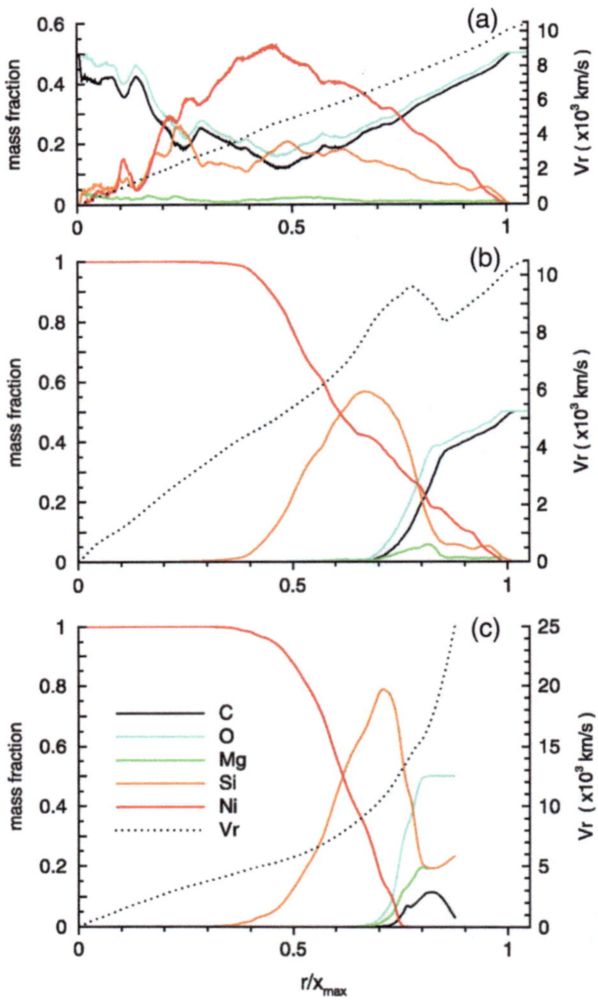

Fig. 22.2 Angle-averaged mass fractions of selected elements and radial velocity plotted against radial coordinate for a 3D deflagration model (panel **a**) and two delayed-detonation models (panels **b** and **c**), about 2 s after explosion. From "Three-dimensional Delayed-detonation Model of Type Ia Supernovae" (Gamezo et al. 2005). © AAS. Reproduced with permission

density of the outer layers. In these layers, the detonation produces IMEs in amounts and ratios close to those demanded by observations (Fig. 20.5). In multidimensional models, the detonation propagates not only outward, but also inward, burning (at least) any topologically connected pockets of fuel that were left by the turbulent deflagration. This inward burning mutes or eliminates the problem of unburned carbon and oxygen at low velocities that burdens multidimensional pure deflagration models.

In 1D, the speed of the centrally initiated deflagration (a small fraction of the local sound speed; Sect. 19.6) and the density at which the DDT occurs must be specified. The DDT takes place simultaneously on a sphere and only propagates outward. The choice of the transition density, the main free parameter of the models, is guided to a certain extent by theory (Niemeyer and Woosley 1997; Khokhlov et al. 1997), and to a certain extent by the roughly congruent values needed to reproduce observed light curves and spectral evolution (Höflich and Khokhlov 1996; Domínguez et al. 2001). The higher the transition density, the sooner the DDT occurs and the more that ^{56}Ni is formed at the expense of IMEs. In models, the outermost ^{56}Ni can range from $\sim$5000 km s^{-1} to nearly 15,000 km s^{-1}, depending on parameters. This difference could have a strong influence on the dark time (Sect. 22.2).

Detailed radiative-transfer calculations of 1D delayed-detonation models having a range of DDT transition densities, and hence a range in M_{Ni}, show broad agreement with many of the observed SN Ia features. The overall shape of the bolometric and broad-band light curves agree with observations. In particular, the models show single peaks in optical bands and double peaks in near-IR bands (Höflich and Khokhlov 1996; Kasen 2006; Jack et al. 2012a). The double peaks are associated with evolution of the ionization state of the IGEs.

This class of models also gives a reasonable representation of the width-luminosity relation (Sect. 20.4). Qualitatively, models with a higher transition density yield more ^{56}Ni, are hotter, and hence have higher opacity. The greater mass of ^{56}Ni yields a brighter peak luminosity and the higher opacity leads to a slower, broader light curve (Höflich et al. 1995). In more detail, the shape and amplitude of the light curve depend on the evolution of the ionization state of the ejecta and the associated redistribution of flux (Kasen and Woosley 2007). The actual dependence of the luminosity evolution on the diffusion and hence opacity per se is a secondary effect. The amplitude and shape of the light curve again depend on the role of the IGEs, especially the ubiquitous lines of Fe II and Co II that blanket the continuum in the blue, restricting the flow of radiation, but that increase the emissivity at longer wavelengths, beyond the R band. Cooler models with less ^{56}Ni undergo earlier recombination of Fe III to Fe II and Co III to Co II. This leads to a more rapid redistribution of the flux from blue to red. The result is a more rapid decline in the blue and enhanced secondary maximum in the red. This physics may apply to other explosion models, but it has been established to operate with reasonable success in delayed-detonation models. Models with lower transition densities and hence with $M_{Ni} \lesssim 0.3\ M_{\odot}$ match some photometric aspects of SN 1991bg-like events (Höflich et al. 1995; Blondin et al. 2013), although these have other observed peculiarities that suggest they could be a physically distinct subgroup of SN Ia (Sect. 20.4).

One-dimensional delayed-detonation models also provide a reasonable representation of the multiwavelength, spectroscopic evolution of SN Ia. Models that represent typical SN Ia have somewhat high kinetic energies, such that the model spectra best match spectroscopically Broad Line events (Blondin et al. 2015; Fig. 22.3). Whether similar models can provide good matches to spectroscopically Core Normal (or LVG) events remains to be seen. Models with small ^{56}Ni mass

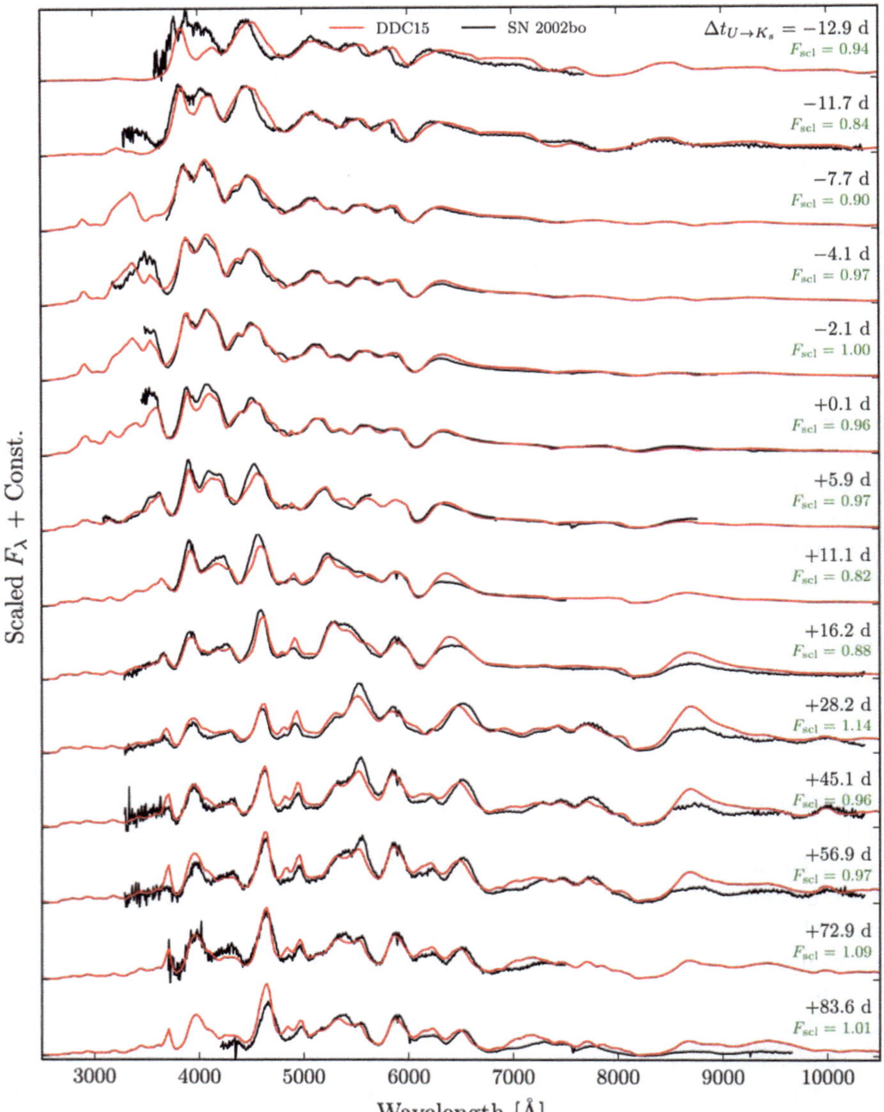

Fig. 22.3 Spectroscopic evolution of SN 2002bo (*red*), a broad-lined SN Ia, and CMFGEN spectra of a delayed-detonation model. Epochs are with respect to bolometric maximum light. A scaling factor (F_{scl}) is applied to the synthetic spectra to reproduce the mean observed flux in the range 5000–6500 Å. Figure 8 "A one-dimensional Chandrasekhar-mass delayed-detonation model for the broad-lined Type Ia supernova 2002bo" from Blondin et al. (2015), by permission of Oxford University Press on behalf of the Royal Astronomical Society

provide spectra (even spectropolarimetry) that are reminiscent of SN 1991bg-like events (Howell et al. 2001; Patat et al. 2012). Models must also be challenged to match UVOIR spectra well before and after maximum light. Early work showed that models chosen to reproduce optical behavior near maximum light also gave a reasonable representation of near-IR spectra with no further tuning of the parameters (Wheeler et al. 1998). Contemporary models provide reasonable results from the near- UV to the near-IR (Blondin et al. 2013).

Overall, 1D delayed-detonation models are as successful as any class of models. They thus represent a standard against which other models need to be compared. As noted by Bravo and García-Senz (2009), it is remarkable that spherically symmetric explosion models with only a single primary free parameter, the DDT transition density, continue to provide the best agreement with observations of SN Ia and even more remarkable that elaborate 3D simulations (discussed below) neither reproduce 1D results nor improve on them.

The mass of unburned carbon in 1D delayed-detonation models, $\sim 10^{-2}$ $M_\odot$, may be consistent with observations, but the carbon is predicted to be ejected only at very high velocity, while observed C II lines form at lower velocities (Sect. 20.4). To a lesser extent, oxygen also does not extend to sufficiently low velocities in the models. The lack of carbon and oxygen at sufficiently low velocities is likely to be due to the limitations of 1D representations of delayed-detonation models. Models of SN 1991bg-likes, with low DDT transition densities and low ^{56}Ni ejection tend to leave more unburned carbon, roughly consistent with the stronger C II lines observed in these events.

Even among typical events, light curves of SN Ia reveal small, second-order departures that reflect variations in the physical conditions in the underlying explosion. In models, variations of the mass and metallicity of the ZAMS progenitor and in the central density at which the explosion is initiated yield small variations in spectral distributions that are reflected in multicolor light curves. Intercomparing very accurate multicolor data can then yield information on these physical parameters. A study of this sort in the context of 1D delayed-detonation models concluded that most SN Ia come from progenitors with ZAMS mass $\gtrsim 4$ $M_\odot$, that the metallicity ranges from 0.03 to 1 times solar, with only a small fraction of events showing supersolar abundances, and that the central density falls in the range 1 to 7×10^9 g cm^{-3} (Höflich et al. 2010). The latter result is consistent with ignition in a near-M_{Ch} white dwarf.

A qualitative difference between the near-M_{Ch} delayed-detonation models and some other classes of models such as sub-Chandrasekhar explosions (Sect. 22.3.8) and violent double-degenerate mergers (Sect. 22.4.2) is that the former explode from rather high density ($\gtrsim 10^9$ g cm^{-3}) and the latter from substantially lower densities ($\gtrsim 10^7$ g cm^{-3}). Merger explosions occur at low density because the white dwarfs individually have less than M_{Ch}. In addition to the light-curve variations mentioned above, some other lines of evidence suggest that the higher densities are preferred (Sect. 21.6.8). During the deflagration phase of near-M_{Ch} delayed-detonation models, high central densities induce electron capture on IGEs. In 1D, these neutron-rich IGEs tend to be centrally concentrated and surrounded by ^{56}Ni that is formed mainly by the detonation, consistent with some evidence that central neutronization does occur in SN Ia. This neutronization is not predicted in sub-Chandrasekhar models or violent mergers (nor is it necessarily predicted in 3D delayed-detonations, barring effects of strong magnetic fields; see below).

Nebular-phase spectroscopy probes the structure in deep layers of the ejecta. The [Fe II] 1.644 μm line is especially useful because it is nearly unblended. The line profile thus yields direct information on the distribution of the iron. At late phases, the ejecta are transparent to both optical photons and to γ-rays, but the positrons may be deposited locally in a tangled magnetic field. If so, the [Fe II] line would be excited only by positron deposition from the decay of coresident ^{56}Co. In some events for which the nebular [Fe II] line profile has been determined, it tends to show a flat top and asymmetric wings (Höflich et al. 2004; Motohara et al. 2006). This suggests that the ionized, excited iron is offset from the geometric center of expansion, and that there is a central cavity in the distribution of the [Fe II] and the ^{56}Co that excites it. Mid-IR nebular-phase spectra of a small sample of SN Ia are consistent with this picture (Gerardy et al. 2007). In 1D delayed-detonation models, the cavity arises because electron capture in the original dense inner parts of the ejecta form neutron-rich isotopes of nickel and iron. With trapping by ambient magnetic fields, positrons from the outer shell of ^{56}Ni and ^{56}Co cannot penetrate to ionize the iron near the center. It will be important to determine how common the flat-topped lines are and how the phenomenon is distributed among the various SN Ia sub-types. Direct imaging of the ejecta of SN 1885A in Andromeda, seen in absorption against the background light of the galactic bulge, suggests that calcium was not mixed into the center of the ejecta (Fesen et al. 2007). It is unlikely, however, that this event was a normal or even an SN 1991bg-like SN Ia (Sects. 7.4.9 and 23.4), so it is not clear that the apparent lack of mixing pertains to SN Ia.

In 2D delayed-detonation models, ignition of deflagrations is restricted neither to the center nor to one location, and the DDT is not restricted to occur simultaneously on a sphere. The number of ignition points, their distance from the center, and the solid angle in which they are distributed all become parameters of the models. Model spectra and light curves depend on viewing angle (Kasen et al. 2009; Blondin et al. 2011), but tend to be in broad agreement with observations of SN Ia. Some of the more asymmetric models fail to match observations. The model rise times to maximum light are too long. Owing to high model kinetic energies, the blueshifts of the Si II λ6355 absorption tend to be too large for Core Normal SN Ia. A caveat is that, because of the unphysical reverse turbulent cascade to larger scales that typifies 2D simulations, 2D calculations may be intrinsically suspect not only compared to 3D models, but even to 1D models.

Models of 3D delayed-detonation explosions show some qualitatively different aspects (Khokhlov 2000; Gamezo et al. 2005; Seitenzahl et al. 2013b). As for 2D, models require a choice of the central density at runaway, the number and distribution of ignition points for the deflagration, and a prescription for the DDT. The distribution of ignition points should be controlled by physics, but in the current state of the art this remains a key part of the parametrization in many models. Recent models have prescriptions based not on a critical density for DDT per se, but on the strength of the turbulence (Sect. 19.8). Models with a suitable range of deflagration ignition points can produce masses of ^{56}Ni ranging from 0.3 to 1.1 $M_{\odot}$ (Nonaka et al. 2012; Seitenzahl et al. 2013b). Although there is no single DDT transition density in 3D delayed-detonation models, the models with fewer deflagration ignition points apparently give a higher effective transition density. Models with only a few ignition points are rather asymmetric and experience weaker deflagrations and slower expansions. With the particular model for DDT based on an estimate of turbulent intensity, models with few ignition points synthesize larger masses of ^{56}Ni, $\sim$1.0 $M_{\odot}$, in the detonations; thus models with few ignition points lead to light curves that are too bright. More centrally concentrated and symmetric distributions of larger numbers of deflagration ignition points are required in order to produce the $\sim$0.6 $M_{\odot}$ that is more typical of SN Ia, but it is not clear that many ignition points are consistent with the systematics of the smoldering phase and the onset of dynamical burning (Sect. 19.2).

In some respects, 3D delayed-detonation models (Figs. 20.5, 22.1, and 22.2) provide reasonable agreement with observations (Röpke et al. 2012; Sim et al. 2013), but discrepancies remain. Comparison with observed spectra again indicates that model kinetic energies tend to be too high. One way to reduce the energy would be to adopt a larger O/C ratio in the white dwarf, but this ratio must be consistent with known nuclear physics and the constraints of stellar evolution. A positive aspect of the 3D models is that carbon and oxygen are typically present down to about 10,000 and 4000 km s^{-1}, respectively, which seems promising to account for the observed C II and O I lines (although excessive nebular [O I] emission may be an issue). The models tend to be too red around maximum light and not to produce satisfactory I-band light curves. It is not clear that current 3D delayed-detonation models robustly account for the variety of properties of SN Ia, including the width-luminosity relation and the various spectral subclasses discussed in Sect. 20.5.

In contrast to 1D models, in 3D the ^{56}Ni, formed by the detonation, tends to be concentrated in the central regions, while most of the neutron-rich IGEs, formed in the deflagration, are carried by buoyant plumes of ash to intermediate velocities, $\sim$3000 to 10,000 km s^{-1} (Seitenzahl et al. 2013b). This behavior would seem to be more physical than that constrained by the limits of spherical symmetry in 1D models, yet some of the evidence outlined above suggests that observed ejecta do have a central cavity in the ^{56}Ni and ^{56}Co distributions. One possibility is that magnetic fields arising in the turbulence of the smoldering phase inhibit the mixing that characterizes 3D models, thus yielding structures closer to the 1D models (Diamond et al. 2015).

Other observations are consistent with delayed-detonation models. In the near-IR, the 1.0927 μm line of Mg II showed a distinct minimum velocity in SN 2011fe (Hsiao et al. 2013), consistent with the magnesium being in an outer layer with a well-defined inner radius. This stratified configuration occurs naturally in 1D delayed-detonation models and is roughly consistent with 3D models. It is not clear that strongly-asymmetric models or violent-merger models (Sect. 22.4.2) can preserve this distinct minimum velocity.

In the nebular phase, emission lines attributed to [Fe II] λ7155 and [Ni II] λ7378, which form mainly in the higher-density, more rapidly-cooling ashes of the deflagration, tend to be substantially redshifted (up to 3500 km s^{-1}) in HVGs and blueshifted or only mildly redshifted in LVGs (Sect. 20.5). The Doppler shifts of λ7155 and λ7378 indicate that the deflagration occurs off-center. This behavior has been interpreted in terms of asymmetric delayed-detonation models (Maeda et al. 2010a,c). When the deflagration is displaced to the near side of the observer's line of sight, the [Fe II] and [Ni II] lines in the cooler deflagration ashes are blueshifted. Models show that the presence of deflagration ashes on the near side inhibits the outward-propagating detonation. This leads to a steep density gradient, confinement of silicon to a rather narrow velocity interval, and thus a slow evolution of the blueshift of the Si II absorption—i.e., to an LVG event. When the deflagration displacement is to the far side, the forbidden lines are redshifted, the detonation is uninhibited, and a shallow density gradient leads to silicon in a broad velocity interval and a rapid evolution of the blueshift of the Si II absorption—i.e, an HVG. Thus SN Ia with different characteristics, HVG versus LVG, could represent similar, but asymmetric, explosions viewed from different perspectives, like identical eggs viewed along different lines of sight. It is not clear whether the required offset of the deflagration is consistent with the need for ignition only mildly off-center (Nonaka et al. 2012; Seitenzahl et al. 2013b). There is additional tension between the asymmetry interpretation of HVGs and LVGs and some observational evidence that LVGs and HVGs have statistically different distributions of host galaxies and locations within galaxies (Sect. 20.9).

A potentially powerful constraint on models is the nature of the late-time trapping of the positrons from ^{56}Co decay. The positrons would freely stream from the ejecta were it not for entanglement with magnetic fields. As long as the positrons are trapped, the UVOIR light curve decays at the ^{56}Co rate. If they escape, there should be cooling and a shift of emission from optical to forbidden iron lines in the mid- and far-IR, with the optical emission sharply declining in an infrared catastrophe (Sect. 5.5.1). In SN 2011fe, there was no evidence for a sharp decline as late as 30 months, owing perhaps to radiative-transfer effects (Sect. 20.2.1).

Late-time spectra may also help to probe, indirectly, the magnetic field. In delayed-detonation models, a tangled magnetic field could inhibit mixing in the ejecta due to Rayleigh–Taylor and other instabilities, and sufficiently strong fields could lead to local positron energy deposition. The latter is involved in models for the suppression of [Fe II] emission in an inner cavity as an explanation for the observed flat-topped profiles discussed earlier in this section. This interpretation of the flat-topped lines requires a magnetic field of order 10^6 G (Höflich et al. 2004; Penney and Höflich 2014). Such fields in the expanded ejecta would imply that original fields in the white dwarf were strong enough to have been dynamically important in the smoldering and early deflagration phases (Sect. 19.2).

22.3.5 Pulsation-Driven Detonations

Several classes of models seek to avoid the necessity of artificially invoking a DDT by arranging for ignition of a *confined* detonation. In the 1D *pulsating delayed detonation* (PDD) model (Khokhlov 1991b; Khokhlov et al. 1993; Dessart et al. 2014a) the deflagration is weak—and the burning is quenched while the white dwarf is still bound. The white dwarf expands, then contracts—i.e., it undergoes a strong pulsation. The ∼10 to 100 s duration of the pulsation may be long enough for Rayleigh–Taylor instabilities to cause turbulent mixing at the intersection of the hot burned and cooler unburned matter. When this volatile mixed layer is compressed at the end of the contraction, a detonation, confined by contraction, is assumed to develop by the Zel'dovich gradient mechanism (Sect. 19.11.1). An important constraint on the PDD model is that the deflagration must burn $\sim 0.3\ M_{\odot}$, not much more and not much less, in order to achieve a large-amplitude, long-duration expansion, yet not unbind the white dwarf (Bravo et al. 2009). Depending on the extent of mixing of burned and unburned matter, PDD models might produce a range of ^{56}Ni masses consistent with the range inferred for SN Ia. This mechanism might produce IMEs only in a relatively narrow velocity interval, perhaps accounting for events that exhibit a "silicon-velocity plateau" (Sect. 20.5). The mechanism might also be more consistent with the early, fast, blue rise of the first-light phase of some SN Ia that is not reproduced in standard delayed-detonation models in the absence of CSI or interaction with a companion (Dessart et al. 2014a).

The *pulsating reverse detonation* (PRD) model is closely related to the PDD model, but simulated in multiple dimensions. In 1D PDD models, burned matter never mixes substantially with outer, unburned fuel. The unburned matter restricts the expansion of the inner material, leading to longer time at high density and more thorough burning. In 3D PRD models, the buoyancy of bubbles and plumes of deflagration ash produces a crucial ingredient that cannot be achieved in 1D: bubbles or plumes that expand from ignition points do penetrate into the outer layers. The associated flows can yield a composition inversion, with unburned C/O in a $\sim 1\ M_{\odot}$ core, surrounded by hot ashes of the deflagration. Deposition of energy from the ashes into the outer layers ensures a large-amplitude expansion even if the deflagration mass is low. In the PRD model (Bravo and García-Senz 2009; Bravo et al. 2009), the constraint on the mass burned by the deflagration is from $\lesssim 0.1$ to $\lesssim 0.3\ M_{\odot}$. In the multidimensional models, expansion tends to be restricted to outer layers rather than involving the overall pulsation and contraction of the star that characterizes PDD models. Nevertheless, the contraction phase in PRD models generates an accretion shock that heats the C/O core and is assumed to ignite a confined detonation that initially propagates inwards (thus a "reverse" detonation).

In common with related multidimensional models, the outcome of PRD models depends on the number and distribution of the sites of initial runaway to deflagration. If the number of ignition sites is too small or the burning rate is too slow, there may be a "bubble catastrophe" in which the bubbles float to the surface before the combustion wave can substantially influence the dynamics. The star may remain bound, and the explosion may fail (Livne et al. 2005; but see Sect. 22.3.6). The result would be a cold C/O core surrounded by hot IGEs. This sets the lower limit on the mass burned in the deflagration phase of successful PRD models and the amount of ^{56}Ni they produce. As for all models of SN Ia, the formation of a detonation is critical and must be assumed, failing a rigorous theory. The deflagration mass must not be $\gtrsim 0.3\ M_{\odot}$, or the energy released by the deflagration would nearly unbind the white dwarf, and the accretion shock would be too inefficient to detonate the C/O core.

PRD models tuned to give about the correct amount of ^{56}Ni for a typical SN Ia presumably also give reasonable bolometric light curves. There have been, as yet, no calculations of broad-band light curves for comparison with data. Spectra of angle-averaged PRD models contained excessively strong iron lines (Baron et al. 2008), but 3D effects may be important for synthetic spectra of these models. The covering fraction of iron in the outer layers of the models where the photosphere forms is less than unity. In 3D spectrum calculations, this would reduce the strength of iron lines, even those of high optical depth (Bravo et al. 2009). Still, because of high ^{56}Ni production, PRD models may be unable to account for SN Ia that are less luminous than average. In addition, the chemical profile of the PRD models lacks the degree of stratification suggested by observations. PRD models do explode from high densities, as some observations require. About 0.1 $M_{\odot}$ of carbon is produced in these models, in the range 6000 to 12,000 km s^{-1}. The degree to which this carbon would show up in spectra as C I and C II is unclear. No rigorous test of these models against polarization data has been attempted.

22.3.6 *Gravitationally-Confined Detonations*

What may be deemed a failure in the PRD model is made a virtue in the *gravitationally confined detonation* (GCD) model (Plewa et al. 2004; Kasen and Plewa 2005; Jordan et al. 2008). In this model (Fig. 22.4), a single bubble of buoyant deflagration ash bursts through the surface of the white dwarf. The resulting eruption drives a "flood" around the white dwarf surface. The flood converges at the antipode, where a detonation is assumed to develop (Sect. 19.9). Once initiated, the detonation propagates inward, producing IGEs in the dense inner layers. During the flood, the reduced mass of the underlying white dwarf causes expansion, so the outer layers are of low enough density to be burned only to IMEs by the detonation. When the flow convergence is sufficient in 3D to initiate a detonation, however, the densities are high enough that more than 1 $M_{\odot}$ of ^{56}Ni is produced, making the model more luminous than most SN Ia.

There are technical issues with this model. The formation and evolution of the plume ignores the convective flows that exist from the prior smoldering phase, no consideration is given to the size and distribution of composition-dependent detonation cells, and the white dwarf rotation has not been taken into account. This GCD model contrasts with the single-plume model of Seitenzahl et al. (2013b),

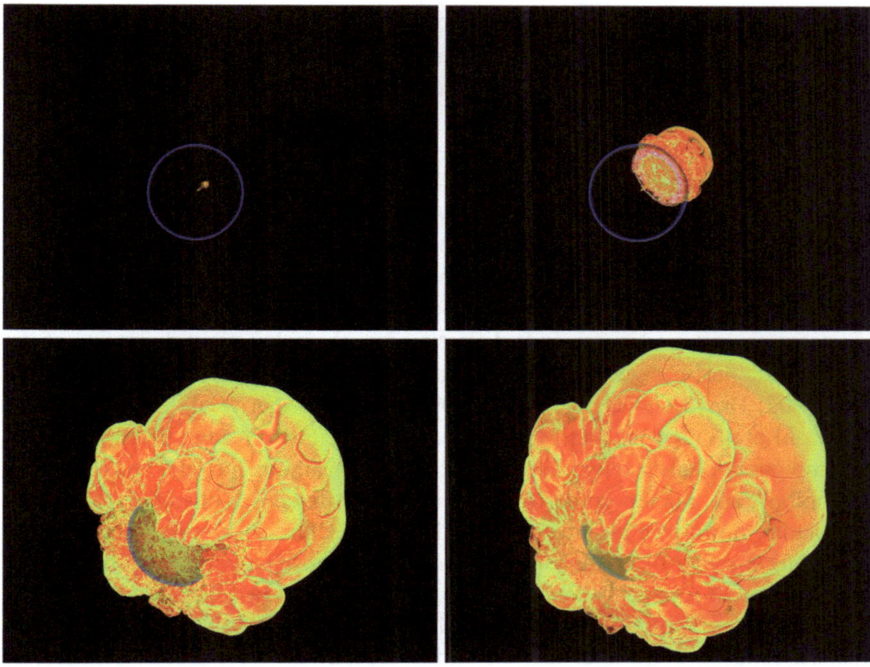

Fig. 22.4 Images from a 3D simulation of the GCD model. *Top left*: 0.5 s after ignition of the deflagration. *Top right*: 1 s, as the deflagration ash bursts through the white dwarf surface. *Lower left*: 1.5 s, as the ash flow over the white dwarf surface. *Lower right*: 1.7 s, just before the ash converges at the antipode. From "Three-dimensional Simulations of the Deflagration Phase of the Gravitationally Confined Detonation Model of Type Ia Supernovae" (Jordan et al. 2008). © AAS. Reproduced with permission

in that the latter model invokes a deflagration-detonation transition based on a turbulence criterion, and a detonation is initiated as the plume first rises toward the surface.

In a variant of the model, the *pulsationally-assisted GCD* (Jordan et al. 2012a), the deflagration releases enough energy, and causes enough expansion, to prevent initiation of a detonation when the flood first converges at the antipode. Instead, the detonation is assumed to be initiated at the antipode by subsequent contraction in a manner somewhat similar to the PDD and PRD models. Lower densities allow less ^{56}Ni to be produced. The detonation produces a smooth interior of NSE material surrounded by a layer of IMEs. The IMEs are, in turn, surrounded by a mixture of IMEs and IGEs from the deflagration.

It is not clear how this and other variations on GCD models reproduce the detailed light curve and spectral evolution of typical SN Ia and satisfy constraints on the carbon, magnesium, and ^{56}Ni distributions. The model demands high-velocity IGEs in the outer layers, which may produce excessively strong iron features. If this outer material contains ^{56}Ni then the photosphere may be too hot and blue.

The model might provide outer clumps of IMEs from the deflagration that could influence the line polarization, but the model might not satisfy the limits on the asymmetry of the bulk ejecta from the continuum polarization. These issues require detailed quantitative work.

An important issue latent in the delayed-detonation, PRD, and GCD models is the number, distribution and dynamics of the original dynamic burning seeds. The first GCD models invoked a single point of runaway that may be consistent with analyses of the end of the smoldering phase (Sect. 19.2). Later GCD models, the PRD models, and delayed-detonation models seem to require more seeds. Despite similar distributions in initiation sites, tens to hundreds of kilometers off-center, there are also apparent differences in the subsequent dynamics of the resulting bubbles, e.g., whether they become toroidal (Zingale et al. 2005; Röpke et al. 2007; Bravo and García-Senz 2009) or not (Jordan et al. 2008). See also Seitenzahl et al. (2013b) and Malone et al. (2014). Few, if any, of the papers on these topics undertake a quantitative intercomparison among models. The dynamics, morphology, and nucleosynthesis, are sensitive to the choice of initial conditions. Does nature uniformly provide just the right conditions, or a range of conditions demanding a range of results? Related issues of the sensitivity to burning seeds arise in Chap. 23 for events that differ from, but appear to be related to, SN Ia.

If carbon burning ignites at a single buoyant point near the center of a near-Chandrasekhar C/O white dwarf, the explosion may lack a vigorous deflagration phase, leading to synthesis of a large mass of ^{56}Ni, $\gtrsim 1$ $M_\odot$ in a subsequent detonation phase. Thus the canonical SD scenario may lead predominantly to GCD and produce only overluminous SN 1991T-like events, not typical SN Ia (Fisher and Jumper 2015). Note, however, that synthetic spectra calculated for detailed 3D hydrodynamical simulations of GCD models were not in good agreement with SN 1991T-like events (Seitenzahl et al. 2016).

22.3.7 Single-Degenerate Super-Chandrasekhar Models

As mentioned in Sect. 22.1 and discussed in more detail in Sect. 21.2.5 , there are classes of spin-up/spin-down models in which mass transfer provides sufficient angular momentum to the white dwarf that it can support more than M_{Ch}. Only after angular momentum is lost or redistributed by various dissipation processes does the white dwarf explode. Depending on conditions at dynamical runaway, these models might produce typical SN Ia as well as, perhaps, the observed class of substantially super-Chandrasekhar events (Sect. 20.7). As for canonical near-Chandrasekhar SD models that approach carbon ignition by accretion from below M_{Ch}, spin-up/spin-down models might undergo a smoldering phase and then proceed to deflagration and delayed-detonation (Chap. 19). There has yet to be any detailed study of the smoldering phase specific to these models. We discuss this class of models here in the context of SD evolution, but variations also apply to certain DD scenarios (Sect. 22.4.1).

Delayed-detonation explosion models of rigidly rotating, moderately super-Chandrasekhar (up to 1.5 $M_{\odot}$) C/O white dwarfs were explored by Domínguez et al. (2006). The calculated light curves and the nucleosynthesis resembled those of near-Chandrasekhar delayed-detonation models, but spectra were not synthesized. Because real white dwarfs are rotating, models such as these should be computed in full detail.

Models of pure deflagration-driven explosions in rapidly rotating white dwarfs of mass well in excess of M_{Ch} do not resemble SN Ia (Pfannes et al. 2010b). The burning front develops strong anisotropy with preferred directions toward the poles. In the central regions and in the equatorial plane, large amounts of C/O are left unburned. These models also depend on the distribution of ignition points. It would be interesting to explore more thoroughly whether the unburned central regions could be eliminated in variations on this model. These models are likely to cause too much distortion from spherical symmetry to agree with the constraints from spectropolarimetry of typical SN Ia.

Rough estimates of the production of IMEs and IGEs in rotating super-Chandrasekhar white dwarfs that experience artificially initiated central detonations might be consistent with the observed substantially super-Chandrasekhar SN Ia (Pfannes et al. 2010a). As discussed in Sect. 22.3.1, central detonations in nonrotating near-M_{Ch} white dwarfs do not resemble SN Ia, and apparently do not occur. It is not clear, then, why central detonations should occur only in rapidly-rotating super-Chandrasekhar white dwarfs. Considering that runaway central carbon burning in nonrotating models is expected to begin as deflagration, not detonation, this seems even more likely in rotating models that ignite carbon at comparable or even lower densities.

The artificial initiation of a delayed detonation in a differentially-rotating super-Chandrasekhar white dwarf yielded velocities of IMEs that were too high (Röpke et al. 2013). These models produced $\gtrsim 10^{-3}$ $M_{\odot}$ of carbon that is likely to be at too high a velocity to account for C II lines in typical SN Ia and to be too small in amount to account for the strong C II lines observed in substantially super-Chandrasekhar SN Ia. This class of models may deserve more study. Once again, the issue of how to initiate a detonation remains problematic and worthy of broader study.

22.3.8 Single-Degenerate Sub-Chandrasekhar Models

When central detonations are artificially initiated in sub-Chandrasekhar C/O white dwarfs of ~ 1 $M_{\odot}$ (Shigeyama et al. 1992; Sim et al. 2010), the resulting light curves and spectra are in reasonable agreement with observations of SN Ia. The low pre-explosion densities of these white dwarfs allow synthesis of IMEs in the outer layers, rather than the IGEs that would be produced by a central detonation in a near-M_{Ch} white dwarf (Sect. 22.3.1). The problem with these models is the artificial initiation of the detonation. These calculations do show, however, that a model that gave an optical display dominated by the ejecta of a detonated $\sim$1 $M_\odot$ C/O white dwarf would be viable (Sim et al. 2010).

In the SD double-detonation model (Sect. 21.2.6), a detonation of a sub-Chandrasekhar C/O white dwarf is triggered by a detonation in a surface helium layer that is accreted from a nondegenerate helium-star companion. Accumulation of helium at a rate of $\sim 10^{-8}$ to $\sim 10^{-7}$ $M_\odot$ y^{-1} can lead to explosive ignition of degenerate helium when the density reaches $\lesssim 10^6$ g cm^{-3} (Woosley and Weaver 1994). The uncertain (see below) minimum mass of the helium layer that will yield dynamical burning during the degenerate helium flash, a necessary condition to produce a detonation, depends inversely on the mass of the C/O core.

If a detonation occurs in the helium layer, nuclear burning provides energy to eject the layer and synthesize heavier elements. If *only* the helium layer detonates, the optical display will not resemble a traditional SN Ia. The effect of a helium detonation on the C/O core depends on, among other factors, the height above the composition boundary at which the helium ignites (Livne and Glasner 1990; García-Senz et al. 1999). In simulations, the helium detonation usually does cause disruption of a C/O core of $\gtrsim$0.8 $M_\odot$ (Livne 1990; Fink et al. 2010; Moll and Woosley 2013; Shen and Bildsten 2014), either by an ingoing detonation starting from the outer edge of the C/O core—an *edge-lit double detonation*—or by compression of the core that initiates an outgoing detonation—a *shock-compression double detonation*. Lower-mass C/O cores and ONeMg cores are much more difficult, perhaps impossible, to disrupt (Shen and Bildsten 2014).

If the C/O core disrupts, the mass of the helium layer at ignition is crucial. Higher-mass helium layers on lower-mass C/O cores make more ^{56}Ni in the outer layers; lower-mass helium layers on higher-mass C/O cores make more IMEs. Excessive production of ^{56}Ni by the helium detonation is deleterious; thus the lower the mass of the helium layer, the smaller its unwanted contribution to the optical display will be. Early work (Woosley and Weaver 1994; Livne and Arnett 1995) inferred that $\lesssim$0.2 $M_\odot$ of helium would be required. With this much helium, the light curves and especially the spectra do not reproduce those of SN Ia (Höflich and Khokhlov 1996; Höflich et al. 1997; Nugent et al. 1997; Fink et al. 2007). The problem of ^{56}Ni production in the helium detonation was discussed in detail by Woosley and Kasen (2011): spectral features of IMEs are too weak because (1) heating from the decay of ^{56}Ni produced by the helium detonation overionizes silicon and sulphur, (2) continuum emission from the hot, helium-burned layer dilutes the spectral features from the underlying core, and (3) the helium-burned layer decelerates the underlying IMEs and confines them to a velocity range that is too narrow.

Recent work suggests that the required mass of helium may be lower, from $\gtrsim$0.1 $M_\odot$ for an 0.8 $M_\odot$ C/O core down to 0.0035 $M_\odot$ for a near-Chandrasekhar C/O core (Bildsten et al. 2007; Fink et al. 2010; Shen and Moore 2014). This raises the possibility that the optical display may be dominated by the underlying detonated C/O core which, as mentioned above, may be favorable for matching the properties of SN Ia. Nucleosynthesis in the helium layer depends on the extent to which carbon is mixed into the layer prior to the detonation (Kromer et al. 2010; Shen and Moore 2014; Piro 2015a). Instead of ^{56}Ni, the helium detonation may produce short-lived radioactivities such as ^{48}Cr and ^{52}Fe, long-lived ^{44}Ti (leading to ^{44}Ca, which is underproduced in supernova models not involving helium detonations), and/or IMEs. The HVFs that are commonly observed in SN Ia spectra (Sect. 20.4) could reflect IMEs in the helium-burned layer (Shen and Moore 2014).

Double-detonation models would have ^{56}Ni in the central regions of the ejecta and hence probably would not show flat-topped nebular-phase [Fe II] profiles, unless gravitational settling of ^{22}Ne in the white dwarf leads to a more neutron-rich central composition. Excessive asymmetry in 3D realizations of double-detonations also may be an issue (Kromer et al. 2010).

As for so many other models, double-detonation models are subject to uncertainty regarding the initiation of detonations in the helium and the C/O (Moll and Woosley 2013). Although not rigorously proven, the high specific energy of helium burning does, however, make the initiation of a detonation in degenerate helium plausible. In the SD context, the white dwarf must be rotating to some degree; this aspect has yet to be thoroughly explored.

22.4 Double-Degenerate Models

The other major progenitor channel under consideration involves the inspiral and merger of two white dwarfs. This channel has the virtue of being trivially in agreement with the failure to find surviving secondary stars and with the general lack of hydrogen. The issue of how mergers of white dwarfs might be brought to the point of explosion has received much attention (Sect. 21.3). Detailed explosion models, which need to be 3D, have received less attention. Some of those that have been calculated treat only the hydrodynamics and nucleosynthesis, a few also present light curves and spectra.

22.4.1 The Canonical Double-Degenerate Model

In the canonical DD model (Sect. 21.3) a super-Chandrasekhar merger remnant spins down over a long timescale until the central density is high enough for carbon ignition. The configuration at explosion would resemble a spun-down super-Chandrasekhar white dwarf in the SD scenario (Sect. 21.2.5). The mildly super-Chandrasekhar explosion models of Domínguez et al. (2006), mentioned in Sect. 22.3.7, could apply to this case. Results of future detailed simulations for spun-down mildly super-Chandrasekhar DD, CD, and SD models may generally resemble each other. They might undergo the standard sequence of carbon ignition, smoldering, deflagration, and DDT of the rather successful near-Chandrasekhar SD models (e.g., delayed-detonation models; Sect. 22.3.4).

22.4.2 Other Double-Degenerate Possibilities

In the DD channel, the alternative to a long spin-down time is explosion on a much shorter timescale (Sect. 21.3.3). In these scenarios, pervasive issues are whether detonations initiate in reality and, if they do, whether the ejecta are excessively asymmetric.

In a simulation of the violent merger of 0.9 and 1.1 $M_\odot$ C/O white dwarfs, in which a detonation was assumed to form and propagate while the merger was in progress (Pakmor et al. 2012), the primary exploded first and wrapped around the secondary before the secondary exploded about a second later. Although the total mass was super-Chandrasekhar, owing to the relatively low central density of the primary the production of ^{56}Ni was relatively modest, about 0.6 $M_\odot$, thus this model was proposed to apply to typical SN Ia rather than to the overluminous, substantially super-Chandrasekhar events (Sect. 20.7). The primary burned at densities that led to production of ^{56}Ni and IMEs, while the lower-density secondary burned only to IMEs. The central regions of the ejecta were occupied by IMEs, surrounded by an asymmetric distribution of ^{56}Ni from the explosion of the primary (Fig. 22.5). Spectra and light curves calculated for this and related models agreed in some respects with observations at some phases (Röpke et al. 2012; Moll et al. 2014), but the large asymmetry and viewing-angle dependence may be inconsistent with the spectroscopic homogeneity and continuum polarization observations of typical SN Ia, and the total ejecta mass may be excessive for typical SN Ia.

Simulations of super-Chandrasekhar mergers of C/O white dwarfs, based on the different assumption that the secondary fully disrupts and forms a disk around the primary before the primary explodes, also were highly asymmetric and the light curves and spectra did not compare favorably with observations of SN Ia (Raskin et al. 2014).

In simulations of the violent mergers of two $\sim$0.9 $M_\odot$ C/O white dwarfs, again assuming the initiation and propagation of a carbon detonation, only small masses of ^{56}Ni, $\sim$0.1 $M_\odot$, were produced. Most of the 1.8 $M_\odot$ of ejecta was composed of IMEs, thus the model was presented as a candidate for SN 1991bg-like explosions (Pakmor et al. 2011) or SN 2010lp-like events (Kromer et al. 2013b). This would be in conflict with suggestions that the ejected masses of rapidly-declining SN Ia are sub-Chandrasekhar (Sect. 20.4).

During a merger of nearly equal-mass C/O white dwarfs of sub-Chandrasekhar total mass, shear and compressional heating might cause central ignition of carbon followed by detonation. The initiation of a detonation at the low central densities ($\lesssim 10^7\,\mathrm{g\,cm^{-3}}$) of the merger and propagation to even lower densities are major issues. Low densities are likely to inhibit the initiation and propagation of a

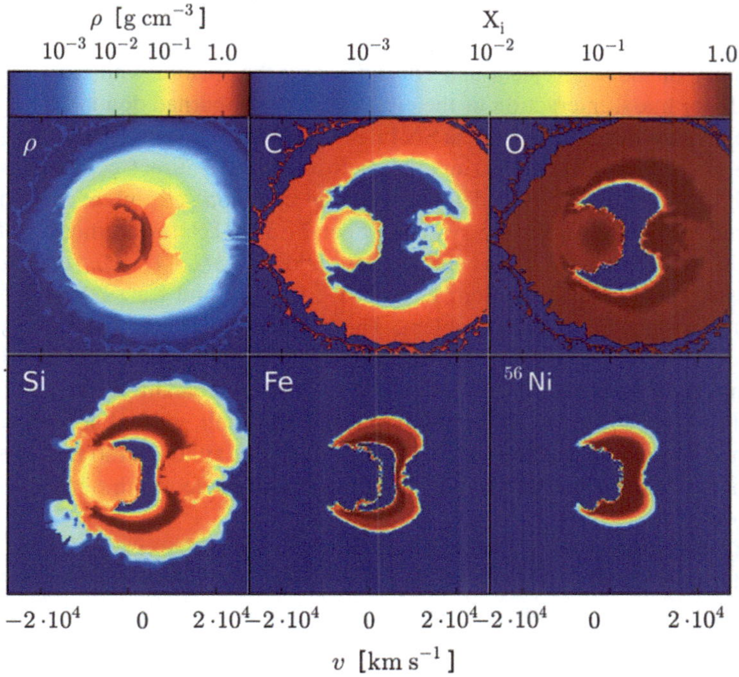

Fig. 22.5 Density and composition of a violent-merger model, 100 s after explosion. From "Normal Type Ia Supernovae from Violent Mergers of White Dwarf Binaries" (Pakmor et al. 2012). © AAS. Reproduced with permission

detonation in C/O and their burning products, since detonation cell sizes will grow to become larger than the scale of the merger.

Ignition of a low-mass ($\sim 10^{-3}$ to 10^{-2} $M_{\odot}$) helium-rich surface layer on one or both of merging C/O white dwarfs, whether of super- or sub-Chandrasekhar total mass, may provide an "initiator" to begin a detonation. No detailed explosion models and radiative transfer calculations have been presented for this case, but this small amount of helium might not have the adverse observational effects of the more massive helium layers that were discussed in the context of SD double-detonation models (Sect. 22.3.8).

Variations on this theme involve a C/O white dwarf accreting from a helium-rich white dwarf. Viscous heating might produce a detonation in the accreted helium layer (Dan et al. 2014). Alternatively, instabilities in the helium accretion stream may produce dense knots of material that strike the surface of the primary, raising the temperature high enough to ignite a detonation in the accreted helium layer (Guillochon et al. 2010). In both cases, ingoing shock waves might then compress the C/O white dwarf, producing a double-detonation supernova in the DD scenario. Because the helium white dwarf would be very close to the accretor, the ejecta are expected to be excessively asymmetric to correspond to typical SN Ia and their

SNRs. If the donor's mass exceeds about 0.4 $M_\odot$, the impact of the ejecta might cause even the donor to explode (a triple-detonation situation), producing excessive asymmetry of the total ejecta and contaminating the spectra with He I lines. For lower donor mass, interaction of the ejecta with the unburned donor may also lead to excessive asymmetry for typical SN Ia as well as their SNRs (Papish et al. 2015).

One way for a prompt DD explosion to occur is a direct collision between two white dwarfs, triggering detonations. The collision rate might be somewhat enhanced in hierarchical triple systems (Sect. 21.3.3). Explosion models of head-on collisions between two C/O white dwarfs yield $0.1 \lesssim M_{Ni} \lesssim 1.1$ $M_\odot$, depending on the masses involved (García-Senz et al. 2013; Kushnir and Katz 2014). Excessive asymmetry is likely to be a problem. The presence of shells of helium of 0.01 to 0.1 $M_\odot$ on one or both of the colliding C/O white dwarfs can influence the outcome (Papish and Perets 2016). In models, the helium had little effect on the C/O detonations, but the ashes of helium burning in the shell, ejected at high velocity, contained substantial fractions of IMEs including ^{44}Ti; although too rare to account for typical SN Ia, such events might be significant for nucleosynthesis and chemical evolution.

Merger remnants that do not explode promptly enter the viscous phase, 10^4 to 10^8 s after merger, during which the disk transforms into a more spherical envelope. If an explosion occurs during this phase it is "tamped" by the envelope (Höflich and Khokhlov 1996). Detailed simulations produce outcomes that are not likely to be consistent with observations of SN Ia (Fryer et al. 2010; Raskin et al. 2014). Emission from shock breakout may be sufficiently extended in time to have been detected, in conflict with observation. The early-time model spectra are unlike those of SN Ia, and the asymmetry is excessive.

22.5 Summary

Accurate radiative transfer calculations of light curves, spectra, and polarization of 3D models will be required for detailed comparison of explosion models with observations of SN Ia.

Expansion from a compact white dwarf state leads to rapid adiabatic cooling to a state of low temperature and low opacity, a dark time. There should be a delay of about a day between the brief breakout emission and the onset of the rise to maximum light powered by deposition of radioactive decay energy. Measuring the delay between breakout and first light, as well as the shape of the early light curve, could prove to be sensitive tests of explosion models.

In terms of accounting for multiwavelength light curves and spectra of SN Ia, 1D delayed-detonations in near-M_{Ch} C/O white dwarfs are the most successful current explosion models. While imperfect, these models represent the standard against which other models must be compared. Delayed-detonation models in 3D tend to produce mixing that does not improve the agreement with observations. There may be processes, for instance the role of strong magnetic fields, that limit the mixing in the deflagration phase. The delayed-detonation model fits most naturally in the SD scenario. A major issue is the lack of evidence for a surviving companion. It is also not clear that this scenario provides explosions at sufficiently large rates or that it can match observed DTDs. It remains possible that delayed-detonation models could arise in variations of DD scenarios.

Many competitive explosion models—delayed detonation, PRD, and GCD—depend on the number, distribution and dynamics of the original dynamic burning seeds that mark the transition from the convective, smoldering, phase to the dynamical phases of deflagration. The dynamics, morphology and nucleosynthesis of a given model are sensitive to this choice of initial conditions. A quantitative intercomparison among models would be beneficial.

Sub-Chandrasekhar explosion models in which a C/O detonation is triggered by a helium-shell detonation can account for a variety of observations. This class of models may be constrained by evidence that the central density of the C/O white dwarf exceeds $10^9\ \mathrm{g\,cm^{-3}}$.

For all the progress on explosion models, there remain aspects that models do not effectively address. An example is the high-velocity features (Sect. 20.4) that are ubiquitous in typical SN Ia, but essentially absent in SN 1991bg-like events. Suggestions have been made that these features could arise in the explosion process or in CSI, but these features have not yet been shown to arise naturally in any explosion model.

While progress has been made toward identifying the nature of the explosion of a typical SN Ia, it remains a challenge to account for the observational properties. A proper explanation of super-Chandrasekhar explosions also remains elusive, a challenge to all models.

Models in the DD scenario may have advantages in accounting for the DTD and rates of SN Ia. DD models naturally account for the absence of a surviving companion star. Unlike the canonical DD model, DD models involving the inspiral of two C/O white dwarfs that merge or violently collide do not seem promising to reproduce the observed light curves, spectra, and polarization of typical SN Ia. It is not clear that invocation of variations such as helium layers or transfer materially improves the situation. The effects of magnetic fields in the progenitor system and ejecta in this context also deserve more study.

All viable models of SN Ia invoke detonations. Calculation of many of the models described here needs to be redone in the context of the unstable flame-brush mechanism (Sect. 19.11.3) and of future insights into the process of deflagration-to-detonation transition.

Chapter 23
Related Explosions

23.1 Introduction

This chapter is concerned with events that are not traditional SN Ia, but that may be related to them, as complete or partial explosions of white dwarfs.

23.2 SN 2002ic-Likes: SN Ia-CSM

The first member of this class to be recognized was SN 2002ic (Hamuy et al. 2003). Early spectra showed lines of Fe III, as in SN 1991T-like SN Ia, but the broad photospheric features were muted and narrow emission lines of H were present (Fig. 23.1). SN 2002ic was overluminous at maximum light and its postmaximum decline was slower than that of typical SN Ia. Two months postmaximum, signatures of strong interaction with circumstellar hydrogen appeared; the early overluminosity and the muting of the spectral features were due to continuum emission from early, milder CSI. Apparently, a SN 1991T-like explosion occurred inside a dense, massive (several $M_{\odot}$), moderately asymmetric (Wang et al. 2004; Chugai and Chevalier 2007), hydrogen-rich circumstellar shell that extended from $\sim 10^{15}$ to $\sim 10^{16}$ cm (Wood-Vasey et al. 2004).

Other well-observed members of the class include SN 2005gj (Aldering et al. 2006; Figs. 23.2 and 23.3), SN 2008J (Taddia et al. 2012a), SN PTF11kx (Dilday et al. 2012), SN 2012ca (Inserra et al. 2014), and SN 2013dn (Fox et al. 2015).

The issue of whether the underlying explosions really were SN Ia appeared to be settled by observations of SN PTF11kx. The early photospheric spectra closely resembled those of SN 1999aa, a spectroscopic Shallow Silicon (Sect. 20.5) that was somewhat similar to but less extreme than SN 1991T-likes. Early spectra of SN PTF11kx also contained narrow H lines with P Cygni profiles, as

D. Branch, J.C. Wheeler, *Supernova Explosions*, Astronomy and Astrophysics Library, DOI 10.1007/978-3-662-55054-0_23

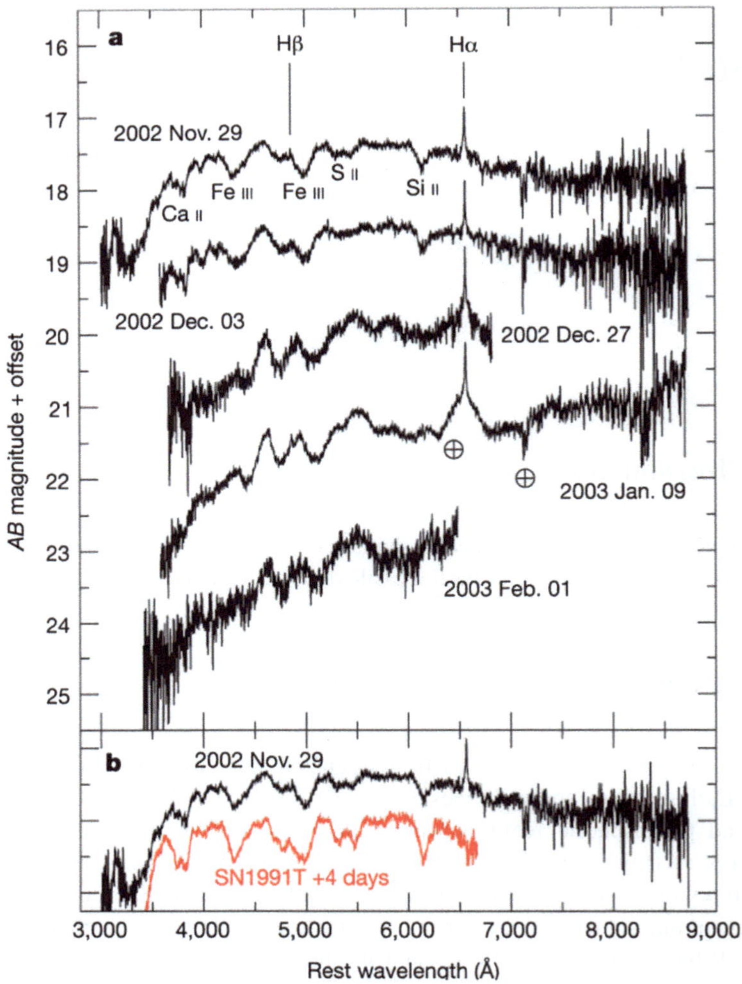

Fig. 23.1 *Top*, (**a**): spectroscopic evolution of SN 2002ic. Maximum light was estimated to be 2002 November 23. *Bottom*, (**b**): the first spectrum of SN 2002ic resembled an early spectrum of the peculiar SN 1991T, except that in SN 2002ic the features were weaker. Reprinted by permission from Macmillan Publishers Ltd.: NATURE (Hamuy et al. 2003)

well as narrow blueshifted absorptions of He I, Na I, Ca II, Ti II, and Fe II. These features, together with postmaximum overluminosity, indicated complex interaction with multiple shells of CSM (Dilday et al. 2012). Late-time spectra resembled those of the previously observed SN 2002ic-likes (Silverman et al. 2013c). SN PTF11kx may constitute a link between SN 2002ic-likes and more typical SN Ia such as SN 2006X (Sect. 21.6.5) that undergo less conspicuous CSI. Alternatively, SN 2002ic-likes may be associated exclusively with SN 1991T-like and SN 1999aa-like events (Fisher and Jumper 2015). Nebular spectra of

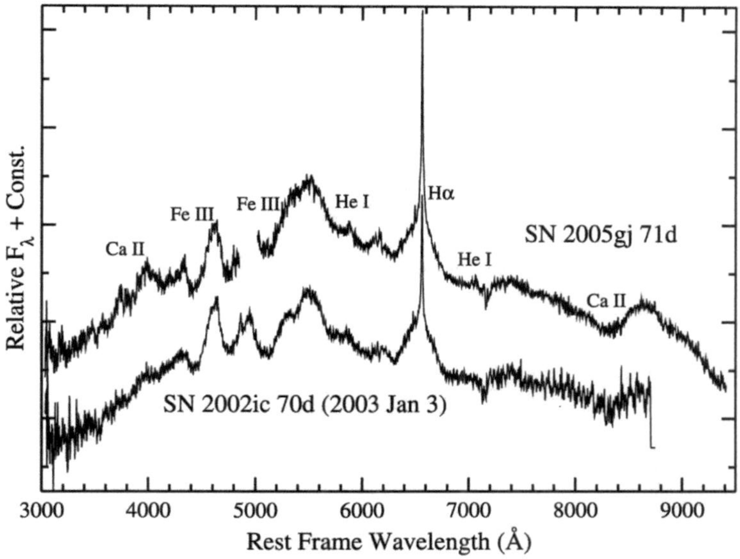

Fig. 23.2 The spectroscopic similarity of SN 2002ic and SN 2005gj after the onset of strong CSI. From "Nearby Supernova Factory Observations of SN 2005gj: Another Type Ia Supernova in a Massive Circumstellar Envelope" (Aldering et al. 2006). © AAS. Reproduced with permission

SN 2012ca and SN 2013dn showed low ratios of [Fe III]/[Fe II] emissions and strong [Ca II] emission (Fox et al. 2015), reminiscent of substantially super-Chandrasekhar SN Ia (Sect. 20.7), in which these properties are attributed to high density of the ejecta. High density is caused by low ejection velocity in the super-Chandrasekhar SN Ia and, presumably, by deceleration of ejecta by CSI in SN 2012ca and SN 2013dn.

Despite the presence of H lines, which traditionally has required a Type II classification, Silverman et al. (2013b) proposed that SN 2002ic-likes be designated SN Ia-CSM to emphasize the connection with traditional SN Ia. The defining properties included bright absolute magnitude at peak, $-21.3 < M_R < -19$; relatively long rise times to peak of 20–40 days; and a high degree of spectroscopic homogeneity after the onset of the strong CSI (Fig. 23.2), with Hα emission of width $\sim$2000 km s^{-1}, possibly due to electron scattering rather than bulk motion (Sect. 4.4.3), as well as narrow, 50–100 km s^{-1}, P Cygni features of Hα. Presupernova mass-loss rates may be $\sim$0.1 M$_\odot$ y^{-1}. The nature of the host galaxies indicate that SN 2002ic-likes are produced by a stellar population that is somewhat young by SN Ia standards. The volumetric rate of SN 2002ic-likes may be $\sim$1% of the SN Ia rate.

Suggestions for the origin of the CSM in the context of the SD channel include accretion from a relatively massive (3–6 M$_\odot$) main-sequence donor that experiences efficient mass stripping by a white dwarf wind (Hachisu et al. 2008); dynamical instability within $\sim 10^4$ years prior to explosion (Han and Podsiadlowski 2006;

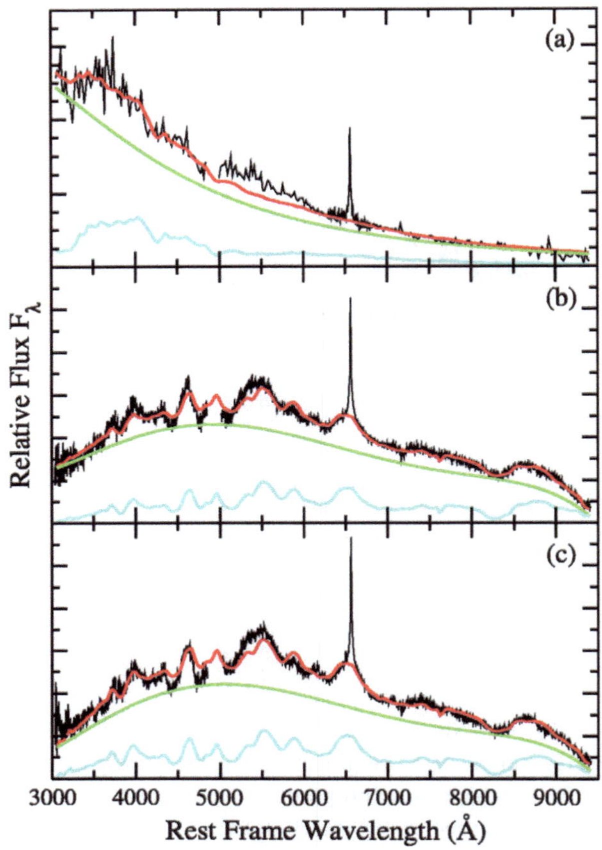

Fig. 23.3 *Top*, (**a**): the spectrum of the SN 2002ic-like SN 2005gj obtained 11 days after explosion (*black*) compared to the sum (*red*) of an 11-day spectrum of SN 1991T (*blue*) and a 12,400 K blackbody component (*green*). *Middle and bottom*, (**b** and **c**): similar, but for 64 and 71 days after explosion and smooth polynomials instead of blackbody components. From "Nearby Supernova Factory Observations of SN 2005gj: Another Type Ia Supernova in a Massive Circumstellar Envelope" (Aldering et al. 2006). © AAS. Reproduced with permission

Meng and Yang 2010a); recurrent novae (Wood-Vasey and Sokoloski 2006); and accretion from the wind of a rotating red giant that is surrounded by a dense, equatorially-enhanced, circumstellar disk (Lü et al. 2009). Livio and Riess (2003) suggested a core-degenerate model (Sect. 21.4).

SN 2002ic-likes share some observational properties with SN IIn. Some events classified as SN IIn may be SN 2002ic-likes in which strong CSI began sufficiently early that an underlying SN Ia spectrum could not be recognized (Silverman et al. 2013b; Sect. 21.6.5).

On the other hand, despite the early spectroscopic resemblance to SN Ia, core-collapse origins of SN 2002ic (Benetti et al. 2006), SN 2005gj (Trundle et al. 2008), and SN 2012ca (Inserra et al. 2014) were suggested. At 2 months postmaximum, the spectra of SN 2002ic were nearly identical to those of SN 1997cy (Germany et al. 2000; Turatto et al. 2000; Deng et al. 2004) and SN 1995E (Rigon et al. 2003), neither of which had been observed near maximum and both of which had been thought to be core-collapse events, tentatively associated with GRBs. That association is unlikely if they were circumstellar-interacting SN Ia. Whether some supernovae that resemble SN 2002ic-likes are core-collapse events remains an open question.

23.3 SN 2002cx-Likes: SN Iax

A new class of transients was recognized with the discovery of SN 2002cx (Li et al. 2003) and its well-observed near twin, SN 2005hk (Phillips et al. 2007; Sahu et al. 2008). The spectroscopic evolution (Fig. 23.4) was unlike anything seen before. The early spectra bore some resemblance to SN 1991T-likes, with conspicuous lines of Fe III and weak lines of Si II and S II, but photospheric velocities were low, $\sim$7000 km s^{-1} (Fig. 23.5). A low level of polarization of SN 2005hk, more like that of traditional SN Ia than like core-collapse supernovae, excluded large asymmetries of the outer layers (Chornock et al. 2006; Maund et al. 2010). Features in postmaximum spectra were like those of SN Ia, but narrower (Fig. 23.6). Late-time spectra, however, were unlike those of SN Ia; they contained a forest of narrow, $\gtrsim$500 km s^{-1}, P Cygni profiles of permitted lines, mostly Fe II, as well as a few forbidden emission lines (Jha et al. 2006; Fig. 23.7; McCully et al. 2014b). Like SN 1991bg-likes, the near-IR light curves lacked double peaks, but unlike SN 1991bg-likes, the light curves were not particularly narrow.

Other well-observed members of the class include SN 2007qd (McClelland et al. 2010); SN 2009ku (Narayan et al. 2011); SN 2010ae (Stritzinger et al. 2014); SN 2012Z (Yamanaka et al. 2015); and SN 2013en (Liu et al. 2015c).

The properties of members of the class, designated SN Iax ("x" for SN 2002cx), were reviewed by Foley et al. (2013). Criteria for membership were (1) no evidence of hydrogen; (2) low ejecta velocity, $\lesssim$8000 km s^{-1}, near maximum light; (3) subluminosity with respect to the SN Ia width-luminosity relation; and (4) spectra resembling those of SN 2002cx and SN 2005hk. Other properties of the class include a large range of absolute magnitudes, $M_V = -14$ to -19, and expansion velocities, 2000–8000 km s^{-1}, at maximum light, and a wide range in estimated ejecta masses, from very low to $\lesssim$1.4 $M_\odot$, but typically about 0.5 $M_\odot$ (Fig. 23.8). Early spectra show the presence of carbon and products of carbon burning.

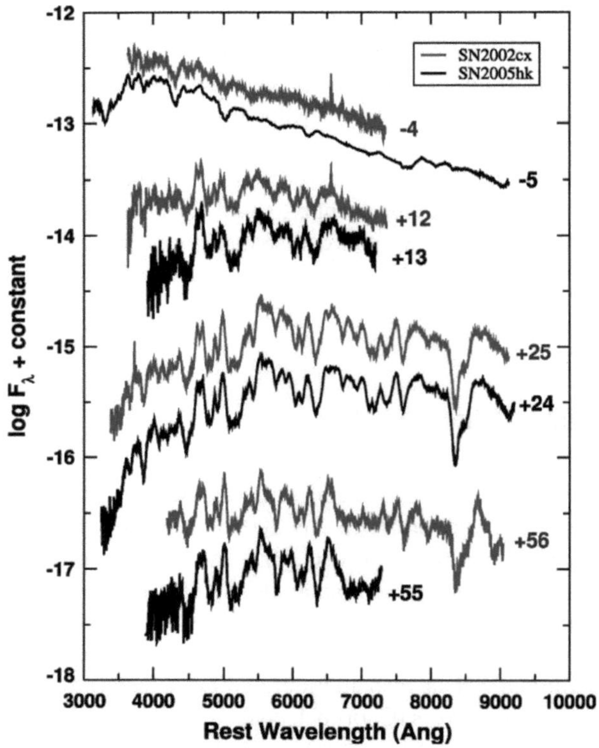

Fig. 23.4 Spectroscopic evolution of SN 2002cx and SN 2005hk. Figure 8 "The Peculiar SN 2005hk: Do Some Type Ia Supernovae Explode as Deflagrations?" from Phillips et al. (2007)

Estimates of the rate of SN 2002cx-likes range from about 6% (White et al. 2015) to about 30% (Foley et al. 2013) of the SN Ia rate. Most SN 2002cx-likes occur in late-type, star-forming galaxies (Foley et al. 2009a; Valenti et al. 2009; Lyman et al. 2013), indicating progenitors that are young by SN Ia standards. Some SN 2002cx-likes, however, appear in populations that lack any signs of recent star formation (Foley et al. 2010b), suggesting a white dwarf origin. White et al. (2015) made a case for dividing the members of the class into those resembling SN 2002cx and those resembling SN 2002es (Ganeshalingam et al. 2012). Among the apparent differences, SN 2002cx-likes preferentially occur in young populations while SN 2002es-likes occur in older populations, and unlike SN 2002cx-likes, spectra of SN 2002es-likes contain Ti II absorption troughs, indicative of low temperature. A strong, early UV pulse from the SN 2002es-like SN iPTF14atg was consistent with interaction between the ejecta and a companion star (Cao et al. 2015).

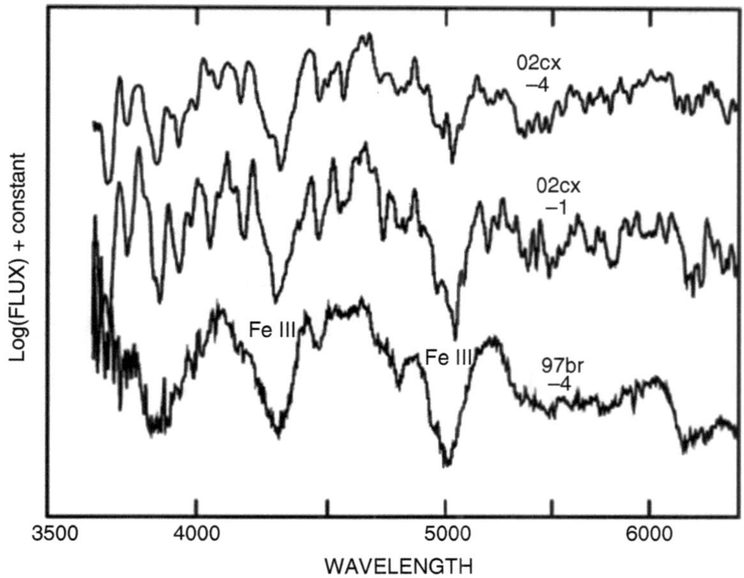

Fig. 23.5 Spectra of SN 2002cx 4 days and 1 day before maximum light compared to an early spectrum of the SN 1991T-like SN 1997br. The SN 1997br spectrum is redshifted by 3000 km s^{-1} to align the Fe III absorptions. Figure 2 "Reading the Spectra of the Most Peculiar Type Ia Supernova 2002cx" from Branch et al. (2004a) ©The Astronomical Society of the Pacific. Reproduced by Permission. All Rights Reserved

Suggestions for explosion mechanisms include "pure" deflagrations, without transition to detonation, of Chandrasekhar or sub-Chandrasekhar C/O white dwarfs (Branch et al. 2004a; Jha et al. 2006; Phillips et al. 2007; Long et al. 2014b); double-detonation explosions (Wang et al. 2013a); explosions of hybrid white dwarfs having C/O cores surrounded by ONeMg (Meng and Podsiadlowski 2014; Wang et al. 2014b; Denissenkov et al. 2015; Kobayashi et al. 2015); PDD (Sect. 22.3.5) explosions of near-M_{Ch} C/O white dwarfs (Stritzinger et al. 2015); and the current front-runner, "failed" deflagrations that only partially disrupt white dwarfs (Livne et al. 2005; Jordan et al. 2012b; Kromer et al. 2013a; Fink et al. 2014; McCully et al. 2014b; Yamanaka et al. 2015).

The presence of He I lines in two SN 2002cx-likes (but see White et al. (2015) on the possibility that these were not SN Iax, but core-collapse events of Type IIb) may indicate that the mass donor was a nondegenerate helium star (Foley et al. 2013). Such explosions would occur in relatively young populations (Liu et al.

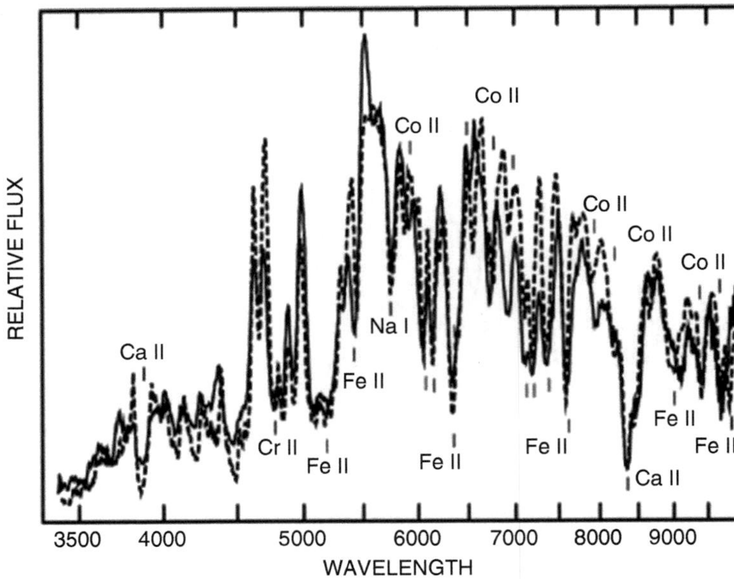

Fig. 23.6 The spectrum of SN 2002cx 25 days after maximum light (*solid line*) compared to a SYNOW synthetic spectrum (*dashed line*) having a photospheric velocity of 5000 km s^{-1}. Figure 6 "Reading the Spectra of the Most Peculiar Type Ia Supernova 2002cx" from Branch et al. (2004a) ©The Astronomical Society of the Pacific. Reproduced by Permission. All Rights Reserved

2015b). (Explosions with helium white dwarfs as donors would occur in older populations; Sects. 23.4 and 23.5.) A luminous ($M_V \simeq -5.3$), blue, probable progenitor system of SN 2012Z was detected in pre-explosion imaging, and further arguments were presented for accretion from a nondegenerate helium star (McCully et al. 2014a).

SN 2008ha, a particularly weak explosion with $M_B \simeq -14$, photospheric velocity near 5000 km s^{-1}, and an ejected mass of only about 0.2 M$_\odot$, was included as a SN 2002cx-like by Foley et al. (2013). It is difficult to explain this event in terms of a failed deflagration and partial disruption of a C/O white dwarf because too much ^{56}Ni is produced (Kromer et al. 2013a). SN 2008ha may have been the result of core collapse (Valenti et al. 2009). Extensive fallback and attendant deceleration of the overlying layers of ejecta might account for the low ejected mass and the low expansion velocities (Moriya et al. 2010a). Analysis of near-IR spectra of the SN 2008ha-like SN 2010ae, however, established the presence of numerous lines of Co II (Stritzinger et al. 2014), which were then recognized to be present in SN 2002cx-likes generally. This appears to provide a link to SN Ia, rather than core collapse, for SN 2008ha and all members of the class. A model of a partial

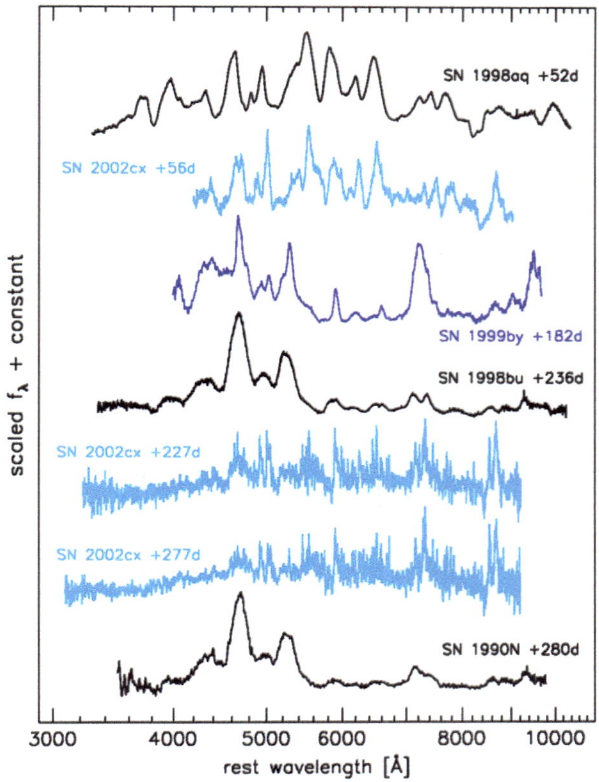

Fig. 23.7 Spectra of SN 2002cx, the typical Type Ia SN 1998aq and SN 1998bu, and the SN 1991bg-like SN 1999by at various epochs. The 56-day spectrum of SN 2002cx resembles the 52-day spectrum of SN 1998bu except that the features are narrower and shifted to the red, but the 227- and 277-day spectra of SN 2002cx do not resemble the late-time spectra of SN Ia. From "Late-Time Spectroscopy of SN 2002cx: The Prototype of a New Subclass of Type Ia Supernovae" (Jha et al. 2006). © AAS. Reproduced with permission

deflagration of a hybrid white dwarf having a C/O core surrounded by a nonburning ONeMg layer that ejected $M_{Ni} \simeq 0.0035$ and $M_{ej} \simeq 0.14$ $M_{\odot}$ with an asymptotic kinetic energy of 0.0018 B showed some of the properties of SN 2008ha (Kromer et al. 2015). Foley et al. (2014b) identified a possible bound remnant or former companion star of SN 2008ha.

Partial disruptions of white dwarfs could leave behind white dwarfs of unusual properties (Jordan et al. 2012b; Kromer et al. 2015), and low ejected mass raises the possibility of recurrent explosions.

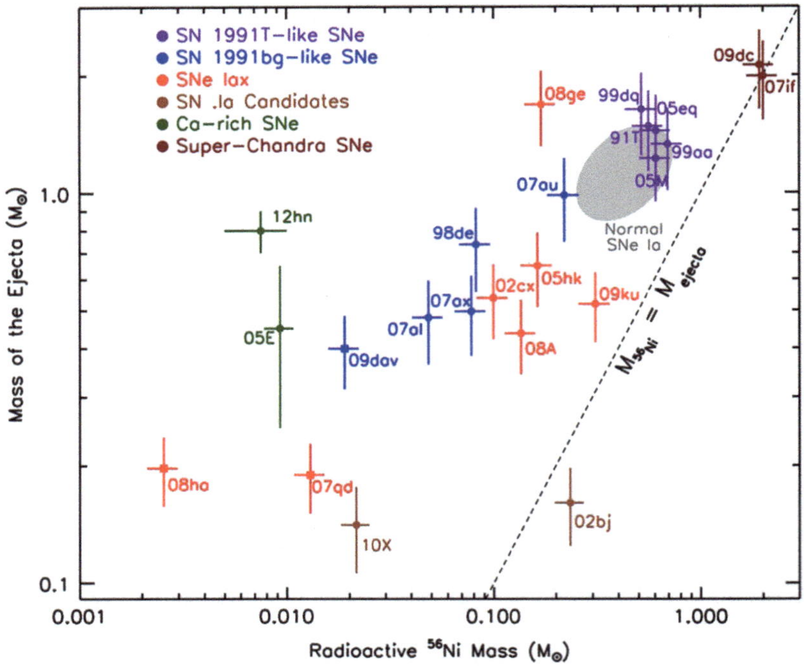

Fig. 23.8 Estimates of ejected mass and M_{Ni} for various supernova subclasses, as indicated in the legend. SN .Ia and Ca-rich events are discussed in Sects. 23.4 and 23.5, respectively. From "Hubble Space Telescope and Ground-based Observations of the Type Iax Supernovae SN 2005hk and SN 2008A" (McCully et al. 2014b). © AAS. Reproduced with permission

23.4 SN 2002bj-Likes: SN .Ia?

The first of these events to be recognized was SN 2002bj (Poznanski et al. 2010a). The well-observed SN 2010X (Kasliwal et al. 2010) was similar in some respects. The most striking characteristic is a very fast light curve (Fig. 23.9): a rapid rise (<7 days) to a moderately bright peak magnitude ($M_R \simeq -18$ for SN 2002bj, $M_R \simeq -17$ for SN 2010X), followed by a very rapid postpeak decline ($\Delta m_{15} \simeq 3.2$ for SN 2002bj). Photospheric-phase spectra are diverse, differing markedly from those of SN Ia. Ejecta velocities tend to be about 10,000 km s^{-1} (Kasliwal et al. 2010). The occurrence rate may be a few percent of the SN Ia rate. SN 1885A in the Andromeda galaxy (Sect. 7.4.9) and SN 1939B in an elliptical galaxy were tentatively associated with this class (Perets et al. 2011); the old populations in which these two events occurred would suggest white dwarf progenitors.

A possible model for SN 2002bj-likes is a helium detonation that does not disrupt an underlying carbon–oxygen or ONeMg white dwarf—the *.Ia* (or *point Ia*) model: about one tenth as bright for about one tenth as long as a typical SN Ia

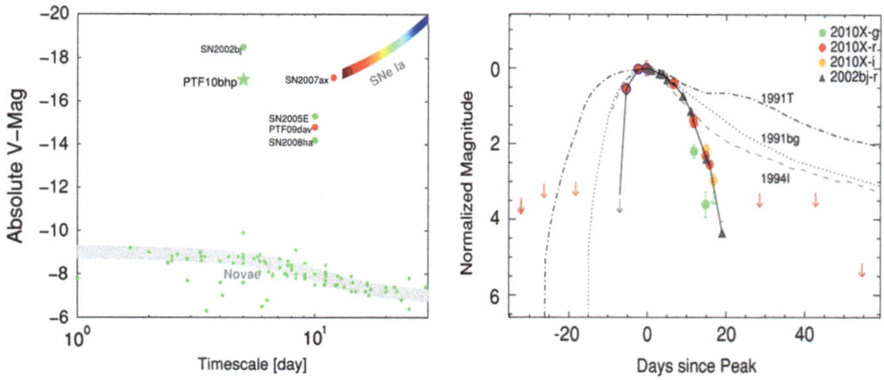

Fig. 23.9 *Left*: M_V plotted against a characteristic timescale for various transients. SN 2002bj and SN PTF10bhp ≡ SN 2010X have short timescales but are moderately luminous. SN 2007ax is an extreme SN 1991bg-like SN Ia. The three events that fall in the luminosity gap between other supernovae and novae are calcium-rich transients (Sect. 23.5). *Right*: optical light curves of SN 2002bj and SN 2010X as well as those of two peculiar SN Ia and the rapidly declining Type Ic SN 1994I. From "Rapidly Decaying Supernova 2010X: A Candidate ".Ia" Explosion" (Kasliwal et al. 2010). © AAS. Reproduced with permission

(Bildsten et al. 2007; Shen et al. 2010b). In this model, the ejected mass is low, $\lesssim 0.2\ M_\odot$, and the kinetic energy is $\gtrsim 0.1$ B. The light curve may be powered primarily by short-lived radioactivities such as ^{48}Cr (half-life 21.6 h) and ^{52}Fe (8.3 h) synthesized during incomplete helium burning (Townsley et al. 2012; Dessart and Hillier 2015) as well as, perhaps, some ^{56}Ni. If, instead, the detonation of the helium shell disrupts the C/O core by launching a prompt ingoing detonation from the outer edge of the core (the *edge-lit* double detonation; Sect. 22.3.8) and the core mass is $\lesssim 0.6\ M_\odot$, the central density will be too low for synthesis of significant amounts of IGEs and ^{56}Ni. Thus the light curve, powered mainly by the radioactivities in the helium shell, might still be as fast as the light curves of SN 2002bj-likes (Sim et al. 2012).[1]

Several helium white dwarfs of extremely low mass ($\lesssim 0.3\ M_\odot$) are known to have massive ($\gtrsim 0.9\ M_\odot$), close, degenerate companions. Some of these systems may evolve to a condition of stable mass transfer from the low-mass white dwarf. In a system in which the massive star is a white dwarf (rather than a neutron-star), a helium detonation might occur in the surface layers of the accretor, producing an SN .Ia (Kilic et al. 2014) if the core does not detonate.

If a deflagration, rather than a detonation, of a surface helium layer of a white dwarf occurred, the core would not be disrupted. Less helium burns in a deflagration than in a detonation, so the result would be a faint, low-velocity

[1] It should be noted, however, that the innermost ejecta of SN 1885A in Andromeda, a possible SN 2002bj-like event, are iron-rich (Sect. 7.4.9).

transient, possibly resembling the weak SN 2008ha-like explosions (Sect. 23.3; Woosley and Kasen 2011).

An entirely different model, that of a SN Ic core-collapse event, was proposed for SN 2010X (Kleiser and Kasen 2014). In this model, owing to extensive fallback possibly driven by a reverse shock from strong CSI, the ejecta contain little or no ^{56}Ni; the light curve is powered only by thermal energy deposited by the shock. Owing to rapid recombination of oxygen and an associated drop in opacity in the surface layers, the model light curve declines rapidly even for an ejected mass of 3 $M_\odot$ and a kinetic energy of 1 B. Whether this model applies to SN 2010X, it provides a valuable illustration of the possible pitfalls of interpreting light curves on the assumption of constant opacity. A similar model might apply to the Type Ic SN 2005ek (Drout et al. 2013; Tauris et al. 2013a), which resembled SN 2010X in some respects, and with adjusted parameters the model might even apply to SN 2002bj itself.

Alternatively, core collapse in an ultra-stripped progenitor, proposed for SN 2005ek (Sect. 17.5.1), could also apply to SN 2002bj and SN 2010X. SN 2010X, SN 2005ek, and SN 2002bj appeared in spiral galaxies. A core-collapse model is not likely to apply to SN 1885A and SN 1939B, which appeared in old populations.

23.5 SN 2005E-Likes: Calcium-Rich Transients

This class is sometimes referred to as SN 2005E-likes, after the first member to be recognized (Perets et al. 2010). Because SN 2005E proved to be not particularly representative of the class, the term "calcium-rich gap transients" (now usually just *calcium-rich transients; CRTs*) was proposed (Kasliwal et al. 2012). These events share five characteristics: rather dim peak absolute magnitudes, typically $M_R \simeq -16$, that fall in the brightness gap between novae and supernovae (Fig. 23.9); moderately fast light curves that rise to maximum light in less than 15 days; photospheric velocities at maximum light of about 10,000 km s^{-1}; early spectroscopic onset of the nebular phase within a few months; and (especially) peculiar nebular spectra dominated by emission lines of [Ca II] $\lambda\lambda$7291,7323 with little or no emission from iron (Fig. 23.10).

CRTs are subluminous with respect to the SN Ia width-luminosity relation. Their photospheric spectra are diverse. Spectra of SN PTF09dav (Sullivan et al. 2011) were reminiscent of SN 1991bg-likes, but with peculiarities such as strong lines of Sc II. Early spectra of SN 2012hn (Valenti et al. 2014) resembled those of SN Ic, but with [Ca II] lines superimposed. Others, including SN 2005E, had He I lines and hence sometimes are referred to as peculiar Type Ib, although a core-collapse origin is improbable. Spectra of the somewhat bright ($M_I \simeq -17.6$) but probable CRT OGLE-2013-SN-079 contained strong lines of Ti II (Inserra et al. 2015).

CRTs occur at several percent of the SN Ia rate. Ejected masses are estimated to be in the range 0.4–1 $M_\odot$, carrying kinetic energies of $\sim$0.1 B. The high calcium

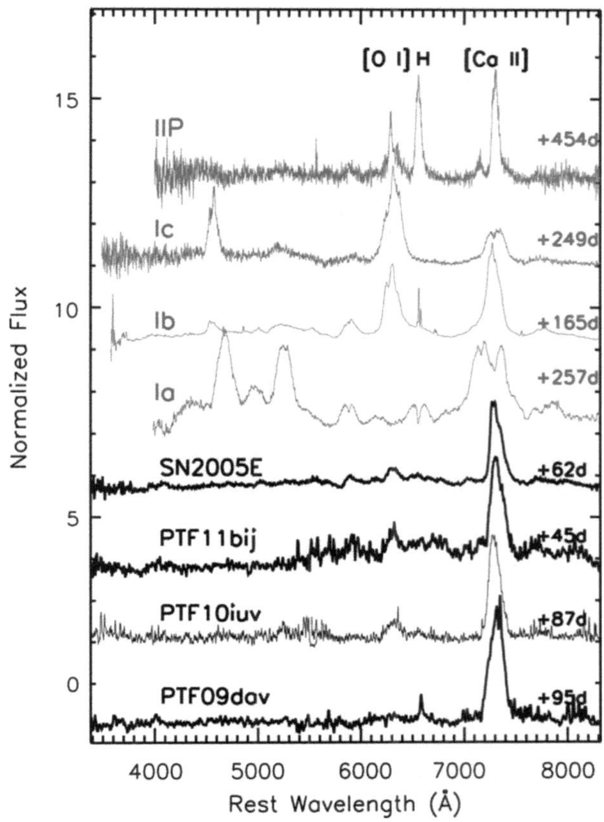

Fig. 23.10 Nebular-phase spectra of four calcium-rich transients (*bottom*) and four supernovae of other types (*top*). From "Calcium-rich Gap Transients in the Remote Outskirts of Galaxies" (Kasliwal et al. 2012). © AAS. Reproduced with permission

abundance suggests explosive helium burning. A possible progenitor system for SN 2005E-likes is a low-mass, $\lesssim 0.6\,M_\odot$, C/O white dwarf accreting from a helium white dwarf (Perets et al. 2010; Waldman et al. 2011; Meng and Han 2015). A detonation in the accreted helium layer might disrupt the C/O core not by an edge-lit detonation as discussed in Sect. 23.4, but by converging shocks that raise the central density before ignition, resulting in the synthesis of the small amount of ^{56}Ni required to power the light curve (Sim et al. 2012). Disrupting a low-mass C/O core by converging shocks appears to be difficult, however (Shen and Bildsten 2014).

CRTs not only appear in old populations, they also have a strong tendency to occur far from the central regions of their host galaxies (Yuan et al. 2013). They might be mergers of neutron stars and white dwarfs in binary systems that received velocity kicks from the supernova that produced the neutron star (Metzger 2012; Lyman et al. 2014b) or mergers of C/O and helium white dwarfs in systems that were ejected from the central regions of their host galaxies (with associated shortening of their orbital periods) by interactions with supermassive black holes (Foley 2015). Tidal detonations of white dwarfs by black holes also has been suggested as a possible origin of at least some CRTs (Sell et al. 2015).

23.6 Summary

The observational diversity of the transients discussed in this chapter is manifest, but the physical origin of the transients is obscure. SN 2002ic-likes appear to be typical SN Ia exploding in CSM, the origin of which is uncertain. SN 2002cx-likes may be partial disruptions, by deflagration, of white dwarfs in SD binary systems, but the nature of these binary systems has not been clarified. Whether SN 2002bj-likes involve white dwarfs or are core-collapse events is an open question. Calcium-rich transients are likely to involve white dwarfs but beyond that, nothing is clear.

Part IV
Consequences, Applications, and Summary

Chapter 24
Consequences of Supernovae

24.1 Introduction

In this chapter, some of the many consequences of supernovae for other areas of astronomy are discussed briefly.

24.2 Compact Remnants

A core-collapse supernova is accompanied by the birth of a neutron star or a black hole. The properties of these compact remnants have the potential to inform us about the supernova progenitors and their explosions. If, for instance, stars of ZAMS mass in the range 8 to about 25 $M_{\odot}$ leave neutron stars while more massive stars leave black holes, then neutron stars outnumber black holes by nearly an order of magnitude. Our Galaxy has produced $\sim 10^9$ compact remnants, with the distances to the nearest neutron star and black hole being $\sim$10 and $\sim$20 pc, respectively. Detections represent only a tiny fraction of the total: $\sim$2500 neutron stars, most as isolated, rotation-powered, radio pulsars and a few hundred in accretion-powered X-ray binaries, and $\sim$50 black holes in X-ray binaries. Consequently, observational selection is an important consideration for investigations of statistical properties of Galactic neutron stars and black holes.

Recent review articles include Miller and Miller (2015) on masses and spins of neutron stars and black holes, Mereghetti et al. (2015) on magnetars, Lattimer and Prakash (2016) on the equation of state of hot, dense matter and neutron stars, and Özel and Freire (2016) on masses, radii, and equation of state of neutron stars. Books include Gosh (2007) on pulsars and Meier (2012) on black hole astrophysics.

D. Branch, J.C. Wheeler, *Supernova Explosions*, Astronomy and Astrophysics Library, DOI 10.1007/978-3-662-55054-0_24

24.2.1 Neutron Stars

The gravitational mass, M_g, of a neutron star at formation depends on the core mass of the immediate stellar progenitor, the amount of ejecta fallback, the equation of state of dense nuclear matter, and the nature of strong-field gravity, which could deviate from general relativity. After formation, some neutron stars increase their mass by accreting from a binary companion. Well-measured values of M_g of neutron stars range from 1.2 to 2.1 $M_{\odot}$, with most in the range 1.3–1.4 $M_{\odot}$. The inferred mass distribution depends on the samples used and the method of statistical analysis (Schwab et al. 2010; Özel et al. 2012).

Measurements of masses of radio pulsars in binaries suggest that the M_g distribution is bimodal (Kiziltan et al. 2013; Fig. 24.1). The M_g distribution of one peak, based on 18 neutron stars in double-neutron star binaries, which should be near their birth masses, is narrow, peaking at 1.33 $M_{\odot}$ with a standard deviation of only $\sim$0.1 $M_{\odot}$. For neutron stars, M_g and baryonic mass, M_b, satisfy $M_b \simeq M_g + 0.075M_g^2$ (Timmes et al. 1996), so the peak is at $M_b = 1.46\ M_{\odot}$. The small dispersion and somewhat high baryon mass are interesting. The high mass presumably is not due to fallback, because a range of fallback masses would be expected to increase the dispersion.

The other peak in the M_g distribution is due to neutron stars with white dwarf companions. These neutron stars are "recycled" to periods of $\lesssim$30 ms owing to accretion of mass and angular momentum. The M_g distribution of 18 such neutron stars peaks at 1.55 $M_{\odot}$ ($M_b = 1.76\ M_{\odot}$) with a standard deviation of $\sim$0.2 $M_{\odot}$ and a high-mass tail extending to $\gtrsim$2 $M_{\odot}$. The masses of these neutron stars, whether formed during normal core-collapse supernovae or by accretion-induced collapse of white dwarfs (Tauris et al. 2013b; Ablimit and Li 2015), are somewhat enhanced ($\lesssim$0.1 $M_{\odot}$) by accretion from their companions. The high masses appear, however,

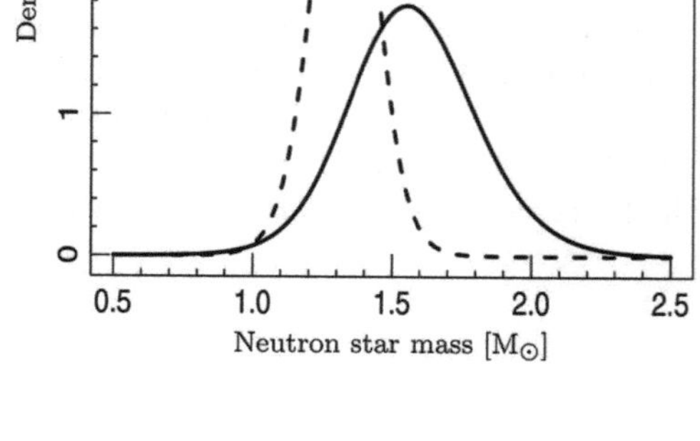

Fig. 24.1 Posterior predictive density estimates for the mass distributions of neutron stars with neutron star (*dashed line*) and white dwarf (*solid line*) companions. From "The Neutron Star Mass Distribution" (Kiziltan et al. 2013). © AAS. Reproduced with permission

to be primarily natal (Antoniadis et al. 2016). Some neutron stars are born with M_b significantly more than 1.46 $M_\odot$.

The highest well-measured masses of neutron stars are about 2 $M_\odot$ (Demorest et al. 2010; Antoniadis et al. 2013). Assuming that general relativity is correct, this mass constrains the equation of state at very high density to be "stiff" (larger Γ_1, pressure sensitive to density) rather than "soft." Associated constraints on the internal composition of neutron stars (purely hadronic? hyperons? free quarks? hybrids? Bose condensates? an admixture of dark matter?) is a matter of much discussion (Hebeler et al. 2013; Orsaria et al. 2014; Drago et al. 2014). Estimated masses of two "black widow" pulsars (pulsars that accrete matter from very low-mass binary companions) are substantially higher than 2 $M_\odot$. Such masses would provide much stronger constraints on the equation of state, but in these cases the uncertainties are such that masses lower than or about 2 $M_\odot$ are not excluded (van Kerkwijk et al. 2011; Romani et al. 2012).

An empirical mass-radius relation could further constrain the equation of state (Lattimer and Prakash 2001), but measuring the radii of neutron stars (Özel et al. 2010a; Steiner et al. 2010; Hebeler et al. 2013) is at present problematic. On the assumption of a unique neutron star radius, independent of mass, an analysis of neutron stars in quiescent low-mass X-ray binaries in globular clusters gave an unexpectedly low $R \simeq 9$ km (Guillot et al. 2013). Other analyses yielded higher values: 11.4–12.8 km (Lattimer and Steiner 2014a) and a causality-based lower limit of 10.7 km (Chen and Piekarewicz 2015) for a 1.4 $M_\odot$ neutron star and 10.8 km for a 1.5 $M_\odot$ neutron star (Özel et al. 2016). Current estimates of radii of neutron stars may be consistent with predictions of the equation of state from nuclear experiment and theory (Lattimer and Steiner 2014b; Heinke et al. 2014).

Considerations of angular momenta of compact remnants invoke the *spin parameter* (or *Kerr parameter*, or simply *spin*), defined as $a \equiv cJ/GM^2$, where J is angular momentum and $0 < a < 1$. The spin of a neutron star near breakup would be $a \simeq 0.7$. Neutron stars in binaries may spin up or down by large factors by accreting matter or by radiating, with associated gains or losses of angular momentum. Thus in general their observed spins do not correspond to their spins at formation. In a small sample of isolated rotation-powered pulsars (Faucher-Giguère and Kaspi 2006), the spins of which are likely to be natal, or nearly so, the spins are low, with a mean value of a in the range 0.02–0.03, corresponding to periods of a few hundred milliseconds (Miller et al. 2011). This is consistent with other evidence, e.g., the moderate spins of pulsars in SNRs of known age (Chevalier 2011; Popov and Turolla 2012). It is possible that these low spins are not truly natal if new-born neutron stars spin down rapidly. Neutron stars may have inherited their spins from the angular momenta of the cores of the supernova progenitors, although in the only known binary radio pulsar, a large misalignment of the spin of one of the neutron stars from the orbital angular momentum vector suggests that this neutron star obtained its spin by hydrodynamical processes during the explosion (Farr et al. 2011). Some of the initial angular momentum of the proto-neutron star at bounce or after deleptonization (with associated contraction and spin-up) could be dissipated in the process of powering a supernova (Sect. 9.5), thus leading to a decrease in rotation rate before the neutron star becomes observable as a pulsar. Thus the final spin for a collapse theorist is the initial spin for a pulsar observer. An issue of whether the observed spins could be generated even if the progenitor did not rotate is debated (Blondin and Mezzacappa 2007; Blondin and Shaw 2007; Rantsiou et al. 2011).

Surface magnetic fields of neutron stars, usually inferred from observed pulsar spin periods and spindown rates, range from 10^8 to 10^9 G for millisecond pulsars (the magnetic fields of which may have been largely "buried" by accretion), through 10^{12} to 10^{13} G for typical pulsars, to $\gtrsim 10^{15}$ G in extreme cases. There is some evidence for a positive correlation between progenitor ZAMS mass and magnetic-field strength (Chevalier 2011). The several dozen neutron stars for which the field is inferred to be $\gtrsim 10^{13}$ G and which occasionally undergo super-Eddington outbursts of X-rays and soft γ-rays are referred to as *magnetars*,[1] also known as soft γ-ray repeaters or anomalous X-ray pulsars. Most (but not all; Davies et al. 2009) magnetars appear to come mainly from stars of ZAMS masses $\gtrsim 30\ M_{\odot}$, indicating that not all stars of ZAMS mass $\gtrsim 30\ M_{\odot}$ become black holes (Muno et al. 2006). Magnetars may be born in close binary systems (Clark et al. 2014). They are X-ray pulsars, and a minority of them also have been detected as radio and/or optical pulsars. They are young, $\lesssim 10^4$ years, and they spin slowly, with periods of 2–12 s. Their persistent radiative output, as high as 10^{35} erg s^{-1} in some cases, far exceeds their spin-down energy losses; this emission is powered by the release of energy stored in their magnetic fields (Mereghetti 2013). In order to generate the magnetic field by a dynamo (Duncan and Thompson 1992) or by hydromagnetic instabilities (Cheng and Yu 2014) they may have spun rapidly, on the order of a millisecond, at birth (but see Vink and Kuiper 2006), with an enormous initial spin-down power possibly leading to long gamma-ray bursts (Bucciantini et al. 2009).

Pulsar space velocities span a wide range but average several hundred km s^{-1} (Hobbs et al. 2005; Faucher-Giguère and Kaspi 2006), which indicates that many neutron stars acquire "kick" velocities from asymmetries in core-collapse explosions. The kicks may be produced by asymmetrical hydrodynamic processes during the explosion (Nordhaus et al. 2010; Wongwathanarat et al. 2013), although it is not clear that this would lead to the observed correlation between the kick directions and the spin orientations for young pulsars (Johnston et al. 2007; Ng and Romani 2007; Noutsos et al. 2013; Rankin 2015). Asymmetry in the neutrino luminosity also may be a source of kicks, e.g., kicks arising from vacuum mixing of active neutrinos to a sterile neutrino with mass of a few kev (Kusenko et al. 2008). A neutrino mechanism might be able to account for the kick-spin correlation. Whatever their origin, the kicks received by neutron stars cause them to emerge from the Galactic disk in 1–10 Myr; the bulk of the Galactic population of neutron stars may be in the halo, with a significant fraction of them escaping from the Galaxy (Sartore et al. 2010; Taani et al. 2012). Neutron stars that result from AIC may, however, receive only

[1]Several objects with magnetic fields $\sim 10^{13}$ G are thought to be relatively old ($\sim 10^5$ to 10^6 years) "fossil" magnetars (Rea et al. 2013).

small kicks; this could account for the presence of pulsars in globular clusters, which have escape velocities of only about 50 km s^{-1} (Tauris et al. 2013b).

Central compact objects (CCOs) are found near the centers of a significant fraction of SNRs of age $\lesssim$7000 years (including Cas A, Sect. 7.4.1), which indicates that they are young and that a significant fraction of new-born neutron stars are CCOs (possibly including an as yet undetected neutron star in SN 1987A; Sect. 11.3). CCOs are apparently isolated neutron stars that emit $\sim 10^{33}$ erg s^{-1} in nonvariable, thermal X-rays (Pavlov et al. 2004; de Luca 2008). No optical or radio counterparts, or pulsar wind nebulae, have been detected. The several CCOs that are observed as X-ray pulsars spin relatively slowly, a few hundred milliseconds—these are their natal spins, given their young ages—and their surface magnetic fields are weak, $\gtrsim 10^{10}$ G (Gotthelf et al. 2013a). Those not observed as pulsars may have even weaker surface magnetic fields. Surface temperature inhomogeneities, however, may imply strong subsurface fields. The initial magnetic fields of CCOs may be typical for neutron stars, but temporarily buried by fallback (Ho 2011; Shabaltas and Lai 2012). Re-emergence of the magnetic fields after $\sim 10^3$ to $\sim 10^5$ years could explain why older descendants of CCOs have not been found (Gotthelf et al. 2013b; Bogdanov et al. 2014). CCOs may even be potential/dormant magnetars, containing very strong buried fields that eventually will re-emerge (de Luca 2008; Viganò and Pons 2012).

24.2.2 Black Holes

One of the enduring issues with the core-collapse problem is which stars form neutron stars and which form black holes (Sect. 9.6). Early work considered the nature of the stars that collapsed to form the detected black holes, with mass typically in the range 5–10 $M_\odot$. Black holes of these masses were unlikely to have formed by binary accretion onto an initial neutron star because of Eddington-limit restrictions. The masses also were too low to be the ZAMS masses of stars that were likely to form black holes (Wheeler and Shields 1976). The obvious suggestion was that the detected black holes arose from the helium cores of stars that had shed their envelopes. The cores of massive stars typically represent about 1/3 of the ZAMS mass, just the right scale if the ZAMS masses exceeded about 30 $M_\odot$. Another factor was that the density gradient between the core and the envelope was thought to get steadily less steep as the mass increased, adding to the overburden, and making a failed explosion and black hole formation more likely.

Observational data that are relevant to this issue have accumulated from the study of neutron stars and black holes in binary stellar systems. These data are subject to the complexities of binary orbital analysis where one can measure a *mass function*, but often not the mass of the compact object directly. The relation between the mass of a compact object to the ZAMS mass of the progenitor also involves complex issues of metallicity, stellar winds, and binary-star evolution that remain uncertain (Sect. 8.8).

Since 2 $M_\odot$ is a lower bound to the maximum mass of neutron stars, 2 $M_\odot$ also is a likely lower bound to the minimum mass of stellar-mass black holes. Mass estimates for several dozen black holes in X-ray binaries range from about 5 to 15 $M_\odot$, with a rather narrow peak near 8 $M_\odot$ (Özel et al. 2010b; Kreidberg et al. 2012). The lack of more massive black holes in the sample could be a reflection of particular binary-formation channels that lead to X-ray binaries. The detection of gravitational waves from the merger of black holes of masses about 36 and 29 $M_\odot$ (Abbott et al. 2016) demonstrated that such massive black holes do form.

If selection effects or errors in the mass estimates are not responsible for the near absence of compact remnants in the range 2 to $\lesssim$5 $M_\odot$, then there is a gap, or at least a deep minimum, in the distribution of the masses of neutron stars and black holes (Farr et al. 2011; Fig. 24.2). A mass gap may contain clues to the physics of supernova explosions and the roles of neutrino-driven convection and fallback (Sect. 9.4.5). An important consideration is that a collapsing iron core of about M_{Ch} becomes sonically disconnected from the outer portions of the star. The inner core must go through the stages of homologous collapse, core bounce, and PNS formation whether or not subsequent fallback leads to the formation of a black hole. The existence of the gap implies that the transition from yielding a neutron star with little fallback to forming a black hole, with or without an attendant explosion, must be abrupt as a function of ZAMS mass. The roles of metallicity, mass loss, rotation, and magnetic fields remain unclear. The diagnostics of just when and how a black hole forms in the wake of PNS formation will probably require observations of the neutrino and gravitational-wave signatures.

There remains to be explained why black holes formed from massive stars usually seem to involve only the mass of the inner core. One possibility is that black holes form from stars that have lost their envelopes to winds, but this would readily

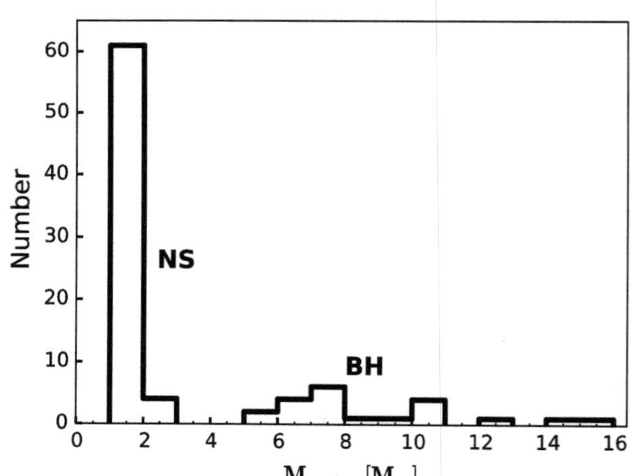

Fig. 24.2 The mass distributions of neutron stars and black holes, illustrating the "mass gap" between the two. Adapted from Belczynski et al. (2012) by Grzegorz Wiktorowicz

work for only the most massive single stars, $\gtrsim$50 M$_\odot$, with associated massive cores, $\gtrsim$15 M$_\odot$. Yet another possibility is that the envelope is lost by mass transfer. All current black hole candidates are perforce in binaries, so a strong selection effect may be at work. If some isolated single stars collapse, taking nearly their whole ZAMS mass with them, we would be hard pressed to detect them. Abbott et al. (2016) may have detected such black holes. Programs to search for the sudden disappearance of stars are relevant here (Kochanek et al. 2008; Gerke et al. 2014).

Yet another explanation of the distribution of black hole masses is that core collapse in red giants descended from stars of ZAMS mass of about 17–25 M$_\odot$ results in black holes (Kochanek 2014, 2015) with expulsion of the outer envelope. The loss of rest mass by neutrino flux when the core collapses reduces the binding energy of any extended envelope. The process of black hole formation could thus be accompanied by a low-energy ($\sim$10^{-4} B) ejection of the hydrogen envelope (Nadyozhin 1980; Lovegrove and Woosley 2013; Piro 2013). The expected distribution of helium-core masses would be broadly consistent with the observational distribution of black hole masses. The detected black hole candidates are, however, all in binary systems with close orbits that could not have harbored an RSG progenitor, so this mechanism does not apply directly to the observed sample.

The maximum spin parameter that a black hole can achieve is nearly unity (Thorne 1974). The spins of about a dozen black holes in X-ray binaries, inferred indirectly from the X-ray properties of their accretion disks (Miller 2007; Reis et al. 2009; Motta et al. 2014), tend to be high, typically about 0.7, much higher than the spins of neutron stars. In several cases, the spins appear to be as high as 0.97 (Gou et al. 2011; Fabian et al. 2012; Miller et al. 2013; Ludlam et al. 2015). Accretion cannot have contributed significantly to these spins because low-mass companions do not have enough mass to give, and high-mass companions do not have enough time. Thus the high spins are largely natal, consistent with the hypothesis that black holes are produced in supernovae with jets (Miller et al. 2011).

The observed spatial distribution of black holes suggests that, on average, their kick velocities are comparable to those of neutron stars (Repetto et al. 2012). If the linear momentum transferred to neutron stars and black holes were equal, as might be expected if kicks were caused by anisotropic emission of neutrinos, then black holes would have lower kick velocities. The black hole kicks may rather be due to large-scale asymmetries in supernova ejecta created by the explosion mechanism (Janka 2013). Black holes that form unaccompanied by a supernova, however, can receive kicks only from anisotropic neutrino emission, possibly leading to a bimodal distribution of black hole kicks.

While firm proof remains elusive, much work has gone into exploring the connection between black hole formation and gamma-ray bursts (Sect. 17.4.3). In the *collapsar* model (Woosley 1993), the explosion may occur by the production of magnetic jets in the vicinity of the black hole or from the propagation of an accretion shock on the rotationally-supported matter that could surround a new-born black hole (Lindner et al. 2010).

24.3 Surviving Companion Stars

Surviving companions of SN Ia were discussed in Sect. 21.6. In the double-degenerate scenario, there are none. In the single-degenerate scenario, the mass donors become single stars that temporarily have unusual properties that may be used to identify them as presupernova donors. Because they are constrained to be of modest mass, all former donors eventually become white dwarfs. Some of these may account for observed samples of white dwarfs having otherwise unexpected properties, e.g., single white dwarfs that have high space velocities yet are not particularly old.

Most massive stars have low space velocities. The $\sim$30% of O- and B-type stars that have velocities in excess of about 40 km s^{-1} are termed *runaway stars*. Some of them probably originated in the *dynamical ejection scenario* (Poveda et al. 1967; Leonard 1991), involving dynamical processes in star clusters, but the majority probably originated in the *binary supernova scenario*, involving ejection of mass from a binary system by a supernova (Blaauw 1961). One runaway B-type star has been found inside an SNR (Dinçel et al. 2015).

Most massive stars are born in binaries (Sana et al. 2012; Chini et al. 2012). For circular orbits and no kicks, the sudden loss of more than half of the mass of the system in a spherical supernova explosion unbinds the binary. Depending strongly on the separation at the time of the explosion, up to 25% of the companion's mass also may be lost by shock heating (Hirai et al. 2014). In most cases, when the first star explodes the binary becomes unbound. The companion leaves the scene at approximately its orbital velocity, while the velocity of the neutron star or black hole is determined by its orbital velocity and whatever kick velocity it receives during the explosion.

When the binary remains bound, which is more likely when mass transfer has made the companion star more massive than the exploding star, the kick momentum is given to the system, perhaps resulting in unusually high system velocity. If subsequent mass transfer to the neutron star or black hole occurs, the system becomes an X-ray binary. Contamination of the surface abundances of the donor star can probe the nature of the supernova progenitor (Suárez-Andrés et al. 2015). Observations can reveal additional properties of the original binary system including its location at the time of explosion (generally in the disk; Napiwotzki and Silva 2012).

If a second supernova occurs, the system is likely to become unbound, resulting in two single compact remnants, the one first formed leaving the scene at its orbital velocity and the second leaving at a velocity determined by the orbital velocity of its progenitor and the kick velocity. In rare cases, owing to judiciously-oriented kicks, the second supernova does not unbind the system, resulting in binary pairs of compact objects.

Population synthesis calculations with binary evolution taken into account predict a velocity distribution of runaway stars that is consistent with observation. Bound systems containing two compact objects are predicted to have sufficiently high space velocities that some of them escape from their host galaxies (Eldridge et al. 2011).

Most of the rare *hypervelocity stars*, those having velocities high enough to lead to escape from the Galaxy ($\gtrsim$550 km s^{-1}, depending on location), were accelerated near the Galactic center, probably due to close approach to the central supermassive black hole (Brown et al. 2012c; Rossi et al. 2014). Other hypervelocity stars appear to have originated in the disk. Some of these may be extreme results of the binary supernova scenario (Przybilla et al. 2008; Tauris 2015; Brown et al. 2015c).

Another possibility, after the first supernova and the birth of a neutron star, is the formation of a Thorne-Żytkow object (TŻO), an RSG with a neutron star core (Thorne and Żytkow 1977). The formation might occur promptly, when the neutron star receives a kick in the direction of its companion (Leonard et al. 1994), or later during common-envelope evolution, when the neutron star spirals into the companion without ejecting its envelope (Terman et al. 1995). A few percent of the RSGs of the Galaxy may be TŻOs (Podsiadlowski et al. 1995), and they may be important sources of p-nuclides (Cannon 1993; Sect. 24.5). The best prospect of identifying a TŻO is from the expected unusual heavy-element surface abundances that result from the fully convective envelope extending down to an extremely hot burning region near the neutron star. A star apparently resident in the Small Magellanic Cloud was proposed as a TŻO candidate (Levesque et al. 2014; Tout et al. 2014; Worley et al. 2016). If the star is a foreground Galactic halo star, however, then it is insufficiently luminous to be a TŻO (Maccarone and de Mink 2016).

24.4 Neutrinos and Gravitational Waves

Unlike photons, neutrinos and gravitational waves promptly emerge from the deep layers of a core-collapse event and therefore have the potential to provide a wealth of information on the core-collapse mechanism.[2]

In Sect. 11.3, we discussed the detection of two dozen (anti)neutrinos from SN 1987A and some of the implications for core collapse and the physics of neutrinos. In recent decades, model predictions of the neutrino emission as well as the capability to detect neutrinos from a core collapse have greatly improved.

A core collapse, whether it results in a neutron star and a supernova, only a neutron star (from AIC; Sect. 9.4.2), or only a black hole, emits $\sim 10^{58}$ neutrinos at characteristic energies of 10 MeV. Experimental techniques for detecting supernova neutrinos were discussed by Scholberg (2012). At present, a Galactic core collapse at a distance of 10 kpc could be detected at several locations on Earth. Super-Kamiokande could detect about 7000 neutrino events, providing excellent statistics; it could detect a few hundred neutrino events from a core collapse in the LMC.

[2]Prospects for future detection of the much weaker neutrino emission from Galactic SN Ia were discussed by Odrzywolek and Plewa (2011) and Wright et al. (2016).

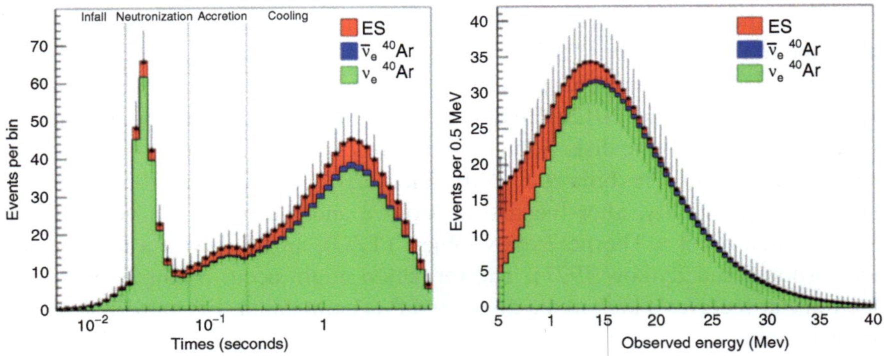

Fig. 24.3 *Left*: predicted time-dependent signal in 40 kilotons of liquid argon for an electron-capture supernova at 10 kpc. *Right*: predicted time-integrated spectrum. Color coding: detections by means of elastic scattering on electrons (*red*) and charged-current interactions with ^{40}Ar of electron antineutrinos (*blue*) and neutrinos (*green*). The latter are dominant. From "Detection of Supernova Neutrinos" (Gil-Botella 2016)

The neutrino emission begins within seconds of the onset of collapse while photons begin to escape only after hours or days, so neutrinos arrive before photons, and the detection of a neutrino burst would signal the imminent arrival of photons, perhaps with some directional information promptly available (Abe et al. 2016). An abrupt interruption of the neutrino signal would mean that a black hole had formed. The SuperNova Early Warning System (SNEWS[3]) is an international network of detectors designed to provide a prompt alert of a core collapse (Antonioli et al. 2004). The Employing Gadolinium to Autonomously Detect Supernovas (EGADS) system is a detector of modest size that is designed to announce a core collapse within 1 s of the arrival of the first neutrino (Adams et al. 2013).[4]

Modeling of neutrino emission from core-collapse supernovae is highly advanced (Nakazato et al. 2013a; Tamborra et al. 2014b; Melson et al. 2015), although predictions of flavor- and time-dependent spectra depend on uncertain properties of neutrinos and details of the explosion dynamics. With good statistics (Fig. 24.3), much could be learned about the core-collapse mechanism. The neutrino signal from the next Galactic core collapse may reveal the primary hydrodynamical instability: e.g., neutrino-driven convection, SASI (Sect. 9.4.7), or some magnetorotational phenomenon (Sect. 9.5). Important constraints on neutrino physics also might be obtained (Sect. 25.5.1).

[3]http://snews.bnl.gov.

[4]During their late, precollapse, neutrino-cooling phases of evolution, massive ($\gtrsim$20 $M_\odot$) stars emit neutrinos at MeV energies. The flux is low compared to that from a core-collapse, but future detection of such neutrinos from a nearby massive star such as Betelgeuse could provide a much earlier alert, *hours, days, or even months* before collapse (Kato et al. 2015; Patton et al. 2017; Asakura et al. 2016).

The cosmic diffuse MeV neutrino background, produced by core-collapse supernovae of various redshifts, was reviewed by Beacom (2010) and Lunardini and Tamborra (2012). The observational upper limit is within a factor of ten of the predicted flux, so detection may occur in the near future. The core-collapse rate at high redshift and the average neutrino emission from core collapse could be constrained. In successful explosion models, the neutrino emission is predicted to increase with the shock revival time, thus it may be possible to infer this crucial parameter of the core-collapse mechanism (Nakazato et al. 2015).

IceCube detected events induced by neutrinos of TeV and PeV energies (Aartsen et al. 2014; Halzen 2014). Liu et al. (2014) and Senno et al. (2015) discussed prospects for the production of GeV to PeV neutrinos by typical and high-energy core-collapse supernovae and their remnants.

Elaborate models predict gravitational-wave emission from core collapse due to the breaking of spherical symmetry during the collapse and bounce phases as well as to hydrodynamical instabilities during the post-bounce phase (Fryer and New 2011; Müller et al. 2013; Abdikamalov et al. 2014; Fuller et al. 2015b). Future gravitational-wave observations will open up a completely new way to probe the quantitative details of core collapse. They may help to constrain the primary hydrodynamical instability and discriminate between some proposed explosion mechanisms: e.g., neutrino-driven, magnetorotational, and acoustic (Sects. 9.4.5 and 9.5). Gravitational waves from fallback accretion onto neutron stars (Piro and Thrane 2012) and hyperaccreting collapsar disks (Kotake et al. 2012) also might be detectable.

Supernovae, both core-collapse and SN Ia, may contribute to the cosmic gravitational-wave background. Uncertainties are large, but the supernova contribution could be comparable to, and difficult to separate from, the contribution predicted by standard inflationary models, thus obscuring this potential probe of the very early Universe (Buonanno et al. 2005).

24.5 Nucleosynthesis

The main goal of nucleosynthesis studies is to identify the physical processes and the astrophysical sites that produce each of the nuclides. Review articles include José and Iliadis (2011) on nucleosynthesis in general, Thielemann et al. (2011b) and Nomoto et al. (2013b) with emphasis on explosive nucleosynthesis in massive stars, Thielemann et al. (2011a) and Qian (2014) with emphasis on the r-process, Jacobson and Frebel (2014) on neutron-capture element abundances in old, metal-poor stars, and Rauscher (2014) on the synthesis of nuclides heavier than iron. The many relevant books include those of Clayton (1984) and Arnett (1996).

Most nuclides of mass number $A < 16$ are produced mainly by nonsupernova sources (big-bang nucleosynthesis, cosmic-ray spallation, novae, AGB stars, winds of Wolf–Rayet stars), although supernovae are important for a few of them. For $16 < A \lesssim 90$, supernovae are the primary nucleosynthesis events. The most accurately known set of relative abundances of the nuclides is that of solar-system material, which reflects 9 Gyr of Galactic nucleosynthesis prior to the formation of the Sun. The relative abundances of most of the nuclides in the range $16 < A \lesssim 90$ can be accounted for reasonably well (Timmes et al. 1995) by combining theoretical nucleosynthetic yields of standard explosion models of (1) core-collapse supernovae (Heger and Woosley 2010; Chieffi and Limongi 2013; Nomoto et al. 2013b), convolved with an initial-mass function for massive stars; and (2) yields of SN Ia models (Iwamoto et al. 1999). Core-collapse supernovae are the dominant source of isotopes in the range $16 < A \lesssim 50$, while SN Ia contribute about half of iron and its neighboring elements. The origin of heavier elements is discussed below.

In a model of a nonrotating, nonmagnetic massive star, hydrostatic burning generates dense shells composed primarily of ^{56}Fe, ^{28}Si, ^{16}O, ^{16}O+^{12}C+^{20}Ne, and ^{12}C+^{4}He, surrounded (if it has not been lost) by an extended, unburned, hydrogen-rich envelope. After collapse of the iron core and successful launch of an outgoing shock, shock heating of the innermost ^{28}Si to $T_9 \gtrsim 5$ ($T_9 \equiv T/10^9\ K$) causes *complete silicon burning* to iron-group elements under conditions of nuclear-statistical-equilibrium (NSE); because the neutron excess of the fuel is small, the primary burning product is ^{56}Ni. At high density and low entropy, complete silicon burning terminates in a *normal freezeout*, where too few light particles are available for capture by heavier ones to alter the NSE composition. For lower density ($\rho \lesssim 10^8$ g cm^{-3}) and higher entropy, complete silicon burning terminates in an *α-rich freezeout*, where capture of abundant alphas does alter the composition; the main product remains ^{56}Ni, but radioactive ^{44}Ti and some other non-NSE nuclides also are produced. The outermost part of the ^{28}Si shell is typically heated to $T_9 \lesssim 5$ and undergoes *incomplete silicon burning* to produce a mixture of iron-group isotopes, including ^{56}Ni as well as ^{28}Si and other nuclei of intermediate mass. The oxygen shell burns in the range $T_9 = 3$–4 to produce mainly ^{28}Si, ^{32}S, ^{36}Ar, and ^{40}Ca. The ^{16}O+^{12}C+^{20}Ne shell burns in the range $T_9 = 2$–3 to produce mainly ^{16}O, ^{20}Ne, ^{24}Mg, and ^{28}Si. The explosion ejects the remaining zones, which include most of the oxygen, unburned. The yields depend on the ZAMS mass and the metallicity. For example, because the mass of ejected oxygen increases strongly with ZAMS mass, the oxygen-to-iron ratio also increases.

The above general picture is secure, but quantitatively many serious uncertainties persist even for a typical core-collapse supernova, not to mention high-energy and very faint core-collapse events. During the presupernova evolution, the treatments of convection, mass loss, and rotation have important consequences for the precollapse structure and composition, and some key nuclear reaction rates remain uncertain, e.g., ^{12}C(α,γ)^{16}O (Ogata et al. 2009). In 1D explosion modeling, important issues include: (1) how to artificially initiate the explosion, e.g., with a piston (Heger and Woosley 2010) or deposition of thermal (Kobayashi et al. 2006) or kinetic (Chieffi and Limongi 2013) energy; (2) the amount of energy to deposit (constrained by observationally-inferred kinetic energies of ejecta); and (3) the placement of the mass cut between matter that escapes and that which falls back (constrained by the need to avoid overproduction of neutron-rich nuclides of the iron group and by neutron star masses). Furthermore, observations of core-collapse supernovae and their remnants show that they are asymmetric, so 1D modeling cannot be expected to give accurate results.

Several radioactive nuclides are of particular interest. The amounts of ejected ^{56}Ni and ^{57}Ni, critical for supernova light curves (Chap. 5), depend in 1D on the placement of the mass cut, but in reality on details of the explosion mechanism. Observations of diffuse γ-ray line emission of ^{26}Al (half-life 7.2×10^5 years) and ^{60}Fe (2.6×10^6 years) provide direct evidence for Galactic nucleosynthesis during the last few million years (Diehl et al. 1995; Wang et al. 2007b; Bouchet et al. 2015). Winds from Wolf–Rayet stars and massive AGB stars, classical novae, and the ejecta of core-collapse supernovae all may be significant for ^{26}Al, the main supernova contribution being explosive burning in the ^{16}O+^{12}C+^{20}Ne shell. Most of the ^{60}Fe may come from a weak s-process (see below) operating during hydrostatic burning in the same shell, followed by ejection by a supernova (Limongi and Chieffi 2006). The discovery of live ^{60}Fe in the Earth and Moon has generated additional interest in the synthesis of this nuclide (Sect. 24.11). The half-life of ^{44}Ti is 60 years, comparable to the mean time between Galactic supernovae, so γ-ray emission from ^{44}Ti should probe individual supernovae. So far, such emission has been detected only from the Cas A and Tycho SNRs (Sect. 7.4) and SN 1987A (Sect. 11.4.2). Synthesis of ^{44}Ti may occur during α-rich freezeout at the termination of complete silicon burning. In principle, the relative and absolute γ-ray intensities of ^{26}Al, ^{60}Fe, and ^{44}Ti constrain supernova models and the Galactic supernova rate, but predictions of their yields are uncertain owing to poorly-known nuclear data (Tur et al. 2010).

About half of the nuclides of $A > 90$ come from a rather well understood nonsupernova source, the main s-process (Käppeler et al. 2011). In AGB stars, a relatively low density of free neutrons ($< 10^8$ cm^{-3}) is released by the ^{13}C(α,n)^{16}O and the ^{22}Ne(α,n)^{25}Mg reactions during helium burning. These neutrons are captured onto a distribution of iron-group "seed nuclei" on a timescale of $\sim$1 year, which is slow compared to beta-decay timescales. The path therefore runs close to the valley of nuclear stability where the required nuclear data are reasonably well known. Winds and planetary nebulae eventually eject the products of the s-process. Supernovae synthesize, or at least eject, most of the other heavy nuclides.

In stars of ZAMS mass $\gtrsim$13 M$_\odot$, the weak s-process (Pignatari et al. 2010) produces some nuclides in the range $56 < A < 90$. Neutrons released by the ^{22}Ne(α,n)^{25}Mg reaction are captured on seed nuclei on a slow time scale in the ^{16}O and ^{16}O+^{12}C+^{20}Ne shells, and the products eventually are ejected by core-collapse supernovae.

The γ-process[5] (Rayet et al. 1990) may produce more than 30 low-abundance nuclides of $A \gtrsim 74$ on the proton-rich side of stability (*p-nuclides*). This spallation-like process invokes photo-neutron emission by (γ,n) reactions in a hot photon environment that partially disintegrates pre-existing, heavy, neutron-rich seed nuclides on a short timescale. Predictions based on simulations of the γ-process in the shocked ^{16}O+^{12}C+^{20}Ne shell of massive stars are consistent with observed

[5] Also known as the *p-process*.

abundances of many of the p-nuclides. Some, however, especially the lighter ones ($A < 120$), are strongly underproduced (Dillmann et al. 2008). The simulations involve substantial uncertainties owing to uncertain nuclear data (Rauscher et al. 2013), and additional origins for p-nuclides are possible, including electron-capture supernovae (Kitaura et al. 2006) and SN Ia (Kusakabe et al. 2011; Travaglio et al. 2015).

Neutrinos emitted by core-collapse supernovae can interact with matter in the expanding ejecta to populate excited nuclear levels that decay via emission of protons, neutrons, and alphas. In the high-temperature environment, these light particles interact with other nuclides. Because the relevant reactions are mediated by a weak interaction, this *ν-process* (Woosley et al. 1990) may be important only for nuclides of low abundance such as ^{11}B and ^{19}F, some of the lighter p-nuclides, and a few heavy nuclides.

After core-collapse and the successful launch of the shock, the neutrino emission drives a continuous flow of protons and neutrons from the surface of the young neutron star. If the initial composition of this *neutrino-driven wind* (Woosley et al. 1994b; Arcones and Thielemann 2013) is proton-rich, conditions in the expanding and cooling wind are favorable for the synthesis of various neutron-deficient nuclides. Neutrons combine with protons to create alphas and an excess of protons, and the alphas then combine to form nuclides made of integer numbers of alphas such as ^{56}Ni ^{60}Zn, and ^{64}Ge. The charged-current reaction $p + \bar{\nu}_e \rightarrow n + e^+$ then produces free neutrons that are captured to form heavier nuclides that are nevertheless neutron-deficient. This *νp-process* (Fröhlich et al. 2006; Pruet et al. 2006; Wanajo 2006) may produce some of the lighter p-nuclides and perhaps other nuclides, but the outcome depends on the detailed conditions in the wind and on uncertain nuclear data (Arcones and Thielemann 2013).

The classical *r-process* (Seeger et al. 1965) produces about half of the nuclides beyond the iron group, including long-lived radioactivities such as ^{232}Th, ^{235}U, and ^{238}U. In general terms, *how* the process works seems reasonably clear. In the presence of a very high density of free neutrons ($\gtrsim 10^{22}$ cm^{-3}) at high temperature ($\gtrsim 10^9$ K) for a short duration (seconds), neutrons are captured onto iron-group seed nuclei on a timescale that is rapid (milliseconds) compared to beta-decay timescales. The nucleosynthesis path is driven far to the neutron-rich side of nuclear stability by the rapid neutron captures, then returns towards stability by beta decays when the neutron density falls, leaving behind an array of r-process nuclides (Fig. 24.4).

Where the r-process occurs is a major issue. Abundances observed in extremely metal-poor stars (Sneden et al. 2008; Roederer et al. 2010; Cowan et al. 2011) indicate that the r-process was operating early in the history of the Galaxy, before the s-process and SN Ia began to contribute to nucleosynthesis. Observations show that the r-process robustly produces heavy ($A \gtrsim 130$) nuclides in nearly the same relative proportions as r-process nuclides in the Sun, suggesting that a unique physical process operates in many generations of stars. If core-collapse supernovae were the main producers, the average mass of r-process nuclides per supernova would need to be $\sim 10^{-4}$ M$_\odot$ (Mathews and Cowan 1990).

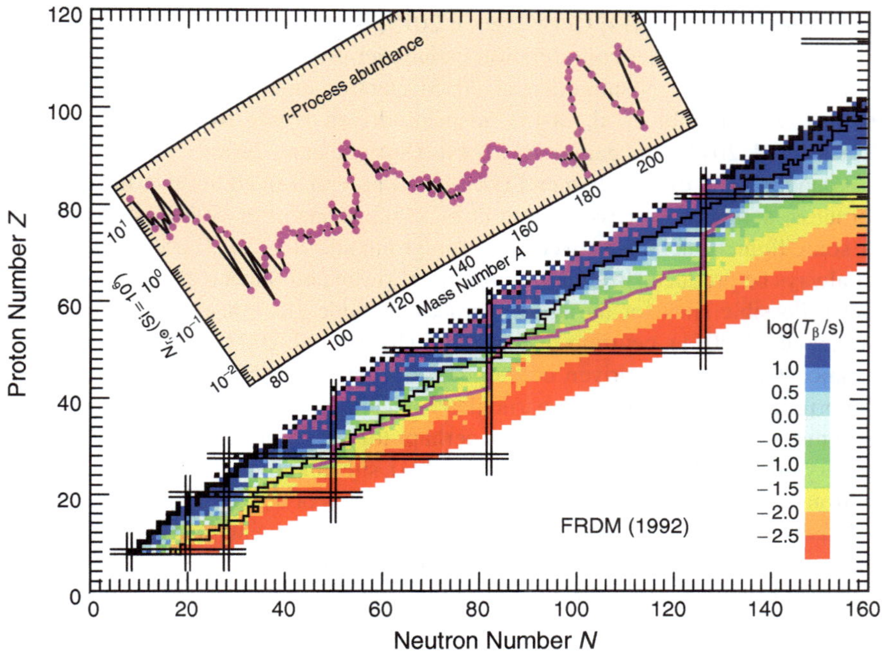

Fig. 24.4 The stable nuclides (*black and magenta boxes*) form the valley of nuclear stability. Unstable neutron-rich nuclides are color-coded on the basis of their mean lifetimes to undergo beta decay, τ_β. The *magenta line* is a typical r-process path. Closed neutron and proton shells are represented by *vertical and horizontal double lines*, respectively. Each r-process path turns vertical at neutron numbers corresponding to closed neutron shells and eventually ends at one of the *magenta squares*, leading to the distribution of r-process nuclides shown in the *inset*. From "Nuclear Properties for Astrophysical and Radioactive Ion Beam Applications" Möller et al. (1997). Reprinted with permission from Elsevier

Numerical simulations of the r-process are uncertain because of poorly known data for nuclides that are far from stability (Mumpower et al. 2015) as well as the unknown physical conditions at the site(s). For some time, a neutron-rich neutrino-driven wind[6] associated with typical core-collapse supernovae was thought to be a promising formation site, but the wind appears to be insufficiently neutron-rich for a robust r-process (Arcones and Thielemann 2013; Wanajo 2013; Martínez-Pinedo et al. 2014). The current favorites are mergers of binary neutron stars or neutron stars and black holes, ejecting up to 10^{-2} $M_\odot$ of r-process enriched material per

[6]As opposed to the proton-rich neutrino-driven wind mentioned above for the νp-process; both could occur in nature, e.g., an evolving neutrino-driven wind could be neutron-rich and proton-rich at different times.

event[7] (Perego et al. 2014; Goriely et al. 2015; Ishimaru et al. 2015; Martin et al. 2015). Other sites under consideration include jet-driven core-collapse supernovae (Winteler et al. 2012; Nishimura et al. 2015), electron-capture supernovae (Wheeler et al. 1998; Pllumbi et al. 2015), neutrino-driven winds from protomagnetars (Vlasov et al. 2014), and gamma-ray bursts (Surman et al. 2006).

Abundances in extremely metal-poor stars imply that in addition to the classical r-process, a *weak r-process* (Truran et al. 2002; Kratz et al. 2007) is responsible for some lighter nuclides between the iron group and $A \lesssim 130$. These nuclides may not always be produced in solar proportions (Cowan and Sneden 2006; Honda et al. 2006). Possible sites for the weak r-process include electron-capture supernovae (Wanajo et al. 2011), neutrino-driven winds (Kratz et al. 2007; Farouqi et al. 2010; Wanajo 2013), and jet-driven core-collapse supernovae (Nishimura et al. 2015). The origin of these nuclides is particularly uncertain, however, and some or all of them may come from other processes, perhaps the νp-process, or a "lighter element primary process" (LEPP; Travaglio et al. 2004), site unknown. The LEPP could be a neutron-capture process related to the weak r-process or a charged-particle process.

Massive stars can produce nonstandard core-collapse events such as high-energy and very faint supernovae, and perhaps PISN. The yields of these events are discussed briefly in Sect. 24.8 in connection with observed element abundances in stars of very low metallicity.

24.6 Dust

In the current Universe, AGB stars such as those with ZAMS masses in the range 3–4 $M_\odot$, which expel dust in winds (Ferrarotti and Gail 2006), may be the main source of dust grains in galaxies. Subsequent grain growth in the ISM may substantially increase the mass of dust (Draine 2009). Most dust destruction probably occurs in SNRs (Raymond et al. 2013; Lakićević et al. 2015; Slavin et al. 2015). Thus the amount of dust in a galaxy is determined by a balance between grain growth in the ISM and destruction in SNRs (Asano et al. 2013).

There is little evidence for dust formation associated with SN Ia. Dust formation associated with core-collapse supernovae and their progenitor stars is, however, an issue of intense interest. Winds of supernova progenitors and common-envelope episodes expel *pre-existing dust* (Sect. 6.6) into the circumstellar environment prior to explosion. Shock-breakout radiation and the forward shock may vaporize pre-existing dust within about 10^{18} cm of the supernova. Supernova radiation can heat surviving dust, leading to mid-IR and/or far-IR echoes as well as scattering echoes of UV and optical radiation. Dust also may form in the heavy-element-rich ejecta, as it did in SN 1987A (Sect. 11.4.3). Ejecta dust, initially warm, may emit in the near-IR

[7]The decay of unstable nuclei in the ejecta may produce a *kilonova* that peaks at 10^{40} to 10^{41} erg s^{-1}, days to weeks postmerger (Metzger et al. 2010; Barnes et al. 2016; Tanaka 2016).

and mid-IR. With its excellent sensitivity and angular resolution, the *Spitzer Space Telescope* permitted detailed investigations of warm dust associated with SN IIP (Sect. 12.8) and SN IIn (Sect. 14.3.4), although even *Spitzer* was insensitive to cold dust that emits in the far-IR and submillimeter ranges.

Several lines of evidence suggest the presence of large amounts of dust ($\gtrsim 10^8$ $M_\odot$) in hyperluminous infrared galaxies and some host galaxies of quasars at redshifts $z > 6$ (Bertoldi et al. 2003; Beelen et al. 2006). At $z = 6$, the age of the Universe was about 1 Gyr and the mean age of the stellar populations in galaxies was less than 0.5 Gyr, perhaps much less. This time, 0.5 Gyr, is the main-sequence lifetime of a low-metallicity star of 3 $M_\odot$, which makes it unlikely that AGB stars can be responsible for so much dust (Michałowski et al. 2010).

Whether supernovae were the main contributors to the high-redshift dust is not clear (Gall et al. 2011). The required amount of surviving dust associated with each core-collapse supernova (both pre-existing and ejecta dust) is $\sim$0.1 to 1 $M_\odot$ (Dwek et al. 2007), consistent with some theoretical predictions of dust production per core-collapse supernova (Nozawa et al. 2003; Bianchi and Schneider 2007). Prior to the advent of the *Herschel Space Observatory* in 2009, most observational estimates of dust masses associated with core-collapse supernovae were only $\sim 10^{-2}$ $M_\odot$ per supernova or less (Kotak et al. 2009; Meikle et al. 2011), although the possibility that additional dust could be hidden, perhaps as cold dust emitting only at far-IR and submillimeter wavelengths, was recognized. Far-IR and submillimeter observations indicated the presence of 1 $M_\odot$ of cold ($\sim$20 K) ejecta dust in Cas A (Sect. 7.4.1) and from 0.5 to 0.8 $M_\odot$ in SN 1987A (Sect. 11.4.3). This supports the notion that core-collapse supernovae could be an important source of high-redshift dust. An important consideration is that in the SNR phase, the reverse shock may destroy much of the ejecta dust by sputtering (Bocchio et al. 2016; Biscaro and Cherchneff 2016). If so, grain growth in the high-density ISM of young galaxies may be required to account for the amount of high-z dust (Michałowski 2015).

24.7 Mechanical and Radiative Feedback

The time-integrated radiative output of a galaxy's supernovae, including the ionizing radiation, is less than that from its stars. Apart from its effects on CSM and nearby ISM (and on whatever life forms may be in the vicinity; Sect. 24.12), the radiation from supernovae themselves is relatively unimportant. Radiation from SNRs is more significant.

Almost all of the initial supernova energy is kinetic (neutrinos aside). If the supernova occurs in a relatively uniform, moderately dense ISM, a classical SNR develops. Some of the kinetic energy is converted into thermal energy of hot ($\gtrsim 10^6$ K) shocked gas and radiation therefrom, and some energy (10–30%) goes into cosmic rays. The remaining kinetic energy, perhaps only a few percent of the original amount, is mechanically deposited into the ISM. Nevertheless, *feedback* of energy and momentum from supernovae (along with stellar winds) plays a major role in determining the nature of the ISM and the evolution of galaxies. An important, complicating issue is how SNRs interact with the multiphase ISM (Martizzi et al. 2015; Kim and Ostriker 2015; Iffrig and Hennebelle 2015; Li et al. 2015).

24.7.1 Current Epoch

A longstanding issue is whether supernova feedback triggers star formation by compressing giant molecular clouds. The answer, once thought to be yes, now appears to be no, or at least not ordinarily. SNRs do tend to be associated with young stars, but it does not necessarily follow that star-formation is triggered by SNRs. The progenitors of core-collapse supernovae are born in regions of recent star formation and they die there, so many SNRs should be found near young stars. Furthermore, the winds of the supernova's progenitor and of nearby stars may trigger star formation. A survey of the 45 known SNRs in the LMC, some of which are apparently associated with young stars and/or giant molecular clouds, produced no evidence of SNR-triggered star formation (Desai et al. 2010). This is consistent with the theoretical expectation that shocks faster than 45 km s^{-1}, such as exist in SNRs that are young enough to be recognizable, disrupt molecular clouds rather than compressing them to initiate star formation (Vanhala and Cameron 1998). Under certain circumstances, however, an individual SNR may trigger some star formation (Cichowolski et al. 2014). The extent to which individual SNRs near molecular clouds eventually trigger star formation after their shocks slow down in 10^5 to 10^6 years is difficult to check because by then the SNRs are not recognizable.

Star formation triggered by overlapping SNRs is a different matter (Sharma et al. 2014). As a consequence of the steep initial-mass function and the dependence of stellar lifetimes on initial mass, OB stars in a rich association or a young cluster can provide a quasi-continuous input of energy to the ISM for $>10^7$ years. The winds of massive stars and the SNRs of multiple supernovae can overlap to form a hot ($>10^6$ K), low-density, weakly-radiating *superbubble*, bounded by a thin, dense *supershell* of cool, compressed ISM. After $\sim 10^6$ years, when a supershell of typical radius $\gtrsim 100$ pc slows to tens of kilometers per second, it may initiate star formation by compressing molecular clouds at sufficiently low velocity. The supershell eventually becomes gravitationally unstable and fragments into clumps that may become molecular and form more stars (McCray and Kafatos 1987). Supershells are observed to be common in star-forming galaxies (Dawson et al. 2013; Ehlerová and Palouŝ 2013), and some shell-like distributions of young stars associated with expanding supershells centered on OB associations or young clusters are observed (Chakraborti and Ray 2011; Roccatagliata et al. 2013). Similarly, multiple generations of superbubbles can form *supergiant shells* of radius ~ 1000 pc that may initiate further star formation (Book et al. 2009).

On galactic scales, star formation in disks is self-regulated by the competition between gravitational instabilities in the cool, dense ISM that drive molecular-cloud coagulation, collapse, and star formation, versus the fraction of the disk volume that the hot gas of stellar winds and overlapping SNRs occupy (Silk 1997; Dobbs et al. 2011; Nath and Shchekinov 2013). In star-forming galaxies, the pressure of hot gas may drive gas out of the disk through "chimneys" to form "fountains," with cooling gas eventually falling back to the disk (Shapiro and Field 1976; Bregman 1980). If the volume filling-factor of the hot gas becomes high enough, a *galactic wind* may develop and expel gas from the galaxy (Larson 1974; Strickland and Heckman 2009; Creasey et al. 2013). The process is self-regulated: if the rate of star formation rises, so does the fraction occupied by hot gas, leaving less cool gas for further star formation. If the rate of star formation drops, so does the hot-gas fraction, leaving more cool gas.

Galaxies like the Milky Way have been forming stars at a relatively constant rate for nearly a Hubble time. If all of the galaxy's ISM had been present in the disk from the beginning, star formation could not have continued so long. Supernova-driven fountains of metal-enriched gas may mix with hot, metal-deficient gas at the base of the galactic corona. With its metallicity thus increased, the hot gas may eventually cool and accrete onto the disk, thus sustaining star formation (Fraternali et al. 2013).

The distribution of the stellar masses of galaxies differs from the distribution of the masses of dark matter halos: the ratio of the masses of stellar to dark matter is lower in less massive galaxies. The relation between the luminosities and maximum rotational velocities of spiral galaxies (Tully and Fisher 1977) changes slope at about $100\,\mathrm{km\,s^{-1}}$. These two characteristics may be due to more efficient regulation of star formation by supernova feedback in the shallower potential wells of less massive galaxies (de Rossi et al. 2010; Puchwein and Springel 2013; Lagos et al. 2013).

Supernovae provide kinetic energy and momentum to interstellar clouds, which can generate turbulence and lead to galactic dynamos and the amplification of galactic magnetic fields (Gent et al. 2013).

During the Sedov phase of an SNR, a substantial fraction (about 40%) of the initial kinetic energy may be converted to ionizing photons that have higher mean energy than those emitted by stars. The higher energy would allow the SNR photons to ionize an extended volume, thus contributing to the maintenance of a warm ($\lesssim 10^4$ K), diffuse component of the ISM (Salpeter 1976). Supernova-driven turbulence produces holes in the ISM that enhance the ability of OB stars to ionize high-altitude gas and contribute to the diffuse component (Wood et al. 2010; Barnes et al. 2014).

In galactic bulges and elliptical galaxies, where mass is continually lost from red giants and planetary nebulae, outflows driven by SN Ia help to keep the density of the ISM low (Tang et al. 2009; Voit et al. 2015).

According to the inflationary ΛCDM paradigm, the density of matter in a dark-matter halo of a dwarf galaxy should increase strongly towards the innermost parsecs to form a density cusp, but observations suggest instead that the density profile in the innermost kiloparsec typically is rather flat, forming a core of uniform density (Oh et al. 2011). One of several possible explanations is that supernova feedback, driving alternating baryonic outflows and inflows from the central regions, may cause (by gravitational attraction) the dark-matter profiles to flatten, transforming cusps into cores (de Souza et al. 2011; Del Popolo and Pace 2016).

24.7.2 Early Universe

The formation and evolution of galaxies is a complex research topic involving a multiphase, turbulent ISM; energy, momentum, and nucleosynthetic input from stars and supernovae; magnetic fields; galaxy rotation and spiral shocks; and much more. The general topic is far beyond the scope of this book. In the remainder of this section, a few of the possible effects of early supernova feedback are mentioned briefly.

It is not clear which radiation sources reionized the Universe at $z > 6$. AGN may have been most important, but supernovae may have contributed in several ways. In the prevalent dwarf galaxies of the early Universe, SNRs may have facilitated the escape of stellar ionizing radiation (Kimm and Cen 2014). SNR photons are less numerous than stellar ionizing photons, but of higher mean energy. The higher energy gives them a higher probability of escaping from the parent galaxy and contributing to reionization (Johnson and Khochfar 2011). Black hole remnants of supernovae, accreting from companions in high-mass X-ray binaries (Mirabel et al. 2011) and/or from high-density ISM (Wheeler and Johnson 2011), also may have provided significant ionizing radiation.

In the early Universe, galactic outflows driven by supernovae, AGNs, and cosmic rays affected dark-matter minihaloes ($\sim 10^6$ M$_\odot$), young galaxies ($\gtrsim 10^8$ M$_\odot$), and the IGM. The removal of gas from a galaxy by outflows reduced the mass of the ISM, tending to impede star formation, while the outflowing gas raised the metallicity of the IGM. The energy of the outflow also may have removed some gas from neighboring protogalactic haloes, impeding their ability to become galaxies. On the other hand, the deposition of metals into such gas increased the cooling rate, tending to promote star formation. The extent to which supernova-driven winds inhibited and regulated star formation in the early Universe remains uncertain (Powell et al. 2011; Wyithe and Loeb 2013; Kim et al. 2015a). Damped Lyman-α absorbers (Sect. 24.8.2) show evidence of strong supernova feedback (Bird et al. 2014).

By infusing the magnetic fields of their progenitor stars into the ISM, supernovae may have provided the initial seeding of magnetic fields in protogalaxies. Subsequent amplification by compression, shear flows, and random motions might then be responsible for the early origin of strong magnetic fields in protogalaxies (Beck et al. 2013). Supernovae also may have contributed to the magnetization of the IGM as early as $z \simeq 20$ (Seifried et al. 2014).

In the disks of young galaxies, supernova feedback generated turbulent viscosity that facilitated outward transport of angular momentum and inward transport of gas (Kumar and Johnson 2010; Yan and Wang 2010). The large specific angular momentum of gas in the galactic bulge presented an obstacle to the accretion of mass onto supermassive black holes at the centers of galaxies. Supernova-driven turbulence in the bulge may have randomized the velocity distribution of the gas, allowing enough gas to accrete (Hobbs et al. 2011; Herrero-Illana et al. 2012). Another possibility is that fallback after explosions of supermassive (10^4 to 10^6 $M_\odot$) Pop III stars was responsible for the origin of the black holes (Whalen et al. 2013b).

24.8 Chemical Enrichment

The observed distributions of element abundances at various locations and epochs constrain cosmic history. Supernovae synthesize and expel into the ISM most of the elements heavier than nitrogen. The abundances at a particular location and a particular time depend not only on the history of star formation at that location, but also on the history of star formation elsewhere; chemical enrichment is closely coupled with mechanical feedback because supernova ejecta are not deposited in situ. Observational constraints on chemical evolution include the relative abundances of elements at various locations at a given time, best determined for the Galaxy (and a few of its nearby, satellite dwarf galaxies; Frebel et al. 2014), and the evolution of abundances with time. The latter can be probed locally by observing stars of a wide range of ages and on a cosmic scale by observing objects at a wide range of redshifts and lookback times. Making inferences from these abundances requires consideration of the histories of the star-formation rate and the IMF, and the transport of gas by means of fountains, inflows, outflows, and radial flows in disks. Chemical and dynamical evolution are inseparable, and supernovae are important for both (Kobayashi and Nakasato 2011). The chemical enrichment of the Universe was reviewed by Frebel and Norris (2013) and Karlsson et al. (2013). Books on chemical evolution include that of (Matteucci 2012).

This section is restricted to a brief outline of the role of supernovae in chemical enrichment, excluding heavy neutron-capture elements, which were discussed in Sect. 24.5. It is useful to note some basic features of the expected nucleosynthetic output of supernovae. The ejecta of core-collapse events of typical energy, about 1 B, have high α/Fe ratios, where alpha elements are those that form by capture of alpha particles: oxygen, neon, magnesium, silicon, sulfur, and calcium. Owing to low neutron excess in the nuclear fuels, core-collapse of metal-deficient stars produces ejecta that have an "odd-even" effect: ratios of abundances of elements of odd atomic number to those of even atomic number are lower than they are in the Sun. The effect diminishes as progenitor metallicity and neutron excess rise. The ejecta of SN Ia have low α/Fe. The ejecta of hypernova models have intermediate α/Fe, comparable to the solar value, and, relative to oxygen, the abundances of silicon and sulfur are high. Relative to iron, the abundances of zinc, cobalt, and vanadium are high while those of manganese and cobalt are low (Tominaga et al. 2007b). Abundances in the ejecta of PISN models are distinctly nonsolar: the odd-even effect is stronger than it is in core-collapse ejecta and relative to iron, the abundances of silicon, sulfur, and calcium are high, that of zinc is low, and that of carbon is approximately solar.

24.8.1 The Galaxy

The history of the Galaxy is written in the evolution of its composition. Metal-poor stars are particularly informative. The first stars, of zero metallicity (*Population III stars*), may have had a top-heavy IMF, composed perhaps only of stars in the range 10 to $\sim$1000 $M_{\odot}$ (Hirano et al. 2014), in which case no first-generation stars survive to be observed today.[8] A limited number of Galactic stars of nonzero but quite low metallicity are observed, and their observed element abundance patterns can be used to constrain the nature of the first stars and their explosions (Nomoto et al. 2013b; Tominaga et al. 2014). Stars of low metallicity are classified (Beers and Christlieb 2005) as *mega-metal-poor* (MMP: $[Fe/H] < -6$),[9] *hyper-metal-poor* (HMP: $-6 < [Fe/H] < -5$), *ultra-metal-poor* (UMP: $-5 < [Fe/H] < -4$), *extremely-metal-poor* (EMP: $-4 < [Fe/H] < -3$), and *very-metal-poor* (VMP: $-3 < [Fe/H] < -2$).

At this writing, the most iron-deficient star known, in which iron lines are not seen at all, has $[Fe/H] < -7.5$. From the limited abundance information available, including extremely high $[C/Fe] > 5$, the star may have been enriched by ejecta of a single, low-energy, black hole-forming explosion of a star having ZAMS mass about 50 $M_{\odot}$ (Bessell et al. 2015). In HMP stars, too, the carbon and heavier elements are likely to have been produced in the explosions of only one or a few first-generation, zero-metallicity stars. The several known HMP stars have peculiar abundance patterns that include high [C/Fe],[10] up to about 4. A phenomenological, 1D, "mixing-fallback" model, in which outward mixing of deep layers of supernova ejecta is followed by fallback of most of the IGEs, can produce these peculiar abundance patterns (Umeda and Nomoto 2003, 2005; Heger and Woosley 2010). This model is intended to mimic what really happens in 3D. Similar abundance patterns may be realized naturally in jet-induced high-energy explosions (Tominaga et al. 2007a).

UMP and EMP stars are likely to have been enriched by only a small number of supernovae. These stars tend to have enhanced C/Fe and distinctly nonsolar

[8] If, however, star formation extended down to 0.8 $M_{\odot}$, some observed low-metallicity stars could be "Population III survivors" that have accreted metals from the ISM (Komiya et al. 2010).

[9] $[Fe/H] \equiv log(Fe/H)_{star} - log(Fe/H)_{Sun}$, where Fe/H is the iron-to hydrogen abundance ratio. Thus, for example, in a star of $[Fe/H] = -6$, the Fe/H ratio is 10^6 times lower than it is in the Sun.

[10] Most very metal-deficient stars are carbon-enhanced (CEMP stars: carbon-enhanced metal-poor stars). Some CEMP stars also have high abundances of s-process elements; many of these stars are known to be in binaries, and they are thought to have acquired their carbon and s-process elements by accretion from a companion AGB star rather than from a supernova (Bonifacio et al. 2015).

relative abundances of iron-group elements, characteristics that have been attributed to jet-induced high-energy explosions (Nomoto et al. 2013b) or the winds of rapidly-rotating massive stars that experience strong mixing and mass loss ("spinstars"; Maeder et al. 2015). Stars having abundances that would implicate PISN ejecta have not been found. The nature of the first generations of stars in the Galaxy and the fraction of them that exploded with kinetic energies near 1 or 10 B remain important unresolved issues.

The rather uniform abundance patterns in VMP stars suggest that they were enriched by the well-mixed ejecta of many supernovae, so signatures of the first stars are washed out. Thus the composition of VMP stars provides an opportunity to compare observations with theoretical yields of typical core-collapse supernovae from stars in the ZAMS mass range 10 to 50 $M_{\odot}$, without the need to consider contributions from SN Ia, which do not begin to occur with great frequency until the mean metallicity rises to $[Fe/H] \simeq -1$. For most elements, the agreement between theory and observation is reasonable (Tominaga et al. 2007b). As expected, the α/Fe ratio in VMP stars is higher than it is in the Sun owing to enrichment by core-collapse events.

Similarly, in the Galaxy's globular clusters (ages about 11–13 Gyr, $[Fe/H] \simeq -0.5$ to -2.5), the α/Fe ratio is higher than the solar ratio (Roediger et al. 2014), indicating that chemical enrichment was primarily by core-collapse supernovae. Following a brief ($\sim$0.1 Gyr?) epoch of star formation and explosions of core-collapse supernovae, the intracluster gas density dropped below a threshold that is required for star formation, so ejecta from the subsequently occurring SN Ia were not incorporated into stars.

In other old components of the Galaxy—the halo, central regions, bulge, and thick disk—the situation is more complex. Suffice to say that as reviewed by Nomoto et al. (2013b) and Feltzing and Chiba (2013) and discussed by many others (Beers et al. 2012; Bensby et al. 2013; Tissera et al. 2014; Grieco et al. 2015), subsets of stars with various spatial distributions, kinematics, and abundance patterns (most but not all subsets having high α/Fe) have been identified, but their origins and chemical-enrichment histories remain unclear.

In the thin disk, where star formation has continued, the α/Fe ratio has gradually decreased as SN Ia have contributed iron-rich ejecta, beginning when the typical $[Fe/H]$ was $\lesssim -1$ (Venn et al. 2004). As mentioned in Sect. 24.5, an appropriate combination of normal core-collapse supernovae of various ZAMS masses, together with SN Ia, can account reasonably well for solar abundances. Chemical enrichment in the disk during the 4.6 Gyr since the formation of the Sun has been modest.

24.8.2 Damped Lyman-α Clouds

Damped Lyman-α absorbers (DLAs)—metal-deficient galaxies and intergalactic gas clouds that imprint absorption lines of neutral and singly-ionized metals on the background light of quasars—provide the best way to determine relative abundances of some elements at high redshift. The majority of the observed DLAs have redshifts between 2 and 4. With few mild exceptions, DLAs up to $z \lesssim 5$ have $[Fe/H] \gtrsim -2.8$ (Rafelski et al. 2012), so they do not probe abundances at extremely low metallicities as some Galactic halo stars do (Sect. 24.8.1). Nevertheless, element abundances in DLAs are of interest because they are independent of uncertainties in the spectroscopic analysis of Galactic halo stars. In addition, of course, there is no guarantee that the early evolution of the Galaxy was typical. Abundances in DLAs appear to be generally consistent with those in Galactic halo stars of comparable metallicity: some metal-deficient DLAs have enhanced α/Fe ratios, some carbon-enhanced DLAs have been found (Cooke et al. 2012b; Salvadori and Ferrara 2012), and DLAs show no signs of enrichment by the ejecta of PISN. Samples of DLAs of lower metallicity, at $z > 5$ (Kulkarni et al. 2013; Rafelski et al. 2014), may prove to be valuable probes of enrichment by the first stars.

24.9 Cosmic Rays

The energy spectrum of hadronic cosmic rays extends from about 10^9 eV (1 Gev) to $\lesssim 10^{21}$ eV $\equiv 10^6$ PeV $\equiv$ 1 ZeV. The overall elemental composition is consistent with solar abundances at the source, modified by propagation effects in a dusty ISM. The origin of cosmic rays was reviewed by Helder et al. (2012) and Blasi (2013).

For a long time, SNRs were suspected to be the main source of cosmic rays having energies at least up to the "knee" of the cosmic-ray spectrum at about 3×10^{15} eV (Baade and Zwicky 1934; Ginzburg and Syrovatskii 1963; Ellison et al. 1997). Cosmic rays of these energies probably originate in the Galaxy. The chief argument for an SNR origin was based on energetics. Starlight, neutrinos, and gravitational waves cannot act as accelerators. SNRs are the only known Galactic sources that can provide enough power, $\sim 10^{41}$ erg s^{-1}, in forms suitable to accelerate particles to relativistic speeds and maintain a cosmic-ray energy density of ~ 1 eV cm^{-3}. This would require that SNRs, on average, put 10–30% of the initial supernova kinetic energy into cosmic rays (Ptuskin et al. 2010). The radio synchrotron radiation of SNRs showed that electrons are accelerated to relativistic speeds, but proof that hadrons also are accelerated was elusive. In one SNR, a postshock temperature lower than expected on the basis of the shock velocity was attributed to cosmic-ray production (Helder et al. 2009). High-energy (TeV) γ-rays coming from SNRs interacting with molecular clouds hinted that acceleration of hadrons to cosmic-ray energies was occurring, but the possibility that the γ-ray emission process is leptonic was still not entirely excluded.

The *Fermi* satellite, launched in 2008, has high sensitivity to γ-rays in the energy range 20 Mev to 300 Gev. *Fermi* observations of the shape of their γ-ray spectra produced strong evidence that at least some SNRs do generate hadronic cosmic rays (Ackermann et al. 2013). The distinctive spectra result from the decay of neutral pions that are produced by hadronic collisions between cosmic-ray nuclei and the ambient plasma. The acceleration process is nonlinear diffusive shock acceleration and magnetic-field amplification at strong collisionless shock fronts propagating into CSM or ISM (Malkov and Drury 2001; Bell et al. 2013).

Many interesting issues remain to be fully resolved. For example, what is the relative importance of SN Ia and core-collapse events in the production of cosmic rays, and the relative importance of acceleration at the forward and reverse shocks (Telezhinsky et al. 2012, 2013)? What fraction of cosmic rays originate in interacting SNRs in superbubbles (Higdon and Lingenfelter 2005; Parizot 2014)?

Ultra-high-energy cosmic rays at energies above the "ankle" of the cosmic-ray spectrum near 3×10^{18} eV have an extragalactic origin because their Larmor radius exceeds the size of the Galaxy. Candidate sources include newly-born pulsars, gamma-ray bursts, neutrino-driven winds of protomagnetars, and AGNs. The extent to which SNRs are a significant source of cosmic rays between the knee and the ankle or even beyond is unclear (Blasi 2014).

24.10 Gamma Rays

Apart from the gamma-ray bursts associated with some SN Ic-bl, detection of γ-rays from individual supernovae has been limited to SN 1987A (Sect. 11.4.2) and the Type Ia SN 2014J (Sect. 20.2.2)

Supernova explosions themselves do not appear to make important contributions to the diffuse γ-ray background from the Galactic disk (Hunter et al. 1997) or the isotropic extragalactic γ-ray background from 200 MeV to 100 GeV (Abdo et al. 2010b). Radioactivity in supernovae produces γ-rays at MeV energies, but in core-collapse events a large fraction of the γ-rays are thermalized in the ejecta. The γ-ray escape fraction from SN Ia is higher, perhaps high enough to make a significant contribution to the cosmic γ-ray background at MeV energies (Ruiz-Lapuente et al. 2016).

Gamma-rays at energies above 1 GeV can be produced by the interaction of cosmic rays with the ambient medium to produce neutral pions, the decay of which produces the γ-rays. This process is especially efficient in SNRs, where the cosmic-ray energy spectrum is much harder than in the Galaxy generally, owing to energy-dependent escape from the Galaxy; thus the resulting γ-ray spectrum of SNRs is much harder than that produced by Galactic cosmic-rays in general. Gamma-rays from the SNRs of core-collapse events (Fields et al. 2010; Senno et al. 2015) and SN Ia (Lien and Fields 2012) may make a contribution to the extragalactic background in the GeV range. The diffuse Galactic γ-ray background, especially above 100 GeV, may come from SN II and SN Ib/c exploding into high-density shells that surround wind bubbles of the massive supernova progenitor stars, and at TeV energies from SNRs that expand into a uniform ISM. In both cases, the shock interaction produces γ-rays by pion decay and by inverse Compton scattering on the background photon field (Berezhko and Völk 2004).

24.11 Effects on the Solar System?

The recognition that short-lived radioactive nuclides such as ^{26}Al (half-life 0.7 Myr) were present in chondritic meteorites when they solidified in the early solar system (Lee et al. 1976) motivated suggestions that a supernova shock triggered the collapse of a molecular cloud from which the solar system then formed (Cameron and Truran 1977; Lattimer et al. 1977; Gritschneder et al. 2012). The discovery that the ratio of radioactive ^{60}Fe (half-life 2.6 Myr) to ^{56}Fe also was elevated in chondrules (Tachibana et al. 2006) intensified interest in this proposition (Boss and Keiser 2012). Subsequent lower measurements of the ^{60}Fe/^{56}Fe ratio appeared to weaken the case for a supernova-triggered origin (Moynier et al. 2011; Tang and Dauphas 2012, 2015). A first generation of supernovae in the molecular cloud in which the Sun formed may have produced the ^{60}Fe, and winds from Wolf–Rayet stars may have provided the ^{26}Al (Gounelle and Meynet 2012). Supernova contamination of forming planetary systems may be a common, universal process, not a special condition of the Sun (Pan et al. 2012b; Jura et al. 2013; Vasileiadis et al. 2013). Further measurements of elevated ^{60}Fe/^{56}Fe ratios and the finding that in chondrules the initial ^{60}Fe/^{56}Fe and ^{26}Al/^{27}Al ratios were well correlated (Mishra and Goswami 2014) suggested, however, co-injection of these two nuclides. This reinvigorated the supernova-trigger hypothesis (Boss and Keiser 2014). The issue is unresolved.

The discovery of live atoms of ^{60}Fe in deep-ocean ferromanganese crusts (Knie et al. 1999, 2004; Fitoussi et al. 2008) and in lunar drill cores (Cook et al. 2009; Fimiani et al. 2016) provided evidence that a relatively nearby supernova occurred about 2 Myr ago (Feige et al. 2012). The most plausible source is a core-collapse supernova, perhaps an electron-capture event (Fry et al. 2015). The concentration of ^{60}Fe suggests a distance from Earth of $\lesssim$100 pc, depending on the amount of ^{60}Fe ejected by the supernova. Because only supernova dust particles, not gas particles, can overcome the ram pressure of the solar wind and the interplanetary magnetic field, the inferred distances also depend on the efficiency of condensation of ^{60}Fe into dust grains (Fields et al. 2005; Athanassiadou and Fields 2011). If at 100 pc, the supernova may have caused changes in climate and increases in cancer and mutation rates (Thomas et al. 2016). Several young neutron stars within 100 pc are candidates to be the compact remnant of the supernova (Tetzlaff et al. 2010). Certain otherwise puzzling features in the observed cosmic-ray spectrum appear to be due to a nearby supernova, perhaps the same supernova, that occurred about 2 Myr ago (Kachelrieß et al. 2015).

24.12 Life

Life as we know it, and even Earth-like planets, would be impossible if it were not for the synthesis of heavy elements in supernovae.

On the other hand, a supernova, with its prompt burst of hard radiation, radioactive ejecta, and eventual generation of cosmic rays during the SNR phase, can be destructive of life in its vicinity. For current life on Earth, the "kill radius," within which a supernova would be catastrophic, is $\sim$10 pc. Gamma-rays and hard X-rays promote the formation of stratospheric nitric oxide, which would in turn initiate catalytic destruction of ozone and allow solar UV radiation to dispose of land life and much of marine life (Ruderman 1974; Gehrels et al. 2003; Melott and Thomas 2011). Coincidentally, the subsequent SNR at a similar distance might compress the heliosphere to within 1 AU and expose Earth to a dangerously increased flux of ambient high-energy cosmic rays (Fields et al. 2008). Land life might be effectively sterilized by a nearby supernova, but surviving bacteria and/or marine life could lead to its eventual re-emergence. At the current epoch, the mean time between supernovae within 10 pc is $\sim$0.1 Gyr (Beech 2011), so Earth (and Mars) probably was subjected to the effects of nearby supernovae in the past.

On statistical grounds, a nearby supernova is more likely to be a core-collapse event rather than an SN Ia because core-collapses occur at a higher rate, and they are more concentrated to the Galactic plane. A census of stars that are currently nearby reveals no progenitors of core-collapse events within 10 pc. The eruption of Betelgeuse as a SN IIP (Chap. 12) some time within the next hundred thousand years will be spectacular, but at a distance of about 200 pc it will not threaten Earth. IK Pegasi, the nearest known SN Ia candidate (Wonnacott et al. 1993), is at a present distance of 46 pc, also too far to be a threat.

An Earth-directed, long-soft gamma-ray burst (the kind of burst associated with SN Ic-bl; Sect. 17.4.3) within about 1 kpc could pose a similar danger to the ozone layer (Piran and Jimenez 2014). In addition, the UV flash at the surface of the Earth that would result from atmospheric processing of the γ-rays (Smith et al. 2004a,b) would inhibit photosynthesis in tens of meters of the water column (Peñate et al. 2010). The burst also might lead to an increase in the flux of high-energy cosmic rays (Dermer and Holmes 2005). At the current epoch, the mean time between such bursts is $\sim$1 Gyr, perhaps longer since bursts tend to occur in star-forming regions of low metallicity. Short-hard gamma-ray bursts also may cause significant ozone depletions (Melott and Thomas 2011).

Estimates of the location and time dependence of a *Galactic habitability zone* within which planets harboring complex life are most likely to exist (Gonzalez et al. 2001) require many assumptions: the kill radius, the spatial distribution and rates of supernovae in the Galaxy, the history of the star-formation rate, the evolving Galactic metallicity distribution (relevant for the formation of Earth-like planets), and the characteristic time for the evolutionary development of complex life. The estimates vary. Lineweaver et al. (2004) suggested that the current zone occupies an annular region between 7 and 9 kpc from the Galactic center that widens with time. Prantzos (2008) countered that the uncertainties are such that the contours of such a zone cannot be even approximately defined, neither in space nor in time, and suggested that the entire Galactic disk may be suitable for life. Gowanlock et al. (2011) suggested that the majority of planets currently capable of supporting complex life may be towards the inner Galaxy, distributed not only near, but also significantly above and below the Galactic plane, where the supernova rate is lower. Spitoni et al. (2014) predicted that the number of habitable planets peaks at a radial distance of 8 kpc, about where we are. It is possible, though, that supernovae do not even determine the location of a GHZ (Filipovic et al. 2013).

Supernovae and gamma-ray bursts that occur beyond the kill radius from Earth, but still close enough to affect the biosphere, are much more frequent. Contemporary literature often regards UV and ionizing radiation from a parent star or Galactic source as primarily destructive; however, the role of radiation may extend far beyond destruction scenarios. Radiation may have been constructively involved in the origin and evolution of life. The point of view that the evolution of terrestrial and extraterrestrial life is driven by radiation sources was first outlined in the classic book by Sagan and Shklovskii (1966). For more recent discussions, see Scalo et al. (2001) and Smith et al. (2004a,b).

Increases in the flux of high-energy cosmic rays due to close approaches of the young Earth to SNRs, which were more prevalent earlier in the history of the Galaxy, may have caused increases in the frequency of lightning strokes, in turn promoting the formation of important complex molecules necessary for the emergence of life (Erlykin and Wolfendale 2010). Approaches to SNRs also may have caused episodes of global cooling, ozone depletion, and mass extinctions, followed by bursts of biodiversity (Svensmark 2012; Kataoka et al. 2013).

Interactions between the spins of neutrinos from core-collapse supernovae with the ^{14}N nuclide could be at the root of chiral selection of left-handed amino acids (Boyd et al. 2012).

Long-soft GRBs that occur in the Galaxy (but not beyond) are capable of providing biologically significant photon jolts to the biosphere. More than 1000 of these probably have occurred during the history of life on Earth. It is not clear, however, that gamma-ray bursts are the most important source of stochastic irradiation events (Scalo and Wheeler 2002).

24.13 Summary

Supernovae are important for the formation of compact objects, emission of neutrinos and gravitational waves, nucleosynthesis, feedback and chemical enrichment, the formation of SNRs and the origin of cosmic rays, and their effects on life in the universe. Thus supernovae deservedly occupy their central place in modern astrophysics.

Chapter 25
Applications of Supernovae to Other Areas of Astrophysics and Physics

25.1 Introduction

While supernovae are intrinsically fascinating events in their own right, they also serve a broader purpose by their applications to other areas of astronomy and physics. In this chapter, some of these applications are mentioned briefly.

25.2 Distances to Type Ia Supernovae

As luminous, rather homogeneous point sources amenable to being "standardized candles" based on the shapes of their light curves and colors, SN Ia are outstanding indicators of extragalactic distances. For some applications (Sects. 25.2.1–25.2.3), only relative distances are needed; for others, absolute distances are required (Sect. 25.2.4).

25.2.1 Accelerating Cosmic Expansion and Dark Energy

The most spectacular application of SN Ia was the discovery of the acceleration of the cosmic expansion, an achievement for which the 2011 Nobel Prize in Physics was awarded to Saul Perlmutter, Adam Riess, and Brian Schmidt. The acceleration is attributed to some form of *dark energy*, perhaps the energy density of the vacuum associated with Einstein's *cosmological constant* or some other entity that could vary slowly with time (Wang 2010). Because numerous articles have discussed in detail the discovery of the acceleration and future attempts to probe the nature of dark energy, our account will be brief and nontechnical.

D. Branch, J.C. Wheeler, *Supernova Explosions*, Astronomy and Astrophysics Library, DOI 10.1007/978-3-662-55054-0_25

The discovery (Riess et al. 1998; Perlmutter et al. 1999) was made by treating SN Ia as standardized candles, using methods to allow for the correlation between peak luminosity and light-curve shape. The "corrected" peak apparent magnitudes of SN Ia were plotted against host-galaxy redshift, for a range of redshifts up to $z \lesssim 1$. Supernova faintness gave (relative) distance and hence travel time of supernova light, and redshift yielded the factor by which the Universe expanded since the supernova occurred; thus the expansion factor as a function of time was revealed. These early studies were reviewed by Riess (2000), Leibundgut (2001), Perlmutter (2003), and Perlmutter and Schmidt (2003).

Subsequent studies of larger samples of SN Ia up to $z \simeq 1.5$ (Astier et al. 2006; Riess et al. 2007; Kessler et al. 2009; Suzuki et al. 2012) used advanced methods for standardizing SN Ia by taking into account correlations of peak luminosity with light-curve shape and colors. As reviewed by Goobar and Leibundgut (2011) and Astier (2012), these second-generation studies confirmed the original discovery of the acceleration.

There is broad agreement that in the context of a homogeneous Friedman–Lemaître–Robertson–Walker cosmology, studies of SN Ia together with other techniques—cosmic microwave background (CMB), baryon acoustic oscillations, weak lensing, growth of clusters of galaxies—demand an accelerating Universe. Observations to date indicate that the Universe is flat, or very nearly so, and that the energy density of the Universe is $\sim$70% dark energy and $\sim$30% matter, mostly nonbaryonic. The observations are consistent with a cosmological constant, Λ (Betoule et al. 2014; Fig. 25.1). Thus the current standard cosmological model, or "concordance model," is ΛCDM (for cold dark matter). This is a book on supernovae, not cosmology, but we acknowledge that there have been dissenting voices. One possibility is that General Relativity is not an exact description of gravity on cosmological scales (Carroll et al. 2004; Jain and Khoury 2010). Another concern is whether averaging over inhomogeneities in terms of spatial curvature, variations in the rate of ticking of clocks, and gravitational energy gradients can cause effects that mimic acceleration (Wiltshire 2013).

An enormous effort is being made to more precisely determine the properties of dark energy, especially whether it is constant or varies with time. In the era of deep, wide-field searches (Sect. 2.4) such as the Palomar Transient Factory, PanSTARRs, the Dark Energy Survey, and eventually the *LSST*, thousands and eventually many thousands of SN Ia will be observed. Techniques are being developed to identify and employ SN Ia using their photometric behavior alone (Wang et al. 2015d), since the flood of discoveries will transcend the capability to do time-consuming spectroscopic followup. Some of the systematic effects that will need to be controlled are discussed in the reviews cited above as well as in many others, including Kim et al. (2015a) and Saunders et al. (2015). The whole field has turned into a mad carnival of statistical analysis, slicing and dicing current samples and much of it dealing in futures.

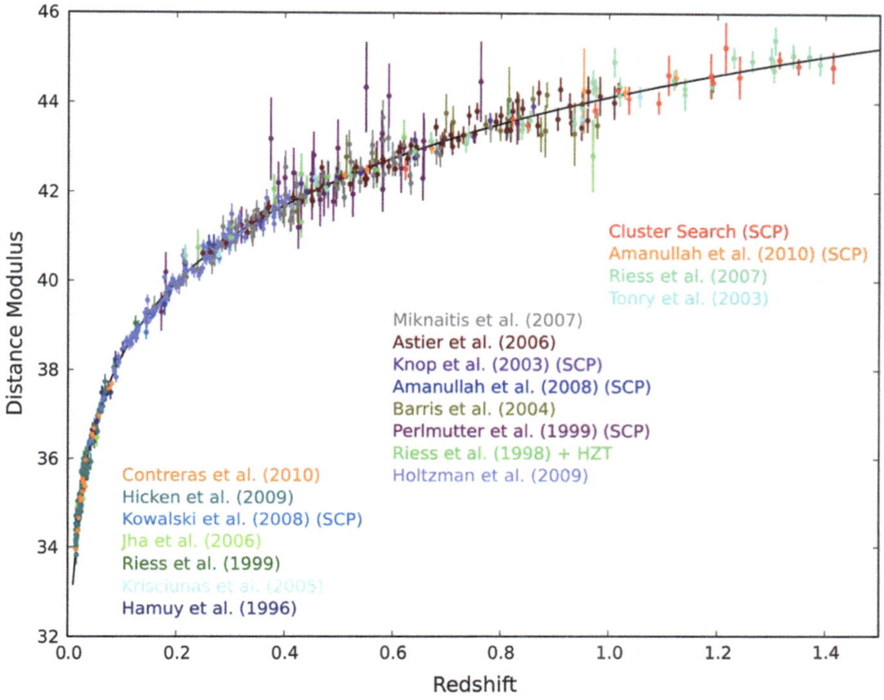

Fig. 25.1 A Hubble diagram for SN Ia (data points). The *solid line* represents a spatially flat ΛCDM cosmology in which the cosmological constant accounts for 73% of the energy density. From "The Hubble Space Telescope Cluster Supernova Survey. V. Improving the Dark-energy Constraints above $z > 1$ and Building an Early-type-hosted Supernova Sample" (Suzuki et al. 2012). © AAS. Reproduced with permission

25.2.2 Flows, Peculiar Velocities, and Anisotropy

Observations of the CMB show that the Local Group of galaxies is moving relative to the CMB by $\gtrsim$600 km s^{-1} (Kogut et al. 1993). The motion is in the general direction of the Shapley supercluster of galaxies, the radial velocity of which is $\sim$13,500 km s^{-1}. Distances and redshifts of SN Ia samples permit the investigation of bulk flows—coherent motions in large volumes—caused by matter inhomogeneities, both overdensities and voids. Studies of SN Ia suggest that the local bulk flow extends beyond the Shapley supercluster, in which case a more distant mass concentration would be indicated (Feindt et al. 2013).

Future large samples of low-z SN Ia will allow an independent measurement of the peculiar-velocity power spectrum, which in turn is directly related to the underlying matter power spectrum. Thus SN Ia will probe the value of the cosmological parameter σ_8, a measure of the amplitude of matter density fluctuations on small scales of $\gtrsim$10 Mpc (Hannestad et al. 2008).

Existing samples of SN Ia indicate that any anisotropic expansion on large scales is likely to be small (Chang et al. 2014; Bengaly et al. 2015).

25.2.3 Cosmic Opacity

Particles of which we are not yet aware might conceivably constitute an interactive medium for photons and hence a form of *cosmic opacity.* One form that was considered as an alternative to cosmic acceleration, but later discarded, was a "gray dust" that would provide a wavelength-independent opacity that could dim distant SN Ia and mimic acceleration. Constraints on cosmic opacity can, in principle, limit physics beyond the standard model and give insight into axions (hypothetical particles with no electrical charge that undergo feeble strong and weak interactions; in the presence of a magnetic field, photons may convert to axions), chameleons (a candidate for dark energy and dark matter that has a nonlinear self-interaction so that the particle has an effective mass that depends on its environment), and mini-charged particles (particles predicted in some extensions of the standard model that have nonquantized electrical charge that is much smaller than the charge of an electron). Constraints on such opacities may be obtained by comparing SN Ia luminosity distances with distances that would not be affected by the opacity, such as angular-diameter distances. Studies to date are consistent with a transparent universe (Liao et al. 2015; Holanda et al. 2016).

25.2.4 The Hubble Constant

The Hubble constant, $H_o = d/(cz)$, is the current value of the redshift-dependent Hubble parameter $H(z)$. Conventional units are km s^{-1} Mpc^{-1}, but the inverse of the Hubble constant has dimensions of time. Together with other cosmological parameters, H_0 gives the expansion age of the Universe. The value of H_0 is obtained from distances and redshifts of galaxies that are in the Hubble flow, $0.03 \lesssim z \lesssim 0.1$, distant enough for the result to be not strongly affected by galaxy peculiar velocities while near enough for the result to be effectively independent of other cosmological parameters.

Among the many ways to obtain estimates of such distances (reviewed by Freedman and Madore 2010), the use of SN Ia as standardized candles is perhaps the most accurate, owing to their near uniformity and their high luminosities. The peak absolute magnitudes of SN Ia in nearby galaxies are calibrated by other high-quality, but much less luminous, distance indicators such as Cepheid variables in young populations and the tip of the red-giant branch in old populations. Calibrated in this way, SN Ia are said to be *secondary* distance indicators.[1] Comparison of the apparent magnitudes and colors of SN Ia in Hubble-flow galaxies with the locally calibrated SN Ia yields the desired distances. A detailed application of this method resulted in $H_o = 73.00 \pm 1.75\,\mathrm{km\,s^{-1}\,Mpc^{-1}}$ (Riess et al. 2016).[2]

25.3 Distances to Core-Collapse Supernovae

Among various ways to determine the distance to SN 1987A in the LMC, the most accurate value probably is that based on the inner ring, 51.4 ± 1.2 kpc (Sect. 11.6).

SN IIP are less luminous than SN Ia, but they are more numerous, especially at high redshift. The ability to infer redshift-independent distances to SN IIP, less accurate than but independent of distances to SN Ia, is valuable. SN IIP are used as both secondary and primary distance indicators.

The *standardized candle* method for SN IIP is based on a correlation, during the plateau phase, between luminosity and photospheric velocity as inferred from the blueshift of spectral features (e.g., Fe II $\lambda5169$ and Hβ; Hamuy and Pinto 2002; Nugent et al. 2006). As for SN Ia, this secondary method is calibrated by high quality, less luminous distance indicators. This technique and generalizations of it (Rodríguez et al. 2014) currently produce distances of about 10% accuracy out to $z \simeq 0.02$ (Poznanski et al. 2010b). Future applications of the method that extend to higher redshift may provide an independent probe of the cosmic expansion history. As for SN Ia (Sect. 25.2.1), techniques based on photometry alone are being developed for SN IIP (de Jaeger et al. 2015b).

Physically-based primary methods for estimating the distances to SN IIP include the expanding photosphere method (EPM: Kirshner and Kwan 1974; Vinkó et al. 2012; Bose and Kumar 2014) and the spectral expanding atmosphere method (SEAM; Baron et al. 2004; see also Dessart et al. 2008). The EPM method equates distance-independent, blackbody photospheric angular diameters inferred from photometry with distance-dependent angular diameters based on photospheric velocities measured from spectra. One problem with the method is uncertain values of photospheric *dilution factors* that take into account deviations from blackbody emission. In SEAM-like methods, spectra of supernova models are calculated with detailed radiation-transfer codes and compared directly, without resorting to dilution factors, to observed SN IIP spectra. The detailed spectral modeling allows the interstellar reddening to be determined; e.g., if a trial value of the reddening is too

[1]Physics-based estimates of SN Ia distances, without the need for local calibration, allow SN Ia to be used also as *primary* distance indicators. Results are consistent with, but less accurate than, those of the standardized candle approach; for a review see Branch (1998).

[2]Analysis of observations of the CMB with the *Planck* satellite, assuming a flat ΛCDM cosmology, resulted in $H_o = 67.8 \pm 0.9\,\mathrm{km\,s^{-1}\,Mpc^{-1}}$ (Planck Collaboration et al. 2016). This provides strong additional motivation for refining the astronomical methods, because a genuine conflict with the *Planck* result would suggest the possibility of new exotic physics, perhaps in the dark sector.

high, the corrected shape of the continuum will require a temperature that is too high to account for the spectral lines. The calculated luminosity of a model, the spectra of which best match the observations, leads to the distance estimate.

Because SN IIP are less luminous than SN Ia, constraining dark energy with SN IIP will require observations with future large telescopes. A *dense shell method*, related to EPM but based on thin, expanding, constant-velocity, circumstellar shells associated with SN IIn, may allow estimates of distances to SN IIn (Potashov et al. 2013), some of which are much more luminous than SN IIP. The even more luminous SLSN I (Sect. 18.3) may prove to be valuable standardizable candles out to $z \gtrsim 3$ (Inserra and Smartt 2014; Scovacricchi et al. 2016).

It may prove feasible to use gamma-ray bursts as distance indicators to extend the Hubble diagram to $z \simeq 10$ (Schaefer 2007; Amati and Valle 2013; Liu and Wei 2015), well beyond the range of SN Ia or SN IIP. Methods restricted to gamma-ray bursts that are associated with supernovae (Pisani et al. 2013) and to supernovae that are associated with gamma-ray bursts (Cano 2014) may be useful.

25.4 Astrophysics

25.4.1 Interstellar Medium

Studies of interstellar extinction of supernovae and light echoes (Crotts 2015) associated with supernovae provide information on the nature and distribution of dust in external galaxies.

Narrow absorption features superimposed on supernova spectra provide a unique way to obtain information on the constituents of interstellar gas in galaxies (Cox and Patat 2014). Because SN 2014J in M82 (Sect. 20.2.2) was nearby and occurred behind a large column density of host-galaxy ISM, it was an exceptional source of information on interstellar atoms and molecules (Ritchey et al. 2015) and on the enigmatic *diffuse interstellar bands, DIBs* (Welty et al. 2014; Fig. 25.2). The spectrum of the Type Ic-bl SN 2012ap (Sect. 17.4.1) contained strong DIBs that varied on short (<30 days) timescales; the supernova appeared to be interacting with DIBs carriers associated with mass loss from the progenitor star (Milisavljevic et al. 2014).

Time-varying strengths of interstellar absorption lines forming in front of expanding supernova photospheres may provide a means to study small-scale ($\sim$100 A.U.) inhomogeneities in the ISM, a scale that is out of reach by other methods (Patat et al. 2010). For a Galactic supernova, sub-A.U. structure could be probed.

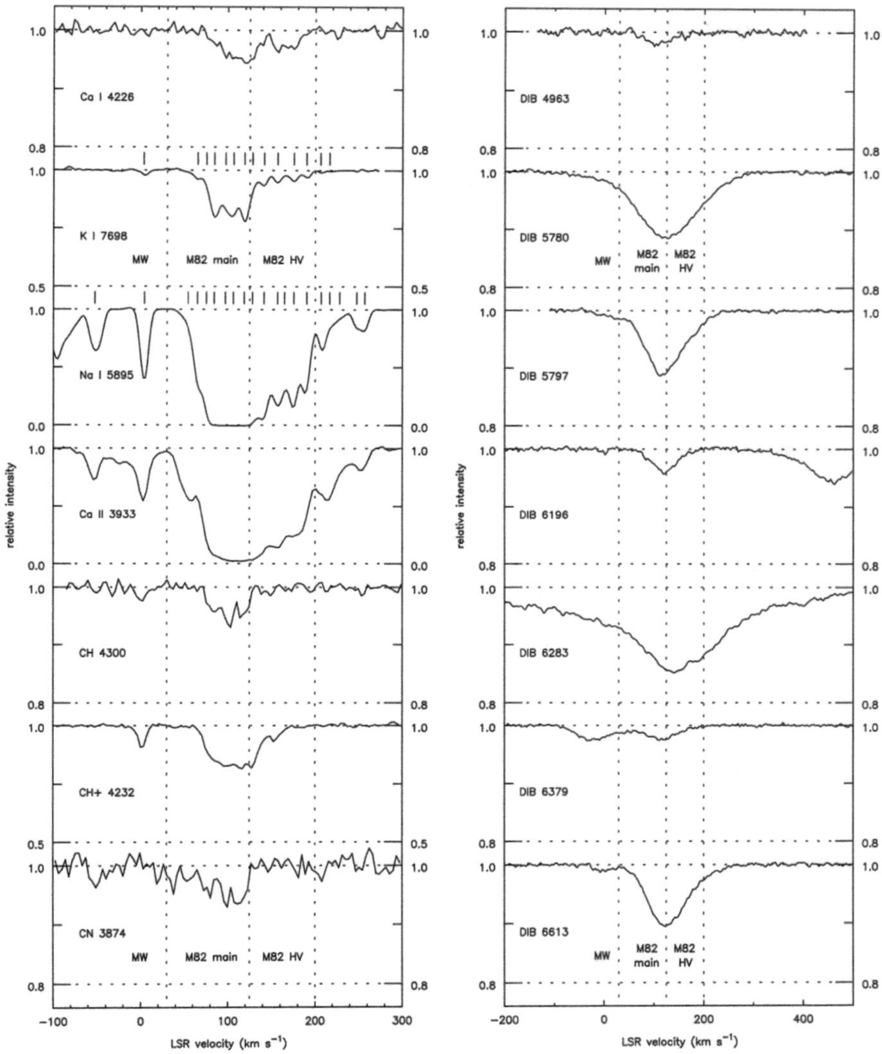

Fig. 25.2 Absorption features of interstellar atomic and molecular gas (*left*) and diffuse interstellar bands (*right*), superimposed on the spectrum of the Type Ia SN 2014J in M82. *Tick marks in the upper left panel* denote individual absorption components used in model fits by Ritchey et al. (2015) to the K I and Na I features. *Vertical lines* refer to three groups of K I absorptions: from left to right, a weak Galactic feature, strong M82 features, and weaker, higher-velocity M82 features. From "Diffuse Interstellar Bands versus Known Atomic and Molecular Species in the Interstellar Medium of M82 toward SN 2014J" (Welty et al. 2014). © AAS. Reproduced with permission

25.4.2 Supernovae from the First Stars

At the end of the cosmic dark ages, the first stars in the Universe (Population III stars) began the ionization and chemical enrichment of galaxies and the intergalactic medium. The formation of the first stars was reviewed by Bromm (2013). Detecting individual Population III stars, at $z \simeq 10$–20, is not currently feasible, but future facilities should be able to detect luminous supernovae produced by Population III stars. These events may include SN IIn (Whalen et al. 2013a), high-energy core-collapse events (Smidt et al. 2014), PISN (Joggerst and Whalen 2011), and more hypothetically, supermassive (10^4 to 10^6 $M_\odot$) thermonuclear explosions (Whalen et al. 2013c). Observations of these explosions would shed light on the nature of the first stars and the primordial IMF (de Souza et al. 2014). Observations of afterglows of gamma-ray bursts also may constrain the environments and properties of early explosions, even at $z \gtrsim 20$ (Mesler et al. 2014; Toma et al. 2016).

25.4.3 Lensing

Weak lensing by inhomogeneities in the cosmic matter distribution is expected to contribute to the observed luminosity dispersion of supernovae. The current upper limit on the lensing contribution to the dispersion of SN Ia luminosities already provides constraints on the cosmic matter power spectrum (Ben-Dayan and Kalaydzhyan 2014; Fedeli and Moscardini 2014).

Galaxies and clusters of galaxies at intermediate redshifts can magnify the flux of distant background sources by large factors. First discoveries included the Type Ia SN 1997ff at $z \simeq 1.7$, magnified by a factor of 1.4 by several foreground lensing galaxies between $z \simeq 0.6$–0.9 (Benítez et al. 2002); an SN Ia at $z \simeq 1.4$, magnified by a factor of $\sim$400 by a lensing galaxy at $z \simeq 1.1$ (Quimby et al. 2014); and an SN Ia at $z \simeq 1.3$, magnified by a factor of two by a lensing galaxy at $z \simeq 0.3$ (Rodney et al. 2015). In November 2015, four images of a SN 1987A-like supernova ("supernova Refsdal") at $z = 1.49$, strongly lensed by a cluster of galaxies at $z = 0.54$, appeared in an "Einstein-cross" configuration (Kelly et al. 2015). In December 2015, a new image of the SN appeared in another of the multiple images of the host galaxy, near the position and time that had been predicted on the basis of lens models of the cluster (Kelly et al. 2016; Fig. 25.3).

Magnification of supernovae that would otherwise be too faint to be detected will bring even more distant events into the realm of quantitative analysis. Studies of lensed SN Ia, for which the intrinsic luminosity and therefore the amount of amplification may be inferred accurately (Patel et al. 2014), may be used to probe the distribution of dark matter in clusters of galaxies (Nordin et al. 2014). Time delays between the light curves of multiple images of supernovae that are strongly lensed by well-modeled clusters of galaxies may provide independent estimates of distances at high redshift (Riehm et al. 2011).

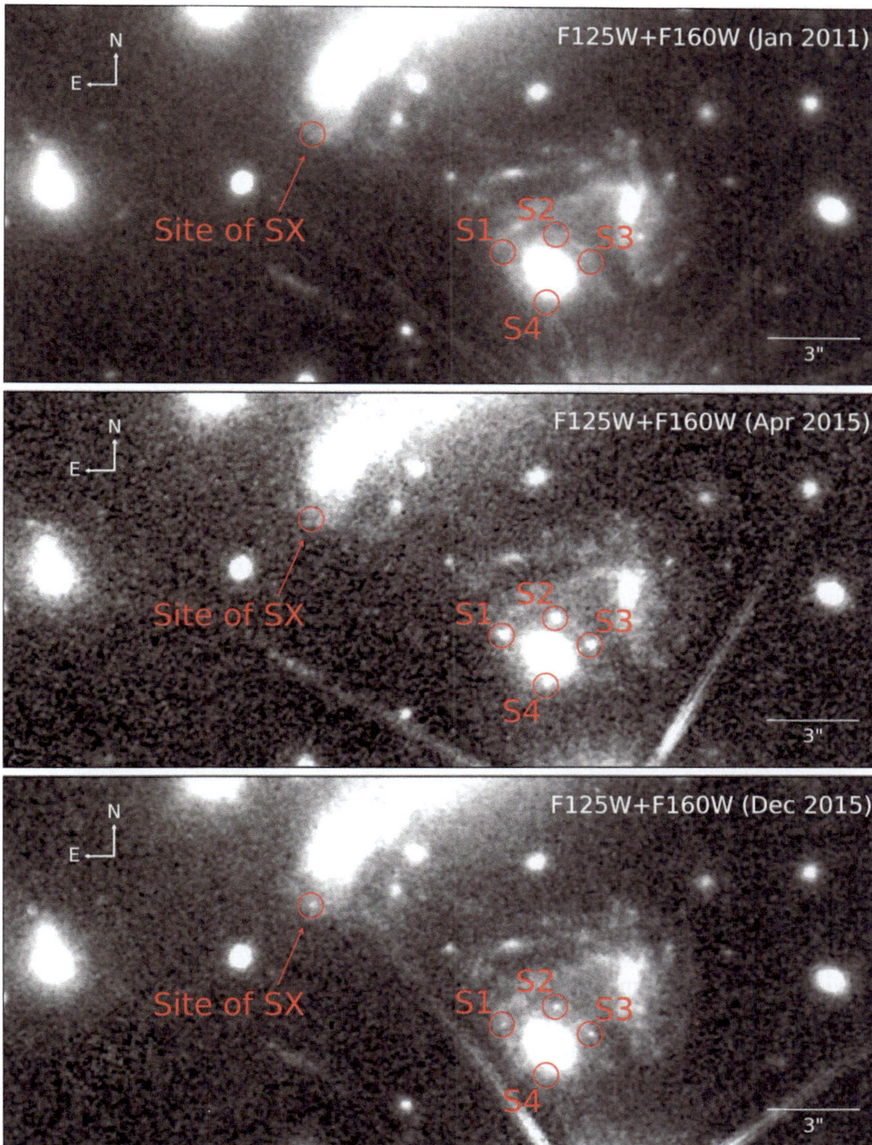

Fig. 25.3 *Top*: An *HST* observation obtained before any images of "supernova Refsdal" had appeared. *Middle*: An observation obtained after four images of the supernova (S1 to S4) had appeared. *Bottom*: An observation obtained after another image of the supernova (SX) had appeared. From "Deja Vu All Over Again: The Reappearance of Supernova Refsdal" (Kelly et al. 2016). © AAS. Reproduced with permission

25.4.4 Time Dilation

Comparing the rates of photometric (Leibundgut et al. 1996; Goldhaber et al. 2001) and spectroscopic (Blondin et al. 2008) evolution of SN Ia at a range of redshifts provides checks on time dilation. To date, the results are consistent with the predictions of general relativity.

25.4.5 Finding Dwarf Galaxies and Intracluster Populations

Deep follow-up of apparently hostless supernovae (both core-collapse events and SN Ia) discovered by future large-scale surveys such as *LSST* out to Gpc distances may allow detection and study of up to 10^4 dwarf galaxies of $\lesssim 10^6\,M_\odot$. Such galaxies play important roles in theories of large-scale structure, and the predicted population of them is enormous ($>1\,\mathrm{Mpc}^{-3}$), but they will remain otherwise difficult to find beyond 1 Mpc (Conroy and Bullock 2015).

Genuinely hostless SN Ia discovered in clusters of galaxies will help to probe the intracluster population of stars stripped from their host galaxies (Graham et al. 2015b).

25.5 Physics

25.5.1 Constraints from Core Collapse

High-statistics observations of the time evolution of the energy spectra of each of the neutrino flavors from a core-collapse event would not only shed light on the collapse process and the equation of state of dense matter (Sect. 24.2.1), but also would provide, in principle, much information on neutrino properties and interactions, as well as other physics constraints. Detecting the nonelectronic flavors, dealing with degeneracies, and disentangling the information will be challenging, however. Fundamental neutrino properties that might be constrained include mass (Ellis et al. 2012b) and the mass hierarchy, i.e., normal: $m_1, m_2 << m_3$, or inverted: $m_1 << m_2, m_3$ (Kotake et al. 2006; Chakraborty et al. 2011; Lai et al. 2016). Other neutrino phenomena that might be probed include flavor oscillations (Dasgupta and Beacom 2011), spectral swaps and splits (Cherry et al. 2012), self interactions (Duan and Kneller 2009), Lorentz invariance (Ellis et al. 2012a), the existence of a hypothetical sterile neutrino (Raffelt and Zhou 2011; Wu et al. 2014; Esmaili et al. 2014), and neutrino-antineutrino transformation (Vlasenkoetal 2014).

Neutrino observations also might provide constraints on exotic physics, such as the production of radions, particles associated with the Randall–Sundrum model of extra dimensions (Das et al. 2013), Goldstone bosons masquerading as fractional cosmic neutrinos (Weinberg 2013; Keung et al. 2014), the existence of an ultra-light gravitino (Catena et al. 2015), and the nature of some hypothesized dark-sector particles (Dreiner et al. 2014).

The nondetection of prompt γ-rays associated with the neutrino burst of SN 1987A constrained the production of axions (Giannotti et al. 2011; Payez et al. 2015) and the properties of a hypothesized dark photon (Kazanas et al. 2015).

25.5.2 *Constraints from SN Ia*

To the best of our knowledge, the values of the physical "constants" are indeed constant, independent of cosmic location and time, but this fundamental assumption is worth checking by all means possible. The light-curve shapes and peak luminosities of SN Ia depend on radioactive decay, thus variations in space and time (redshift) can be used to attempt to constrain weak-interaction physics. The homogeneity of (standardized) SN Ia light curves indicates that any rate of change of the Fermi constant, G_F, is no more than one part in a billion per year (Ferrero and Altschul 2010). With certain assumptions, SN Ia can constrain variations in the gravitational constant, G, over the last nine Gyrs (Mould and Uddin 2014) and constrain certain decaying dark matter scenarios (Blackadder and Koushiappas 2014). Current SN Ia data cannot significantly constrain space or time variations in the fine-structure constant (Kraiselburd et al. 2015) or variable speed-of-light cosmologies (Zhang and Meng 2014; Qi et al. 2014).

25.6 Summary

Because they are luminous point sources, the distances to which can be estimated, supernovae have many applications to astrophysics and cosmology. In view of the extreme physical conditions encountered, especially in core collapse, supernovae also provide useful constraints for physics.

Chapter 26
Summary and Prospects

26.1 Introduction

Throughout this book, we have pointed to open issues and challenges, both observational and theoretical. In this brief chapter, we mention some of the major uncertainties and prospects for progress.

26.2 Massive-Star Supernovae

Our understanding of the late stages of single-star evolution near the low-mass limit (ZAMS mass $\sim$8 $M_\odot$) of "massive stars" is seriously incomplete. We need to better quantify the metallicity-dependent ZAMS-mass boundaries between stars that become white dwarfs, those (if any) that proceed to degenerate carbon ignition, those that are subject to collapse by electron capture in degenerate cores, and those that proceed to iron-core collapse.

We have a rudimentary theoretical understanding of core collapse in massive stars, but that understanding is not yet highly advanced. Multidimensional effects, rotation, and magnetic fields will be important to this complex topic. We are unsure about which stars leave behind typical neutron stars or magnetars, which stars explode while leaving black holes, and which stars completely collapse to black holes with no explosion. According to models, a very massive star that has an oxygen core in the range 60–130 $M_\odot$ undergoes pair instability and a powerful explosion that leaves no compact remnant. It is not clear that a supernova corresponding to such a model has been observed.

The best-observed supernova is the peculiar SN 1987A. Its progenitor was a star of ZAMS mass about 18 $M_\odot$ that was a blue supergiant when it exploded. The neutrino detections confirmed the basic theory of core collapse. Extensive multi-wavelength observations allowed the ejecta and the CSI to be well characterized,

D. Branch, J.C. Wheeler, *Supernova Explosions*, Astronomy and Astrophysics Library, DOI 10.1007/978-3-662-55054-0_26

and the transition to a SNR is being watched closely. Nevertheless, much remains to be learned. We are unsure about the presupernova evolution that produced the blue supergiant and the processes that shaped the rich, asymmetric CSM. We would like to know whether the compact remnant is now a neutron star or a black hole.

The best-understood and most common supernovae are SN IIP. Their progenitors are effectively single stars that have ZAMS masses in the range of about 8–16 $M_{\odot}$. These stars evolve to become red supergiants, the cores of which collapse, probably to become neutron stars. A typical SN IIP ejects $\sim$10 $M_{\odot}$, mostly hydrogen, and the kinetic energy is $\sim$1 B. Outstanding questions include why the kinetic energy appears to increase monotonically with ejected mass, why some SN IIP are substantially dimmer and weaker ($\sim$0.1 B) than typical ones, and why stars of somewhat higher ZAMS mass do not become SN IIP (the red-supergiant problem; Sect. 3.2).

There are major gaps in our understanding of other massive-star supernovae. SN IIL, the progenitors of which may have ZAMS masses somewhat higher than those of SN IIP, eject only $\sim$1 $M_{\odot}$ of hydrogen. When and how most of their envelopes were lost during the presupernova evolution is not clear. SN IIn, which may come from a wide range of ZAMS masses, undergo strong CSI. The degree to which the CSM, the structure of which varies greatly among SN IIn, comes from distinct and perhaps multiple episodes of heavy mass loss or from approximately steady-state winds is not well understood.

Before they explode, the progenitors of SN IIb, SN Ib, SN Ibn, and SN Ic lose much of their envelopes, from almost all of their hydrogen (SN IIb) to most or all of their helium (SN Ibn and SN Ic). For these stripped-envelope events, the outstanding issue is in which cases the mass loss is by winds of stars of high ZAMS mass and in which cases the explosions are of stars of somewhat lower ZAMS mass that lose their envelopes by means of binary-star evolution. Inferences about the mass ranges of various subtypes of core-collapse supernovae from their occurrence rates need to be reconciled with evidence from their locations in their host galaxies.

For a SN Ic-bl, not only the mass-loss process but also the origin of the high-velocity ejecta and the apparently high total kinetic energy ($\sim$10 B, as in hypernova models) is not clear. The extraordinarily high luminosity of a hydrogen-rich SLSN-II is due to CSI, but for hydrogen-poor SLSN-I, the relative importance of energy input from CSI, magnetars, and radioactivity is debated.

26.3 Type Ia Supernovae

It is highly likely that SN Ia are explosions of C/O white dwarfs in binary systems, but even for typical events the presupernova evolution is uncertain and the observational diversity is not well understood. In the traditional SD model, the binary contains a C/O white dwarf that accretes from a nondegenerate companion until it approaches the Chandrasekhar mass, ignites carbon deep in its interior, and explodes, perhaps by means of a deflagration that undergoes a transition to a detonation. For most typical SN Ia, however, observational evidence for a companion and the associated CSM that might be expected in the SD scenario is scant, and some explosions appear to be significantly sub-Chandrasekhar while a small fraction are substantially super-Chandrasekhar.

In the DD model, which is favored according to most binary population-synthesis calculations, two white dwarfs in a binary merge and explode, perhaps by detonation. The total ejected mass could range from super-Chandrasekhar to, conceivably, sub-Chandrasekhar. In the sub-Chandrasekhar double-detonation model, which could be SD or DD, ignition in an accreted helium layer provokes ignition of the underlying C/O core. A potential problem with DD mergers and perhaps double-detonation models is that they may be too asymmetric to be consistent with observations.

Other supernova subtypes that may be explosions of or on C/O white dwarfs include SN Ia-CSM, SN Iax, SN 2002bj-likes, and calcium-rich transients. A major issue for all models of SN Ia and these subtypes concerns the ignition and propagation of the nuclear burning. Conditions required for initiating and quenching a detonation are particularly uncertain.

26.4 Prospects

26.4.1 Observations

There is a burgeoning literature on the science potential of current (Chap. 2), developing, and planned optical and near-IR supernova searches from ground and space. Some searches are aimed at increasing the discovery rate of supernovae in nearby galaxies—those that can be observed in most detail and for the longest times (*SkyMapper*,[1] Keller et al. 2007; La Silla-QUEST Supernova Survey *LSQ*, Walker et al. 2015; *Gaia*,[2] Altavilla et al. 2012; *ZTF*,[3] Smith et al. 2014b; *ATLAS*[4]; *Dome A in Antarctica*, Kim et al. 2010; Fig. 26.1). Some are high-cadence searches aimed at observing very early light curves, including shock breakout of core-collapse events (*KISS*, Morokuma et al. 2014; *SHOOT*, Tanaka et al. 2016). Deep searches to high redshift are aimed at improving the use of supernovae for cosmology and discovering explosions of the first stars (*LSST*,[5] Young et al. 2008; Lien and Fields 2009; *JWST*,[6] Gardner et al. 2006; *Euclid*,[7] Laureijs et al. 2011; *E-ELT*, Tamai

[1]http://msowww.anu.edu.au/skymapper.

[2]http://gaia.esa.int.

[3]http://ptf.caltech.

[4]http://www.fallingstar.com.

[5]http://www.lsst.org.

[6]http://www.jwst.nasa.gov.

[7]http://sci.esa.int/euclid.

Fig. 26.1 To observe supernovae, astronomers will go to the ends of the Earth ... Courtesy of F. Du and Z. Li, Nanjing Institute of Astronomical Optics

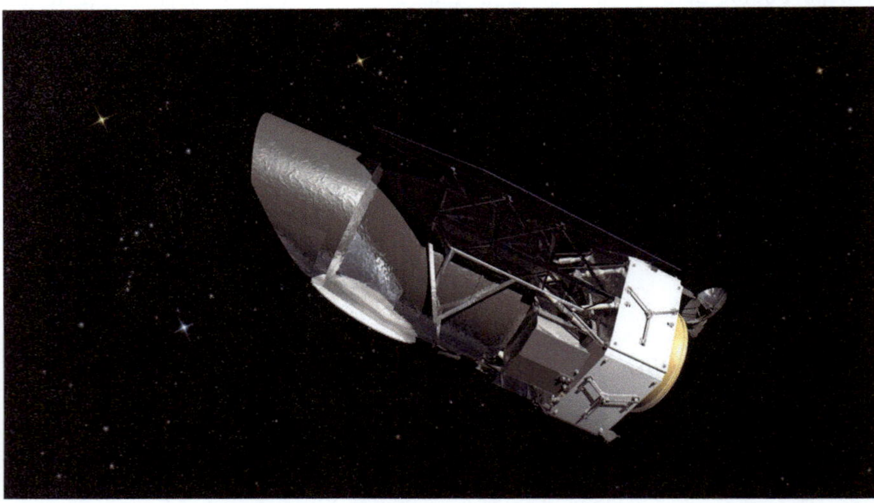

Fig. 26.2 ... and beyond. Artist's rendering of the Wide Field Infrared Survey Telescope, scheduled to be launched in the mid-2020s. Credit: NASA/Goddard Space Flight Center/Conceptual Image Lab

and Spyromilio 2014; *WFIRST*,[8] Spergel et al. 2015, Fig. 26.2; *WISH*,[9] Yamada et al. 2012). These searches and others will discover many thousands of supernovae in the coming years, perhaps millions in the coming decades. Some SN Ia may be observed to $z \simeq 4$ and some overluminous core-collapse events and/or PISN to $z \simeq 15$. For most high-redshift discoveries, spectroscopic followup will not be possible. Accurate techniques for photometric classification and characterization of supernovae will be required.

Multiwavelength observations across the electromagnetic spectrum will become increasingly valuable. Potential new facilities relevant to supernovae include the

[8]http://www.stsci.edu/wfirst.

[9]http://www.wishmission.org/en/index.html.

Space Variable Objects Monitor *SVOM*[10] (γ-rays), Advanced Telescope for High Energy Astrophysics *ATHENA*[11] (X-rays; O'Brien 2016), Ultraviolet Transient Astronomy Satellite *ULTRASAT*[12] (UV; Ganot et al. 2016), and Square Kilometer Array *SKA*[13] (radio: Perez-Torres et al. 2015; Wang et al. 2015b).

Multimessenger observations of supernovae will include neutrinos and eventually gravitational waves. Neutrinos will signal the occurrence of supernovae in nearby galaxies (*Hyper-K*,[14] Abe et al. 2011). The next core-collapse event in the Galaxy almost certainly will be first detected in neutrinos (Jiangmen Underground Neutrino Observatory *JUNO*,[15] Li 2014; Deep Underground Neutrino Experiment *DUNE*,[16] Acciarri et al. 2015) and perhaps in gravitational waves (Advanced Laser Interferometer Gravitational-wave Observatory *aLIGO*[17]; Advanced Virgo *adVIRGO*[18]; Kamioka Gravitational wave detector *KAGRA*,[19] Nakamura et al. 2016). In spite of extinction by dust in the ISM, a Galactic supernova very likely also will be detectable in the near-IR and even the optical (Adams et al. 2013).

26.4.2 Modeling

Observations of young SNRs and spectropolarimetry of supernovae establish that neither core-collapse events nor SN Ia are spherical. Photometric, spectroscopic, and spectropolarimetric data have not yet been properly integrated with 3D modeling of stellar evolution, supernova dynamics, and radiative transfer to take full advantage of the tools we now have to more deeply understand the nature of supernovae. In the absence of such thorough, self-consistent analysis, it remains appropriate to be cautious of the quantitative results of current modeling.

For modeling of supernovae and their presupernova evolution, most of the required physics is known, although improvements in nuclear reaction rates and atomic and molecular data for opacity computations will be helpful. A deeper understanding of the physics of turbulence and some of the dense-matter physics involved in core collapse is desired.

The major requirement for more advanced modeling is increased computational capacity. With future generations of computers, increasingly realistic 3D numerical

---

[10] www.svom.fr/svom.html.

[11] sci.esa.int/cosmic-visions/54517-athena.

[12] www.ultrasat.org.

[13] www.skatelescope.org.

[14] www.hyperk.org.

[15] english.ihep.cas.cn/rs/fs/juno0815/.

[16] www.dunescience.org.

[17] https://www.advancedligo.mit.edu.

[18] http://public.virgo-gw.eu/advanced-virgo/.

[19] http://gwcenter.icrr.u-tokyo.ac.jp/en.

simulations of, e.g., the merger scenario for the progenitor of SN 1987A, CSI with asymmetric CSM, nuclear burning fronts in SN Ia, core collapse in massive stars, and the presupernova evolution and explosion of PISN and PPISN, will become possible.

26.5 Conclusion

Future supernova studies are sure to advance our knowledge not only of the explosions themselves but also, by exploiting supernovae as cosmic probes, of astrophysics and cosmology more generally.

Appendix: Abbreviations

1D	one dimensional
2D	two dimensional
3D	three dimensional
AGB	asymptotic giant branch
AIC	accretion-induced collapse
ALMA	Atacama Large Millimeter/Submillimeter Array
ASAS-SN	All-Sky Automated Survey for Supernova
ASCA	Advanced Satellite for Cosmology and Astrophysics
C/O	carbon/oxygen
CANDELS	Cosmic Assembly near-IR Deep Extragalactic Legacy Survey
CCO	central compact object
CD	core degenerate
CDM	cold dark matter
CDS	cool dense shell
CE	common envelope
CEMP	carbon-enhanced metal-poor stars
CHASE	CHilean Automatic Supernova sEarch
CJD	Chapman–Jouguet detonation
CLASH	Cluster Lensing and Supernova survey with Hubble
CRT	calcium-rich transient
CRTS	Catalina Real-time Transient Survey
CSI	circumstellar interaction
CSM	circumstellar matter
CSS	Catalina Sky Survey
CTIO	Cerro Tololo Inter-American Observatory
DD	double degenerate
DES	dark energy survey
DIB	diffuse interstellar bands
DDT	deflagration-to-detonation transition

D. Branch, J.C. Wheeler, *Supernova Explosions*, Astronomy and Astrophysics Library, DOI 10.1007/978-3-662-55054-0

DTD	delay-time distribution
E-AGB	early asymptotic giant branch
EELG	extreme emission-line galaxy
EMP	extremely metal-poor
EOS	equation of state
EPM	expanding photosphere method
FMK	fast-moving knots
FWHM	full width half maximum
FOV	field of view
GCD	gravitationally confined detonation
HMP	hyper-metal-poor
HRD	Hertzsprung–Russell diagram
HST	Hubble Space Telescope
HVF	high velocity feature
HVG	high velocity gradient
IGE	iron-group elements
IGM	intergalactic medium
IME	intermediate-mass elements
IMFM	initial-mass final-mass relation
INTEGRAL	International Gamma Ray Astrophysical Laboratory
iPTF	intermediate Palomar Transient Factory
IR	infrared
IRC	infrared catastrophe
ISM	interstellar matter
ISP	interstellar polarization
IUE	International Ultraviolet Explorer
JWST	James Webb Space Telescope
LEPP	light element primary process
LVG	low velocity gradient
LBV	luminous blue variable
LMC	Large Magellanic Cloud
LOSS	Lick Observatory Supernova Search
LSST	Large Synoptic Survey Telescope
LTE	local thermodynamic equilibrium
MGFLD	multi-energy group flux-limited diffusion
MRI	magnetorotational instability
NAXI	nonaxisymmetric instability
NLTE	nonlocal thermodynamic equilibrium
NSE	nuclear statistical equilibrium
NSF	Nearby Supernova Factory
NuSTAR	Nuclear Spectroscopic Telescope Array
O/Ne/Mg	oxygen–neon–magnesium
OSSE	Oriented Scintillation Spectrometer Experiment
PDD	pulsating delayed detonation
PRD	pulsating reverse detonation

PISN	pair-instability supernova
PNS	protoneutron star
PPISN	pulsational pair-instability supernova
PTF	Palomar Transient Factory
PWN	pulsar wind nebula
RLOF	Roche lobe overflow
RN	recurrent nova
ROSAT	Röentgen Satellite
ROTSE	Robotic Optical Transient Source Experiment
RSG	red supergiant
RSVP	ROTSE supernova verification project
SAGB	super-asymptotic giant branch
SASI	standing accretion shock instability
SCP	Supernova Cosmology Project
SD	single degenerate
SDSS	Sloan Digital Sky Survey
SEAM	spectral expanding atmosphere method
SED	spectral energy distribution
SFR	star formation rate
SLSN	superluminous supernova
SMC	Small Magellanic Cloud
SNLS	Supernova Legacy Survey
SNR	supernova remnant
SP	spectropolarimetry type
TIGRE	Telescopio Internacional de Guanajuato Robótico Espectróscopico
TP-AGB	thermally pulsing asymptotic giant branch
TSS	Texas Supernova Search
UMP	ultra-metal-poor
UV	ultraviolet
VLBI	very long baseline interferometry
VMP	very metal-poor
ZAMS	zero-age main sequence
ZND	Zel'dovich von Neumann–Doering detonation

References

Aartsen, M. G., Ackermann, M., Adams, J., et al. 2014, PRL, 113, 101101
Abbott, B. P., Abbott, R., Abbott, T. D., et al. 2016, PRL, 116, 061102
Abbott, D. C., & Lucy, L. B. 1985, ApJ, 288, 679
Abdikamalov, E., Burrows, A., Ott, C. D., et al. 2012, ApJ, 755, 111
Abdikamalov, E., Gossan, S., DeMaio, A. M., & Ott, C. D. 2014, Physical Review D, 90, 044001
Abdikamalov, E. B., Ott, C. D., Rezzolla, L., et al. 2010, Physical Review D, 81, 044012
Abdo, A. A., Ackermann, M., Ajello, M., et al. 2010a, ApJ, 708, 1254
Abdo, A. A., Ackermann, M., Ajello, M., et al. 2010b, ApJ, 720, 435
Abe, K., Abe, T., Aihara, H., et al. 2011, arXiv:1109.3262
Abe, K., Haga, Y., Hayato, Y., et al. 2016, Astroparticle Physics, 81, 39
Abel, T., Anninos, P., Norman, M. L., & Zhang, Y. 1998, ApJ, 508, 518
Abel, T., Bryan, G. L., & Norman, M. L. 2000, ApJ, 540, 39
Ablimit, I., & Li, X.-D. 2015, ApJ, 800, 98
Ablimit, I., Xu, X.-j., & Li, X.-D. 2014, ApJ, 780, 80
Acciarri, R., Acero, M. A., Adamowski, M., et al. 2015, arXiv:1512.06148
Acheson, D. J. 1978, RSLPT, 289, 459
Ackermann, M., Ajello, M., Allafort, A., et al. 2013, Science, 339, 807
Adams, S. M., & Kochanek, C. S. 2015, MNRAS, 452, 2195
Adams, S. M., Kochanek, C. S., Beacom, J. F., et al. 2013, ApJ, 778, 164
Aerts, C. 2015, AN, 336, 477
Aglietta, M., Badino, G., Bologna, G., et al. 1987, EPL (Europhysics Letters), 3, 1315
Aitken, D. K., Smith, C. H., James, S. D., et al. 1988, MNRAS, 231, 7P
Akerlof, C. W., Kehoe, R. L., McKay, T. A., et al. 2003, PASP, 115, 132
Akiyama, S., Wheeler, J. C., Meier, D. L., & Lichtenstadt, I. 2003, ApJ, 584, 954
Aldering, G., Adam, G., Antilogus, P., et al. 2002, Proceedings of the SPIE, 4836, 61
Aldering, G., Antilogus, P., Bailey, S., et al. 2006, ApJ, 650, 510
Aldering, G., Humphreys, R. M., & Richmond, M. 1994, AJ, 107, 662
Alexeyev, E. N., Alexeyeva, L. N., Krivosheina, I. V., & Volchenko, V. I. 1988, Physics Letters B, 205, 209
Altavilla, G., Botticella, M. T., Cappellaro, E., & Turatto, M. 2012, ApSpSci, 341, 163
Altavilla, G., Stehle, M., Ruiz-Lapuente, P., et al. 2007, A&A, 475, 585
Amanullah, R., Goobar, A., Johansson, J., et al. 2014, ApJL, 788, LL21
Amanullah, R., Johansson, J., Goobar, A., et al. 2015, MNRAS, 453, 3300
Amati, L., & Valle, M. D. 2013, International Journal of Modern Physics D, 22, 30028
Ambwani, K., & Sutherland, P. 1988, ApJ, 325, 820

D. Branch, J.C. Wheeler, *Supernova Explosions*, Astronomy and Astrophysics Library, DOI 10.1007/978-3-662-55054-0

Anderson, J. P., Dessart, L., Gutierrez, C. P., et al. 2014a, MNRAS, 441, 671
Anderson, J. P., González-Gaitán, S., Hamuy, M., et al. 2014b, ApJ, 786, 67
Anderson, J. P., Habergham, S. M., James, P. A., & Hamuy, M. 2012, MNRAS, 424, 1372
Anderson, J. P., & James, P. A. 2008, MNRAS, 390, 1527
Anderson, J. P., & James, P. A. 2009, MNRAS, 399, 559
Anderson, J. P., James, P. A., Förster, F., et al. 2015a, MNRAS, 448, 732
Anderson, J. P., James, P. A., Habergham, S. M., et al. 2015b, Publications of the Astron. Soc. of Australia, 32, 19
Andersson, N. 1998, ApJ, 502, 708
Andrews, J. E., Clayton, G. C., Wesson, R., et al. 2011a, AJ, 142, 45
Andrews, J. E., Gallagher, J. S., Clayton, G. C., et al. 2010, ApJ, 715, 541
Andrews, J. E., Smith, N., & Mauerhan, J. C. 2015, MNRAS, 451, 1413
Andrews, J. E., Sugerman, B. E. K., Clayton, G. C., et al. 2011b, ApJ, 731, 47
Angel, J. R. P. 1977, ApJ, 216, 1
Antoniadis, J., Freire, P. C. C., Wex, N., et al. 2013, Science, 340, 448
Antoniadis, J., Tauris, T. M., Ozel, F., et al. 2016, arXiv:1605.01665
Antonioli, P., Tresch Fienberg, R., Fleurot, R., et al. 2004, New Journal of Physics, 6, 114
Anupama, G. C., Sahu, D. K., Gurugubelli, U. K., et al. 2009, MNRAS, 392, 894
Appenzeller, I. 1970, A&A, 9, 216
Arcavi, I., Gal-Yam, A., Kasliwal, M. M., et al. 2010, ApJ, 721, 777
Arcavi, I., Gal-Yam, A., Yaron, O., et al. 2011, ApJL, 742, L18
Arcones, A., & Thielemann, F.-K. 2013, Journal of Physics G Nuclear Physics, 40, 013201
Aretxaga, I., Benetti, S., Terlevich, R. J., et al. 1999, MNRAS, 309, 343
Arnett, W. D. 1969, ApSpSci, 5, 180
Arnett, W. D. 1982, ApJ, 253, 785
Arnett, D. 1996, Supernovae and Nucleosynthesis: an Investigation of the History of Matter from the Big Bang to the Present (Princeton Un. Press)
Arnett, W. D., Bahcall, J., Kirshner, R. P., Woosley, S.E., 1989, A&A, 27
Arnett, W. D., & Fu, A. 1989, ApJ, 340, 396
Arnett, W. D., & Meakin, C. 2011a, ApJ, 733, 7
Arnett, W. D., & Meakin, C. 2011b, ApJ, 741, 33
Arnett, D., Meakin, C., & Young, P. A. 2009, ApJ, 690, 1715
Asakura, K., Gando, A., Gando, Y., et al. 2016, ApJ, 818, 91
Asano, R. S., Takeuchi, T. T., Hirashita, H., & Nozawa, T. 2013, MNRAS, 432, 637
Ashall, C., Mazzali, P., Bersier, D., et al. 2014, MNRAS, 445, 4424
Ashworth, W. B., Jr. 1980, Journal for the History of Astronomy, 11, 1
Aspden, A. J., Bell, J. B., Day, M. S., et al. 2008, ApJ, 689, 1173
Aspden, A. J., Bell, J. B., Dong, S., & Woosley, S. E. 2011, ApJ, 738, 94
Aspden, A. J., Bell, J. B., & Woosley, S. E. 2010, ApJ, 710, 1654
Astier, P. 2012, Rep. Prog. Phys., 75, 116901
Astier, P., Guy, J., Regnault, N., et al. 2006, A&A, 447, 31
Athanassiadou, T., & Fields, B. D. 2011, Proceedings of the National Academy, 16, 229
Auer, L. H., & van Blerkom, D. 1972, ApJ, 178, 175
Axelrod, T. S. 1980, Ph.D. Thesis, Un. California, Santa Cruz
Aznar-Siguán, G., García-Berro, E., Lorén-Aguilar, P. et al. 2015, MNRAS, 450, 2948
Baade, W., & Zwicky, F. 1934, Proc. NAS, 20, 259
Badenes, C., Borkowski, K. J., Hughes, J. P., et al. 2006, ApJ, 645, 1373
Badenes, C., Hughes, J. P., Bravo, E., & Langer, N. 2007, ApJ, 662, 472
Badenes, C., & Maoz, D. 2012, ApJL, 749, L11
Badenes, C., Maoz, D., & Draine, B. T. 2010, MNRAS, 407, 1301
Badenes, C., Mullally, F., Thompson, S. E., & Lupton, R. H. 2009, ApJ, 707, 971
Bailey, J. 1988, PASA, 7, 405
Baklanov, P. V., Blinnikov, S. I., & Pavlyuk, N. N. 2005, Astronomy Letters, 31, 429
Baklanov, P. V., Sorokina, E. I., & Blinnikov, S. I. 2015, Astronomy Letters, 41, 95

Balberg, S., & Loeb, A. 2011, MNRAS, 414, 1715
Balbus, S. A., & Hawley, J. F. 1998, RMP, 70, 1
Ball, L., Campbell-Wilson, D., Crawford, D. F., Turtle, A. J. 1995, ApJ, 453, 864
Bandiera, R., & Petruk, O. 2010, A&A, 509, A34
Baraffe, I., Heger, A., & Woosley, S. E. 2001, ApJ, 550, 890
Barbarino, C., Dall'Ora, M., Botticella, M. T., et al. 2015, MNRAS, 448, 2312
Barbary, K., Aldering, G., Amanullah, R., et al. 2012, ApJ, 745, 32
Barbary, K., Dawson, K. S., Tokita, K., et al. 2009, ApJ, 690, 1358
Barbieri, C., Bonoli, C., & Cristiani, S. 1982, A&A, 114, 216
Barbon, R., Buondí, V., Cappellaro, E., & Turatto, M. 1999, A&A Supplement, 139, 531
Barbon, R., Ciatti, F., & Rosino, L. 1973, A&A, 29, 57
Barbon, R., Ciatti, F., & Rosino, L. 1979, A&A, 72, 287
Barbon, R., Ciatti, F., Rosino, L., et al. 1982, A&A, 116, 43
Barkat, Z. 1975, ARAA, 13, 45
Barkat, Z., & Marom, A. 1990, in Supernovae, ed. J. C. Wheeler, T. Piran, & S. Weinberg (World Scientific), 95
Barkat, Z., Rakavy, G., & Sack, N. 1967, PRL, 18, 379
Barkat, Z., Reiss, Y., & Rakavy, G. 1974, ApJL, 193, L21
Barna, B., Vinkó, J., Silverman, J. M., Marion, G. H., & Wheeler, J. C. 2016, MNRAS, 457, 3225
Barnes, J., Kasen, D., Wu, M.-R., & Martínez-Pinedo, G. 2016, ApJ, 829, 110
Barnes, J. E., Wood, K., Hill, A. S., & Haffner, L. M. 2014, MNRAS, 440, 3027
Baron, E., Bongard, S., Branch, D., & Hauschildt, P. H. 2006, ApJ, 645, 480
Baron, E., Branch, D., Hauschildt, P. H., 1999, ApJ, 527, 739
Baron, E., Branch, D., Hauschildt, P. H., et al. 2000, ApJ, 545, 444
Baron, E., Branch, D., & Hauschildt, P. H. 2007, ApJ, 662, 1148
Baron, E., Cooperstein, J., & Kahana, S. 1987, ApJ, 320, 300
Baron, E., Hauschildt, P. H., Branch, D., et al. 1993, ApJL, 416, L21
Baron, E., Hauschildt, P. H., & Branch, D. 1994, ApJ, 426, 334
Baron, E., Hauschildt, P. H., Branch, D., et al. 1995, ApJ, 441, 170
Baron, E., Hauschildt, P. H., & Chen, B. 2009, A&A, 498, 987
Baron, E., Höflich, P., Krisciunas, K., et al. 2012, ApJ, 753, 105
Baron, E., Jeffery, D. J., Branch, D., et al. 2008, ApJ, 672, 1038
Baron, E., Nugent, P. E., Branch, D., & Hauschildt, P. H. 2004, ApJL, 616, L91
Barone-Nugent, R. L., Lidman, C., Wyithe, J. S. B., et al. 2012, MNRAS, 425, 1007
Bartel, N., & Bietenholz, M. F. 2008, ApJ, 682, 1065
Bartel, N., Bietenholz, M. F., Rupen, M. P., & Dwarkadas, V. V. 2007, ApJ, 668, 924
Bartel, N., Bietenholz, M. F., Rupen, M. P., et al. 1994, Nature, 368, 610
Beacom, J. F. 2010, ARN&PS, 60, 439
Beck, A. M., Dolag, K., Lesch, H., & Kronberg, P. P. 2013, MNRAS, 435, 3575
Becker, S. A., & Iben, I., Jr. 1979, ApJ, 232, 831
Becker, S. A., & Iben, I., Jr. 1980, ApJ, 237, 111
Bedin, L. R., Ruiz-Lapuente, P., González Hernández, J. I., et al. 2014, MNRAS, 439, 354
Beech, M. 2011, ApSpSci, 336, 287
Beelen, A., Cox, P., Benford, D. J., et al. 2006, ApJ, 642, 694
Beers, T. C., Carollo, D., Ivezić, Ž., et al. 2012, ApJ, 746, 34
Beers, T. C., & Christlieb, N. 2005, ARAA, 43, 531
Begelman, M. C., & Sarazin, C. L. 1986, ApJL, 302, L59
Belczynski, K., Kalogera, V., Rasio, F. A., et al. 2008, ApJS, 174, 223
Belczynski, K., Wiktorowicz, G., Fryer, C. L., et al. 2012, ApJ, 757, 91
Bell, A. R. 1978, MNRAS, 182, 443
Bell, A. R., Schure, K. M., Reville, B., & Giacinti, G. 2013, MNRAS, 431, 415
Ben-Dayan, I., & Kalaydzhyan, T. 2014, Physical Review D, 90, 083509
Ben-Ami, S., Gal-Yam, A., Mazzali, P. A., et al. 2014, ApJ, 785, 37
Ben-Ami, S., Hachinger, S., Gal-Yam, A., et al. 2015, ApJ, 803, 40

Benetti, S., Bufano, F., Vinkó, J., et al. 2010, CBET, 2536, 1
Benetti, S., Cappellaro, E., Danziger, I. J., et al. 1998, MNRAS, 294, 448
Benetti, S., Cappellaro, E., Mazzali, P. A., et al. 2005, ApJ, 623, 1011
Benetti, S., Cappellaro, E., Turatto, M., et al. 1994a, A&A, 285, 147
Benetti, S., Cappellaro, E., Turatto, M., et al. 2006, ApJL, 653, L129
Benetti, S., Nicholl, M., Cappellaro, E., et al. 2014, MNRAS, 441, 289
Benetti, S., Patat, F., Turatto, M., et al. 1994b, A&A, 285, L13
Benetti, S., Turatto, M., Cappellaro, E., et al. 1999, MNRAS, 305, 811
Benetti, S., Turatto, M., Valenti, S., et al. 2011, MNRAS, 411, 2726
Bengaly, C. A. P., Jr., Bernui, A., & Alcaniz, J. S. 2015, ApJ, 808, 39
Benítez, N., Riess, A., Nugent, P., et al. 2002, ApJL, 577, L1
Bensby, T., Yee, J. C., Feltzing, S., et al. 2013, A&A, 549, A147
Benvenuto, O. G., Bersten, M. C., & Nomoto, K. 2013, ApJ, 762, 74
Berezhko, E. G., & Völk, H. J. 2004, ApJ, 611, 12
Berger, E., Chornock, R., Lunnan, R., et al. 2012, ApJL, 755, L29
Berger, E., Kulkarni, S. R., & Chevalier, R. A. 2002, ApJL, 577, L5
Berger, E., & Soderberg, A. M. 2008, GCN, 7159, 1
Bernstein, J. P., Kessler, R., Kuhlmann, S., et al. 2012, ApJ, 753, 152
Bersten, M. C., Benvenuto, O. G., Folatelli, G., et al. 2014, AJ, 148, 68
Bersten, M. C., Benvenuto, O., & Hamuy, M. 2011, ApJ, 729, 61
Bersten, M. C., Benvenuto, O. G., Nomoto, K., et al. 2012, ApJ, 757, 31
Bersten, M. C., & Hamuy, M. 2009, ApJ, 701, 200
Bersten, M. C., Tanaka, M., Tominaga, N., et al. 2013, ApJ, 767, 143
Bertelli, G., Bressan, A. G., & Chiosi, C. 1985, A&A, 150, 33
Bertola, F. 1964, Annales d'Astrophysique, 27, 319
Bertola, F. Mammano, A. & Perinotto, M. 1965, Contr. Asiago Obs. No. 174
Bertoldi, F., Carilli, C. L., Cox, P., et al. 2003, A&A, 406, L55
Bessell, M. S., Collet, R., Keller, S. C., et al. 2015, ApJL, 806, L16
Bethe, H. A. 1990, RMP, 62, 801
Bethe, H. A., Brown, G. E., Appvlegate, J., & Lattimer, J. M. 1979, Nuclear Physics A, 324, 487
Bethe, H. A., & Wilson, J. R. 1985, ApJ, 295, 14
Betoule, M., Kessler, R., Guy, J., et al. 2014, A&A, 568, AA22
Bevan, A., Barlow, M. J., & Milisavljevic, D. 2017, MNRAS, 465, 4044
Bianchi, S., & Schneider, R. 2007, MNRAS, 378, 973
Bianco, F. B., Howell, D. A., Sullivan, M., et al. 2011, ApJ, 741, 20
Bianco, F. B., Modjaz, M., Hicken, M., et al. 2014, ApJS, 213, 19
Bietenholz, M. F., Brunthaler, A., Soderberg, A. M., et al. 2012, ApJ, 751, 125
Bietenholz, M. F., Soderberg, A. M., & Bartel, N. 2009, ApJL, 694, L6
Bildsten, L., Shen, K. J., Weinberg, N. N., & Nelemans, G. 2007, ApJL, 662, L95
Bionta, R. M., Blewitt, G., Bratton, C. B., et al. 1987, PRL, 58, 1494
Bird, S., Vogelsberger, M., Haehnelt, M., et al. 2014, MNRAS, 445, 2313
Biscaro, C., & Cherchneff, I. 2016, A&A, 589, A132
Bisnovatyi-Kogan, G. S. 1971, Soviet Astronomy, 14, 652
Bisnovatyi-Kogan, G. S. 2011, Stellar Physics (Springer)
Björnsson, C.-I. 2015, ApJ, 813, 43
Björnsson, C.-I., & Lundqvist, P. 2014, ApJ, 787, 143
Blaauw, A. 1961, Bulletin Astronomical Institute of the Netherlands, 15, 265
Black, C. S., & Fesen, R. A. 2015, MNRAS, 447, 2540
Black, C. S., Fesen, R. A., & Parrent, J. T. 2016, MNRAS, 462, 649
Blackadder, G., & Koushiappas, S. M. 2014, Physical Review D, 90, 103527
Blair, W. P., Chandar, R., Dopita, M. A., et al. 2014, ApJ, 788, 55
Blanco, V. M., Blanco, B. M., & McCarthy, M. F. 1980, ApJ, 242, 938
Blanco, V. M., Gregory, B., Hamuy, M., et al. 1987, ApJ, 320, 589
Blandford, R. D., & Ostriker, J. P. 1978, ApJL, 221, L29

Blandford, R. D., & Payne, D. G. 1982, MNRAS, 199, 883
Blasi, P. 2013, The Astronomy and Astrophysics Review, 21, 70
Blasi, P. 2014, Comptes Rendus Physique, 15, 329
Blinnikov, S., Lundqvist, P., Bartunov, O., et al. 2000, ApJ, 532, 1132
Blinnikov, S. I., & Bartunov, O. S. 1993, A&A, 273, 106
Blinnikov, S. I., Eastman, R., Bartunov, O. S., et al. 1998, ApJ, 496, 454
Blondin, J. & Lundqvist, P. 1993, ApJ, 405, 1
Blondin, J. M., & Mezzacappa, A. 2006, ApJ, 642, 401
Blondin, J. M., & Mezzacappa, A. 2007, Nature, 445, 58
Blondin, J. M., Mezzacappa, A., & DeMarino, C. 2003, ApJ, 584, 971
Blondin, J. M., & Shaw, S. 2007, ApJ, 656, 366
Blondin, S., Davis, T. M., Krisciunas, K., et al. 2008, ApJ, 682, 724
Blondin, S., Dessart, L., & Hillier, D. J. 2015, MNRAS, 448, 2766
Blondin, S., Dessart, L., Hillier, D. J., & Khokhlov, A. M. 2013, MNRAS, 429, 2127
Blondin, S., Kasen, D., Röpke, F. K., et al. 2011, MNRAS, 417, 1280
Blondin, S., Matheson, T., Kirshner, R. P., et al. 2012, AJ, 143, 126
Bloom, J. S., Kasen, D., Shen, K. J., et al. 2012, ApJL, 744, L17
Bludman, S. A., & Cernohorsky, J. 1995, Physics Reports, 256, 37
Bocchio, M., Marassi, S., Schneider, R., et al. 2016, A&A, 587, A157
Bogdanov, S., Ng, C.-Y., & Kaspi, V. M. 2014, ApJL, 792, LL36
Boggs, S. E., Harrison, F. A., Miyasaka, H., et al. 2015, Science, 348, 670
Bogomazov, A. I., & Tutukov, A. V. 2009, Astronomy Reports, 53, 214
Boisseau, J. R., Wheeler, J. C., Oran, E. S., & Khokhlov, A. M. 1996, ApJL, 471, L99
Bonanno, A., Baldo, M., Burgio, G. F., & Urpin, V. 2014, A&A, 561, L5
Bond, J. R., Arnett, W. D., & Carr, B. J. 1984, ApJ, 280, 825
Bonifacio, P., Caffau, E., Spite, M., et al. 2015, A&A, 579, A28
Book, L. G., Chu, Y.-H., Gruendl, R. A., & Fukui, Y. 2009, AJ, 137, 3599
Borish, H. J., Huang, C., Chevalier, R. A., et al. 2015, ApJ, 801, 7
Borkowski, K. J., Reynolds, S. P., Green, D. A., et al. 2010, ApJL, 724, L161
Borkowski, K. J., Reynolds, S. P., Green, D. A., et al. 2014, ApJL, 790, L18
Borkowski, K. J., Reynolds, S. P., Hwang, U., et al. 2013, ApJL, 771, L9
Bose, S., & Kumar, B. 2014, ApJ, 782, 98
Bose, S., Kumar, B., Sutaria, F., et al. 2013, MNRAS, 433, 1871
Bose, S., Valenti, S., Misra, K., et al. 2015, MNRAS, 450, 2373
Boss, A. P., & Keiser, S. A. 2012, ApJL, 756, L9
Boss, A. P., & Keiser, S. A. 2014, ApJ, 788, 20
Botticella, M. T., Riello, M., Cappellaro, E., et al. 2008, A&A, 479, 49
Botticella, M. T., Smartt, S. J., Kennicutt, R. C., et al. 2012, A&A, 537, A132
Botticella, M. T., Trundle, C., Pastorello, A., et al. 2010, ApJL, 717, L52
Bouchet, L., Jourdain, E., & Roques, J.-P. 2015, ApJ, 801, 142
Bouchet, P., Dwek, E., Danziger, J., et al. 2006, ApJ, 650, 212
Bouchet, P., Phillips, M. M., Suntzeff, N. B., et al. 1991, A&A, 245, 490
Bours, M. C. P., Toonen, S., & Nelemans, G. 2013, A&A, 552, A24
Bowen, D. V., Roth, K. C., Meyer, D. M., & Blades, J. C. 2000, ApJ, 536, 225
Bowers, E. J. C., Meikle, W. P. S., Geballe, T. R., et al. 1997, MNRAS, 290, 663
Boyd, R. N., Kajino, T., & Onaka, T. 2012, JP Conf. Series, 403, 012032
Boyles, J., Lorimer, D. R., Turk, P. J., et al. 2011, ApJ, 742, 51
Brachwitz, F., Dean, D. J., Hix, W. R., et al. 2000, ApJ, 536, 934
Bramante, J. 2015, PRL, 115, 141301
Branch, D. 1987, ApJL, 320, L121
Branch, D. 1990, in Supernovae, ed. A. G. Petschek (Springer), 30
Branch, D. 1998, ARAA, 36, 17
Branch, D., Baron, E., Hall, N., et al. 2005, PASP, 117, 545
Branch, D., Baron, E., Thomas, R. C., et al. 2004a, PASP, 116, 903

Branch, D., Benetti, S., Kasen, D., et al. 2002, ApJ, 566, 1005
Branch, D., & Cowan, J. J. 1985, ApJL, 297, L33
Branch, D., Dang, L. C., & Baron, E. 2009, PASP, 121, 238
Branch, D., Dang, L. C., Hall, N., et al. 2006a, PASP, 118, 560
Branch, D., Doggett, J. B., Nomoto, K., & Thielemann, F.-K. 1985, ApJ, 294, 619
Branch, D., Drucker, W., & Jeffery, D. J. 1988, ApJL, 330, L117
Branch, D., Falk, S. W., Uomoto, A. K., et al. 1981, ApJ, 244, 780
Branch, D., Garnavich, P., Matheson, T., et al. 2003, AJ, 126, 1489
Branch, D., & Greenstein, J. L. 1971, ApJ, 167, 89
Branch, D., Jeffery, D. J., Blaylock, M., & Hatano, K. 2000, PASP, 112, 217
Branch, D., Jeffery, D. J., Parrent, J., et al. 2008, PASP, 120, 135
Branch, D., Jeffery, D. J., Young, T. R., & Baron, E. 2006b, PASP, 118, 791
Branch, D., Lacy, C. H., McCall, M. L., et al. 1983, ApJ, 270, 123
Branch, D., Romanishin, W., & Baron, E. 1996, ApJ, 467, 473
Branch, D., & Tammann, G. A. 1992, ARAA, 30, 359
Branch, D., Thomas, R. C., Baron, E., et al. 2004b, ApJ, 606, 413
Branch, D., & van den Bergh, S. 1993, AJ, 105, 2231
Brandt, T. D., Tojeiro, R., Aubourg, É., et al. 2010, AJ, 140, 804
Bravo, E., Althaus, L. G., García-Berro, E., & Domínguez, I. 2011, A&A, 526, A26
Bravo, E., & García-Senz, D. 2009, ApJ, 695, 1244
Bravo, E., García-Senz, D., Cabezón, R. M., & Domínguez, I. 2009, ApJ, 695, 1257
Bregman, J. N. 1980, ApJ, 236, 577
Brinkworth, C. S., Burleigh, M. R., Lawrie, K., et al. 2013, ApJ, 773, 47
Broersen, S., Chiotellis, A., Vink, J., & Bamba, A. 2014, MNRAS, 441, 3040
Bromm, V. 2013, Reports on Progress in Physics, 76, 112901
Bromm, V., Coppi, P. S., & Larson, R. B. 2002, ApJ, 564, 23
Bromm, V., & Larson, R. B. 2004, ARAA, 42, 79
Brott, I., de Mink, S. E., Cantiello, M., et al. 2011a, A&A, 530, A115
Brott, I., Evans, C. J., Hunter, I., et al. 2011b, A&A, 530, A116
Brown, G. E., Lee, C.-H., Wijers, R. A. M. J., et al. 2000, Proceedings of the National Academy, 5, 191
Brown, P. J. 2014, ApJL, 796, LL18
Brown, P. J., Baron, E., Milne, P., et al. 2015a, ApJ, 809, 37
Brown, P. J., Dawson, K. S., de Pasquale, M., et al. 2012a, ApJ, 753, 22
Brown, P. J., Dawson, K. S., Harris, D. W., et al. 2012b, ApJ, 749, 18
Brown, P. J., Kuin, P., Scalzo, R., et al. 2014, ApJ, 787, 29
Brown, P. J., Smitka, M. T., Wang, L., et al. 2015b, ApJ, 805, 74
Brown, W. R., Anderson, J., Gnedin, O. Y., et al. 2015c, ApJ, 804, 49
Brown, W. R., Cohen, J. G., Geller, M. J., & Kenyon, S. J. 2012c, ApJL, 754, L2
Brown, W. R., Kilic, M., Allende Prieto, C., et al. 2013, ApJ, 769, 66
Bruenn, S. W. 1973, ApJL, 183, L125
Bruenn, S. W. 1985, ApJS, 58, 771
Bruenn, S. W., & Dineva, T. 1996, ApJL, 458, L71
Bruenn, S. W., Mezzacappa, A., Hix, W. R., et al. 2013, ApJL, 767, L6
Bucciantini, N., Metzger, B. D., Thompson, T. A., & Quataert, E. 2012, MNRAS, 419, 1537
Bucciantini, N., Quataert, E., Arons, J., et al. 2008, MNRAS, 383, L25
Bucciantini, N., Quataert, E., Metzger, B. D., et al. 2009, MNRAS, 396, 2038
Buchler, J. R., Wheeler, J. C., & Barkat, Z. 1972, ApJ, 172, 713
Bufano, F., Pian, E., Sollerman, J., et al. 2012, ApJ, 753, 67
Bufano, F., Pignata, G., Bersten, M., et al. 2014, MNRAS, 439, 1807
Bulla, M., Sim, S. A., & Kromer, M. 2015, MNRAS, 450, 967
Buonanno, A., Sigl, G., Raffelt, G. G., Janka, H.-T., & Müller, E. 2005, Physical Review D, 72, 084001
Burkey, M. T., Reynolds, S. P., Borkowski, K. J., & Blondin, J. M. 2013, ApJ, 764, 63

Burns, C. R., Stritzinger, M., Phillips, M. M., et al. 2014, ApJ, 789, 32
Burrows, A. 2013, RMP, 85, 245
Burrows, A., Dessart, L., Livne, E., et al. 2007, ApJ, 664, 416
Burrows, A., Dolence, J. C., & Murphy, J. W. 2012, ApJ, 759, 5
Burrows, A., & Goshy, J. 1993, ApJL, 416, L75
Burrows, A., & Lattimer, J. M. 1984, ApJ, 285, 294
Burrows, A., & Lattimer, J. M. 1985, ApJL, 299, L19
Burrows, A., & Lattimer, J. M. 1986, ApJ, 307, 178
Burrows, A., Livne, E., Dessart, L., et al. 2006, ApJ, 640, 878
Burrows, C. J., Krist, J., Hester, J. J., et al. 1995, ApJ, 452, 680
Burrows, D. N., Michael, E., Hwang, U., et al. 2000, ApJ, 543, 2
Buta, R. J. 1982, PASP, 94, 578
Cabot, W. H., & Cook, A. W. 2006, Nature Physics, 2, 562
Cameron, A. G. W., & Truran, J. W. 1977, Icarus, 30, 447
Campana, S., & Immler, S. 2012, MNRAS, 427, L70
Canizares, C. R., Kriss, G. A., & Feigelson, E. D. 1982, ApJL, 253, L17
Cannon, R. C. 1993, MNRAS, 263, 817
Cano, Z. 2013, MNRAS, 434, 1098
Cano, Z. 2014, ApJ, 794, 121
Cano, Z., Maeda, K., & Schulze, S. 2014, MNRAS, 438, 2924
Cantiello, M., Yoon, S.-C., Langer, N., & Livio, M. 2007, A&A, 465, L29
Cao, Y., Kasliwal, M. M., Arcavi, I., et al. 2013, ApJL, 775, L7
Cao, Y., Kulkarni, S. R., Howell, D. A., et al. 2015, Nature, 521, 328
Cappellaro, E., Danziger, I. J., della Valle, M., et al. 1995, A&A, 293, 723
Cappellaro, E., della Valle, M., Iijima, T., & Turatto, M. 1990, A&A, 228, 61
Cappellaro, E., Mazzali, P. A., Benetti, S., et al. 1997, A&A, 328, 203
Cardall, C. Y., Endeve, E., & Mezzacappa, A. 2013, Physical Review D, 88, 023011
Carlton, A. K., Borkowski, K. J., Reynolds, S. P., et al. 2011, ApJL, 737, L22
Carroll, S. M., Duvvuri, V., Trodden, M., & Turner, M. S. 2004, Physical Review D, 70, 043528
Cartier, R., Hamuy, M., Pignata, G., et al. 2014, ApJ, 789, 89
Castellani, V., Chieffi, A., Tornambe, A., & Pulone, L. 1985, ApJ, 296, 204
Castor, J. I. 1970, MNRAS, 149, 111
Castor, J. I. 1972, ApJ, 178, 779
Castor, J. I., Abbott, D. C., & Klein, R. I. 1975, ApJ, 195, 157
Catchpole, R. M., Whitelock, P. A., Feast, M. W., et al. 1988, MNRAS, 231, 75p
Catchpole, R. M., Whitelock, P. A., Menzies, J. W., et al. 1989, MNRAS, 237, 55P
Catena, R., Covi, L., & Emken, T. 2015, Physical Review D, 91, 123524
Chakraborti, S., & Ray, A. 2011, ApJ, 728, 24
Chakraborty, S., Fischer, T., Mirizzi, A., et al. 2011, PRL, 107, 151101
Chan, K.-W., & Lingenfelter, R. E. 1993, ApJ, 405, 614
Chan, T. C., Cheng, K. S., & Harko, T. 2009, ApJ, 695, 1
Chandra, P., Chevalier, R. A., Chugai, N., et al. 2012a, ApJ, 755, 110
Chandra, P., Chevalier, R. A., Irwin, C. M., et al. 2012b, ApJL, 750, L2
Chandra, P., Dwarkadas, V. V., Ray, A., et al. 2009a, ApJ, 699, 388
Chandra, P., Stockdale, C. J., Chevalier, R. A., et al. 2009b, ApJ, 690, 1839
Chandrasekhar, S. 1960, Radiative Transfer (Dover)
Chandrasekhar, S. 1961, Hydrodynamic and Hydromagnetic Stability (Clarendon Press)
Chang, P., Bildsten, L., & Arras, P. 2010, ApJ, 723, 719
Chang, Z., Li, X., Lin, H.-N., & Wang, S. 2014, European Physical Journal C, 74, 2821
Chaplin, W. J., & Miglio, A. 2013, ARAA, 51, 353
Charlebois, M., Drissen, L., Bernier, A.-P., et al. 2010, AJ, 139, 2083
Chatzopoulos, E., & Wheeler, J. C. 2012a, ApJ, 748, 42
Chatzopoulos, E., & Wheeler, J. C. 2012b, ApJ, 760, 154
Chatzopoulos, E., Wheeler, J. C., & Couch, S. M. 2013a, ApJ, 776, 129

Chatzopoulos, E., Wheeler, J. C., & Vinkó, J. 2012, ApJ, 746, 121
Chatzopoulos, E., Wheeler, J. C., Vinkó, J., et al. 2011, ApJ, 729, 143
Chatzopoulos, E., Wheeler, J. C., Vinkó, J., et al. 2013b, ApJ, 773, 76
Chatzopoulos, E., van Rossum, D. R., Wheeler, J. C., et al. 2015, ApJ, 799, 18
Chen, H.-L., Woods, T. E., Yungelson, L. R., et al. 2014a, MNRAS, 445, 1912
Chen, J., Wang, X., Ganeshalingam, M., et al. 2014b, ApJ, 790, 120
Chen, K.-J. 2015, Modern Physics Letters A, 30, 1530002
Chen, M. C., Herwig, F., Denissenkov, P. A., & Paxton, B. 2014c, MNRAS, 440, 1274
Chen, T.-W., Smartt, S. J., Bresolin, F., et al. 2013, ApJL, 763, L28
Chen, T.-W., Smartt, S. J., Jerkstrand, A., et al. 2015, MNRAS, 452, 1567
Chen, W.-C., & Li, X.-D. 2009, ApJ, 702, 686
Chen, W.-C., & Piekarewicz, J. 2015, PRL, 115, 161101
Chen, X., Han, Z., & Tout, C. A. 2011, ApJL, 735, L31
Chen, X., Jeffery, C. S., Zhang, X., & Han, Z. 2012, ApJL, 755, L9
Cheng, Q., & Yu, Y.-W. 2014, ApJL, 786, L13
Cherepashchuk, A. M., & Karetnikov, V. G. 2003, Astronomy Reports, 47, 38
Cherry, J. F., Wu, M.-R., Carlson, J., et al. 2012, Physical Review D, 85, 125010
Chevalier, R. A. 1982a, ApJ, 258, 790
Chevalier, R. A. 1982b, ApJ, 259, 302
Chevalier, R. A. 1997, Science, 276, 1374
Chevalier, R. A. 1998, ApJ, 499, 810
Chevalier, R. A. 2005, ApJ, 619, 839
Chevalier, R. A. 2011, in Astrophysics of Neutron Stars, AIP Conf. Series, 1379, 5
Chevalier, R. A. 2012, ApJL, 752, LL2
Chevalier, R. A., Blondin, J. M., & Emmering, R. T. 1992, ApJ, 392, 118
Chevalier, R. A., Dwarkadas, V. V. 1995, ApJ letters, 452, 45
Chevalier, R. A., & Fransson, C. 1994, ApJ, 420, 268
Chevalier, R. A., & Fransson, C. 2003, in Supernovae and Gamma-Ray Bursters, ed. K. W. Weiler (Springer), 171
Chevalier, R. A., & Fransson, C. 2008, ApJL, 683, L135
Chevalier, R. A., Fransson, C., & Nymark, T. K. 2006, ApJ, 641, 1029
Chevalier, R. A., & Irwin, C. M. 2011, ApJL, 729, L6
Chevalier, R. A., & Irwin, C. M. 2012, ApJL, 747, L17
Chevalier, R. A., & Oishi, J. 2003, ApJL, 593, L23
Chevalier, R. A., & Raymond, J. C. 1978, ApJL, 225, L27
Chevalier, R. A., & Soderberg, A. M. 2010, ApJL, 711, L40
Chieffi, A., & Limongi, M. 2013, ApJ, 764, 21
Childress, M., Aldering, G., Antilogus, P., et al. 2013a, ApJ, 770, 107
Childress, M. J., Filippenko, A. V., Ganeshalingam, M., & Schmidt, B. P. 2014a, MNRAS, 437, 338
Childress, M. J., Hillier, D. J., Seitenzahl, I., et al. 2015, MNRAS, 454, 3816
Childress, M. J., Scalzo, R. A., Sim, S. A., et al. 2013b, ApJ, 770, 29
Childress, M. J., Wolf, C., & Zahid, H. J. 2014b, MNRAS, 445, 1898
Chini, R., Hoffmeister, V. H., Nasseri, A., et al. 2012, MNRAS, 424, 1925
Chiosi, E., Chiosi, C., Trevisan, P., Piovan, L., & Orio, M. 2015, MNRAS, 448, 2100
Chiotellis, A., Kosenko, D., Schure, K. M., et al. 2013, MNRAS, 435, 1659
Chiotellis, A., Schure, K. M., & Vink, J. 2012, A&A, 537, A139
Chita, S. M., Langer, N., van Marle, A. J., et al. 2008, A&A, 488, L37
Chomiuk, L. 2013, Publications of the Astron. Soc. of Australia, 30, 46
Chomiuk, L., Chornock, R., Soderberg, A. M., et al. 2011, ApJ, 743, 114
Chomiuk, L., Soderberg, A. M., Moe, M., et al. 2012, ApJ, 750, 164
Chornock, R., Berger, E., Rest, A., et al. 2013, ApJ, 767, 162
Chornock, R., Filippenko, A. V., Branch, D., et al. 2006, PASP, 118, 722
Chornock, R., Filippenko, A. V., Li, W., & Silverman, J. M. 2010, ApJ, 713, 1363

Chornock, R., Filippenko, A. V., Li, W., et al. 2011, ApJ, 739, 41
Chugai, N. N. 1980, Soviet Astronomy Letters, 6, 266
Chugai, N. N. 1986, Soviet Astronomy Letters, 12, 192
Chugai, N. N. 1990, Soviet Astronomy Letters, 16, 457
Chugai, N. N. 1992a, Soviet Astronomy, 36, 63
Chugai, N. N. 1992b, Soviet Astronomy Letters, 18, 168
Chugai, N. N. 1993, ApJL, 414, L101
Chugai, N. N. 2001, MNRAS, 326, 1448
Chugai, N. N. 2006, Astronomy Letters, 32, 739
Chugai, N. N. 2009, MNRAS, 400, 866
Chugai, N. N., Blinnikov, S. I., Cumming, R. J., et al. 2004, MNRAS, 352, 1213
Chugai, N. N., & Chevalier, R. A. 2006, ApJ, 641, 1051
Chugai, N. N., & Chevalier, R. A. 2007, ApJ, 657, 378
Chugai, N. N., Chevalier, R. A., & Utrobin, V. P. 2007, ApJ, 662, 1136
Chugai, N. N., & Danziger, I. J. 1994, MNRAS, 268, 173
Chugai, N. N., & Utrobin, V. P. 2000, A&A, 354, 557
Chugai, N. N., & Utrobin, V. P. 2008, Astronomy Letters, 34, 589
Chugai, N. N., & Utrobin, V. P. 2014, Astronomy Letters, 40, 291
Churazov, E., Sunyaev, R., Isern, J., et al. 2014, Nature, 512, 406
Churazov, E., Sunyaev, R., Isern, J., et al. 2015, ApJ, 812, 62
Cichowolski, S., Pineault, S., Gamen, R., et al. 2014, MNRAS, 438, 1089
Claeys, J. S. W., de Mink, S. E., Pols, O. R., et al. 2011, A&A, 528, A131
Claeys, J. S. W., Pols, O. R., Izzard, R. G., et al. 2014, A&A, 563, AA83
Claret, A. 2014, A&A, 562, A31
Clark, J. S., Ritchie, B. W., Najarro, F., et al. 2014, A&A, 565, A90
Clayton, D. D. 1984, Principles of stellar evolution and nucleosynthesis. D. D. Clayton.2nd edition. The University of Chicago Press, Chicago - London.
Clifford, F. E., & Tayler, R. J. 1965, Memoirs of the RAS, 69, 21
Clocchiatti, A., Suntzeff, N. B., Covarrubias, R., & Candia, P. 2011, AJ, 141, 163
Clocchiatti, A., Suntzeff, N. B., Phillips, M. M., et al. 2001, ApJ, 553, 886
Clocchiatti, A., & Wheeler, J. C. 1997, ApJ, 491, 375
Clocchiatti, A., Wheeler, J. C., Benetti, S., & Frueh, M. 1996a, ApJ, 459, 547
Clocchiatti, A., Wheeler, J. C., Brotherton, M. S., et al. 1996b, ApJ, 462, 462
Clocchiatti, A., Wheeler, J. C., Kirshner, R. P., et al. 2008, PASP, 120, 290
Colgate, S. A., Petschek, A. G., & Kriese, J. T. 1980, ApJL, 237, L81
Colgate, S. A., & White, R. H. 1966, ApJ, 143, 626
Conley, A., Guy, J., Sullivan, M., et al. 2011, ApJS, 192, 1
Conroy, C., & Bullock, J. S. 2015, ApJL, 805, L2
Cook, D. L., Berger, E., Faestermann, T., et al. 2009, in Lunar and Planetary Science Conference, 40, 1129
Cooke, J. 2008, ApJ, 677, 137
Cooke, J., Ellis, R. S., Sullivan, M., et al. 2011, ApJL, 727, L35
Cooke, J., Sullivan, M., Barton, E. J., et al. 2009, Nature, 460, 237
Cooke, J., Sullivan, M., Gal-Yam, A., et al. 2012a, Nature, 491, 228
Cooke, R., Pettini, M., & Murphy, M. T. 2012b, MNRAS, 425, 347
Copin, Y., Blanc, N., Bongard, S., et al. 2006, New Astron Rev, 50, 436
Corsi, A., Ofek, E. O., Frail, D. A., et al. 2011, ApJ, 741, 76
Couch, S. M. 2013a, ApJ, 765, 29
Couch, S. M. 2013b, ApJ, 775, 35
Couch, S. M., Chatzopoulos, E., Arnett, W. D., & Timmes, F. X. 2015, ApJL, 808, L21
Couch, S. M., & O'Connor, E. P. 2014, ApJ, 785, 123
Couch, S. M., & Ott, C. D. 2013, ApJL, 778, L7
Couch, S. M., & Ott, C. D. 2015, ApJ, 799, 5
Couch, S. M., Pooley, D., Wheeler, J. C., & Milosavljević, M. 2011, ApJ, 727, 104

Couch, S. M., Wheeler, J. C., & Milosavljević, M. 2009, ApJ, 696, 953
Courant, R., & Friedrichs, K. O. 1948, Pure and Applied Mathematics (Interscience)
Cowan, J. J., Roederer, I. U., Sneden, C., & Lawler, J. E. 2011, in RR Lyrae Stars, Metal-Poor Stars, and the Galaxy (Carnegie Institution) 223
Cowan, J. J., & Sneden, C. 2006, Nature, 440, 1151
Cox, D. P., & Smith, B. W. 1974, ApJL, 189, L105
Cox, J. P., & Giuli, R. T. 1968, Principles of Stellar Structure (Gordon and Breach)
Cox, N. L. J., & Patat, F. 2014, A&A, 565, AA61
Creasey, P., Theuns, T., & Bower, R. G. 2013, MNRAS, 429, 1922
Crockett, R. M., Smartt, S. J., Eldridge, J. J., et al. 2007, MNRAS, 381, 835
Crockett, R. M., Eldridge, J. J., Smartt, S. J., et al. 2008, MNRAS, 391, L5
Crockett, R. M., Smartt, S. J., Pastorello, A., et al. 2011, MNRAS, 410, 2767
Cropper, M., Bailey, J., McCowage, J., et al. 1988, MNRAS, 231, 695
Crotts, A. P. S. 2015, ApJL, 804, L37
Crotts, A. P. S. & Heathcote, S. R. 2000, ApJ, 528, 1
Crotts, A. P. S., Kunkel, W. E., & McCarthy, P. J. 1989, ApJL, 347, L61
Crowther, P. A., Schnurr, O., Hirschi, R., et al. 2010, MNRAS, 408, 731
Culhane, M., & McCray, R. 1995, ApJ, 455, 335
D'Andrea, C. 2014, AAS Meeting 223, abstract #141.05
D'Ercole, A., D'Antona, F., Carini, R., et al. 2012, MNRAS, 2921
Dahlen, T., Strolger, L.-G., Riess, A. G., et al. 2012, ApJ, 757, 70
Dall'Ora, M., Botticella, M. T., Pumo, M. L., et al. 2014, ApJ, 787, 139
Damköhler, G. 1940, Electrochemie und Angewandte Physikalische Chemie, 46, 601, English translation NASA TM No. 1112 (1947).
Dan, M., Rosswog, S., Brüggen, M., & Podsiadlowski, P. 2014, MNRAS, 438, 14
Dan, M., Rosswog, S., Guillochon, J., & Ramirez-Ruiz, E. 2011, ApJ, 737, 89
Dan, M., Rosswog, S., Guillochon, J., & Ramirez-Ruiz, E. 2012, MNRAS, 422, 2417
Darbha, S., Metzger, B. D., Quataert, E., et al. 2010, MNRAS, 409, 846
Darnley, M. J., Henze, M., Steele, I. A., et al. 2015, A&A, 580, A45
Darrieus, G. 1938. "Propagation d'un front de flamme," La Technique Moderne and Congrés de Mécanique Appliquée, Paris
Das, P. K., Selvaganapathy, J., Sharma, C., et al. 2013, International Journal of Modern Physics A, 28, 50152
Dasgupta, B., & Beacom, J. F. 2011, Physical Review D, 83, 113006
Davidson, K., & Fesen, R. A. 1985, ARAA, 23, 119
Davies, B., Figer, D. F., Kudritzki, R.-P., et al. 2009, ApJ, 707, 844
Dawson, J. R., McClure-Griffiths, N. M., Wong, T., et al. 2013, ApJ, 763, 56
Dawson, K. S., Aldering, G., Amanullah, R., et al. 2009, AJ, 138, 1271
De Donder, E., & Vanbeveren, D. 1998, A&A, 333, 557
de Jaeger, T., Anderson, J. P., Pignata, G., et al. 2015a, ApJ, 807, 63
de Jaeger, T., González-Gaitán, S., Anderson, J. P., et al. 2015b, ApJ, 815, 121
de Jager, C., Nieuwenhuijzen, H., & van der Hucht, K. A. 1988, A&A Supplement, 72, 259
De Looze, I., Barlow, M. J., Swinyard, B. M., et al. 2017, MSRAS, 465, 3309
de Luca, A. 2008, 40 Years of Pulsars: Millisecond Pulsars, Magnetars and More, 983, 311
de Mink, S. E., Cantiello, M., Langer, N., et al. 2009, A&A, 497, 243
de Mink, S. E., Sana, H., Langer, N., Izzard, R. G., & Schneider, F. R. N. 2014, ApJ, 782, 7
de Rossi, M. E., Tissera, P. B., & Pedrosa, S. E. 2010, A&A, 519, A89
de Rújula, A. 1987, Physics Letters B, 193, 514
De Smedt, K., Van Winckel, H., Karakas, A. I., et al. 2012, A&A, 541, A67
de Souza, R. S., Ishida, E. E. O., Whalen, D. J., et al. 2014, MNRAS, 442, 1640
de Souza, R. S., Rodrigues, L. F. S., Ishida, E. E. O., & Opher, R. 2011, MNRAS, 415, 2969
de Vaucouleurs, G., & Corwin, H. G., Jr. 1985, ApJ, 295, 287
de Vaucouleurs, G., de Vaucouleurs, A., Buta, R., et al. 1981, PASP, 93, 36
Dearborn, D. S. P., Lattanzio, J. C., & Eggleton, P. P. 2006, ApJ, 639, 405

Del Popolo, A., & Pace, F. 2016, ApSpSci, 361, 162
DeLaney, T., Kassim, N. E., Rudnick, L., & Perley, R. A. 2014, ApJ, 785, 7
DeLaney, T., Rudnick, L., Stage, M. D., et al. 2010, ApJ, 725, 2038
Deloye, C. J., & Taam, R. E. 2010, ApJL, 719, L28
Demorest, P. B., Pennucci, T., Ransom, S. M., et al. 2010, Nature, 467, 1081
Deng, J., Kawabata, K. S., Ohyama, Y., et al. 2004, ApJL, 605, L37
Denisov Yu. N. & Troshin, Ya. K. 1959, Dokl. Akad. Nauk SSSR (Phys-Chem. Sec), 125, 110
Denissenkov, P. A., Herwig, F., Truran, J. W., & Paxton, B. 2013, ApJ, 772, 37
Denissenkov, P. A., Truran, J. W., Herwig, F., et al. 2015, MNRAS, 447, 2696
Dermer, C. D., & Holmes, J. M. 2005, ApJL, 628, L21
Desai, K. M., Chu, Y.-H., Gruendl, R. A., et al. 2010, AJ, 140, 584
Dessart, L., Blondin, S., Brown, P. J., et al. 2008, ApJ, 675, 644
Dessart, L., Blondin, S., Hillier, D. J., & Khokhlov, A. 2014a, MNRAS, 441, 532
Dessart, L., Burrows, A., Livne, E., & Ott, C. D. 2007, ApJ, 669, 585
Dessart, L., Gutierrez, C. P., Hamuy, M., et al. 2014b, MNRAS, 440, 1856
Dessart, L., & Hillier, D. J. 2005, A&A, 437, 667
Dessart, L., & Hillier, D. J. 2006, A&A, 447, 691
Dessart, L., & Hillier, D. J. 2008, MNRAS, 383, 57
Dessart, L., & Hillier, D. J. 2010, MNRAS, 405, 2141
Dessart, L., & Hillier, D. J. 2011a, MNRAS, 410, 1739
Dessart, L., & Hillier, D. J. 2011b, MNRAS, 415, 3497
Dessart, L., & Hillier, D. J. 2015, MNRAS, 447, 1370
Dessart, L., Hillier, D. J., Gezari, S., et al. 2009, MNRAS, 394, 21
Dessart, L., Hillier, D. J., Li, C., & Woosley, S. 2012a, MNRAS, 424, 2139
Dessart, L., Hillier, D. J., Livne, E., et al. 2011, MNRAS, 414, 2985
Dessart, L., Hillier, D. J., Waldman, R., et al. 2012b, MNRAS, 426, L76
Dessart, L., Hillier, D. J., Waldman, R., & Livne, E. 2013a, MNRAS, 433, 1745
Dessart, L., Livne, E., & Waldman, R. 2010a, MNRAS, 405, 2113
Dessart, L., Livne, E., & Waldman, R. 2010b, MNRAS, 408, 827
Dessart, L., Waldman, R., Livne, E., et al. 2013b, MNRAS, 428, 3227
Detmers, R. G., Langer, N., Podsiadlowski, P., & Izzard, R. G. 2008, A&A, 484, 831
Dewey, D., Dwarkadas, V. V., Haberl, F., et al. 2012, ApJ, 752, 103
Dewi, J. D. M., Pols, O. R., Savonije, G. J., & van den Heuvel, E. P. J. 2002, MNRAS, 331, 1027
Dexter, J., & Kasen, D. 2013, ApJ, 772, 30
Dhawan, S., Leibundgut, B., Spyromilio, J., & Maguire, K. 2015, MNRAS, 448, 1345
Di Carlo, E., Corsi, C., Arkharov, A. A., et al. 2008, ApJ, 684, 471
Di Carlo, E., Massi, F., Valentini, G., et al. 2002, ApJ, 573, 144
Di Stefano, R. 2010a, ApJ, 712, 728
Di Stefano, R. 2010b, ApJ, 719, 474
Di Stefano, R., Fisher, R., Guillochon, J., & Steiner, J. F. 2015, arXiv:1501.07837
Di Stefano, R., & Kilic, M. 2012, ApJ, 759, 56
Di Stefano, R., Voss, R., & Claeys, J. S. W. 2011, ApJL, 738, L1
Diamond, T. R., Höflich, P., & Gerardy, C. L. 2015, ApJ, 806, 107
Dicus, D. A., Kolb, E. W., Schramm, D. N., & Tubbs, D. L. 1976, ApJ, 210, 481
Diehl, R., Dupraz, C., Bennett, K., et al. 1995, A&A, 298, 445
Diehl, R., Halloin, H., Kretschmer, K., et al. 2006, Nature, 439, 45
Diehl, R., Siegert, T., Hillebrandt, W., et al. 2014, Science, 345, 1162
Diehl, R., Siegert, T., Hillebrandt, W., et al. 2015, A&A, 574, AA72
Dilday, B., Howell, D. A., Cenko, S. B., et al. 2012, Science, 337, 942
Dillmann, I., Rauscher, T., Heil, M., et al. 2008, Journal of Physics G Nuclear Physics, 35, 014029
Dimitriadis, G., Chiotellis, A., & Vink, J. 2014, MNRAS, 443, 1370
Dinçel, B., Nevusäuser, R., Yerli, S. K., et al. 2015, MNRAS, 448, 3196
Dittmann, J. A., Soderberg, A. M., Chomiuk, L., et al. 2014, ApJ, 788, 38
Dobbie, P. D., Napiwotzki, R., Burleigh, M. R., et al. 2006, MNRAS, 369, 383

Dobbs, C. L., Burkert, A., & Pringle, J. E. 2011, MNRAS, 417, 1318
Doggett, J. B., & Branch, D. 1985, AJ, 90, 2303
Doherty, C. L., Gil-Pons, P., Siess, L., Lattanzio, J. C., & Lau, H. H. B. 2015, MNRAS, 446, 2599
Dolence, J. C., Burrows, A., Murphy, J. W., & Nordhaus, J. 2013, ApJ, 765, 110
Domínguez, I., Höflich, P., & Straniero, O. 2001, ApJ, 557, 279
Domínguez, I., & Khokhlov, A. 2011, ApJ, 730, 87
Domínguez, I., Piersanti, L., Bravo, E., et al. 2006, ApJ, 644, 21
Dong, S., Shappee, B. J., Prieto, J. L., et al. 2016, Science, 351, 257
Dopita, M. A., Cohen, M., Schwartz, R. D., & Evans, R. 1984, ApJL, 287, L69
Doroshenko, V. T., Efimov, Y. S., & Shakhovskoi, N. M. 1995, Astron Let, 21, 513
Dotani, T., Hayashida, K., Inoue, H., et al. 1987, Nature, 330, 230
Drago, A., Lavagno, A., & Pagliara, G. 2014, Physical Review D, 89, 043014
Dragulin, P., & Höeflich, P. 2016, ApJ, 818, 26
Draine, B. T. 2009, in Cosmic Dust – Near and Far, ASP Conf. Series, 414, 453
Draine, B. T. 2011, Physics of the Interstellar and Intergalactic Medium (Princeton Un. Press)
Drake, A. J., Djorgovski, S. G., Mahabal, A., et al. 2009, ApJ, 696, 870
Drake, A. J., Djorgovski, S. G., Mahabal, A., et al. 2012, IAU Symposium 285, New Horizons in Time-domain Astronomy, 306
Dreiner, H. K., Fortin, J.-F., Hanhart, C., & Ubaldi, L. 2014, Physical Review D, 89, 105015
Drout, M. R., Soderberg, A. M., Gal-Yam, A., et al. 2011, ApJ, 741, 97
Drout, M. R., Soderberg, A. M., Mazzali, P. A., et al. 2013, ApJ, 774, 58
Duan, H., & Kneller, J. P. 2009, Journal of Physics G Nuclear Physics, 36, 113201
Duncan, R. C., & Thompson, C. 1992, ApJL, 392, L9
Dunstall, P. R., Dufton, P. L., Sana, H., et al. 2015, A&A, 580, A93
Dwarkadas, V. V. 2005, ApJ, 630, 892
Dwarkadas, V. V. 2007, ApSpSci, 307, 153
Dwarkadas, V. V. 2011, MNRAS, 412, 1639
Dwarkadas, V. V. 2014, MNRAS, 440, 1917
Dwek, E. 1983, ApJ, 274, 175
Dwek, E. 1985, ApJ, 297, 719
Dwek, E., A'Hearn, M. F., Becklin, E. E., et al. 1983, ApJ, 274, 168
Dwek, E., Arendt, R. G., Bouchet, P., et al. 2008, ApJ, 676, 1029
Dwek, E., Arendt, R. G., Bouchet, P., et al. 2010, ApJ, 722, 425
Dwek, E., Galliano, F., & Jones, A. P. 2007, ApJ, 662, 927
Edwards, Z. I., Pagnotta, A., & Schaefer, B. E. 2012, ApJL, 747, L19
Eggleton, P. P. 1973, MNRAS, 163, 279
Eggleton, P. P. 1983, ApJ, 268, 368
Ehlerová, S., & Palouŝ, J. 2013, A&A, 550, A23
Ekström, S., Georgy, C., Eggenberger, P., et al. 2012, A&A, 537, A146
Ekström, S., Meynet, G., Chiappini, C., et al. 2008, A&A, 489, 685
Eldridge, J. J., Fraser, M., Maund, J. R., & Smartt, S. J. 2015, MNRAS, 446, 2689
Eldridge, J. J., Fraser, M., Smartt, S. J., et al. 2013, MNRAS, 436, 774
Eldridge, J. J., Izzard, R. G., & Tout, C. A. 2008, MNRAS, 384, 1109
Eldridge, J. J., Langer, N., & Tout, C. A. 2011, MNRAS, 414, 3501
Eldridge, J. J., Mattila, S., & Smartt, S. J. 2007, MNRAS, 376, L52
Eldridge, J. J., & Tout, C. A. 2004, MNRAS, 353, 87
Elias, J. H., Matthews, K., Neugebauer, G., & Persson, S. E. 1985, ApJ, 296, 379
Elias-Rosa, N., Van Dyk, S. D., Li, W., et al. 2010, ApJL, 714, L254
Elias-Rosa, N., Van Dyk, S. D., Li, W., et al. 2011, ApJ, 742, 6
Ellis, J., Janka, H.-T., Mavromatos, N. E., et al. 2012a, Physical Review D, 85, 045032
Ellis, J., Janka, H.-T., Mavromatos, N. E., et al. 2012b, Physical Review D, 85, 105028
Ellison, D. C., Drury, L. O'C., & Meyer, J.-P. 1997, ApJ, 487, 197
Elmhamdi, A., Danziger, I. J., Branch, D., et al. 2006, A&A, 450, 305
Elmhamdi, A., Danziger, I. J., Cappellaro, E., et al. 2004, A&A, 426, 963

Elmhamdi, A., Danziger, I. J., Chugai, N., et al. 2003, MNRAS, 338, 939
Elshamouty, K. G., Heinke, C. O., Sivakoff, G. R., et al. 2013, ApJ, 777, 22
Endal, A. S., & Sofia, S. 1976, ApJ, 210, 184
Endal, A. S., & Sofia, S. 1978, ApJ, 220, 279
Endeve, E., Cardall, C. Y., Budiardja, R. D., et al. 2012, ApJ, 751, 26
Ensman, L., & Burrows, A. 1992, ApJ, 393, 742
Ensman, L. M., & Woosley, S. E. 1988, ApJ, 333, 754
Epstein, R. I. 1979, MNRAS, 188, 305
Epstein, R. I., & Pethick, C. J. 1981, ApJ, 243, 1003
Ergma, E., & Paczyński, B. 1974, Acta Astron, 24, 1
Ergon, M., Jerkstrand, A., Sollerman, J., et al. 2015, A&A, 580, A142
Ergon, M., Sollerman, J., Fraser, M., et al. 2014, A&A, 562, A17
Eriksen, K. A., Arnett, D., McCarthy, D. W., & Young, P. 2009, ApJ, 697, 29
Erlykin, A. D., & Wolfendale, A. W. 2010, Surveys in Geophysics, 31, 383
Esmaili, A., Peres, O. L. G., & Serpico, P. D. 2014, Physical Review D, 90, 033013
Evans, R. 1994, PASA, 11, 7
Fabbri, J., Otsuka, M., Barlow, M. J., et al. 2011, MNRAS, 418, 1285
Fabian, A. C., & Terlevich, R. 1996, MNRAS, 280, L5
Fabian, A. C., Wilkins, D. R., Miller, J. M., et al. 2012, MNRAS, 424, 217
Falk, S. W., & Arnett, W. D. 1977, ApJS, 33, 515
Fano, U. 1963, Ann. Rev. Nucl. Sci., 13, 1
Faran, T., Poznanski, D., Filippenko, A. V., et al. 2014a, MNRAS, 442, 844
Faran, T., Poznanski, D., Filippenko, A. V., et al. 2014b, MNRAS, 445, 554
Farouqi, K., Kratz, K.-L., Pfeiffer, B., et al. 2010, ApJ, 712, 1359
Farr, W. M., Kremer, K., Lyutikov, M., & Kalogera, V. 2011, ApJ, 742, 81
Fassia, A., Meikle, W. P. S., Chugai, N., et al. 2001, MNRAS, 325, 907
Fassia, A., Meikle, W. P. S., Vacca, W. D., et al. 2000, MNRAS, 318, 1093
Faucher-Giguère, C.-A., & Kaspi, V. M. 2006, ApJ, 643, 332
Fedeli, C., & Moscardini, L. 2014, MNRAS, 442, 2659
Feige, J., Wallner, A., Winkler, S. R., et al. 2012, Publications of the Astron. Soc. of Australia, 29, 109
Feindt, U., Kerschhaggl, M., Kowalski, M., et al. 2013, A&A, 560, A90
Felten, J. E., Dwek, E., & Viegas-Aldrovandi, S. M. 1989, ApJ, 340, 943
Feltzing, S., & Chiba, M. 2013, New Astron Rev, 57, 80
Fenech, D. M., Muxlow, T. W. B., Beswick, R. J., et al. 2008, MNRAS, 391, 1384
Fermi, E. 1949, Physical Review, 75, 1169
Fernández, R. 2015, MNRAS, 452, 2071
Fernández, R., Müller, B., Foglizzo, T., & Janka, H.-T. 2014, MNRAS, 440, 2763
Ferrarotti, A. S., & Gail, H.-P. 2006, A&A, 447, 553
Ferrero, A., & Altschul, B. 2010, Physical Review D, 82, 123002
Fesen, R. A., & Becker, R. H. 1990, ApJ, 351, 437
Fesen, R. A., Gerardy, C. L., Filippenko, A. V., et al. 1999a, AJ, 117, 725
Fesen, R. A., Gerardy, C. L., McLin, K. M., & Hamilton, A. J. S. 1999b, ApJ, 514, 195
Fesen, R. A., Hammell, M. C., Morse, J., et al. 2006, ApJ, 645, 283
Fesen, R. A., Höflich, P. A., & Hamilton, A. J. S. 2015, ApJ, 804, 140
Fesen, R. A., Höflich, P. A., Hamilton, A. J. S., et al. 2007, ApJ, 658, 396
Fesen, R. A., Hurford, A. P., & Matonick, D. M. 1995, AJ, 109, 2608
Fesen, R. A., Kremer, R., Patnaude, D., & Milisavljevic, D. 2012, AJ, 143, 27
Fesen, R. A., & Matonick, D. M. 1993, ApJ, 407, 110
Fesen, R., Rudie, G., Hurford, A., & Soto, A. 2008, ApJS, 174, 379
Fesen, R. A., Saken, J. M., & Hamilton, A. J. S. 1989, ApJL, 341, L55
Fesen, R. A., Shull, J. M., & Hurford, A. P. 1997, AJ, 113, 354
Fickett, W., & Davis, C. 1979, Los Alamos Series in Basic and Applied Sciences (Un. California Press)

Fields, B. D., Athanassiadou, T., & Johnson, S. R. 2008, ApJ, 678, 549
Fields, B. D., Hochmuth, K. A., & Ellis, J. 2005, ApJ, 621, 902
Fields, B. D., Pavlidou, V., & Prodanović, T. 2010, ApJL, 722, L199
Filipovic, M. D., Horner, J., Crawford, E. J., et al. 2013, Serbian Astronomical Journal, 187, 43
Filippenko, A. V. 1988, AJ, 96, 1941
Filippenko, A. V. 1992, ApJL, 384, L37
Filippenko, A. V. 1997, ARAA, 35, 309
Filippenko, A. V., Barth, A. J., Bower, G. C., et al. 1995a, AJ, 110, 2261
Filippenko, A. V., Barth, A. J., Matheson, T., et al. 1995b, ApJL, 450, L11
Filippenko, A. V., Li, W. D., Treffers, R. R., & Modjaz, M. 2001, in Small-telescope Astronomy on Global Scales, ASP Conf. Series, 246, 121
Filippenko, A. V., Matheson, T., & Ho, L. C. 1993, ApJL, 415, L103
Filippenko, A. V., Porter, A. C., & Sargent, W. L. W. 1990, AJ, 100, 1575
Filippenko, A. V., Richmond, M. W., Branch, D., et al. 1992a, AJ, 104, 1543
Filippenko, A. V., Richmond, M. W., Matheson, T., et al. 1992b, ApJL, 384, L15
Filippenko, A. V., & Sargent, W. L. W. 1986, AJ, 91, 691
Fimiani, L., Cook, D. L., Faestermann, T., et al. 2016, PRL, 116, 151104
Fink, M., Hillebrandt, W., & Röpke, F. K. 2007, A&A, 476, 1133
Fink, M., Kromer, M., Seitenzahl, I. R., et al. 2014, MNRAS, 438, 1762
Fink, M., Röpke, F. K., Hillebrandt, W., et al. 2010, A&A, 514, A53
Firth, R. E., Sullivan, M., Gal-Yam, A., et al. 2015, MNRAS, 446, 3895
Fischer, T., Sagert, I., Pagliara, G., et al. 2011, ApJS, 194, 39
Fisher, A. K. 2000, Ph.D. Thesis, Un. Oklahoma
Fisher, R., & Jumper, K. 2015, ApJ, 805, 150
Fitoussi, C., Raisbeck, G. M., Knie, K., et al. 2008, Physical Review Letters, 101, 121101
Fitzpatrick, E. L., & Garmany, C. D. 1990, ApJ, 363, 119
Flannery, B. P. 1976, ApJ, 205, 217
Foglizzo, T., Galletti, P., Scheck, L., & Janka, H.-T. 2007, ApJ, 654, 1006
Folatelli, G., Bersten, M. C., Benvenuto, O. G., et al. 2014a, ApJL, 793, LL22
Folatelli, G., Bersten, M. C., Kuncarayakti, H., et al. 2014b, ApJ, 792, 7
Folatelli, G., Contreras, C., Phillips, M. M., et al. 2006, ApJ, 641, 1039
Folatelli, G., Morrell, N., Phillips, M. M., et al. 2013, ApJ, 773, 53
Folatelli, G., Phillips, M. M., Burns, C. R., et al. 2010, AJ, 139, 120
Folatelli, G., Phillips, M. M., Morrell, N., et al. 2012, ApJ, 745, 74
Foley, R. J. 2012, ApJ, 748, 127
Foley, R. J. 2015, MNRAS, 452, 2463
Foley, R. J., Berger, E., Fox, O., et al. 2011, ApJ, 732, 32
Foley, R. J., Challis, P. J., Chornock, R., et al. 2013, ApJ, 767, 57
Foley, R. J., Chornock, R., Filippenko, A. V., et al. 2009a, AJ, 138, 376
Foley, R. J., Fox, O. D., McCully, C., et al. 2014a, MNRAS, 443, 2887
Foley, R. J., & Kasen, D. 2011, ApJ, 729, 55
Foley, R. J., Matheson, T., Blondin, S., et al. 2009b, AJ, 137, 3731
Foley, R. J., McCully, C., Jha, S. W., et al. 2014b, ApJ, 792, 29
Foley, R. J., Narayan, G., Challis, P. J., et al. 2010a, ApJ, 708, 1748
Foley, R. J., Papenkova, M. S., Swift, B. J., et al. 2003, PASP, 115, 1220
Foley, R. J., Rest, A., Stritzinger, M., et al. 2010b, AJ, 140, 1321
Foley, R. J., Simon, J. D., Burns, C. R., et al. 2012, ApJ, 752, 101
Foley, R. J., Smith, N., Ganeshalingam, M., et al. 2007, ApJL, 657, L105
Font, J. A., Cerdá-Durán, P., Gabler, M., et al. 2011, JP Conf. Series, 283, 012011
Förster, F., & Schawinski, K. 2008, MNRAS, 388, L74
Fossey, J., Cooke, B., Pollack, G., et al. 2014, CBET, 3792, 1
Fowler, W. A. 1966, ApJ, 144, 180
Fowler, W. A., & Hoyle, F. 1964, ApJS, 9, 201
Fox, D. W., Lewin, W. H. G., Fabian, A., et al. 2000, MNRAS, 319, 1154

Fox, O. D., Azalee Bostroem, K., Van Dyk, S. D., et al. 2014, ApJ, 790, 17
Fox, O. D., Chevalier, R. A., Dwek, E., et al. 2010, ApJ, 725, 1768
Fox, O. D., Chevalier, R. A., Skrutskie, M. F., et al. 2011, ApJ, 741, 7
Fox, O. D., Silverman, J. M., Filippenko, A. V., et al. 2015, MNRAS, 447, 772
Fraley, G. S. 1968, ApSpSci, 2, 96
France, K., McCray, R., Heng, K., et al. 2010, Science, 329, 1624
France, K., McCray, R., Penton, S. V., et al. 2011, ApJ, 743, 186
Fransson, C. 1982, A&A, 111, 140
Fransson, C. 1984, A&A, 132, 115
Fransson, C., & Jerkstrand, A. 2015, ApJL, 814, L2
Fransson, C., Benvenuti, P., Wamsteker, W., et al. 1984, A&A, 132, 1
Fransson, C., & Björnsson, C.-I. 1998, ApJ, 509, 861
Fransson, C., Cassatella, A., Gilmozzi, R., et al. 1989, ApJ, 336, 429
Fransson, C., Challis, P. M., Chevalier, R. A., et al. 2005, ApJ, 622, 991
Fransson, C., & Chevalier, R. A. 1987, ApJL, 322, L15
Fransson, C., & Chevalier, R. A. 1989, ApJ, 343, 323
Fransson, C., Chevalier, R. A., Filippenko, A. V., et al. 2002, ApJ, 572, 350
Fransson, C., Ergon, M., Challis, P. J., et al. 2014, ApJ, 797, 118
Fransson, C., & Kozma, C. 1993, ApJL, 408, L25
Fransson, C., Larsson, J., Migotto, K., et al. 2015, ApJL, 806, L19
Fransson, C., Larsson, J., Spyromilio, J., et al. 2013, ApJ, 768, 88
Fransson, C., Larsson, J., Spyromilio, J., et al. 2016, ApJL, 821, L5
Fraser, M., Ergon, M., Eldridge, J. J., et al. 2011, MNRAS, 417, 1417
Fraser, M., Kotak, R., Pastorello, A., et al. 2015, MNRAS, 453, 3886
Fraser, M., Maund, J. R., Smartt, S. J., et al. 2012, ApJL, 759, L13
Fraser, M., Maund, J. R., Smartt, S. J., et al. 2014, MNRAS, 439, L56
Fraser, M., Takáts, K., Pastorello, A., et al. 2010, ApJL, 714, L280
Fraternali, F., Marasco, A., Marinacci, F., & Binney, J. 2013, ApJL, 764, L21
Frebel, A., & Norris, J. E. 2013, in Planets, Stars and Stellar Systems. Volume 5: Galactic Structure and Stellar Populations, 55
Frebel, A., Simon, J. D., & Kirby, E. N. 2014, ApJ, 786, 74
Freedman, D. Z. 1974, Physical Review D, 9, 1389
Freedman, W. L., & Madore, B. F. 2010, ARAA, 48, 673
Freese, K., Bodenheimer, P., Spolyar, D., & Gondolo, P. 2008, ApJL, 685, L101
Fremling, C., Sollerman, J., Taddia, F., et al. 2014, A&A, 565, A114
Frey, L. H., Even, W., Whalen, D. J., et al. 2013, ApJS, 204, 16
Friedman, A. S., Wood-Vasey, W. M., Marion, G. H., et al. 2015, ApJS, 220, 9
Frieman, J. A., Bassett, B., Becker, A., et al. 2008, AJ, 135, 338
Friesen, B., Baron, E., Branch, D., et al. 2012, ApJS, 203, 12
Friesen, B., Baron, E., Wisniewski, J. P., et al. 2014, ApJ, 792, 120
Friesen, B., Baron, E., Parrent, J. T., et al. 2017, MNRAS, 467, 2392
Fröhlich, C., Martínez-Pinedo, G., Liebendörfer, M., et al. 2006, PRL, 96, 142502
Fry, B. J., Fields, B. D., & Ellis, J. R. 2015, ApJ, 800, 71
Fryer, C. L. 2009, ApJ, 699, 409
Fryer, C. L., Belczynski, K., Wiktorowicz, G., et al. 2012, ApJ, 749, 91
Fryer, C. L., Colgate, S. A., & Pinto, P. A. 1999, ApJ, 511, 2
Fryer, C. L., & New, K. C. B. 2011, Living Reviews in Relativity, 14, 1
Fryer, C. L., Ruiter, A. J., Belczynski, K., et al. 2010, ApJ, 725, 296
Fryer, C. L., Woosley, S. E., & Heger, A. 2001, ApJ, 550, 372
Fryer, C. L., & Young, P. A. 2007, ApJ, 659, 1438
Fu, W., & Lai, D. 2011, MNRAS, 413, 2207
Fuller, G. M., Fowler, W. A., & Newman, M. J. 1982, ApJS, 48, 279
Fuller, J., Cantiello, M., Stello, D., Garcia, R. A., & Bildsten, L. 2015a, Science, 350, 423
Fuller, J., Klion, H., Abdikamalov, E., & Ott, C. D. 2015b, MNRAS, 450, 414

Fuller, J., & Lai, D. 2012, MNRAS, 421, 426
Gaensler, B. M., Manchester, R. N., Staveley-Smith, L., et al. 1997, ApJ, 479, 845
Galbany, L., Moreno-Raya, M. E., Ruiz-Lapuente, P., et al. 2016, MNRAS, 457, 525
Gall, C., Hjorth, J., & Andersen, A. C. 2011, The Astronomy and Astrophysics Review, 19, 43
Gall, C., Hjorth, J., Watson, D., et al. 2014, Nature, 511, 326
Gall, E., Taubenberger, S., Kromer, M., et al. 2012, MNRAS, 427, 994
Gall, E. E. E., Polshaw, J., Kotak, R., et al. 2015, A&A, 582, A3
Gallagher, J. S., Garnavich, P. M., Caldwell, N., et al. 2008, ApJ, 685, 752
Gal-Yam, A. 2012, Science, 337, 927
Gal-Yam, A., Arcavi, I., Ofek, E. O., et al. 2014, Nature, 509, 471
Gal-Yam, A., & Leonard, D. C. 2009, Nature, 458, 865
Gal-Yam, A., Leonard, D. C., Fox, D. B., et al. 2007, ApJ, 656, 372
Gal-Yam, A., & Maoz, D. 2004, MNRAS, 347, 942
Gal-Yam, A., Mazzali, P. A., Manulis, I., & Bishop, D. 2013, PASP, 125, 749
Gal-Yam, A., Mazzali, P., Ofek, E. O., et al. 2009, Nature, 462, 624
Gamezo, V. N., Khokhlov, A. M., & Oran, E. S. 2005, ApJ, 623, 337
Gamezo, V. N., Khokhlov, A. M., Oran, E. S., et al. 2003, Science, 299, 77
Gamezo, V. N. & Oran, E. S. 2007 Paths to Exploding Stars: Accretion and Eruption, Kavli Institute for Theoretical Physics, (http://online.kitp.ucsb.edu/online/snovae_c07/gamezo/)
Gamezo, V. N., Poludnenko, A. Y., Oran, E. S., & Williams, F. A. 2014, Combust. & Flame, 161, 950
Gamezo, V. N., Wheeler, J. C., Khokhlov, A. M., & Oran, E. S. 1999, ApJ, 512, 82
Gandhi, P., Yamanaka, M., Tanaka, M., et al. 2013, ApJ, 767, 166
Ganeshalingam, M., Li, W., & Filippenko, A. V. 2011, MNRAS, 416, 2607
Ganeshalingam, M., Li, W., Filippenko, A. V., et al. 2012, ApJ, 751, 142
Ganot, N., Gal-Yam, A., Ofek, E. O., et al. 2016, ApJ, 820, 57
Gao, Y., & Pritchet, C. 2013, AJ, 145, 83
Garcia-Berro, E., & Iben, I. 1994, ApJ, 434, 306
García-Hernández, D. A., García-Lario, P., Plez, B., et al. 2006, Science, 314, 1751
García-Senz, D., & Bravo, E. 2005, A&A, 430, 585
García-Senz, D., Bravo, E., & Woosley, S. E. 1999, A&A, 349, 177
García-Senz, D., Cabezón, R. M., Arcones, A., et al. 2013, MNRAS, 436, 3413
García-Senz, D., Cabezón, R. M., Domínguez, I., & Thielemann, F. K. 2016, ApJ, 819, 132
Gardner, J. P., Mather, J. C., Clampin, M., et al. 2006, Space Science Reviews, 123, 485
Gaskell, C. M., Cappellaro, E., Dinerstein, H. L., et al. 1986, ApJL, 306, L77
Gehrels, N., Laird, C. M., Jackman, C. H., et al. 2003, ApJ, 585, 1169
Geier, S., Fürst, F., Ziegerer, E., et al. 2015, Science, 347, 1126
Geier, S., Marsh, T. R., Wang, B., et al. 2013, A&A, 554, A54
Geier, S., Nesslinger, S., Heber, U., et al. 2007, 15th European Workshop on White Dwarfs, 372, 393
Gent, F. A., Shukurov, A., Sarson, G. R., et al. 2013, MNRAS, 430, L40
Georgy, C., Ekström, S., Meynet, G., et al. 2012, A&A, 542, A29
Georgy, C., Meynet, G., Walder, R., et al. 2009, A&A, 502, 611
Gerardy, C. L., Fesen, R. A., Höflich, P., & Wheeler, J. C. 2000, AJ, 119, 2968
Gerardy, C. L., Fesen, R. A., Marion, G. H., et al. 2004a, in Cosmic Explosions in Three Dimensions, ed. P. Höflich, P. Kumar, & J. C. Wheeler (Cambridge University Press) 57
Gerardy, C. L., Fesen, R. A., Nomoto, K., et al. 2002, ApJ, 575, 1007
Gerardy, C. L., Höflich, P., Fesen, R. A., et al. 2004b, ApJ, 607, 391
Gerardy, C. L., Meikle, W. P. S., Kotak, R., et al. 2007, ApJ, 661, 995
Gerke, J., Kochanek, C. S., & Stanek, K. Z. 2014, AAS Meeting 223, #235.01
Gerke, J. R., Kochanek, C. S., & Stanek, K. Z. 2015, MNRAS, 450, 3289
Germany, L. M., Reiss, D. J., Sadler, E. M., et al. 2000, ApJ, 533, 320
Gezari, S., Basa, S., Martin, D. C., et al. 2008a, ApJ, 676, 944
Gezari, S., Dessart, L., Basa, S., et al. 2008b, ApJL, 683, L131

Gezari, S., Halpern, J. P., Grupe, D., et al. 2009, ApJ, 690, 1313
Gezari, S., Jones, D. O., Sanders, N. E., et al. 2015, ApJ, 804, 28
Gezari, S., Rest, A., Huber, M. E., et al. 2010, ApJL, 720, L77
Ghosh, P. 2007, Rotation and Accretion Powered Pulsars, World Scientific
Giannotti, M., Duffy, L. D., & Nita, R. 2011, Journal of Cosmology and Astroparticle Physics, 1, 15
Gil-Botella, I. 2016, arXiv:1605.02204
Gilfanov, M., & Bogdán, Á. 2010, Nature, 463, 924
Gilmozzi, R. & Panagia, N. 1999, Mem. Soc. Astron. Ital., 70, 583
Ginzburg, S., & Balberg, S. 2012, ApJ, 757, 178
Ginzburg, V. L., & Syrovatskii, S. I. 1963, Soviet Astronomy, 7, 357
Godon, P., Sion, E. M., Starrfield, S., et al. 2014, ApJL, 784, LL33
Goldhaber, G., Groom, D. E., Kim, A., et al. 2001, ApJ, 558, 359
Goldreich, P., & Weber, S. V. 1980, ApJ, 238, 991
Gomez, G., & Lopez, R. 1994, AJ, 108, 195
Gomez, H. L., Clark, C. J. R., Nozawa, T., et al. 2012a, MNRAS, 420, 3557
Gomez, H. L., Krause, O., Barlow, M. J., et al. 2012b, ApJ, 760, 96
Gonzalez, G., Brownlee, D., & Ward, P. 2001, Icarus, 152, 185
González-Gaitán, S., Tominaga, N., Molina, J., et al. 2015, MNRAS, 451, 2212
González Hernández, J. I., Ruiz-Lapuente, P., Tabernero, H. M., et al. 2012, Nature, 489, 533
Goobar, A. 2008, ApJL, 686, L103
Goobar, A., Johansson, J., Amanullah, R., et al. 2014, ApJL, 784, LL12
Goobar, A., Kromer, M., Siverd, R., et al. 2015, ApJ, 799, 106
Goobar, A., & Leibundgut, B. 2011, ARN&PS, 61, 251
Goodrich, R. W., Stringfellow, G. S., Penrod, G. D., & Filippenko, A. V. 1989, ApJ, 342, 908
Gorbikov, E., Gal-Yam, A., Ofek, E. O., et al. 2014, MNRAS, 443, 671
Goriely, S., Bauswein, A., Just, O., Pllumbi, E., & Janka, H.-T. 2015, MNRAS, 452, 3894
Gorosabel, J., Larionov, V., Castro-Tirado, A. J., et al. 2006, A&A, 459, L33
Gotthelf, E. V., Halpern, J. P., & Alford, J. 2013a, ApJ, 765, 58
Gotthelf, E. V., Halpern, J. P., Allen, B., & Knispel, B. 2013b, ApJ, 773, 141
Gotthelf, E. V., Helfand, D. J., & Newburgh, L. 2007, ApJ, 654, 267
Gou, L., McClintock, J. E., Reid, M. J., et al. 2011, ApJ, 742, 85
Gould, R. J. 1972, Physics, 60, 145
Gounelle, M., & Meynet, G. 2012, A&A, 545, A4
Gowanlock, M. G., Patton, D. R., & McConnell, S. M. 2011, Astrobiology, 11, 855
Goyal, A., Dutta, S., Choudhury, S. R. 1995, Physics Letters B, 346, 312–316
Graham, J. R., Meikle, W. P. S., Allen, D. A., et al. 1986, MNRAS, 218, 93
Graham, M. L., Nugent, P. E., Sullivan, M., et al. 2015a, MNRAS, 454, 1948
Graham, M. L., Sand, D. J., Valenti, S., et al. 2014, ApJ, 787, 163
Graham, M. L., Sand, D. J., Zaritsky, D., & Pritchet, C. J. 2015b, ApJ, 807, 83
Graham, M. L., Valenti, S., Fulton, B. J., et al. 2015c, ApJ, 801, 136
Graham, P. W., Rajendran, S., & Varela, J. 2015d, Physical Review D, 92, 063007
Graur, O., Bianco, F. B., & Modjaz, M. 2015, MNRAS, 450, 905
Graur, O., Maoz, D., & Shara, M. M. 2014a, MNRAS, 442, L28
Graur, O., Rodney, S. A., Maoz, D., et al. 2014b, ApJ, 783, 28
Graur, O., Zurek, D., Shara, M. M., et al. 2016, ApJ, 819, 31
Graves, G. J. M., Challis, P. M., Chevalier, R. A., et al. 2005, ApJ, 629, 944
Green, D. A. 2014a, BASI, 42, 47
Green, D. A. 2014b, in Supernova Environmental Impacts, ed. A. Ray & R. A. McCray (Cambridge Un. Press) 188
Grefenstette, B. W., Harrison, F. A., Boggs, S. E., et al. 2014, Nature, 506, 339
Greif, T. H., Springel, V., White, S. D. M., et al. 2011, ApJ, 737, 75
Grieco, V., Matteucci, F., Ryde, N., et al. 2015, MNRAS, 450, 2094
Gritschneder, M., Lin, D. N. C., Murray, S. D., et al. 2012, ApJ, 745, 22

Groenewegen, M. A. T., Sloan, G. C., Soszyński, I., & Petersen, E. A. 2009, A&A, 506, 1277
Grogin, N. A., Kocevski, D. D., Faber, S. M., et al. 2011, ApJS, 197, 35
Groh, J. H. 2014, A&A, 572, L11
Groh, J. H., Meynet, G., & Ekström, S. 2013, A&A, 550, L7
Guilet, J., Sato, J., & Foglizzo, T. 2010, ApJ, 713, 1350
Guillochon, J., Dan, M., Ramirez-Ruiz, E., & Rosswog, S. 2010, ApJL, 709, L64
Guillochon, J., Parrent, J., Kelley, L. Z., & Margutti, R. 2017, ApJ, 835, 64
Guillot, S., Servillat, M., Webb, N. A., & Rutledge, R. E. 2013, ApJ, 772, 7
Gullikson, K., Kraus, A., & Dodson-Robinson, S. 2016, AJ, 152, 40
Gunn, J. E., & Ostriker, J. P. 1969, Nature, 221, 454
Gutiérrez, C. P., Anderson, J. P., Hamuy, M., et al. 2014, ApJL, 786, L15
Guzik, J. A., & Lovekin, C. C. 2012, The Astronomical Review, 7, 030000
Haas, M. R., Erickson, E. F., Lord, Steven D., et al. 1990, ApJ, 360, 257
Habergham, S. M., Anderson, J. P., James, P. A., & Lyman, J. D. 2014, MNRAS, 441, 2230
Hachinger, S., Mazzali, P. A., Sullivan, M., et al. 2013, MNRAS, 429, 2228
Hachinger, S., Mazzali, P. A., Taubenberger, S., et al. 2012a, MNRAS, 422, 70
Hachinger, S., Mazzali, P. A., Taubenberger, S., et al. 2012b, MNRAS, 427, 2057
Hachisu, I., & Kato, M. 2003, ApJ, 598, 527
Hachisu, I., Kato, M., & Luna, G. J. M. 2007, ApJL, 659, L153
Hachisu, I., Kato, M., & Nomoto, K. 1999, ApJ, 522, 487
Hachisu, I., Kato, M., & Nomoto, K. 2008, ApJ, 679, 1390
Hachisu, I., Kato, M., & Nomoto, K. 2010, ApJL, 724, L212
Hachisu, I., Kato, M., & Nomoto, K. 2012, ApJL, 756, L4
Hakobyan, A. A., Adibekyan, V. Z., Aramyan, L. S., et al. 2012, A&A, 544, A81
Halzen, F. 2014, AN, 335
Hamers, A. S., Pols, O. R., Claeys, J. S. W., & Nelemans, G. 2013, MNRAS, 430, 2262
Hamilton, A. J. S., Fesen, R. A., & Blair, W. P. 2007, MNRAS, 381, 771
Hamilton, A. J. S., Fesen, R. A., Wu, C.-C., et al. 1997, ApJ, 481, 838
Hamlington, P. E., Poludnenko, A. Y., & Oran, E. S. 2011, Physics of Fluids, 23, 125111
Hamlington, P. E., Poludnenko, A. Y., & Oran, E. S. 2012, Physics of Fluids, 24, 075111
Hammer, N. J., Janka, H.-T., Müller, E. 2010, ApJ, 714, 1371
Hamuy, M. 2003, ApJ, 582, 905
Hamuy, M., Folatelli, G., Morrell, N. I., et al. 2006, PASP, 118, 2
Hamuy, M., Maza, J., Phillips, M. M., et al. 1993, AJ, 106, 2392
Hamuy, M., Maza, J., Pinto, P. A., et al. 2002, AJ, 124, 417
Hamuy, M., Phillips, M. M., Suntzeff, N. B., et al. 1996a, AJ, 112, 2391
Hamuy, M., Phillips, M. M., Suntzeff, N. B., et al. 1996b, AJ, 112, 2408
Hamuy, M., Phillips, M. M., Suntzeff, N. B., et al. 2003, Nature, 424, 651
Hamuy, M., & Pinto, P. A. 2002, ApJL, 566, L63
Hamuy, M., Pinto, P. A., Maza, J., et al. 2001, ApJ, 558, 615
Hamuy, M., & Suntzeff, N. B. 2015, in Fifty Years of Wide-field Studies in the Southern Hemisphere, ASP Conf. Series, 491, 203
Hamuy, M., Suntzeff, N. B., Gonzalez, R., & Martin, G. 1988, AJ, 95, 63
Han, Z., & Podsiadlowski, P. 2006, MNRAS, 368, 1095
Hancock, P. J., Gaensler, B. M., & Murphy, T. 2011, ApJL, 735, L35
Hanke, F., Müller, B., Wongwathanarat, A., et al. 2013, ApJ, 770, 66
Hannestad, S., Haugbølle, T., & Thomsen, B. 2008, Journal of Cosmology and Astroparticle Physics, 2, 22
Hansen, B. M. S. 2003, ApJ, 582, 915
Hansen, C. J., Kawaler, S. D., & Trimble, V. 2004, Stellar Interiors (Springer)
Hansen, C. J., & Wheeler, J. C., 1969, Ap&SS, 3, 464
Hanuschik, R. W., & Dachs, J. 1988, AAP, 205, 135
Hanuschik, R. W., & Thimm, G. J. 1990, A&A, 231, 77
Harkness, R. P., & Wheeler, J. C. 1990, in Supernovae, ed. A. G. Petschek (Springer) 1

Harkness, R. P., Wheeler, J. C., Margon, B., et al. 1987, ApJ, 317, 355
Harris, C. E., Nugent, P. E., & Kasen, D. N. 2016, ApJ, 823, 100
Hashimoto, M., Nomoto, K., & Shigeyama, T. 1989, ApJ, 210, L5
Hasinger, G., Aschenbach, B., Truemper, J., et al. 1996, A&A, 312, L9
Hatano, K., Branch, D., Fisher, A., et al. 1999a, ApJ, 525, 881
Hatano, K., Branch, D., Fisher, A., et al. 1999b, ApJS, 121, 233
Hauschildt, P. H., & Baron, E. 1999, Journal of Computational and Applied Mathematics, 109, 41
Hayashi, C. 1961, Publications of the ASJ, 13, 450
Hayato, A., Yamaguchi, H., Tamagawa, T., et al. 2010, ApJ, 725, 894
Hayden, B. T., Garnavich, P. M., Kasen, D., et al. 2010a, ApJ, 722, 1691
Hayden, B. T., Garnavich, P. M., Kessler, R., et al. 2010b, ApJ, 712, 350
He, L. & Lee, H. S. 1995, Physics of Fluids, 7, 1151
Hebeler, K., Lattimer, J. M., Pethick, C. J., & Schwenk, A. 2013, ApJ, 773, 11
Heger, A., Fryer, C. L., Woosley, S. E., et al. 2003, ApJ, 591, 288
Heger, A., Langer, N., & Woosley, S. E. 2000, ApJ, 528, 368
Heger, A., & Woosley, S. E. 2002, ApJ, 567, 532
Heger, A., & Woosley, S. E. 2010, ApJ, 724, 341
Heger, A., Woosley, S. E., & Spruit, H. C. 2005, ApJ, 626, 350
Heinke, C. O., Cohn, H. N., Lugger, P. M., et al. 2014, MNRAS, 444, 443
Heinke, C. O., & Ho, W. C. G. 2010, ApJL, 719, L167
Helder, E. A., Broos, P. S., Dewey, D., et al. 2013a, ApJ, 764, 11
Helder, E. A., Vink, J., Bamba, A., et al. 2013b, MNRAS, 435, 910
Helder, E. A., Vink, J., Bassa, C. G., et al. 2009, Science, 325, 719
Helder, E. A., Vink, J., Bykov, A. M., et al. 2012, Space Science Reviews, 173, 369
Helfand, D. J., Velusamy, T., Becker, R. H., & Lockman, F. J. 1989, ApJ, 341, 151
Heng, K. 2010, Publications of the Astron. Soc. of Australia, 27, 23
Heng, K., Haberl, F., Aschenbach, B., & Hasinger, G. 2008, ApJ, 676, 361
Heng, K., McCray, R., Zhekov, S. A., et al. 2006, ApJ, 644, 959
Henry, R. B. C., & Branch, D. 1987, PASP, 99, 112
Henze, M., Ness, J.-U., Darnley, M. J., et al. 2015, A&A, 580, A46
Herant, M., Benz, W., Hix, W. R., et al. 1994, ApJ, 435, 339
Hernandez, M., Meikle, W. P. S., Aparicio, A., et al. 2000, MNRAS, 319, 223
Herrero-Illana, R., Pérez-Torres, M. Á., & Alberdi, A. 2012, A&A, 540, L5
Herwig, F. 2005, ARAA, 43, 435
Hester, J. J. 2008, ARAA, 46, 127
Hicken, M., Garnavich, P. M., Prieto, J. L., et al. 2007, ApJL, 669, L17
Hicken, M., Wood-Vasey, W. M., Blondin, S., et al. 2009, ApJ, 700, 1097
Higdon, J. C., & Lingenfelter, R. E. 2005, ApJ, 628, 738
Hillebrandt, W., Hoflich, P., 1989, Progress Reports in Physics, 52, 11
Hillebrandt, W., Kromer, M., Röpke, F. K., & Ruiter, A. J. 2013, Frontiers Phys., 8, 116
Hillebrandt, W., & Niemeyer, J. C. 2000, ARAA, 38, 191
Hillebrandt, W., Nomoto, K., & Wolff, R. G. 1984, A&A, 133, 175
Hillier, D. J. 1991, A&A, 247, 455
Hillier, D. J. 2011, ApSpSci, 336, 87
Hillier, D. J., & Dessart, L. 2012, MNRAS, 424, 252
Hillman, Y., Prialnik, D., Kovetz, A., & Shara, M. M. 2015, MNRAS, 446, 1924
Hinkley, S., Kraus, A. L., Ireland, M. J., et al. 2015, ApJL, 806, L9
Hirai, R., Sawai, H., & Yamada, S. 2014, ApJ, 792, 66
Hirano, S., Hosokawa, T., Yoshida, N., et al. 2014, ApJ, 781, 60
Hirata, K., Kajita, T., Koshiba, M., et al. 1987, PRL, 58, 1490
Hirschi, R., Meynet, G., & Maeder, A. 2004, A&A, 425, 649
Ho, W. C. G. 2011, MNRAS, 414, 2567
Hobbs, A., Nayakshin, S., Power, C., & King, A. 2011, MNRAS, 413, 2633
Hobbs, G., Lorimer, D. R., Lyne, A. G., & Kramer, M. 2005, MNRAS, 360, 974

Höflich, P., Dragolin, P. Mitchell, et al. 2013, Frontiers of Physics, 8, 144
Hoffman, J. L., Leonard, D. C., Chornock, R., et al. 2008, ApJ, 688, 1186
Höflich, P. 1991, A&A, 246, 481
Höflich, P. 2005, ApSpSci, 298, 87
Höflich, P. 2009, in Recent Directions in Astrophysical Quantitative Spectroscopy and Radiation Hydrodynamics, AIP Conf. Series, 1171, 161
Höflich, P., Gerardy, C. L., Fesen, R. A., & Sakai, S. 2002, ApJ, 568, 791
Höflich, P., Gerardy, C. L., Marion, H., & Quimby, R. 2006, New Astron Rev, 50, 470
Höflich, P., Gerardy, C. L., Nomoto, K., et al. 2004, ApJ, 617, 1258
Höflich, P., & Khokhlov, A. 1996, ApJ, 457, 500
Höflich, P., Khokhlov, A. M., & Müller, E. in Thermonuclear Supernovae, ed. P. Ruiz-Lapuente, R. Canal, & J. Isern, 1997 (NATO ASIC) 681
Höflich, P., Khokhlov, A. M., & Wheeler, J. C. 1995, ApJ, 444, 831
Höflich, P., Krisciunas, K., Khokhlov, A. M., et al. 2010, ApJ, 710, 444
Hoflich, P., Langer, N., & Duschinger, M. 1993, A&A, 275, L29
Höflich, P., & Schaefer, B. E. 2009, ApJ, 705, 483
Höflich, P., & Stein, J. 2002, ApJ, 568, 779
Höflich, P., Wheeler, J. C., Hines, D. C., & Trammell, S. R. 1996, ApJ, 459, 307
Höflich, P., Wheeler, J. C., & Wang, L. 1999, ApJ, 521, 179
Holanda, R. F. L., Jesus, J. F., & Dantas, M. A. 2016, arXiv:1605.01342
Hole, K. T., Kasen, D., & Nordsieck, K. H. 2010, ApJ, 720, 1500
Holoien, T. W.-S., Prieto, J. L., Pejcha, O., et al. 2016, ACTAA, 66, 219
Honda, S., Aoki, W., Ishimaru, Y., et al. 2006, ApJ, 643, 1180
Hopkins, A. M., & Beacom, J. F. 2006, ApJ, 651, 142
Horesh, A., Kulkarni, S. R., Fox, D. B., et al. 2012, ApJ, 746, 21
Horiuchi, S., & Beacom, J. F. 2010, ApJ, 723, 329
Horiuchi, S., Beacom, J. F., Kochanek, C. S., et al. 2011, ApJ, 738, 154
Horiuchi, S., Nakamura, K., Takiwaki, T., et al. 2014, MNRAS, 445, L99
Houck, J. C., & Fransson, C. 1996, ApJ, 456, 811
Howe, R. 2009, LRSP, 6, 1
Howell, D. A. 2001, ApJL, 554, L193
Howell, D. A. 2011, Nature Communications, 2, 350
Howell, D. A., Höflich, P., Wang, L., & Wheeler, J. C. 2001, ApJ, 556, 302
Howell, D. A., Kasen, D., Lidman, C., et al. 2013, ApJ, 779, 98
Howell, D. A., Sullivan, M., Brown, E. F., et al. 2009, ApJ, 691, 661
Howell, D. A., Sullivan, M., Conley, A., & Carlberg, R. 2007, ApJL, 667, L37
Howell, D. A., Sullivan, M., Nugent, P. E., et al. 2006, Nature, 443, 308
Hoyle, F. 1946, MNRAS, 106, 343
Hoyle, F., & Fowler, W. A. 1960, ApJ, 132, 565
Hsiao, E. Y., Burns, C. R., Contreras, C., et al. 2015, A&A, 578, A9
Hsiao, E. Y., Marion, G. H., Phillips, M. M., et al. 2013, ApJ, 766, 72
Hughes, J. P., Rakowski, C. E., Burrows, D. N., & Slane, P. O. 2000, ApJL, 528, L109
Hummel, J. A., Pawlik, A. H., Milosavljević, M., & Bromm, V. 2012, ApJ, 755, 72
Hummer, D. G. 1968, MNRAS, 141, 479
Humphreys, R. M., & Davidson, K. 1979, ApJ, 232, 409
Humphreys, R. M., & Davidson, K. 1994, PASP, 106, 1025
Humphreys, R. M., Davidson, K., Jones, T. J., et al. 2012, ApJ, 760, 93
Humphreys, R. M., & McElroy, D. B. 1984, ApJ, 284, 565
Hungerford, A. L., Fryer, C. L., & Rockefeller, G. 2005, ApJ, 635, 487
Hunter, D. A., Shaya, E. J., Holtzman, J. A., et al. 1995, ApJ, 448, 179
Hunter, D. J., Valenti, S., Kotak, R., et al. 2009, A&A, 508, 371
Hunter, S. D., Bertsch, D. L., Catelli, J. R., et al. 1997, ApJ, 481, 205
Hurley, J. R., Tout, C. A., & Pols, O. R. 2002, MNRAS, 329, 897
Hutton, S., Ferreras, I., & Yershov, V. 2015, MNRAS, 452, 1412

Hwang, U., & Laming, J. M. 2012, ApJ, 746, 130
Hwang, U., Laming, J. M., Badenes, C., et al. 2004, ApJL, 615, L117
Iben, I., Jr. 1975, ApJ, 196, 525
Iben, I., Jr. 1978, ApJ, 219, 213
Iben, I., Jr. 1982, ApJ, 253, 248
Iben, I., Jr. 2013, Stellar Evolution Physics (Cambridge Un. Press)
Iben, I., Jr., & Renzini, A. 1983, ARAA, 21, 271
Iben, I., Jr., & Tutukov, A. V. 1984, ApJS, 54, 335
Iben, I., Jr., & Tutukov, A. V. 1985, ApJS, 58, 661
Idan, I., Shaviv, N. J., & Shaviv, G. 2013, MNRAS, 433, 2884
Iffrig, O., & Hennebelle, P. 2015, A&A, 576, A95
Ilkov, M., & Soker, N. 2012, MNRAS, 419, 1695
Ilkov, M., & Soker, N. 2013, MNRAS, 428, 579
Immler, S., Brown, P. J., Milne, P., et al. 2006, ApJL, 648, L119
Immler, S., Brown, P. J., Milne, P., et al. 2007, ApJ, 664, 435
Immler, S., Modjaz, M., Landsman, W., et al. 2008, ApJL, 674, L85
Immler, S., Pietsch, W., & Aschenbach, B. 1998, A&A, 336, L1
Immler, S., Wilson, A. S., & Terashima, Y. 2002, ApJL, 573, L27
Immler, S., Milne, P., & Pooley, D. 2010, The Astronomer's Telegram, 3012
Imshennik, V. S. Nadezhin, D. K., 1989, SSR, 8, 1
Imshennik, V. S., & Ryazhskaya, O. G. 2004, Astronomy Letters, 30, 14
Indebetouw, R., Matsuura, M., Dwek, E., et al. 2014, ApJL, 782, L2
Inoue, H., Hayashida, K., Itoh, M., et al. 1991, ApJ, 43, 213
Inserra, C., Baron, E., & Turatto, M. 2012a, MNRAS, 422, 1178
Inserra, C., Sim, S. A., Wyrzykowski, L., et al. 2015, ApJL, 799, LL2
Inserra, C., & Smartt, S. J. 2014, ApJ, 796, 87
Inserra, C., Smartt, S. J., Jerkstrand, A., et al. 2013, ApJ, 770, 128
Inserra, C., Smartt, S. J., Scalzo, R., et al. 2014, MNRAS, 437, L51
Inserra, C., Turatto, M., Pastorello, A., et al. 2011, MNRAS, 417, 261
Inserra, C., Turatto, M., Pastorello, A., et al. 2012b, MNRAS, 422, 1122
Isensee, K., Olmschenk, G., Rudnick, L., et al. 2012, ApJ, 757, 126
Isern, J., Jean, P., Bravo, E., et al. 2016, A&A, 588, A67
Ishimaru, Y., Wanajo, S., & Prantzos, N. 2015, ApJL, 804, L35
Ivanova, N., Belczynski, K., Kalogera, V., et al. 2003, ApJ, 592, 475
Ivanova, N., Justham, S., Chen, X., et al. 2013, The Astronomy and Astrophysics Review, 21, 59
Iwakami, W., Kotake, K., Ohnishi, N., et al. 2009, ApJ, 700, 232
Iwamoto, K., Brachwitz, F., Nomoto, K., et al. 1999, ApJS, 125, 439
Iwamoto, K., Nakamura, T., Nomoto, K., et al. 2000, ApJ, 534, 660
Iwamoto, K., Nomoto, K., Hoflich, P., et al. 1994, ApJL, 437, L115
Izzard, R. G., Ramirez-Ruiz, E., & Tout, C. A. 2004, MNRAS, 348, 1215
Jack, D., Baron, E., & Hauschildt, P. H. 2015a, MNRAS, 449, 3581
Jack, D., Hauschildt, P. H., & Baron, E. 2012a, A&A, 538, AA132
Jack, D., Hauschildt, P. H., & Baron, E. 2012b, A&A, 546, A39
Jack, D., Mittag, M., Schröder, K.-P., et al. 2015b, MNRAS, 451, 4104
Jacobson, H. R., & Frebel, A. 2014, Journal of Physics G Nuclear Physics, 41, 044001
Jain, B., & Khoury, J. 2010, Annals of Physics, 325, 1479
Jakobsen, P., Albrecht, R., Barbieri, C., et al. 1991, ApJL, 369, L63
Jakobsen, P., Macchetto, F., & Panagia, N. 1993, ApJ, 403, 2
James, S., & Baron, E. 2010, ApJ, 718, 957
Janka, H.-T. 1992, A&A, 256, 452
Janka, H.-T. 2012, AR&PS, 62, 407
Janka, H.-T. 2013, MNRAS, 434, 1355
Jeffery, D. J. 1987, Nature, 329, 419
Jeffery, D. J. 1991, ApJ, 375, 264

Jeffery, D. J., & Branch, D. 1990, in Supernovae, ed. J. C. Wheeler, T. Piran, & S. Weinberg (World Scientific), 149
Jeffery, D. J., Kirshner, R. P., Challis, P. M., et al. 1994, ApJL, 421, L27
Jencson, J. E., Prieto, J. L., Kochanek, C. S., et al. 2016, MNRAS, 456, 2622
Jennings, Z. G., Williams, B. F., Murphy, J. W., et al. 2014, ApJ, 795, 170
Jerkstrand, A., Ergon, M., Smartt, S. J., et al. 2015a, A&A, 573, A12
Jerkstrand, A., Fransson, C., & Kozma, C. 2011, A&A, 530 A45
Jerkstrand, A., Fransson, C., Maguire, K., et al. 2012, A&A, 546, A28
Jerkstrand, A., Smartt, S. J., Fraser, M., et al. 2014, MNRAS, 439, 3694
Jerkstrand, A., Smartt, S. J., Sollerman, J., et al. 2015b, MNRAS, 448, 2482
Jerkstrand, A., Timmes, F. X., Magkotsios, G., et al. 2015c, ApJ, 807, 110
Jha, S., Branch, D., Chornock, R., et al. 2006, AJ, 132, 189
Ji, S., Fisher, R. T., García-Berro, E., et al. 2013, ApJ, 773, 136
Joggerst, C. C., & Whalen, D. J. 2011, ApJ, 728, 129
Johansson, J., Thomas, D., Pforr, J., et al. 2013, MNRAS, 435, 1680
Johansson, J., Goobar, A., Kasliwal, M. M., et al. 2017, MNRAS, 466, 3442
Johansson, J., Woods, T. E., Gilfanov, M., et al. 2014, MNRAS, 442, 1079
Johnson, J. L., & Khochfar, S. 2011, ApJ, 743, 126
Johnston, M. D., & Yahil, A. 1984, ApJ, 285, 587
Johnston, S., Kramer, M., Karastergiou, A., et al. 2007, MNRAS, 381, 1625
Jones, D. O., Riess, A. G., & Scolnic, D. M. 2015, ApJ, 812, 31
Jones, S., Hirschi, R., Nomoto, K., et al. 2013, ApJ, 772, 150
Jordan, G. C., IV, Fisher, R. T., Townsley, D. M., et al. 2008, ApJ, 681, 1448
Jordan, G. C., IV, Graziani, C., Fisher, R. T., et al. 2012a, ApJ, 759, 53
Jordan, G. C., IV, Perets, H. B., Fisher, R. T., & van Rossum, D. R. 2012b, ApJL, 761, L23
Jose, J. 2016, Stellar Explosions: Hydrodynamics and Nucleosynthesis (CRC Press)
José, J., Hernanz, M., & Isern, J. 1993, A&A, 269, 291
José, J., & Iliadis, C. 2011, Reports on Progress in Physics, 74, 096901
Joss, P. C., Salpeter, E. E., & Ostriker, J. P. 1973, ApJ, 181, 429
Jura, M., Xu, S., & Young, E. D. 2013, ApJL, 775, L41
Justham, S. 2011, ApJL, 730, L34
Justham, S., Wolf, C., Podsiadlowski, P., & Han, Z. 2009, A&A, 493, 1081
Kachelrieß, M., Neronov, A., & Semikoz, D. V. 2015, PRL, 115, 181103
Kafka, S., Honeycutt, R. K., & Williams, R. 2012, MNRAS, 425, 1585 34
Kaiser, N., Aussel, H., Burke, B. E., et al. 2002, Proceedings of the SPIE, 4836, 154
Kaiser, N., Burgett, W., Chambers, K., et al. 2010, Proceedings of the SPIE, 7733,
Kalirai, J. S., Hansen, B. M. S., Kelson, D. D., et al. 2008, ApJ, 676, 594
Kalirai, J. S., Marigo, P., & Tremblay, P.-E. 2014, ApJ, 782, 17
Kamble, A., Margutti, R., Soderberg, A. M., et al. 2016, ApJ, 818, 111
Kamenetzky, J., McCray, R., Indebetouw, R., et al. 2013, ApJL, 773, L34
Kangas, T., Mattila, S., Kankare, E., et al. 2013, MNRAS, 436, 3464
Kankare, E., Ergon, M., Bufano, F., et al. 2012, MNRAS, 424, 855
Kankare, E., Kotak, R., Pastorello, A., et al. 2015, A&A, 581, L4
Kankare, E., Mattila, S., Ryder, S., et al. 2014, MNRAS, 440, 1052
Karakas, A. I., García-Hernández, D. A., & Lugaro, M. 2012, ApJ, 751, 8
Karlsson, T., Bromm, V., & Bland-Hawthorn, J. 2013, RMP, 85, 809
Kasen, D. 2006, ApJ, 649, 939
Kasen, D. 2010, ApJ, 708, 1025
Kasen, D., & Bildsten, L. 2010, ApJ, 717, 245
Kasen, D., & Nugent, P. 2013, ARN&PS, 63, 153
Kasen, D., Nugent, P., Wang, L., et al. 2003, ApJ, 593, 788
Kasen, D., & Plewa, T. 2005, ApJL, 622, L41
Kasen, D., & Plewa, T. 2007, ApJ, 662, 459
Kasen, D., Röpke, F. K., & Woosley, S. E. 2009, Nature, 460, 869

Kasen, D., Thomas, R. C., & Nugent, P. 2006, ApJ, 651, 366
Kasen, D., & Woosley, S. E. 2007, ApJ, 656, 661
Kasen, D., & Woosley, S. E. 2009, ApJ, 703, 2205
Kasen, D., Woosley, S. E., & Heger, A. 2011, ApJ, 734, 102
Kashi, A., & Soker, N. 2011, MNRAS, 417, 1466
Kashyap, R., Fisher, R., García-Berro, E., et al. 2015, ApJL, 800, LL7
Kasliwal, M. M., Kulkarni, S. R., Gal-Yam, A., et al. 2010, ApJL, 723, L98
Kasliwal, M. M., Kulkarni, S. R., Gal-Yam, A., et al. 2012, ApJ, 755, 161
Kataoka, R., Ebisuzaki, T., Miyahara, H., & Maruyama, S. 2013, Proceedings of the National Academy, 21, 50
Kato, M. 2013, IAU Symposium, 281, 145
Kato, C., Delfan Azari, M., Yamada, S., et al. 2015, ApJ, 808, 168
Kato, M., & Hachisu, I. 2004, ApJL, 613, L129
Kato, M., & Hachisu, I. 2012, BASI, 40, 393
Kato, M., Saio, H., Hachisu, I., & Nomoto, K. 2014, ApJ, 793, 136
Katsuda, S., Maeda, K., Nozawa, T., et al. 2014, ApJ, 780, 184
Katsuda, S., Mori, K., Maeda, K., et al. 2015, ApJ, 808, 49
Katsuda, S., Petre, R., Hughes, J. P., et al. 2010, ApJ, 709, 1387
Katsuda, S., Petre, R., Long, K. S., et al. 2009, ApJL, 692, L105
Kattner, S., Leonard, D. C., Burns, C. R., et al. 2012, PASP, 124, 114
Katz, B., Sapir, N., & Waxman, E. 2012, ApJ, 747, 147
Kawabata, K. S., Deng, J., Wang, L., et al. 2003, ApJL, 593, L19
Kawabata, K. S., Akitaya, H., Yamanaka, M., et al. 2014, ApJL, 795, LL4
Kawabata, K. S., Jeffery, D. J., Iye, M., et al. 2002, ApJL, 580, L39
Kawaler, S. D. 2004, Stellar Rotation, IAU Symp. 215, 561
Kazanas, D., Mohapatra, R. N., Nussinov, S., et al. 2015, Nuclear Physics B, 890, 17
Keller, S. C., Schmidt, B. P., Bessell, M. S., et al. 2007, Publications of the Astron. Soc. of Australia, 24, 1
Kelly, P. L., Filippenko, A. V., Modjaz, M., & Kocevski, D. 2014a, ApJ, 789, 23
Kelly, P. L., Fox, O. D., Filippenko, A. V., et al. 2014b, ApJ, 790, 3
Kelly, P. L., Hicken, M., Burke, D. L., et al. 2010, ApJ, 715, 743
Kelly, P. L., & Kirshner, R. P. 2012, ApJ, 759, 107
Kelly, P. L., Rodney, S. A., Treu, T., et al. 2015, Science, 347, 1123
Kelly, P. L., Rodney, S. A., Treu, T., et al. 2016, ApJL, 819, L8
Kepler, S. O., Kleinman, S. J., Nitta, A., et al. 2007, MNRAS, 375, 1315
Kerzendorf, W. E., Yong, D., Schmidt, B. P., et al. 2013, ApJ, 774, 99
Kerzendorf, W. E., Childress, M., Scharwächter, J., et al. 2014a, ApJ, 782, 27
Kerzendorf, W. E., Schmidt, B. P., Laird, J. B., et al. 2012, ApJ, 759, 7
Kerzendorf, W. E., & Sim, S. A. 2014, MNRAS, 440, 387
Kerzendorf, W. E., Taubenberger, S., Seitenzahl, I. R., & Ruiter, A. J. 2014b, ApJL, 796, LL26
Kessler, R., Becker, A. C., Cinabro, D., et al. 2009, ApJS, 185, 32
Keung, W.-Y., Ng, K.-W., Tu, H., & Yuan, T.-C. 2014, Physical Review D, 90, 075014
Khan, R., Stanek, K. Z., Stoll, R., & Prieto, J. L. 2011, ApJL, 737, L24
Khokhlov, A. M. 1991a, A&A, 245, 114
Khokhlov, A. M. 1991b, A&A, 245, L25
Khokhlov, A. M. 1993, ApJ, 419, 200
Khokhlov, A. M. 1995, ApJ, 449, 695
Khokhlov, A. M. 2000, arXiv:astro-ph/0008463
Khokhlov, A., Müller, E., & Höflich, P. 1993, A&A, 270, 223
Khokhlov, A. M., Oran, E. S., & Wheeler, J. C. 1997, ApJ, 478, 678
Kiewe, M., Gal-Yam, A., Arcavi, I., et al. 2012, ApJ, 744, 10
Kifonidis, K., Plewa, T., Scheck, L., et al. 2006, A&A, 453, 661
Kilic, M., Hermes, J. J., Gianninas, A., et al. 2014, MNRAS, 438, L26
Kim, A., Bonissent, A., Christiansen, J. L., et al. 2010, Astroparticle Physics, 33, 248

Kim, A. G., Padmanabhan, N., Aldering, G., et al. 2015a, Astroparticle Physics, 63, 2
Kim, C.-G., & Ostriker, E. C. 2015, ApJ, 815, 67
Kim, H.-J., Yoon, S.-C., & Koo, B.-C. 2015b, ApJ, 809, 131
Kimm, T., & Cen, R. 2014, ApJ, 788, 121
King, A. R., Rolfe, D. J., & Schenker, K. 2003, MNRAS, 341, L35
Kippenhahn, R., Thomas, H. C., & Weigert, A. 1965, Zeitschrift fuer Astrophysik, 61, 241
Kippenhahn, R., & Weigert, A. 1994, Stellar Structure and Evolution (Springer)
Kippenhahn, R., Weigert, A., & Weiss, A. 2012, Stellar Structure and Evolution (Springer-Verlag)
Kirshner, R. P., & Kwan, J. 1974, ApJ, 193, 27
Kirshner, R. P., & Oke, J. B. 1975, ApJ, 200, 574
Kirshner, R. P., Sonneborn, G., Crenshaw, D. M., Nassiopoulos, G. E. 1987, AJ, 320, 602
Kistler, M. D., Yüksel, H., Ando, S., et al. 2011, Physical Review D, 83, 123008
Kitaura, F. S., Janka, H.-T., & Hillebrandt, W. 2006, A&A, 450, 345
Kiziltan, B., Kottas, A., De Yoreo, M., & Thorsett, S. E. 2013, ApJ, 778, 66
Kjær, K., Leibundgut, B., Fransson, C., et al. 2010, A&A, 517, A51
Klein, R. I., & Chevalier, R. A. 1978, ApJL, 223, L109
Klein, R. I., McKee, C. F., & Colella, P. 1994, ApJ, 420, 213
Kleiser, I. K. W., & Kasen, D. 2014, MNRAS, 438, 318
Kluźniak, W., & Ruderman, M. 1998, ApJL, 505, L113
Knie, K., Korschinek, G., Faestermann, T., et al. 1999, PRL, 83, 18
Knie, K., Korschinek, G., Faestermann, T., et al. 2004, PRL, 93, 171103
Kobayashi, C., & Nakasato, N. 2011, ApJ, 729, 16
Kobayashi, C., Nomoto, K., & Hachisu, I. 2015, ApJL, 804, L24
Kobayashi, C., Umeda, H., Nomoto, K., et al. 2006, ApJ, 653, 1145
Kochanek, C. S. 2009, ApJ, 707, 1578
Kochanek, C. S. 2014, ApJ, 785, 28
Kochanek, C. S. 2015, MNRAS, 446, 1213
Kochanek, C. S., Beacom, J. F., Kistler, M. D., et al. 2008, ApJ, 684, 1336
Kochanek, C. S., Khan, R., & Dai, X. 2012a, ApJ, 759, 20
Kochanek, C. S., Szczygiel, D. M., & Stanek, K. Z. 2011, ApJ, 737, 76
Kochanek, C. S., Szczygieł, D. M., & Stanek, K. Z. 2012b, ApJ, 758, 142
Koekemoer, A. M., Faber, S. M., Ferguson, H. C., et al. 2011, ApJS, 197, 36
Kogut, A., Lineweaver, C., Smoot, G. F., et al. 1993, ApJ, 419, 1
Kohmura, Y., Inoue, H., Aoki, T., et al. 1994, PASJ, 46, L157
Komiya, Y., Habe, A., Suda, T., & Fujimoto, M. Y. 2010, ApJ, 717, 542
Korolev, V. V., Vasiliev, E. O., Kovalenko, I. G., & Shchekinov, Y. A. 2015, Astronomy Reports, 59, 690
Kotak, R., Meikle, W. P. S., Farrah, D., et al. 2009, ApJ, 704, 306
Kotake, K., Sato, K., & Takahashi, K. 2006, Reports on Progress in Physics, 69, 971
Kotake, K., Takiwaki, T., & Harikae, S. 2012, ApJ, 755, 84
Kothes, R. 2010, in The Dynamic Interstellar Medium, ASP Conf. Series, 438, 347
Kozai, Y. 1962, AJ, 67, 591
Kozasa, T., Nozawa, T., Tominaga, N., et al. 2009, in Cosmic Dust - Near and Far, ASP Conf. Series, 414, 43
Kozma, C. & Fransson, C. 1998, ApJ, 496, 946
Kozma, C., & Fransson, C. 1992, ApJ, 390, 602
Kozma, C., Fransson, C., Hillebrandt, W., et al. 2005, A&A, 437, 983
Kozyreva, A., Blinnikov, S., Langer, N., & Yoon, S.-C. 2014a, A&A, 565, A70
Kozyreva, A., Yoon, S.-C., & Langer, N. 2014b, A&A, 566, A146
Kraiselburd, L., Landau, S. J., Negrelli, C., & García-Berro, E. 2015, ApSpSci, 357, 4
Kratz, K.-L., Farouqi, K., Pfeiffer, B., et al. 2007, ApJ, 662, 39
Krause, O., Birkmann, S. M., Usuda, T., et al. 2008a, Science, 320, 1195
Krause, O., Tanaka, M., Usuda, T., et al. 2008b, Nature, 456, 617
Krauss, M. I., Soderberg, A. M., Chomiuk, L., et al. 2012, ApJL, 750, L40

Kreidberg, L., Bailyn, C. D., Farr, W. M., & Kalogera, V. 2012, ApJ, 757, 36
Krisciunas, K., Li, W., Matheson, T., et al. 2011, AJ, 142, 74
Kromer, M., Fink, M., Stanishev, V., et al. 2013a, MNRAS, 429, 2287
Kromer, M., Ohlmann, S. T., Pakmor, R., et al. 2015, MNRAS, 450, 3045
Kromer, M., Pakmor, R., Taubenberger, S., et al. 2013b, ApJL, 778, L18
Kromer, M., & Sim, S. A. 2009, MNRAS, 398, 1809
Kromer, M., Sim, S. A., Fink, M., et al. 2010, ApJ, 719, 1067
Kroupa, P., Tout, C. A., & Gilmore, G. 1993, MNRAS, 262, 545
Kuhlen, M., Woosley, S. E., & Glatzmaier, G. A. 2006, ApJ, 640, 407
Kulkarni, G., Rollinde, E., Hennawi, J. F., & Vangioni, E. 2013, ApJ, 772, 93
Kulkarni, S. R. 2012, IAU Symposium 285, New Horizons in Time-domain Astronomy, 55
Kumagai, S., Shigeyama, T., Nomoto, K., et al. 1989, ApJ, 345, 412
Kumar, B., Pandey, S. B., Sahu, D. K., et al. 2013, MNRAS, 431, 308
Kumar, P., & Johnson, J. L. 2010, MNRAS, 404, 2170
Kuncarayakti, H., Doi, M., Aldering, G., et al. 2013a, AJ, 146, 31
Kuncarayakti, H., Doi, M., Aldering, G., et al. 2013b, AJ, 146, 30
Kuncarayakti, H., Maeda, K., Bersten, M. C., et al. 2015, A&A, 579, A95
Kundt, W. 1976, Nature, 261, 673
Kunkel, W., Madore, B., Shelton, I., et al. 1987, IAU Circ., 4316, 1
Kuroda, T., Kotake, K., & Takiwaki, T. 2012, ApJ, 755, 11
Kusakabe, M., Iwamoto, N., & Nomoto, K. 2011, ApJ, 726, 25
Kusenko, A., Mandal, B. P., & Mukherjee, A. 2008, Physical Review D, 77, 123009
Kushnir, D., & Katz, B. 2014, ApJ, 785, 124
Kushnir, D., Katz, B., Dong, S., et al. 2013, ApJL, 778, L37
Kutsuna, M., & Shigeyama, T. 2015, Publications of the ASJ, 67, 54
Käppeler, F., Gallino, R., Bisterzo, S., & Aoki, W. 2011, RMP, 83, 157
Lagos, C. d. P., Lacey, C. G., & Baugh, C. M. 2013, MNRAS, 436, 1787
Lai, K.-C., Lee, F.-F., Lee, F.-S., et al. 2016, Journal of Cosmology and Astroparticle Physics, 7, 039
Lakićević, M., van Loon, J. T., Meixner, M., et al. 2015, ApJ, 799, 50
Lamb, D. Q., Lattimer, J. M., Pethick, C. J., & Ravenhall, D. G. 1978, PRL, 41, 1623
Lamb, D. Q., Lattimer, J. M., Pethick, C. J., & Ravenhall, D. G. 1981, Nuclear Physics A, 360, 459
Landau, F. D. & Lifshitz, E. M. 1959 Fluid mechanics (Addison-Wesley)
Landau, L. D. 1944, Zh. Eksper. Teoret. Fiz., 14, 240
Langer, N. 2012, ARAA, 50, 107
Langer, N., Norman, C. A., de Koter, A., et al. 2007, A&A, 475, L19
Laor, A. 2006, ApJ, 643, 112
Larson, R. B. 1974, MNRAS, 169, 229
Larsson, J., Fransson, C., Kjaer, K., et al. 2013, ApJ, 768, 89
Larsson, J., Fransson, C., Östlin, G., et al. 2011, Nature, 747, 7352
Lattanzio, J. C. 1986, ApJ, 311, 708
Lattimer, J. M., & Prakash, M. 2000, Physics Reports, 333, 121
Lattimer, J. M., & Prakash, M. 2001, ApJ, 550, 426
Lattimer, J. M., & Prakash, M. 2016, Physics Reports, 621, 127
Lattimer, J. M., Schramm, D. N., & Grossman, L. 1977, Nature, 269, 116
Lattimer, J. M., & Steiner, A. W. 2014a, ApJ, 784, 123
Lattimer, J. M., & Steiner, A. W. 2014b, European Physical Journal A, 50, 40
Lattimer, J. M., & Swesty, F. 1991, Nuclear Physics A, 535, 331
Lau, H. H. B., Gil-Pons, P., Doherty, C., & Lattanzio, J. 2012, A&A, 542, A1
Laureijs, R., Amiaux, J., Arduini, S., et al. 2011, arXiv:1110.3193
Law, N. M., Kulkarni, S. R., Dekany, R. G., et al. 2009, PASP, 121, 1395
Lawrence, S. S., Sugerman, B. E., Bouchet, P. 2000, ApJ, 537, 2
Lazzati, D., Morsony, B. J., Blackwell, C. H., & Begelman, M. C. 2012, ApJ, 750, 68
Leaman, J., Li, W., Chornock, R., & Filippenko, A. V. 2011, MNRAS, 412, 1419

LeBlanc, J. M., & Wilson, J. R. 1970, ApJ, 161, 541
Lee, J.-J., Park, S., Hughes, J. P., & Slane, P. O. 2014, ApJ, 789, 7
Lee, J. H., & Lee, M. G. 2014a, ApJ, 786, 130
Lee, J. H., & Lee, M. G. 2014b, ApJ, 793, 134
Lee, T., Papanastassiou, D. A., & Wasserburg, G. J. 1976, Geophysics Research Letters, 3, 41
Leibundgut, B. 2001, ARAA, 39, 67
Leibundgut, B., Kirshner, R. P., Phillips, M. M., et al. 1993, AJ, 105, 301
Leibundgut, B., Schommer, R., Phillips, M., et al. 1996, ApJL, 466, L21
Leibundgut, B., & Suntzeff, N. B. 2003, in Supernovae and Gamma-Ray Bursters, ed. K. W. Weiler (Springer), 77
Leloudas, G., Chatzopoulos, E., Dilday, B., et al. 2012, A&A, 541, A129
Leloudas, G., Schulze, S., Krühler, T., et al. 2015, MNRAS, 449, 917
Lennarz, D., Altmann, D., & Wiebusch, C. 2012, A&A, 538, A120
Lentz, E. J., Baron, E., Branch, D., et al. 2000, ApJ, 530, 966
Lentz, E. J., Baron, E., Lundqvist, P., et al. 2001, ApJ, 547, 406
Lentz, E. J., Bruenn, S. W., Hix, W. R., et al. 2015, ApJL, 807, L31
Lentz, E. J., Mezzacappa, A., Bronson Messer, O. E., et al. 2012, ApJ, 747, 73
Leonard, D. C. 2007, ApJ, 670, 1275
Leonard, D. C. 2011, ApSpSci, 336, 117
Leonard, D. C., Dessart, L., Hillier, D. J., & Pignata, G. 2012, AIP Conf. Series, 1429, 204
Leonard, D. C., Filippenko, A. V., Barth, A. J., & Matheson, T. 2000, ApJ, 536, 239
Leonard, D. C., Filippenko, A. V., Chornock, R., & Foley, R. J. 2002a, PASP, 114, 1333
Leonard, D. C., Filippenko, A. V., Ganeshalingam, M., et al. 2006, Nature, 440, 505
Leonard, D. C., Filippenko, A. V., Gates, E. L., et al. 2002b, PASP, 114, 35
Leonard, D. C., Filippenko, A. V., Li, W., et al. 2002c, AJ, 124, 2490
Leonard, P. J. T. 1991, AJ, 101, 562
Leonard, P. J. T., Hills, J. G., & Dewey, R. J. 1994, ApJL, 423, L19
Leonard, D. C., Filippenko, A. V., Ardila, D. R., & Brotherton, M. S. 2001, ApJ, 553, 861
Lepp, S., Dalgarno, A., & McCray, R. 1990, ApJ, 358, 262
Levanon, N., Soker, N., & García-Berro, E. 2015, MNRAS, 447, 2803
Levermore, C. D., & Pomraning, G. C. 1981, ApJ, 248, 321
Levesque, E. M., Massey, P., Żytkow, A. N., & Morrell, N. 2014, MNRAS, 443, L94
Lewis, J. R., Walton, N. A., Meikle, W. P. S., et al. 1994, MNRAS, 266, L27
Li, C., Hillier, D. J., & Dessart, L. 2012, MNRAS, 426, 1671
Li, H., McCray, R., & Sunyaev, R. A. 1993, ApJ, 419, 824
Li, L.-X. 2008, MNRAS, 388, 603
Li, M., Ostriker, J. P., Cen, R., Bryan, G. L., & Naab, T. 2015, ApJ, 814, 4
Li, W., Bloom, J. S., Podsiadlowski, P., et al. 2011a, Nature, 480, 348
Li, W., Chornock, R., Leaman, J., et al. 2011b, MNRAS, 412, 1473
Li, W., Filippenko, A. V., Chornock, R., et al. 2003, PASP, 115, 453
Li, W., Filippenko, A. V., Gates, E., et al. 2001, PASP, 113, 1178
Li, W., Leaman, J., Chornock, R., et al. 2011c, MNRAS, 412, 1441
Li, W., Van Dyk, S. D., Filippenko, A. V., et al. 2006, ApJ, 641, 1060
Li, W. D. 1999, IAU Circ., 7294, 1
Li, W.-D., Li, C., Filippenko, A. V., & Moran, E. C. 1998, IAU Circ., 6829, 1
Li, Y.-F. 2014, IJMP Conf. Series, 31, 1460300
Li, Z., Wheeler, J. C., Bash, F. N., & Jefferys, W. H. 1991, ApJ, 378, 93
Liao, K., Avgoustidis, A., & Li, Z. 2015, Physical Review D, 92, 123539
Lidov, M. L. 1962, Planetary Space Science, 9, 719
Liebendörfer, M. 2005, ApJ, 633, 1042
Liebendörfer, M., Rampp, M., Janka, H.-T., & Mezzacappa, A. 2005, ApJ, 620, 840
Lien, A., & Fields, B. D. 2009, Journal of Cosmology and Astroparticle Physics, 1, 047
Lien, A., & Fields, B. D. 2012, ApJ, 747, 120
Limongi, M., & Chieffi, A. 2006, ApJ, 647, 483

Lin, D. N. C., Faulkner, J., & Papaloizou, J. 1985, MNRAS, 212, 105
Lindner, C. C., Milosavljević, M., Couch, S. M., & Kumar, P. 2010, ApJ, 713, 800
Lineweaver, C. H., Fenner, Y., & Gibson, B. K. 2004, Science, 303, 59
Lira, P. 1995, Masters Thesis, Un. Chile
Lira, P., Suntzeff, N. B., Phillips, M. M., et al. 1998, AJ, 115, 234
Liu, D.-D., Zhou, W.-H., Wu, C.-Y., & Wang, B. 2015a, Research in Astronomy and Astrophysics, 15, 1813
Liu, J., Di Stefano, R., Wang, T., & Moe, M. 2012a, ApJ, 749, 141
Liu, J., & Wei, H. 2015, General Relativity and Gravitation, 47, 141
Liu, R.-Y., Wang, X.-Y., Inoue, S., et al. 2014, Physical Review D, 89, 083004
Liu, W.-M., Chen, W.-C., Wang, B., & Han, Z. W. 2010, A&A, 523, A3
Liu, X. W., Liang, J. D., Xu, & R. X. 2012b, MNRAS, 424, 4
Liu, Z.-W., Moriya, T. J., Stancliffe, R. J., & Wang, B. 2015b, A&A, 574, AA12
Liu, Z.-W., Pakmor, R., Röpke, F. K., et al. 2012c, A&A, 548, A2
Liu, Z.-W., Pakmor, R., Röpke, F. K., et al. 2013, A&A, 554, A109
Liu, Z.-W., Zhang, J.-J., Ciabattari, F., et al. 2015c, MNRAS, 452, 838
Livio, M., & Riess, A. G. 2003, ApJL, 594, L93
Livio, M., & Truran, J. W. 1992, ApJ, 389, 695
Livne, E. 1990, ApJL, 354, L53
Livne, E., & Arnett, D. 1995, ApJ, 452, 62
Livne, E., Asida, S. M., & Höflich, P. 2005, ApJ, 632, 443
Livne, E., & Glasner, A. S. 1990, ApJ, 361, 244
Loll, A. M., Desch, S. J., Scowen, P. A., & Foy, J. P. 2013, ApJ, 765, 152
Long, K. S., Blair, W. P., Godfrey, L. E. H., et al. 2012, ApJ, 756, 18
Long, K. S., Blair, W. P., Winkler, P. F., et al. 2010, ApJS, 187, 495
Long, K. S., Kuntz, K. D., Blair, W. P., et al. 2014a, ApJS, 212, 21
Long, M., Jordan, G. C., IV, van Rossum, D. R., et al. 2014b, ApJ, 789, 103
Longo, M. J. 1987, Physical Review D, 36, 3276
Lopez, L. A., Ramirez-Ruiz, E., Huppenkothen, D., et al. 2011, ApJ, 732, 114
Lorén-Aguilar, P., Isern, J., & García-Berro, E. 2009, A&A, 500, 1193
Lorén-Aguilar, P., Isern, J., & García-Berro, E. 2010, MNRAS, 406, 2749
Lovegrove, E., & Woosley, S. E. 2013, ApJ, 769, 109
Lozinskaya, T. A. 1992, Supernovae and Stellar Winds in the Interstellar Medium (AIP)
Lu, F. J., Wang, Q. D., Ge, M. Y., et al. 2011, ApJ, 732, 11
Lucy, L. B. 1971, ApJ, 163, 95
Lucy, L. B. 1991, ApJ, 383, 308
Lucy, L. B. 1999, A&A, 345, 211
Lucy, L. B., Danziger, I. J., Gouiffes, C., & Bouchet, P. 1989, IAU Colloq. 120: Structure and Dynamics of the Interstellar Medium, 164
Ludlam, R. M., Miller, J. M., & Cackett, E. M. 2015, ApJ, 806, 262
Lunardini, C., & Tamborra, I. 2012, Journal of Cosmology and Astroparticle Physics, 7, 12
Lundqvist, P., & Fransson, C. 1991, ApJ, 380, 575
Lundqvist, P., & Fransson, C. 1996, ApJ, 464, 924
Lundqvist, P., Mattila, S., Sollerman, J., et al. 2013, MNRAS, 435, 329
Lundqvist, P., Nyholm, A., Taddia, F., et al. 2015, A&A, 577, A39
Lundqvist, P., & Tziamtzis, A. 2012, MNRAS, 423, 1571
Lunnan, R., Chornock, R., Berger, E., et al. 2014, ApJ, 787, 138
Luo, D., & McCray, R. 1991, ApJ, 379, 659
Lyman, J. D., Bersier, D., & James, P. A. 2014a, MNRAS, 437, 3848
Lyman, J. D., Bersier, D., James, P. A., et al. 2016, MNRAS, 457, 328
Lyman, J. D., James, P. A., Perets, H. B., et al. 2013, MNRAS, 434, 527
Lyman, J. D., Levan, A. J., Church, R. P., et al. 2014b, MNRAS, 444, 2157
Lü, G., Zhu, C., Wang, Z., & Wang, N. 2009, MNRAS, 396, 1086
Ma, H., Woosley, S. E., Malone, C. M., Almgren, A., & Bell, J. 2013a, ApJ, 771, 58

Ma, X., Chen, X., Chen, H.-l., Denissenkov, P. A., & Han, Z. 2013b, ApJL, 778, LL32
Maccarone, T. J., & de Mink, S. E. 2016, MNRAS, 458, L1
MacAlpine, G. M., & Satterfield, T. J. 2008, AJ, 136, 2152
Mackey, J., Mohamed, S., Gvaramadze, V. V., et al. 2014, Nature, 512, 282
Madgwick, D. S., Hewett, P. C., Mortlock, D. J., & Wang, L. 2003, ApJL, 599, L33
Maeda, K. 2013a, ApJ, 762, 14
Maeda, K. 2013b, ApJL, 762, L24
Maeda, K., Benetti, S., Stritzinger, M., et al. 2010a, Nature, 466, 82
Maeda, K., Hattori, T., Milisavljevic, D., et al. 2015, ApJ, 807, 35
Maeda, K., Katsuda, S., Bamba, A., et al. 2014a, ApJ, 785, 95
Maeda, K., Kawabata, K., Li, W., et al. 2009, ApJ, 690, 1745
Maeda, K., Kawabata, K., Mazzali, P. A., et al. 2008, Science, 319, 1220
Maeda, K., Kutsuna, M., & Shigeyama, T. 2014b, ApJ, 794, 37
Maeda, K., Leloudas, G., Taubenberger, S., et al. 2011, MNRAS, 413, 3075
Maeda, K., Mazzali, P. A., Deng, J., et al. 2003, ApJ, 593, 931
Maeda, K., Nakamura, T., Nomoto, K., et al. 2002, ApJ, 565, 405
Maeda, K., Röpke, F. K., Fink, M., et al. 2010b, ApJ, 712, 624
Maeda, K., Tajitsu, A., Kawabata, K. S., et al. 2016, ApJ, 816, 57
Maeda, K., Tanaka, Nomoto, K., et al. 2007, ApJ, 666, 1069
Maeda, K., Taubenberger, S., Sollerman, J., et al. 2010c, ApJ, 708, 1703
Maeder, A. 1987, A&A, 178, 159
Maeder, A. 2000, New Astron Rev, 44, 291
Maeder, A. 2009, Physics, Formation, and Evolution of Rotating Stars (Springer)
Maeder, A., & Meynet, G. 1987, A&A, 182, 243
Maeder, A., & Meynet, G. 1989, A&A, 210, 155
Maeder, A., Meynet, G., & Chiappini, C. 2015, A&A, 576, A56
Maeder, A., Meynet, G., Lagarde, N., & Charbonnel, C. 2013, A&A, 553, A1
Maggi, P., Haberl, F., Sturm, R., & Dewey, D. 2012, A&A, 548, L3
Maguire, K., Di Carlo, E., Smartt, S. J., et al. 2010a, MNRAS, 404, 981
Maguire, K., Jerkstrand, A., Smartt, S. J., et al. 2012a, MNRAS, 420, 3451
Maguire, K., Kotak, R., Smartt, S. J., et al. 2010b, MNRAS, 403, L11
Maguire, K., Sullivan, M., Ellis, R. S., et al. 2012b, MNRAS, 426, 2359
Maguire, K., Sullivan, M., Pan, Y.-C., et al. 2014, MNRAS, 444, 3258
Maguire, K., Sullivan, M., Patat, F., et al. 2013, MNRAS, 436, 222
Maguire, K., Sullivan, M., Thomas, R. C., et al. 2011, MNRAS, 418, 747
Mahabal, A. A., Djorgovski, S. G., Drake, A. J., et al. 2011, BASI, 39, 387
Mahabal, A. A., Drake, A. J., Djorgovski, S. G., et al. 2010, ATel, 2490
Malesani, D., Fynbo, J. P. U., Hjorth, J., et al. 2009, ApJL, 692, L84 M
Malkov, M. A., & Drury, L. O'C. 2001, Reports on Progress in Physics, 64, 429
Malone, C. M., Nonaka, A., Woosley, S. E., et al. 2014, ApJ, 782, 11
Mandel, K. S., Foley, R. J., & Kirshner, R. P. 2014, ApJ, 797, 75
Mannarelli, M., Pagliaroli, G., Parisi, A., & Pilo, L. 2014, Physical Review D, 89, 103014
Mannucci, F., Della Valle, M., Panagia, N., et al. 2005, A&A, 433, 807
Maoz, D., & Badenes, C. 2010, MNRAS, 407, 1314
Maoz, D., & Mannucci, F. 2008, MNRAS, 388, 421
Maoz, D., & Mannucci, F. 2012, Publications of the Astron. Soc. of Australia, 29, 447
Maoz, D., Mannucci, F., & Brandt, T. D. 2012, MNRAS, 426, 3282
Maoz, D., Mannucci, F., Li, W., et al. 2011, MNRAS, 412, 1508
Maoz, D., Mannucci, F., & Nelemans, G. 2014, ARAA, 52, 107
Maran, S. P., Sonneborn, G., Pun, C. S. J., et al. 2000, ApJ, 545, 1
Marcaide, J. M., Martí-Vidal, I., Alberdi, A., et al. 2009a, A&A, 505, 927
Marcaide, J. M., Martí-Vidal, I., Perez-Torres, M. A., et al. 2009b, A&A, 503, 869
Margutti, R., Kamble, A., Milisavljevic, D., et al. 2017, ApJ, 835, 140
Margutti, R., Milisavljevic, D., Soderberg, A. M., et al. 2014a, ApJ, 780, 21

Margutti, R., Parrent, J., Kamble, A., et al. 2014b, ApJ, 790, 52
Margutti, R., Soderberg, A. M., Chomiuk, L., et al. 2012, ApJ, 751, 134
Marietta, E., Burrows, A., & Fryxell, B. 2000, ApJS, 128, 615
Marigo, P., & Girardi, L. 2007, A&A, 469, 239
Marion, G. H., Brown, P. J., Vinkó, J., et al. 2016, ApJ, 820, 92
Marion, G. H., Höflich, P., Gerardy, C. L., et al. 2009, AJ, 138, 727
Marion, G. H., Höflich, P., Wheeler, J. C., et al. 2006, ApJ, 645, 1392
Marion, G. H., Sand, D. J., Hsiao, E. Y., et al. 2015, ApJ, 798, 39
Marion, G. H., Vinkó, J., Kirshner, R. P., et al. 2014, ApJ, 781, 69
Marion, G. H., Vinkó, J., Wheeler, J. C., et al. 2013, ApJ, 777, 40
Martin, D., Perego, A., Arcones, A., et al. 2015, ApJ, 813, 2
Martizzi, D., Faucher-Giguère, C.-A., & Quataert, E. 2015, MNRAS, 450, 504
Martí-Vidal, I., Marcaide, J. M., Alberdi, A., et al. 2011a, A&A, 526, A142
Martí-Vidal, I., Marcaide, J. M., Alberdi, A., et al. 2011b, A&A, 526, A143
Martí-Vidal, I., Tudose, V., Paragi, Z., et al. 2011c, A&A, 535, L10
Martínez-Pinedo, G., Fischer, T., & Huther, L. 2014, Journal of Physics G Nuclear Physics, 41, 044008
Masada, Y., Takiwaki, T., Kotake, K., & Sano, T. 2012, ApJ, 759, 110
Mason, E. 2013, A&A, 556, 2
Mason, E., Ederoclite, A., Williams, R. E., et al. 2012, A&A, 544, A149
Matheson, T., Filippenko, A. V., Barth, A. J., et al. 2000a, AJ, 120, 1487
Matheson, T., Filippenko, A. V., Chornock, R., et al. 2000b, AJ, 119, 2303
Matheson, T., Filippenko, A. V., Ho, L. C., et al. 2000c, AJ, 120, 1499
Matheson, T., Filippenko, A. V., Li, W., et al. 2001, AJ, 121, 1648
Matheson, T., Garnavich, P. M., Stanek, K. Z., et al. 2003, ApJ, 599, 394
Matheson, T., Joyce, R. R., Allen, L. E., et al. 2012, ApJ, 754, 19
Mathews, G. J., & Cowan, J. J. 1990, Nature, 345, 491
Mathews, G. J., Meixner, M., Olson, J. P., et al. 2013, JP Conf. Series, 445, 012023
Mathewson, D. S., & Clarke, J. N. 1973, ApJ, 180, 725
Matsuura, M., Dwek, E., Barlow, M. J., et al. 2015, ApJ, 800, 50
Matsuura, M., Dwek, E., Meixner, M., et al. 2011, Science, 333, 6047
Matsuura, M., Indebetouw, R., Woosley, S., et al. 2017, arXiv:1704.02324
Mattei, J., Johnson, G. E., Rosino, L., et al. 1979, IAU Circ., 3348, 1
Matteucci, F. 2012, Chemical Evolution of Galaxies (Springer)
Mattila, S., Lundqvist, P., Gröningsson, P., et al. 2010, ApJ, 717, 2
Mattila, S., Lundqvist, P., Sollerman, J., et al. 2005, A&A, 443, 649
Mattila, S., Meikle, W. P. S., Lundqvist, P., et al. 2008, MNRAS, 389, 141
Matzner, C. D., Levin, Y., & Ro, S. 2013, ApJ, 779, 60
Matzner, C. D., & McKee, C. F. 1999, ApJ, 510, 379
Mauerhan, J., & Smith, N. 2012, MNRAS,424,2659
Mauerhan, J., Williams, G. G., Smith, N., et al. 2014, MNRAS, 442, 1166
Mauerhan, J. C., Smith, N., Filippenko, A. V., et al. 2013a, MNRAS, 430, 1801
Mauerhan, J. C., Smith, N., Silverman, J. M., et al. 2013b, MNRAS, 431, 2599
Mauerhan, J. C., Van Dyk, S. D., Graham, M. L., et al. 2015a, MNRAS, 447, 1922
Mauerhan, J. C., Williams, G. G., Leonard, D. C., et al. 2015b, MNRAS, 453, 4467
Maund, J. R., Arcavi, I., Ergon, M., et al. 2015, MNRAS, 454, 2580
Maund, J. R., Fraser, M., Ergon, M., et al. 2011, ApJL, 739, L37
Maund, J. R., Fraser, M., Smartt, S. J., et al. 2013a, MNRAS, 431, L102
Maund, J. R., Mattila, S., Ramirez-Ruiz, E., & Eldridge, J. J. 2014a, MNRAS, 438, 1577
Maund, J. R., Reilly, E., & Mattila, S. 2014b, MNRAS, 438, 938
Maund, J. R., & Smartt, S. J. 2009, Science, 324, 486
Maund, J. R., Smartt, S. J., & Danziger, I. J. 2005, MNRAS, 364, L33
Maund, J. R., Smartt, S. J., Kudritzki, R. P., et al. 2004, Nature, 427, 129
Maund, J. R., Spyromilio, J., Höflich, P. A., et al. 2013b, MNRAS, 433, L20

Maund, J. R., Wheeler, J. C., Baade, D., et al. 2009, ApJ, 705, 1139
Maund, J. R., Wheeler, J. C., Patat, F., et al. 2007a, ApJ, 671, 1944
Maund, J. R., Wheeler, J. C., Patat, F., et al. 2007b, A&A, 475, L1
Maund, J. R., Wheeler, J. C., Patat, F., et al. 2007c, MNRAS, 381, 201
Maund, J. R., Wheeler, J. C., Wang, L., et al. 2010, ApJ, 722, 1162
Maurer, I., Jerkstrand, A., Mazzali, P. A., et al. 2011, MNRAS, 418, 1517
Maurer, I., Mazzali, P. A., Taubenberger, S., & Hachinger, S. 2010a, MNRAS, 409, 1441
Maurer, J. I., Mazzali, P. A., Deng, J., et al. 2010b, MNRAS, 402, 161
Maxted, P. F. L., Marsh, T. R., & North, R. C. 2000, MNRAS, 317, L41
Mayle, R., & Wilson, J. R. 1988, ApJ, 334, 909
Mazurek, T. J. 1973, ApSpSci, 23, 365
Mazurek, T. J. 1974, Nature, 252, 287
Mazurek, T. J., Meier, D. L., & Wheeler, J. C. 1977, ApJ, 213, 518
Mazzali, P. A. 2000, A&A, 363, 705
Mazzali, P. A., Benetti, S., Altavilla, G., et al. 2005a, ApJL, 623, L37
Mazzali, P. A., Chugai, N., Turatto, M., et al. 1997, MNRAS, 284, 151
Mazzali, P. A., Deng, J., Maeda, K., et al. 2002, ApJL, 572, L61
Mazzali, P. A., Deng, J., Maeda, K., et al. 2004, ApJ, 614, 858
Mazzali, P. A., Iwamoto, K., & Nomoto, K. 2000, ApJ, 545, 407
Mazzali, P. A., Kawabata, K. S., Maeda, K., et al. 2005b, Science, 308, 1284
Mazzali, P. A., Kawabata, K. S., Maeda, K., et al. 2007, ApJ, 670, 592
Mazzali, P. A., & Lucy, L. B. 1993, A&A, 279, 447
Mazzali, P. A., Lucy, L. B., Danziger, I. J., et al. 1993, A&A, 269, 423
Mazzali, P. A., Maurer, I., Valenti, S., et al. 2010, MNRAS, 408, 87
Mazzali, P. A., Sullivan, M., Filippenko, A. V., et al. 2015, MNRAS, 450, 2631
Mazzali, P. A., Sullivan, M., Hachinger, S., et al. 2014, MNRAS, 439, 1959
Mazzali, P. A., Valenti, S., Della Valle, M., et al. 2008, Science, 321, 1185
Mazzali, P. A., Walker, E. S., Pian, E., et al. 2013, MNRAS, 432, 2463
McCall, M. L. 1984, MNRAS, 210, 829
McClelland, C. M., Garnavich, P. M., Galbany, L., et al. 2010, ApJ, 720, 704
McClelland, C. M., Garnavich, P. M., Milne, P. A., et al. 2013, ApJ, 767, 119
McCray, R. 1993, ARAA, 31, 175
McCray, R., & Kafatos, M. 1987, ApJ, 317, 190
McCrum, M., Smartt, S. J., Kotak, R., et al. 2014, MNRAS, 437, 656
McCrum, M., Smartt, S. J., Rest, A., et al. 2015, MNRAS, 448, 1206
McCully, C., Jha, S. W., Foley, R. J., et al. 2014a, Nature, 512, 54
McCully, C., Jha, S. W., Foley, R. J., et al. 2014b, ApJ, 786, 134
McKee, C. F., & Ostriker, J. P. 1977, ApJ, 218, 148
Mcley, L., & Soker, N. 2014, MNRAS, 445, 2492
Meier, D. L. 2012, Black Hole Astrophysics: The Engine Paradigm (Springer)
Meier, D. L., Epstein, R. I., Arnett, W. D., & Schramm, D. N. 1976, ApJ, 204, 869
Meikle, W. P. S., Kotak, R., Farrah, D., et al. 2011, ApJ, 732, 109
Meikle, W. P. S., Matcher, S. J., & Morgan, B. L. 1987, Nature, 329, 608
Melott, A. L., & Thomas, B. C. 2011, Astrobiology, 11, 343
Melson, T., Janka, H.-T., & Marek, A. 2015, ApJL, 801, LL24
Mendez, M., Clocchiatti, A., Benvenuto, O. G., et al. 1988, ApJ, 334, 295
Meng, X., & Han, Z. 2015, A&A, 573, AA57
Meng, X., & Podsiadlowski, P. 2014, ApJL, 789, L45
Meng, X., & Yang, W. 2010a, ApSpSci, 329, 287
Meng, X., & Yang, W. 2010b, ApJ, 710, 1310
Meng, X., & Yang, W. 2010c, MNRAS, 401, 1118
Meng, X., & Yang, W. 2012, A&A, 543, A137
Mennekens, N., Vanbeveren, D., De Greve, J. P., & De Donder, E. 2010, A&A, 515, A89
Menou, K., Balbus, S. A., & Spruit, H. C. 2004, ApJ, 607, 564

Menzies, J. W., Catchpole, R. M., van Vuuren, G., et al. 1987, MNRAS, 227, 39P
Mereghetti, S. 2013, Brazilian Journal of Physics, 43, 356
Mereghetti, S., La Palombara, N., Tiengo, A., et al. 2013, IAU Symposium, 281, 68
Mereghetti, S., Tiengo, A., Esposito, P., et al. 2009, Science, 325, 1222
Mereghetti, S., Pons, J. A., & Melatos, A. 2015, SSR, 191, 315
Mesler, R. A., Whalen, D. J., Smidt, J., et al. 2014, ApJ, 787, 91
Metzger, B. D. 2010, MNRAS, 409, 284
Metzger, B. D. 2012, MNRAS, 419, 827
Metzger, B. D., Giannios, D., Thompson, T. A., et al. 2011, MNRAS, 413, 2031
Metzger, B. D., Martínez-Pinedo, G., Darbha, S., et al. 2010, MNRAS, 406, 2650
Metzger, B. D., Quataert, E., & Thompson, T. A. 2008, MNRAS, 385, 1455
Metzger, B. D., Vurm, I., Hascoët, R., & Beloborodov, A. M. 2014, MNRAS, 437, 703
Meyers, J., Aldering, G., Barbary, K., et al. 2012, ApJ, 750, 1
Meynet, G., & Maeder, A. 2000, A&A, 361, 101
Meynet, G., & Maeder, A. 2005, A&A, 429, 581
Mezzacappa, A. 2005, ARN&PS, 55, 467
Miceli, M., Sciortino, S., Troja, E., & Orlando, S. 2015, ApJ, 805, 120
Michael, E., McCray, R., Chevalier, R., et al. 2003, ApJ, 593, 809
Michael, E., McCray, R., Pun, C. S. J., et al. 2000, ApJL, 542, L53
Michałowski, M. J. 2015, A&A, 577, A80
Michałowski, M. J., Watson, D., & Hjorth, J. 2010, ApJ, 712, 942
Mihalas, D. & Mihalas, B. 1984, Foundations of Radiation Hydrodynamics (Oxford Un. Press)
Mihalas, D., & Weaver, R. 1982, Journal of Quantitative Spectroscopy and Radiative Transfer, 28, 213
Miknaitis, G., Pignata, G., Rest, A., et al. 2007, ApJ, 666, 674
Mikołajewska, J. 2013, IAU Symposium 281, Binary Paths to Type Ia Supernova Explosions, 162
Milisavljevic, D., & Fesen, R. A. 2013, ApJ, 772, 134
Milisavljevic, D., & Fesen, R. A. 2015, Science, 347, 526
Milisavljevic, D., Fesen, R. A., Chevalier, R. A., et al. 2012, ApJ, 751, 25
Milisavljevic, D., Fesen, R. A., Gerardy, C. L., et al. 2010, ApJ, 709, 1343
Milisavljevic, D., Fesen, R. A., Kirshner, R. P., & Challis, P. 2009, ApJ, 692, 839
Milisavljevic, D., Fesen, R. A., Leibundgut, B., & Kirshner, R. P. 2008, ApJ, 684, 1170
Milisavljevic, D., Margutti, R., Crabtree, K. N., et al. 2014, ApJL, 782, L5
Milisavljevic, D., Margutti, R., Parrent, J. T., et al. 2015, ApJ, 799, 51
Millard, J., Branch, D., Baron, E., et al. 1999, ApJ, 527, 746
Miller, A. A., Silverman, J. M., Butler, N. R., et al. 2010a, MNRAS, 404, 305
Miller, A. A., Smith, N., Li, W., et al. 2010b, AJ, 139, 2218
Miller, J. M. 2007, ARAA, 45, 441
Miller, J. M., Miller, M. C., & Reynolds, C. S. 2011, ApJL, 731, L5
Miller, J. M., Parker, M. L., Fuerst, F., et al. 2013, ApJL, 775, L45
Miller, M. C., & Miller, J. M. 2015, Physics Reports, 548, 1
Milne, P. A., Brown, P. J., Roming, P. W. A. et al. 2013, ApJ, 779, 23
Milne, P. A., Foley, R. J., Brown, P. J., & Narayan, G. 2015, ApJ, 803, 20
Milne, P. A., The, L.-S., & Leising, M. D. 1999, ApJS, 124, 503
Milne, P. A., Williams, G. G., Porter, A., et al. 2017, ApJ, 835, 100
Minkowski, R. 1939, ApJ, 89, 156
Minkowski, R. 1941, PASP, 53, 224
Mirabel, I. F., Dijkstra, M., Laurent, P., et al. 2011, A&A, 528, A149
Mishra, R. K., & Goswami, J. N. 2014, Geochimica Cosmochimica Acta, 132, 440
Mocák, M., Campbell, S. W., Müller, E., & Kifonidis, K. 2010, A&A, 520, A114
Modjaz, M. 2012, in Death of Massive Stars, ed. P. Roming, N. Kawai, & E. Pian (Cambridge Un. Press), 207
Modjaz, M., Blondin, S., Kirshner, R. P., et al. 2014, AJ, 147, 99
Modjaz, M., Kirshner, R. P., Blondin, S., et al. 2008, ApJL, 687, L9

Modjaz, M., Li, W., Butler, N., et al. 2009, ApJ, 702, 226
Moll, R., Raskin, C., Kasen, D., & Woosley, S. E. 2014, ApJ, 785, 105
Moll, R., & Woosley, S. E. 2013, ApJ, 774, 137
Möller, P., Nix, J. R., & Kratz, K. L. 1997, Atomic Data and Nuclear Data Tables, 66, 131
Montes, M. J., van Dyk, S. D., Weiler, K. W., et al. 1998, ApJ, 506, 874
Montes, M. J., Weiler, K. W., Van Dyk, S. D., et al. 2000, ApJ, 532, 1124
Moore, K., & Bildsten, L. 2012, ApJ, 761, 182
Morales-Garoffolo, A., Elias-Rosa, N., Benetti, S., et al. 2014, MNRAS, 445, 1647
Morales-Garoffolo, A., Elias-Rosa, N., Bersten, M., et al. 2015, MNRAS, 454, 95
Moriya, T., Tominaga, N., Blinnikov, S. I., et al. 2011, MNRAS, 415, 199
Moriya, T., Tominaga, N., Tanaka, M., et al. 2010a, ApJ, 719, 1445
Moriya, T., Tominaga, N., Tanaka, M., et al. 2010b, ApJL, 717, L83
Moriya, T. J. 2015, ApJL, 803, L26
Moriya, T. J., Blinnikov, S. I., Tominaga, N., et al. 2013, MNRAS, 428, 1020
Moriya, T. J., & Langer, N. 2015, A&A, 573, A18
Moriya, T. J., & Maeda, K. 2012, ApJL, 756, L22
Moriya, T. J., & Maeda, K. 2014, ApJL, 790, LL16
Moriya, T. J., Tominaga, N., Langer, N., et al. 2014a, A&A, 569, A57
Moriya, T. J., Maeda, K., Taddia, F., et al. 2014b, MNRAS, 439, 2917
Morokuma, T., Tominaga, N., Tanaka, M., et al. 2014, Publications of the ASJ, 66, 114
Morris, T., & Podsiadlowski, P. 2007 Science, 315, 5815
Morris, T., & Podsiadlowski, Ph. 2009, MNRAS, 339, 2
Moseley, S. H., Dwek, E., Glaccum, W., et al., 1989, Nature, 340, 697
Mösta, P., Ott, C. D., Radice, D., et al. 2015, Nature, 528, 376
Mösta, P., Richers, S., Ott, C. D., et al. 2014, ApJL, 785, L29
Motohara, K., Maeda, K., Gerardy, C. L., et al. 2006, ApJL, 652, L101
Motta, S. E., Muñoz-Darias, T., Sanna, A., et al. 2014, MNRAS, 439, L65
Mould, J., & Reid, N. 1987, ApJ, 321, 156
Mould, J., & Uddin, S. A. 2014, Publications of the Astron. Soc. of Australia, 31, 15
Moynier, F., Blichert-Toft, J., Wang, K., et al. 2011, ApJ, 741, 71
Müller, E., & Hillebrandt, W. 1979, A&A, 80, 147
Mumpower, M. R., Surman, R., Fang, D.-L., et al. 2015, Physical Review C, 92, 035807
Munari, U., Barbon, R., Piemonte, A., et al. 1998, A&A, 333, 159
Munari, U., Henden, A., Belligoli, R., et al. 2013, Proceedings of the National Academy, 20, 30
Muno, M. P., Clark, J. S., Crowther, P. A., et al. 2006, ApJL, 636, L41
Murphy, J. W., Dolence, J. C., & Burrows, A. 2013, ApJ, 771, 52
Müller, B., Janka, H.-T., & Dimmelmeier, H. 2010, ApJS, 189, 104
Müller, B., Janka, H.-T., & Marek, A. 2013, ApJ, 766, 43
Münch, G. 1948, ApJ, 108, 116
Nadyozhin, D. K. 1980, ApSpSci, 69, 115
Nadyozhin, D. K. 1994, ApJS, 92, 527
Nagy, A. P., Ordasi, A., Vinkó, J., & Wheeler, J. C. 2014, A&A, 571, A77
Nagy, A. P., & Vinkó, J. 2016, AAP, 589, A53
Nakamura, K., Horiuchi, S., Tanaka, M., et al. 2016, MNRAS, 461, 3296
Nakar, E. 2015, ApJ, 807, 172
Nakar, E., Poznanski, D., & Katz, B. 2016, ApJ, 823, 127
Nakar, E., & Sari, R. 2010, ApJ, 725, 904
Nakazato, K., Mochida, E., Niino, Y., & Suzuki, H. 2015, ApJ, 804, 75
Nakazato, K., Sumiyoshi, K., Suzuki, H., et al. 2013a, ApJS, 205, 2
Nakazato, K., Sumiyoshi, K., & Yamada, S. 2013b, A&A, 558, A50
Napiwotzki, R., Christlieb, N., Drechsel, H., et al. 2001, Astr. Nachr., 322, 411
Napiwotzki, R., & Silva, M. D. V. 2012, Mem. Societa Astronomica Italiana, 83, 272
Narayan, G., Foley, R. J., Berger, E., et al. 2011, ApJL, 731, L11
Nath, B. B., & Shchekinov, Y. 2013, ApJL, 777, L12

Neill, J. D., Sullivan, M., Gal-Yam, A., et al. 2011, ApJ, 727, 15
Neill, J. D., Sullivan, M., Howell, D. A., et al. 2009, ApJ, 707, 1449
Nelemans, G., Verbunt, F., Yungelson, L. R., & Portegies Zwart, S. F. 2000, A&A, 360, 1011
Nelemans, G., Voss, R., Nielsen, M. T. B., & Roelofs, G. 2010, MNRAS, 405, L71
Ng, C.-Y., Gaensler, B. M., & Murray, S. S. 2009, ApJ Letters, 706, 1
Ng, C.-Y., Potter, T. M., Staveley-Smith, L, et al. 2011 ApJ Letters, 728, 1
Ng, C.-Y., & Romani, R. W. 2007, ApJ, 660, 1357
Nicholl, M., & Smartt, S. J. 2016, MNRAS, 457, L79
Nicholl, M., Smartt, S. J., Jerkstrand, A., et al. 2013, Nature, 502, 346
Nicholl, M., Smartt, S. J., Jerkstrand, A., et al. 2014, MNRAS, 444, 2096
Nicholl, M., Smartt, S. J., Jerkstrand, A., et al. 2015a, ApJL, 807, L18
Nicholl, M., Smartt, S. J., Jerkstrand, A., et al. 2015b, MNRAS, 452, 3869
Nielsen, M. T. B., Dominik, C., Nelemans, G., & Voss, R. 2013, A&A, 549, A32
Nielsen, M. T. B., Gilfanov, M., Bogdán, Á., et al. 2014a, MNRAS, 442, 3400
Nielsen, M. T. B., Nelemans, G., Voss, R., & Toonen, S. 2014b, A&A, 563, A16
Nielsen, M. T. B., Voss, R., & Nelemans, G. 2012, MNRAS, 426, 2668
Niemeyer, J. C., & Woosley, S. E. 1997, ApJ, 475, 740
Niino, Y., Nagamine, K., & Zhang, B. 2015, MNRAS, 449, 2706
Nisenson, P., & Papaliolios, C. 1999, ApJ, 518, 1
Nisenson, P., Papaliolios, C., Karovska, M., & Noyes, R. 1987, ApJL, 320, L15
Nishimura, N., Fischer, T., Thielemann, F.-K., et al. 2012, ApJ, 758, 9
Nishimura, N., Takiwaki, T., & Thielemann, F.-K. 2015, ApJ, 810, 109
Nomoto, K. 1982, ApJ, 253, 798
Nomoto, K. 1984, ApJ, 277, 791
Nomoto, K. 1987, ApJ, 322, 206
Nomoto, K., & Hashimoto, M. 1988, Physics Reports, 163, 13
Nomoto, K., Iwamoto, K., & Suzuki, T. 1995, Physics Reports, 256, 173
Nomoto, K., Kamiya, Y., & Nakasato, N. 2013a, IAU Symposium, 281, 253
Nomoto, K., Kobayashi, C., & Tominaga, N. 2013b, ARAA, 51, 457
Nomoto, K., & Kondo, Y. 1991, ApJL, 367, L19
Nomoto, K., & Sugimoto, D. 1977, Publications of the ASJ, 29, 765
Nomoto, K., Sugimoto, D., Sparks, W. M., et al. 1982, Nature, 299, 803
Nomoto, K., Suzuki, T., Shigeyama, T., et al. 1993, Nature, 364, 507
Nomoto, K., Thielemann, F.-K., & Yokoi, K. 1984, ApJ, 286, 644
Nomoto, K., Yamaoka, H., Pols, O. R., et al. 1994, Nature, 371, 227
Nonaka, A., Aspden, A. J., Zingale, M., et al. 2012, ApJ, 745, 73
Nordhaus, J., Brandt, T. D., Burrows, A., & Almgren, A. 2012, MNRAS, 423, 1805
Nordhaus, J., Brandt, T. D., Burrows, A., et al. 2010, Physical Review D, 82, 103016
Nordin, J., Rubin, D., Richard, J., et al. 2014, MNRAS, 440, 2742
Norgaard-Nielsen, H. U., Hansen, L., Jorgensen, H. E., et al. 1989, Nature, 339, 523
Noutsos, A., Schnitzeler, D. H. F. M., Keane, E. F., et al. 2013, MNRAS, 430, 2281
Nozawa, T., Kozasa, T., Umeda, H., et al. 2003, ApJ, 598, 785
Nugent, P., Baron, E., Branch, D., et al. 1997, ApJ, 485, 812
Nugent, P., Phillips, M., Baron, E., et al. 1995, ApJL, 455, L147
Nugent, P., Sullivan, M., Bersier, D., et al. 2011a, ATel, 3581, 1
Nugent, P. E., Sullivan, M., Cenko, S. B., et al. 2011b, Nature, 480, 344
Nugent, P., Sullivan, M., Ellis, R., et al. 2006, ApJ, 645, 841
Nugent, P., Thomas, R., & Aldering, G. 2008, JP Conf. Series, 125, 012011
O'Brien, P. T. 2016, AAS/High Energy Astrophysics Division, 15, 112.03
O'Connor, E., & Ott, C. D. 2011, ApJ, 730, 70
O'Connor, E., & Ott, C. D. 2013, ApJ, 762, 126
Ober, W. W., El Eid, M. F., & Fricke, K. J. 1983, AAP, 119, 61
Obergaulinger, M., Aloy, M. A., Dimmelmeier, H., Müller, E. 2006, A&A, 457, 209
Obergaulinger, M., Janka, H.-T., & Aloy, M. A. 2014, MNRAS, 445, 3169

Odrzywolek, A., & Plewa, T. 2011, A&A, 529, A156
Oemler, A., Jr., & Tinsley, B. M. 1979, AJ, 84, 985
Ofek, E. O., Fox, D., Cenko, S. B., et al. 2013, ApJ, 763, 42
Ofek, E. O., Rabinak, I., Neill, J. D., et al. 2010, ApJ, 724, 1396
Ofek, E. O., Sullivan, M., Shaviv, N. J., et al. 2014a, ApJ, 789, 104
Ofek, E. O., Zoglauer, A., Boggs, S. E., et al. 2014b, ApJ, 781, 42
Ogata, K., Kan, M., & Kamimura, M. 2009, Progress of Theoretical Physics, 122, 1055
Ögelman, H., & Alpar, M. A. 2004, ApJL, 603, L33
Oh, S.-H., Brook, C., Governato, F., et al. 2011, AJ, 142, 24
Ohkubo, T., Nomoto, K., Umeda, et al. 2009, ApJ, 706, 1184
Ohkubo, T., Umeda, H., Maeda, K., et al. 2006, ApJ, 645, 1352
Ohkubo, T., Umeda, H., & Nomoto, K. 2003, Nuclear Physics A, 718, 632
Okumura, J. E., Ihara, Y., Doi, M., et al. 2014, Publications of the ASJ, 66, 49
Oliva, E. 1987, ApJL, 321, L45
Olling, R. P., Mushotzky, R., Shaya, E. J., et al. 2015, Nature, 521, 332
Oran, E. S. & Boris, J. P. 2005 Numerical Simulation of Reactive Flow, Cambridge: Cambridge University Press
Orsaria, M., Rodrigues, H., Weber, F., & Contrera, G. A. 2014, Physical Review C, 89, 015806
Osborne, J. P., Page, K. L., Beardmore, A. P., et al. 2011, ApJ, 727, 124
Ostriker, J. P., & Bodenheimer, P. 1968, ApJ, 151, 1089
Ostriker, J. P., & Gunn, J. E. 1971, ApJL, 164, L95
Ott, C. D., Abdikamalov, E., Mösta, P., et al. 2013, ApJ, 768, 115
Ott, C. D., Burrows, A., Thompson, T. A., et al. 2006, ApJS, 164, 130
Ott, C. D., Ou, S., Tohline, J. E., & Burrows, A. 2005, ApJL, 625, L119
Ou, S., & Tohline, J. E. 2006, ApJ, 651, 1068
Ou, S., Tohline, J. E., & Lindblom, L. 2004, ApJ, 617, 490
Ouyed, R., Dey, J., & Dey, M. 2002, A&A, 390, L39
Ouyed, R., Kostka, M., Koning, N., et al. 2012, MNRAS, 423, 1652
Ouyed, R., Leahy, D., & Koning, N. 2015a, MNRAS, 454, 2353
Ouyed, R., Leahy, D., & Koning, N. 2015b, Research in Astronomy and Astrophysics, 15, 483
Owen, P. J., & Barlow, M. J. 2015, ApJ, 801, 141
Owocki, S. P. 2015, Astrophysics and Space Science Library, 412, 113
Özel, F., Baym, G., & Güver, T. 2010a, Physical Review D, 82, 101301
Özel, F., & Freire, P. 2016, ARAA, 54, 401
Özel, F., Psaltis, D., Güver, T., et al. 2016, ApJ, 820, 28
Özel, F., Psaltis, D., Narayan, R., & McClintock, J. E. 2010b, ApJ, 725, 1918
Özel, F., Psaltis, D., Narayan, R., & Santos Villarreal, A. 2012, ApJ, 757, 55
Paczyński, B. 1970, Acta Astron, 20, 47
Paczyński, B. 1971, ARAA, 9, 183
Paczyński, B. 1972, Astrophysics Letters, 11, 53
Page, D., Prakash, M., Lattimer, J. M., & Steiner, A. W. 2011, PRL, 106, 081101
Pagliaroli, G., Rossi-Torres, F., & Vissani, F. 2010, Astroparticle Physics, 33, 287
Pagliaroli, G., Vissani, F., Costantini, M. L., & Ianni, A. 2009, Astropart. Phys. 31, 3
Pagnotta, A., & Schaefer, B. E. 2014, ApJ, 788, 164
Pagnotta, A., Schaefer, B. E., Clem, J. L., et al. 2015, ApJ, 811, 32
Pakmor, R., Hachinger, S., Röpke, F. K., & Hillebrandt, W. 2011, A&A, 528, A117
Pakmor, R., Kromer, M., Taubenberger, S., & Springel, V. 2013, ApJL, 770, L8
Pakmor, R., Kromer, M., Taubenberger, S., et al. 2012, ApJL, 747, L10
Pan, K.-C., Ricker, P. M., & Taam, R. E. 2012a, ApJ, 760, 21
Pan, K.-C., Ricker, P. M., & Taam, R. E. 2013, ApJ, 773, 49
Pan, L., Desch, S. J., Scannapieco, E., & Timmes, F. X. 2012b, ApJ, 756, 102
Pan, L., Wheeler, J. C., & Scalo, J. 2008, ApJ, 681, 470
Pan, T., Kasen, D., & Loeb, A. 2012c, MNRAS, 422, 2701
Pan, T., Loeb, A., & Kasen, D. 2012d, MNRAS, 423, 2203

Pan, Y.-C., Foley, R. J., Kromer, M., et al. 2015a, MNRAS, 452, 4307
Pan, Y.-C., Sullivan, M., Maguire, K., et al. 2014, MNRAS, 438, 1391
Pan, Y.-C., Sullivan, M., Maguire, K., et al. 2015b, MNRAS, 446, 354
Panagia, N. 2005, in IAU Colloq. 192, Cosmic Explosions, 585
Panagia, N., Sramek, R. A., & Weiler, K. W. 1986, ApJL, 300, L55
Panagia, N., Tanzi, E. G., & Tarenghi, M. 1983, ApJ, 272, 123
Panagia, N., Vettolani, G., Boksenberg, A., et al. 1980, MNRAS, 192, 861
Panagia, N., Van Dyk, S. D., Weiler, K. W., et al. 2006, ApJ, 646, 369
Pankey, T. 1963, Nuclear Science Abstracts, 17, No. 4, 1395
Papadopoulos, A., D'Andrea, C. B., Sullivan, M., et al. 2015, MNRAS, 449, 1215
Papaliolios, C., Krasovska, M., Koechlin, L., et al. 1989, Nature, 338, 565
Papish, O., & Perets, H. B. 2016, ApJ, 822, 19
Papish, O., Soker, N., García-Berro, E., & Aznar-Siguán, G. 2015, MNRAS, 449, 942
Parizot, E. 2014, Nuclear Physics B Proc. Suppl., 256, 197
Park, S., Zhekov, S. A., Burrows, D. N., et al. 2005, Adv. Space Res., 35, 6
Park, S., Zhekov, S. A., Burrows, D. N., et al. 2006, ApJ, 646, 1001
Park, S., Zhekov, S. A., Burrows, D. N., Garmire, G. P., & McCray, R. 2004, ApJ, 610, 275
Parrent, J., Branch, D., Troxel, M. A., et al. 2007, PASP, 119, 135
Parrent, J., Friesen, B., & Parthasarathy, M. 2014, ApSpSci, 351, 1
Parrent, J. T., Howell, D. A., Friesen, B., et al. 2012, ApJL, 752, L26
Parrent, J. T., Milisavljevic, D., Soderberg, A. M., & Parthasarathy, M. 2016, ApJ, 820, 75
Parthasarathy, M., Branch, D., Jeffery, D. J., & Baron, E. 2007, New Astron Rev, 51, 524
Pastorello, A., Baron, E., Branch, D., et al. 2005, MNRAS, 360, 950
Pastorello, A., Benetti, S., Brown, P. J., et al. 2015a, MNRAS, 449, 1921
Pastorello, A., Cappellaro, E., Inserra, C., et al. 2013, ApJ, 767, 1
Pastorello, A., Hadjiyska, E., Rabinowitz, D., et al. 2015b, MNRAS, 449, 1954
Pastorello, A., Kasliwal, M. M., Crockett, R. M., et al. 2008a, MNRAS, 389, 955
Pastorello, A., Mattila, S., Zampieri, L., et al. 2008b, MNRAS, 389, 113
Pastorello, A., Prieto, J. L., Elias-Rosa, N., et al. 2015c, MNRAS, 453, 3649
Pastorello, A., Pumo, M. L., Navasardyan, H., et al. 2012, A&A, 537, 141
Pastorello, A., Quimby, R. M., Smartt, S. J., et al. 2008c, MNRAS, 389, 131
Pastorello, A., Sauer, D., Taubenberger, S., et al. 2006, MNRAS, 370, 1752
Pastorello, A., Smartt, S. J., Botticella, M. T., et al. 2010, ApJL, 724, L16
Pastorello, A., Smartt, S. J., Mattila, S., et al. 2007, Nature, 447, 829
Pastorello, A., Turatto, M., Benetti, S., et al. 2002, MNRAS, 333, 27
Pastorello, A., Valenti, S., Zampieri, L., et al. 2009, MNRAS, 394, 2266
Pastorello, A., Wang, X.-F., Ciabattari, F., et al. 2016, MNRAS, 456, 853
Pastorello, A., Wyrzykowski, Ł., Valenti, S., et al. 2015d, MNRAS, 449, 1941
Pastorello, A., Zampieri, L., Turatto, M., et al. 2004, MNRAS, 347, 74
Patat, F., Baade, D., Höflich, P., et al. 2009, A&A, 508, 229
Patat, F., Barbon, R., Cappellaro, E., & Turatto, M. 1993, A&A Supplement, 98, 443
Patat, F., Barbon, R., Cappellaro, E., & Turatto, M. 1994, A&A, 282, 731
Patat, F., Benetti, S., Cappellaro, E., et al. 1996, MNRAS, 278, 111
Patat, F., Cappellaro, E., Danziger, J., et al. 2001, ApJ, 555, 900
Patat, F., Chandra, P., Chevalier, R., et al. 2007, Science, 317, 924
Patat, F., Chugai, N. N., Podsiadlowski, P., et al. 2011a, A&A, 530, A63
Patat, F., Cordiner, M. A., Cox, N. L. J., et al. 2013, A&A, 549, A62
Patat, F., Cox, N. L. J., Parrent, J., & Branch, D. 2010, A&A, 514, A78
Patat, F., Höflich, P., Baade, D., et al. 2012, A&A, 545, A7
Patat, F., & Romaniello, M. 2006, PASP, 118, 146
Patat, F., Taubenberger, S., Benetti, S., et al. 2011b, A&A, 527, L6
Patat, F., Taubenberger, S., Cox, N. L. J., et al. 2015, A&A, 577, A53
Patel, B., McCully, C., Jha, S. W., et al. 2014, ApJ, 786, 9
Patnaude, D. J., Badenes, C., Park, S., & Laming, J. M. 2012, ApJ, 756, 6

Patnaude, D. J., & Fesen, R. A. 2014, ApJ, 789, 138
Patnaude, D. J., Loeb, A., & Jones, C. 2011, Proceedings of the National Academy, 16, 187
Patterson, J., Oksanen, A., Monard, B., et al. 2014, in Stella Novae: Past and Future Decades, ASP Conf. Series, 490, 35
Pavlov, G. G., Sanwal, D., & Teter, M. A. 2004, IAU Symposium 218, Young Neutron Stars and Their Environments, 239
Patton, K. M., Lunardini, C., & Farmer, R. J. 2017, ApJ, 840, 2
Pavlović, M. Z., Urošević, D., Vukotić, B., et al. 2013, ApJS, 204, 4
Paxton, B., Bildsten, L., Dotter, A., et al. 2011, ApJS, 192, 3
Paxton, B., Cantiello, M., Arras, P., et al. 2013, ApJS, 208, 4
Paxton, B., Marchant, P., Schwab, J., et al. 2015, ApJS, 220, 15
Payez, A., Evoli, C., Fischer, T., et al. 2015, Journal of Cosmology and Astroparticle Physics, 2, 006
Pejcha, O., & Prieto, J. L. 2015, ApJ, 806, 225
Pejcha, O., & Thompson, T. A. 2015, ApJ, 801, 90
Penney, R., & Höflich, P. 2014, ApJ, 795, 84
Perego, A., Rosswog, S., Cabezón, R. M., et al. 2014, MNRAS, 443, 3134
Pereira, R., Thomas, R. C., Aldering, G., et al. 2013, A&A, 554, A27
Perets, H. B., Badenes, C., Arcavi, I., et al. 2011, ApJ, 730, 89
Perets, H. B., Gal-Yam, A., Mazzali, P. A., et al. 2010, Nature, 465, 322
Perez-Torres, M., Alberdi, A., Beswick, R. J., et al. 2015, in Advancing Astrophysics with the Square Kilometre Array (AASKA14), 60
Perlmutter, S. 2003, Physics Today, 56, 040000
Perlmutter, S., Aldering, G., Goldhaber, G., et al. 1999, ApJ, 517, 565
Perlmutter, S., Deustua, S., Gabi, S., et al. 1997, in Thermonuclear Supernovae, ed. P. Ruiz–Lapuente, R. Canal, & J. Isern (Kluwer), 749
Perlmutter, S., & Schmidt, B. P. 2003, in Supernovae and Gamma-Ray Bursters, ed. K. Weiler, 195
Perrett, K., Sullivan, M., Conley, A., et al. 2012, AJ, 144, 59
Peters, N. 2000, Turbulent Combustion, Cambridge: Cambridge University Press
Pethick, C. J., & Ravenhall, D. G. 1995, ARN&PS, 45, 429
Petrovic, J., Langer, N., & van der Hucht, K. A. 2005, A&A, 435, 1013
Peñate, L., Martín, O., Cárdenas, R., & Agustí, S. 2010, ApSpSci, 330, 211
Pfannes, J. M. M., Niemeyer, J. C., & Schmidt, W. 2010a, A&A, 509, A75
Pfannes, J. M. M., Niemeyer, J. C., Schmidt, W., & Klingenberg, C. 2010b, A&A, 509, A74
Phillips, M. M. 1993, ApJL, 413, L105
Phillips, M. M. 2012, Publications of the Astron. Soc. of Australia, 29, 434
Phillips, M. M., & Heathcote, S. R. 1989, PASP, 101, 137
Phillips, M. M., Li, W., Frieman, J. A., et al. 2007, PASP, 119, 360
Phillips, M. M., Lira, P., Suntzeff, N. B., et al. 1999, AJ, 118, 1766
Phillips, M. M., Phillips, A. C., Heathcote, S. R., et al. 1987, PASP, 99, 592
Phillips, M. M., Simon, J. D., Morrell, N., et al. 2013, ApJ, 779, 38
Phillips, M. M., Wells, L. A., Suntzeff, N. B., et al. 1992, AJ, 103, 1632
Piersanti, L., Tornambé, A., Yungelson, L., & Straniero, O. 2013, IAU Symposium, 281, 209
Pignata, G., Maza, J., Antezana, R., et al. 2009, in Probing Stellar Populations Out to the Distant Universe, AIP Conf. Series, 1111, 551
Pignata, G., Stritzinger, M., Soderberg, A., et al. 2011, ApJ, 728, 14
Pignatari, M., Gallino, R., Heil, M., et al. 2010, ApJ, 710, 1557
Pinto, P. A., & Eastman, R. G. 2000, ApJ, 530, 757
Pinto, P. A. & Woosley, S. E. 1988, ApJ, 329, 820
Piran, T., & Jimenez, R. 2014, PRL, 113, 231102
Piro, A. L. 2008, ApJ, 679, 616
Piro, A. L. 2011, ApJL, 738, L5
Piro, A. L. 2013, ApJL, 768, L14
Piro, A. L. 2015a, ApJ, 801, 137

Piro, A. L. 2015b, ApJL, 808, L51
Piro, A. L., Chang, P., & Weinberg, N. N. 2010, ApJ, 708, 598
Piro, A. L., & Morozova, V. S. 2014, ApJL, 792, L11
Piro, A. L., & Nakar, E. 2013, ApJ, 769, 67
Piro, A. L., & Ott, C. D. 2011, ApJ, 736, 108
Piro, A. L., Thompson, T. A., & Kochanek, C. S. 2014, MNRAS, 438, 3456
Piro, A. L., & Thrane, E. 2012, ApJ, 761, 63
Pisani, G. B., Izzo, L., Ruffini, R., et al. 2013, A&A, 552, L5
Plait, P. C., Lundqvist, P., Chevalier, R. A., et al. 1995, ApJ, 439, 730
Planck Collaboration, Ade, P. A. R., Aghanim, N., et al. 2016, A&A, 594, A13
Plewa, T., Calder, A. C., & Lamb, D. Q. 2004, ApJL, 612, L37
Pllumbi, E., Tamborra, I., Wanajo, S., et al. 2015, ApJ, 808, 188
Podsiadlowski, P. 1992, PASP, 104, 717
Podsiadlowski, P., Cannon, R. C., & Rees, M. J. 1995, MNRAS, 274, 485
Podsiadlowski, P., Hsu, J. J. L., Joss, P. C., & Ross, R. R. 1993, Nature, 364, 509
Podsiadlowski, P., Ivanova, N., Justham, S., & Rappaport, S. 2010, MNRAS, 406, 840
Podsiadlowski, P., Joss, P. C., & Hsu, J. J. L. 1992, ApJ, 391, 246
Poelarends, A. J. T., Herwig, F., Langer, N., & Heger, A. 2008, ApJ, 675, 614
Pollas, C., Cappellaro, E., Turatto, M., & Candeo, G. 1988, IAU Circ., 4691, 1
Poludnenko, A. Y. 2015, Physics of Fluids, 27, 014106
Poludnenko, A. Y., Gardiner, T. A., & Oran, E. S. 2011, PRL, 107, 054501
Poludnenko, A. Y., & Oran, E. S. 2010, Combust. & Flame, 157, 995
Poludnenko, A. Y., & Oran, E. S. 2011, Combust. & Flame, 158, 301
Pooley, D., & Lewin, W. H. G. 2004, IAU Circ., 8323, 2
Pooley, D., Lewin, W. H. G., Fox, D. W., et al. 2002, ApJ, 572, 932
Popov, S. B., & Turolla, R. 2012, ApSpSci, 341, 457
Posselt, B., Pavlov, G. G., Suleimanov, V., & Kargaltsev, O. 2013, ApJ, 779, 186
Postman, M., Coe, D., Benítez, N., et al. 2012, ApJS, 199, 25
Postnov, K. A., & Yungelson, L. R. 2014, Living Reviews in Relativity, 17, 3
Potashov, M., Blinnikov, S., Baklanov, P., & Dolgov, A. 2013, MNRAS, 431, L98
Potter, T. M., Staveley-Smith, L., Ng, C.-Y., et al. 2009, ApJ, 705, 1
Poveda, A., Ruiz, J., & Allen, C. 1967, Boletin de los Observatorios Tonantzintla y Tacubaya, 4, 86
Powell, L. C., Slyz, A., & Devriendt, J. 2011, MNRAS, 414, 3671
Poznanski, D. 2013, MNRAS, 436, 3224
Poznanski, D., Chornock, R., Nugent, P. E., et al. 2010a, Science, 327, 58
Poznanski, D., Nugent, P. E., & Filippenko, A. V. 2010b, ApJ, 721, 956
Pozzo, M., Meikle, W. P. S., Fassia, A., et al. 2004, MNRAS, 352, 457
Prantzos, N. 2008, Space Science Reviews, 135, 313
Prieto, J. L., Kistler, M. D., Thompson, T. A., et al. 2008, ApJL, 681, L9
Pruet, J., Hoffman, R. D., Woosley, S. E., et al. 2006, ApJ, 644, 1028
Pskovskii, I. P. 1977, Soviet Astronomy, 21, 675
Ptuskin, V., Zirakashvili, V., & Seo, E.-S. 2010, ApJ, 718, 31
Puchwein, E., & Springel, V. 2013, MNRAS, 428, 2966
Puckett, T., Armstrong, J., Johnson, W., et al. 1994, IAU Circ., 5961, 1
Pumo, M. L., & Zampieri, L. 2011, ApJ, 741, 41
Pun, C. S. J., Michael, E., Zhekov, S. A., et al. 2002, ApJ, 572, 906
Pzrybilla, N., Nieva, M. F., Heber, U., & Butler, K. 2008, ApJL, 684, L103
Pérez-Torres, M. A., Lundqvist, P., Beswick, R. J., et al. 2014, ApJ, 792, 38
Qi, J.-Z., Zhang, M.-J., & Liu, W.-B. 2014, Physical Review D, 90, 063526
Qian, Y.-Z. 2014, Journal of Physics G Nuclear Physics, 41, 044002
Qiu, Y., Li, W., Qiao, Q., & Hu, J. 1999, AJ, 117, 736
Quimby, R. M., Aldering, G., Wheeler, J. C., et al. 2007a, ApJL, 668, L99
Quimby, R. M., Castro, F., Gerardy, C. L., et al. 2005, BAAS, 37, #171.02
Quimby, R., Höflich, P., & Wheeler, J. C. 2007b, ApJ, 666, 1083

Quimby, R., Höflich, P., Kannappan, S. J., et al. 2006, ApJ, 636, 400
Quimby, R. M., Kulkarni, S. R., Kasliwal, M. M., et al. 2011, Nature, 474, 487
Quimby, R. M., Kulkarni, S. R., Ofek, E., et al. 2010, The Astronomer's Telegram, 2492
Quimby, R. M., Oguri, M., More, A., et al. 2014, Science, 344, 396
Quimby, R. M., Wheeler, J. C., Höflich, P., et al. 2007c, ApJ, 666, 1093
Quimby, R. M., Yuan, F., Akerlof, C., & Wheeler, J. C. 2013, MNRAS, 431, 912
Quiroga, C., Mikołajewska, J., Brandi, E., et al. 2002, A&A, 387, 139
Rabinak, I., Livne, E., & Waxman, E. 2012, ApJ, 757, 35
Racusin, J. L., Park, S., Zhekov, S., et al. 2009, ApJ, 703, 2
Rafelski, M., Neeleman, M., Fumagalli, M., et al. 2014, ApJL, 782, L29
Rafelski, M., Wolfe, A. M., Prochaska, J. X., et al. 2012, ApJ, 755, 89
Raffelt, G. G., & Zhou, S. 2011, Physical Review D, 83, 093014
Rakavy, G., & Shaviv, G. 1967, ApJ, 148, 803
Ramirez-Ruiz, E., & Serenelli, A. M. 2004, MNRAS, 353, L7
Rankin, J. M. 2015, ApJ, 804, 112
Rantsiou, E., Burrows, A., Nordhaus, J., & Almgren, A. 2011, ApJ, 732, 57
Raskin, C., & Kasen, D. 2013, ApJ, 772, 1
Raskin, C., Kasen, D., Moll, R., et al. 2014, ApJ, 788, 75
Raskin, C., Scannapieco, E., Rockefeller, G., et al. 2010, ApJ, 724, 111
Rau, A., Kulkarni, S. R., Law, N. M., et al. 2009, PASP, 121, 1334
Rauscher, T. 2014, AIP Advances, 4, 041012
Rauscher, T., Dauphas, N., Dillmann, I., et al. 2013, Reports on Progress in Physics, 76, 066201
Ray, A., Singh, K. P., & Sutaria, F. K. 1993, J Astrophys Astron, 14, 53
Rayet, M., Arnould, M., & Prantzos, N. 1990, A&A, 227, 271
Raymond, J. C., Ghavamian, P., Williams, B. J., et al. 2013, ApJ, 778, 161
Rea, N., Israel, G. L., Pons, J. A., et al. 2013, ApJ, 770, 65
Reed, J. E., Hester, J. J., Fabian, A. C., & Winkler, P. F. 1995, ApJ, 440, 706
Reilly, E. 2015, PhD dissertation, University of Sheffield
Reilly, E., Maund, J. R., Baade, D., et al. 2016, MNRAS, 457, 288
Reimers, D. 1975, MSRSL, 8, 369
Reis, R. C., Fabian, A. C., Ross, R. R., & Miller, J. M. 2009, MNRAS, 395, 1257
Renzini, A. 1981, Physical Processes in Red Giants, 88, 431
Repetto, S., Davies, M. B., & Sigurdsson, S. 2012, MNRAS, 425, 2799
Rest, A., Foley, R. J., Gezari, S., et al. 2011a, ApJ, 729, 88
Rest, A., Foley, R. J., Sinnott, B., et al. 2011b, ApJ, 732, 3
Rest, A., Matheson, T., Blondin, S., et al. 2008, ApJ, 680, 1137
Rest, A., Scolnic, D., Foley, R. J., et al. 2014, ApJ, 795, 44
Rest, A., Suntzeff, N. B., Olsen, K., et al. 2005, Nature, 438, 1132
Reynolds, S. P., Borkowski, K. J., Green, D. A., et al. 2008, ApJL, 680, L41
Reynolds, S. P., Borkowski, K. J., Green, D. A., et al. 2009, ApJL, 695, L149
Reynolds, S. P., Borkowski, K. J., Hwang, U., et al. 2007, ApJL, 668, L135
Rezzolla, L., Lamb, F. K., Marković, D., & Shapiro, S. L. 2001a, Physical Review D, 64, 104013
Rezzolla, L., Lamb, F. K., Marković, D., & Shapiro, S. L. 2001b, Physical Review D, 64, 104014
Richardson, D., Jenkins, R. L., III, Wright, J., & Maddox, L. 2014, AJ, 147, 118
Richmond, M. W., & Smith, H. A. 2012, JAAVSO, 40, 872
Richmond, M. W., Treffers, R. R., Filippenko, A. V., & Paik, Y. 1996a, AJ, 112, 732
Richmond, M. W., Treffers, R. R., Filippenko, A. V., et al. 1994, AJ, 107, 1022
Richmond, M. W., van Dyk, S. D., Ho, W., et al. 1996b, AJ, 111, 327
Richtler, T., & Sadler, E. M. 1983, A&A, 128, L3
Riehm, T., Mörtsell, E., Goobar, A., et al. 2011, A&A, 536, A94
Riess, A. G. 2000, PASP, 112, 1284
Riess, A. G., Filippenko, A. V., Challis, P., et al. 1998, AJ, 116, 1009
Riess, A. G., Macri, L. M., Hoffmann, S. L., et al. 2016, ApJ, 826, 56
Riess, A. G., Strolger, L.-G., Casertano, S., et al. 2007, ApJ, 659, 98

Rigault, M., Aldering, G., Kowalski, M., et al. 2015, ApJ, 802, 20
Rigault, M., Copin, Y., Aldering, G., et al. 2013, A&A, 560, A66
Rigon, L., Turatto, M., Benetti, S., et al. 2003, MNRAS, 340, 191
Ripero, J., Garcia, F., Rodriguez, D., et al. 1993, IAU Circ., 5731, 1
Ritchey, A. M., Welty, D. E., Dahlstrom, J. A., & York, D. G. 2015, ApJ, 799, 197
Ritossa, C., Garcia-Berro, E., & Iben, I., Jr. 1996, ApJ, 460, 489
Roccatagliata, V., Preibisch, T., Ratzka, T., & Gaczkowski, B. 2013, A&A, 554, A6
Roche, P. F., Aitken, D. K., Smith, C. H., et al. 1991, MNRAS, 252, 39P
Rodney, S. A., Patel, B., Scolnic, D., et al. 2015, ApJ, 811, 70
Rodney, S. A., Riess, A. G., Strolger, L.-G., et al. 2014, AJ, 148, 13
Rodríguez, Ó., Clocchiatti, A., & Hamuy, M. 2014, AJ, 148, 107
Roederer, I. U., Cowan, J. J., Karakas, A. I., et al. 2010, ApJ, 724, 975
Roediger, J. C., Courteau, S., Graves, G., & Schiavon, R. P. 2014, ApJS, 210, 10
Roelofs, G., Bassa, C., Voss, R., & Nelemans, G. 2008, MNRAS, 391, 290
Romani, R. W., Filippenko, A. V., Silverman, J. M., et al. 2012, ApJL, 760, L36
Roming, P. W. A., Pritchard, T. A., Brown, P. J., et al. 2009, ApJL, 704, L118
Roming, P. W. A., Pritchard, T. A., Prieto, J. L., et al. 2012, ApJ, 751, 92
Röpke, F. K., Kromer, M., Seitenzahl, I. R., et al. 2012, ApJL, 750, L19
Röpke, F. K., & Niemeyer, J. C. 2007, A&A, 464, 683
Röpke, F. K., Sim, S. A., Fink, M., et al. 2013, IAU Symposium, Binary Paths to Type Ia Supernova Explosions, 281, 261
Röpke, F. K., Woosley, S. E., & Hillebrandt, W. 2007, ApJ, 660, 1344 Niemeyer, J. C. 2007, A&A, 464, 683
Rossi, E. M., Kobayashi, S., & Sari, R. 2014, ApJ, 795, 125
Rosswog, S., Kasen, D., Guillochon, J., & Ramirez-Ruiz, E. 2009, ApJL, 705, L128
Roy, R., Kumar, B., Benetti, S., et al. 2011, ApJ, 736, 76
Roy, R., Kumar, B., Maund, J. R., et al. 2013, MNRAS, 434, 2032
Roy, S., & Pal, S. 2013, ApJ, 774, 150
Rubin, K. H. R., Williams, K. A., Bolte, M., & Koester, D. 2008, AJ, 135, 2163
Ruderman, M. A. 1974, Science, 184, 1079
Ruiter, A. J., Belczynski, K., & Fryer, C. 2009, ApJ, 699, 2026
Ruiter, A. J., Belczynski, K., Sim, S. A., et al. 2014, MNRAS, 440, L101
Ruiz-Lapuente, P. 2004, ApJ, 612, 357
Ruiz-Lapuente, P. 2014, New Astron Rev, 62, 15
Ruiz-Lapuente, P., Comeron, F., Méndez, J., et al. 2004, Nature, 431, 1069
Ruiz-Lapuente, P., The, L.-S., Hartmann, D. H., et al. 2016, ApJ, 820, 142
Rupen, M. P., van Gorkom, J. H., Knapp, G. R., et al. 1987, AJ, 94, 61
Russell, B. R., & Immler, S. 2012, ApJL, 748, L29
Sagan, C. & Shklovskii 1966, Intelligent Life in the Universe (Holden–Day)
Sagert, I., Fischer, T., Hempel, M., et al. 2009, PRL, 102, 081101
Sahman, D. I., Dhillon, V. S., Marsh, T. R., et al. 2013, MNRAS, 433, 1588
Sahu, D. K., Anupama, G. C., & Chakradhari, N. K. 2013, MNRAS, 433, 2
Sahu, D. K., Anupama, G. C., Srividya, S., & Muneer, S. 2006, MNRAS, 372, 1315
Sahu, D. K., Gurugubelli, U. K., Anupama, G. C., & Nomoto, K. 2011, MNRAS, 413, 2583
Sahu, D. K., Tanaka, M., Anupama, G. C., et al. 2008, ApJ, 680, 580
Sahu, D. K., Tanaka, M., Anupama, G. C., et al. 2009, ApJ, 697, 67
Saio, H., & Nomoto, K. 1985, A&A, 150, L21
Saio, H., & Nomoto, K. 2004, ApJ, 615, 444
Sakon, I., Onaka, T., Wada, T., et al. 2009, ApJ, 692, 546
Salpeter, E. E. 1955, ApJ, 121, 161
Salpeter, E. E. 1976, ApJ, 206, 673
Salvadori, S., & Ferrara, A. 2012, MNRAS, 421, L29
Salvo, M. E., Cappellaro, E., Mazzali, P. A., et al. 2001, MNRAS, 321, 254
Sana, H., de Koter, A., de Mink, S. E., et al. 2013, A&A, 550, A107

Sana, H., de Mink, S. E., de Koter, A., et al. 2012, Science, 337, 444
Sand, D. J., Graham, M. L., Bildfell, C., et al. 2012, ApJ, 746, 163
Sanders, N. E., Soderberg, A. M., Foley, R. J., et al. 2013, ApJ, 769, 39
Sanders, N. E., Soderberg, A. M., Gezari, S., et al. 2015, ApJ, 799, 208
Sanduleak, N. 1969, AJ, 74, 877
Sarangi, A., & Cherchneff, I. 2015, A&A, 575, A95
Sarikas, S., Tamborra, I., Raffelt, G., et al. 2012, Physical Review D, 85, 113007
Sartore, N., Ripamonti, E., Treves, A., & Turolla, R. 2010, A&A, 510, A23
Saruwatari, M., & Hashimoto, M. 2010, Progress of Theoretical Physics, 124, 925
Sasdelli, M., Mazzali, P. A., Pian, E., et al. 2014, MNRAS, 445, 711
Sato, K. 1975, Progress of Theoretical Physics, 53, 595
Sato, Y., Nakasato, N., Tanikawa, A., et al. 2015, ApJ, 807, 105
Satterfield, T. J., Katz, A. M., Sibley, A. R., et al. 2012, AJ, 144, 27
Sauer, D. N., Mazzali, P. A., Blondin, S., et al. 2008, MNRAS, 391, 1605
Sauer, D. N., Mazzali, P. A., Deng, J., et al. 2006, MNRAS, 369, 1939
Saunders, C., Aldering, G., Antilogus, P., et al. 2015, ApJ, 800, 57
Sawai, H., & Yamada, S. 2014, ApJL, 784, L10
Scalo, J. M. 1973, ApJ, 184, 801
Scalo, J. M. 1986, Fundamental Cosmic Physics, 11, 1
Scalo, J. M., Despain, K. H., & Ulrich, R. K. 1975, ApJ, 196, 805
Scalo, J., & Wheeler, J. C. 2002, ApJ, 566, 723
Scalo, J. M., Wheeler, J. C., & Williams, P. 2001, arXiv:astro-ph/0104209; in Frontiers of Life; 12th Rencontres de Blois, ed. L. M. Celnikier
Scalzo, R., Aldering, G., Antilogus, P., et al. 2012, ApJ, 757, 12
Scalzo, R. A., Aldering, G., Antilogus, P., et al. 2010, ApJ, 713, 1073
Scalzo, R., Aldering, G., Antilogus, P., et al. 2014a, MNRAS, 440, 1498
Scalzo, R. A., Childress, M., Tucker, B., et al. 2014b, MNRAS, 445, 30
Scalzo, R. A., Ruiter, A. J., & Sim, S. A. 2014c, MNRAS, 445, 2535
Scannapieco, E., & Bildsten, L. 2005, ApJL, 629, L85
Scannapieco, E., Ferrara, A., Heger, A., et al. 2006, New Astron Rev, 50, 89
Schaefer, B. E. 1993, PASP, 105, 1238
Schaefer, B. E. 2007, ApJ, 660, 16
Schaefer, B. E., & Girard, T. M. 1999, in Anni Mirabiles, ed. A. G. Davis Phillip, W. F. Van Altena, A. R. Upgren, 69
Schaefer, B. E., & Pagnotta, A. 2012, Nature, 481, 164
Schawinski, K. 2009, MNRAS, 397, 717
Schawinski, K., Justham, S., Wolf, C., et al. 2008, Science, 321, 223
Schinzel, F. K., Taylor, G. B., Stockdale, C. J., et al. 2009, ApJ, 691, 1380
Schlegel, E. M. 1990, MNRAS, 244, 269
Schlegel, E. M. 1994, AJ, 108, 1893
Schlegel, E. M. 1996, AJ, 111, 1660
Schlegel, E. M., & Kirshner, R. P. 1989, AJ, 98, 577
Schlegel, E. M., & Petre, R. 2006, ApJ, 646, 378
Schmidt, B. P., Kirshner, R. P., Eastman, R. G., et al. 1993, Nature, 364, 600
Schmidt, B. P., Suntzeff, N. B., Phillips, M. M., et al. 1998, ApJ, 507, 46
Schmidt, W., Ciaraldi-Schoolmann, F., Niemeyer, J. C., et al. 2010, ApJ, 710, 1683
Scholberg, K. 2012, ARN&PS, 62, 81
Schwab, J., Podsiadlowski, P., & Rappaport, S. 2010, ApJ, 719, 722
Schwab, J., Quataert, E., & Bildsten, L. 2015, MNRAS, 453, 1910
Schwab, J., Shen, K. J., Quataert, E., et al. 2012, MNRAS, 427, 190
Schwarz, H. E. 1987, A&A, 177, L4
Schwarzschild, M. 1958 Structure and Evolution of the Stars (Princeton Un. Press)
Schwarzschild, M., & Härm, R. 1965, ApJ, 142, 855
Schweizer, F., & Middleditch, J. 1980, ApJ, 241, 1039

Scovacricchi, D., Nichol, R. C., Bacon, D., et al. 2016, MNRAS, 456, 1700
Sedov, L. 1959, Similarity and Dimensional Methods in Mechanics (Academic Press)
Seeger, P. A., Fowler, W. A., & Clayton, D. D. 1965, ApJS, 11, 121
Seifried, D., Banerjee, R., & Schleicher, D. 2014, MNRAS, 440, 24
Seitenzahl, I. R., Cescutti, G., Röpke, F. K., et al. 2013a, A&A, 559, L5
Seitenzahl, I. R., Ciaraldi-Schoolmann, F., Röpke, F. K., et al. 2013b, MNRAS, 429, 1156
Seitenzahl, I. R., Kromer, M., Ohlmann, S. T., et al. 2016, A&A, 592, A57
Seitenzahl, I. R., Taubenberger, S., & Sim, S. A. 2009, MNRAS, 400, 531
Seitenzahl, I. R., Timmes, F. X., & Magkotsios, G. 2014, ApJ, 792, 10
Sekiguchi, Y., & Shibata, M. 2011, ApJ, 737, 6
Sell, P. H., Maccarone, T. J., Kotak, R., et al. 2015, MNRAS, 450, 4198
Senno, N., Mészáros, P., Murase, K., et al. 2015, ApJ, 806, 24
Seok, J. Y., Koo, B.-C., & Onaka, T. 2013, ApJ, 779, 134
Shabaltas, N., & Lai, D. 2012, ApJ, 748, 148
Shafter, A. W., Henze, M., Rector, T. A., et al. 2015, ApJS, 216, 34
Shapiro, P. R., & Field, G. B. 1976, ApJ, 205, 762
Shapiro, S. L., & Teukolsky, S. A. 1986, Black Holes, White Dwarfs and Neutron Stars: The Physics of Compact Objects (Wiley)
Shappee, B. J., Kochanek, C. S., & Stanek, K. Z. 2013a, ApJ, 765, 150
Shappee, B., Prieto, J., Stanek, K. Z., et al. 2014, AAS Meeting 223, #236.03
Shappee, B. J., Stanek, K. Z., Pogge, R. W., & Garnavich, P. M. 2013b, ApJL, 762, L5
Shara, M. M., & Prialnik, D. 1994, AJ, 107, 1542
Sharma, P., Roy, A., Nath, B. B., & Shchekinov, Y. 2014, MNRAS, 443, 3463
Shen, G., Horowitz, C. J., & Teige, S. 2010a, Physical Review C, 82, 045802
Shen, K. J., & Bildsten, L. 2014, ApJ, 785, 61
Shen, K. J., Bildsten, L., Kasen, D., & Quataert, E. 2012, ApJ, 748, 35
Shen, K. J., Guillochon, J., & Foley, R. J. 2013, ApJL, 770, L35
Shen, K. J., Kasen, D., Weinberg, N. N., et al. 2010b, ApJ, 715, 767
Shen, K. J., & Moore, K. 2014, ApJ, 797, 46
Shi, G., Qian, S.-B., & Fernández Lajús, E. 2014, Publications of the ASJ, 66, 8
Shibata, M., Karino, S., & Eriguchi, Y. 2003, MNRAS, 343, 619
Shibata, M., & Sekiguchi, Y.–I. 2005, Physical Review D, 71, 024014
Shigeyama, T., & Nomoto, K. 1990, ApJ, 360, 242
Shigeyama, T., Nomoto, K., Tsujimoto, T., & Hashimoto, M.-A. 1990, ApJL, 361, L23
Shigeyama, T., Nomoto, K., Yamaoka, H., & Thielemann, F.-K. 1992, ApJL, 386, L13
Shigeyama, T., Suzuki, T., Kumagai, S., et al. 1994, ApJ, 420, 341
Shiode, J. H., & Quataert, E. 2014, ApJ, 780, 96
Shivvers, I., Groh, J. H., Mauerhan, J. C., et al. 2015, ApJ, 806, 213
Shternin, P. S., Yakovlev, D. G., Heinke, C. O., et al. 2011, MNRAS, 412, L108
Shtykovskiy, P. E., Lutovinov, A. A., Gilfanov, M. R., & Sunyaev, R. A. 2005, Astronomy Letters, 31, 258
Siegert, T., Diehl, R., Krause, M. G. H., & Greiner, J. 2015, A&A, 579, A124
Siess, L. 2007, A&A, 476, 893
Siess, L. 2010, A&A, 512, A10
Silk, J. 1997, ApJ, 481, 703
Silverman, J. M., & Filippenko, A. V. 2012, MNRAS, 425, 1917
Silverman, J. M., Ganeshalingam, M., Cenko, S. B., et al. 2012a, ApJL, 756, L7
Silverman, J. M., Ganeshalingam, M., & Filippenko, A. V. 2013a, MNRAS, 430, 1030
Silverman, J. M., Ganeshalingam, M., Li, W., & Filippenko, A. V. 2012b, MNRAS, 425, 1889
Silverman, J. M., Ganeshalingam, M., Li, W., et al. 2011, MNRAS, 410, 585
Silverman, J. M., Kong, J. J., & Filippenko, A. V. 2012c, MNRAS, 425, 1819
Silverman, J. M., Nugent, P. E., Gal-Yam, A., et al. 2013b, ApJS, 207, 3
Silverman, J. M., Nugent, P. E., Gal-Yam, A., et al. 2013c, ApJ, 772, 125
Silverman, J. M., Vinkó, J., Kasliwal, M. M., et al. 2013d, MNRAS, 436, 1225

Silverman, J. M., Vinkó, J., Marion, G. H., et al. 2015, MNRAS, 451, 1973
Sim, S. A., Fink, M., Kromer, M., et al. 2012, MNRAS, 420, 3003
Sim, S. A., Röpke, F. K., Hillebrandt, W., et al. 2010, ApJL, 714, L52
Sim, S. A., Seitenzahl, I. R., Kromer, M., et al. 2013, MNRAS, 436, 333
Simon, J. D., Gal-Yam, A., Gnat, O., et al. 2009, ApJ, 702, 1157
Sinnott, B., Welch, D. L., Rest, A., Sutherland, P. G., & Bergmann, M. 2013, ApJ, 767, 45
Siverd, R. J., Goobar, A., Stassun, K. G., & Pepper, J. 2015, ApJ, 799, 105
Skopal, A. 2015, Proceedings of the National Academy, 34, 123
Slavin, J. D., Dwek, E., & Jones, A. P. 2015, ApJ, 803, 7
Smartt, S. J. 2009, ARAA, 47, 63
Smartt, S. J. 2015, Publications of the Astron. Soc. of Australia, 32, 16
Smartt, S. J., Eldridge, J. J., Crockett, R. M., & Maund, J. R. 2009, MNRAS, 395, 1409
Smartt, S. J., Valenti, S., Fraser, M., et al. 2015, A&A, 579, A40
Smidt, J., Whalen, D. J., Chatzopoulos, E., et al. 2015, ApJ, 805, 44
Smidt, J., Whalen, D. J., Wiggins, B. K., et al. 2014, ApJ, 797, 97
Smith, D. S., Scalo, J., & Wheeler, J. C. 2004a, Origins of Life and Evolution of the Biosphere, 34, 513
Smith, D. S., Scalo, J., & Wheeler, J. C. 2004b, Icarus, 171, 229
Smith, J. A., Tucker, D. L., Kent, S., et al. 2002, AJ, 123, 2121
Smith, N. 2007, AIP Conf. Proc., 937, 163
Smith, N. 2013, MNRAS, 434, 102
Smith, N. 2014, ARAA, 52, 487
Smith, N., Foley, R. J., Bloom, J. S., et al. 2010a, ApJ, 686, 485–491
Smith, N., Arnett, W. D., Bally, J., et al. 2013a, MNRAS, 429, 1324
Smith, N., Chornock, R., Li, W., et al. 2010b, ApJ, 686, 467–484
Smith, N., Chornock, R., Silverman, J. M., et al. 2010c, ApJ, 709, 856
Smith, N., Foley, R. J., & Filippenko, A. V. 2008, ApJ, 680, 568
Smith, N., Ganeshalingam, M., Chornock, R., et al. 2009a, ApJL, 697, L49
Smith, N., Li, W., Filippenko, A. V., & Chornock, R. 2011a, MNRAS, 412, 1522
Smith, N., Li, W., Foley, R. J., et al. 2007, ApJ, 666, 1116
Smith, N., Li, W., Miller, A. A., et al. 2011b, ApJ, 732, 63
Smith, N., Li, W., Silverman, J. M., et al. 2011c, MNRAS, 415, 773
Smith, N., Mauerhan, J. C., Cenko, S. B., et al. 2015, MNRAS, 449, 1876
Smith, N., Mauerhan, J. C., Kasliwal, M. M., & Burgasser, A. J. 2013b, MNRAS, 434, 2721
Smith, N., Mauerhan, J. C., & Prieto, J. L. 2014a, MNRAS, 438, 1191
Smith, N., Mauerhan, J. C., Silverman, J. M., et al. 2012a, MNRAS, 426, 1905
Smith, N., & McCray, R. 2007, ApJL, 671, L17
Smith, N., Miller, A., Li, W., et al. 2010d, AJ, 139, 1451
Smith, N., & Owocki, S. P. 2006, ApJL, 645, L45
Smith, N., Silverman, J. M., Chornock, R., et al. 2009b, ApJ, 695, 1334
Smith, N., Silverman, J. M., Filippenko, A. V., et al. 2012b, AJ, 143, 17
Smith, R. M., Dekany, R. G., Bebek, C., et al. 2014b, Proceedings of the SPIE, 9147, 914779
Smith, V. V., & Lambert, D. L. 1990, ApJS, 72, 387
Sneden, C., Cowan, J. J., & Gallino, R. 2008, ARAA, 46, 241
Sobolev, V. V. 1960, Moving Envelopes of Stars (Harvard Un. Press)
Soderberg, A. M., Berger, E., Page, K. L., et al. 2008, Nature, 453, 469
Soderberg, A. M., Chevalier, R. A., Kulkarni, S. R., & Frail, D. A. 2006a, ApJ, 651, 1005
Soderberg, A. M., Gal-Yam, A., & Kulkarni, S. R. 2004, GCN, 2586, 1
Soderberg, A. M., Kulkarni, S. R., Nakar, E., et al. 2006b, Nature, 442, 1014
Soderberg, A. M., Margutti, R., Zauderer, B. A., et al. 2012, ApJ, 752, 78
Soker, N. 2015, MNRAS, 450, 1333
Soker, N., Kashi, A., García-Berro, E., et al. 2013, MNRAS, 431, 1541
Sollerman, J., Cumming, R. J., & Lundqvist, P. 1998, ApJ, 493, 933
Sollerman, J., Jaunsen, A. O., Fynbo, J. P. U., et al. 2006, A&A, 454, 503

Sollerman, J., Kozma, C., & Lundqvist, P. 2001, A&A, 366, 197
Song, H. F., Maeder, A., Meynet, G., et al. 2013, A&A, 556, A100
Sonneborn, G., Altner, B., & Kirshner, R. P. 1987, ApJL, 323, L35
Soraisam, M. D., & Gilfanov, M. 2015, A&A, 583, A140
Soria, R., Balestrieri, R., & Ohtsuka, Y. 2013, Publications of the Astron. Soc. of Australia, 30, 28
Sorokina, E., Blinnikov, S., Nomoto, K., et al. 2016, ApJ, 829, 17
Spergel, D., Gehrels, N., Baltay, C., et al. 2015, arXiv:1503.03757
Spiro, S., Pastorello, A., Pumo, M. L., et al. 2014, MNRAS, 439, 2873
Spitoni, E., Matteucci, F., & Sozzetti, A. 2014, MNRAS, 440, 2588
Spitzer, L. 1998, Physical Processes in the Interstellar Medium (Wiley)
Spruit, H. C. 1998, A&A, 333, 603
Spruit, H. C. 1999, A&A, 349, 189
Spruit, H. C. 2002, A&A, 381, 923
Spyromilio, J., Meikle, W.P.S., & Allen, D. A. 1990, MNRAS, 242, 669
Spyromilio, J., Leibundgut, B., & Gilmozzi, R. 2001, A&A, 376, 188
Sramek, R. A., Panagia, N., & Weiler, K. W. 1984, ApJL, 285, L59
Sramek, R. A., & Weiler, K. W. 2003, in Supernovae and Gamma-Ray Bursters, ed. K. W. Weiler (Springer), 145
Srivastav, S., Anupama, G. C., & Sahu, D. K. 2014, MNRAS, 445, 1932
Srivastav, S., Ninan, J. P., Kumar, B., et al. 2016, MNRAS, 457, 1000
Stacy, A., Greif, T. H., & Bromm, V. 2010, MNRAS, 403, 45
Stancliffe, R. J., & Eldridge, J. J. 2009, MNRAS, 396, 1699
Stanishev, V., Goobar, A., Amanullah, R., et al. 2015, arXiv:1505.07707
Stanishev, V., Goobar, A., Benetti, S., et al. 2007, A&A, 469, 645
Starrfield, S. 2015, arXiv:1502.02665
Starrfield, S., Sparks, W. M., & Truran, J. W. 1985, ApJ, 291, 136
Starrfield, S., Timmes, F. X., Hix, W. R., et al. 2013, IAU Symposium, 281, 166
Stathakis, R. A., & Sadler, E. M. 1991, MNRAS, 250, 786
Stein, J., Barkat, Z., & Wheeler, J. C. 1999, ApJ, 523, 381
Steiner, A. W., Lattimer, J. M., & Brown, E. F. 2010, ApJ, 722, 33
Stephenson, F. R. 2010, Astronomy and Geophysics, 51, 27
Stephenson, F. R., & Green, D. A. 2002, Historical Supernovae and their Remnants (Clarendon Press)
Sternberg, A., Gal-Yam, A., Simon, J. D., et al. 2011, Science, 333, 856
Sternberg, A., Gal-Yam, A., Simon, J. D., et al. 2014, MNRAS, 443, 1849
Stevenson, D. J. 1977, PASA, 3, 167
Stockdale, C. J., Rupen, M. P., Cowan, J. J., et al. 2001, AJ, 122, 283
Stockdale, C. J., Van Dyk, S. D., Weiler, K. W., et al. 2004, BAAS, 36, 1464
Stodolsky, L. 1988, Physics Letters B, 201, 353
Stoll, R., Prieto, J. L., Stanek, K. Z., et al. 2011, ApJ, 730, 34
Strickland, D. K., & Heckman, T. M. 2009, ApJ, 697, 2030
Stritzinger, M., Hamuy, M., Suntzeff, N. B., et al. 2002, AJ, 124, 2100
Stritzinger, M. D., Hsiao, E., Valenti, S., et al. 2014, A&A, 561, A146
Stritzinger, M., Leibundgut, B., Walch, S., & Contardo, G. 2006, A&A, 450, 241
Stritzinger, M., Mazzali, P., Phillips, M. M., et al. 2009, ApJ, 696, 713
Stritzinger, M. D., Phillips, M. M., Boldt, L. N., et al. 2011, AJ, 142, 156
Stritzinger, M., Taddia, F., Fransson, C., et al. 2012, ApJ, 756, 173
Stritzinger, M. D., Valenti, S., Höeflich, P., et al. 2015, A&A, 573, AA2
Strolger, L.-G., Riess, A. G., Dahlen, T., et al. 2004, ApJ, 613, 200
Sugerman, B. E. K., Andrews, J. E., Barlow, M. J., et al. 2012, ApJ, 749, 170
Suh, H., Yoon, S.-c., Jeong, H., & Yi, S. K. 2011, ApJ, 730, 110
Sukhbold, T., & Woosley, S. E. 2014, ApJ, 783, 10
Suleimanov, V. F., Klochkov, D., Pavlov, G. G., & Werner, K. 2014, ApJS, 210, 13
Sullivan, M., Conley, A., Howell, D. A., et al. 2010, MNRAS, 406, 782

Sullivan, M., Ellis, R. S., Howell, D. A., et al. 2009, ApJL, 693, L76
Sullivan, M., Kasliwal, M. M., Nugent, P. E., et al. 2011, ApJ, 732, 118
Sumiyoshi, K., Takiwaki, T., Matsufuru, H., & Yamada, S. 2015, ApJS, 216, 5
Sumiyoshi, K., & Yamada, S. 2012, ApJS, 199, 17
Suntzeff, N., & Bouchet, P. 1990, AJ, 650
Suntzeff, N. B., Hamuy, M., Martin, G., et al. 1988, AJ, 96, 1864
Sunyaev, R. A., Kaniovskii, A., Efremov, V., et al. 1987, Soviet Astronomy Letters, 13, 431
Sunyaev, R. A.; Kaniovskii, A. S.; Efremov, V. V., et al. 1989, Soviet Astronomy Letters, 15, 125
Surman, R., McLaughlin, G. C., & Hix, W. R. 2006, ApJ, 643, 1057
Sutantyo, W. 1978, ApSpSci, 54, 479
Sutherland, P. G., & Wheeler, J. C. 1984, ApJ, 280, 282
Suwa, Y., Kotake, K., Takiwaki, T., et al. 2011, ApJ, 738, 165
Suwa, Y., Yoshida, T., Shibata, M., et al. 2015, MNRAS, 454, 3073
Suzuki, A., & Shigeyama, T. 2010, ApJL, 717, L154
Suzuki, N., Rubin, D., Lidman, C., et al. 2012, ApJ, 746, 85
Suzuki, T. K., Sumiyoshi, K., & Yamada, S. 2008, ApJ, 678, 1200
Suárez-Andrés, L., González Hernández, J. I., Israelian, G., Casares, J., & Rebolo, R. 2015, MNRAS, 447, 2261
Svensmark, H. 2012, MNRAS, 423, 1234
Svirski, G., & Nakar, E. 2014a, ApJ, 788, 113
Svirski, G., & Nakar, E. 2014b, ApJL, 788, L14
Svirski, G., Nakar, E., & Sari, R. 2012, ApJ, 759, 108
Swartz, D. A. 1991, ApJ, 373, 604
Swartz, D. A. 1994, ApJ, 428, 267
Swartz, D. A., Clocchiatti, A., Benjamin, R., et al. 1993a, Nature, 365, 232
Swartz, D. A., Filippenko, A. V., Nomoto, K., & Wheeler, J. C. 1993b, ApJ, 411, 313
Swartz, D. A., Sutherland, P. G., & Harkness, R. P. 1995, ApJ, 446, 766
Swartz, D. A., & Wheeler, J. C. 1991, ApJL, 379, L13
Symbalisty, E. M. D. 1984, ApJ, 285, 729
Szalai, T., Vinkó, J., Balog, Z., et al. 2011, A&A, 527, A61
Szalai, T., Vinkó, J., Nagy, A. P., et al. 2016, MNRAS, 460, 200
Taam, R. E., & Ricker, P. M. 2010, New Astron Rev, 54, 65
Taani, A., Naso, L., Wei, Y., et al. 2012, ApSpSci, 341, 601
Tachibana, S., Huss, G. R., Kita, N. T., et al. 2006, ApJL, 639, L87
Taddia, F., Sollerman, J., Fremling, C., et al. 2015a, A&A, 580, A131
Taddia, F., Sollerman, J., Leloudas, G., et al. 2015b, A&A, 574, A60
Taddia, F., Sollerman, J., Razza, A., et al. 2013, A&A, 558, A143
Taddia, F., Stritzinger, M. D., Phillips, M. M., et al. 2012a, A&A, 545, L7
Taddia, F., Stritzinger, M. D., Sollerman, J., et al. 2012b, A&A, 537, A140
Takahashi, K., Yoshida, T., & Umeda, H. 2013, ApJ, 771, 28
Takiwaki, T., Kotake, K., & Suwa, Y. 2012, ApJ, 749, 98
Takáts, K., Pignata, G., Pumo, M. L., et al. 2015, MNRAS, 450, 3137
Takáts, K., Pumo, M. L., Elias-Rosa, N., et al. 2014, MNRAS, 438, 368
Takáts, K., & Vinkó, J. 2006, MNRAS, 372, 1735
Tamai, R., & Spyromilio, J. 2014, Proceedings of the SPIE, 9145, 91451E
Tamborra, I., Hanke, F., Janka, H.-T., et al. 2014a, ApJ, 792, 96
Tamborra, I., Raffelt, G., Hanke, et al. 2014b, Physical Review D, 90, 045032
Tanaka, M. 2016, Advances in Astronomy, 2016, 634197
Tanaka, M., Kawabata, K. S., Hattori, T., et al. 2012a, ApJ, 754, 63
Tanaka, M., Kawabata, K. S., Maeda, K., et al. 2008, ApJ, 689, 1191
Tanaka, M., Kawabata, K. S., Maeda, K., et al. 2009a, ApJ, 699, 1119
Tanaka, M., Kawabata, K. S., Yamanaka, M., et al. 2010, ApJ, 714, 1209
Tanaka, M., Mazzali, P. A., Maeda, K., & Nomoto, K. 2006, ApJ, 645, 470

Tanaka, M., Nozawa, T., Sakon, I., et al. 2012b, ApJ, 749, 173
Tanaka, M., Tominaga, N., Morokuma, T., et al. 2016, ApJ, 819, 5
Tanaka, M., Yamanaka, M., Maeda, K., et al. 2009b, ApJ, 700, 1680
Tanaka, Y., & Lewin, W. H. G. 1995, X-ray Binaries, 126
Tananbaum, H. 1999, IAU Circ., 7246, 1
Tang, H., & Dauphas, N. 2012, Earth and Planetary Science Letters, 359, 248
Tang, H., & Dauphas, N. 2015, ApJ, 802, 22
Tang, S., Bildsten, L., Wolf, W. M., et al. 2014, ApJ, 786, 61
Tang, S., Wang, Q. D., Mac Low, M.-M., & Joung, M. R. 2009, MNRAS, 398, 1468
Tanikawa, A., Nakasato, N., Sato, Y., et al. 2015, ApJ, 807, 40
Taubenberger, S., Benetti, S., Childress, M., et al. 2011a, MNRAS, 412, 2735
Taubenberger, S., Elias-Rosa, N., Kerzendorf, W. E., et al. 2015, MNRAS, 448, L48
Taubenberger, S., Hachinger, S., Pignata, G., et al. 2008, MNRAS, 385, 75
Taubenberger, S., Kromer, M., Hachinger, S., et al. 2013a, MNRAS, 432, 3117
Taubenberger, S., Kromer, M., Pakmor, R., et al. 2013b, ApJL, 775, L43
Taubenberger, S., Navasardyan, H., Maurer, J. I., et al. 2011b, MNRAS, 413, 214
Taubenberger, S., Pastorello, A., Mazzali, P. A., et al. 2006, MNRAS, 371, 1459
Taubenberger, S., Valenti, S., Benetti, S., et al. 2009, MNRAS, 397, 677
Tauris, T. M. 2015, MNRAS, 448, L6
Tauris, T. M., Langer, N., Moriya, T. J., et al. 2013a, ApJL, 778, L23
Tauris, T. M., Sanyal, D., Yoon, S.-C., & Langer, N. 2013b, A&A, 558, A39
Tayar, J., & Pinsonneault, M. H. 2013, ApJL, 775, L1
Telesco, C. M., Höflich, P., Li, D., et al. 2015, ApJ, 798, 93
Telezhinsky, I., Dwarkadas, V. V., & Pohl, M. 2012, Astroparticle Physics, 35, 300
Telezhinsky, I., Dwarkadas, V. V., & Pohl, M. 2013, A&A, 552, A102
Temim, T., & Dwek, E. 2013, ApJ, 774, 8
Temim, T., Sonneborn, G., Dwek, E., et al. 2012, ApJ, 753, 72
Terman, J. L., Taam, R. E., & Hernquist, L. 1995, ApJ, 445, 367
Tetzlaff, N., Neuhäuser, R., Hohle, M. M., & Maciejewski, G. 2010, MNRAS, 402, 2369
The, L.-S., & Burrows, A. 2014, ApJ, 786, 141
Thielemann, F.-K., Arcones, A., Käppeli, R., et al. 2011a, Progress in Particle and Nuclear Physics, 66, 346
Thielemann, F.-K., Argast, D., Brachwitz, F., et al. 2003, Nuclear Physics A, 718, 139
Thielemann, F.-K., Hashimoto, M.-A., Nomoto, K., et al. 1990, ApJ, 349, 222
Thielemann, F.-K., Hirschi, R., Liebendörfer, M., & Diehl, R. 2011b, Lecture Notes in Physics, 812, 153
Thielemann, F.-K., Nomoto, K., & Hashimoto, M.-A. 1996, ApJ, 460, 408
Thomas, B. C., Engler, E. E., Kachelrieß, M., et al. 2016, ApJL, 826, L3
Thomas, R. C., Branch, D., Baron, E., et al. 2004, ApJ, 601, 1019
Thomas, R. C., Nugent, P. E., & Meza, J. C. 2011, PASP, 123, 237
Thompson, G. D., Evans, R. O., Hers, J., et al. 1983, IAU Circ., 3835, 1
Thompson, L. A. 1982, ApJL, 257, L63
Thompson, T. A. 2011, ApJ, 741, 82
Thompson, T. A., Chang, P., & Quataert, E. 2004, ApJ, 611, 380
Thompson, T. A., Prieto, J. L., Stanek, K. Z., Kistler, M. D., et al. 2009, ApJ, 705, 1364
Thompson, T. A., Quataert, E., & Burrows, A. 2005, ApJ, 620, 861
Thomson, M. G., & Chary, R. R. 2011, ApJ, 731, 72
Thöne, C. C., de Ugarte Postigo, A., García-Benito, R., et al. 2015, MNRAS, 451, L65
Thöne, C. C., Michałowski, M. J., Leloudas, G., et al. 2009, ApJ, 698, 1307
Thorne, K. S. 1974, ApJ, 191, 507
Thorne, K. S., & Żytkow, A. N. 1977, ApJ, 212, 832
Thoroughgood, T. D., Dhillon, V. S., Littlefair, S. P., et al. 2001, MNRAS, 327, 1323
Timmes, F. X. 2000, ApJ, 528, 913
Timmes, F. X., Brown, E. F., & Truran, J. W. 2003, ApJL, 590, L83

Timmes, F. X., & Woosley, S. E. 1992, ApJ, 396, 649
Timmes, F. X., Woosley, S. E., & Weaver, T. A. 1995, ApJS, 98, 617
Timmes, F. X., Woosley, S. E., & Weaver, T. A. 1996, ApJ, 457, 834
Tissera, P. B., Beers, T. C., Carollo, D., & Scannapieco, C. 2014, MNRAS, 439, 3128
Toma, K., Yoon, S.-C., & Bromm, V. 2016, SSR, 202, 159
Tomasella , L., Benetti, S., Cappellaro, E., et al. 2014, AN, 335, 84
Tomasella, L., Cappellaro, E., Fraser, M., et al. 2013, MNRAS, 434, 1636
Tominaga, N., Limongi, M., Suzuki, T., et al. 2008, ApJ, 687, 1208
Tominaga, N., Blinnikov, S., Baklanov, P., et al. 2009, ApJL, 705, L10
Tominaga, N., Blinnikov, S. I., & Nomoto, K. 2013, ApJL, 771, L12
Tominaga, N., Iwamoto, N., & Nomoto, K. 2014, ApJ, 785, 98
Tominaga, N., Maeda, K., Umeda, H., et al. 2007a, ApJL, 657, L77
Tominaga, N., Morokuma, T., Blinnikov, S. I., et al. 2011, ApJS, 193, 20
Tominaga, N., Tanaka, M., Nomoto, K., et al. 2005, ApJL, 633, L97
Tominaga, N., Umeda, H., & Nomoto, K. 2007b, ApJ, 660, 516
Tomita, H., Deng, J., Maeda, K., et al. 2006, ApJ, 644, 400
Toonen, S., Nelemans, G., & Portegies Zwart, S. 2012, A&A, 546, A70
Toonen, S., Voss, R., & Knigge, C. 2014, MNRAS, 441, 354
Tornambé, A., & Piersanti, L. 2013, MNRAS, 431, 1812
Totani, T., Morokuma, T., Oda, T., et al. 2008, Publications of the ASJ, 60, 1327
Tout, C. A., Żytkow, A. N., Church, R. P., et al. 2014, MNRAS, 445, L36
Townsley, D. M., & Bildsten, L. 2005, ApJ, 628, 395
Townsley, D. M., Moore, K., & Bildsten, L. 2012, ApJ, 755, 4
Trammell, S. R., Hines, D. C., & Wheeler, J. C. 1993, ApJL, 414, L21
Tran, H. D., Filippenko, A. V., Schmidt, G. D., et al. 1997, PASP, 109, 489
Travaglio, C., Gallino, R., Arnone, E., et al. 2004, ApJ, 601, 864
Travaglio, C., Gallino, R., Rauscher, T., et al. 2015, ApJ, 799, 54
Trimble, V. 1973, PASP, 85, 579
Trimble, V. 1982, RMP, 54, 1183
Tripp, R. 1998, A&A, 331, 815
Troja, E., Segreto, A., La Parola, V., et al. 2014, ApJL, 797, L6
Trundle, C., Kotak, R., Vink, J. S., & Meikle, W. P. S. 2008, A&A, 483, L47
Trundle, C., Pastorello, A., Benetti, S., et al. 2009, A&A, 504, 945
Truran, J. W., Cowan, J. J., Pilachowski, C. A., & Sneden, C. 2002, PASP, 114, 1293
Truran, J. W., & Livio, M. 1986, ApJ, 308, 721
Tsebrenko, D., & Soker, N. 2015, MNRAS, 447, 2568
Tsvetkov, D. Y., Metlov, V. G., Shugarov, S. Y., et al. 2014, Contr. Astron. Obs. Skalnate Pleso, 44, 67
Tsvetkov, D. Y., Shugarov, S. Y., Volkov, I. M., et al. 2013, Contr. Astron. Obs. Skalnate Pleso, 43, 94
Tsvetkov, D. Y., Volkov, I. M., Sorokina, E., et al. 2012, P Zvezdy, 32, 6
Tueller, J., Barthelmy, S., Gehrels, N., et al. 1990, ApJL, 351, L41
Tully, R. B., & Fisher, J. R. 1977, A&A, 54, 661
Tur, C., Heger, A., & Austin, S. M. 2010, ApJ, 718, 357
Turatto, M., Cappellaro, E., Danziger, I. J., et al. 1993, MNRAS, 262, 128
Turatto, M., Mazzali, P. A., Young, T. R., et al. 1998, ApJL, 498, L129
Turatto, M., Suzuki, T., Mazzali, P. A., et al. 2000, ApJL, 534, L57
Turtle, A. J., Campbell-Wilson, D., Bunton, J. D., et al. 1987, Nature, 327, 38
Uchida, H., Yamaguchi, H., & Koyama, K. 2013, ApJ, 771, 56
Ugliano, M., Janka, H.–T., Marek, A., & Arcones, A. 2012, ApJ, 757, 69
Umeda, H., & Nomoto, K. 2003, Nature, 422, 871
Umeda, H., & Nomoto, K. 2005, ApJ, 619, 427
Umeda, H., & Nomoto, K. 2008, ApJ, 673, 1014
Umeda, H., Nomoto, K., Yamaoka, H., & Wanajo, S. 1999, ApJ, 513, 861

Uno, S., Mitsuda, K., Inoue, H., et al. 2002, ApJ, 565, 419
Uomoto, A. 1991, AJ, 101, 1275
Uomoto, A., & Kirshner, R. P. 1985, A&A, 149, L7
Uomoto, A., & Kirshner, R. P. 1986, ApJ, 308, 685
Usov, V. V. 1992, Nature, 357, 472
Utrobin, V. P. 1994, A&A, 281, L89
Utrobin, V. P. 1996, A&A, 306, 219
Utrobin, V. P. 2004, Astronomy Letters, 30, 293
Utrobin, V. P. 2007a, A&A, 461, 233
Utrobin, V. P. 2007b, in Supernova 1987A: 20 Years After: Supernovae and Gamma-Ray Bursters, AIP Conf. Series, 937, 25
Utrobin, V. P., & Chugai, N. N. 2005, A&A, 441, 271
Utrobin, V. P., & Chugai, N. N. 2008, A&A, 491, 507
Utrobin, V. P., & Chugai, N. N. 2009, A&A, 506, 829
Utrobin, V. P., & Chugai, N. N. 2011, A&A, 532, A100
Utrobin, V. P., & Chugai, N. N. 2013, A&A, 555, A145
Utrobin, V. P., & Chugai, N. N. 2015, A&A, 575, A100
Utrobin, V. P., Chugai, N. N., & Andronova, A.A. 1995, A&A, 295, 129
Utrobin, V. P., Chugai, N. N., & Botticella, M. T. 2010, ApJL, 723, L89
Utrobin, V. P., Wongwathanarat, A., Janka, H.-T., Müller, E. 2015, A&A, 581, A40
Vacca, W. D., Hamilton, R. T., Savage, M., et al. 2015, ApJ, 804, 66
Valenti, S., Benetti, S., Cappellaro, E., et al. 2008a, MNRAS, 383, 1485
Valenti, S., Elias-Rosa, N., Taubenberger, S., et al. 2008b, ApJL, 673, L155
Valenti, S., Fraser, M., Benetti, S., et al. 2011, MNRAS, 416, 3138
Valenti, S., Pastorello, A., Cappellaro, E., et al. 2009, Nature, 459, 674
Valenti, S., Sand, D., Stritzinger, M., et al. 2015, MNRAS, 448, 2608
Valenti, S., Taubenberger, S., Pastorello, A., et al. 2012, ApJL, 749, L28
Valenti, S., Yuan, F., Taubenberger, S., et al. 2014, MNRAS, 437, 1519
van Aarle, E., van Winckel, H., Lloyd Evans, T., et al. 2011, A&A, 530, A90
van den Bergh, S. 1994, ApJ, 424, 345
van den Bergh, S., Li, W., & Filippenko, A. V. 2005, PASP, 117, 773
van den Bergh, S., & Tammann, G. A. 1991, ARAA, 29, 363
van den Heuvel, E. P. J., & Portegies Zwart, S. F. 2013, ApJ, 779, 114
van den Heuvel, E. P. J., & Yoon, S.-C. 2007, ApSpSci, 311, 177
van der Horst, A. J., Kamble, A. P., Paragi, Z., et al. 2011, ApJ, 726, 99
Van Dyk, S. D., Cenko, S. B., Poznanski, D., et al. 2012, ApJ, 756, 131
Van Dyk, S. D., Lee, J. C., Anderson, J., et al. 2015, ApJ, 806, 195
Van Dyk, S. D., & Matheson, T. 2012, ApJ, 746, 179
Van Dyk, S. D., Peng, C. Y., Barth, A. J., et al. 1999, PASP, 111, 313
Van Dyk, S. D., Peng, C. Y., King, J. Y., et al. 2000, PASP, 112, 1532
Van Dyk, S. D., Weiler, K. W., Sramek, R. A., et al. 1994, ApJL, 432, L115
Van Dyk, S. D., Weiler, K. W., Sramek, R. A., et al. 1996, AJ, 111, 1271
Van Dyk, S. D., Weiler, K. W., Sramek, R. A., & Panagia, N. 1993, ApJL, 419, L69
Van Dyk, S. D. 2013, AJ, 145, 118
Van Dyk, S. D., Zheng, W., Clubb, K. I., et al. 2013, ApJL, 772, L32
Van Dyk, S. D., Zheng, W., Fox, O. D., et al. 2014, AJ, 147, 37
van Horn, H. M. 1968, ApJ, 151, 227
van Kerkwijk, M. H., Breton, R. P., & Kulkarni, S. R. 2011, ApJ, 728, 95
van Kerkwijk, M. H., Chang, P., & Justham, S. 2010, ApJL, 722, L157
van Marle, A. J., Smith, N., Owocki, S. P., & van Veelen, B. 2010, MNRAS, 407, 2305
van Putten, M. H. P. M., Lee, G. M., Della Valle, M., et al. 2014, MNRAS, 444, L58
Vanhala, H. A. T., & Cameron, A. G. W. 1998, ApJ, 508, 291
Vasileiadis, A., Nordlund, Å., & Bizzarro, M. 2013, ApJL, 769, L8
Vassiliadis, E., & Wood, P. R. 1993, ApJ, 413, 641

Venn, K. A., Irwin, M., Shetrone, M. D., et al. 2004, AJ, 128, 1177
Ventura, P., & D'Antona, F. 2011, MNRAS, 410, 2760
Ventura, P., D'Antona, F., & Mazzitelli, I. 2000, A&A, 363, 605
Ventura, P., di Criscienzo, M., Schneider, R., et al. 2012, MNRAS, 420, 1442
Viganò, D., & Pons, J. A. 2012, MNRAS, 425, 2487
Viganò, D., Rea, N., Pons, J. A., et al. 2013, MNRAS, 434, 123
Vink, J., Bleeker, J., van der Heyden, K., et al. 2006, ApJL, 648, L33
Vink, J. S., de Koter, A., & Lamers, H. J. G. L. M. 2001, A&A, 369, 574
Vink, J., & Kuiper, L. 2006, MNRAS, 370, L14
Vink, J. S., Muijres, L. E., Anthonisse, B., et al. 2011, A&A, 531, A132
Vinkó, J., Blake, R. M., Sárneczky, K., et al. 2004, A&A, 427, 453
Vinkó, J., Takáts, K., Szalai, T., et al. 2012, A&A, 540, A93
Vissani, F. 2015, Journal of Physics G Nuclear Physics, 42, 013001
Vlasenko, A., Fuller, G. M., & Cirigliano, V. 2014, Physical Review D, 89, 105004
Vlasov, A. D., Metzger, B. D., & Thompson, T. A. 2014, MNRAS, 444, 3537
Vogt, F., & Dopita, M. A. 2010, ApJ, 721, 597
Vogt, F., & Dopita, M. A. 2011, ApSpSci, 331, 521
Voit, G. M., Donahue, M., O'Shea, B. W., et al. 2015, ApJL, 803, L21
Voitsekhovsky, B. V. Mitrofanov, V. V. & Topchian, M. E. 1963, Otdel Akad. Nauk SSSR
Voss, R., & Nelemans, G. 2008, Nature, 451, 802
Vreeswijk, P. M., Savaglio, S., Gal-Yam, A., et al. 2014, ApJ, 797, 24
Vučetić, M. M., Arbutina, B., & Urošević, D. 2015, MNRAS, 446, 943
Wagenhuber, J., & Groenewegen, M. A. T. 1998, A&A, 340, 183
Walborn, N. R., Prevot, M. L., Prevot, L., et al. 1989, A&A, 219, 229
Waldman, R., Sauer, D., Livne, E., et al. 2011, ApJ, 738, 21
Walker, E. S., Baltay, C., Campillay, A., et al. 2015, ApJS, 219, 13
Walker, E. S., Hachinger, S., Mazzali, P. A., et al. 2012, MNRAS, 427, 103
Walker, E. S., Hook, I. M., Sullivan, M., et al. 2011, MNRAS, 410, 1262
Walker, E. S., Mazzali, P. A., Pian, E., et al. 2014, MNRAS, 442, 2768
Walmswell, J. J., & Eldridge, J. J. 2012, MNRAS, 419, 2054
Wampler, E. J. & Richichi, A. 1989, ApJ, 217, 31
Wampler, E. J., Wang, L., Baade, D., et al. 1990, ApJ Letters, 362, L13
Wanajo, S. 2006, ApJ, 647, 1323
Wanajo, S. 2013, ApJL, 770, L22
Wanajo, S., Janka, H.-T., & Müller, B. 2011, ApJL, 726, L15
Wanajo, S., Nomoto, K., Janka, H.-T., et al. 2009, ApJ, 695, 208
Wang, B., Chen, X., Meng, X., & Han, Z. 2009a, ApJ, 701, 1540
Wang, B., & Han, Z. 2009, A&A, 508, L27
Wang, B., & Han, Z.-W. 2010, Research in Astronomy and Astrophysics, 10, 681
Wang, B., & Han, Z. 2012, New Astron Rev, 56, 122
Wang, B., Justham, S., & Han, Z. 2013a, A&A, 559, A94
Wang, B., Justham, S., Liu, Z.-W., et al. 2014a, MNRAS, 445, 2340
Wang, B., Li, X.-D., & Han, Z.-W. 2010a, MNRAS, 401, 2729
Wang, B., Li, Y., Ma, X., et al. 2015a, A&A, 584, A37
Wang, B., Liu, Z., Han, Y., et al. 2010b, Science China Physics, Mechanics, and Astronomy, 53, 586
Wang, B., Meng, X., Liu, D.-D., Liu, Z.-W., & Han, Z. 2014b, ApJL, 794, LL28
Wang, F. Y., & Dai, Z. G. 2014, ApJS, 213, 15
Wang, L. 2005, ApJL, 635, L33
Wang, L., Baade, D., Höflich, P., et al. 2004, ApJL, 604, L53
Wang, L., Baade, D., Höflich, P., et al. 2006, ApJ, 653, 490
Wang, L., Baade, D., Höflich, P., & Wheeler, J. C. 2003, ApJ, 592, 457
Wang, L., Baade, D., & Patat, F. 2007a, Science, 315, 212

Wang, L., Cui, X., Zhu, H., & Tian, W. 2015b, Advancing Astrophysics with the Square Kilometre Array (AASKA14), 64
Wang, L., Howell, D. A., Höflich, P., & Wheeler, J. C. 2001, ApJ, 550, 1030
Wang, L., Mazzali, P. A. 1992, Nature, 355, 58
Wang, L., & Wheeler, J. C. 2008, ARAA, 46, 433
Wang, L., Wheeler, J. C., Höflich, P., et al. 2002, ApJ, 579, 2
Wang, L., Wheeler, J. C., Li, Z., & Clocchiatti, A. 1996, ApJ, 467, 435
Wang, S. Q., Wang, L. J., Dai, Z. G., & Wu, X. F. 2015c, ApJ, 799, 107
Wang, W., Harris, M. J., Diehl, R., et al. 2007b, A&A, 469, 1005
Wang, X., Filippenko, A. V., Ganeshalingam, M., et al. 2009b, ApJL, 699, L139
Wang, X., Li, W., Filippenko, A. V., et al. 2009c, ApJ, 697, 380
Wang, X., Wang, L., Filippenko, A. V., et al. 2013b, Science, 340, 170
Wang, Y. 2010, Dark Energy (Wiley)
Wang, Y., Gjergo, E., & Kuhlmann, S. 2015d, MNRAS, 451, 1955
Warner, B. 1995, Cambridge Astrophysics Series, 28
Warren, D. C., & Blondin, J. M. 2013, MNRAS, 429, 3099
Washabaugh, P. C., & Bregman, J. N. 2013, ApJ, 762, 1
Weaver, T. A., & Woosley, S. E. 1993, Physics Reports, 227, 65
Webbink, R. F. 1976, ApJ, 209, 829
Webbink, R. F. 1984, ApJ, 277, 355
Weidemann, V., & Koester, D. 1983, A&A, 121, 77
Weigert, A. 1966, Zeitschrift fuer Astrophysik, 64, 395
Weiler, K. W., Panagia, N., & Sramek, R. A. 1990, ApJ, 364, 611
Weiler, K. W., Panagia, N., Stockdale, C., et al. 2011, ApJ, 740, 79
Weiler, K. W., Sramek, R. A., Panagia, N., et al. 1986, ApJ, 301, 790
Weiler, K. W., van Dyk, S. D., Discenna, J. L., et al. 1991, ApJ, 380, 161
Weiler, K. W., van Dyk, S. D., Panagia, N., & Sramek, R. A. 1992, ApJ, 398, 248
Weiler, K. W., Williams, C. L., Panagia, N., et al. 2007, ApJ, 671, 1959
Weinberg, N. N., & Quataert, E. 2008, MNRAS, 387, L64
Weinberg, S. 2013, PRL, 110, 241301
Wells, L. A., Phillips, M. M., Suntzeff, B., et al. 1994, AJ, 108, 2233
Wellstein, S., Langer, N., & Braun, H. 2001, A&A, 369, 939
Welty, D. E., Ritchey, A. M., Dahlstrom, J. A., & York, D. G. 2014, ApJ, 792, 106
Wesson, R., Barlow, M. J., Matsuura, M., & Ercolano, B. 2015, MNRAS, 446, 2089
Whalen, D. J., Even, W., Lovekin, C. C., et al. 2013a, ApJ, 768, 195
Whalen, D. J., Even, W., Smidt, J., et al. 2013b, ApJ, 778, 17
Whalen, D. J., Johnson, J. L., Smidt, J., et al. 2013c, ApJ, 777, 99
Whalen, D. J., Smidt, J., Even, W., et al. 2014a, ApJ, 781, 106
Whalen, D. J., Smidt, J., Heger, A., et al. 2014b, ApJ, 797, 9
Wheeler, J. C. 1977, ApSpSci, 50, 125
Wheeler, J. C. 1990, in Supernovae, ed. J. C. Wheeler, T. Piran, & S. Weinberg, (World Scientific), 1
Wheeler, J. C. 2012, ApJ, 758, 123
Wheeler, J. C., & Akiyama, S. 2007, ApJ, 654, 429
Wheeler, J. C., Barker, E., Benjamin, R., et al. 1993, ApJL, 417, L71
Wheeler, J. C., Cowan, J. J., & Hillebrandt, W. 1998, ApJL, 493, L101
Wheeler, J. C., & Filippenko, A. V. 1996, IAU Colloq. 145, Supernovae and Supernova Remnants, 241
Wheeler, J. C., & Hansen, C. J. 1971, ApSpSci, 11, 373
Wheeler, J. C., & Harkness, R. P. 1986, in NATO ASIC Proc. 180, Galaxy Distances and Deviations from Universal Expansion, 45
Wheeler, J. C., & Harkness, R. P. 1990, Reports on Progress in Physics, 53, 1467
Wheeler, J. C., Harkness, R. P., Barker, E. S., et al. 1987, ApJL, 313, L69
Wheeler, J. C., Harkness, R. P., Clocchiatti, A., et al. 1994, ApJL, 436, L135

Wheeler, J. C., & Johnson, V. 2011, ApJ, 738, 163
Wheeler, J. C., Johnson, V., & Clocchiatti, A. 2015a, MNRAS, 450, 1295
Wheeler, J. C., Kagan, D., & Chatzopoulos, E. 2015b, ApJ, 799, 85
Wheeler, J. C., Lecar, M., & McKee, C. F. 1975, ApJ, 200, 145
Wheeler, J. C., & Levreault, R. 1985, ApJL, 294, L17
Wheeler, J. C., Meier, D. L., & Wilson, J. R. 2002, ApJ, 568, 807
Wheeler, J. C., & Pooley, D. 2013, ApJ, 762, 75
Wheeler, J. C., & Shields, G. A. 1976, Nature, 259, 642
Wheeler, J. C., Yi, I., Höflich, P., & Wang, L. 2000, ApJ, 537, 810
Whelan, J., & Iben, I., Jr. 1973, ApJ, 186, 1007
White, C. J., Kasliwal, M. M., Nugent, P. E., et al. 2015, ApJ, 799, 52
White, G. L., Malin, D. F. 1987, Nature, 327, 36
Whitelock, P. A., Catchpole, R. M., Menzies, J. W., et al. 1989, MNRAS, 240, 7P
Wild, P., & Barbon, R. 1980, IAU Circ., 3532, 1
Williams, B. J., Blair, W. P., Blondin, J. M., et al. 2011, ApJ, 741, 96
Williams, B. J., Borkowski, K. J., Reynolds, S. P., et al. 2012, ApJ, 755, 3
Williams, C. L., Panagia, N., Van Dyk, S. D., et al. 2002, ApJ, 581, 396
Williams, F. 1985, Combustion Theory: the Fundamental Theory of Chemically Reacting Flow Systems (Benjamin/Cummings)
Williams, K. A., Bolte, M., & Koester, D. 2009, ApJ, 693, 355
Williams, S. C., Darnley, M. J., Bode, M. F., & Shafter, A. W. 2014, in Stella Novae: Past and Future Decades, 490, 85
Willson, L. A. 2000, ARAA, 38, 573
Wilson, J. R., & Mayle, R. W. 1988, Physics Reports, 163, 63
Wiltshire, D. L. 2013, arXiv:1311.3787
Winget, D. E., Kepler, S. O., Campos, F., et al. 2009, ApJL, 693, L6
Winkler, P. F., Gupta, G., & Long, K. S. 2003, ApJ, 585, 324
Winkler, P. F., Hamilton, A. J. S., Long, K. S., & Fesen, R. A. 2011, ApJ, 742, 80
Winkler, P. F., Long, K. S., Hamilton, A. J. S., & Fesen, R. A. 2005, ApJ, 624, 189
Winkler, P. F., Twelker, K., Reith, C. N., & Long, K. S. 2009, ApJ, 692, 1489
Winkler, P. F., Williams, B. J., Reynolds, S. P., et al. 2014, ApJ, 781, 65
Winteler, C., Käppeli, R., Perego, A., et al. 2012, ApJL, 750, L22
Wolf, W. M., Bildsten, L., Brooks, J., & Paxton, B. 2013, ApJ, 777, 136
Woltjer, L. 1972, ARAA, 10, 129
Wong, T.-W., Fryer, C. L., Ellinger, C. I., et al. 2014, arXiv:1401.3032
Wongwathanarat, A., Janka, H.-T., Müller, E. 2010, ApJL, 725, L106
Wongwathanarat, A., Janka, H.-T., Müller, E. 2013, A&A, 552, A126
Wongwathanarat, A., Müller, E., & Janka, H.-T. 2015, A&A, 577, A48
Wonnacott, D., Kellett, B. J., & Stickland, D. J. 1993, MNRAS, 262, 277
Wood, K., Hill, A. S., Joung, M. R., et al. 2010, ApJ, 721, 1397
Wood, P. R., Bessell, M. S., & Fox, M. W. 1983, ApJ, 272, 99
Wood, P. R., & Faulkner, D. J. 1986, ApJ, 307, 659
Wood, P. R., Whiteoak, J. B., Hughes, S. M. G., et al. 1992, ApJ, 397, 552
Wood-Vasey, W. M., & Sokoloski, J. L. 2006, ApJL, 645, L53
Wood-Vasey, W. M., Wang, L., & Aldering, G. 2004, ApJ, 616, 339
Wooden, D. H., Rank, D. M., Bregman, J. D., et al. 1993, ApJS, 88, 477
Woods, T. E., & Gilfanov, M. 2014, MNRAS, 439, 2351
Woosley, S. E. 1988, ApJ, 330, 218
Woosley, S. E. 1993, ApJ, 405, 273
Woosley, S. E. 2010, ApJL, 719, L204
Woosley, S. E., & Baron, E. 1992, ApJ, 391, 228
Woosley, S. E., Blinnikov, S., & Heger, A. 2007, Nature, 450, 390
Woosley, S. E., & Bloom, J. S. 2006, ARAA, 44, 507
Woosley, S. E., Eastman, R. G., Weaver, T. A., & Pinto, P. A. 1994a, ApJ, 429, 300

Woosley, S. E., Hartmann, D. H., Hoffman, R. D., & Haxton, W. C. 1990, ApJ, 356, 272
Woosley, S. E., & Heger, A. 2006, ApJ, 637, 914
Woosley, S. E., & Heger, A. 2015, ApJ, 810, 34
Woosley, S. E., Heger, A., & Weaver, T. A. 2002, RMP, 74, 1015
Woosley, S. E., & Kasen, D. 2011, ApJ, 734, 38
Woosley, S. E., Langer, N., & Weaver, T. A. 1995, ApJ, 448, 315
Woosley, S. E., Pinto, P. A., & Ensman, L. 1988, ApJ, 324, 466
Woosley, S. E., & Weaver, T. A. 1988, Physics Reports, 163, 79
Woosley, S. E., & Weaver, T. A. 1994, ApJ, 423, 371
Woosley, S. E., & Weaver, T. A. 1995, ApJS, 101, 181
Woosley, S. E., Weaver, T. A., & Taam, R. E. 1980, in Texas Workshop on Type I Supernovae, ed. J. C. Wheeler (Un. Texas), 96
Woosley, S. E., Wilson, J. R., Mathews, G. J., et al. 1994b, ApJ, 433, 229
Worley, C. C., Irwin, M. J., Tout, C. A., et al. 2016, MNRAS, 459, L31
Woudt, P. A., Steeghs, D., Karovska, M., et al. 2009, ApJ, 706, 738
Wright, W. P., Nagaraj, G., Kneller, J. P., et al. 2016, Physical Review D, 94, 025026
Wu, M.-R., Fischer, T., Huther, L., et al. 2014, Physical Review D, 89, 061303
Wyithe, J. S. B., & Loeb, A. 2013, MNRAS, 428, 2741
Wynn, G. 2008, in RS Ophiuchi (2006) and the Recurrent Nova Phenomenon, 401, 73
Xiao, L., & Eldridge, J. J. 2015, MNRAS, 452, 2597
Ya'Ari, A., & Tuchman, Y. 1996, ApJ, 456, 350
Ya'Ari, A., & Tuchman, Y. 1999, ApJL, 514, L35
Yahil, A. 1983, ApJ, 265, 1047
Yamada, T., Iwata, I., Ando, M., et al. 2012, Proceedings of the SPIE, 8442, 84421A
Yamaguchi, H., Badenes, C., Foster, A. R., et al. 2015, ApJL, 801, L31
Yamaguchi, H., Badenes, C., Petre, R., et al. 2014, ApJL, 785, L27
Yamaguchi, H., Koyama, K., & Uchida, H. 2011, Publications of the ASJ, 63, 837
Yamanaka, M., Kawabata, K. S., Kinugasa, K., et al. 2009, ApJL, 707, L118
Yamanaka, M., Maeda, K., Kawabata, K. S., et al. 2015, ApJ, 806, 191
Yan, C.-S., & Wang, J.-M. 2010, ApJ, 725, 2359
Yan, L., Quimby, R., Ofek, E., et al. 2015, ApJ, 814, 108
Yang, H., & Chevalier, R. A. 2015, ApJ, 806, 153
Yaron, O., Prialnik, D., Shara, M. M., & Kovetz, A. 2005, ApJ, 623, 398
Yasuda, N., & Fukugita, M. 2010, AJ, 139, 39
Yoon, S.-C. 2004, Ph.D. Thesis, Utrecht Un.
Yoon, S.-C. 2015, Publications of the Astron. Soc. of Australia, 32, e015
Yoon, S.-C., & Cantiello, M. 2010, ApJL, 717, L62
Yoon, S.-C., Dierks, A., & Langer, N. 2012a, A&A, 542, A113
Yoon, S.-C., Gräfener, G., Vink, J. S., et al. 2012b, A&A, 544, LL11
Yoon, S.-C., & Langer, N. 2004a, A&A, 419, 623
Yoon, S.-C., & Langer, N. 2004b, A&A, 419, 645
Yoon, S.-C., & Langer, N. 2005a, A&A, 435, 967
Yoon, S.-C., & Langer, N. 2005b, A&A, 443, 643
Yoon, S.-C., Langer, N., & van der Sluys, M. 2004, A&A, 425, 207
Yoon, S.-C., Podsiadlowski, P., & Rosswog, S. 2007, MNRAS, 380, 933
Yoon, S.-C., Woosley, S. E., & Langer, N. 2010, ApJ, 725, 940
Yoshida, T., Okita, S., & Umeda, H. 2014, MNRAS, 438, 3119
Young, D. R., Smartt, S. J., Mattila, S., et al. 2008, A&A, 489, 359
Young, D. R., Smartt, S. J., Valenti, S., et al. 2010, A&A, 512, A70
Young, P. A., Fryer, C. L., Hungerford, A., et al. 2006, ApJ, 640, 891
Young, T. R., Smith, D., & Johnson, T. A. 2005, ApJL, 625, L87
Yuan, F., Akerlof, C., Quimby, R., et al. 2007, BAAS, 39, #105.05
Yuan, F., Kobayashi, C., Schmidt, B. P., et al. 2013, MNRAS, 432, 1680
Yuan, F., Quimby, R. M., Wheeler, J. C., et al. 2010, ApJ, 715, 1338

Yungelson, L. R. 2010, Astronomy Letters, 36, 780
Yungelson, L. R., van den Heuvel, E. P. J., Vink, J. S., et al. 2008, A&A, 477, 223
Yusof, N., Hirschi, R., Meynet, G., et al. 2013, MNRAS, 433, 1114
Zampieri, L., Pastorello, A., Turatto, M., et al. 2003, MNRAS, 338, 711
Zampieri, L., Shapiro, S. L., & Colpi, M. 1998, ApJL, 502, L149
Zanardo, G., Staveley-Smith, L., Ball, L., et al. 2010, ApJ, 710, 2
Zanardo, G., Staveley-Smith, L., Indebetouw, R., et al. 2014, ApJ, 796, 82
Zanardo, G., Staveley-Smith, L., Ng, C.-Y., et al. 2013, ApJ, 767, 98
Zel'Dovich, Y. B. 1966, Journal of Applied Mechanics and Technical Physics, 7, 68
Zel'dovich, Y. B., Librovich, V. B., Makhviladze, G. M., & Sivashinsky, G. I. 1970, Acta Astron., 15, 313
Zel'dovich, Y. B., & Raizer, Y. P. 1967, Physics of Shock Waves and High-temperature Hydrodynamic Phenomena, (Academic Press)
Zel'dovich, Y. B., & Raizer, Y. P. 1969, Elements of Gas Dynamics and the Classical Theory of Shock Waves (Academic Press)
Zelaya, P., Quinn, J. R., Baade, D., et al. 2013, AJ, 145, 27
Zhang, J., Wang, X., Mazzali, P. A., et al. 2014a, ApJ, 797, 5
Zhang, J.-J., Wang, X.-F., Bai, J.-M., et al. 2014b, AJ, 148, 1
Zhang, P., & Meng, X. 2014, Modern Physics Letters A, 29, 1450103
Zhang, T., Wang, X., Wu, C., et al. 2012, AJ, 144, 131
Zhang, T., Wang, X., Zhou, X., et al. 2004, AJ, 128, 1857
Zhang, Y., & Burrows, A. 2013, Physical Review D, 88, 105009
Zhekov, S. A., Park, S., McCray, R., et al. 2010, MNRAS, 407, 1157
Zheng, W., Shivvers, I., Filippenko, A. V., et al. 2014, ApJL, 783, LL24
Zheng, W., Silverman, J. M., Filippenko, A. V., et al. 2013, ApJL, 778, L15
Zhou, F., Strom, R. G., & Jiang, S. 2006, Chin. J. A&A, 6, 635
Zhou, W.-H., Wang, B., Meng, X.-C., et al. 2015, Research in Astronomy and Astrophysics, 15, 1701
Zhu, C., Chang, P., van Kerkwijk, M. H., & Wadsley, J. 2013, ApJ, 767, 164
Zhu, C., Pakmor, R., van Kerkwijk, M. H., & Chang, P. 2015, ApJL, 806, L1
Ziebarth, K. 1970, ApJ, 162, 947
Zimmermann, H.-U., Lewin, W., Predehl, P., et al. 1994, Nature, 367, 621
Zingale, M., Almgren, A. S., Bell, J. B., et al. 2009, ApJ, 704, 196
Zingale, M., & Dursi, L. J. 2007, ApJ, 656, 333
Zingale, M., Woosley, S. E., Bell, J. B., et al. 2005, JP Conf. Series, 16, 405
Zingale, M., Nonaka, A., Almgren, A. S., et al. 2011, ApJ, 740, 8
Zinn, P.-C., Stritzinger, M., Braithwaite, J., et al. 2012, A&A, 538, A30
Zwicky, F. 1938, PASP, 50, 215
Zwicky, F. 1942, ApJ, 96, 28
Zwicky, F. 1964, ApJ, 139, 514
Zwicky, F. 1974, in Supernovae and Supernova Remnants, ed. C. B. Cosmovici (Reidel), 1

Index

D. Branch, J.C. Wheeler, *Supernova Explosions*, Astronomy and Astrophysics Library, DOI 10.1007/978-3-662-55054-0

MIX
Papier aus verantwortungsvollen Quellen
Paper from responsible sources
FSC® C105338

If you have any concerns about our products,
you can contact us on
ProductSafety@springernature.com

In case Publisher is established outside the EU,
the EU authorized representative is:
Springer Nature Customer Service Center GmbH
Europaplatz 3, 69115 Heidelberg, Germany

Printed by Libri Plureos GmbH
in Hamburg, Germany